Phase Transitions in Materials

Second Edition

The new edition of this popular textbook provides a fundamental approach to phase transformations and thermodynamics of materials. Explanations are emphasized at the level of atoms and electrons, and it comprehensively covers the classical topics from classical metallurgy to nanoscience and magnetic phase transitions. The book has three parts, covering the fundamentals of phase transformations, the origins of the Gibbs free energy, and the major phase transformations in materials science. A fourth part on advanced topics is available online. Much of the content from the first edition has been expanded, notably precipitation transformations in solids, heterogeneous nucleation, and energy, entropy, and pressure. Three new chapters have been added to cover interactions within microstructures, surfaces, and solidification. Containing over 170 end-of-chapter problems, it is a valuable companion to graduate students and researchers in materials science, engineering, and applied physics.

Brent Fultz is the Rawn Professor of Materials Science and Applied Physics at the California Institute of Technology. His awards include the 2016 William Hume-Rothery Award of The Minerals, Metals and Materials Society (TMS). He is a fellow of the American Physical Society (APS), TMS, and the Neutron Scattering Society of America (NSSA).

Phase Transitions in Materials

Second Edition

BRENT FULTZ

California Institute of Technology

CAMBRIDGE
UNIVERSITY PRESS

University Printing House, Cambridge CB2 8BS, United Kingdom

One Liberty Plaza, 20th Floor, New York, NY 10006, USA

477 Williamstown Road, Port Melbourne, VIC 3207, Australia

314–321, 3rd Floor, Plot 3, Splendor Forum, Jasola District Centre, New Delhi – 110025, India

79 Anson Road, #06–04/06, Singapore 079906

Cambridge University Press is part of the University of Cambridge.

It furthers the University's mission by disseminating knowledge in the pursuit of
education, learning, and research at the highest international levels of excellence.

www.cambridge.org
Information on this title: www.cambridge.org/9781108485784
DOI: 10.1017/9781108641449

First published 2014
Second edition 2020

Printed in the United Kingdom by TJ International Ltd, Padstow Cornwall

A catalogue record for this publication is available from the British Library.

Library of Congress Cataloging-in-Publication Data
Names: Fultz, B. (Brent), author.
Title: Phase transitions in materials / Brent Fultz.
Description: Second edition. | Cambridge ; New York, NY : Cambridge
University Press, 2020. | Includes bibliographical references and index.
Identifiers: LCCN 2019043465 (print) | LCCN 2019043466 (ebook) |
ISBN 9781108485784 (hardback) | ISBN 9781108641449 (epub)
Subjects: LCSH: Phase transformations (Statistical physics)–Textbooks. |
Thermodynamics–Textbooks. | Materials–Thermal properties–Textbooks. |
Statistical mechanics–Textbooks.
Classification: LCC QC175.16.P5 F86 2020 (print) | LCC QC175.16.P5
(ebook) | DDC 530.4/74–dc23
LC record available at https://lccn.loc.gov/2019043465
LC ebook record available at https://lccn.loc.gov/2019043466

ISBN 978-1-108-48578-4 Hardback

Additional resources for this publication at www.cambridge.org/fultz2

Contents

Preface

Content

This book explains the thermodynamics and kinetics of most of the important phase transitions in materials science. It is a textbook, so the emphasis is on explanations of phenomena rather than a scholarly assessment of their origins. The goal is explanations that are concise, clear, and reasonably complete. The level and detail are appropriate for upper division undergraduate students and graduate students in materials science and materials physics. The book should also be useful for researchers who are not specialists in these fields. The book is organized for approximately sequential coverage in a graduate-level course. The four parts of the book serve different purposes, however, and should be approached differently.

Part I presents topics that all graduate students in materials science must know.[1] After a general overview of phase transitions, temperature–composition phase diagrams are explained from classical thermodynamics and from the statistical mechanics of Ising lattices. Diffusion, equilibration, and nucleation are then covered, and general aspects of diffusion and nucleation are used with T–c phase diagrams to explain the rates of some phase transformations.

Part II addresses the origins of materials thermodynamics and kinetics at the level of atoms and electrons. Electronic and elastic energy are covered at the level needed in some of the later chapters. The physical origins of entropy (a topic that receives scant coverage in other texts) are presented in the context of phase transitions on Ising lattices. Effects of pressure, combined with temperature, are explained with a few concepts of chemical bonding and antibonding. The thermodynamics of real materials typically involves minimizing a free energy with multiple degrees of freedom, and Chapter 9 shows directions beyond one variable. Chapter 10 on kinetics emphasizes atom movements for diffusion in solids, especially features of atom–vacancy interchanges.

Part III is the longest. It describes important phase transformations in materials, with their underlying concepts. Topics include surface phenomena, melting, solidification, nucleation and growth in solids, spinodal decomposition, phase field theory, continuous ordering, martensitic transformations, phenomena in nanomaterials, and phase transitions involving electrons or spins. Many topics from metallurgy and ceramic engineering are

---

[1] The author asks graduate students to explain some of the key concepts at a blackboard during their Ph.D. candidacy examinations.

covered, although the connection between processing and properties is less emphasized, allowing for a more concise presentation than in traditional texts.

The online Advanced Topics present modern topics that have proved their importance. These chapters are available online at doi:10.7907/05BY-QX43 and can be downloaded at no cost from https://www.library.caltech.edu. The chapters cover low- and high-temperature treatments of the partition function, nonequilibrium states in crystalline alloys, a k-space formulation of elastic energy, fluctuations and how they are measured, high-temperature thermodynamics, the renormalization group, scaling theory, and an introduction to quantum phase transitions. The topics are explained at a fundamental level, but unlike Parts I through III, for conciseness there are more omissions of methods and steps.

Many topics in phase transitions and related phenomena are not covered in this text. These include: polymer flow and dynamics including reptation, phase transitions in fluid systems including phenomena near the critical temperature, crystallographic symmetry in displacive transformations, and massive transformations. Also beyond the scope of the book are computational methods that are increasingly important for studies of phase transformations in materials, including: Monte Carlo methods, molecular dynamics methods (classical and quantum), and density functional theory with time or ensemble averages for materials at finite temperatures.

The field of phase transitions is huge, and continues to grow. This text is a snapshot of the field taken from the viewpoint of the author near the year 2020. Impressively, this field continues to offer a rich source of new ideas and results for both fundamental and applied research, and parts of it will look different in a decade or so. I expect, however, that the core will remain the same – the free energy of materials will be at the center, surrounded by issues of kinetics.

Teaching

I use this text in a course for Ph.D. students in both materials science and in applied physics at the California Institute of Technology. The 10-week course is offered in the third academic quarter as part of a one-year sequence. The first two quarters in this sequence cover thermodynamics and statistical mechanics, so the students are already familiar with using a partition function to obtain thermodynamic quantities. Familiarity with some concepts from solid-state physics and chemistry is certainly helpful, but the text develops many of the important concepts as needed.

In the one-quarter course at Caltech, I cover most topics in Parts I and II, moving in sequence through the chapters. Time limitations force a selection of topics from Part III and Advanced Topics. For example, I tend to cover Chapters 12, 16, 18, and parts of 14, 19, 20 (although sometimes these later parts are replaced by an advanced chapter, such as 25). It is unrealistic to cover the entire content of the book in one course, even with a 15-week semester. An instructor will use discretion in selecting topics for the second half of his or her course.

The problems at the end of each chapter were used for weekly student assignments, and this helped to refine their wording and content. The majority of these problems are based on concepts explained in the text, sometimes filling in explanations or extending the analyses. Other problems, less popular with students, develop new concepts not described in the chapter. These problems usually include longer explanations and hints that may be worth reading even without working the problem. None of the problems are intended to be particularly difficult, and some can be answered quickly with one main idea. For homework, I assign five or six of these problems every week during the term. In their reviews of the course, most students reportedly spend 6–8 hours per week outside the classroom completing these problem sets and reading the text. An online solutions manual is available to course instructors whose identity can be verified. Please ask me for further information.

Acknowledgments

I thank J.J. Hoyt for collaborating with me on a book chapter about phase equilibria and phase transformations that prompted me to get started on the first edition of this book. The development of the topic of vibrational entropy would not have been possible without the contributions of my junior collaborators at Caltech, especially L. Anthony, L.J. Nagel, H.N. Frase, M.E. Manley, J.Y.Y. Lin, T.L. Swan-Wood, A.B. Papandrew, O. Delaire, M.S. Lucas, M.G. Kresch, M.L. Winterrose, J. Purewal, C.W. Li, T. Lan, H.L. Smith, L. Mauger, S.J. Tracy, D.S. Kim, and N. Weadock. Several of them are taking this field into new directions.

Important ideas have come from stimulating conversations over the years with O. Hellman, A. van de Walle, V. Ozolins, G. Ceder, M. Asta, L.-Q. Chen, D.D. Johnson, E.E. Alp, R. Hemley, J. Neugebauer, B. Grabowski, M. Sluiter, F. Körmann, D. de Fontaine, A.G. Khachaturyan, I. Abrikosov, A. Zunger, P. Rez, K. Samwer, and W.L. Johnson.

Brent Fultz

Notation

a	lattice parameter
A	area
$\vec{A}$	vector potential of magnetic field
A-atom	generic chemical element
APDB	antiphase domain boundary
α	coefficient of linear thermal expansion
α	critical exponent for heat capacity
α-phase	generic phase
α-sublattice	a lattice of like atoms within an ordered structure
α_i	root of Bessel function
α^2	electron–phonon coupling factor
$\vec{b}$	Burgers vector of dislocation
b_A	coherent neutron scattering length of isotope A
$b(\vec{k})$	Fourier transform of pairwise energy for two concentration waves
B	bulk modulus
$\vec{B}$	magnetic field
B-atom	generic chemical element
$B(\vec{R})$	pairwise energy between atoms
β	coefficient of volume thermal expansion
β	critical exponent for density
β-phase	generic phase
β-sublattice	a lattice of like atoms within an ordered structure
c	chemical composition (atomic fraction)
c_l^*	chemical composition of liquid at liquid–solid interface
c_s^*	chemical composition of solid at liquid–solid interface
c	speed of sound or light
c_A	concentration of A-atoms
c_A	weight of atomic wavefunction on atom A in a molecular wavefunction
C_{el}	electronic heat capacity
$C_P(T)$	heat capacity at constant pressure
$C_V(T)$	heat capacity at constant volume
C_{ij}, C_{ijlm}	elastic constant

D	diffusion coefficient
D	deformation potential
D_h	thermal (heat) diffusion coefficient
$\vec{D}$	electric polarization
D_0	prefactor for exponential form of diffusion coefficient
$\tilde{D}(c)$	interdiffusion coefficient
$\underline{D}(\vec{k}), D_{ij}(\vec{k})$	dynamical matrix, element of
δ	fractional change in volume (of misfitting sphere)
ΔG_V	change in Gibbs free energy per unit volume
ΔG^*	activation barrier for nucleation
$\Delta(\vec{r})$	static wave of chemical concentration
e	charge of electron
e_A	energy of an A-atom on a crystal site
e_{AB}	energy of a pair (bond) between an A- and a B-atom
e_R, e_W	energy of two atoms, A and B, on their right or wrong sublattices
$\vec{e}_{\kappa j}(\vec{k})$	polarization for atom of basis index κ in phonon of $\vec{k}$ in branch j
$\mathrm{erf}(z)$	error function
E	energy, thermodynamic energy
$\vec{E}$	electric field
E_{el}	elastic energy
E_{elec}	electrostatic energy
ε	energy, energy of phonon
ϵ	energy, energy of electron
ϵ	fractional difference in T from T_c
ϵ_F	Fermi energy
$\epsilon_j, \epsilon_{ij}$	strain
η	fractional change of lattice parameter with composition
η	order parameter
f	correlation factor
f_α	(atomic) fraction of α-phase
f_j	interaction free energy
$f(c)$	free energy per unit volume
F	Helmholtz free energy
$\mathcal{F}$	force
$F_\xi(c, T)$	free energy for phase ξ with composition c at temperature T
$g(\varepsilon)$	phonon density of states
$\vec{g}$	reciprocal lattice vector
$\mathbf{grad}(c)$ or $\overrightarrow{\nabla c}$	gradient (of concentration)
G	Gibbs free energy
$G(\vec{r}, t)$	Van Hove space-time correlation function

$\mathcal{G}$	temperature gradient dT/dx
γ	coefficient for linear electronic heat capacity vs. T
γ	Grüneisen parameter
γ_j	Grüneisen parameter for phonon mode j
γ_{xy}	shear strain
Γ	atomic jump frequency
Γ	point at origin of reciprocal lattice
h	bond integral
$\hbar$	Planck constant divided by 2π
H	Hamiltonian
$\vec{H}$	magnetic field
$\vec{j}$	flux
$J_0(x), J_1(x)$	Bessel functions of zero and first order
J_n	number of clusters per unit time that change from n to $n+1$
J_{ss}	steady-state flux in number space of cluster sizes
J_{hs}, J_{hl}	heat flux in solid and liquid (1D)
$\vec{J}_A$	flux of A-atoms
$J(\vec{r}_1 - \vec{r}_j)$	magnetic exchange energy
k	partitioning ratio $k = c_s/c_l$
$\vec{k}$	wavevector
k_B	Boltzmann constant
κ	coefficient for square gradient energy
κ	Ginzburg–Landau parameter
L	latent heat
L	long-range order parameter
$L(\tau E_0/k_B T)$	Langevin function
LHS	left-hand side
λ	wavelength
λ	electron–phonon coupling parameter
m	mass
m	slope of liquidus curve on phase diagram dT_l/dc
M	mobility
$\vec{M}$	magnetization
$\mathcal{M}$	Mendeleev number
μ	chemical potential
μ	shear modulus
$\vec{\mu}$	magnetic moment

$n(\varepsilon_i, T)$	Planck distribution
N	number (of atoms)
N_A^α	number of A-atoms on α-sublattice (point variable)
$N_{AB}^{\alpha\beta}$	number of A–B pairs with A on α and B on β (pair variable)
$N(k)$	number of quantum states with wavevector less than k
$\underset{\sim}{N}(t)$	vector of number occupancies of states at time t
ν	frequency
ν	Poisson ratio
ν	critical exponent for correlation length
ω	angular frequency
Ω	number of states accessible to the system
Ω	atomic volume
Ω_j	configurations of a system with energy j
p_i	probability of a state
$\vec{p}$	momentum
p_A	partial pressure of vapor of element A
P_A^α	probability of A-atom on α-sublattice (point variable)
$P_{AB}^{\alpha\beta}$	probability of A–B pair with A on α and B on β (pair variable)
P	pressure
P_{th}	thermal pressure (from expansion against a bulk modulus)
$\mathcal{P}$	Péclet number
$\Phi(r)$	interatomic, central-force potential
$\Phi_M(r), \Phi_{L\text{-}J}(r)$	Morse potential, Lennard-Jones potential
Φ_0	quantum of magnetic flux $hc/2e$
Q	compositional wavevector $2\pi/\lambda$
Q	total electrostatic charge
$\vec{Q}$	momentum transfer in scattering
Q	quality factor of damped harmonic oscillator
r_B	Bohr radius $r_B = \hbar^2/(m_e e^2)$
r_{WS}	Wigner–Seitz radius
$\vec{r}_l$	position of unit cell
$\vec{r}_k$	basis vector within unit cell
R	number of right atoms on a sublattice of an ordered structure
$R(Q)$	growth rate for compositional wavevector Q
R^*	critical radius for nucleation
$\vec{R}$	position of atom center
$\vec{R}_n$	displacement after n jumps
$\mathcal{R}$	number of atoms in unit cell
RHS	right-hand side

ρ	density, e.g., [atoms cm^{-3}]
$\rho(\epsilon)$	electronic density of states
$\rho(\epsilon_F)$	electronic density of states at the Fermi energy
$\vec{s}_i$	electronic spin at site i
S	entropy
S	overlap integral
S_{conf}	configurational entropy
S_{vib}	vibrational entropy
S_h	harmonic entropy
S_{qh}	entropy contribution from quasiharmonicity
S_{anh}	entropy contribution from anharmonicity
S_{el}	electronic entropy
S_{epi}	entropy contribution from electron–phonon interaction
S_{mag}	magnetic entropy
$S(\vec{Q}, \omega)$	scattering function
σ	surface energy per unit area
σ	electrical conductivity
σ	spin number (± 1)
σ_{gb}	energy per unit area of grain boundary
σ_{ij}	stress
t	time
T	temperature
T_c	critical temperature
T_C	Curie temperature
T_m	melting temperature
T_N	Néel temperature
$T_1, T_2, \ldots$	sequence of temperatures such that $T_2 > T_1$
$\vec{T}$	translation vector of real space lattice
τ	characteristic time (e.g., for diffusion)
$\vec{\tau}$	electrostatic dipole moment
$\theta(\vec{r})$	Heaviside function, 1 in the region, 0 outside
$\theta(\vec{r}, t)$	phase of wavefunction in space and time
Θ_D	Debye temperature
$\vec{u}(x, y, z)$	displacement vector
U	difference in chemical preferences of A- and B-atoms $U = (e_{AA} - e_{BB})/4V$
U	Coulomb energy penalty for placing a second electron on a site in Hubbard model
Υ_j	Grüneisen parameter for energy of electronic state j

$\vec{v}$	velocity
V	interchange energy $V = (e_{AA} + e_{BB} - 2e_{AB})/4$
V	volume
$V(\vec{r})$	potential energy
V_Q	quantum volume, related to cube of de Broglie wavelength
v_0	volume per atom

W	number of wrong atoms on a sublattice of an ordered structure
W_{ij}	transition rate from state j to state i
$W^{\uparrow}_{\beta A \alpha}$	rate of increase of LRO parameter by jump of A from β- to α-sublattice
$\underset{\approx}{W}(\Delta t)$	transition matrix for time interval Δt

ξ	correlation function
ξ	length
$\{\chi_i\}$	reaction coordinates
χ	susceptibility

Y	Young's modulus
$\psi(\vec{r})$	wavefunction

z	coordination number of lattice
z	partition function of subsystem
Z	partition function
$\mathcal{Z}$	Zeldovich factor

PART I

BASIC THERMODYNAMICS AND KINETICS OF PHASE TRANSFORMATIONS

The field of phase transitions is rich and vast, and continues to grow. This text covers parts of the field relevant to materials physics, but many concepts and tools of phase transitions in materials are used elsewhere in the larger field of phase transitions. Likewise, new methods from the larger field are now being applied to studies of materials.

Part I of the book covers essential topics of free energy, phase diagrams, diffusion, nucleation, and a few classic phase transformations that have been part of the historical core of materials science. In essence, the topics in Part I are the thermodynamics of how atoms prefer to be arranged when brought together at various temperatures, and how the processes of atom movements control the rates and even the structures that are formed during phase transformations. The topics in Part I are largely traditional ones, but formulating the development in terms of statistical mechanics allows more rigor for some topics, and makes it easier to incorporate some deeper concepts from Part II into descriptions of phase transitions in Part III and the online Advanced Topics.

1 Introduction

1.1 What Is a Phase Transition?

A phase transition is an abrupt change in a system that occurs over a small range of a control variable. For thermodynamic phase transitions, typical control variables are the "intensive variables" of temperature, pressure, or magnetic field. Thermodynamic phase transitions in materials and condensed matter, the subject of this book, occur when there is a singularity in the free energy function of the material, or in one of the derivatives of the free energy function.[1] Accompanying a phase transition are changes in some physical properties and structure of the material, and changes in properties or structure are the usual way that a phase transition is discovered. There is a very broad range of systems that can exhibit phase transitions, including atomic nuclei, traffic flow, and social networks. For many systems it is a challenge to find reliable models of the free energy, however, so thermodynamic analyses are not available.

Interacting Components

Our focus is on thermodynamic phase transitions in assemblages of many atoms. How and why do these groups of atoms undergo changes in their structures with temperature and pressure? It is often useful to consider separately the components of the atoms:

- nuclei, which have charges that define the chemical elements,
- nuclear spins and their orientations,
- electrons that occupy states around the nuclei, and
- electron spins, which may have preferred orientations with respect to other spins.

Sometimes a phase transition involves only one of these components. For example, at low temperatures (microkelvin), the weak energy of interaction between spins at different nuclei can lead to a widespread alignment of nuclear spins. An ordered array of aligned nuclear spins may be favored thermodynamically at extremely low temperatures, but thermodynamics favors a disordered nuclear magnetic structure at modest temperatures. Order–disorder phase transitions have spawned several creative methods to understand how an order parameter, energy, and entropy depend on temperature.

---

[1] A brief review of free energy is given in Section 1.6.2.

Sometimes phase transitions involve multiple physical components. Electrons of opposite spin can be coupled together by a wave of nuclear vibration (a phonon). These Cooper pairs can condense into a superconducting state at low temperatures. Perhaps electron charge or spin fluctuations couple the electrons in high-temperature superconductors, although the mechanism is not fully understood today. Much of the fascination with phase transitions such as superconductivity is with the insight they give into the interactions between the electrons and phonons, for example. While these are important subjects for study, they are to some extent diversions from the main topic of phase transitions. Likewise, delving deeper into the first example of nuclear spin alignments at low temperatures reveals that the information about the alignment of one nucleus is carried to a nearby nucleus by the conduction electrons, and these hyperfine interactions between nuclei and electrons are an interesting topic in their own right.[2]

Emergence of Macroscopic Behavior from the Atomistic

In a study of phase transitions, it is easy to lose track of the forest if we focus on the interesting trees. Throughout much of this text, the detailed interactions between the components of matter are replaced with simplifying assumptions that allow for straightforward analysis. Sometimes the essence of the phase transition is captured well with such a simple model. Other times the discrepancies prove interesting in themselves. Perhaps surprisingly, the same mathematical model reappears in explanations of phase transitions involving very different aspects of materials. The ordering of nuclear spins has a natural parallel with the ordering of electron spins in a ferromagnetic material, although the origin of the magnetic moments, their coupling, and the temperature of ordering are completely different. More surprisingly, there is a clear parallel between these spin-ordering problems and chemical ordering in an alloy, where a random distribution of two chemical elements on a crystal lattice evolves into a pattern such as a chessboard with decreasing temperature.

A phase transition is an "emergent phenomenon," meaning that its large-scale features emerge from interactions between numerous individual components. Philosophers classify a phase transition as a type of "weak emergence" because the large-scale properties can be predicted from the interactions of the individual components.[3] How emergence occurs is a topic in itself. When a change in temperature or pressure favors a new phase, it can appear abruptly with macroscopic dimensions, or it can grow continuously from the atomic scale to the macroscopic. Although the atomic-scale processes are statistical, the averaged macroscopic behavior is quite consistent for the same material under the same conditions.

---

[2] At familiar temperatures the nuclear spins are completely disordered, and do not change in a way that affects the thermodynamics of the material. On the other hand, we might expect a coupling between chemical order and magnetic order if the energy scales of their internal interactions are comparable.
[3] "Strong emergence," which cannot be predicted this way, may underlie the origin of consciousness or the soul.

The macroscopic behavior is usually predicted by assuming a material of infinite size, since Avogadro's number can often be approximated adequately as infinity.

1.2 Atoms and Materials

An interaction between atoms is a precondition for a phase transition in a material (and, in fact, for having a material in the first place). Atoms interact in interesting ways when they are brought together. In condensed matter there are liquids of varying density, and numerous types of crystal structures. Magnetic moments form structures of their own, and the electron density can have spatial patterns. In general, chemical bonds are formed when atoms are brought together. The energy of interatomic interactions is dominated by the energy of the electrons, which are usually assumed to adapt instantaneously to the positions of the nuclei. The nuclei, in turn, tend to position themselves to allow the lowest energy of the material, which means that nuclei move around to let the electrons find low-energy states. Once the electronic structure of a material is known, it is often possible to understand many properties of a material, especially its chemical, electronic, magnetic, and optical properties.

For many materials, accurate calculations of electronic structure have arrived. Many reliable quantum mechanical calculations are now routine, and more will be commonplace soon. Electronic structure calculations are an important but large topic that extends beyond the scope of this text. Some aspects of electronic energy are explained in Chapter 6, and other aspects are developed as needed to explain specific phase transformations in Part III. (Fortunately, there are many excellent references on electronic energy and chemical bonding of materials.) Entropy is the other thermodynamic function essential for understanding most phase transitions in materials. Both the energy and entropy of materials

---

Box 1.1   **Microstructure**

Materials are made of atoms, but the structural arrangements of atoms are usually described by a hierarchy of features, each with its own characteristics. Mixtures of crystals and phases, with interfaces between them and defects inside them, are the "microstructure" of a material. One viewpoint is that a complete description of the structural features of microstructure is a full definition of the material. A second viewpoint adds excitations involving electrons, nuclei, or microstructure to the description of a material. The first viewpoint considers only matter, the second viewpoint adds energy to the definition of a material.

Control of microstructure is the means for controlling properties of a material – this is the central paradigm of materials science and engineering (see Fig. 1.1). Designing a microstructure is distinctly different from "Edisonian testing," which is another way to find materials with appropriate properties. Edisonian testing ignores the essence of a material, however, and is not materials science.

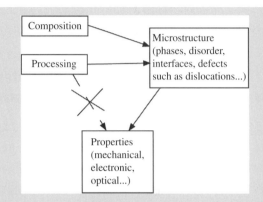

Paradigm of materials science. A direct processing-to-properties relationship, as Edison pursued when finding filament materials for incandescent light bulbs, is not materials science.

depend on the types of atoms and their mutual arrangements, parameterized as "state variables." Careful selections of state variables are critical for developing predictive theories of phase transitions.

There are opportunities to control the states of matter through both thermodynamics and kinetics. Thermodynamic control tends to be the most reliable, at least when the atom motions are fast enough so that equilibrium can be approached in a reasonable time. Thermodynamic control involves selecting the chemical composition, and adjusting the intrinsic variables of temperature, pressure, and external fields. Control of temperature is usually the most accessible way to set the state of equilibrium, and has served us well through the bronze, iron, and silicon ages of humankind. Most of this book is concerned with phase transitions that are driven by temperature.

1.3 Pure Elements

1.3.1 Melting: A Discontinuous Phase Transition

A liquid and a crystal are fundamentally different owing to the symmetry of their atom arrangements. In Chapter 18 when the Landau–Lifshitz criterion for second-order phase transitions is developed, it is proved that melting must involve a discontinuity in the first derivative of the free energy, dF/dT, at the melting temperature. For now, please accept that it is not appropriate to use the same free energy function

$$F(T) = E - TS \qquad (1.1)$$

for both the liquid and the solid phases. Instead, Fig. 1.2 shows two separate curves, $F_s(T)$ and $F_l(T)$, for the solid and liquid phases of a pure element. The curve $F_s(T)$ for the solid lies below that of the liquid at $T = 0$ because the energy of the solid is lower than that of

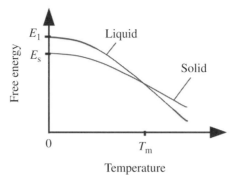

Figure 1.2   Free energy curves of a pure element when its atoms are arranged as a crystalline solid, $F_s(T)$, or as a liquid, $F_l(T)$.

the liquid.[4] As shown on the y-axis, $E_s < E_l$, and at $T = 0$ there is no entropy contribution to the free energy. The free energy of the liquid decreases faster with temperature than that of the solid because $S_s < S_l$.[5] The two curves $F_s(T)$ and $F_l(T)$ cross at the melting temperature, T_m.

In equilibrium, an elemental material follows the solid curve of Fig. 1.2 at low temperatures, and switches to the liquid curve above T_m. At T_m there is a discontinuity in the first derivative of the free energy. A "first-order" phase transition occurs. This is rather catastrophic behavior, with the material changing from all liquid to all solid over an infinitesimal range of temperature across T_m.

Pure elements have well-defined melting temperatures that are set by the equality of the solid and liquid free energies, $F_s(T_m) = F_l(T_m)$. For constant E and S, a consequence is obtained quickly

$$F_s(T_m) = F_l(T_m), \tag{1.2}$$

$$E_s - T_m S_s = E_l - T_m S_l, \tag{1.3}$$

$$S_l - S_s = \frac{E_l - E_s}{T_m} = \frac{L}{T_m}, \tag{1.4}$$

where the latent heat, L, a positive quantity, is defined with the difference in entropy at T_m

$$L \equiv \left[S_l(T_m) - S_s(T_m) \right] T_m. \tag{1.5}$$

The latent heat is absorbed at T_m during melting, and released during solidification.

Equation 1.4 for melting ignores the temperature dependences of E and S, which are important over a range of T. Nevertheless, if E and S in the solid and liquid vary slowly around T_m,[6] the two curves in Fig. 1.2 can be approximated as straight lines. At

---

[4] Perhaps this is intuitive – the atoms in a crystalline solid have optimized their positions with respect to their neighbors, and all atoms are in such optimal positions. The liquid has bond distances and angles that are not at all uniform, meaning that some atoms are in configurations that are less favorable energetically.

[5] Again, perhaps this is intuitive – there are more equivalent ways Ω of arranging the atoms in the liquid than in a crystalline solid, so the entropy, $S = k_B \ln \Omega$, is larger for the liquid.

[6] It is actually the differences, $S_l - S_s$ and $E_l - E_s$, that should vary slowly with T, and this is more plausible.

temperatures very close to T_m, the difference in free energy of the liquid and solid is proportional to the undercooling

$$F_l(T) - F_s(T) = \frac{L(T_m - T)}{T_m} = \frac{L\Delta T}{T_m}, \tag{1.6}$$

with the undercooling defined as $\Delta T \equiv T_m - T$ (see Problem 1.3). The sign is correct in Eq. 1.6 – when $T < T_m$, the $F_s(T)$ is more negative (favorable) than $F_l(T)$.

The thermodynamics of melting (or solidification) illustrates some general truths:

- The low-energy phase is favored at low temperatures.
- The high-entropy phase is favored at high temperatures.
- If the low-energy phase has a lower entropy than the other phase, there will be a phase transition at a finite temperature.

We now take a short digression into kinetics. Although a liquid will eventually solidify at any temperature below T_m, the rate of solidification depends strongly on ΔT. Equation 1.6 is useful for understanding the kinetics of solidification because it relates the undercooling below T_m to the difference in free energy of the liquid and solid. This difference in free energy is available to do work, such as overcoming any potential energy barriers that impede solidification. A larger undercooling makes it more probable for a small region of the liquid to overcome a nucleation barrier and become a solid,[7] and solidification speeds up considerably with undercooling. Equation 1.6 is based on thermodynamics, however. Kinetics requires additional information about the phase transformation. For example, Eq. 1.6 shows that the thermodynamics is symmetrical around T_m for solidification and melting (i.e., $F_l - F_s \propto \Delta T$ for both positive and negative ΔT), but the kinetics is not symmetrical. The enhanced kinetics of solidification with undercooling does not correspond to a more rapid melting with superheating. Melting does not have such a nucleation barrier.

1.3.2 Structural Symmetry and Continuous Phase Transitions

When the high-temperature phase and low-temperature phase have crystal structures with compatible symmetries, the phase transition can be continuous. Figure 1.3 is a map of the phases of cerium metal, charted in a space spanned by T and P, known as a "T–P phase diagram." Upon heating cerium at ambient pressure (0.0001 GPa), it transforms between four different crystalline phases before melting. Our present interest is in the phase transition between two of them, the α- and γ-phases.[8] Curiously, both have the fcc crystal structure, but they differ in volume by about 17% at ambient pressure. Choose a pressure of 1 GPa (to avoid the β-phase), and follow a vertical path in Fig. 1.3b that starts at low temperature, with cerium in the α-phase. Upon heating to a temperature near room

---

[7] Chapter 4 discusses how the nucleation barrier originates from unfavorable surface energy and the large surface-to-volume ratio of small particles.

[8] Solid phases, typically with different crystal structures, are designated by lower case Greek letters. The sequence of letters tends to follow their appearance in a phase diagram, or their sequence of discovery.

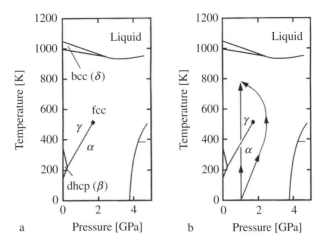

Figure 1.3 (**a**) Low-pressure region of the cerium T–P phase diagram. The solid phases at zero pressure are fcc α, double hcp β, fcc γ, bcc δ. (**b**) Two paths through the phase diagram of part a. One path has a discontinuous expansion of the fcc unit cell, the other a continuous expansion.

temperature, the α-phase undergoes a sudden expansion of its fcc unit cell as it transforms into the γ-phase. The how and why of this phase transition is not fully understood today, but its existence is not in doubt.[9]

Obviously the crystallographic symmetries of two fcc phases are the same, even if they differ in the sizes of their unit cells. With such a special relationship, we might ask if it is possible to go from one to the other in a continuous way, without a discontinuous change in volume. It turns out that this is indeed possible for cerium beyond a pressure of 2 GPa and a temperature of 500 K. The T–P phase diagram of cerium metal has a "critical point," beyond which the two fcc phases are indistinguishable. It is possible to change from a first-order discontinuous phase transition at lower pressures (or temperatures) to a second-order continuous phase transition by taking a path around the critical point in the phase diagram. If we start at $T = 0$ K, $P = 1$ GPa, and go up in temperature and pressure along a curved path that goes to the right of the critical point in Fig. 1.3b, the volume of the fcc unit cell will expand continuously.

It is perhaps better known that the T–P phase diagram of water has a critical point, beyond which the liquid and gas phases become indistinguishable.[10] Evidently there can be a symmetry relationship between atom arrangements in gases and liquids that allows such continuous transitions.

---

[9] Cerium is the first element on the periodic table with an f-electron, and its electronic structure is a challenge to understand today. Its fcc–fcc transformation has attracted considerable attention, in part because of suggestions that its outer electrons become delocalized when the γ-phase collapses into the high-density α-phase.

[10] Carbon dioxide also has a well-known critical point of its liquid and gas phases. When pushed beyond the critical point, "supercritical" carbon dioxide is an effective agent for dry cleaning clothes.

1.4 Alloys: Unmixing and Ordering

An "alloy," which is a combination of two or more chemical elements, brings additional degrees of freedom to arrange the atoms, and more possibilities for energies and entropies. For millennia there has been interest in using temperature to alter the states of alloys. Today the equilibrium states of alloys are mapped on a chart of temperature versus composition called a "T–c phase diagram." Transitions between phases occur at specific temperatures and compositions, and knowing the boundaries of these phase transitions is of both fundamental and practical interest. Understanding T–c phase diagrams is an important goal for this text.

Figure 1.4 shows two basic types of transitions from a high-temperature phase (at top) to a low-temperature phase or phases (at bottom). For clarity, all three atom configurations in Fig. 1.4 are based on an underlying square lattice. The problem is reduced to one of atom positions on an "Ising lattice," where an Ising lattice is a graph of nodes connected by first-neighbor links. Each node (or site) is occupied by one atom of either species, with a probability dependent on the concentration. The links between nodes are chemical bonds, which have energies that depend on the specific atoms in each pair.

The high-temperature phase in Fig. 1.4 is a random (or nearly random) distribution of atoms called a "solid solution." The two species of atoms are mixed together, or dissolved

Solid solution, disordered

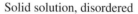

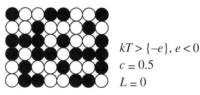

$$kT > \{-e\}, \, e < 0$$
$$c = 0.5$$
$$L = 0$$

High T

Unmixed (two phases)   Ordered
$e_{AA} + e_{BB} < 2e_{AB}$ $2e_{AB} < e_{AA} + e_{BB}$

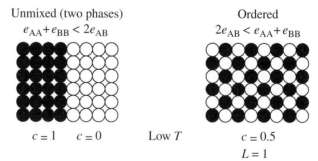

$c = 1$ $c = 0$ Low T $c = 0.5$
$L = 1$

Figure 1.4   Three alloy states. At top is the expected high-temperature phase, a disordered solid solution. This phase is unstable at low temperatures when chemical bond energies overpower the disorder caused by thermal energy. Two low-temperature states are possible – an unmixed state of two phases (left), or an ordered compound (right). Chemical preferences usually select only one of these states. (L is the long-range order parameter, defined later.)

in each other. Some A–B alloys mix better than others, much like "liquid solutions" of ethanol and water, as opposed to unmixing systems such as oil and water. With many equivalent configurations, a solid solution phase has a high configurational entropy, S_{conf}, favoring it at high temperatures.

At low temperatures, two states of low S_{conf} are possible:

- The lower left shows an unmixed alloy, where the B-atoms separate from the A-atoms. Unmixing is expected if atoms prefer their same species as neighbors more than they prefer unlike atoms as neighbors. At lower solute concentrations, unmixing may produce small solute-rich regions called "precipitates."[11]
- The lower right shows an ordered structure. Such ordering is expected if unlike atoms are preferred as neighbors because this structure maximizes the numbers of unlike neighbor pairs.

These unmixing and ordering transitions are usually more complicated than the extremes depicted in Fig. 1.4. At intermediate temperatures there is often partial unmixing, or partial ordering.[12] It is also possible, for example, for a solid solution to be in equilibrium with a partially ordered phase. Nevertheless, Fig. 1.4 depicts the two essential phase transitions that occur with temperature.

It turns out that a simplified two-dimensional square lattice with fixed first-neighbor interactions is the only ordering problem with an exact solution for the free energy through the critical temperature. The solution was reported by Onsager in 1944 [1], and is an impressive achievement, well beyond the scope of this text. Unfortunately, the Onsager solution seems to be a special case, and efforts to extend it beyond the square lattice have not been successful. The difficulty is that the problem is a multibody one, where the energies of all atoms depend on the configurations of all others. Although the free energy of the fully ordered state and the fully disordered solid solutions are reasonably straightforward to understand, the intermediate states near the critical temperature are not. As disorder begins to develop in an ordered structure, the neighborhoods of individual atoms differ. The energy of each atom depends on its bonds to the surrounding atoms, but the types of these atoms change with disorder. Sadly, it is not enough to consider only an average neighborhood about a typical atom. Likewise, the neighbors of the neighbors (some of which are the original atom and its neighbors) influence the types of atoms in neighboring positions.

The interaction of an atom with its neighbors, whose identities depend on how they interact with their own neighbors, leads to cooperative behavior in a phase transition. In essence, with increasing temperature the alloy tends to hold its order locally, but the quality of order over longer distances deteriorates more rapidly. At the critical temperature the long-range order is lost, but the short-range order is not, and persists to higher temperatures. Short-range order is often best treated with its own, independent order parameters. A consistent treatment of short-, intermediate-, and long-range order

---

[11] This is much like a precipitation reaction in a liquid solution, but of course the precipitates in solids cannot fall to the bottom of a test tube under gravity.

[12] An "order parameter" describes how the quality of order changes with temperature.

parameters is a daunting task, however. Approximation methods have been developed, and their strengths and weaknesses are important topics. The more abstract methods of Chapter 27 (online) treat cooperative behavior more properly, but atoms may move too sluggishly to achieve equilibrium across the large spatial distances that become relevant near a critical temperature.

1.5 What Is a Phase Transformation?

We use the words "phase transformation" to mean the life cycle of a "phase transition," but usage in the literature is less consistent. A phase transition originates with the singular character of the free energy, but the free energy does not fully control a phase transformation. The effects of kinetics and mechanisms of atom movements are important for understanding phase transformations, often giving them unique character. For studies of phase transitions, on the other hand, kinetic phenomena are either unimportant, or are considered impediments to understanding. In materials there are many types of phase transformations, and major ones are discussed in Part III of this book. Section 1.5 presents some concepts important for phase transformations.

1.5.1 Diffusional and Diffusionless

One distinction between phase transformations in solids is if diffusive atom motions are involved as in Fig. 1.5a, or if there is a change in structure by a cooperative displacement of all atoms, such as by the shear shown in Fig. 1.5b. Most chemical phase transformations in

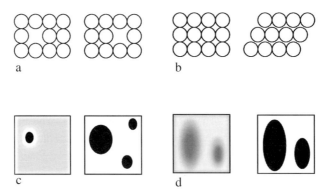

Figure 1.5 Illustrations of four processes in time. In each pair the configuration before is to the left, and after is to the right. (**a**) Atom motion by diffusion, in this case by one atom exchanging sites with a vacancy. The vacancy remains to jump again, likely in a different direction. (**b**) Atom motion by cooperative shear, changing a square crystal into a rhombohedral one. (**c**) Phase transformation by nucleation and growth, where a new phase forms discontinuously from the parent phase. (**d**) Phase transformation by continuous unmixing (ordering is possible too), where a gradual modulation grows to high amplitude. This is typical of spinodal decomposition.

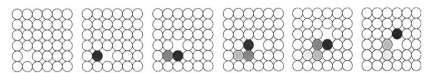

Figure 1.6 Series of vacancy jumps in a square lattice, from left to right. All atoms are the same species, but the darkness of shading indicates how recently the atom has interchanged sites with the vacancy.

solids involve diffusive motions of atoms. For example, the chemical unmixing or ordering transformations of Fig. 1.4 require diffusion because atoms must move individually over modest distances to optimize their local chemical environments. In contrast, "martensitic transformations" described in Chapter 19 occur by cooperative shear motions of all atoms.

Vacancies

Two or more atoms cannot occupy the same site on a crystal lattice.[13] If every site is occupied by an atom, however, it is difficult to understand how atoms can move around. Adjacent atoms do not move by exchanging positions because the energy barrier is too high. The diffusion of atoms in crystals usually occurs by a "vacancy mechanism," as illustrated in Figs. 1.5a and 1.6. A vacancy can exchange sites with a neighboring atom, usually a first-nearest neighbor, of which there are four in these figures. It is important to understand that if a neighboring atom jumps into a vacancy, the vacancy is not annihilated, unless it moves to a surface. Within the bulk of a crystal, the vacancy moves from site to site, each time exchanging sites with a neighboring atom. The vacancy mechanism is the usual way that atoms move in metals, and it is a common mechanism for covalently and ionically bonded materials, too. The concentrations of vacancies and the activation energies for moving them are important parameters in atomistic theories of diffusion.

Interstitials

Interstitial atoms are usually small atoms that fit in the gaps or "interstices" between the larger atoms in a crystal. For example, small carbon atoms fit into interstices between iron atoms on bcc or fcc crystal structures. Interstitial atoms can move rapidly from site to site, in part because the interstitial sites are mostly vacant, but also because the energy barriers are usually low. Interstitial hydrogen atoms move particularly rapidly in metals, for example.

Sometimes an atom can be an interstitial species in its own crystal lattice. The diamond-cubic lattice of silicon is fairly open, and silicon atoms can move between crystal sites and interstitial sites. Diffusive atom movements by such an "interstitialcy mechanism" occur in a number of materials. When an interstitialcy mechanism is active, diffusion is often "anomalously" fast.

---

[13] There are cases of "dumbbells" of two atoms on a site, but these are rare.

1.5.2 Continuous and Discontinuous Transformations

On the mesoscale of the microstructure, diffusional phase transformations can proceed in two essential ways. Figure 1.5c shows the nucleation of a new phase (on left), where embryos appear and then grow in size (Chapter 14). As they grow, the regions of new phase have essentially the same chemical composition and the same crystal structure at all sizes. Figure 1.5d depicts an alternative, continuous process where the unmixing occurs gradually, and differences in composition evolve continuously. This occurs in "spinodal decomposition," for example (Chapter 16). Phase transformations that occur by nucleation and growth seem much more common than those that occur in a continuous process, in part because they are easier to observe.

Differences between continuous and discontinuous unmixing transformations are seen in the experimental images of Fig. 1.7, obtained by position-sensitive atom probe microscopy. Each dot marks the detection of one atom of a specific element. The dots in Fig. 1.7a are cobalt atoms in a material originally prepared as a solid solution of Cu–1%Co. Cobalt is largely insoluble in Cu at low temperatures, and the unmixing transformation is seen in the time sequence of images of Fig. 1.7a. Even in the earliest stages there is clustering of cobalt atoms into distinct zones – cobalt unmixes from copper by a discontinuous "nucleation and growth" transformation. There is an energy associated with the surface around the cobalt atoms, and the minimization of the surface energy causes the cobalt-rich

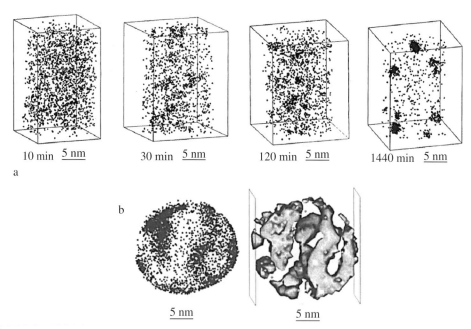

10 min 5 nm   30 min 5 nm   120 min 5 nm   1440 min 5 nm

a

b

5 nm   5 nm

Figure 1.7   Results from position-sensitive atom probe microscopy measurements on unmixing in two alloys. Both alloys were prepared as random solid solutions by rapidly cooling from a high temperature, and heated at low temperatures where the atoms had some diffusive mobility. (Each image is from a different part of the material.) (**a**) Unmixing of Co atoms from a Cu-rich phase during heating at 723 K for times as indicated. (**b**) Unmixing of Cr atoms in an alloy of Fe–Cr–Ni after heating at 673 K for 10 000 hours. Image at right was constructed from the three-dimensional data by sampling the atoms in neighboring voxels (i.e., 3D cubic pixels). Reprinted, with permission, from [2].

precipitates to have approximately spherical shapes. On the other hand, Fig. 1.7b shows a continuous unmixing process called "spinodal decomposition." Chromium atoms have partially separated from the surrounding iron-rich phase in a stainless steel. Notice how the chromium atoms are in diffuse zones without sharp boundaries. The image at right in Fig. 1.7b was obtained by defining a local composition of 30% to help identify the shapes of these zones. The zones are sinuous with a periodicity of 10 nm or so. They are often described as wavelike composition modulations through the alloy. In Chapter 16, the diffuse boundaries of the chromium-rich zones will be characterized by an energy that is not a surface energy, but an energy proportional to the square of the composition gradient.

1.6 Brief Review of Thermodynamics and Kinetics

1.6.1 Partition Function

Throughout the text a physical system is modeled with "state variables" that parameterize essential features of the internal structure in a material. These state variables (simple ones are a chemical composition or an order parameter) are selected so the energies of the system $\{\varepsilon_i\}$ can be calculated for each state i. With these energies, a partition function, Z, is constructed as

$$Z \equiv \sum_i e^{-\varepsilon_i/k_B T}. \tag{1.7}$$

For a particular temperature, the partition function depends on how probability is distributed ("partitioned") between states of different energy. All states of the same energy make the same contribution to the partition function, and if there are more states of a particular energy, this energy carries more total weight in determining thermodynamic properties. Importantly, the partition function contains all thermodynamic information about the state variables of our model system, and therefore determines all thermodynamic properties of the model that gives $\{\varepsilon_i\}$.

Since each state with the same energy ε_j has the same Boltzmann factor, $e^{-\varepsilon_j/k_B T}$, it is often convenient to rewrite the partition function as

$$Z = \sum_j \Omega_j e^{-\varepsilon_j/k_B T}, \tag{1.8}$$

where Ω_j is the number of states having the energy ε_j. For a system with a fixed energy ε_j, the probability of each state with this energy is $p_j = 1/\Omega_j$ and the entropy is[14]

$$S_j = k_B \ln \Omega_j. \tag{1.9}$$

---

[14] This probability is an average over numerous equivalent systems comprising a "microcanonical ensemble." Equation 1.9 is consistent with the expression for the total entropy,

$$S_j = -k_B \sum_{j'}^{\Omega_j} p_{j'} \ln p_{j'}, \tag{1.10}$$

when each $p_{j'}$ is equal to $1/\Omega_j$. Equal probability for states of equal energy is a central tenet of statistical mechanics, and is expected for a system in thermodynamic equilibrium.

We will usually consider systems where the number of atoms is conserved, and our physical system is in thermal contact with a large heat bath of temperature T. Consider the probability of our system being in state i with energy ε_i, or in state i' with energy $\varepsilon_{i'}$. The entropy of the system plus heat bath is *dominated by the entropy of the heat bath*. The entropy of the heat bath, designated by the states j, changes when energy is moved into our small system of interest

$$S_j(E - \varepsilon_i) = S(E) - \varepsilon_i \frac{\partial S}{\partial E}, \tag{1.11}$$

$$S_{j'}(E - \varepsilon_{i'}) = S(E) - \varepsilon_{i'} \frac{\partial S}{\partial E}. \tag{1.12}$$

With the entropy expression Eq. 1.9, recognizing $\partial S/\partial E = 1/T$, and using the fact that the probability of a state p_i is proportional to the number Ω_j for the dominating heat bath

$$\frac{p_i}{p_{i'}} = \frac{\Omega_j}{\Omega_{j'}}, \tag{1.13}$$

$$\frac{p_i}{p_{i'}} = \frac{e^{S_j/k_B}}{e^{S_{j'}/k_B}}, \tag{1.14}$$

$$\frac{p_i}{p_{i'}} = \frac{e^{S/k_B}}{e^{S/k_B}} \frac{e^{-\varepsilon_i/k_B T}}{e^{-\varepsilon_{i'}/k_B T}} = \frac{e^{-\varepsilon_i/k_B T}}{e^{-\varepsilon_{i'}/k_B T}}. \tag{1.15}$$

The Boltzmann factor for an individual state i can be normalized by the sum of all other Boltzmann factors (Eq. 1.7) to give the probability of the state i

$$p_i = \frac{e^{-\varepsilon_i/k_B T}}{Z}. \tag{1.16}$$

With the thermodynamic probabilities of all states, it is straightforward to calculate thermodynamic averages, such as for the energy

$$\langle \varepsilon \rangle = \sum_i p_i \varepsilon_i, \tag{1.17}$$

$$\langle \varepsilon \rangle = k_B T^2 \frac{\partial \ln Z}{\partial T}. \tag{1.18}$$

Finally, the Helmholtz free energy, $F = E - TS$, can be obtained from the partition function as

$$F = -k_B T \ln Z. \tag{1.19}$$

Box 1.2 **Helmholtz or Gibbs?**

An expression for $E(S, V, N)$ is the fundamental equation of thermodynamics. Minimizing E with respect to S is a way to find conditions of thermal equilibrium, but the role of temperature appears through the functional relationship between E and S, i.e., $T = (\partial E/\partial S)_{V,N}$. It is usually easier to work with T directly, and T is a natural variable for the Helmholtz or Gibbs free energies, $F(T, V, N)$ or $G(T, P, N)$. In general, the Gibbs free energy is the most convenient thermodynamic potential if we have control over T and P. If G does not depend significantly on P, however, it is simpler to ignore effects of pressure and focus on the role of T by using the

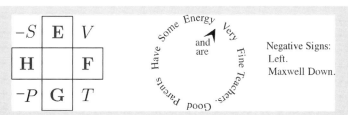

Figure.1.8 Thermodynamic square with mnemonics.

Helmholtz free energy F. We will usually use F because, as explained in Chapter 8, for most solids a change of 100 K in T causes approximately the same change in G as does 5000 atmospheres in P, which is less common in practice.

   The thermodynamic square of Fig. 1.8 can guide the selection of a thermodynamic potential, and it has other uses.

- The thermodynamic potentials are in the four boxes, with their natural variables to their sides.
- Computing total differentials such as dF of Eq. 1.22 uses differentials of its natural variables (with positive signs), and quantities in the opposite corner, e.g., $dF = -SdT - PdV$. The term μdN can be added later.
- Maxwell relations are formed by going down the sides with a $-$ sign, $(\partial S/\partial P)_T = -(\partial V/\partial T)_P$, or across the rows with a $+$ sign, $(\partial P/\partial T)_V = (\partial S/\partial V)_T$.

1.6.2 Free Energy

The historical, but still correct, meaning of "free energy" is the amount of energy available in a system to do external work such as pushing a piston in a steam engine. The Helmholtz free energy, $F \equiv E - TS$, is the internal energy minus the entropic contribution that cannot perform work. When two parts of a system are brought together and interact at the same temperature, one part will do work on the other until the total free energy is a minimum. Minimizing free energy functions will be our primary method for finding states of thermodynamic equilibrium.

   The total differential of the Helmholtz free energy, $F = E - TS$, is

$$dF = d(E - TS) = dE - TdS - SdT. \tag{1.20}$$

The thermodynamic identity[15] is

$$dE = \mu dN - PdV + TdS. \tag{1.21}$$

Substituting Eq. 1.21 into 1.20

$$dF = \mu dN - PdV - SdT. \tag{1.22}$$

---

[15] Obtained by differentiating the fundamental relationship for $E(S, V, N)$, and recognizing (or defining) $(\partial E/\partial S)_{V,N} = T$, $(\partial E/\partial V)_{S,N} = -P$, $(\partial E/\partial N)_{S,V} = \mu$.

Here μ, P, S are equilibrium values. The variations of $F(N, V, T)$ are expressed with variations of its independent variables N, V, and T

$$dF = \left(\frac{\partial F}{\partial N}\right)_{V,T} dN + \left(\frac{\partial F}{\partial V}\right)_{T,N} dV + \left(\frac{\partial F}{\partial T}\right)_{N,V} dT. \tag{1.23}$$

A comparison of Eqs. 1.22 and 1.23 provides these valuable relationships

$$\mu = \left(\frac{\partial F}{\partial N}\right)_{V,T}, \quad P = -\left(\frac{\partial F}{\partial V}\right)_{T,N}, \quad S = -\left(\frac{\partial F}{\partial T}\right)_{N,V}. \tag{1.24}$$

We will often use the first expression to treat the chemical potential as the free energy per atom;[16] the second is used in Chapter 8 to study pressure, and the third expression is handy for obtaining the entropy from Eq. 1.19.

1.6.3 Kinetic Master Equation

The "kinetic master equation" is the most general kinetic equation. The name is grand, and the equation is worthy of respect, but it is quite natural to understand and often easy to use. The ideas of the kinetic master equation are illustrated with Fig. 1.9. Figure 1.9a shows a row of bins (or containers) for entities, in our case atoms. Beneath them are written bin numbers $\ldots, n - 1, n, n + 1, \ldots$ Their contents are listed above as $\ldots, N_{n-1}, N_n, N_{n+1}, \ldots$ For kinetics, we allow the jumping of atoms between adjacent bins. Two types of jumps, left and right, are depicted between bins n and $n + 1$, with definitions in Fig. 1.9b.

The kinetic master equation states that if more atoms enter a bin than leave it in a time Δt, the bin will gain ΔN_n atoms as given by Eq. 1.25. The outflow of atoms from the nth bin is the product of the number of atoms in the bin at the time, and the probability that an individual atom leaves the bin in the time interval Δt. Likewise, the inflow of atoms into the bin n is the product of the number of atoms in the surrounding bins times the probability

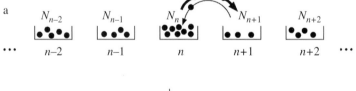

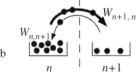

Figure 1.9 (**a**) A row of bins for atoms, numbered at bottom, and with contents at top. (**b**) Flow of atoms between two adjacent bins, with definitions of rates.

---

[16] This is true if F scales with N, but Eq. 1.24 must be used more carefully if surface energy is important, for example.

that each atom goes from the bin n' to n. Using the definitions in Fig. 1.9, the kinetic master equation is

$$\Delta N_n(\Delta t) = \sum_{n'} \Big\{ + W_{n,n'}(\Delta t)\, N_{n'}(t) - W_{n',n}(\Delta t)\, N_n(t) \Big\}, \tag{1.25}$$

where the $W_{n,n'}(\Delta t)$ is proportional to Δt when Δt is small.

Problems

1.1 Explain the basic paradigm of materials science.

1.2 Suppose the internal energies of the solid and liquid phases vary with temperature linearly as $E(T) = E^0 + T\, dE/dT$, where E^0 is the internal energy at $T = 0$, and dE/dT is a constant for a particular phase.
(**a**) How does this modify Eq. 1.4, i.e., $L = E_l - E_s$?
(**b**) Show that Eq. 1.4 is recovered if dE/dT is equal for the solid and liquid phases.

1.3 (**a**) Derive Eq. 1.6 in the text, i.e., $F_l(T) - F_s(T) = \frac{L(T_m - T)}{T_m}$.
(**b**) Assume that the driving potential for a phase transition is the difference in free energy between the two phases. Explain why the driving potential for a first-order phase transition would be proportional to the difference in temperature from the critical temperature, and explain how the difference in entropy of the two phases affects the driving potential.

1.4 In one paragraph, do your best to explain why it is difficult to parameterize the state of order in an alloy with partial order. (Credit will be given generously, but please think about this problem before you write.)

1.5 Suppose an alloy of composition 0.5 undergoes chemical unmixing by a continuous process, and develops concentration modulations of the form

$$c(x) = 0.5 + 0.5\,\sin(kx), \tag{1.26}$$

where the wavelength of the modulation is λ, and $k = 2\pi/\lambda$. Assume that λ is much smaller than the size of the alloy.
(**a**) If the local free energy depends only on composition, i.e., $F = F(c)$, does the total free energy depend on λ?
(**b**) Suppose λ becomes quite short, $\lambda = 2a$, where a is the interatomic spacing. Is the alloy better described as an ordered structure or a chemically unmixed structure? (*Hint*: Draw it.) If the result is an ordered alloy, why is this inconsistent with an unmixing tendency?
(**c**) Suggest a correction to the assumption that $F = F(c)$.

1.6 Draw several frames of a cartoon showing how a vacancy mechanism with first-neighbor jumps can cause neighboring atoms 1 and 2 on a simple cubic lattice to change from an orientation with 1 on the left of 2 to an orientation with 2 on the left of 1.

1.7 (**a**) Is it possible for chemical ordering to occur by a shear transformation (i.e., can the order parameter change during a simple shear transformation)?

(**b**) Does the configurational entropy change during a shear transformation?

1.8 Prove Eq. 1.19.

Temperature–Composition Phase Diagrams

Chapter 2 explains the concepts behind T–c phase diagrams, which are maps of equilibrium alloy phases within a space spanned by temperature T and chemical composition c.[1] When we know T and c, a T–c phase diagram is valuable for finding the equilibrium fraction of each phase and its chemical composition.

Our emphasis is on deriving T–c phase diagrams by minimizing the total free energy, given free energy functions $F(c, T)$ of different phases. The amount of each phase is constrained by solute conservation, giving the straightforward "lever rule." The minimization of the total free energy leads to the more subtle "common tangent construction," which selects the equilibrium phases at T from the $F(c)$ curves of the different phases. For binary alloys, the shapes of $F(c)$ curves and their dependence on temperature are used to deduce eutectic, peritectic, and continuous solid solubility phase diagrams. Some features of ternary alloy phase diagrams are also discussed.

If atoms are confined to sites on a crystal lattice, free energy functions can be calculated with a minimum set of assumptions about the energies of different atomic configurations. Because the key features of phase diagrams can be obtained with general types of interactions between atoms, systems with very different chemical bonds, e.g., both oil in water and iron in copper, can show similar phase transitions. In these unmixing cases, the individual atoms or molecules prefer the same chemical species as neighbors (see Fig. 1.4, lower left), and an unmixing phase diagram is obtained. The opposite case of a preference for unlike neighbors leads to a chemical ordering phase diagram, and additional information about how an order parameter varies with temperature. These generalizations of chemical interactions are useful for identifying phenomena common to many phase transitions. Behavior that emerges on a large length scale can be missed if there is too much emphasis on the details of chemical bonding and atom vibrations. Nevertheless, we must be wise enough to know the predictive power available at different levels of approximation.

2.1 Intuition and Expectations about Alloy Thermodynamics

2.1.1 Overview of the Approach

Temperature promotes disorder in a material, favoring higher-entropy phases such as liquids, but chemical bond energy favors crystals at low temperatures. Melting (or

---

[1] Pressure is assumed constant, or negligible.

solidification) of an elemental material is then understood as occurring at the temperature T_m where the free energies of the solid and liquid are equal. This is similar for a binary alloy of two chemical elements, but the binary system has an additional degree of freedom – the equilibrium solid and liquid phases usually differ in their chemical compositions.

The free energy of an alloy depends on the spatial arrangement of its atoms.[2] In what follows, minimalist models are used to calculate thermodynamic functions and predict the equilibrium phases in alloys at different combinations of T and chemical composition c. Such models predict the main features of T–c phase diagrams for binary A–B alloys (A and B are the two chemical species of atoms). Convenience for calculation is a virtue and a priority for this chapter, but we must be aware of two types of risks:

1. A good parameterization of atom configurations may require more detail than is possible with a simple model.
2. Even if the parameterizations of atom positions are excellent, calculating accurate energies or entropies from these parameterizations requires more sophisticated methods.

Part II of this book addresses these two issues in more detail. Nevertheless, the simple models developed in this chapter are often useful semiquantitatively, and serve as benchmarks for assessing the value added by more sophisticated treatments.

For alloy thermodynamics we need expressions for E and S, for which the Helmholtz free energy is

$$F(c, T) = E(c, T) - TS(c, T). \tag{2.1}$$

In a first approximation, both the internal energy $E(c)$ and entropy $S(c)$ are assumed independent of temperature, and depend only on the types of atoms and their nearest neighbors. For solid solution phases, it is also convenient to assume statistical randomness of atom occupancies of the sites on the lattice. With these assumptions, the two general risks listed above bring these more specific concerns:

1. Chapter 7 goes beyond the assumed random environment of an *average* atom, i.e., the "point approximation." One approach is to use probabilities of local "clusters" of atoms, working up systematically from chemical composition (point), to numbers of atom pairs (pair), then tetrahedra, for example. Usually the energy and entropy of materials originate from short-range interatomic interactions, so this approach has had much success in practice.
2. How accurately can we account for the energy and entropy of the material? For example, the electronic energy, which provides the thermodynamic E, is not rigorously constant for a nearest-neighbor pair of atoms when there are changes in state variables. At first, we also assume that the entropy of an alloy originates entirely from the number of ways that its atoms can be configured on the sites of a crystal. This gives the "configurational entropy," $S_{conf}(c)$. Conveniently, $S_{conf}(c)$ depends only on c and not on T. We first ignore the entropy caused by the thermal vibrations of atoms, $S_{vib}(c, T)$, the entropy

---

[2] This restates the paradigm that the atom arrangements in a material determine its properties, including thermodynamic properties.

from disorder in magnetic spins, $S_{\mathrm{mag}}(c, T)$, and the entropy from disorder in electron state occupancies, $S_{\mathrm{el}}(c, T)$. At modest temperatures, $S_{\mathrm{vib}} > S_{\mathrm{conf}}$, however, so these contributions are assessed later.

Two important T–c phase diagrams, unmixing and ordering, can be obtained readily by minimizing one free energy function on a single lattice (Sections 2.7–2.9). More generally, there is competition between different phases with different crystal structures, so each phase ξ has a different $F_\xi(c, T)$. Separate $F_\xi(c, T)$ curves are used to obtain three other types of phase diagrams: continuous solid solubility, eutectic, and peritectic (Sections 2.4 and 2.5). All these five essential T–c phase diagrams, summarized in Box 2.10, are obtained by minimizing the total free energy with the "common tangent construction," explained in Section 2.3.

2.1.2 Free Energies of Alloy Phases

Alloys transform to different phases at different temperatures. From high to low temperature, here are some typical phases that are found for an alloy system ("system" meaning a range of compositions for specified chemical elements), with considerations of the Helmholtz free energy, $F(c) = E(c) - TS(c)$, for each phase:

- The phase of maximum entropy dominates at the highest temperatures. For most materials this is a gas of isolated atoms, but at lower temperatures most alloys form a liquid phase with continuous solubility of A- and B-atoms.
- At low-to-intermediate temperatures, the equilibrium phases and their chemical compositions depend in detail on the free energy versus composition curves $F_\xi(c)$ for each phase ξ at the temperature of interest. Usually there are chemical unmixings, and the different chemical compositions frequently prefer different crystal structures. The chemical unmixings may not require precise stoichiometries (e.g., a precise composition of A_2B_3) because some spread of compositions may provide entropy to favor off-stoichiometric compositions (e.g., a composition $A_{2-\delta}B_{3+\delta}$, where $\delta \neq 0$).
- At the lowest temperatures, the equilibrium state for a general chemical composition is a combination of crystalline phases with precise stoichiometries and a high degree of long-range order.[3]

Box 2.1	Configurational and Dynamical Sources of Entropy

We take the statistical mechanics approach to entropy, and count the configurations of a system with equivalent macrostates, giving the Ω for Eq. 1.9. For counting:

- Procedures to count atom configurations are independent of temperature.
- Procedures to count states explored with thermal excitations, such as atom vibrations or electronic excitations, depend on temperature.

---

[3] At low temperatures, however, atoms usually do not have enough mobility to form these precise structures, so some chemical disorder is often observed in practice.

This separation of configurational entropy from entropies of excitations works well for most temperatures of solids, where timescales for electrons, spins, and vibrations are much shorter than the timescales for atom diffusion (diffusion is required to change atom configurations).

The reader should know that at a temperature of 1000 K or so, S_{vib} is typically an order-of-magnitude larger than S_{conf} (compare the maximum entropy in Fig. 2.16a to that in Fig. 12.4). It is possible for S_{vib} to be similar for different phases, however, so the difference in entropy caused by vibrations is not always dominant in a phase transition. Nevertheless, S_{vib} usually does depend on atom configurations, and this is developed for independent harmonic oscillations in Section 7.6. At high temperatures, however, the statistical mechanics of harmonic vibrations needs significant modifications (Chapter 26, online).

2.1.3 Thermodynamics of Solutions: Qualitative

Assuming good atomic mixing of two chemical elements in an alloy, Fig. 2.1 shows a typical curve of Helmholtz free energy, $F = E - TS$, versus composition c. Some salient points are:

- There are different bonding energies for the pure elements A and B, so different values of F at $c = 0$ and $c = 1$. The pure elements A and B have different melting temperatures, for example.
- When c is near 0, there are only a few B-atoms in the alloy, and likewise few A-atoms when c is near 1. In such dilute alloys, we expect the solute atoms to be isolated from each other, so each solute atom is in the same chemical environment – it is fully surrounded by solvent atoms. Each solute atom therefore has the same bonding energy. Likewise, the number of solvent atoms with solute neighbors increases linearly with c, and the overall bonding energy increases linearly with the solute concentration.[4]
- From the entropy expression $S = k_B \ln \Omega$, we expect the entropy to be highly sensitive to composition when c is near 0 or 1 because in a nearly pure material there are numerous

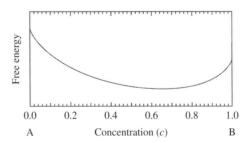

Figure 2.1   Typical $F(c)$ curve for a solution, with features described in the text.

---

[4] In the case of nonlocal "metallic bonds," described in Section 6.4.1, the donation of an electron to the band structure also affects the energy linearly with solute concentration in the dilute limit.

possible sites for the few solute atoms. There is always a downward slope of $F(c)$ near $c = 0$ and $c = 1$ owing to $S_{conf}(c)$.[5]

• Somewhere in the middle of the composition range, $F(c)$ should no longer change with composition (i.e., $dF/dc = 0$). One reason is that $F(c)$ must turn around to accommodate its opposite slopes near $c = 0$ and $c = 1$. Another reason is that at $c = 0.5$ there is a maximum number of unlike pairs, and this number decreases smoothly away from the equiatomic composition.

2.1.4 Hume-Rothery rules

Hume-Rothery rules give conditions for solid solubility and for compound formation. They came from efforts by William Hume-Rothery to reconcile the relative importance of concepts about local covalent bonding from chemistry, and nonlocal free electrons from physics. These rules are intuitively reasonable, and were used as predictive tools for metallic alloys in the mid- to late-twentieth century [3–5]. They use three properties of individual atoms: (1) electronegativity, (2) metallic radius, and (3) electron-to-atom ratio, with the first two factors being the most successful. Although atoms retain some of their identity when assembled into a solid, the outer bonding electrons have different energies and spatial distributions in different crystals. We therefore cannot expect too much accuracy from rules based on atomic properties when predicting alloy phases. Nevertheless, in more detail these atomic properties are:

• Electronegativity. This is a measure of the tendency for electron transfer to an atom during chemical bonding, which provides information on the strength of the bond (see Section 6.2.4).[6]
• Metallic radius. Inserting a larger solute atom into a crystal of smaller atoms causes stresses in the host crystal.[7]
• Electron-to-atom ratio. This accounts for nonlocal Fermi surface effects in metals (Section 6.5.2). The e/a ratio is expected to be a useful parameter when the rigid band model (Section 6.5.3) is appropriate.

The Hume-Rothery rule for solid solutions is reasonably useful. It sets conditions for having "extended solid solubility," meaning something like 5% solubility at 1000 °C:

• The Pauling electronegativity of the elements shall not differ by more than 0.4.
• The difference in metallic radius of the atoms shall not differ by more than 15%.

This rule for solid solubility can be used conveniently by plotting the elements of interest on a map of electronegativity versus metallic radius [5], as shown in Fig. 2.2 for the

---

[5] Equation 2.31 shows that, for small c, the configurational entropy per atom is $S_{conf}(c) \simeq -k_B c \ln c$, which dominates over the proportionality to c that is followed by the energy (and vibrational entropy).

[6] In metals this is only a part of the picture, because conduction electrons are highly effective in screening the charge disturbances that occur with electron transfer, but changes in the screening may scale with differences in electronegativity.

[7] Distortions of electron wavefunctions can raise their kinetic energy, so Hume-Rothery and others treated this effect as an elastic strain energy.

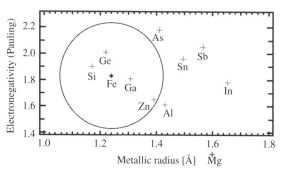

Figure 2.2 "Darken–Gurry plot" for extended solid solubility of some main group elements in iron.

solubility of some main group elements in iron. Instead of a rectangle of fixed width in electronegativity and metallic radius, Darken and Gurry suggested drawing an ellipse with an extent of 0.4 in electronegativity and 15% in radius. The elements within it satisfy the Hume-Rothery rule for good solid solubility, and tend to have higher solubilities than the elements outside it. For example, magnesium has a notoriously low solubility in iron, consistent with Fig. 2.2.

For its ease of use, the Hume-Rothery rule for solid solubility is a best buy.[8] Of course, phase diagrams of most binary alloys are now known, and have much more reliable information about solid solubility, including its temperature dependence. The real value of the Hume-Rothery rules is that they give us intuition about the importance of electronegativity and metallic radius in the thermodynamics of alloying.

Box 2.2   Other Rules of Thumb and Empirical Trends

A novel idea for understanding compound formation was developed by David Pettifor [6], who boldly proposed a new classification scheme for all elements in the periodic table. The ordering of the elements begins with He at 1, and the other inert gases through 6 at Rn, jumping to the alkali elements with 7 at Fr through Li at 12. In general, the sequence runs up and down the columns of the periodic table, with isoelectronic elements placed near each other in the sequence. Approximately, the more electronegative elements (F at 102) are higher in this sequence of "Mendeleev numbers," $\mathcal{M}$. Elements with similar chemical properties have similar values of $\mathcal{M}$.

The interest in this scheme is from the "structure maps" that it generates naturally. If $\mathcal{M}$ is plotted on the x-axis, and $\mathcal{M}$ is also plotted on the y-axis, all binary combinations of elements are available as points on an x–y grid. There is a tendency for specific types of compounds, such as the NaCl structure for an equiatomic AB alloy, to be grouped in specific regions on this grid. There was hope that these structure maps would assist efforts to design alloys by chemical modifications, but this has proved only partly successful. Nevertheless, the structure maps for AB and A_3B alloys do offer an interesting framework to consider chemical effects on crystal structures and bonding. They also show barren regions – for example, there are no ordered compounds for AB alloys if both elements have $\mathcal{M} < 50$.

[8] The Hume-Rothery rules for compound formation are less successful than for solid solubility, however.

There are a number of semiempirical approaches to estimate heats of formation of compounds from their chemical elements. One in widespread use is that of A.R. Miedema, *et al.*, who formulated a heat of formation in terms of an electron density at the outer parts of an atom in a solid, a work function to parameterize how easily the outer electrons could be removed from the atom, and information on the atomic size to account for the overlaps of atoms in a solid [7]. The theoretical justification for this approach is not rigorous. Nevertheless, after a number of years of tuning the parameters in the model, the Miedema heats of formation are now fairly reliable. Tabulations of Miedema heats of formation of different alloy compositions are available, and can be useful for estimating solubility tendencies and the stabilities of compounds.

For compositions and temperatures where solid solubility is not favorable, we expect either an ordered compound near a stoichiometric composition, or a mixture of phases, such as a compound plus a solid solution of different composition. Stoichiometric compounds are especially common when the chemical bonding is strongly covalent or ionic. Materials with ions of specific valence are less tolerant of chemical disorder and deviations from stoichiometry than are metallic alloys. For covalently bonded crystals such as Si or GaAs, the unit cell is understood by the geometry of the bonding orbitals, in this case the tetrahedral sp^3-hybrid orbitals that are compatible with the diamond cubic or zincblende crystal structures. For ionically bonded crystals, the close packing of ions of different radii can be used to explain the preference of NaCl for the B1 structure, and CsCl for the B2 structure – an explanation with kinship to the Hume-Rothery size rule for metallic radii.

2.2 Free Energy Curves, Solute Conservation, and the Lever Rule

2.2.1 Free Energy versus Composition

Free energy versus composition curves, $F(c)$, are essential for understanding T–c phase diagrams. Furthermore, modifications of these analyses are useful for systems out of equilibrium, constrained by kinetic processes.

A simple but important phase diagram is for a binary alloy with only two phases: a liquid phase of continuous solubility, and one crystalline solid phase of continuous solid solubility. The two free energy versus composition curves, $F_s(c)$ and $F_l(c)$, are similar to the curve of Fig. 2.1, but they differ in detail. Figure 2.3 shows a typical case. At the highest temperature, T_4, the liquid curve in Fig. 2.3 has the lowest free energy for any composition. The alloy is a liquid at all compositions, as expected because the liquid has the higher entropy. The $-TS$ part of the free energy for the liquid, $F_l(c)$, changes quickly with temperature, however, because S_l is large. With decreasing temperature, the free energy curve for the liquid rises relative to the solid curve (and both may flatten somewhat). At the lowest temperature, T_1, the crystalline solid curve has the lowest free energy, F, for any composition, consistent with its lower energy, E_s. At the temperature T_1, the alloy is a solid at all compositions.

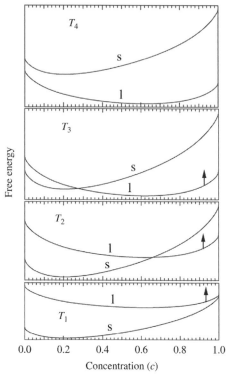

Free energy

Concentration (c)

Figure 2.3   Free energy versus composition curves for solid and liquid phases at four temperatures $T_4 > T_3 > T_2 > T_1$.

The intermediate temperatures T_2 and T_3 of Fig. 2.3 require further analysis. It is not simply a matter of picking the lower free energy curve at a particular composition because the alloy has the freedom to unmix chemically, forming a liquid enriched in B-atoms and a solid depleted in B-atoms, as explained below. The free energy is minimized by selecting an optimal fraction of liquid and solid phases with different chemical compositions.

2.2.2 Conservation of Solute and the Lever Rule

The conservation of solute atoms forces a relationship between the fractional amounts of phases and the chemical compositions of the phases. Consider the material depicted in Fig. 2.4a, which has an overall composition of $c_0 = 0.3$. If it forms a mixture of two phases (here denoted α and β) having different compositions, conservation of solute requires that one of the phases has more solute than c_0, and the other less. The composition deviations from c_0, indicated by the gray zones in Fig. 2.4a, must average to zero. The condition is

$$(c_\alpha - c_0)f_\alpha + (c_\beta - c_0)f_\beta = 0, \tag{2.2}$$

$$\frac{c_0 - c_\alpha}{f_\beta} = \frac{c_\beta - c_0}{f_\alpha}, \tag{2.3}$$

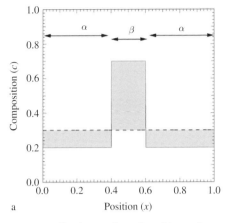

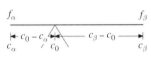

a

b

Figure 2.4 (**a**) Composition profile of a two-phase alloy with overall composition $c_0 = 0.3$. The area of the upper gray band is $(0.7 - 0.3) \times 0.2 = 0.8$, whereas the areas of the two lower gray bands are each $(0.2 - 0.3) \times 0.4 = -0.4$. (**b**) Analogy to a balanced lever, with weights and lever arms as labeled.

where the f_α and f_β are the mole fractions of the α- and β-phases. Because $f_\alpha + f_\beta = 1$,

$$f_\alpha = \frac{c_\beta - c_0}{c_\beta - c_\alpha},\tag{2.4}$$

$$f_\beta = \frac{c_0 - c_\alpha}{c_\beta - c_\alpha}.\tag{2.5}$$

Figure 2.4b shows the analogy to balancing a lever, where the heavier mass (larger f) placed on the shorter arm of the lever (smaller Δc) balances a smaller mass on the longer lever arm. Equation 2.3 is therefore known as the "lever rule." The lever rule is a consequence only of solute conservation, and is useful when working either with curves of free energy versus composition, or with T–c phase diagrams themselves.

2.3 Common Tangent Construction

One more tool is needed to work out phase diagrams from curves of free energy versus composition – the "common tangent construction." It is based on the reasonable expectation that the total free energy, F_T, is the sum of contributions from the fractions of the α- and β-phases:

$$F_T(f_\alpha, c_\alpha, f_\beta, c_\beta) = f_\alpha F_\alpha(c_\alpha) + f_\beta F_\beta(c_\beta),\tag{2.6}$$

where the fractions are normalized to 1 as $f_\alpha + f_\beta = 1$. Using $f_\alpha = 1 - f_\beta$,

$$F_T(c_\alpha, c_\beta, f_\beta) = (1 - f_\beta)F_\alpha(c_\alpha) + f_\beta F_\beta(c_\beta),\tag{2.7}$$

$$F_T(c_\alpha, c_\beta, f_\beta) = F_\alpha(c_\alpha) + f_\beta \left(F_\beta(c_\beta) - F_\alpha(c_\alpha) \right).\tag{2.8}$$

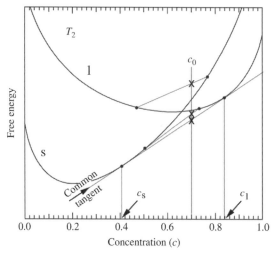

Concentration (c)

Figure 2.5 Free energy versus composition curves for solid (s) and liquid (l) phases at the temperatures T_2. For a selected c_0, the total free energy is obtained at $\times$ on the straight line with endpoints on the two free energy curves. The minimum free energy (lowest $\times$) is found on the tangent line common to the two curves.

Conservation of solute (with Eq. 2.5 from the lever rule) allows transformation from phase fraction to chemical composition:

$$F_T(c_\alpha, c_\beta, c_0) = F_\alpha(c_\alpha) + \left(\frac{c_0 - c_\alpha}{c_\beta - c_\alpha}\right)\left(F_\beta(c_\beta) - F_\alpha(c_\alpha)\right). \tag{2.9}$$

The result, Eq. 2.9, is the equation of a straight line of free energy versus the composition c_0 in the range $c_\alpha < c_0 < c_\beta$. The F_T is at c_0 on a straight line between its values at the endpoints, $F_\alpha(c_\alpha)$ and $F_\beta(c_\beta)$.

This linear relationship of Eq. 2.9 is used to minimize the total free energy of a mixture of two phases by adjusting their chemical compositions. One of the panels of Fig. 2.3 (for temperature T_2) is redrawn in Fig. 2.5, now with three sloping straight lines between pairs of endpoints on the curves F_l and F_s. An overall chemical composition of $c_0 = 0.7$ is picked for illustration. The total free energy F_T is minimized with a geometrical algorithm. Move the endpoints of the sloping lines in Fig. 2.5, demanding that they stay on the curves of the two allowable phases. From the linear relationship of Eq. 2.9, we know that F_T will be at the intersections of the sloping lines with the vertical line at c_0, marked by the crosses "$\times$," in Fig. 2.5. We see that the lowest possible position for the $\times$ is found on the "common tangent" between the two curves of free energy versus composition.

The common tangent construction gives the minimum F_T by identifying the compositions c'_α and c'_β of the two phases in thermodynamic equilibrium.[9] These special compositions c'_α and c'_β from the common tangent rule satisfy two equalities: first, the

---

[9] The fractions of these two phases will be obtained from the lever rule. Note that the lever rule is built into Eq. 2.9, using the three compositions $\{c_0, c'_\alpha, c'_\beta\}$ and Eq. 2.5.

slopes at $F_\alpha(c'_\alpha)$ and $F_\beta(c'_\beta)$ are equal, and second, these two slopes equal the slope of the line between these two special compositions

$$\left.\frac{\partial F_\alpha(c)}{\partial c}\right|_{c'_\alpha} = \left.\frac{\partial F_\beta(c)}{\partial c}\right|_{c'_\beta} = \frac{F_\beta(c'_\beta) - F_\alpha(c'_\alpha)}{c'_\beta - c'_\alpha}. \tag{2.10}$$

The equilibrium compositions of the solid and liquid phases obtained from the common tangent construction are labeled "c_l" and "c_s" in Fig. 2.5.

Box 2.3   **Chemical Potentials**

The common tangent, $F^{c.t.}(c)$, is a straight line that can extend from $0 \leq c \leq 1$, and we find its slope

$$F^{c.t.}(c) = F^{c.t.}(c = 0) + c\left(F^{c.t.}(c = 1) - F^{c.t.}(c = 0)\right), \tag{2.11}$$

$$\frac{\partial F^{c.t.}(c)}{\partial c} = F^{c.t.}(c = 1) - F^{c.t.}(c = 0). \tag{2.12}$$

To change composition on a lattice of N sites, it is necessary to remove one A-atom when we add one B-atom. With a change in composition we therefore obtain the difference in chemical potentials from Eq. 2.12

$$\mu_B - \mu_A = \frac{F^{c.t.}(c = 1)}{N} - \frac{F^{c.t.}(c = 0)}{N}. \tag{2.13}$$

The first term on the RHS is dominated by the B-atoms, and the second by the A. Neglecting any constant shift, these two terms are identified with the two chemical potentials

$$\mu_A = \frac{F^{c.t.}(c = 0)}{N}, \quad \mu_B = \frac{F^{c.t.}(c = 1)}{N}. \tag{2.14}$$

These μ_A and μ_B are marked at the edges of Fig. 2.6. In equilibrium, the μ_B is the same in both phases, or B-atoms would move from one phase to the other.[a] For a composition c_0 between c_s and c_l, the average chemical potential of an atom is the composition-weighted average of μ_A and μ_B. (Outside this range, including compositions near $c_0 = 0$ and $c_0 = 1$, the chemical potential rises above the common tangent as it tracks the individual curves for the solid or liquid.)

When a material is out of equilibrium, as when a phase transformation is underway, the new phases may not yet have the compositions predicted by the common tangent rule. There is, however, an expectation that **local equilibrium** occurs near the interface between the two phases because atoms can quickly jump across the interface from one phase to the other. During equilibration, any change in composition by moving atoms between the two phases involves replacing one species of atom with the other. A requirement of local equilibrium, then, is an equal *difference* in chemical potentials of A and B atoms in both phases. The condition on the compositions of the two phases, c'_α and c'_β, is that they give equal slopes on the free energy curves at c'_α and c'_β (i.e., $dF_\alpha/dc\,|_{c'_\alpha} = dF_\beta/dc\,|_{c'_\beta}$). This is the first equality of Eq. 2.10. When c'_α and c'_β differ from the compositions of the common tangent, however, the material evolves towards equilibrium to lower the total F_T. In doing so, local equilibrium will enforce equal slopes, which eventually become the common tangent.

---

[a] In problems where an A-atom and a B-atom are swapped in position, it is the change in energy from the swap that is important, not the energy of an individual A or B-atom.

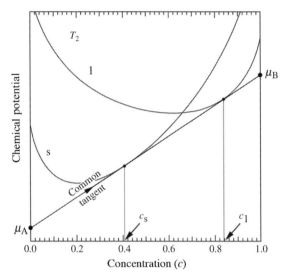

Figure 2.6 Locating the chemical potentials of A- and B-atoms, μ_A and μ_B, for the equilibrium common tangent construction of Fig. 2.5.

It can be shown that the first equality in Eq. 2.10 ensures equal differences in the chemical potentials of the two atom species in the two phases

$$\mu_\alpha^A(T) - \mu_\alpha^B(T) = \mu_\beta^A(T) - \mu_\beta^B(T), \qquad (2.15)$$

but not necessarily the minimum of F_T – equal slopes can be found at two points on the right or left sides of the minima of the two $F(c)$ curves. The second equality in Eq. 2.10 is required to ensure the minimum (so the chemical potential of an atom is the same in either phase).

2.4 Continuous Solid Solubility Phase Diagram

We now construct the phase diagram for the alloy with the free energy versus composition curves of Fig. 2.3, and then show how to use this phase diagram. First note from Fig. 2.5 that the compositions labeled "c_l" and "c_s" from the common tangent construction serve to minimize the free energy of a mixture of solid and liquid phases for any intermediate composition c_0 between 0.4 and 0.84 at this temperature T_2. These two points are transferred to Fig. 2.7. With $F(c)$ curves at different temperatures as in the four panels of Fig. 2.3, it is a straightforward exercise to construct common tangents and identify more compositions at the phase boundaries. Results are presented in Fig. 2.7, which is a T–c phase diagram.

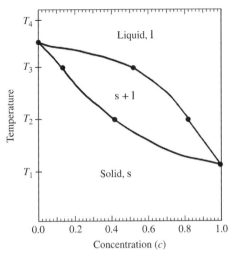

Figure 2.7   Continuous solid solubility phase diagram derived from curves of Fig. 2.3. Dots mark compositions obtained by common tangent constructions on the panels of Fig. 2.3.

The dots on the T–c phase diagram of Fig. 2.7 were obtained directly from the common tangents of the $F(c)$ curves of Fig. 2.3 (as with Fig. 2.5). Some discretion was exercised in interpolating between the points. Between the temperatures T_3 and T_4, perhaps halfway between them, Fig. 2.3 shows that the last of the A-rich solid phase has melted. The pure element A has a single melting temperature, of course, and this is also true for the element B, so the two curves must come together at $c = 0$ and $c = 1$. The phase diagram shows that the A–B alloy does not have a single melting temperature, however. In equilibrium at temperatures above the "solidus" (the lower curve in this phase diagram), the alloy tends to form a little liquid that is enriched in the element B. Just below the "liquidus" (the upper curve in this phase diagram), the alloy is mostly liquid with a small amount of solid phase enriched in the element A.

For any point within the "two-phase region," marked "s+l" in the phase diagram of Fig. 2.7, Fig. 2.8 shows how the equilibrium phases are obtained:

- Draw a horizontal line across the diagram at the temperature of interest. From the intersections with the solidus and liquidus curves, read off the compositions c_s and c_l. These are the equilibrium compositions of the solid and liquid phases at the temperature of interest. Such curves of $c_s(T)$ and $c_l(T)$ are shown on the right of Fig. 2.8.
- To obtain the fractions of the solid and liquid phases, note the composition of the alloy c_0, and use the lever rule with the compositions $c_s < c_0 < c_l$ in Eqs. 2.4 and 2.5. This is illustrated with Fig. 2.8 for $c_0 = 0.6$ over a range of temperatures. To quickly sketch graphs of $f_s(T)$ and $f_l(T)$, it is handy to find the temperature where a horizontal line is equidistant between the solidus and liquidus, and $f_s = f_l = 0.5$. This helps determine the curvature of the $f_s(T)$ and $f_l(T)$. Another example is given in Section 5.2.1.

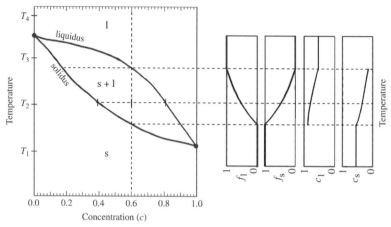

Figure 2.8   Use of phase diagram of Fig. 2.7 to predict equilibrium phase fractions and compositions of an alloy with composition $c_0 = 0.6$ at different T (four graphs at right, rotated 90°). Upper and lower horizontal dashed lines are liquidus and solidus at c_0, respectively. The middle dashed line was chosen so that c_0 is the same horizontal distance between the liquidus and solidus, so the phase fractions of liquid and solid are both 0.5.

2.5 Eutectic and Peritectic Phase Diagrams

It is possible to have equilibrium between three phases in a binary A–B alloy at a fixed temperature and pressure. We consider a liquid phase in equilibrium with two solid phases. For example, Pb dissolves in Sb in the liquid phase, but at low temperatures the Pb-rich phase is fcc, whereas the Sb-rich phase is hexagonal. The free energy is minimized at low temperatures when the material is a heterogeneous mixture of a Pb-rich fcc phase and a Sn-rich hexagonal phase. To find the chemical compositions of the equilibrium phases, each phase needs a distinct free energy versus composition curve.

There are two possible types of phase diagrams with a high-temperature liquid phase and two low-temperature solid phases – "eutectic" and "peritectic" phase diagrams. The type depends on the compositional sequence of the minima of the three $F_\xi(c)$ curves, as shown in Fig. 2.9.

2.5.1 Eutectic Phase Diagram

The eutectic case occurs when the minimum of the $F_l(c)$ curve of the liquid phase lies between the minima of the curves for the two solid phases α and β, as shown in Fig. 2.9a (and in Fig. 2.10a). Figure 2.10a shows four liquid free energy curves for four temperatures. For simplicity, the solid phase curves are assumed constant with temperature, since they have the lower entropy. With increasing temperature, the liquid free energy curve falls relative to the solid curves owing to the large $-TS_l$ term in the liquid free energy. Three common tangent constructions are shown in Fig. 2.10a. The most interesting one is at the temperature T_2, where the common tangent touches all three phases, with the liquid in the middle. This temperature T_2 is the "eutectic temperature." These three points of contact

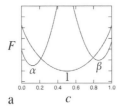
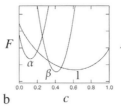
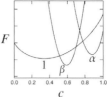

Figure 2.9   (**a**) Free energy curves for two solid phases, α and β, and a liquid phase with minima positioned to give a eutectic phase diagram. (**b**) Same phases as panel a, but minima positioned for peritectic phase diagrams.

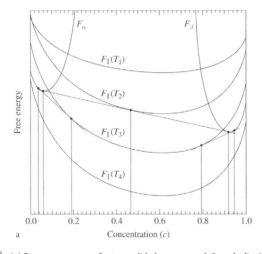
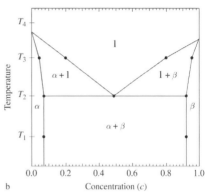

Figure 2.10   (**a**) Free energy curves for two solid phases, α and β, and a liquid phase drawn for four temperatures $T_1 < T_2 < T_3 < T_4$. Three common tangent constructions are shown, from which seven compositions are found. (**b**) Eutectic phase diagram derived from the curves of part (a) – the nine dots mark the seven compositions from common tangent constructions.

are shown at T_2 in Fig. 2.10b, which is a complete eutectic phase diagram. Please compare Figs. 2.10a and 2.10b to check the correspondence between the compositions from the common tangent constructions, and the points on the phase diagram at T_1, at T_2 , at T_3.

2.5.2 Peritectic Phase Diagram

The other possible compositional sequence for three free energy curves occurs when the liquid curve has its minimum on one side of the minima of the two curves for the solids (Fig. 2.9b). An example is shown in Fig. 2.11a, with its corresponding "peritectic" phase diagram in Fig. 2.11b. This phase diagram can be constructed by the same common tangent exercise as was used for the eutectic case in Fig. 2.10 – lowering the $F_1(c)$ curve relative to $F_\alpha(c)$ and $F_\beta(c)$ with increasing temperature. Peritectic phase diagrams are common when one crystalline phase, here the β-phase, has a much lower melting temperature than the α-phase. If the liquid free energy curve were to lie to the left of the two solid curves, the common tangent construction would produce a peritectic phase diagram with the higher melting temperatures to the right.

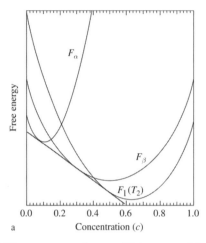 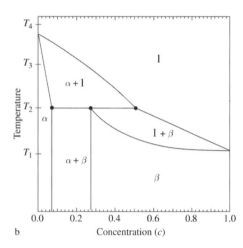

Figure 2.11 (**a**) Free energy curves for two solid phases, α and β, and a liquid phase drawn for the peritectic temperature T_2. (**b**) Peritectic phase diagram estimated from curves of part (a).

2.5.3 More Complex Phase Diagrams

Eutectic, peritectic, and continuous solid solubility phase diagrams can be assembled into more complex diagrams with multiple phases and two-phase regions (see [8], for example). For example, two eutectic phase diagrams over composition ranges from $0 < c < 0.5$ and $0.5 < c < 1.0$ could be attached together at the composition $c = 0.5$ as in Fig. 9.1b where a third solid phase (γ in Fig. 9.1b), probably an ordered compound, is stable. Alternatively, a peritectic phase diagram could be attached to one side of a eutectic diagram. In another variation, the liquid phase in a eutectic diagram could be replaced by a random solid solution, perhaps the disordered solid solution in the phase diagram of Fig. 2.7. (When a solid solution replaces the liquid region in Figs. 2.10b or 2.11b, the diagram is called a "eutectoid" or "peritectoid" diagram, respectively.) In various combinations, the forms of Figs. 2.7, 2.10b, 2.11b, plus the ordering and unmixing diagrams discussed below, account for almost all features found in real binary alloy phase diagrams.

Box 2.4   **Cooling Curves**

Cooling curves are measured to find temperatures of phase boundaries in T–c phase diagrams. A thermocouple in a sample that is cooling in a tranquil environment shows a monotonic decrease of temperature with time, perhaps like an exponential decay. If there is a phase transition in the material and a release of latent heat during cooling, however, this smooth cooling curve is interrupted. For modest rates of cooling, the release of latent heat keeps the temperature stable at the transformation temperature for a "thermal arrest." The temperature decreases again once the transformation is complete. The duration of the thermal arrest hints at the volume fraction of the phase transformation, but the temperature of the arrest is quantitative. By measuring cooling curves on alloys with a sequence of chemical compositions, points at various T, c are acquired for sketching out a phase diagram.

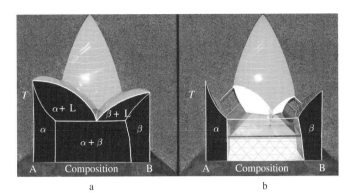

Figure 2.12   (**a**) A ternary phase diagram of elements A, B, and C (to back, with region of C-rich γ-phase, not labeled). (**b**) Sections removed to show that the ternary eutectic composition has a lower melting temperature than any of the three binaries. The triangular grid at the bottom depicts the alloy composition as in Fig. 2.13a.

2.6 Ternary Phase Diagrams

A binary alloy has one independent composition variable, but a ternary alloy has two. A two-dimensional T–c diagram is no longer sufficient. Figure 2.12 shows an image of a compound eutectic diagram for a ternary alloy (a special case where each pair of elements, A–B, B–C, and A–C, has a eutectic phase diagram as labeled in Fig. 2.12a, and there are no other phases). Temperature is vertical, and the concentration of the third element, C, increases with distance from the front to the back of the model. The Gibbs phase rule states that adding a third component (the element C) gives an additional degree of freedom in the intensive variables of temperature, pressure or mole fractions. The two-phase areas of the binary T–c eutectic diagram become volumes in the ternary diagram. Two of these volumes are removed in Fig. 2.12b, showing more clearly the solidus surfaces and the low-temperature ternary eutectic point. The eutectic point in this ternary diagram has three solid phases and a liquid phase in equilibrium (compared with two solid phases and one liquid phase for a binary alloy).

Box 2.5	Gibbs Phase Rule

This rule is [9]

$$P + F = C + 2, \qquad (2.16)$$

where P is the number of phases, F is the number of degrees of freedom, and C is the number of components (chemical species, e.g., 2 for a binary A–B alloy).

The T–p phase diagram of cerium (Fig. 1.3) is simple to interpret. With one component (Ce), it is possible to have three phases in equilibrium at one point (zero degrees of freedom), such as the α, β, and γ at $T = 230$ K and $p = 0.3$ GPa. Equation 2.16 becomes $3 + 0 = 1 + 2$.

The degrees of freedom must be handled with care for T–c phase diagrams, since Eq. 2.16 allows freedom of either temperature or pressure. Furthermore, the chemical composition can be used as an intensive variable, giving the set $\{T, p, c\}$. Here is an interpretation for the eutectic point in a eutectic T–c phase diagram. For a binary alloy there are two components, A and B, so $C = 2$, and Eq. 2.16 becomes $3 + 1 = 2 + 2$ at the eutectic point. The one degree of freedom refers to pressure, which is not shown in a eutectic T–c phase diagram. At different pressures, we expect the eutectic point to move in T and c, tracing a line with one degree of freedom. A line in a T–c phase diagram, such as a liquidus line, gives us one more degree of freedom (compared to a point), and now two phases are in equilibrium at the liquidus. The situation is more subtle for a two-phase region of the diagram, such as the $\alpha + \beta$ region in Fig. 2.10b. Although this region occupies a two-dimensional area in the diagram, the amounts of the two phases are constrained by the lever rule, so one degree of freedom is lost and we do not gain a degree of freedom over the case of the liquidus line. (Yes, this is a tricky feature of applying the phase rule.) Within a single-phase region (e.g., "α") we have a real degree of freedom to change c, but of course only one phase in equilibrium.

If pressure is kept constant, as in a T–c phase diagram, it is appropriate to rewrite the phase rule as

$$P + F = C + 1, \tag{2.17}$$

but this is unconventional and may be declared incorrect by a casual observer. Beware, however, if the magnetic properties of the phases are significant and different, and a magnetic field is added to give the set $\{T, p, B, c\}$, it is necessary to write

$$P + F = C + 3. \tag{2.18}$$

The essential idea behind the Gibbs phase rule is that adding a new phase puts another constraint on the chemical potentials through expressions like Eq. 2.15, causing a loss of a degree of freedom for the intensive variables.

Handbooks of ternary phase diagrams usually do not show three-dimensional data structures as in Fig. 2.12, but instead present various cuts through them. Horizontal cuts as in Fig. 2.13 are "isothermal sections." For any point inside Fig. 2.13a, the compositions $\{c_A, c_B, c_C\}$ are marked at the ends of the three intersecting grid lines. The isothermal section of Fig. 2.13b is for the alloy of Fig. 2.12 at a low temperature. A vertical stack of isothermal sections for different temperatures can depict the full ternary diagram. Compared with a binary alloy, adding a third component gives another degree of freedom by Eq. 2.16, imposing a new feature in the two-phase regions of ternary phase diagrams. In a two-phase region of a ternary alloy, it is necessary to know the "tie-lines" that identify the compositions of the two phases in equilibrium.

Tie-lines affect the utility of vertical sections of Fig. 2.12 (which show all temperatures, but cannot include all compositions). Vertical sections, as in the plane of the paper of Fig. 2.12 (with constant c_C) are called "isopleths." Isopleths have the same usage as binary phase diagrams for locating single-phase regions. For a two-phase region in an isopleth, the two phases generally do not have the same composition c_C, however, and generally lie out of the plane of the isopleth (unless the tie-line is along constant c_C, as for a binary diagram with $c_C = 0$).

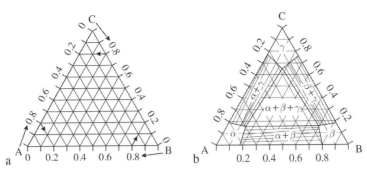

(**a**) Grid for chemical compositions of ternary alloy. Arrows show how numbers at edges represent compositions of elements at corners. At any point in the field, the sum of compositions of all three elements equals 1.0. (**b**) Schematic isothermal section of a ternary phase diagram (for the low-temperature part of Fig. 2.12). The three phases α, β, and γ are rich in the elements A, B, C, respectively. The two-phase regions are shown with thin gray tie-lines indicating the specific compositions in equilibrium with each other. The central triangle is a region of three-phase equilibrium.

Three-phase triangular regions are often handy structural units for understanding isothermal sections, such as the region "$\alpha + \beta + \gamma$" in Fig. 2.13. Around each triangular three-phase region there must be two-phase regions, and single-phase regions at their ends. Within the three-phase region, the phases have compositions given by the corners of the triangle,[10] and their fractions are set by a pair of lever rules for the distances from the alloy composition to the compositions of these three corners. Figures 2.12 and 2.13 are actually a rather simple example. Ternary alloys allow for phases and microstructures beyond those possible for binary alloys, but today our knowledge of ternary phase diagrams is much less complete.[11]

2.7 Free Energy of a Solid Solution

Considerable insight into phase transitions is provided by a model with atoms placed on the fixed sites of an "Ising lattice," which also gives the connections between the sites. The Ising model is natural for obtaining E for an alloy by summing pairwise interaction energies.[12] The entropy of an alloy, S, also depends on the configurations of atoms. This is natural for the configurational entropy, S_{conf}, but it is not too difficult for the vibrational entropy, S_{vib}, as shown in Chapter 7.

---

[10] At the low temperature T_1, Fig. 2.10a shows a common tangent between two minima of the free energy curves of phases α and β. For our ternary eutectic system at low temperature, a tangent plane will touch near the minima of three bowl-shaped $F_\xi(c)$ curves. If Fig. 2.13b corresponds to a temperature just below the existence of the liquid, the contact points are at the corners of the triangular three-phase region in Fig. 2.13b.

[11] Today many ternary diagrams are missing phases, temperatures, or tie-lines, and a substantial effort continues to obtain this important information by experiment and theory [10].

[12] Historically, the energies of atom pairs ("bonds") were adjusted as parameters to account for phase transitions in specific materials. Today E is found by electronic structure calculations, a large topic with books of its own.

2.7.1 Parameters for Atom Configurations and Bonds

Consider a binary alloy with chemical species A and B placed at random on the sites of a lattice. This is a model for a solid solution.

- The concentration of B-atoms is c, and by using mole fractions (or atomic fractions) the concentration c varies from 0 to 1. When $c = 0$ the alloy is pure element A; when $c = 1$ the alloy is pure element B.
- The concentration of A-atoms is $1 - c$.
- The crystal has N sites, and therefore a number cN of B-atoms and $(1 - c)N$ of A-atoms.
- Each site is surrounded by z sites as first-nearest neighbors (1nn).
- For a random solid solution, on average each A-atom is surrounded by zc B-atoms, and its other $(1 - c)z$ neighbors are A-atoms.
- The alloy is assumed to be completely random, so the species of atom on any lattice site does not depend on the species of atoms occupying its neighboring sites (see point 1 of Section 2.1.1).
- All the energy of atom configurations depends on 1nn (first-nearest-neighbor) pairs of atoms[13] (see point 2 of Section 2.1.1).

  - Each A–A pair has energy e_{AA},
  - Each A–B pair has energy e_{AB},
  - Each B–B pair has energy e_{BB}.
  - To obtain the total energy of an alloy configuration, simply count the number of pairs with each energy.

Consider an A-atom on a lattice site. Its energy depends on the number of its A-atom neighbors, $z(1-c)$, and the number of its B-atom neighbors, zc. The energy for this average A-atom, and likewise for the average B-atom, is

$$e_A = \frac{z}{2}(1 - c)\,e_{AA} + \frac{z}{2}c\,e_{AB}, \tag{2.19}$$

$$e_B = \frac{z}{2}(1 - c)\,e_{AB} + \frac{z}{2}c\,e_{BB}. \tag{2.20}$$

The division by 2 corrects for a double-counting of bonds when summing them as pairs over all atoms in the crystal – the bond from each neighbor atom back to the original atom of the pair gets counted a second time. A pictorial justification for using $z/2$ bonds per atom is shown in Fig. 2.14b.

2.7.2 Partition Function and Free Energy

We begin by writing the partition function for a single site, Z_{1site}, which contains either an A-atom or a B-atom

$$Z_{1site} = e^{-\beta e_A} + e^{-\beta e_B}, \tag{2.21}$$

[13] One way to justify this assumption is to argue that most of the chemical bonding energy is associated with 1nn pairs, which is often approximately correct. Sometimes, however, a 1nn model can average successfully over other interactions.

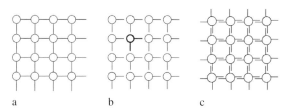

Figure 2.14 (**a**) Square lattice with coordination number $z = 4$, and four neighbors about each atom. (**b**) Tiling of a square lattice with one atom per site, and two bonds per site, $z/2$, to avoid overcounting the bonds. (**c**) Double-counting of bonds with z bonds per site.

where $\beta \equiv 1/(k_B T)$. For a random solid solution, all site occupancies are independent, so the total partition function of the alloy with N sites is

$$Z_N = Z_{1\text{site}}^N, \tag{2.22}$$

$$Z_N = \left(e^{-\beta e_A} + e^{-\beta e_B}\right)^N. \tag{2.23}$$

We can evaluate Z_N with the binomial expansion,

$$Z_N = \sum_{n=0}^{N} \left(\frac{N!}{(N-n)!\,n!}\right)\left(e^{-\beta e_A}\right)^{N-n}\left(e^{-\beta e_B}\right)^n. \tag{2.24}$$

For a fixed concentration c (one value of n, where $n = cN$), we use only one term from the sum[14] over n

$$Z_N(c) \simeq \left(\frac{N!}{(N-n)!\,n!}\right)\left(e^{-\beta e_A}\right)^{N-n}\left(e^{-\beta e_B}\right)^n. \tag{2.25}$$

Obtaining the Helmholtz free energy from the expression $F = -k_B T \ln Z$, and using the Stirling approximation $\ln(x!) \simeq x \ln x - x$

$$F(c) = \left[(N-n)e_A + ne_B\right]$$
$$- k_B T \left[N \ln N - n \ln n - (N-n)\ln(N-n)\right], \tag{2.26}$$

$$F(c) = \left[e_{AA} + 2c(e_{AB} - e_{AA}) + c^2(4V)\right](zN/2)$$
$$- k_B T \left[\ln N - c \ln n - (1-c)\ln(N-n)\right]N, \tag{2.27}$$

which can be written compactly as

$$F(c) = E_{\text{conf}}(c) - TS_{\text{conf}}(c), \tag{2.28}$$

by using the definitions

$$E_{\text{conf}}(c) = \left[e_{AA} + 2c(e_{AB} - e_{AA}) + c^2(4V)\right]\frac{z}{2}N, \tag{2.29}$$

$$E_{\text{conf}}(c) = \left[(1-c)e_{AA} + c\,e_{BB} - c(1-c)4V\right]\frac{z}{2}N, \tag{2.30}$$

$$S_{\text{conf}}(c) = -k_B\left[c \ln c + (1-c)\ln(1-c)\right]N. \tag{2.31}$$

[14] Also, when N is very large, if the equilibrium concentration is c, Z_N becomes a sharply peaked function with a maximum at $n = cN$. It is then standard practice in statistical mechanics to replace Z_N with its maximum value (see [11]).

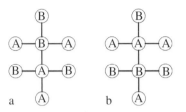

(**a**) Initial configuration around central A- and B-atoms. (**b**) New configuration after exchange of the central A–B pair.

Here $E_{conf}(c)$ and $S_{conf}(c)$ are the configurational energy and configurational entropy of the alloy.[15] Section 7.1.2 presents a physical derivation of $S_{conf}(c)$ of Eq. 2.31. Equation 2.27 defines implicitly an important parameter of the energy

$$V \equiv (e_{AA} + e_{BB} - 2e_{AB})/4. \tag{2.32}$$

This parameter V is sometimes called the "interchange energy." Its physical meaning is made clear with the change in configuration of an A- and B-atom pair in Fig. 2.15. For the initial configuration of the central A- and B-atoms in Fig. 2.15a, the figure shows a total of five A–B pairs, one A–A pair, and one B–B pair. After interchange of only the central A- and B-atoms,[16] there are three A–B pairs, two A–A pairs, and two B–B pairs (Fig. 2.15b). After the interchange in the figure, the change in pair energy is precisely $4V = e_{AA} + e_{BB} - 2e_{AB}$. For other interchanges, or for multiple interchanges, the change in pair energy will always be an integer multiple of $\pm 4V$.

Box 2.6 — **Pairwise Energy in the Point Approximation**

The $E(c)$ can be calculated by summing the number of each chemical species times its average energy

$$E(n) = (N - n)e_A + ne_B. \tag{2.33}$$

With Eqs. 2.19 and 2.20, and noting that $n = Nc$ and $N - n = N(1 - c)$

$$E(c) = \frac{zN}{2}(1 - c)\big[(1 - c)e_{AA} + c\,e_{AB}\big] + \frac{zN}{2}c\big[(1 - c)e_{AB} + c\,e_{BB}\big], \tag{2.34}$$

$$E(c) = \frac{zN}{2}\big[(1 - c)^2 e_{AA} + c^2 e_{BB} + 2c(1 - c)\,e_{AB}\big]. \tag{2.35}$$

Equation 2.35 shows that all the energy from A–B pairs scales as $2c(1 - c)$, which is the probability that a 1nn pair on the lattice is an A–B or B–A.

We now seek terms proportional to c. These give a linear interpolation of the energy from pure A to pure B. Deviations from this straight line arise from the chemical preference for like or unlike pairs of atoms.

[15] The word "configurational" refers to arrangements of atoms on the sites of the Ising lattice.
[16] Note that this interchange conserves the total number of atoms of each type. Also note that an interchange of like pairs, A–A or B–B, has no effect on the total number of like or unlike pairs.

Arranging $(1 - c)^2 e_{AA} = (1 - 2c + c^2) e_{AA} = [(1 - c) - c(1 - c)] e_{AA},$ (2.36)

and $c^2 e_{BB} = (c - c + c^2) e_{BB} = [c - c(1 - c)] e_{BB},$ (2.37)

gives the linear terms $(1 - c) e_{AA}$ and $c\, e_{BB}$. Substituting Eqs. 2.36 and 2.37 into 2.35

$$E(c) = \frac{zN}{2} \left[[1 - c - c(1 - c)] e_{AA} + [c - c(1 - c)] e_{BB} + 2c(1 - c) e_{AB} \right],$$ (2.38)

$$E(c) = \frac{zN}{2} \left[e_{AA} + c(e_{BB} - e_{AA}) + c(1 - c)(-e_{AA} - e_{BB} + 2e_{AB}) \right],$$ (2.39)

$$E(c) = \frac{zN}{2} \left[e_{AA} + c(e_{BB} - e_{AA}) - c(1 - c)\, 4V \right],$$ (2.40)

which is the same as Eq. 2.30.

| Box 2.7 | Configurational Entropy in the Point Approximation |

The partition function of Eq. 2.25 has a combinatorial prefactor that becomes the configurational entropy when the free energy is calculated as $F = -k_B T \ln Z$

$$S_{conf} = k_B \ln \left(\frac{N!}{(N - n)!\, n!} \right).$$ (2.41)

Section 7.1.2 with Fig. 7.2a presents a more detailed derivation of Eq. 2.41. Using the Stirling approximation, $\ln x! = x \ln x - x$

$$S_{conf} = k_B \left[N \ln N - N - (N - n) \ln (N - n) + (N - n) - n \ln n + n \right].$$ (2.42)

There is a cancellation of the terms $-N + (N - n) + n$, and we use the expressions $n = Nc$ and $N - n = N(1 - c)$

$$S_{conf} = k_B N \left[\ln N - (1 - c) \ln \left[N(1 - c) \right] - c \ln \left[Nc \right] \right].$$ (2.43)

There is a cancellation of the terms $\ln N - (1 - c) \ln N - c \ln N$ giving

$$S_{conf} = -k_B N \left[(1 - c) \ln (1 - c) + c \ln c \right],$$ (2.44)

which is the same as Eq. 2.31. Because c and $1 - c$ are less than 1, the two logarithm functions are negative, and S_{conf} is positive. It has a maximum value of $0.69\, k_B N$ at $c = 0.5$, and is zero at $c = 0$ and $c = 1$.

Logarithm functions have singularities as $c \to 0$ or $c \to 1$. These are slow divergences, and the factors c and $1 - c$ have stronger dependences on c, so the functions $c \ln c$ and $(1 - c) \ln (1 - c)$ are both well behaved in Eq. 2.44 at the concentration limits. ($S_{conf}(c)$ does have singularities in its derivatives as $c \to 0$ or $c \to 1$.) Nevertheless, as $c \to 0$ the function $-c \ln c$ becomes larger than $c\, \varepsilon$, where ε is a constant, such as the energy of an isolated defect or impurity atom. The configurational entropy therefore dominates as $c \to 0$ and $c \to 1$, so mixing a small amount of impurity is always favorable in the dilute limits at finite T, even if chemical bonding does not favor it.

2.8 Unmixing Phase Diagram

2.8.1 Instability against Unmixing

First suppose that $V = 0$, so $E_{conf}(c)$ of Eq. 2.30 has only the first two terms with coefficients $1 - c$ and c. The energy is therefore linear in the alloy composition c from pure A to pure B. Furthermore, when $V = 0$, the pairwise energy is insensitive to atom interchanges as in Fig. 2.15 – replacing like pairs with unlike pairs has no effect on $E_{conf}(c)$ if $V = 0$. With no energetic tendency for unmixing (or ordering), but with an entropy that favors mixing, the alloy will remain a solid solution at all temperatures. It is called an "ideal solution."

Unmixing is controlled by the term $c(1-c)\,4V$ in Eq. 2.30. Consider first the case where $V < 0$, meaning that the average of A–A and B–B pairs is more favorable than A–B pairs. This gives the alloy a tendency to unmix chemically.[17] The energy and entropy terms of Eqs. 2.30 and 2.31 are graphed in Fig. 2.16a. When $-TS_{conf}(c)$ is added to $E_{conf}(c)$ to obtain the free energy curves in Fig. 2.16b, the configurational entropy has logarithmic singularities in its composition derivatives that always force a reduction in F from its local maxima at $c = 0$ and $c = 1$. The energy term of Eq. 2.30 or 2.40, positive in this case with $V < 0$, can dominate at intermediate concentrations, however, especially at low temperatures.

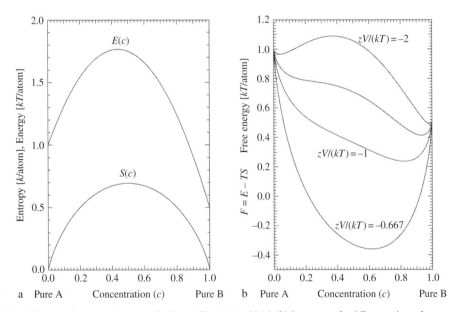

Figure 2.16   (a) Configurational energy and entropy for $V < 0$ (Eqs. 2.29 and 2.31). (b) Free energy for different values of $zV/(kT)$ as labeled.

---

[17] The opposite sign of $V > 0$ means that at low temperatures the alloy will tend to maximize the number of its A–B pairs. It does so by developing chemical order, as in Section 2.9.

For small magnitudes of zV/k_BT (high temperatures), a random solid solution offers the minimum in free energy for any composition. For the small $zV/k_BT = -0.667$ in Fig. 2.16b, the curve $F_{conf}(c)$ is concave upwards for all c. On the other hand, Fig. 2.16b shows that the free energy $F_{conf}(c)$ has peculiar curvature when the ratio zV/k_BT is more negative than -1. When $F_{conf}(c)$ has a curvature that is concave downwards over some range of c, a common tangent can touch $F_{conf}(c)$ at two compositions. This is consistent with chemical unmixing.

We obtain the critical temperature for unmixing by first writing $F_{conf}(c)$, on a per atom basis with Eqs. 2.30 and 2.31

$$F_{conf}(c) = \frac{z}{2}\Big[(1-c)\,e_{AA} + c\,e_{BB} - 4Vc(1-c)\Big] + k_BT\Big[c\ln c + (1-c)\ln(1-c)\Big]. \quad (2.45)$$

Setting its second derivative equal to zero (which gives the boundary between positive and negative curvature)[18]

$$\frac{d^2 F_{conf}(c)}{dc^2} = 4zV + \frac{k_BT}{c\,(1-c)} = 0, \quad (2.46)$$

gives the special temperature

$$T_{sp} = -\frac{4zV}{k_B}\,c\,(1-c). \quad (2.47)$$

The highest temperature for unmixing occurs when $c = 1/2 = 1 - c$

$$T_c = -\frac{zV}{k_B}. \quad (2.48)$$

This T_c is the "critical temperature" for unmixing at the composition $c = 1/2$. Further interpretation of these curves, the subject of the next section, leads to the unmixing phase diagram of Fig. 2.17.

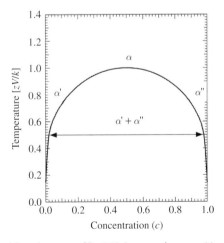

Figure 2.17 Unmixing phase diagram derived from the curves of Fig. 2.16. Arrow marks composition limits obtained by common tangent construction from top curve of Fig. 2.16b.

[18] The reader is encouraged to work all steps for obtaining T_{sp} by calculating the first and second derivatives of $F_{conf}(c)$; see Problem 2.5.

Our $E_{conf}(c)$ and $S_{conf}(c)$ were derived for a random arrangement of atoms. If this is true, as it could be at sufficiently high temperature, the probabilities for an A- or B-atom on a site (($1 - c$) and c) are accurate, as are the probabilities for the pairs of the different species ($zc^2/2$ for a B–B pair, for example). At lower temperatures the equilibrium site occupancies are not random, however, owing to the energetics of chemical preferences. Both $E_{conf}(c)$ and $S_{conf}(c)$ need to be modified when there is a tendency for A-atoms to prefer A-atom neighbors at lower temperatures. We ignore this in the next sections, but here is an important consequence.

At temperatures somewhat above the critical temperature for unmixing, the alloy is not random. Over short distances there is a tendency for more A–A and B–B pairs because thermal energy does not entirely disrupt the chemical preferences. One consequence is that the configurational entropy is not so large as for a random solid solution. Likewise, the hump in the curve of $E(c)$ in Fig. 2.16a is also suppressed. The overall result turns out to be a significant reduction of the hump in the curve of $F(c)$ in Fig. 2.16b at low temperatures. Unfortunately our point approximation, which has no independent variables to give information on pairs or larger clusters, cannot account for these effects of short-range order. Even the pair approximation becomes inadequate as we cool closer to T_c (and below T_c). Closer to T_c, the spatial range of the order increases exponentially, and diverges to infinity at T_c itself. This is the topic of Chapter 27 (online).

2.8.2 Features of the Unmixing Phase Diagram

Using the common tangent construction to span between the two local minima in the $F(c)$ curves at low T in Fig. 2.16b (upper curves in the figure),[19] the unmixing phase diagram of Fig. 2.17 is constructed. It shows a solid solution at high temperatures, denoted the "α-phase." At low temperatures this solution becomes unstable against chemical unmixing, and forms an A-rich "α'-phase" and a B-rich "α''-phase." In the two-phase region beneath the arching curve in Fig. 2.17, the chemical compositions of the two phases can be read directly from the curve itself for any temperature.[20] The horizontal double arrow points to these two compositions for a normalized temperature of 0.5 (which is consistent with the common tangent applied to the top free energy curve in Fig. 2.16b). In this unmixing phase diagram of Fig. 2.17, the composition segregation is larger at lower temperatures, and is infinitesimally small at $c = 0.5$ at the critical temperature $T_c = z|V|/k_B$.

Figure 2.18a shows a typical $F(c)$ curve for $T < T_c$. Its curvature is negative at intermediate values of c, between approximately $c = 0.26$ and $c = 0.74$ as indicated by the vertical dashed lines. Sometimes an additional "spinodal curve" is drawn on the unmixing phase diagram to bound the region where $F(c)$ has negative curvature. On Fig. 2.18b, the dashed spinodal curve reaches up to the critical temperature at $c = 0.5$, and has the parabolic shape in composition of the form $c(1-c)$. The spinodal curve is given by Eq. 2.47, which was calculated from the condition $d^2F/dc^2 = 0$.

[19] See also Box 2.9.
[20] The fractional amounts of the phases are set by the lever rule or Eqs. 2.4 and 2.5.

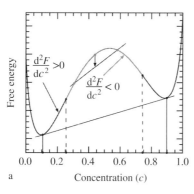
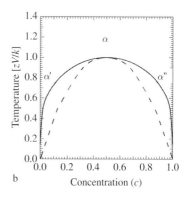

Figure 2.18 (**a**) Typical free energy versus composition curve for an unmixing alloy at low temperature. The curvature of the function is negative for $0.26 < c < 0.74$, but positive for all other compositions. The inflection points of $d^2F/dc^2 = 0$ are marked with vertical dashed lines. (**b**) Unmixing phase diagram with dashed "spinodal" from inflection points of curve in panel a.

Within the composition range of negative curvature (where $F(c)$ is shown as gray in Fig. 2.18a), any line that intersects the two sides of the free energy curve will lie below $F(c)$ (between the points of intersection). A typical line is shown below the composition $c = 0.44$, with an arrow showing how an alloy of this composition can reduce its total free energy by separating into the two compositions at the ends of the line. (This follows from the same algorithm illustrated in Fig. 2.5.) In this region of negative curvature, $d^2F/dc^2 < 0$, the free energy is unstable against even infinitesimal changes in composition – we could have drawn the straight line only slightly below the gray curve, and still have unmixing. When infinitesimally small composition changes are favorable, nucleation of a distinct zone is not needed to start the unmixing process, and unmixing can start with small fluctuations in composition. This is called "spinodal decomposition," and is shown in Fig. 1.5d. On the other hand, nucleation and growth is required between the spinodal curve and the solid boundary lines of Fig. 2.18b.

Beware, though, a free energy curve with $d^2F/dc^2 < 0$ is so unstable that it is not useful for equilibrium thermodynamics, including the development of phase diagrams. Spinodal decomposition does occur in nature, and it is a continuous process. Spinodal decomposition is explained in Chapter 16 as a kinetic phenomenon based on diffusion, however, rather than a thermodynamic concept based on an alloy free energy function. Obtaining a spinodal curve with $d^2F/dc^2 < 0$ as in Fig. 2.18a is not rigorous.

2.9 Order–Disorder Phase Diagram

2.9.1 Long-Range Order Parameterization

The chemical preferences of atoms in an A–B binary alloy are specified as (Eq. 2.32)

$$V \equiv \frac{e_{AA} + e_{BB} - 2e_{AB}}{4}. \tag{2.49}$$

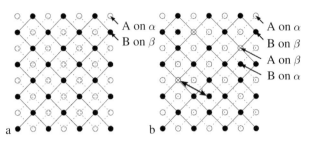

Figure 2.19 (a) Ordered square lattice with two species {A, B} and two interpenetrating square sublattices {α, β}. (b) Misplaced atoms A and B (antisites) on well-ordered sublattices. The double arrow indicates how a duo of A and B atoms can be exchanged to eliminate antisites.

An alloy with an ordering tendency, $V > 0$, maximizes the number of its A–B pairs at low temperatures.[21] At high temperatures, alloys with either unmixing or ordering tendencies will be random solid solutions if configurational entropy is dominant, but their phases at low temperatures are fundamentally different (see Fig. 1.4).

For illustration of an ordering transformation, we use an equiatomic A–B alloy on a square lattice, as shown in Fig. 2.19. For $V > 0$, we expect the structure of Fig. 2.19a at $T = 0$. The solid and dashed lines in Fig. 2.19a show two interpenetrating square sublattices, α and β. Each is occupied by one type of atom; the A are on the α-sublattice, and B are on the β. For a crystal with N sites, each of these sublattices contains $N/2$ sites.

Finite temperatures favor some disorder in the structure – putting a few atoms on the wrong sublattice gives a big increase in the configurational entropy,[22] driving some mixing of A-atoms onto the β-sublattice (Fig. 2.19b). By conservation of atoms and conservation of sublattice sites, an equal number of B-atoms must move onto the α-sublattice. The number of these "wrong" atoms (or "antisites") on each sublattice is W, and the number of "right" atoms is R. The sublattice concentrations are used to define a "long-range order" (LRO) parameter, L, for the alloy [12]

$$L \equiv \frac{R - W}{N/2}. \tag{2.50}$$

This L ranges over $-1 < L < 1$. The case of $L = 1$ corresponds to that of Fig. 2.19a, where $R = N/2$ and $W = 0$. The disordered equiatomic alloy has as many right as wrong atoms on each sublattice, i.e., $R = W$, so $L = 0$ for a disordered alloy. For the alloy of Fig. 2.19b, $W = 3$, $R = 25$, and $L = 0.79$.

Figure 2.20a depicts the general case of an alloy with two identical sublattices, but an overall chemical composition $c < 0.5$. We want the order parameter to quantify the difference in solute concentrations on the two sublattices. The L of Eq. 2.50 is still useful if R and W are averaged for the two sublattices.[23] Figure 2.20b shows how the sublattice

---

[21] Section 2.8 analyzed $V < 0$, where like atom pairs are preferred to unlike pairs, and unmixing occurs at low temperatures.

[22] Near $L = 1$ the derivative of the entropy with respect to L has a logarithmic singularity in sublattice concentrations like the concentration dependence of Eq. 2.31 near $c = 1$.

[23] In general, the right atoms on the two sublattices are unequal, $R^\alpha \neq R^\beta$, unless $c = 0.5$.

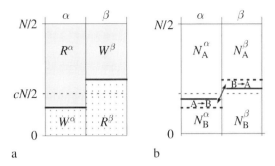

Figure 2.20 (**a**) Sublattice concentrations for an alloy of overall composition c. The distances between the heavy and dashed horizontal lines are equal for both sublattices α and β. (**b**) Switching A-atoms to B on the α-sublattice requires a compensating change of B-atoms to A on the β-sublattice, so the difference $R - W$ undergoes the same change for both sublattices.

populations change as L is changed (reduced in this case). Owing to the conservation of sublattice sites and the conservation of atoms, the transfer of B-atoms onto the α-sublattice must be exactly compensated by the transfer of A-atoms onto the β-sublattice.

2.9.2 Temperature Dependence of Long-Range Order

We now calculate the temperature dependence of the long-range order parameter, $L(T)$, for the case of an equiatomic AB alloy. The energy is obtained by counting the number of A–A, A–B, and B–B pairs of atoms as in Section 2.7. As atoms are moved between sublattices, we must be sure to conserve the number of each species.

The essential trick is to consider two atoms at a time – we transform between a "right duo" that comprises an A-atom and a B-atom on their proper sublattices, and a "wrong duo" that comprises an A-atom and a B-atom on the wrong sublattices (antisites). This corresponds to swapping the two atoms at the ends of the double arrow in Fig. 2.19b. To calculate the equilibrium value of the LRO parameter, we seek the ratio of right and wrong duos as a function of temperature. The energy of the two atoms in the wrong duo (a *wrong A* plus a *wrong B*) depends on the starting atom and its neighbors

$$e_{\mathrm{W}} = \frac{zR}{N/2}e_{\mathrm{AA}} + \frac{zR}{N/2}e_{\mathrm{BB}} + 2\frac{zW}{N/2}e_{\mathrm{AB}}. \qquad (2.51)$$

$$\text{starting wrong atom : wrong A} \quad \text{wrong B} \quad \text{wrong A (B)}$$
$$\text{neighbor : right A} \quad \text{right B} \quad \text{wrong B (A)}$$

To understand Eq. 2.51, remember that the neighbors of an atom are on the other sublattice, so the first term is from the wrong A in the duo that has right A-atoms as neighbors, the second term is from the wrong B in the duo that has right B-atoms as neighbors, and the third term (two terms really) is from the B-neighbors of the wrong A that are themselves wrong (plus the wrong A-neighbors of the wrong B, giving the factor of 2).

Exchanging the positions of the wrong A- and B-atoms in the duo (as indicated by the double arrow in Fig. 2.19) puts the two atoms on their proper sublattices. The analogous expression for the pair energy of the "right" duo is

$$e_R = \frac{zW}{N/2}e_{AA} + \frac{zW}{N/2}e_{BB} + 2\frac{zR}{N/2}e_{AB}, \tag{2.52}$$

where the A-neighbors of the right A are wrong, the B-neighbors of the right B are wrong, but the B (A)-neighbors of the right A (B) are right. (Notice how the swap of the variables R and W in Eqs. 2.51 and 2.52 follows the swap of the atoms in the duo.)

The energy of a single atom will be half that of Eq. 2.51 or Eq. 2.52 for the reason illustrated in Fig. 2.14b. We calculate the ratio of right to wrong atoms, R/W, as the ratio of their Boltzmann factors

$$\frac{R}{W} = \frac{e^{-\beta e_R/2}}{e^{-\beta e_W/2}}, \tag{2.53}$$

$$\frac{R}{W} = \exp\left[-\beta\frac{1}{2}(R - W)(2e_{AB} - e_{AA} - e_{BB})\frac{z}{N/2}\right]. \tag{2.54}$$

With Eq. 2.50 and noting that $R + W = N/2$, it is found that $R = (1 + L)N/4$ and $W = (1 - L)N/4$, so Eq. 2.54 becomes

$$\frac{1 + L}{1 - L} = \exp\left[\frac{L2Vz}{k_B T}\right], \tag{2.55}$$

where again $V \equiv (e_{AA} + e_{BB} - 2e_{AB})/4$. Equation 2.55 can be solved numerically, and the result for $L(T)$ is shown in Fig. 2.21.

Figure 2.21 shows that the LRO is largest at low temperatures. As the order decreases, there is also a decrease in the energy penalty of creating a wrong duo of atoms. At higher temperatures, an increase in temperature becomes increasingly effective in creating more wrong duos, causing L to fall catastrophically. The value of L becomes zero at a critical temperature, T_c.

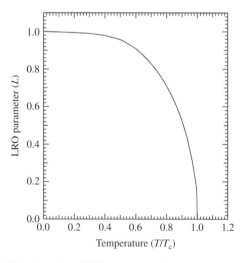

Figure 2.21 Long-range order versus T in the Gorsky–Bragg–Williams approximation.

To obtain the critical temperature T_c, we can solve Eq. 2.55 analytically with a trick of linearization. The T_c is found by taking L as infinitesimally small, for which Eq. 2.55 becomes

$$\ln(1 + 2L) = \frac{L2Vz}{k_B T_c}, \tag{2.56}$$

$$2L = \frac{L2Vz}{k_B T_c}, \tag{2.57}$$

$$T_c = \frac{zV}{k_B}. \tag{2.58}$$

Please compare this result to the critical temperature for the unmixing alloy with $c = 0.5$, Eq. 2.48.

The theory presented in this section is called the "Gorsky–Bragg–Williams"[24] or "mean field" theory of ordering [13–15]. Equation 2.51 makes a simplifying assumption about counting pairs – if we pick an A-atom, its neighbors on the adjacent sublattice sites will be determined only by the overall concentration on the sublattice. Because the Gorsky–Bragg–Williams theory keeps track only of the average sublattice concentrations, there is no spatial scale. The sublattices are infinite in size. It is therefore appropriate to call L a long-range order parameter.

The Gorsky–Bragg–Williams approach is a theory in the "point approximation" because the theory uses only a simple sublattice concentration for each site (or "point") in the structure. The occupation probabilities for both a site and its neighbors are assumed to be the sublattice concentrations, and the types of neighbors are the same when the site is occupied by either an A-atom or a B-atom. (For example, there is no allowance for a preference for the first-nearest neighbors of a wrong A-atom to be B-atoms.) In nature, however, atoms tend to satisfy their chemical preferences over short distances, and alloys exhibit "short-range order," even at temperatures above the critical temperature. This short-range order reduces the configurational entropy of the alloy as it makes the bond energy more favorable. A better thermodynamic theory is possible by defining "pair variables" to account for the short-range preference of neighboring atoms [16]. This is developed further in Section 7.2.

Box 2.9   **A Correspondence between Unmixing and Ordering**

There is a fundamental similarity between the unmixing and ordering phase diagrams in the point approximation. For convenience, here again is the free energy per atom of an unmixing alloy (from Eqs. 2.30 and 2.31):

$$F(c) = \frac{z}{2}\left[e_{AA} + c(e_{BB} - e_{AA}) - c(1 - c)4V\right] + k_B T\left[(1 - c)\ln(1 - c) + c\ln c\right]. \tag{2.59}$$

---

[24] It is often called "Bragg–Williams theory" but Gorsky was first to publish the correct mean field theory for chemical ordering in CuAu, although in a peculiar form [13]. Unfortunately, he was arrested by Stalin's secret police and executed shortly after he refused to testify that Lev Landau used excessively tough standards for grading students.

Somewhat hidden in Eq. 2.59 is a symmetry about $c = 1/2$, consistent with the symmetry of the phase diagram of Fig. 2.18b. To bring this out, transform the variable c to a new variable L that is zero at $c = 1/2$, and use it to rewrite Eq. 2.59

$$L \equiv 2c - 1, \quad c = \frac{1+L}{2}, \quad (1-c) = \frac{1-L}{2}, \tag{2.60}$$

$$F(L) = \frac{z}{2} \left[e_{AA} + \left(\frac{1+L}{2} \right) (e_{BB} - e_{AA}) - \left(\frac{1-L^2}{4} \right) 4V \right]$$

$$+ k_B T \left[\left(\frac{1-L}{2} \right) \ln \left(\frac{1-L}{2} \right) + \left(\frac{1+L}{2} \right) \ln \left(\frac{1+L}{2} \right) \right]. \tag{2.61}$$

For the equilibrium phase diagram, we need the common tangent construction using the first derivative, $dF/dc = 2dF/dL$

$$2\frac{dF}{dL} = \frac{z}{2} (e_{BB} - e_{AA}) + zL2V + k_B T \ln \left(\frac{1+L}{1-L} \right), \tag{2.62}$$

$$2\frac{dF}{dL} - \frac{z}{2} (e_{BB} - e_{AA}) = zL2V + k_B T \ln \left(\frac{1+L}{1-L} \right). \tag{2.63}$$

The RHS of Eq. 2.63 is antisymmetric in L, changing sign when L is changed to $-L$. Along any common tangent, however, the compositional derivatives of F must be equal, and must be equal for Eq. 2.62 at $+L$ and $-L$. This can be true only if the RHS of Eq. 2.63 is zero.

Two equations result when the RHS $= 0$. One is the slope for the common tangent

$$\frac{dF}{dc} = \frac{z}{2} (e_{BB} - e_{AA}), \tag{2.64}$$

caused by different energies of A–A and B–B pairs. The more interesting equation from the condition RHS $= 0$ is

$$\ln \left(\frac{1+L}{1-L} \right) = -\frac{L2Vz}{k_B T}. \tag{2.65}$$

Equation 2.65, obtained from the common tangent condition for the unmixing problem, has the same form as Eq. 2.55 for the Gorsky–Bragg–Williams model of ordering (with a sign change because $V < 0$ for unmixing). It is now easy to see that the temperature dependence of the order parameter, $L(T)$, has a shape identical to the phase boundary of the unmixing diagram. Rotate the curve $L(T)$ of Fig. 2.21 by 90° counterclockwise, reflect it about $L = 0$, and compare it to Fig. 2.18b.

Does the condition for absolute instability against unmixing, $d^2F/dc^2 = 0$ as shown in Fig. 2.18a, pertain to the ordering problem? Making a 90° clockwise rotation of the dashed curve in the unmixing phase diagram of Fig. 2.18b, and again using Eq. 2.60, a new curve $L_{sp}(T)$ is obtained that generally lies below the $L(T)$ of Fig. 2.21. This new curve bounds a region where the high-temperature disordered phase is absolutely unstable against the formation of chemical order. Perhaps the region beneath $L_{sp}(T)$ could be called a region of "spinodal ordering," in a natural analogy to chemical unmixing by spinodal decomposition [17].

2.10 Alloy Phase Diagrams

Box 2.10   **Summary of T–c Phase Diagrams**

The five major T–c phase diagrams are:

- Continuous solid solubility. One crystal structure accommodates all chemical compositions from pure B to pure A. Between the solidus and liquidus lines, the compositions of the solid and liquid phases depend on temperature.
- Eutectic phase diagram. The liquid free energy curve has a minimum at a composition between those of the two solid phases. The lowest temperature of liquid stability is at an intermediate composition.
- Peritectic phase diagram. One of the solid phases has a lower melting temperature than the other, and does not exist above the peritectic temperature. Just below this temperature, the composition of the liquid phase lies outside the range of the two solid phases.
- Unmixing phase diagram. The species A and B prefer like neighbors more than unlike, so $V < 0$ and unmixing occurs at low temperatures and intermediate compositions. Spinodal decomposition is possible when the solid solution is unstable, and composition fluctuations evolve continuously into chemical segregations (Chapter 16).
- Ordering phase diagram. The species A and B prefer unlike A–B pairs more than like, so $V > 0$. At low temperatures, a periodic ordered structure is formed to maximize the number of unlike pairs. In cases allowed by symmetry, spinodal ordering is possible, where the solid solution is unstable and the order parameter evolves continuously (Section 18.4).

Compendia of binary and ternary alloy phase diagrams from experimental results and theoretical assessments have been assembled as handbooks and texts [8, 10, 18]. The details of these real phase diagrams are useful for practical heat treatments of materials, but they are also fascinating for what they reveal about the bonding and entropy of alloys and compounds. The reader is encouraged to peruse them to appreciate the depth of information that is available today. For example, an internet search shows that the Cu–Sn phase diagram is not so simple as may be assumed, considering the ancient human knowledge of bronze. A surprisingly large number of different crystalline phases can be found on a typical alloy phase diagram. Some phases have stoichiometric compositions. Others exist over a range of compositions, where the composition deviations are accommodated by antisite atoms, interstitials, vacancies, or combinations of these. What makes the understanding of phase diagrams even more complicated is that the composition ranges for phases depend on the other phases with which they are in equilibrium, and these phases may have disorder, too.

Minimization of the free energy of the material is the underlying rule for phase diagrams, and obtaining the free energy from fundamentals of electronic structure and bonding is a forefront research topic in materials science. For more practical work, polynomial models of free energy functions are used for calculating phase diagrams in the CALPHAD approach [19–21], which is often used for multicomponent alloys for which experimental phase diagrams are not available.

The models for free energy presented in Chapter 2 offer valuable conceptual understanding, but are insufficient for accurate predictions of real phase diagrams. Quantitative predictions require additional physical concepts and complexity for obtaining free energy functions of good quality, such as:

- *Enthalpy.* At low temperatures the enthalpy originates with static structures of the nuclei and electrons. Both the bonding energy and the effects of pressure can be obtained from the electronic structure of the alloy. Most quantitative electronic structure calculations of today are based on the approximation that the electron exchange and correlation energy depend on the density of electrons. The reliability of these methods is now known, and they are often satisfactory.
- *Entropy.* The configurational entropy can be calculated more accurately than with Eq. 2.31. The problem with Eq. 2.31 is that at intermediate temperatures it is not correct to assume the occupancy of a crystal site is independent of the occupancy of neighboring sites. There are short-range correlations between atom positions, and these reduce the configurational entropy. A better expression is obtained with the pair approximation Eqs. 7.15–7.16. (Short-range correlations also alter the enthalpy.) The pair approximation provides a significant improvement in calculating the configurational thermodynamics of an alloy, but higher-order cluster approximations are practical, too.

Configurational entropy is not the only source of entropy. Crystals of pure elements have no configurational entropy, but they have solid-state phase transitions, too. For shear transformations there is also no change in the configurational entropy. Atomic vibrations are the primary source of heat in materials, so different alloy phases can differ in vibrational entropy if their atomic vibrations differ in frequency and amplitude. At elevated temperature the vibrations may no longer be harmonic, and this adds considerable complexity to the problem, as discussed in Section 12.3 and Chapter 26 (online).

Motions of magnetic spins and electronic excitations can make important contributions to the entropy, too. Metals always have some contribution from electronic entropy, where temperature causes disorder in the occupancy of electronic states near the Fermi level (Section 6.4.2).

Box 2.11	Convex Hull of Energies

Which phases appear on a phase diagram? For the most part, these phases are the ones observed experimentally, but phases occurring for kinetic reasons are not placed on equilibrium phase diagrams. A big challenge is to calculate the equilibrium phases in an alloy system. Today the computational approach is to consider all plausible crystal structures, maybe hundreds of them, populate them with atoms at an appropriate stoichiometry, and calculate their energies of formation. There are usually many different crystal structures that can accommodate a particular stoichiometry, and these give vertical columns of crosses in Fig. 2.22.

The reference energies of Au and Na are set at zero in Fig. 2.22, even though these metals have favorable (negative) cohesive energies. There are six crosses in Fig. 2.22 that lie above the zero line, so these structures are quite unlikely to be found on a phase diagram. The energies of formation of most compounds lie below zero and are favorable. The most stable compounds lie at the bottoms of the vertical columns of crosses. With a

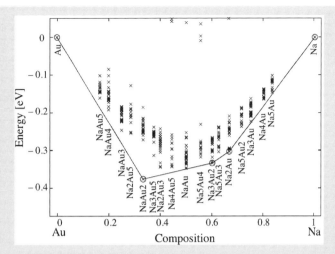

Figure.2.22 Energies per atom of a few hundred possible compounds of Au–Na with 17 different stoichiometries. The convex hull is the lower bound between the compounds of lowest energy, which are circled. From [22].

compositional degree of freedom, however, many of these low-lying compounds could be replaced by pairs of compounds of neighboring compositions with a lower average energy. The problem is much like the common tangent construction of Fig. 2.5, but in Fig. 2.22 a low-lying cross must fall below any line between compounds of neighboring compositions. Geometrically, this constructs a "convex hull," shown in Fig. 2.22 by the solid lines.[a] The compounds on the convex hull are those expected on the phase diagram at low temperatures.

The convex hull is constructed from energies at zero temperature. There are uncertainties in the computed energies – the energies of many compounds differ by only a few meV per atom, which may be an optimistic accuracy in 2020. There is also a role for temperature. We know that some phases appear only at high temperatures (the liquid is an obvious example), and these are favored by the $-TS$ contribution to the free energy. When using results like Fig. 2.22 for predicting the equilibrium phases, it is appropriate to look for candidate compounds within an energy range of, say, 25 meV above the convex hull. This does not alter the result for $NaAu_2$, which indeed appears on the experimental phase diagram, but a phase of composition AuNa also appears on the experimental phase diagram although it is calculated to lie slightly above the convex hull.

[a] In mathematical usage, "convex hull" means a bounding surface connecting the outermost points in a set, so that all other points are contained within the bounding surface.

Problems

2.1   Use the Gibbs phase rule, Eq. 2.16, to explain your answers to these questions about the ternary eutectic diagram shown in Fig. 2.12.

   (a) How many phases are in equilibrium at the eutectic point?

   (b) Why is the β–liquid boundary a surface in this model?

(c) How many phases are in equilibrium at the composition $A_{0.33}B_{0.33}C_{0.34}$ for the lowest temperature shown in the model?

(d) If pressure were applied to this system, what would happen to the liquidus surfaces and the eutectic point? Assume that the liquid has a larger specific volume than any of the solid phases. Assume also that the three solid phases have identical specific volumes, and equal compressibilities for all compositions.

2.2 Suppose the energy cost of an impurity atom (or a point defect like a vacancy) is ε, and the energy cost for n of them (on a lattice of N sites) is $n\varepsilon$. In the dilute limit of the impurity concentration, $c \to 0$, show that the impurity concentration $c(T)$ at temperature T, where $c = n/N$, is

$$c(T) = \exp\left(-\varepsilon/k_B T\right). \tag{2.66}$$

(*Hint*: Use $S_{conf}(c)$ for a solid solution, but explain why you can ignore the term $(1 - c)\ln(1 - c)$. Set $dF/dc = 0$ to find the minimum of $F(c)$.)

2.3 Random solid solutions

(a) Use the algorithm of the binomial expansion of $(A + B)^N$ to justify why the number of arrangements of n equivalent atoms on N sites is $N! / [(N - n)! \, n! \,]$.

(b) Derive the configurational entropy of mixing from part a using the Stirling approximation.

(c) An atom in a bcc crystal has eight nearest neighbors (coordination number $z = 8$). If the chemical composition of the alloy is $c = 0.5$, what is the numerical probability that a given atom in a bcc A–B alloy has n B-atoms as nearest neighbors, where n ranges from 0 to 8? Make a graph of $P(8, n, 0.5)$ versus n.

2.4 A "high-entropy alloy" is a multicomponent alloy that could have a higher configurational entropy than a binary alloy.

(a) Show that the configurational entropy of an n-component alloy can be as large as $S_{max} = k_B \ln n$.
(*Hint*: Use the entropy expression $S = -k_B \sum_j^n c_j \ln c_j$, and argue why it is a maximum when $c_j = 1/n$ for all j.)

(b) Make a plot of S_{max} versus n for $2 \le n \le 10$.

(c) How much does the configurational entropy change by adding a third component to a binary alloy, and by adding an eighth component to a seven-component alloy?

(d) What is the effect of chemical short-range order on the result of part c?

(e) For metallic alloys with larger n, what other sources of entropy might be comparable with the effect of part c?

2.5 Calculate the critical temperature for unmixing for a binary alloy with composition $c = 0.5$. Use the energy and configurational entropy of the mean field theory (for example Eqs. 2.30 and 2.31). The curvature of $F(c)$ is zero at the critical temperature. Explain why this sets the critical condition, and explain why this approach is irrelevant when the interchange energy $V > 0$.

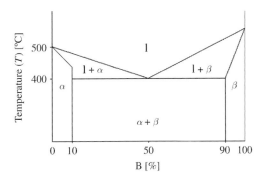

Simple eutectic phase diagram constructed from straight lines.

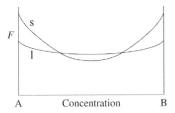

$F(c)$ curves for solid (s) and liquid (l) at an intermediate temperature.

2.6   Figure 2.23 consists of phase boundaries made from straight lines. Draw sketches of free energy versus composition curves, with common tangents, for the phases α, β, and *liquid* at temperatures of 300 °C, 400 °C, and 500 °C that are consistent with this phase diagram. Show how the positions of these $F(c)$ curves shift with temperature.

2.7   Using free energy versus composition curves like those of Fig. 2.11a, graph them at different temperatures and construct a peritectic phase diagram like that of Fig. 2.11b. You may need to widen the $F(c)$ curve of the β-phase.

2.8   Figure 11.16 shows free energy versus composition curves at a low temperature for a liquid and three solid phases. (Here, please ignore the kinetic effects in Section 11.6.2 that are used to discuss Fig. 11.16.)
   (a) Draw two more figures with free energy versus composition for higher temperatures (these should help you with part b).
   (b) With your $F(c)$ curves of part a, draw the phase diagram for this A–B alloy.

2.9   Figure 2.24 shows free energy versus composition curves for a solid phase and a liquid phase having a phase diagram with continuous solid solubility. The shape of the phase diagram will be different from that of Fig. 2.7, however. For example, notice that two common tangent curves can be drawn for the positions of the two curves in Fig. 2.24.
   (a) Assuming as usual that the entropy of the liquid is much higher than that of the solid, construct two or three more pairs of $F(c)$ curves for this system that are useful for constructing a phase diagram.
   (b) Draw the phase diagram consistent with your curves in part a.

(c) What is the curvature of the solidus curve around $c = 0.5$, which is a special point in the diagram?

(d) Suppose the $F(c)$ curves for the solid and liquid were interchanged in Fig. 2.24. Draw the phase diagram for this case.

2.10 Figure 2.22 shows only energies of Au–Na compounds, and not their free energies. Assume the entropy of the compounds can differ by $\pm 0.2\,k_B$/atom. In addition to the five compounds that appear on the convex hull of Fig. 2.22, approximately how many more compounds would be candidate equilibrium phases at a temperature of 10 K? At 100 K? At 1000 K?

Diffusion

Inside solids, atoms move by diffusion. The vacancy mechanism for diffusion in crystals was presented in Section 1.5.1 and illustrated with Fig. 1.6. Mention was made of interstitial diffusion and interstitialcy diffusion. This chapter gives more information about these diffusion mechanisms, with emphasis on how their rates depend on temperature. Mass transport in glasses and liquids can also occur by atomic-level diffusion, but for gases or fluids of low viscosity there are larger-scale convective currents with dynamics unlike diffusion.[1]

The main part of Chapter 3 derives the "diffusion equation," and gives some of its solutions for the chemical composition $c(\vec{r}, t)$ at positions $\vec{r}$ and at times t. The diffusion equation has the same mathematical form as the equation for heat conduction, if solute concentration is replaced by heat or by temperature. The heat equation has been known for centuries, and methods for its solution have a long history in mathematical physics. Some solutions are standard for diffusion in materials, such as Gaussian functions and error functions for one-dimensional problems. Chapter 3 also presents the method of separation of variables for three-dimensional problems in Cartesian and cylindrical coordinates. The Laplacian, ∇^2, is separable in nine other coordinate systems, each with its own special functions and orthogonality relationships, but these are beyond the scope of this book. For problems in ellipsoidal coordinates, for example, the reader may consult classic texts in mathematical physics (e.g., [23]). Today, numerical methods are practical for solving diffusion problems, and can prove more efficient than analytical methods.

The analytical power to solve the diffusion equation does not necessarily transfer to a predictive power for understanding phase transformations in materials. Because diffusion depends on atomic-scale processes, local changes in chemical composition or atomic structure during the phase transformation can depreciate the value of the diffusion equation because the "diffusion constant," D, is not constant. A partly successful approach to problems with a varying D is presented in Section 10.3.3. The diffusion equation is best used for systems in steady state, however, and should be used with caution, or with extensions, when describing the kinetics of a phase transformation.[2]

---

[1] Convective currents can be driven by differences in density, such as the rising of a hot liquid in a gravitational field.

[2] Chapter 23 (online) uses the kinetic master equation first to derive the diffusion equation, and then replaces the assumption of random atom jumps with chemically biased jumps.

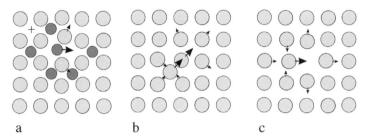

Figure 3.1 (**a**) Interstitial mechanism for diffusive jump of a (dark) interstitial atom. (**b**) Interstitialcy mechanism for self diffusion. (**c**) Vacancy mechanism. For all images, atoms are drawn near their initial positions. Large arrows are the displacements for the main jump; small arrows are displacements of neighboring atoms at some stage in the process.

3.1 Processes of Atom Movements in Crystals

3.1.1 Mechanisms

Three mechanisms for atom diffusion in crystals are illustrated in Fig. 3.1.

- Figure 3.1a shows an interstitial site, marked by a cross, that can be vacant or occupied by a small interstitial atom. The large arrow shows an interstitial atom jumping into a neighboring, unoccupied interstitial site. This is the elementary kinetic step of the "interstitial mechanism."
- Figure 3.1b shows atoms of the same chemical species, but one is in an interstitial site. This "self-interstitial" atom can move by pushing a neighboring atom off its lattice site and into an interstitial site. This is the "interstitialcy mechanism."
- Figure 3.1c illustrates the "vacancy mechanism," which was discussed in Section 1.5.1. Like the other two mechanisms, the vacancy mechanism requires displacements of surrounding atoms and thermal activation.
- Older literature mentions a "ring mechanism," where, for example, a square cluster of four atoms on a square lattice makes a rotation of 90°. In a crystal, the energy required for this mechanism is too high to activate at reasonable temperatures. Such cooperative mechanisms of diffusion are not ruled out for glasses and liquids, however.

3.1.2 Temperature Dependence of Diffusion

Diffusion coefficients for interstitial, interstitialcy, and vacancy diffusion are

$$D_i(T) = f_i\, \Gamma_i\, \frac{a_i^2}{6}\, c_i(1 - c_i)\, e^{-\Delta G_{m,i}/k_B T}, \tag{3.1}$$

$$D_{int}(T) = \Gamma_{int}\, \frac{a_{int}^2}{6}\, c_{int}(T)\, e^{-\Delta G_{m,int}/k_B T}, \tag{3.2}$$

$$D_v(T) = f_v\, \Gamma_v\, \frac{a_v^2}{6}\, c_v(T)\, e^{-\Delta G_{m,v}/k_B T}, \tag{3.3}$$

where the factors are defined and justified as follows. The jump lengths a for the different processes are in three dimensions, and the factor $a^2/6$ is discussed with Eq. 3.18. All three processes have "attempt frequencies," Γ, as the atom moves towards its future site in a local vibration. (The vibrational modes differ for the three mechanisms, however.) Work is required to displace surrounding atoms and reconfigure the chemical bond energy to achieve the activated state of the diffusive jump. All three processes therefore have Boltzmann probabilities with activation energies ΔG_m, where the m denotes "migration."

The other factors in Eqs. 3.1–3.3 are specialized to the mechanism. For interstitial diffusion, the overall jump rate is proportional to the concentration of interstitial atoms c_i, but also to the concentration of vacant neighboring sites $1 - c_i$, assuming a random distribution of interstitial atoms.[3] The total jump rates for interstitialcy and vacancy diffusion are proportional to the concentrations of self-interstitial atoms c_{int} and vacancies c_v, which are small.

The correlation factors f are subtle. In the interstitialcy mechanism, the self-interstitial atom can make a random walk, jumping in any direction, and its $f = 1$. For the vacancy mechanism and the interstitial mechanism, the jump of an atom requires a neighboring vacancy. If an atom exchanges sites with a vacancy, the next move of the atom is biased to move back into this adjacent vacancy. Atoms that move by the vacancy mechanism or interstitial mechanism do not perform a random walk, and the fraction of their jumps that contribute to diffusion is f, which is less than one. Section 10.1 discusses this in detail.

The concentrations of self-interstitials and vacancies, $c_{int}(T)$ and $c_v(T)$, depend on temperature. It takes thermal activation to create these defects, and typically their concentrations are very low. At low concentrations these defects do not interact, so the probabilities of any site being occupied by a self-interstitial or a vacancy are

$$c_{int}(T) = e^{-\Delta G_{f,int}/k_B T}, \qquad (3.4)$$

$$c_v(T) = e^{-\Delta G_{f,v}/k_B T}, \qquad (3.5)$$

where the subscript f denotes "formation." The interstitial mechanism does not have such a factor, since c_i is set by the alloy composition.

Typical values are $\Delta G_f = 1.5\,\text{eV}$ and $\Delta G_m = 0.5\,\text{eV}$. These values are by no means universal (for example, close-packed structures tend to have values of ΔG_f and ΔG_m that are nearly the same). Nevertheless, this helps explain why interstitial diffusion is generally much faster than vacancy diffusion. The vacancy formation energy can suppress $c_v(T)$ to low values such as 10^{-9} at modest temperatures, but this does not occur for interstitial diffusion.[4]

---

[3] This is rarely a good assumption, since ordering or clustering of occupied and unoccupied interstitial sites is generally expected.
[4] Sometimes it is hard to understand how vacancy diffusion occurs at all, given the low equilibrium values of $c_v(T)$ from Eq. 3.5. Often the vacancy concentration is not in equilibrium, and vacancies were trapped in the material after cooling from a higher temperature where c_v was larger.

Box 3.1   Simmons–Balluffi Experiments

These experiments have determined vacancy concentrations by measuring the change in length of crystals at high temperature, T. Most of the change in length is from thermal expansion, and this needs to be known. Thermal expansion can be assessed from the temperature dependence of the lattice parameter, a

$$a(T) = a_0 + a_0 \int_0^T \alpha(T')\, dT', \tag{3.6}$$

where $\alpha(T')$ is the coefficient of linear thermal expansion. The lattice parameter $a(T)$ is referenced to its value a_0 at low temperature.

The length of a piece of material also follows the trend of Eq. 3.6, with a critical difference. The atomic fraction of vacancies, $c_v(T)$, is the same as the fraction of new sites added to the crystal, expanding its volume by the same fraction.[a] The fractional change of length is 1/3 of this, so at temperature T

$$\frac{l(T)}{l_0} = 1 + \int_0^T \alpha(T')\, dT' + \frac{c_v(T)}{3}, \tag{3.7}$$

where the length is referenced to l_0, the length at low temperature.

After dividing Eq. 3.6 by a_0 and subtracting it from Eq. 3.7, the lengthening from thermal expansion is removed, giving the lengthening from vacancies alone

$$c_v(T) = 3\left[\frac{l(T)}{l_0} - \frac{a(T)}{a_0}\right]. \tag{3.8}$$

To determine $c_v(T)$ for aluminum, Simmons and Balluffi used X-ray diffractometry to measure $a(T)$ accurately, and $l(T)$ was measured simultaneously with a micrometer microscope [24]. Similar measurements were done on Au, Cu, and Pb [25–27].

A laser interferometer offers an improvement in the dilatometry, and has led to a new technique that does not require a separate measurement of lattice parameter. Schaefer et al. [28] studied the time dependence of the vacancy concentration in Fe–Al and Ni–Al by increasing the temperature, and measuring how the sample had a quick thermal expansion, followed by a slower lengthening as more vacancies migrated from the surface into the sample. Both the vacancy formation enthalpy and vacancy migration enthalpy were determined this way.

[a] There is a subtlety about the volume of a vacancy Ω_v, which is typically smaller than the volume of an atom Ω_{at} because neighboring atoms relax into the empty site. Fortunately, any effects of this local structure on the overall volume are the same as their effects on the lattice parameter measured by X-ray diffractometry, which are subtracted in Eq. 3.8.

The interstitialcy mechanism appears to be less common in crystalline materials, but a type of it is responsible for much of the self-diffusion in silicon, especially at higher temperatures.[5] Other elements from Groups III and V can diffuse by the interstitialcy mechanism when dissolved in Si. Some solutes in metallic alloys, such as Au in Pb, have an anomalously high diffusivity. In this case a Au atom jumps off a lattice site (leaving a

[5] Note that Si has a rather open diamond cubic structure.

vacancy behind), and diffuses interstitially through the Pb-rich matrix until it finds a vacant site. This process is called "anomalous" diffusion.[6]

Sometimes the details of the prefactors involving a, Γ, f, and the entropic part of ΔG are ignored, since they are expected to vary slowly over a range of temperatures. This way the mechanism itself can be ignored, and the $D(T)$ for any diffusion mechanism is summarized as

$$D(T) = D_0 \, e^{-Q/k_B T}. \tag{3.9}$$

Handbooks of diffusion data present pairs of constants $\{D_0, Q\}$ for different materials over different ranges of temperature.

3.1.3 Ionic Crystals

In an ionic crystal, the transfer of an electron from one atom to a more electronegative atom causes both atoms to change their sizes significantly. The negative anion tends to be much larger than the positive cation, and sometimes the large negative anions are essentially immobile. To make vacancies for cation diffusion, it is necessary to respect charge neutrality by removing equal numbers of cations and anions. Two types of defect pairs are common:

- A Schottky defect pair, as in NaCl, where vacancies are formed on both the Na^+ and Cl^- sublattices. The ionic radius of Na^+ is about half that of the Cl^-, and the Na^+ proves to be the mobile species.
- A Frenkel defect pair, as in AgBr. In this case no atoms are removed, and no defects are introduced into the Br^- sublattice. An Ag^+ ion moves into an interstitial site, creating a vacancy on the Ag^+ sublattice. The Ag^+ can move by either the vacancy or interstitialcy mechanism, and the contributions of the two vary with temperature (the interstitialcy mechanism is more important at low temperatures).

One consequence of creating defects in pairs is that the energy of formation is the energy for making two point defects. Even if only one of them is involved in diffusion, its concentration, proportional to $\exp(-\Delta G_{f,\,pair}/k_B T)$, depends on making the other defect, too. The close proximity of local charges causes the defects and diffusion to be strongly altered by Coulombic interactions. The Coulomb energies are much larger than thermal energies.

Electrons can move by a type of diffusion in mixed-valent materials, such as a material containing a mixture of Fe^{2+} and Fe^{3+} ions. Electrical conductivity involves an electron on an Fe^{2+} cation hopping to a neighboring Fe^{3+}, transforming it to Fe^{2+}. This process requires a thermal activation of the surrounding anions. Anions such as O^{2-} are pulled significantly closer to an Fe^{3+} cation than to an Fe^{2+}. The electron cannot move from an Fe^{2+} to a neighboring Fe^{3+} until thermal vibrations temporarily increase the distances to its neighboring O^{2-} anions. It is useful to consider the electron with the distortion of its neighboring anions as a composite charge carrier called a "small polaron," as described in

---

[6] "Anomalous diffusion" has a number of other meanings, however.

Section 9.2. Temperature facilitates the hopping of small polarons from site to site, and the electrical conductivity increases with thermal activation.

3.2 The Diffusion Equation

3.2.1 Atom Jumps, Fluxes, and Fick's First Law

Consider the flow of atoms between two adjacent atomic planes of crystal, separated by a, as shown in Fig. 3.2. The key assumptions are:

- Each atom is independent, and does not interact with the other atoms.
- Each atom jumps off its plane with a rate Γ [jumps/(unit time)].
- Each jump off a plane goes into an adjacent plane, but the choice is random, so the jump rate to a neighboring plane is $\Gamma/2$ (and $\Gamma/2$ to the plane on the other side that is neglected for now).

The overall rate of atom flow between the two planes, J [atoms/(unit time)], has a positive contribution J_+ from the atoms moving from n to $n+1$, and a negative contribution J_- from the jumps from $n + 1$ to n (x increases from left to right)

$$J = J_+ - J_-. \tag{3.10}$$

These two flows are the atom jump rates left or right times the number of atoms in each plane N_n and N_{n+1}

$$J = \frac{\Gamma}{2}N_n - \frac{\Gamma}{2}N_{n+1}, \quad \text{or} \tag{3.11}$$

$$J = \frac{\Gamma}{2}Vc_n - \frac{\Gamma}{2}Vc_{n+1}. \tag{3.12}$$

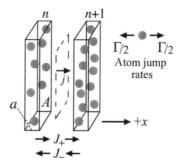

Figure 3.2 Exploded view of two adjacent planes of material, showing physical parameters in the derivation of Fick's first law in the text. Dashed rectangle is an imaginary plane between the planes that is useful for quantifying the flux, j.

Here $N_n = Vc_n$, where c_n is the concentration, and the volume $V = Aa$ (see Fig. 3.2). Equation 3.12 becomes

$$J = \frac{\Gamma}{2}Aa\left(c_n - c_{n+1}\right). \tag{3.13}$$

The flux of atoms between the planes is $j = J/A$, with units [atoms (unit time)$^{-1}$ (unit area)$^{-1}$]. Divide Eq. 3.13 by A, and multiply by a/a

$$j = -\frac{\Gamma}{2}a^2\frac{c_{n+1} - c_n}{a}. \tag{3.14}$$

The fraction on the RHS is identified as a spatial derivative, dc/dx. With the definition of the diffusion constant, D,

$$D \equiv \Gamma a^2/2, \tag{3.15}$$

Eq. 3.14 becomes "Fick's first law" of diffusion

$$j = -D\frac{dc}{dx}. \tag{3.16}$$

In three dimensions it is

$$\vec{j} = -D\vec{\nabla}c, \tag{3.17}$$

where[7]

$$D = \Gamma a^2/6. \tag{3.18}$$

Fick's first law states that the flux of atoms moves down the concentration gradient (hence the minus sign), and is proportional to the gradient. The constant of proportionality is D, which has dimensions [length^2/time].

As a practical matter, fluxes between planes are usually harder to measure than spatial profiles of concentration. Also, when there are fluxes between planes, the concentrations in the planes change with time. For this reason, Fick's first law is less useful than Fick's second law, which is explained next. In fact, Fick's second law is called "The Diffusion Equation."

Box 3.2 **Scaling of Distance with Time**

For Eq. 3.14 we use a as the distance of an atomic jump. The diffusion equation was known long before atom jump distances and frequencies were understood, however. If we worked with slabs of thickness Δ instead of atomic planes of width a in Fig. 3.2, an atom must make more jumps before leaving its slab, so its jump frequency out of the slab, Γ_Δ, is smaller. This approach must predict the same flux j in Fig. 3.2 even if the slabs are wider, because the atoms at the surfaces of the slabs are still jumping at the same rates. Assuming the same physical concentration gradient,[a] dc/dx, the effective Γ_Δ in Eq. 3.14 must therefore be reduced for larger Δ as Δ^{-2}. Suppose we define a characteristic time $\tau = 1/\Gamma_\Delta$, and define a characteristic distance

[7] For example, in Cartesian coordinates the components of atom jumps along the three axes are independent, giving a vector sum of fluxes along all three axes. If no axis is preferred, each jump has components along all three axes, and the RHS of 3.15 must be divided by 3.

$\bar{x} = \Delta$. Evidently then, $\bar{x} \propto \sqrt{\tau}$. This dependence of a characteristic distance on the square root of time is a general feature of diffusion, and is discussed later with Eq. 3.45.

[a] In a concentration gradient, the difference in concentration, $c_{n+1} - c_n$, grows in proportion to the distance between the centers of the slabs, so with increasing Δ, the increase in the numerator in Eq. 3.14 cancels the increase in the denominator (here Δ is replacing a).

3.2.2 The Diffusion Equation

The conservation of atoms links Fick's first law (Eq. 3.16) to "the Diffusion Equation" (Eq. 3.24 below). Figure 3.3 depicts a concentration profile that is highest in the central bin, which we assume to be an atomic plane. Consequently, the flow of atoms out of the nth plane is larger than the flow into it from its neighbors. With the conservation of atoms, N_n decreases with time, and the peaked concentration profile in Fig. 3.3 levels out. We seek a time-dependent concentration profile, $c(x, t)$, that describes the evolution of the system. First, we derive the diffusion equation that $c(x, t)$ obeys.

Consider the four arrows at the top of Fig. 3.3, giving the four flows of atoms on and off plane n. The (positive or negative) accumulation on plane n is the sum of the atom flows [atoms per unit time] from planes $n - 1$ and $n + 1$ onto plane n, so we use the flow equation (Eq. 3.11) twice. The total atom flow onto plane n, $J_{\to n}$, has a contribution along $+x$ from flow from $n - 1$ to n, and a second contribution along $-x$ from flow from n to $n + 1$

$$J_{\to n} = \left(\frac{\Gamma}{2}N_{n-1} - \frac{\Gamma}{2}N_n\right) - \left(\frac{\Gamma}{2}N_n - \frac{\Gamma}{2}N_{n+1}\right), \tag{3.19}$$

$$J_{\to n} = \frac{\Gamma}{2}\left(N_{n-1} + N_{n+1} - 2N_n\right). \tag{3.20}$$

Multiplying by a^2/a^2, and recognizing the diffusion constant D from Eq. 3.15

$$J_{\to n} = D\frac{N_{n-1} + N_{n+1} - 2N_n}{a^2}. \tag{3.21}$$

Dividing Eq. 3.21 by the volume of the plane, V, gives concentrations such as $c_n = N_n/V$ on the RHS. On the LHS, $J_{\to n}$, the rate of flow of atoms onto plane n, when divided by V, is the rate of change of concentration on the plane n

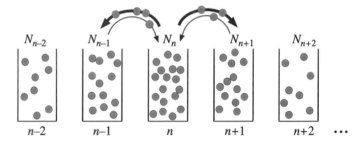

Figure 3.3 Physical parameters defined in the text to derive the diffusion equation. There will be a greater flow of atoms off the plane n because it has the highest concentration. (Compare to Fig. 1.9.)

$$\frac{dc_n}{dt} = D\, \frac{c_{n-1} + c_{n+1} - 2c_n}{a^2}. \tag{3.22}$$

Consistent with Eq. 3.14, the fraction in Eq. 3.22 is a second derivative with respect to position

$$\frac{dc}{dt} = D\, \frac{d^2c}{dx^2}. \tag{3.23}$$

Equation 3.23 is true at any time or position, so we change the notation to accommodate a broad range of times and positions for $c(x, t)$

$$\frac{\partial}{\partial t}\, c(x, t) = D\, \frac{\partial^2}{\partial x^2}\, c(x, t). \tag{3.24}$$

Equation 3.24 is "The Diffusion Equation," (sometimes called "Fick's second law"). It is a partial differential equation in the independent variables for position and time. Solving it gives $c(x, t)$. One solution to Eq. 3.24 is a uniform chemical composition c_0 for all (x, t), since the derivatives on both the LHS and RHS are zero. More interesting solutions are obtained in Section 3.3.

In three dimensions, the derivation from Eq. 3.17 uses the "divergence theorem," which ensures the conservation of atoms when the flux $\vec{j}$ varies throughout the material

$$\vec{\nabla} \cdot \vec{j} = -\frac{\partial c}{\partial t}. \tag{3.25}$$

Equation 3.25 states that a flux leaving a small volume causes a decrease in the concentration in that volume – a consequence of the conservation of atoms. Applying Eq. 3.25 to Fick's first law (Eq. 3.17)

$$-\frac{\partial c}{\partial t} = -\vec{\nabla} \cdot [D\,\vec{\nabla} c]. \tag{3.26}$$

Assuming again that D is a constant, and using a standard vector identity to obtain the Laplacian

$$\frac{\partial c}{\partial t} = D\,\vec{\nabla} \cdot \vec{\nabla} c, \tag{3.27}$$

$$\frac{\partial c}{\partial t} = D\,\nabla^2 c\,, \tag{3.28}$$

which is the diffusion equation in three dimensions (cf., Eq. 3.24). Methods of solving the diffusion equation for $c(\vec{r}, t)$ are celebrated in mathematical physics, and some of them are described below.

Box 3.3   **Radiotracer Experiments**

These experiments are arguably the most important methods for measuring diffusion coefficients. Most elements have radioisotopes suitable for these experiments. The steps for measuring the "self-diffusion" coefficient of Co in Co, for example, may be: (1) depositing some radioactive ^{57}Co on the surface of natural Co metal, (2) heating for a fixed time and temperature to cause diffusion of the ^{57}Co into the bulk Co, (3) precisely removing layers of material (called "sectioning"),[a] and (4) measuring the radioactivity of the material in the removed sections. The quantities of radioisotope are usually in the dilute limit – in this case the fraction of radioactive atoms need be only 10^{-12}. The initial profile can be approximated as a δ-function, and the distribution of radioactivity can be fit to a Gaussian function to obtain the self-diffusion coefficient. These "radiotracer" experiments are also used to measure the "tracer" diffusion coefficients for radioactive elements in other elements or alloys.

Many lighter elements, including O and Si, do not have suitable radioisotopes. For O, "activation analysis" can be used. Most O nuclei are ^{16}O, and the isotope ^{18}O has a low natural abundance of about 0.2 %. It is not a radioisotope, but it can be used in a similar way because under a proton beam ^{18}O transmutes to ^{18}F (whereas ^{16}O does not). The ^{18}F is a short-lived radioisotope which reveals the positions of ^{18}O.

^{a} This has been done by cutting off surface layers by machining, grinding on abrasive papers to various depths, or by sputtering by bombardment of Ar$^+$ ions, for example. Collecting the removed material and knowing its depth and mass are essential parts of sectioning protocols.

3.3 Gaussian and Error Functions in One Dimension

3.3.1 Gaussian Solution

One type of solution to the one-dimensional diffusion equation (Eq. 3.16) is a Gaussian function that depends on distance x and time t

$$c(x, t) = \frac{c'}{\sqrt{4\pi Dt}} e^{-\frac{x^2}{4Dt}}. \tag{3.29}$$

The partial derivatives of $c(x, t)$ are

$$\frac{\partial}{\partial t} c(x, t) = \frac{c'}{\sqrt{4\pi Dt}} e^{-\frac{x^2}{4Dt}} \left[-\frac{1}{2t} + \frac{x^2}{4Dt^2} \right], \tag{3.30}$$

$$\frac{\partial}{\partial x} c(x, t) = \frac{c'}{\sqrt{4\pi Dt}} e^{-\frac{x^2}{4Dt}} \left[-\frac{2x}{4Dt} \right], \tag{3.31}$$

$$\frac{\partial^2}{\partial x^2} c(x, t) = \frac{c'}{\sqrt{4\pi Dt}} e^{-\frac{x^2}{4Dt}} \left[-\frac{2}{4Dt} + \left(\frac{2x}{4Dt} \right)^2 \right]. \tag{3.32}$$

Substitution of these partial derivatives into the one-dimensional diffusion equation, Eq. 3.24, demonstrates the validity of the Gaussian solution, Eq. 3.29.

Figure 3.4 shows the Gaussian concentration profile $c(x, t)$ and its spatial derivatives. All have physical importance. From Fick's first law (Eq. 3.16), the first spatial derivative of $c(x, t)$ is proportional to the negative of the flux

$$\frac{\partial}{\partial x} c(x, t) = -\frac{j}{D}, \tag{3.33}$$

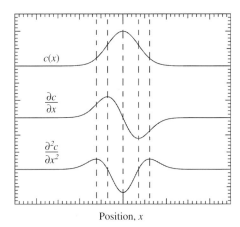

Position, x

Figure 3.4 A Gaussian function at top, and its derivatives below. The Gaussian represents concentration, the first derivative is the negative of the flux, and the second derivative is the rate of accumulation. Inflection points and zeros are indicated with vertical dashed lines.

and from the diffusion equation 3.24, the second spatial derivative of $c(x,t)$ is proportional to the accumulation of solute

$$\frac{\partial^2}{\partial x^2} c(x,t) = \frac{\frac{\partial}{\partial t} c(x,t)}{D}. \tag{3.34}$$

The Gaussian function solution becomes a Dirac δ-function when $t \to 0$.[8] The Gaussian solution is used when a very thin layer of diffusing species is placed between two thick blocks. It is also useful when a thin, nonvolatile, layer is deposited on the surface of a solid, and diffuses into the solid along positive x. With the surface at $x = 0$, the concentration profile for $x < 0$ is inverted ($c(-x,t) \to c(x,t)$) and added to the profile at $x > 0$. Adding these halves of the solution is possible if the atoms move independently (without interactions, doubling the concentration profile at $x > 0$ does not alter the time evolution).

Box 3.4   **Concentration Profiles during Diffusion**

"Random jumps by independent atoms" was our main assumption about how atoms move by diffusion – each atom moves independently of the others. Now use this assumption to follow the time evolution of a concentration profile $c(x,t)$ by focusing on the bin n in Fig. 1.9a.

- A flat concentration, $c(x) = c_0$, has the same concentration in bins $n-1, n, n+1$. With random jumps between bins, there is no net flux between any pair of adjacent bins, and no change of concentration in bin n.
- A concentration profile with a linear variation with position, $\partial c/\partial x = c_{x0}$, has more atoms jumping from bins of higher n than lower n, and a net flux of atoms to lower n (assuming $c_{x0} > 0$; see Eq. 3.33). The flux

---

[8] The peak concentration of the species decreases with time as $1/\sqrt{t}$, and the concentration profile increases in breadth as $\sqrt{t}$, preserving the integral of the concentration. Atoms are also conserved, although their density becomes unphysical as $t \to 0$.

is proportional to the difference in concentrations of two adjacent bins, so for a constant slope c_{x0} the flux of atoms into bin n from $n + 1$ equals the flux out from n to $n - 1$. The concentration of bin n remains unchanged if the concentration gradient is constant.

- A concentration profile having curvature with position, $\partial^2 c / \partial x^2 = c_{xx0}$, has a difference in concentration between bins n and $n+1$ that is larger than the difference between n and $n-1$ (for $c_{xx0} > 0$). The net flux into bin n is therefore greater than its loss, and its concentration increases with time, i.e., $\partial c / \partial t = D\, c_{xx0}$, where the constant of proportionality is D (Eq. 3.34).

These points are illustrated with the Gaussian function in Fig. 3.4. Its first spatial derivative gives the (negative) flux at each point x, as expected from Fick's first law $\partial c / \partial x = -j/D$. The flux is zero where $c(x)$ is flat, and is high at the sloping parts. The second spatial derivative gives the rate of accumulation at each point x. It is zero at inflection points (with zero curvature). There is concentration loss where the curvature is negative, and concentration gain where the curvature is positive.

3.3.2 Green's Function Solution

Consider a one-dimensional concentration profile along a row of atomic planes. We can center a narrow Gaussian function at each plane (a narrow width can confine it to one plane). Along 1000 planes we can center 1000 Gaussian functions, for example, to describe a 1D composition profile. This is the situation at the time $t = 0$. The basic assumption of independent atom motions then allows adding the concentration profiles that evolve with time around each plane. Overlaps of the Gaussian profiles are no problem for diffusion behavior, assuming independent atom movements.

The idea is illustrated in Fig. 3.5a, where an arbitrary function can be approximated as a sum of narrow Gaussians, taken to the limit as δ-functions. Figure 3.5b illustrates the "error function solution" obtained below for a flat initial composition profile with a sharp interface at $x = 0$. Here all the initially narrow Gaussians have the same amplitude for $x \leq 0$, but have zero amplitude for $x > 0$. The individual Gaussians evolve in time, becoming wider and shorter, but with unchanged area (so atoms are conserved for each Gaussian). Far from

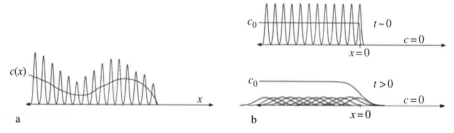

Figure 3.5 (**a**) A set of evenly spaced Gaussian functions with amplitudes set so their sum approximates the function $c(x)$. (**b**) A set of evenly spaced Gaussian functions with amplitudes set so that their sum matches a Heaviside function. Top shows narrow Gaussians near $t = 0$, and a sharp drop in $c(x)$ at $x = 0$. Bottom shows broadening of all Gaussian functions at a later time, and how their sum develops a rounded corner near $x = 0$.

$x = 0$ their sum remains constant, again because their areas are conserved (for $x \ll 0$ the sum is c_0, and for $x \gg 0$ the sum is 0). Near $x = 0$ the sum of Gaussians develops a broader interface at longer times. Our goal is to calculate how this interface broadens with time.

Technically, what we will do is perform a convolution. Recall that the convolution of a Dirac δ-function with any function returns the function. For the concentration profile at $t = 0$ (such as the one in Fig. 3.5a)

$$c(x, 0) = \int_{-\infty}^{\infty} c(x', 0)\, \delta(x - x')\mathrm{d}x'. \tag{3.35}$$

This integral amounts to summing a large number of δ-functions, each with a different center x', and weighted by the concentration at x', which is $c(x', 0)$. For these δ-functions we use normalized Gaussian functions of Eq. 3.29 at $t = \tau$, where τ is very small[9]

$$c(x, 0) = \lim_{\tau \to 0} \int_{-\infty}^{\infty} c(x', 0)\, \frac{1}{\sqrt{4\pi D\tau}} e^{-\frac{(x-x')^2}{4D\tau}}\, \mathrm{d}x'. \tag{3.36}$$

The advantage of Eq. 3.36 is that all the Gaussian functions broaden independently with time, so the same Eq. 3.36 gives the concentration profile at later times

$$c(x, t) = \int_{-\infty}^{\infty} c(x', 0)\, \frac{1}{\sqrt{4\pi Dt}} e^{-\frac{(x-x')^2}{4Dt}}\, \mathrm{d}x'. \tag{3.37}$$

The disadvantage of Eq. 3.37 is that the integral is usually difficult to perform for an arbitrary initial profile $c(x, 0)$.

We now calculate $c(x, t)$ for the initial profile $c(x, 0)$ with the sharp edge illustrated at the top of Fig. 3.5b. This $c(x, 0) = c_0\, \overline{\theta}(x)$, where $\overline{\theta}(x)$ is the downstep function with $\overline{\theta}(x) = 1$ for $x \leq 0$ and $\overline{\theta}(x) = 0$ for $x > 0$

$$c(x, t) = \int_{-\infty}^{\infty} c_0\, \overline{\theta}(x')\, \frac{1}{\sqrt{4\pi Dt}} e^{-\frac{(x-x')^2}{4Dt}}\, \mathrm{d}x'. \tag{3.38}$$

The same initial concentration profile can be constructed by setting the concentration to a constant c_0, but with an upper cutoff for the positions of the Gaussian centers

$$c(x, t) = \int_{-\infty}^{0} c_0\, \frac{1}{\sqrt{4\pi Dt}} e^{-\frac{(x-x')^2}{4Dt}}\, \mathrm{d}x', \tag{3.39}$$

$$c(x, t) = c_0\, \frac{1}{\sqrt{4\pi Dt}} \int_{-\infty}^{0} e^{-\frac{(x-x')^2}{4Dt}}\, \mathrm{d}x'. \tag{3.40}$$

The ratio $(x - x')^2/(4Dt)$ is dimensionless, and reappears many times when working with diffusion equations. We make the standard substitution

$$\eta \equiv \frac{x - x'}{\sqrt{4Dt}} \tag{3.41}$$

---

[9] The Gaussian, which solves the problem for an initial point of concentration, is therefore a Green's function that is convoluted with a general initial profile to give the profile at later times.

and note that the limits change from $\{-\infty, 0\}$ to $\{\infty, x/\sqrt{4Dt}\}$. The dx' changes to $-\sqrt{4Dt}\, d\eta$ (allowing us to switch the limits of integration)

$$c(x, t) = c_0 \frac{1}{\sqrt{\pi}} \int_{\frac{x}{\sqrt{4Dt}}}^{\infty} e^{-\eta^2} d\eta. \tag{3.42}$$

Equation 3.42 is as far as we can go in the analytical reduction, but it is traditional to write it somewhat differently.

The "error function" is defined as

$$\text{erf}(z) \equiv \frac{2}{\sqrt{\pi}} \int_0^z e^{-\chi^2} d\chi, \tag{3.43}$$

and it is known that $\text{erf}(\infty) = 1$ (cf. Eq. 9.52). Equation 3.42 can be rewritten in terms of the error function as

$$c(x, t) = \frac{c_0}{2} \left[1 - \text{erf}\left(\frac{x}{\sqrt{4Dt}}\right) \right]. \tag{3.44}$$

It is useful to know that $\text{erf}(0) = 0$ and $\text{erf}(-\infty) = -1$, so for all times, $c(-\infty, t) = c_0$ and $c(0, t) = c_0/2$ and $c(\infty, t) = 0$. The breadth of the composition spread near $x = 0$ increases as $\sqrt{t}$, but far from the interface the composition remains unchanged for all time.

Box 3.5   **Characteristic Distance**

A characteristic distance of diffusion can be obtained by a scaling argument. Consider again the Gaussian solution, and how a sharp Gaussian evolves with time into broader Gaussians. Each Gaussian is specified completely by three parameters – area, center, width. By normalization (conservation of atoms) all these Gaussians have the same area, and their centers do not move (unbiased diffusion jumps). This leaves one parameter, in this case the width, to rescale each Gaussian for a different time. Evidently the value of the function $c(x, t)$ (with two independent variables) is determined by a single parameter through a relationship between x and t. This relationship is $\eta = x/\sqrt{t}$. For the same η, the concentration is the same. To be used as an argument in a function, the dimensionless form is D/η^2. For the same initial and boundary conditions, different materials have the same concentration at the same value of $D/\eta^2 = D\,t/x^2$. This pertains to the error function solution, too, since its component Gaussian functions evolve independently in time. In fact, it applies to all solutions to the diffusion equation, as can be seen in the time dependence of Eqs. 3.70 and 3.80 below. This scaling behavior of the concentration profile during diffusion gives a handy estimate of a characteristic distance of diffusion in the time t,

$$x \propto \sqrt{Dt}, \tag{3.45}$$

although numerical factors differ from problem to problem.

This error function solution is useful for an infinite bar of material, initially with an abrupt change in composition in the middle. If there are ends to the bar, sometimes it is possible to use the trick described for the Gaussian solution – when the solute is not volatile, it reflects from the ends, and the tail of the composition profile can be inverted and superimposed on the main profile. For a short bar, however, this needs to be done multiple times, which is tedious. The next method to solve the diffusion equation handles short bars or long bars quite nicely.

3.4 Fourier Series Solutions to the Diffusion Equation

3.4.1 Separation of Variables in Cartesian Coordinates

The diffusion equation 3.28 is a partial differential equation for composition, which depends on both position and time, i.e., $c(\vec{r}, t)$. The previous analysis initially assigned a sharp Gaussian function to each point in space, and then let all the Gaussians broaden in time. The concentration profile at a later time is the sum of these independent Gaussians. This is one type of systematic mathematical approach to solving partial differential equations. The other involves selecting an appropriate coordinate system, and using a sum of special functions of the coordinates. These special functions form a complete set that can describe any reasonable function of composition in the coordinate system. The special functions are all "orthogonal" to each other, and evolve independently in time. They are added to give the total profile at later times.

This method of solution is called "separation of variables," because it involves rearranging the diffusion equation (more than once) so all terms with one coordinate appear on one side of an equation. For our first example in Cartesian coordinates, where the special functions are sines and cosines, the result is a transformation to simple ordinary differential equations for the x, y, z, and t dependences of $c(x, y, z, t)$. Boundary conditions and initial conditions then specify the problem precisely, giving weights and wavelengths to the independent sine functions.

The method of separation of variables differs in detail for different coordinate systems, but is simplest for Cartesian coordinates. We select a problem in two dimensions, with no variation of the concentration along the direction $\hat{z}$ (where $\hat{z}$ is a unit vector in the z-direction). The diffusion equation 3.28 is[10]

$$\frac{\partial c}{\partial t} = D \left[\frac{\partial^2 c}{\partial x^2} + \frac{\partial^2 c}{\partial y^2} \right]. \tag{3.46}$$

Fundamental to the method of separation of variables is the assumption (which is not true for all partial differential equations but works here) that the concentration can be expressed as a product of three functions, each involving only one variable. Give these three functions obvious symbols

$$c(x, y, t) = X(x)\, Y(y)\, T(t). \tag{3.47}$$

Substituting Eq. 3.47 into 3.46 and differentiating the relevant variables

$$X(x)Y(y)\frac{\partial T(t)}{\partial t} = D \left[\frac{\partial^2 X(x)}{\partial x^2} Y(y)T(t) + X(x)\frac{\partial^2 Y(y)}{\partial y^2} T(t) \right], \tag{3.48}$$

---

[10] With no variation of the concentration along $\hat{z}$, the term in the Laplacian $\partial^2 c/\partial z^2 = 0$.

and dividing all terms by the product XYT gives

$$\frac{\frac{\partial T(t)}{\partial t}}{T(t)} = D\left[\frac{\frac{\partial^2 X(x)}{\partial x^2}}{X(x)} + \frac{\frac{\partial^2 Y(y)}{\partial y^2}}{Y(y)}\right] = -k_1^2. \tag{3.49}$$

We now make a key observation about Eq. 3.49. The variable t appears only on the LHS and not the right. Likewise, the spatial variables x and y appear only on the right, and not on the left. The spatial variables cannot account for any time dependence, and the time variable cannot account for any spatial dependence. Equality in Eq. 3.49 is therefore possible only if both sides equal a constant. For later convenience, call it $-k_1^2$, which is our first constant of separation. This leads to a simple solution for $T(t)$

$$\frac{\partial T(t)}{\partial t} = -k_1^2\, T(t), \tag{3.50}$$

$$T(t) = T_0\, e^{-k_1^2 t}. \tag{3.51}$$

The spatial variables are separated further

$$-k_1^2 = D\left[\frac{\frac{\partial^2 X(x)}{\partial x^2}}{X(x)} + \frac{\frac{\partial^2 Y(y)}{\partial y^2}}{Y(y)}\right], \tag{3.52}$$

$$D\frac{\frac{\partial^2 X(x)}{\partial x^2}}{X(x)} = -D\frac{\frac{\partial^2 Y(y)}{\partial y^2}}{Y(y)} - k_1^2 = -k_2^2, \tag{3.53}$$

where all terms with x or y were separated to opposite sides of the equation. The independence of x and y requires that the two sides of Eq. 3.53 equal a second constant of separation, $-k_2^2$.

The ordinary differential equations for $X(x)$ and $Y(y)$ in Eq. 3.53 have standard harmonic solutions, with coefficients to be determined later

$$X_{k_2}(x) = A_x\, \sin\left(\frac{k_2}{\sqrt{D}}x\right) + B_x\, \cos\left(\frac{k_2}{\sqrt{D}}x\right), \tag{3.54}$$

$$Y_{k_3}(y) = A_y\, \sin\left(\frac{k_3}{\sqrt{D}}y\right) + B_y\, \cos\left(\frac{k_3}{\sqrt{D}}y\right), \tag{3.55}$$

where the two constants of separation are combined into one for clarity

$$k_3^2 \equiv k_1^2 - k_2^2. \tag{3.56}$$

A more complete solution, with time dependence, is

$$c(x, y, t) = e^{-(k_2^2 + k_3^2)t}\, X_{k_2}(x)\, Y_{k_3}(y). \tag{3.57}$$

3.4.2 Boundary Conditions and Initial Conditions

Now specify the shape of the material and its initial concentration profile, that is, the boundary conditions and the initial conditions. The shape is a rectangular bar, long in the z-direction. We also need to specify what happens when the diffusing atoms reach the surface. At the surface, we assume the atoms are lost.[11] This describes, for example, atoms of hydrogen diffusing out of a solid and leaving as a gas of low density. For initial conditions, we choose a uniform composition throughout the bar. With time, the loss of solute from the surface causes a decrease of composition near the surfaces, and later a loss of solute throughout the bar.

The problem we are considering has no variation of composition along $\hat{z}$. Our solution could pertain to an infinite bar[12] oriented so its long direction is along $\hat{z}$. Intersecting the x–y-plane, one corner is at the origin, and the opposite corner is at $x = L_x$ and $y = L_y$. Since solute is lost at the surface, the surface concentration is 0. The boundary conditions involve only the spatial coordinates, and are valid for all time

$$c(0, y, t) = 0, \ c(x, 0, t) = 0, \ c(L_x, y, t) = 0, \ c(x, L_y, t) = 0. \tag{3.58}$$

Evidently the cosine solutions of Eqs. 3.54 and 3.55 are inappropriate, since they cannot be zero at the origin, and cannot satisfy the first two boundary conditions in Eq. 3.58. With $B_x = 0$ and $B_y = 0$, Eqs. 3.54 and 3.55 give

$$c(x, y, t) = A_x A_y \, e^{-(k_2^2 + k_3^2)t} \, \sin\left(\frac{k_2}{\sqrt{D}} x\right) \sin\left(\frac{k_3}{\sqrt{D}} y\right). \tag{3.59}$$

The constants k_2 and k_3 are obtained by imposing the other two boundary conditions of zero concentration at $x = L_x$ and $y = L_y$. This is achieved when the arguments of the sine functions are multiples of π at L_x and L_y

$$c(x, y, t) = A_{mn} \, e^{-\left(\frac{m^2}{L_x^2} + \frac{n^2}{L_y^2}\right)\pi^2 D t} \, \sin\left(\frac{\pi m}{L_x} x\right) \sin\left(\frac{\pi n}{L_y} y\right), \tag{3.60}$$

so the boundary conditions give

$$k_2 = \frac{\pi \sqrt{D}}{L_x} \, m, \text{ where } m \text{ is an integer,} \tag{3.61}$$

$$k_3 = \frac{\pi \sqrt{D}}{L_y} \, n, \text{ where } n \text{ is an integer.} \tag{3.62}$$

3.4.3 Orthogonality and Initial Conditions

Finally we impose the initial condition of uniform starting concentration, c', at $t = 0$. We need a sum of sine functions along x and y that matches this flat concentration profile. (Once set, each sine function will evolve independently in time.) Noting that the exponential factor with the time dependence in Eq. 3.60 equals 1 at $t = 0$, the initial profile is

$$c' = \sum_m \sum_n A_{mn} \, \sin\left(\frac{\pi m}{L_x} x\right) \sin\left(\frac{\pi n}{L_y} y\right). \tag{3.63}$$

---

[11] If they were not, they would return to the bar, and there would be no change in the flat composition profile.
[12] A shorter bar or even a plate could also be appropriate, but only if solute cannot exit through the surfaces normal to $\hat{z}$.

To solve for the set of constants $\{A_{mn}\}$, we use orthogonality relations of the sine functions. This requires multiplying both sides of Eq. 3.63 by specific sine functions (denoted by m' and n'), and integrating over the intervals from 0 to L_x and 0 to L_y. Equation 3.63 becomes

$$\text{LHS} = \text{RHS}, \tag{3.64}$$

$$\text{LHS} = \int_0^{L_x} \int_0^{L_y} c' \sin\left(\frac{\pi m'}{L_x}x\right) \sin\left(\frac{\pi n'}{L_y}y\right) dx\, dy, \tag{3.65}$$

$$\text{RHS} = \sum_{m,n} A_{mn} \int_0^{L_x} \int_0^{L_y} \sin\left(\frac{\pi m'}{L_x}x\right) \sin\left(\frac{\pi m}{L_x}x\right) \sin\left(\frac{\pi n'}{L_y}y\right) \sin\left(\frac{\pi n}{L_y}y\right) dx\, dy. \tag{3.66}$$

The integrals of Eq. 3.66 are zero unless $m = m'$ and $n = n'$, owing to the orthogonality of sine functions over their interval of periodicity. This is a standard result, but is illustrated for an example in Fig. 3.6.

Orthogonality simplifies Eq. 3.66 – when $m = m'$, the x-part of the integrand becomes $\sin^2(\pi mx/L_x)$, which integrates to $L_x/2$ over the x-interval, and likewise for the y-integral

$$\text{RHS} = A_{m'n'} \frac{L_x}{2} \frac{L_y}{2}. \tag{3.67}$$

The integrals on the LHS (Eq. 3.65) give cosines evaluated between limits

$$\text{LHS} = \frac{c' L_x L_y}{\pi^2 m' n'} \cos\left(\frac{\pi m'}{L_x}x\right) \cos\left(\frac{\pi n'}{L_y}y\right)\Bigg|_{x=0,\ y=0}^{x=L_x,\ y=L_y}. \tag{3.68}$$

At the lower limits, the cosines evaluate to +1. At the upper limits, the cosines evaluate to +1 for even integers, and −1 for odd. Subtracting the results from the two limits gives zero when m' and n' are even, so fitting to the initial conditions forces m' and n' to be odd. Equating the LHS and RHS of Eqs. 3.67 and 3.68, and eliminating the primes for clarity, the Fourier coefficients $A_{m,n}$ are

$$A_{mn} = \frac{16\, c'}{\pi^2\, m\, n}, \tag{3.69}$$

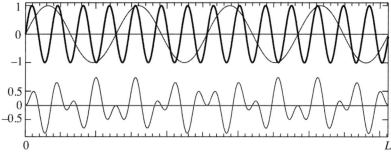

Figure 3.6 Top: two sine waves commensurate with the interval $0 - L$, but with different periodicities. Bottom: product of the two sine waves, showing equal areas of negative and positive values. The net integral is zero.

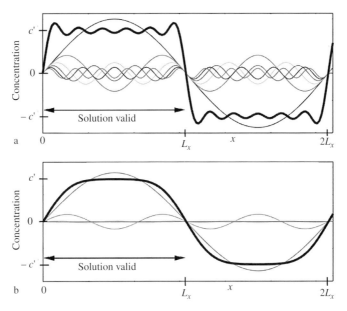

Figure 3.7 Concentration profile along x-axis (black) and individual terms with $1 \leq m = \leq 11$. (**a**) Initial condition at $t = 0$; (**b**) profile after $t = 0.1 L_x^2/(\pi^2 D)$.

and the full solution is

$$c(x, y, t) = c' \sum_{m,n} \frac{16}{\pi^2 \, m \, n} e^{-\left(\frac{m^2}{L_x^2} + \frac{n^2}{L_y^2}\right)\pi^2 D t} \sin\left(\frac{\pi m}{L_x}x\right) \sin\left(\frac{\pi n}{L_y}y\right), \qquad (3.70)$$

$$m = 1, \, 3, \, 5, \, 7, \ldots \qquad (3.71)$$

$$n = 1, \, 3, \, 5, \, 7, \ldots \qquad (3.72)$$

The solution for $c(x, y, t)$ in Eq. 3.70 is a Fourier series in x and y with coefficients that depend on time. When $t = 0$, the exponential is $+1$, and the Fourier series describes a rectangular composition profile of composition c', with sharp edges that drop to zero at the edge of the bar.[13] The time dependences of the Fourier coefficients do not depend on composition, only on the order of the sine waves. Suppose for a moment that the bar has a square cross-section, so $L_x = L_y$. The exponential in Eq. 3.70 is then $\exp(-\pi^2 D(m^2 + n^2)/L_x^2 \, t)$. The rate of decay of a Fourier coefficient goes as $m^2 + n^2$, meaning that the higher-order coefficients decay first. The higher-order Fourier coefficients are the ones needed to give the composition profile its sharp edge, so after only short times the edges broaden, much as shown in Fig. 3.5b. At late times, the higher-order coefficients can be neglected, and the composition profile can be approximated with $m = n = 1$, giving

$$c(x, y, t) = c' e^{-2\pi^2 D t/L_x^2} \sin\left(\frac{\pi}{L_x}x\right) \sin\left(\frac{\pi}{L_y}y\right), \qquad (3.73)$$

[13] This solution can be readily modified by adding a background concentration, instead of having zero concentration outside the bar.

for a square bar at late times. This late-stage composition profile is a smooth mound of considerably less height than c', and decreases uniformly in height with time. Even for shorter times the composition profile is soon dominated by the lowest terms in the Fourier series, as shown in Fig. 3.7b.

3.5 Bessel Functions and Other Special Function Solutions

3.5.1 General Approach for Special Function Solutions

Solving the diffusion equation by the separation of variables is possible for other coordinate systems. It is always straightforward to separate the time dependence from the spatial coordinates, as in Eq. 3.49. The important consideration for other coordinate systems is whether the Laplacian ∇^2 is separable. It turns out that there are 13 coordinate systems for which the Laplacian is separable,[14] so the concentration can be expressed as a product of spatial functions (times exponential functions for time). The solution to the diffusion equation involves these mathematical steps:

- separation of the diffusion equation to obtain a set of constants of separation;
- use of boundary conditions to pick acceptable families of the special functions characteristic of the coordinate system;
- expanding the initial concentration profile as a sum of special functions for that coordinate system;
- using the orthogonality relationships of the special functions to determine the weights of the different functions in the expansion of the initial profile;
- assigning a different time dependence to each term in the expansion, based on the constants of separation.

3.5.2 Bessel Function Solutions to the Diffusion Equation

Here we sketch the approach for cylindrical coordinates, which shows similarities and differences to the Fourier series approach for the Cartesian system. The diffusion equation (Eq. 3.28) in cylindrical coordinates is

$$\frac{\partial c}{\partial t} = D \left[\frac{\partial^2 c}{\partial r^2} + \frac{1}{r} \frac{\partial c}{\partial r} + \frac{1}{r^2} \frac{\partial^2 c}{\partial \phi^2} + \frac{\partial^2 c}{\partial z^2} \right], \tag{3.74}$$

where the concentration is $c(r, \phi, z, t)$. As for our problem in the Cartesian coordinate system, the z-dependence is ignored by assuming an infinitely long bar along $\hat{z}$. Anticipating separation in the coordinates r, ϕ, t, the concentration is written as the product of three functions

---

[14] Rectangular, cylindrical, elliptic cylindrical, parabolic cylindrical, spherical, conical, parabolic, prolate spheroidal, oblate spheroidal, ellipsoidal, paraboloidal, bispherical, toroidal (although the last two require a somewhat different approach).

$$c(r, \phi, t) = e^{-D\lambda^2 t} R(r) \, e^{im\phi}. \tag{3.75}$$

After substitution into Eq. 3.74 and some algebra

$$0 = \frac{d^2 R}{dr^2} + \frac{1}{r}\frac{dR}{dr} + \left(\lambda^2 - \frac{m^2}{r^2}\right)R. \tag{3.76}$$

Equation 3.76 is Bessel's equation. We pick a simple case where the concentration profile is uniform inside the bar, so $m = 0$. The solutions are Bessel functions of zero order, $J_0(\lambda r)$, and its derivatives are

$$J_0(\lambda r) = 1 - \frac{\lambda^2 r^2}{2^2} + \frac{\lambda^4 r^4}{2^2\,4^2} - \frac{\lambda^6 r^6}{2^2\,4^2\,6^2} + \cdots, \tag{3.77}$$

$$\frac{\partial}{\partial r} J_0(\lambda r) = -\frac{\lambda^2 r}{2^1} + \frac{\lambda^4 r^3}{2^2\,4^1} - \frac{\lambda^6 r^5}{2^2\,4^2\,6^1} + \cdots, \tag{3.78}$$

$$\frac{\partial^2}{\partial r^2} J_0(\lambda r) = -\frac{\lambda^2}{2^1} + \frac{3\lambda^4 r^2}{2^2\,4^1} - \frac{5\lambda^6 r^4}{2^2\,4^2\,6^1} + \cdots. \tag{3.79}$$

Substituting the Bessel function and its derivatives into Eq. 3.76, by grouping terms in powers of r it can be confirmed that all vanish as required.

3.5.3 Boundary Conditions, Initial Conditions, Orthogonality

We select a boundary condition for our cylindrical rod of radius a so that the concentration at a is 0. Inside the rod, the initial concentration is c'. A Bessel function is a smooth function with oscillations that swing through zero. One Bessel function cannot reproduce our initial condition with a sharp cutoff of concentration at $r = a$. The parameter λ rescales the Bessel function, however. Just like the case for a Fourier expansion, the solution is a sum of Bessel functions, scaled so their oscillations go to zero precisely at $r = 0$. This is known as a "Fourier–Bessel expansion"

$$c(r, t) = \sum_{\alpha_i} A_i \, e^{-\frac{\alpha_i^2}{a^2}Dt} J_0\left(\frac{\alpha_i r}{a}\right), \tag{3.80}$$

where the different values of λ are set to α_i/a. The α_i are special numbers – the "roots of the Bessel function," corresponding to where the Bessel function crosses zero. A Bessel function with the argument $\alpha_i r/a$ will therefore be zero at $r = a$.

To fit the initial concentration profile at $r < a$, we need the orthogonality relationship

$$\int_0^1 r J_0\left(\frac{\alpha_i r}{a}\right) J_0\left(\frac{\alpha_j r}{a}\right) dr = 0 \quad \text{unless } i = j, \tag{3.81}$$

where r in the integrand gives added weight to the outer parts of the Bessel functions (unlike the orthogonality for Fourier series solutions). The $\{\alpha_i\}$, the roots of J_0, are $\{2.405, 5.52, 8.65, 11.79, \ldots\}$. To match the initial profile at $t = 0$, the LHS of Eq. 3.80 is set equal to c', the exponential is equal to 1, both sides are multiplied by $J_0(\frac{\alpha_1 r}{a})$, and integrated

$$c' \int_0^a r J_0\left(\frac{\alpha_1 r}{a}\right) dr = \sum_{\alpha_i} A_i \int_0^a r J_0\left(\frac{\alpha_1 r}{a}\right) J_0\left(\frac{\alpha_i r}{a}\right) dr. \tag{3.82}$$

The RHS of Eq. 3.82 is zero for all terms in the sum, except when $\alpha_i = \alpha_1$. Here the integrand is $rJ_0^2(\frac{\alpha_1 r}{a})$. These expressions are also needed:

$$\int_0^1 xJ_0(\alpha_i x)\,dx = \frac{1}{\alpha_i}J_1(\alpha_i), \tag{3.83}$$

$$\int_0^1 xJ_0^2(\alpha_i x)\,dx = \frac{1}{2}J_1^2(\alpha_i), \tag{3.84}$$

where

$$\int_0^1 xJ_1(x)\,dx = \frac{x}{2} - \frac{x^3}{2^2 4} + \frac{x^5}{2^2 4^2 6} - \cdots, \tag{3.85}$$

(which also equals $-dJ_0(x)/dx$). The coefficients of the Fourier–Bessel expansion are

$$A_1 = \frac{2c'}{\alpha_1 J_1(\alpha_1)}, \tag{3.86}$$

$$A_2 = \frac{2c'}{\alpha_2 J_1(\alpha_2)}, \tag{3.87}$$

$$\cdots$$

The first terms in the Fourier–Bessel expansion for $c(r, t)$ are shown in Fig. 3.8 for $t = 0$. All have the same characteristic shape of J_0, but are rescaled in height and width. Although there are sign oscillations in A_i from the values of $J_1(\alpha_i)$, all the Bessel functions in the

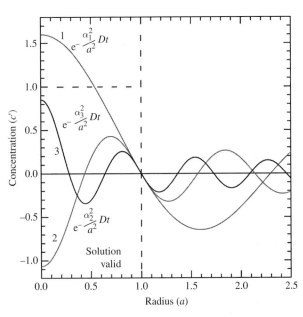

Figure 3.8 The first three functions in the Fourier–Bessel expansion ($i = 1, 2, 3$) for a uniform concentration in a cylindrical bar of radius a, labeled with characteristic decay behaviors.

series make positive contributions to the concentration just below $r = a$. This sharp edge
continues through to $r > a$, but the solution has no meaning beyond $r = a$ because this is
outside the cylindrical bar. The rates of decay increase as the square of α_i, so the higher-
order curves decay much faster than low-order ones. The values of α_i^2 are 5.78, 30.5,
and 74.8 for $i = 1, 2, 3$, and these rapidly increasing α_i^2 are in the exponential of the
time function.

Problems

3.1 (a) For diffusion in one dimension, show by substitution that the partial derivatives
in Eqs. 3.30, 3.31, and 3.32 satisfy the diffusion equation, Eq. 3.24.
(b) Explain why it is physically reasonable that the first and second spatial derivatives of $c(x, t)$ are odd and even functions of x.

3.2 Suppose two diffusion profiles have been measured in two identical samples heated
for the same time of 1000 s, but at two different temperatures. The characteristic
diffusion length at 900 °C was found to be 1.0 μm, and 2.7 μm at 1000 °C.
(a) Calculate the activation energy for this diffusion process.
(b) Suppose the diffusion coefficient is of the form of Eq. 3.9 and the diffusion
distances are precisely $x = \sqrt{4Dt}$. What is D_0?

3.3 Solve the diffusion equation for a long rectangular bar, much as was done in
Section 3.4, but with its cross-section centered about the origin. The initial condition
is still a uniform concentration of c', but the boundary conditions are now

$$c(-L_x/2, y, t) = 0, \ c(x, -L_y/2, t) = 0, c(L_x/2, y, t) = 0, \ c(x, L_y/2, t) = 0. \quad (3.88)$$

3.4 Consider the diffusion problem in cylindrical coordinates, with the initial and
boundary conditions of Section 3.5. Graph the approximate radial concentration
profile in the cylinder for the times $t' = 0$ (easy), $t' = 0.05$, $t' = 0.10$, and $t' = 0.5$,
where t' is in units of a^2/D.

3.5 Consider the diffusion of hydrogen through a thick-walled tube having an inner
diameter, b, that is small compared to the outer radius at $r = a$. The concentration of
hydrogen at $r = 0$ is always c', and the concentration is zero at $r = a$ as it escapes
into the atmosphere. After a long time, what is the concentration profile of hydrogen
at different r through the thick wall of the tube? Graph it.

3.6 Suppose atoms in a particular material move by a vacancy mechanism. Now suppose
there is a background concentration of vacancies in this material at all temperatures,
even at low temperatures, of $c_{v,e} = 10^{-10}$. These are called "extrinsic" vacancies.
(Sometimes they could be associated with chemical impurities, for example.) Added
to this background concentration of extrinsic vacancies are additional vacancies that
have an energy of formation of 1.0 eV/atom, so their population depends strongly on
temperature. Both types of vacancies have an energy of migration of 0.5 eV/atom.

(a) Over the temperature range 300 to 1000 K, make a semi-log plot of the concentrations of both intrinsic and extrinsic vacancies (i.e., plot two curves of c vs. T, using a logarithmic axis for c). At what temperature are the two concentrations equal? At what temperature will the number of intrinsic vacancies be zero (i.e., less than one vacancy) in a piece of material with 10^{23} atoms?

(b) Again on one semi-log plot, for each type of vacancy make a graph of the diffusion coefficient $D(T)$ versus $1/T$ over the range $1/1000$ to $1/300\,\mathrm{K}^{-1}$. (Temperature-independent prefactors involving a, Γ, f, and the entropic part of ΔG can be collected into a constant D_0, which might be $1\,\mathrm{cm}^2/\mathrm{s}$, for example.) What is the temperature where the two curves cross? Why does this relate to a result of part a?

(c) On the same graph as part b, plot the total $D(T)$ from both intrinsic and extrinsic vacancies. (i) Why is this curve well approximated as two straight lines? (ii) If the purity of the material were improved and $c_{v,e} = 10^{-11}$, approximately how much lower in temperature will the intrinsic behavior extend?

Nucleation

A phase transformation can start continuously or discontinuously (Section 1.5.2). The widespread discontinuous case is initiated by a small but distinct volume of material taking a structure and composition that differ from those of the parent phase.[1] A discontinuous transition can be forced by symmetry, as formalized in Section 18.4. There is no continuous way to rearrange the atoms of a liquid into a crystal, for example. A new crystal must appear in miniature in the liquid in a process called "nucleation." If the nucleation event is successful, this crystal will consume more liquid phase as it grows. Nucleation has many variations, but some key concepts can be appreciated immediately:

- Because the new phase and the parent phase have different structures, there must be an interface between them. The atom bonding across this interface is not optimal,[2] so the interfacial energy must be positive. This surface energy is most significant when the new phase is small, because a larger fraction of its atoms are at the interface. Surface energy plays a key role in nucleation.
- For nucleation of a new phase within a solid, the new phase generally differs in shape or density from the parent phase. The mismatch creates an elastic field that costs energy. This differs for nucleation in a liquid or gas, where the surrounding atoms can flow out of the way.
- When the nucleus has a different chemical composition than the parent phase, another issue is the time required for atoms to diffuse to incipient nuclei (sometimes called "embryos"). The addition of atoms to embryos is largely a kinetic phenomenon, although the tendency of atoms to remain on the embryos is a thermodynamic one.

Chapter 4 begins with the thermodynamics of forming "critical" nuclei that can grow. The chapter ends with a discussion of the transient time after a quench from high temperature when the distribution of solute relaxes towards the equilibrium distribution for steady-state nucleation.

4.1 Nucleation Phenomena and Terminology

Nucleation can occur without a change in crystal structure. Consider an A-rich A–B alloy having the α-phase at high temperature, as shown in the unmixing phase diagram

---

[1] This nucleus may or may not have the structure and composition of the final phase because the transformation may occur in stages.

[2] If the structure of the interfacial atoms and bonds were favorable, the new phase would take this local atomic structure.

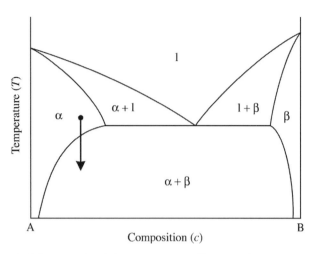

Figure 4.1   Binary phase diagram depicting a quench path from a temperature with pure α-phase to a temperature where some β-phase will nucleate.

of Fig. 2.17. Suppose the alloy is quenched (cooled quickly) to a temperature such as $0.4zV/k_B$, where the equilibrium state would be mostly A-rich α'-phase, plus some B-rich α''-phase. For some compositions, the B-rich α''-phase may nucleate as small zones or "precipitates" in an A-rich matrix.[3] In this case, the underlying crystal lattice remains the same while the solute atoms coalesce. A different case is shown by the arrow in Fig. 4.1. Here the precipitation of β-phase in an alloy quenched from the α-phase requires both a redistribution of chemical elements and a different crystal structure. In both these examples of nucleation, the parent phase is "supersaturated" immediately after the quench, and is unstable against forming the new phase.

"Homogeneous" nucleation occurs when nuclei form randomly throughout the bulk material; i.e., without preference for location. "Heterogeneous" nucleation refers to the formation of nuclei at specific sites. In solid $\rightarrow$ solid transformations, heterogeneous nucleation occurs on grain boundaries, dislocation lines, stacking faults, or other defects or heterogeneities. When freezing a liquid, the wall of the container is the usual site for heterogeneous nucleation. For the nucleation of solid phases, heterogeneous nucleation is more common than homogeneous.

The precipitate phase can be "coherent" or "incoherent" with the surrounding matrix. Figure 4.2a illustrates an incoherent nucleus. The precipitating β-phase has a crystal structure different from the parent α-phase. There is poor registry of atoms across the interface, so atoms at the interface have a higher energy from poor chemical bonding. In contrast, the coherent nucleus in Fig. 4.2b has crystal planes that match up with planes in the α-matrix. Heterogeneous, incoherent nucleation is often stated to be the most common type of precipitation reaction in solids. However, some atomic-level coherence would be expected in the very early stages of nucleation, and may go undetected. Homogeneous, coherent nucleation is well documented in some systems, however, including the case

---

[3] A "matrix" is the surrounding environment in which a new phase develops.

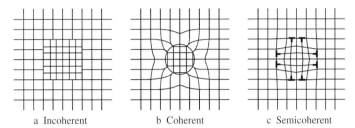

a Incoherent   b Coherent   c Semicoherent

Figure 4.2   Schematic representation of (**a**) an incoherent precipitate, (**b**) a coherent precipitate, and (**c**) a semicoherent precipitate interface.

of Co-rich precipitates in the Cu–Co system shown in Fig. 1.7. To date, there have been no reports of homogeneous, incoherent nucleation in any crystalline solid as depicted in Fig. 4.2a. Incoherent interfaces may form later during precipitate growth, however.

An important feature of the nucleation of a new solid phase within a solid is the elastic distortion both inside and outside the new particle. The relative importance of elastic energy changes as the precipitate grows. In the earliest stages when the precipitate is small and has a high surface-to-volume ratio, the effects of surface energy are dominant. The elastic energy depends on volume, and becomes more important than surface energy as the precipitate grows larger. Some reduction in elastic energy occurs if the interface loses coherency, even if this causes the surface energy to increase. To help relieve the elastic strain, an array of dislocations may form on the boundary, as shown in Fig. 4.2c. An interface that is coherent except for a periodic sequence of dislocations is "semicoherent." With a greater mismatch of lattice parameters across the interface, the spacing between dislocations decreases and the surface energy increases.

The full picture of early and late stages of phase transformations has many variants that are described in Part III of the book. This chapter addresses the earliest stage of a first-order phase transformation driven by a discontinuity in dF/dT. A small, distinct nucleus of the new phase is formed, and a nucleation rate is calculated by considering the probability of finding a nucleus of a particular size, and by considering its tendency to grow. The surface energy plays a key role and is considered first, but the behavior can be modified significantly by elastic energy.

4.2 Critical Nucleus

4.2.1 Fluctuations

In a liquid there are dynamical rearrangements of atoms, and transient appearances and disappearances of different atomic configurations. Some of these have similarities to crystals, or distorted crystals, and these crystal-like fluctuations are expected to become more frequent as the liquid is cooled towards the melting temperature, T_m. Crystalline regions are unstable above T_m, of course, but at temperatures slightly above T_m their penalty in free energy is expected to be small by Eq. 1.6, so crystal-like fluctuations may be larger and more persistent.

As the temperature falls below T_m, the growth of crystalline regions is expected. Infinitesimally below T_m, however, the difference in free energy of the liquid and solid is infinitesimally small. The slightest energy penalty, in particular a surface energy, suppresses nucleation because the crystal-like fluctuations are small in size, and surface energy affects a larger fraction of atoms in small particles.

Now consider a liquid that is quenched rapidly to a temperature moderately below T_m. At first the atom configurations in the liquid are appropriate for a high temperature, but are considerably out of equilibrium at the new low temperature. The local configurations in the undercooled liquid evolve over time, and eventually a steady-state distribution of local structures is achieved. After a time delay, called an "incubation time," a steady-state distribution is attained where some local regions are large enough to continue growing as crystals by adding more atoms to their surfaces. The phase transformation is then well underway.

4.2.2 Surface Energy

Consider the formation of a nucleus as a fluctuation that rearranges atoms in a small volume of the parent phase. The probability, $\mathcal{P}$, that a group of atoms is configured as a nucleus is

$$\mathcal{P}(T) = \frac{\exp\left(-\frac{\Delta G}{k_B T}\right)}{Z}, \tag{4.1}$$

where ΔG is the change in free energy when the group of atoms is configured as the new phase (compared with the parent phase), and Z is a partition function comprising the sum of Gibbs factors for all possible configurations, so Z normalizes the probability. The ΔG includes a term $A\sigma$, the area of interface between the two phases times the specific surface energy. This is a positive quantity, i.e., an energy penalty. The change in Gibbs free energy upon forming the nucleus is

$$\Delta G = V \Delta G_V + A \sigma, \tag{4.2}$$

where ΔG_V is the Gibbs free energy per unit volume of forming the new phase from the parent phase, and V is the volume of the new phase. In what follows, some of the most important results are obtained because A is not proportional to V.

Assume the surface energy is isotropic, so σ is independent of crystallographic direction.[4] The equilibrium shape that minimizes the surface area-to-volume ratio is a sphere of radius R

$$\Delta G(R) = \frac{4}{3}\pi R^3 \Delta G_V + 4\pi R^2 \sigma. \tag{4.3}$$

---

[4] This is not expected for crystalline solids. Usually low-index planes such as {110} or {111} have the lowest surface energy, and the particle is shaped to maximize these surfaces. The particle will have a larger surface area than a sphere, but will have a lower surface energy. Determining this shape requires the "Wulff construction" discussed in Section 11.5.

When $\Delta G_V > 0$, ΔG is never negative (because $\sigma > 0$), so nucleation cannot occur for any R. When $G_V < 0$, nucleation of the new phase is possible for large R.

4.2.3 Critical Radius

Figure 4.3 shows the two terms of Eq. 4.3. It is drawn for $\Delta G_V < 0$, where the volume energy (varying as R^3) is negative and surface energy (varying as R^2) is positive. The surface energy dominates at small R, but the volume term dominates for large R. This trend reflects the relations: $R^2 > R^3$ when $R \rightarrow 0$, and $R^3 > R^2$ when $R \rightarrow \infty$. The slope of $\Delta G(R)$ therefore turns over, so $\Delta G(R)$ has a maximum value, ΔG^*, at R^*, which is found by differentiating Eq. 4.3

$$\frac{d\Delta G}{dR}\bigg|_{R=R^*} = 4\pi R^{*2} \Delta G_V + 8\pi R^* = 0 \quad \Rightarrow \quad R^* = -\frac{2\sigma}{\Delta G_V}, \tag{4.4}$$

and by substitution of R^* into Eq. 4.3

$$\Delta G^* \equiv \Delta G(R^*) = \frac{4}{3}\pi \left(-\frac{2\sigma}{\Delta G_V}\right)^3 \Delta G_V + 4\pi \left(-\frac{2\sigma}{\Delta G_V}\right)^2 \sigma \tag{4.5}$$

$$\Rightarrow \Delta G^* = \frac{16\pi}{3} \frac{\sigma^3}{(\Delta G_V)^2}. \tag{4.6}$$

Approximately, the critical radius R^* separates two types of behavior. For sizes greater than R^*, the particle tends to grow rather than shrink – Fig. 4.3 shows that, for $R > R^*$, particle growth causes $\Delta G(R)$ to decrease. On the other hand, for sizes smaller than R^*,

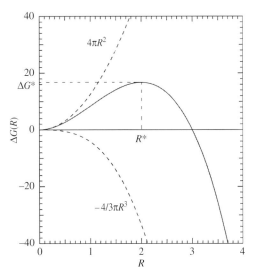

Figure 4.3   Graph of $\Delta G(R) = 4\pi R^2 - 4/3\pi R^3$, with individual terms as dashed lines. The critical radius, R^*, and critical free energy of formation, ΔG^*, are labeled for this case of $\Delta G_V = -1$ and $\sigma = 1$. (Equivalently, the horizontal axis has units of $-\sigma/\Delta G_V$, and the vertical has units of $\sigma^3/\Delta G_V^2$.)

the particle tends to shrink rather than grow because shrinkage causes $\Delta G(R)$ to decrease. The critical radius R^* and the critical free energy of formation ΔG^* are key quantities for nucleation.

At R^* the variation of free energy with respect to R is zero, so a nucleus of radius R^* is in an "unstable equilibrium" with respect to the parent phase. A chemical potential can be defined as in Eq. 1.24, and this is the chemical potential for atoms both in and near the critical nucleus because they are in local equilibrium. The cost in free energy to add an atom to this region is large for a small nucleus with a high surface-to-volume ratio. This effect of surface area on chemical potential is known as the "Gibbs–Thomson effect," and is described in more detail in Section 20.2.[5]

4.2.4 Temperature Dependence

We continue with the thermodynamics of nucleation by returning to Eq. 4.1 for $\mathcal{P}(T)$. Making these approximations:

- The only activation energy ΔG of interest is the activation to the critical radius, ΔG^* in Eq. 4.6.
- The specific surface energy, σ, has no significant temperature dependence near the critical temperature.
- The change in volume free energy between solid and liquid is that of Eq. 1.6, i.e., $\Delta G_V \simeq -L\,\Delta T/T_c$ for a small or modest undercooling $\Delta T = T_c - T$. (Here L is the latent heat of the phase transformation.)
- The partition function Z is not known, but is assumed to be constant because it is dominated by states that depend weakly on ΔT compared to the state at the critical radius.

With these approximations, Eq. 4.6 becomes

$$\Delta G^* = \frac{16\pi}{3} \frac{\sigma^3\,T_c^{\,2}}{L^2\,\Delta T^2},\tag{4.7}$$

so the probability of a critical fluctuation has the temperature dependence

$$\mathcal{P}(T) = \frac{1}{Z}\exp\left[-\frac{16\pi}{3}\frac{\sigma^3}{L^2}\left(\frac{T_c}{\Delta T}\right)^2\frac{1}{k_B T}\right].\tag{4.8}$$

All the factors in Eq. 4.8 depend weakly or modestly on temperature around T_c, except the factor involving the undercooling itself, $1/\Delta T$. This factor diverges at T_c, is squared, and is in an exponential. The undercooling is therefore expected to dominate the probability of a critical fluctuation, at least when the undercooling is small. Although this is a thermodynamic analysis, it points out why the kinetics of nucleation below T_c depend strongly on temperature. In practice it may be useful to assign a specific temperature for the onset of nucleation, since the behavior is so sensitive to undercooling. More quantitative statements require an analysis of the kinetics of nucleation, however.

---

[5] Note that R^* in Fig. 4.3 is at $R = 2$, not at $R = 3$ where the surface and volume energies are equal.

4.3 Heterogeneous Nucleation

4.3.1 General Features of Heterogeneous Nucleation

Homogeneous nucleation in solid materials is not rare, but heterogeneous nucleation is more common. Here the second phase nucleates at dislocations, grain boundaries, or other particles, for example. The size, shape, and free energy of forming a critical nucleus are changed in the presence of a defect. The case of grain boundary nucleation is now treated in some detail.

For heterogeneous nucleation on a planar interface such as a grain boundary, a new feature of nucleation is the elimination of some of the grain boundary area (dashed line in Fig. 4.4), which is replaced by the new particle. For the previous analysis of homogeneous nucleation, a sphere is a natural shape because it minimizes a surface energy for a given volume. For heterogeneous nucleation on a grain boundary, the new shape of the nucleus accommodates these characteristics:

- In contact with the parent phase, the precipitate surface is spherical because the local atom structure is unaware of the grain boundary.
- The precipitate widens to eliminate more of the grain boundary.
- The precipitate is shaped as a double convex lens, or two spherical caps that make contact at the grain boundary.

4.3.2 Activation Barrier for Heterogeneous Nucleation

The surface area of a double spherical cap having a contact angle θ (defined in Fig. 4.4a) is $4\pi(1 - \cos\theta)R^2$, and its volume is $2\pi(2 - 3\cos\theta + \cos^3\theta)R^3/3$, where R is the (constant)

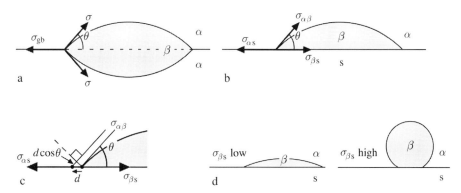

Figure 4.4 (**a**) A grain boundary β-phase nucleus as a double spherical cap. The dashed line shows the region of pre-existing grain boundary that is eliminated by the nucleus. (**b**) Nucleation of β-phase on a substrate, s. (**c**) A shift of d lengthens the α-β interface of panel b by $d\cos\theta$. (**d**) Contact angles of heterogeneous nucleation of β-phase on a substrate with low (left) and high (right) interface energy between particle and substrate.

radius of curvature. With these expressions, the free energy of formation as a function of R can be calculated in the same way as Eqs. 4.2 and 4.3

$$\Delta G = V \Delta G_V + A \sigma - A_{gb} \sigma_{gb}, \tag{4.9}$$

$$\Delta G = \frac{2\pi}{3} \left(2 - 3\cos\theta + \cos^3\theta \right) R^3 \Delta G_V + 4\pi (1 - \cos\theta) R^2 \sigma$$
$$- \pi R^2 \sin^2\theta \, \sigma_{gb}. \tag{4.10}$$

The last term on the RHS of Eq. 4.9 contains the grain boundary surface energy, σ_{gb}, and has a negative sign because a circular area of pre-existing α–α interface is eliminated by the new phase (dashed line in Fig. 4.4a). The energy $-\pi R^2 \sin^2\theta \, \sigma_{gb}$ decreases the free energy of formation. Since this decrease is not realized for homogeneous nucleation in the matrix, heterogeneous nucleation is preferred energetically.

Critical quantities can again be found by taking the derivative of ΔG with respect to the radius of curvature and setting the result equal to zero. It is interesting that the value of R^* is identical to that derived in the homogeneous case

$$R^* = -\frac{2\sigma}{\Delta G_V}, \tag{4.11}$$

whereas the critical free energy of formation becomes

$$\Delta G^* = \frac{16\pi\sigma^3}{3(\Delta G_V)^2} \frac{2 - 3\cos\theta + \cos^3\theta}{2}$$

$$\Delta G^* = \Delta G^*_{hom} \frac{2 - 3\cos\theta + \cos^3\theta}{2}. \tag{4.12}$$

Here ΔG^*_{hom} is the critical free energy of formation for homogeneous nucleation (Eq. 4.6). The term $(2 - 3\cos\theta + \cos^3\theta)/2$ can range from zero to one, so the free energy of formation of a heterogeneous grain boundary nucleus is less than that of a homogeneous nucleus.

4.3.3 Surface Energies in Heterogeneous Nucleation

Nucleation on the surface of a substrate is similar,[6] but as shown in Fig. 4.4b there are three phases, $\{\alpha, \beta, s\}$ (where s is the substrate), and between all pairs of phases there are interfaces with different surface energies, $\{\sigma_{\alpha\beta}, \sigma_{\alpha s}, \sigma_{\beta s}\}$. A feature of heterogeneous nucleation is the angle θ at the point of three-phase contact, which is set by the relative strengths of the three surface energies. If this point moves a small distance d to the left along the substrate surface in Fig. 4.4c, the β–s surface will increase at the direct expense of the α–s surface. Simultaneously, the amount of α–β surface increases in proportion to $\cos\theta$. With the equilibrium angle θ, the net change in surface energy is zero, so

$$\sigma_{\beta s} + \sigma_{\alpha\beta} \cos\theta - \sigma_{\alpha s} = 0. \tag{4.13}$$

The equilibrium contact angle can be understood as balancing the horizontal forces on the point of three-phase contact, where the net force is the change with d of the three surface

---

[6] The elimination of grain boundary energy by the growing β-phase is similar to the elimination of α–s surface, but the area and volume of the β phase are half as large.

energies. Suppose the β–s surface energy $\sigma_{\beta s}$ is very low. The angle θ must then be very small for force balance, meaning that the β phase spreads out over the substrate surface, preferentially wetting it. The opposite case where $\sigma_{\beta s}$ is very high gives a large θ (near 180°) and minimum contact between the β-phase and the substrate. The β-phase will form as spheres over the substrate, with energetics comparable to the homogeneous nucleation of β-phase in α.

Figure 4.4 was drawn assuming an isotropic surface energy for the α–β interface. Such a double spherical cap morphology is unlikely because a faceted nucleus is expected [29]. Different crystallographic interfaces can have substantial differences in their specific energies. Section 11.5 describes how equilibrium shapes of crystals are controlled by differences in surface energies, and similar considerations should apply to nuclei of new phases. Figure 14.1 shows a chemical zone in an early state of a precipitation reaction. This zone is planar and aligned along a $\langle 100 \rangle$ direction. Figure 4.5 shows such a faceted precipitate, called a "grain boundary allotriomorph." The image shows a precipitate of lead at the grain boundary between two aluminum crystals (the Al–Al grain boundary is about a quarter of the way up from the bottom of the image). The checked pattern at the top of the lead particle indicates regular atom matching across low-index crystal planes between the lead and aluminum. The bottom of the crystal is less well oriented for such matching, and the particle minimizes its surface area against the lower aluminum crystal by making a spherical surface with a large radius of curvature. Notice that

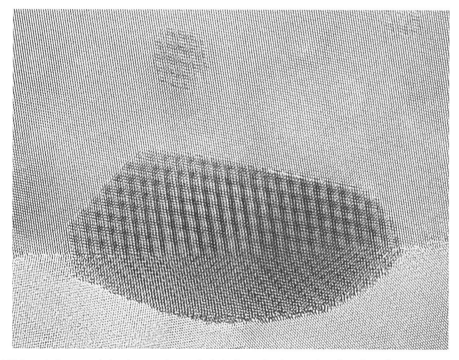

Figure 4.5   A high-resolution transmission electron micrograph of a lead crystal at the grain boundary of two aluminum crystals, and a second lead crystal above it. Figure courtesy of Uli Dahmen (see also [30]).

there are more lead atoms on the top side of the precipitate that has good atomic registry with the aluminum matrix.

The elimination of area of the Al–Al grain boundary favored the nucleation of the lead crystal at a grain boundary in Fig. 4.5, even though homogeneous nucleation also occurred in this material – notice the small lead crystal near the top of the image. Grain boundaries form a three-dimensional network through a polycrystalline material, and this network also offers edges and corners for nucleation. In general, more grain boundary area can be eliminated per unit volume of precipitate if nucleation occurs at an edge, and even more when nucleation is at a corner of three or more crystals. There are fewer of these nucleation sites, however.

4.4 Free Energy Curves and Nucleation

4.4.1 Shifts of Free Energy Curves

The critical nucleus is in a state of equilibrium, with the condition $\mathrm{d}\Delta G/\mathrm{d}R = 0$ at R^*. This is an unstable equilibrium, but an energy and entropy can be defined, and free energy curves prove useful for understanding the local equilibrium. Consider the nucleation of a β-phase precipitate in a matrix of α-phase, which we approach by putting the extra energy for nucleation into the curve $F_\beta(c)$ for the β-phase. We simply elevate the curve $F_\beta(c)$ to account for activating the nucleation of a small particle of radius R^*.[7] This energy includes the activation free energy $\Delta G^* = 16\pi\sigma^3/3(\Delta G_V)^2$ of Eq. 4.6. For solid precipitates that form in a solid matrix, any size or shape misfit generates elastic strains in both the precipitate and the surrounding matrix. The elastic energy, which is always positive,[8] adds to the free energy of nucleation and suppresses nucleation. The general idea is symbolized by adding the term $\Delta G^* + PdV$ to shift the curve $F_\beta(c)$ in Fig. 4.6a, but the elastic energy of a misfitting precipitate depends strongly on its shape, too, as discussed in Section 4.4.2.[9] The ΔG^* also depends strongly on undercooling owing to $\Delta G_V(T)$, causing the shift of $F_\beta(c)$ to decrease with undercooling. An appropriate shift of the curve $F_\beta(c)$ may be difficult to know.

When accounting for the activation barrier of nucleation, Fig. 4.6a shows how the common tangent line intersects the free energy curve of the α-phase at a chemical composition shifted upwards by the amount Δc. This is physically reasonable – with more free energy needed to nucleate the β-phase, the α-phase is supersaturated with more B-atoms. For local equilibrium evaluated by the common tangent, the β-phase is also enriched in B-atoms. These effects are often difficult to measure in small, isolated nuclei,

---

[7] Assume no strong dependence of the critical free energy of formation on chemical composition.

[8] Without the precipitate, the parent phase is assumed free of internal stresses, so precipitate formation generates strain energy. If there are pre-existing internal stresses in the parent phase, however, precipitates may nucleate at locations that reduce the strain energy.

[9] A different aspect of PdV pertains to nucleation in a material under applied pressure P, where pressures of several GPa can alter nucleation rates, depending on the change in volume of precipitation dV.

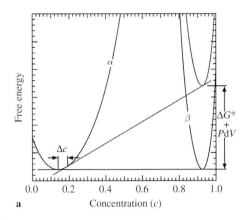
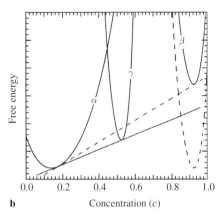

a   Concentration (c)   b   Concentration (c)

Figure 4.6   (**a**) Effect of surface energy and misfit elastic strains on the relative free energy of a β-phase nucleus.
(**b**) Opportunity for an intermediate phase to form in the case shown in panel a.

but they can also account for observable shifts of phase boundaries for nanoparticles. For nanoparticles of uniform size, the analysis can be made quantitative.[10]

Another effect associated with the energy barrier of nucleation is described with Fig. 4.6b. Although the γ-phase is not expected when the β-phase particles are large and $F_\beta(c)$ follows the dashed curve, the γ-phase may form if $F_\beta(c)$ is raised by a nucleation barrier (assuming the γ-phase does not have a high nucleation barrier, too). In many alloy systems there are several crystal structures of intermediate compositions having similar free energies, but different nucleation barriers. When differences in nucleation barriers are comparable to the free energy differences between phases, phases like the γ-phase may form in "pre-precipitation" processes. For example, if elastic energy dominates the problem, and if a nucleus of γ-phase has a small misfit, the γ-phase may be easier to nucleate than the β-phase. Often in these cases, the shapes of the γ-phase particles are flat plates, which may be taken as evidence of strain energy dominance as discussed below in Section 4.4.2. A Guinier–Preston zone shown in Fig. 14.1 and described in Section 14.1 is such a transient structure. There can even be a sequence of transient structures before the equilibrium phase nucleates.

4.4.2 Effects of Elastic Energy on Nucleation

To include the elastic energy from misfit stresses in and around a precipitate that nucleates in a solid, we need a result from Section 6.8. We pick the simple result of Eq. 6.142 for the total elastic energy, E_{tot}, of an ellipsoidal precipitate. This is an approximate result, based on the assumption that the precipitate itself is incompressible, and has no internal elastic energy. The precipitate is parameterized by its fractional volume change with respect to the volume of the same atoms in the parent phase, $\delta v/v$. This change in specific volume generates shear strains in the surrounding matrix, so the elastic energy scales with the shear

---

[10] For nanoparticles, surface energy effects are often called the "Gibbs–Thomson" effect of Section 20.2.

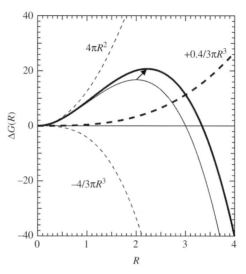

Figure 4.7 Graph of $\Delta G(R) = 4\pi R^2 - 4/3\pi R^3 + 0.4/3\pi R^3$, with individual terms as dashed lines. With this additional, positive strain energy, the critical radius, R^*, and critical free energy of formation, ΔG^*, shift as shown by the arrow. (As in Fig. 4.3, $\Delta G_V = -1$, $\sigma = 1$, and now $6\mu\delta^2 E(y/R) = 0.1$.)

modulus, μ. We also add the factor $E(y/R)$ (discussed after Eq. 6.142), to account for the aspect ratio of the ellipsoidal precipitate. It ranges from 0 for flat plates to 1 for a sphere. Elastic energy often favors precipitates shaped as flat plates (Sections 6.8.5 and 24.3.2 discuss the reasons). Flat precipitates have a high surface area per unit volume, however, so the actual shape also depends on the surface energy.

Adding the elastic energy to Eq. 4.3 gives the free energy change for forming the nucleus out of a volume $v = 4/3\,\pi\,R^3$ of the parent phase. With Eq. 6.142

$$\Delta G = \frac{4}{3}\pi R^3\,\Delta G_V + \frac{2}{3}\mu\frac{(\delta v)^2}{v}E(y/R) + 4\pi R^2\,\sigma. \tag{4.14}$$

For an approximately spherical precipitate, the ratio $\delta v/v$ can be expressed in terms of the relative change in radius $\delta \equiv \delta r/r$, which is $\delta = \delta v/3v$

$$\Delta G = \frac{4}{3}\pi R^3\left(\Delta G_V + 6\mu\delta^2 E(y/R)\right) + 4\pi R^2\,\sigma. \tag{4.15}$$

Figure 4.7 shows how the positive elastic energy counteracts the negative chemical free energy to make nucleation more difficult. Both the critical free energy of formation and the critical radius are increased. The total elastic energy and the change in volume free energy both scale as R^3, so the surface energy is again dominant at small R as seen in Fig. 4.7.

Because the elastic energy has the opposite sign as ΔG_V, a larger undercooling is required for nucleation. For an undercooling ΔT that gives a ΔG^* by Eq. 4.7, with elastic energy the same ΔG^* (and R^*) are obtained[11] at the undercooling $\Delta T + \delta T$ where

[11] Set to zero the derivative $d\Delta G/dR$ from Eq. 4.15 to find R^*, and substitute $\Delta G_V \simeq -L\,(\Delta T + \delta T)/T_c$. Then calculate R^* the same way with $\delta^2 = 0$ using $\Delta G_V \simeq -L\,\Delta T/T_c$. Equate them to find δT.

$$\delta T \simeq \frac{6\mu\delta^2 E(y/R)}{L} \, T_c = \frac{E_{el}}{L} \, T_c. \tag{4.16}$$

It is interesting to ask if elastic energy can suppress a phase transformation entirely, which would be the case if the elastic energy per unit volume of precipitate were greater than or equal to the chemical free energy per unit volume, i.e., $6\mu\delta^2 E(y/R) \geq -\Delta G_V$ in Eq. 4.15. What this would mean is that the phase transformation could occur when the transforming particle is in free space, but not when constrained by the solid matrix. There are in fact cases where elastic strains in the matrix suppress phase transformations by hundreds of degrees kelvin. Complete suppression seems to be unlikely. To help minimize the elastic energy, there are degrees of freedom in the shape and orientation of a precipitate, and $E(y/R)$ approaches zero in the case of flat plates. If homogeneous precipitates are observed as flat plates, or are shaped as lenses, the reason is often that these shapes and orientations minimize the elastic energy, whereas precipitates having more equal dimensions (termed "equiaxed morphologies") are suppressed by elastic energy. In short, elastic energy in solids often alters the shapes of precipitates, their energies, and their nucleation rate. A more detailed analysis of the evolving shape of a growing precipitate is given in Section 15.4. The chemical free energy tends to win in the end, however.

4.5 The Nucleation Rate

The thermodynamic concept of the critical nucleus of Section 4.2 is due to Gibbs [31]. It introduces the key concept of metastable equilibrium at a critical radius, which is useful for obtaining a general temperature dependence of nucleation. A thermodynamic theory is not intended for calculating the kinetics of nucleation, however. The Gibbs droplet picture of nucleation was extended to include kinetics by Volmer and Weber [32], Farkas [33], Becker and Döring [34], and Zeldovich [35] in the 1920s and 1940s. The rest of this chapter presents some of these kinetic analyses, and ends with an introduction to more modern concepts of time-dependent nucleation and nucleation in multicomponent systems.

4.5.1 Clusters of Atoms

All our information about the ΔG for nucleation is parameterized by the size of the nucleus (we assume that its shape is constant). For atomistic kinetics it is more convenient to work with the number of atoms in the cluster, n, instead of the radius, R. The equivalent to Eq. 4.3 is

$$\Delta G(n) = \Delta G_V \, \Omega \, n + \sigma \, s_1 \, n^{2/3}, \tag{4.17}$$

where Ω is the atomic volume, and s_1 is the atomic surface area. Figure 4.8 shows $\Delta G(n)$ for parameters in correspondence to those of Fig. 4.3. The curve is skewed to the right, since the addition of an atom to a large cluster gives a small contribution to the radius.

To describe the curve $\Delta G(n)$ of Fig. 4.8, two parameters are certainly useful – the critical nucleus size, n^*, and the critical free energy of formation, ΔG^*. These give the peak of

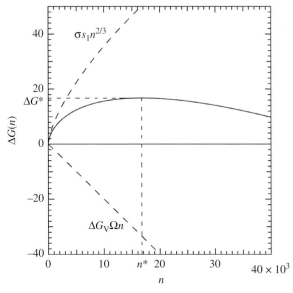

Figure 4.8 Graph of $\Delta G(n) = \Delta G_V \Omega n + \sigma s_1 n^{2/3}$, with individual terms as dashed lines. This graph corresponds to that of Fig. 4.3, with $R = 1$ corresponding to a 10-atom radius of a simple cubic crystal (4,188 atoms). Conversion gives $\Delta G_V \Omega = -0.002$, $\sigma s_1 = 0.07677$. The critical cluster size, $n^* = 16,752$, and critical free energy of formation, $\Delta G^* = 16.75$, are labeled.

$\Delta G(n)$ (the point at the intersection of the vertical and horizontal dashed lines in Fig. 4.8). To more completely specify the critical free energy of formation, we add a third parameter, the negative curvature of $\Delta G(n)$ at n^*. The Taylor expansion near n^* is

$$\Delta G(n^* + \delta n) = \Delta G(n^*) + \left.\frac{\partial \Delta G}{\partial n}\right|_{n^*} \delta n + \frac{1}{2} \left.\frac{\partial^2 \Delta G}{\partial n^2}\right|_{n^*} (\delta n)^2 + \cdots, \quad (4.18)$$

$$\Delta G(n^*) - \Delta G(n^* + \delta n) \simeq -\frac{1}{2} \left.\frac{\partial^2 \Delta G}{\partial n^2}\right|_{n^*} (\delta n)^2, \quad (4.19)$$

where the second term on the RHS of Eq. 4.18 is zero because the function $\Delta G(n)$ is a maximum at n^*.

4.5.2 Atom Additions and Subtractions

Consider the different ways a cluster can change its size. A cluster containing 5 atoms can, by fluctuation processes, combine with 1 atom to produce a cluster of size 6, an $n = 2$ cluster can attach to an $n = 7$ cluster creating a size $n = 9$, etc. The rates of cluster growth depend in part on how many clusters of each type are present to come together. In a vapor especially, it is easy to imagine collisions that make a cluster of n atoms by combining a cluster of m atoms with a cluster of $n - m$ atoms.[12] The rate $R(n, m)$ depends on mobilities

[12] In a hot vapor, when the kinetic energy of a cluster is comparable to the binding energy of an atom, the conservation of momentum may require interactions with a third cluster to carry momentum away from the collision.

and sticking coefficients, but is also proportional to the product of concentrations, $c(m) \times c(n-m)$

$$R(n,m) = c(m)c(n-m), \tag{4.20}$$

$$R(n,m) = c_0\, e^{-[\Delta G_V \Omega m + \sigma\, s_1\, m^{2/3}]/k_B T}\, c_0\, e^{-[\Delta G_V \Omega (n-m) + \sigma\, s_1\, (n-m)^{2/3}]/k_B T}, \tag{4.21}$$

$$R(n,m) = c_0^2\, e^{-\Delta G_V \Omega n/k_B T}\, e^{-\sigma\, s_1\, [m^{2/3} + (n-m)^{2/3}]/k_B T}. \tag{4.22}$$

To find the maximum in $R(n,m)$, examine $[m^{2/3} + (n-m)^{2/3}]$, a factor in the exponent of Eq. 4.22. This factor is positive, so if it is a minimum, the rate $R(n,m)$ is maximized. We seek an extremum as

$$\frac{\partial}{\partial m}[m^{2/3} + (n-m)^{2/3}] = 0, \tag{4.23}$$

$$\frac{2}{3}m^{-1/3} - \frac{2}{3}(n-m)^{-1/3} = 0, \tag{4.24}$$

$$m = \frac{n}{2}. \tag{4.25}$$

Note, however, that the curvature of this factor $[m^{2/3} + (n-m)^{2/3}]$ is negative

$$\frac{\partial^2}{\partial m^2}[m^{2/3} + (n-m)^{2/3}] = -\frac{2}{9}m^{-4/3} - \frac{2}{9}(n-m)^{-4/3}, \tag{4.26}$$

$$\frac{\partial^2}{\partial m^2}[m^{2/3} + (n-m)^{2/3}] < 0. \tag{4.27}$$

Therefore the factor $[m^{2/3} + (n-m)^{2/3}]$ is a maximum at $m = n/2$, meaning that the rate $R(n,n/2)$ is a minimum. The maximum values of $R(n,m)$ are when $m = 1$ or $m = n - 1$. In other words, the most rapid growth of clusters, based on the types of clusters available to combine, occurs by adding one atom at a time. Since single atoms tend to be most mobile, there are also kinetic reasons why single atom additions are faster. We believe that cluster growth mostly occurs by adding one atom at a time, a process termed "monomer attachment."

4.5.3 Steady-State Nucleation Rate

Flux through a Distribution of Cluster Sizes

To calculate a nucleation rate, we return to the kinetic master equation as illustrated in Fig. 1.9. Instead of moving atoms between adjacent bins, however, we move clusters of atoms. This is not a physical process, but it is a good way to track the time evolution of the "cluster size distribution," $\{N_n\}$, which specifies the number of clusters containing n atoms. In a material quenched from high temperature we assume that all clusters contain one atom, so $N_1 = N$, and $N_n = 0$ for $n > 1$. Very quickly, however, we expect the formation of clusters with $n = 2$, and eventually there will be clusters with a very large n that may be better described as crystals. Our concern, however, is for clusters with sizes around the critical n^*, since these are the ones most important for nucleation. In what follows we seek a steady-state condition over the time period of nucleation, where a large supply of single-atom clusters ($N_1 \simeq N$) contributes a flow of clusters from bin to bin of increasingly larger n.

Assuming the clusters grow or shrink by one atom at a time (as argued in the previous section), the movement of clusters occurs only between adjacent bins in Fig. 1.9. The flux of clusters depends on the numbers in the bin and their transition rates, as specified by the kinetic master equation. By analogy to Eq. 1.25

$$J_n = W_n[N_n - N_{n+1}], \tag{4.28}$$

where W_n is a transition rate that accounts for how fast atoms depart from a cluster of size n. The difference in W for a cluster of size n and $n+1$ is assumed small, so W_n is used for both. The flux J_n is the number of clusters per unit time that change from n to $n+1$. Here J_n is positive if more clusters move from n to $n+1$ than the reverse.

Assume there is a cluster distribution in thermodynamic equilibrium that can be known, perhaps with Eq. 4.1. It proves useful to normalize Eq. 4.28 with the equilibrium cluster numbers N_n^{eq}

$$\frac{J_n}{N_n^{\text{eq}}} = \frac{W_n}{N_n^{\text{eq}}} \left[N_n - N_{n+1} \right]. \tag{4.29}$$

For the equilibrium distribution, the net flux between adjacent cluster numbers must be zero (as can be shown with an approach like that of Eq. 5.19, et seq.)

$$W_n N_n^{\text{eq}} - W_{n+1} N_{n+1}^{\text{eq}} = 0, \tag{4.30}$$

$$N_{n+1}^{\text{eq}} = \frac{W_n N_n^{\text{eq}}}{W_{n+1}}, \tag{4.31}$$

and again assuming $W_n = W_{n+1}$,

$$\frac{J_n}{N_n^{\text{eq}}} = W_n \left[\frac{N_n}{N_n^{\text{eq}}} - \frac{N_{n+1}}{N_{n+1}^{\text{eq}}} \right]. \tag{4.32}$$

Steady-State Cluster Size Distribution

To track all cluster sizes $\{N_n\}$, we need equations like Eq. 4.32 for all n. This is generally quite complicated because a high flux into one cluster size can alter the fluxes to clusters of similar size at later times. Fortunately, it is not difficult to analyze the steady-state condition, where the cluster distribution $\{N_n\}$ does not change with time. In steady state

$$J_n = J_m \equiv J_{\text{ss}}, \tag{4.33}$$

because this ensures that the net flow of clusters into a bin n in Fig. 1.9 equals the flow out of the bin to the next bin $n+1$ in the chain. With equal flows in and out, the values of N_n remain constant.[13]

---

[13] This is indeed steady state, although the steady-state cluster size distribution is not necessarily in thermodynamic equilibrium.

The differences between N_n and N_{n+1} are not expected to be large, making it tricky to work with Eq. 4.32 directly. Using the steady-state condition of Eq. 4.33, Eq. 4.32 gives the same cluster flux for all n. Summing the J_n

$$\sum_{n=1}^{m-1} \frac{J_n}{N_n^{\text{eq}}} = W_n \left[\left(\frac{N_1}{N_1^{\text{eq}}} - \frac{N_2}{N_2^{\text{eq}}} \right) + \left(\frac{N_2}{N_2^{\text{eq}}} - \frac{N_3}{N_3^{\text{eq}}} \right) + \left(\frac{N_3}{N_3^{\text{eq}}} - \frac{N_4}{N_4^{\text{eq}}} \right) \right.$$
$$\left. + \cdots \left(\frac{N_{m-1}}{N_{m-1}^{\text{eq}}} - \frac{N_m}{N_m^{\text{eq}}} \right) \right], \tag{4.34}$$

and after numerous cancellations on the RHS

$$J_{\text{ss}} \sum_{n=1}^{m-1} \frac{1}{N_n^{\text{eq}}} = W_n \left[\frac{N_1}{N_1^{\text{eq}}} - \frac{N_m}{N_m^{\text{eq}}} \right]. \tag{4.35}$$

Now we impose physically reasonable boundary conditions. Assume the value of N_1 has its equilibrium value, N_1^{eq}, because small cluster sizes equilibrate most quickly:

$$(\text{B.C. 1)} \ \ N_1/N_1^{\text{eq}} = 1.$$

For steady state during the times of nucleation we need to assume an abundant supply of single atom clusters.[14] On the other hand, we expect few clusters of very large size, so $N_m \simeq 0$ for large m:

$$(\text{B.C. 2)} N_m/N_m^{\text{eq}} = 0.$$

These two boundary conditions simplify Eq. 4.35

$$J_{\text{ss}} = \frac{W_n}{\sum\limits_{n=1}^{m-1} \frac{1}{N_n^{\text{eq}}}} \equiv \frac{W_n}{Z'}. \tag{4.36}$$

The next step is to evaluate the denominator of Eq. 4.36, called Z', using Eq. 4.1

$$Z' = \sum_{n=1}^{m-1} \frac{1}{N_n^{\text{eq}}}, \tag{4.37}$$

$$Z' = \sum_{n=1}^{m-1} \frac{Z}{\exp\left(-\frac{\Delta G(n)}{k_{\text{B}} T}\right)}, \tag{4.38}$$

$$Z' = \sum_{n=1}^{m-1} Z \exp\left(\frac{\Delta G(n)}{k_{\text{B}} T}\right). \tag{4.39}$$

Flux of Clusters near the Critical Size

Clusters near the critical size n^* are the most important for nucleation. Because ΔG^* is the largest value of ΔG, these clusters make the largest contribution to Z'. Equation 4.19 provides useful information about how $\Delta G(n)$ varies with n near n^*

[14] This is not true much later in the post-nucleation stage when there is a significant fraction of new phase.

$$Z' = \sum_{n=1}^{m-1} Z \exp\left(\frac{\Delta G^*}{k_{\mathrm{B}}T}\right) \exp\left(\frac{\left.\frac{\partial^2 \Delta G}{\partial n^2}\right|_{n^*}(\delta n)^2}{2k_{\mathrm{B}}T}\right), \tag{4.40}$$

which is converted to a continuous variable

$$Z' = Z \exp\left(\frac{\Delta G^*}{k_{\mathrm{B}}T}\right) \int_{n=1}^{\infty} \exp\left(\frac{\left.\frac{\partial^2 \Delta G}{\partial n^2}\right|_{n^*}(\delta n)^2}{2k_{\mathrm{B}}T}\right) dn. \tag{4.41}$$

The sign of $\partial^2 \Delta G/\partial n^2|_{n^*}$ is negative, and this factor is a constant, so the integrand is recognized as a Gaussian function

$$Z' = Z \exp\left(\frac{\Delta G^*}{k_{\mathrm{B}}T}\right) \int_{-\infty}^{\infty} \exp\left(-an^2\right) dn, \tag{4.42}$$

where a is defined as

$$a \equiv -\frac{\left.\frac{\partial^2 \Delta G}{\partial n^2}\right|_{n^*}}{2k_{\mathrm{B}}T}. \tag{4.43}$$

There was also a change of the lower limit of integration in going from Eq. 4.41 to 4.42. In Eq. 4.41 the deviations δn are centered around the peak at n^*. Because $\delta n = n - n^*$, at $n = 1$ the absolute value $|\delta n| = n^* - 1$ is large, and the integrand is small. The lower limit can then be extended to $-\infty$. With these limits, the integration gives the same result for δn centered about n^* as for n centered about zero in Eq. 4.42. Recognizing the standard Gaussian integral of Eq. 9.52

$$Z' = Z \exp\left(\frac{\Delta G^*}{k_{\mathrm{B}}T}\right) \sqrt{\frac{\pi}{a}}. \tag{4.44}$$

Using this result in Eq. 4.36, the final expression for the steady-state flux of clusters across the critical cluster size is

$$J_{\mathrm{ss}} = \frac{W_n \mathcal{Z}}{Z} e^{-\frac{\Delta G^*}{k_{\mathrm{B}}T}}, \tag{4.45}$$

where $\mathcal{Z}$ is the "Zeldovich factor"

$$\mathcal{Z} = \sqrt{-\frac{1}{2\pi k_{\mathrm{B}}T} \left.\frac{\partial^2 \Delta G}{\partial n^2}\right|_{n^*}}. \tag{4.46}$$

The Zeldovich factor brings in the curvature of the free energy of cluster formation $\Delta G(n)$. The flux of clusters across the critical size is slower if this curvature is smaller, altering the rate of nucleation as discussed in Section 4.5.4.

Nonthermodynamic Behavior

The nucleation process was decomposed into several scales of length and time, i.e., atom attachment, fluxes through a cluster size distribution, and a macroscopic fraction of new phase. The critical temperature is set by the equality of free energies of the parent phase and the new phase, so below the critical temperature the formation of the new phase is downhill in free energy. Nevertheless, if we look closely at the processes on different length scales, we do find a peculiar behavior where subcritical clusters somehow go uphill in free energy to reach the critical size. A straightforward explanation is that nucleation includes a fluctuation in a large system of atoms, where some unlikely fluctuations do occur over time.

Our analysis of the cluster fluxes, however, did not take fluctuations into account – a global, steady-state flux through all cluster sizes was assumed. We assumed a cluster size distribution that was biased strongly towards smaller clusters, and the kinetic master equation predicted a flux of clusters that grew across the critical size. The cluster size distribution was not in equilibrium, certainly not at first when we assumed the existence of monomers alone. Non-thermodynamic states are unsurprising in the incubation phase as this initial cluster size distribution evolves into a steady-state cluster size distribution. Nevertheless, even in steady state the number of clusters decreases monotonically with cluster size, allowing the kinetic master equation to predict a steady flux to larger cluster sizes. This analysis does not predict a minimum in the cluster size distribution at the critical size, where the free energy is highest. In fact, the cluster size distribution never reaches thermodynamic equilibrium during the nucleation stage of a phase transformation. After all, the phase transformation is actively underway.

4.5.4 Time for a Random Walk across the Critical Size

Equation 4.45 for J_{ss} gives the steady-state flux of clusters from one size to the next. An individual nucleus, however, does not move continuously from small n to large n. A nucleus undergoes fluctuations in its size, and even nuclei that have achieved $n > n^*$ can shrink and disappear. We treat the halting progress over the nucleation barrier as a random walk problem. The idea is illustrated in Fig. 4.9. For simplicity we treat all clusters within $k_B T$ of the top of the barrier as equivalent – there is no energetic bias to their growth or

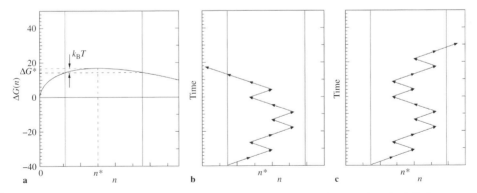

Figure 4.9   (**a**) Energy of formation of a nucleus, following Fig. 4.8, with markers indicating the range of n within $k_B T$ of the top of the barrier. (**b**) Random additions and subtractions of atoms from a subsize cluster, resulting in disappearance of the cluster. (**c**) Random additions and subtractions of atoms from a subsize cluster, resulting in successful nucleation.

shrinkage. The width of this region in cluster size is obtained with Eq. 4.19 by equating the difference in free energy to $k_\mathrm{B}T$

$$k_\mathrm{B}T = -\frac{1}{2}\frac{\partial^2 \Delta G}{\partial n^2}\bigg|_{n^*}(\delta n)^2. \tag{4.47}$$

$$\delta n = \sqrt{-\frac{2k_\mathrm{B}T}{\frac{\partial^2 \Delta G}{\partial n^2}\big|_{n^*}}}, \tag{4.48}$$

which is half the width between the vertical lines in Fig. 4.9a–c.

Figures 4.9b,c depict the fate of a cluster at the smallest size of this range. At any time it may add an atom and grow, or lose an atom and shrink. Without an energetic bias, these events occur at random, and the cluster executes a one-dimensional random walk along the axis of n. The analysis of a random walk is given in Section 10.1. In short, Eq. 10.6 gives a mean-squared displacement after n_j jumps

$$R^2 = n_j\,r^2. \tag{4.49}$$

The distance (in number of atoms) across the nucleation barrier is δn, and the individual jump sizes (the r) are 1 on the cluster axis. For our random walk of Eq. 4.49, n_j is the number of vectors in Fig. 4.9b,c, which is proportional to the time required to cross the barrier. The net time factor for the random walk across the barrier, τ, is therefore proportional to the square of δn of Eq. 4.48. The unit of time $(1/W_n)$ is the time for one atom to attach or detach from the cluster

$$\tau = \frac{(\delta n)^2}{W_n}, \tag{4.50}$$

$$\tau = \frac{1}{W_n}\left(-\frac{2k_\mathrm{B}T}{\frac{\partial^2 \Delta G}{\partial n^2}\big|_{n^*}}\right) = \frac{1}{\pi W_n \mathcal{Z}^2}, \tag{4.51}$$

where $\mathcal{Z}$ is given in Eq. 4.46. With this analysis up to Eq. 4.51 we can estimate the density of nuclei that will grow, and how long it takes for this to happen.

4.5.5 Temperature Dependence of Nucleation Rate

A word of caution – the results in this chapter have not proved useful for predicting absolute numbers. What they can do is predict temperature dependences of nucleation densities and rates, or evaluate quantities such as surface energies from these rates.[15] For example, if we solve Eq. 4.17 for n^* and again use $\Delta G_\mathrm{V} = -L\,\Delta T/T_\mathrm{c}$ (as for Eqs. 4.7 and 4.16), we can calculate the curvature $\mathrm{d}^2\Delta G/\mathrm{d}n^2$ at n^* in terms of the undercooling, ΔT. This curvature of ΔG, when substituted into Eq. 4.51, gives a time for an embryo to cross the critical region, $\tau \propto (\Delta T)^{-4}$. The crossing of the nucleation barrier slows dramatically near T_c, owing to the large critical cluster number $n^* \propto (\Delta T)^{-3}$ and the large number of kinetic steps (atom additions or subtractions) that occur when crossing the nucleation barrier.

[15] Of course, with additional effects from elastic energy even these goals may be challenging.

The nucleation rate of Eq. 4.51 is a strong function of temperature. With Eq. 4.7 for ΔG^* substituted into Eq. 4.45, the nucleation rate is

$$J_{ss} = J_0 \exp\left(-\frac{16\pi\,\sigma^3 T_{\mathrm{m}}^2}{3L^2\,k_{\mathrm{B}}T}\,\frac{1}{(\Delta T)^2}\right), \tag{4.52}$$

where all pre-exponential factors are now in J_0, estimated to be $10^{40}/(\mathrm{m}^3\,\mathrm{s})$ and independent of temperature. Consider the freezing of a liquid with $T_{\mathrm{m}} = 1000\,\mathrm{K}$, $L = 1.0 \times 10^9\,\mathrm{J/m}^3$, and $\sigma = 1\,\mathrm{J/m}^2$. At an undercooling of 126 K below the melting point, after 1 second the liquid will contain 100 nuclei per cubic meter, a number so low that the crystalline phase is undetectable. Just 4 degrees lower in temperature, $\Delta T = 130\,\mathrm{K}$, the nucleation rate jumps by a factor of 100. The extreme sensitivity to temperature suggests that for practical purposes a specific temperature can be identified as the onset of nucleation. The ΔT for this onset is called the "critical undercooling."

4.6 Time-Dependent Nucleation

Consider again the quench of Fig. 4.1. Immediately after the quench there is a metastable α-phase matrix with B-atoms arranged more or less randomly on the lattice (a "supersaturated solid solution"). There are few B-rich clusters of sizes greater than $n = 1$. For a nucleation event to occur at some time after $t = 0$, a critical nucleus must form by atom transport through the α-crystal. Some time will pass before a cluster of critical size can form by fluctuations, and before the steady-state nucleation rate occurs. For times less than this "incubation time" or "nucleation time lag," the nucleation rate is very low. After this time, the nucleation rate is approximately the steady-state value.

4.6.1 Numerical Example of Time-Dependent Nucleation

The kinetic master equation for transitions between the bins of different n is

$$\frac{\partial N(n,t)}{\partial t} = N(n-1,t)W_{n-1}^{\uparrow} + N(n+1,t)W_{n+1}^{\downarrow}$$
$$- N(n,t)\left[W_n^{\uparrow} + W_n^{\downarrow}\right]. \tag{4.53}$$

This master equation allows changes only by the addition or loss of single atoms from a cluster.[16] The expressions for the W are specific for nucleation in a specific material at a specific temperature. Kelton and Greer [36] provide a numerical example for oxide precipitation in silicon at a temperature of $800\,^\circ\mathrm{C}$.[17] The time dependence of the different cluster sizes $N(n,t)$ were found by numerical integration of the cluster size distribution

---

[16] The first two terms in Eq. 4.53 account for the increase in the number of clusters of n atoms by adding an atom to clusters of size $n-1$, and by removing an atom from clusters of size $n+1$. The loss of clusters of size n by the complementary processes is given by the last term of Eq. 4.53.

[17] The details require extensive discussion, but the rates $W^{\downarrow}$ decrease with cluster size, whereas $W^{\uparrow}$ increases with cluster size. At the critical cluster size of 28 atoms for this simulation, $W_{28}^{\uparrow} = W_{28}^{\downarrow}$.

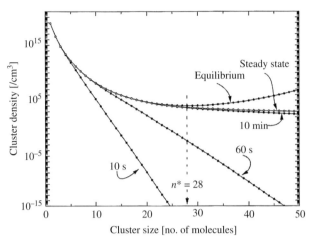

The evolution of cluster sizes with time. The logarithmic scale shows an enormous range in cluster size, and the graph shows enormous changes in cluster sizes over time. For scale, the cluster density in the steady-state solution is exactly half that for the equilibrium solution at the critical size $n = 28$. This image was published in [36], p. 37, copyright Elsevier (2010).

$$N(n, t + \delta t) = N(n, t) + \delta t \frac{\partial N(n, t)}{\partial t}. \tag{4.54}$$

Results are shown in Fig. 4.10. At $t = 0$, the initial cluster size distribution has all atoms in single-atom clusters ($n = 1$), which is representative of infinite temperature. With time, the cluster size distribution curves rotate counterclockwise in Fig. 4.10. The appearance of large clusters takes some minutes in this case. In equilibrium the cluster density has a minimum at the critical size of $n^* = 28$. Nevertheless, because clusters of size greater than $n^* = 28$ continue to grow, the steady-state cluster distribution never rises above this minimum for large n. A minimum in cluster size is expected in thermodynamic equilibrium from the Gibbs droplet analysis, but there is no such minimum when nucleation and growth occurs at a steady state.

4.6.2 Analytical Results for Time-Dependent Nucleation

An analytical calculation of the time dependence of nucleation requires the Fokker–Planck equation. Obtaining its analytical solution is a formidable task, beyond the scope of this text. Two approximate solutions exist, using the initial condition where only clusters of size $n = 1$ exist at $t = 0$. Trinkaus and Yoo [37] used a Green's function approach to arrive at the following time-dependent nucleation rate

$$J(t) = J_{\text{ss}} \frac{1}{\sqrt{1 - E^2}} \exp\left(-\frac{\pi Z^2 E^2 (n^* - 1)^2}{1 - E^2}\right), \tag{4.55}$$

where

$$E = \exp\left(-\frac{t}{\tau}\right), \tag{4.56}$$

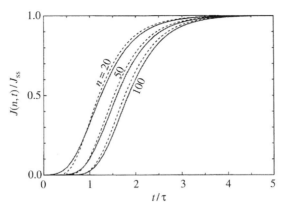

Figure 4.11   Time-dependent nucleation rates normalized by the steady-state value, labeled by the cluster size *n*. The Trinkaus and Yoo [37] solutions are the dashed lines, and the Shi *et al.* [38] solutions are solid curves, with labels for three critical sizes. Reproduced, with permission, from [38]. Copyright (1990) by the American Physical Society.

and τ represents a natural timescale for the kinetic problem

$$\tau = \frac{1}{2\pi \mathcal{Z}^2 \,\mathcal{B}(n^*)}, \tag{4.57}$$

where $\mathcal{B}$ is the rate of monomer attachment.

Shi *et al.* [38] obtained an analytical solution by employing the technique of singular perturbation. Their solution is

$$J(t) = J_{ss} \,\exp\left\{-\left[\exp\left(-2\frac{t - \lambda\tau}{\tau}\right)\right]\right\}, \tag{4.58}$$

where

$$\lambda = (n^*)^{-1/3} - 1 + \ln\left\{3 \sqrt{\pi} \,\mathcal{Z} \,n^*[1 - (n^*)^{-1/3}]\right\}. \tag{4.59}$$

Although Eqs. 4.55 and 4.58 have different mathematical forms, the two approximate solutions are quite similar (see Fig. 4.11). These two predictions of the time-dependent nucleation rate are also consistent with the qualitative picture described above – the rate is quite low for a time of τ, and rises rapidly to the steady-state value at later times. From the definition of τ it is clear that the incubation time is a function of both the critical size and the attachment rate term $\mathcal{B}(n^*)$.

4.7 Nucleation in Multicomponent Systems

All consideration so far has been for nucleation in a one-component system, e.g., a pure element. The extension to nucleation in alloys and compounds is arguably more important, but brings complexities that are not fully understood. For simplicity, first consider the nucleation of an equiatomic AB compound, where the ratio of chemical species remains fixed, even for the smallest clusters. This simple case reduces to a one-component system, where AB dimers are added or removed from clusters to alter the cluster size distribution. The nucleation of solid solution phases cannot be so simple. Although the new phase may eventually be a solid solution with an average composition, we expect different critical nuclei with different chemical compositions. Furthermore, the critical nuclei may not have chemical homogeneity – the segregation of one species to a surface may lower the surface energy for nucleation, favoring the formation of small clusters with surface segregation.

Our treatment of nucleation in a single-component system used the number of atoms in the cluster, n, as the coordinate for tracking the distribution of clusters and their energies. A more complete set of coordinates $\{\eta_i\}$ for nucleation in a binary system may include:

- numbers of A- and B-atoms in a cluster η_1, η_2;
- surface segregation of one of the species η_3;
- shape variations (that alter the specific surface energies) η_4, maybe more;
- variations in elastic energy (inside both the cluster and the parent phase) η_5, maybe more.

(The coordinates η_4, η_5 are expected in single-component systems, too.)

Owing to statistical differences in local compositions of small regions in the parent alloy, and owing to variations in kinetic processes that bring the different atomic species to the growing clusters, we expect that multiple combinations of these coordinates $\{\eta_i\}$ could produce critical nuclei. We therefore expect many sets of critical coordinates $\{\eta_i^*\}$ producing multiple critical $\Delta G^*(\{\eta_i^*\})$ at states of unstable equilibrium. The locus of these critical coordinates $\{\eta_i^*\}$ will likely be a piecewise continuous hypersurface in a space of dimension one less than the total number of coordinates.[18]

It is possible to reduce the two composition coordinates, η_1 and η_2, to a single coordinate, simplifying the flux of clusters through the two-dimensional space of A- and B-atoms to a flux through one coordinate. This new coordinate η_t is not simply the sum of η_1 and η_2 because it must be unique for all combinations of A- and B-atoms. Dimensional reduction from η_1 and η_2 to the single η_t brings a complexity where each η_t' is connected by monomer addition to multiple other η_t, so the sequence of clusters does not make a change of ± 1 in η_t, as was the case for monomer additions or subtractions in a single-component system. Nevertheless, with this reduction of dimension in composition space, the problem of multicomponent nucleation can be solved analytically, much as was done for single-component systems [39]. This approach presumes that all contributions to the $\Delta G(\{\eta_i\})$ of the nucleation problem can be parameterized by composition alone, which would be true if shapes, strains, surface energies, etc., are determined only by the numbers and types of atoms in the cluster. The extent to which this is useful is a topic of investigation today.

---

[18] The flux of clusters over this critical hypersurface may have some similarities to the activation process of atom diffusion described in Section 10.7, but it is not obvious if saddle points play the same role in nucleation as they do in atomistic diffusion.

Problems

4.1 Phase transitions and dimensionality

Assume a temperature where a second phase has a lower volume free energy than its parent phase. Consider the difference in free energy between (1) a parent phase, and (2) a parent phase plus a nucleus of a second phase.

Assume that this difference contains a favorable contribution from the volume free energy of the second phase, which scales as, for example, $-Pr^3$ in three dimensions (where P is a constant, perhaps involving π). A positive surface energy is also needed.

Calculate the critical radius in one, two, and three dimensions. Is the nucleus unstable at the critical radius?

4.2 (**a**) Perform the differentiations and substitutions to obtain R^* and ΔG^* from Eq. 4.3.
(**b**) Starting with Eq. 1.6, derive Eq. 4.7 for $\Delta G^*(\Delta T)$.

4.3 Suppose that nuclei formed in a phase transformation are cubic in shape, rather than spheres. If a is the cube edge length, σ is the specific surface energy, and ΔG_V is the change in volume free energy upon forming the new phase (assume $\Delta G_V < 0$):
(**a**) Calculate the critical size of the nucleus, a^*, and the critical free energy of formation, ΔG^*, in terms of σ and ΔG_V.
(**b**) Explain why these differ from R^* and ΔG^* for a spherical nucleus. (Not difficult.)

4.4 Suppose an $\alpha \to \beta$ phase transition could be second order in free energy. In other words, near T_c the two phases have the free energies

$$G_\alpha(T_c - \Delta T) = G_0(T_c) + G'_0 \Delta T + \frac{1}{2}G''_\alpha(\Delta T)^2 + \cdots \qquad (4.60)$$

$$G_\beta(T_c - \Delta T) = G_0(T_c) + G'_0 \Delta T + \frac{1}{2}G''_\beta(\Delta T)^2 + \cdots \qquad (4.61)$$

where the first two terms on the right-hand sides are equal, and the prime ($\prime$) denotes a derivative with respect to T, evaluated at T_c.
(**a**) If a nucleation and growth transformation were to occur for this system, calculate the critical radius for nucleation, $R^*(\Delta T)$, and the critical free energy of formation, $G^*(\Delta T)$, assuming a spherical nucleus of β-phase.
(**b**) A second-order phase transition can occur continuously without a distinct nucleation event, but this continuous process competes with a process of nucleation and growth. The most expedient is expected to win.

Discuss the kinetic issues favoring or disfavoring the continuous process at small ΔT. Comment on the rate of cooling needed to achieve a large ΔT, and how the relative kinetics of the two processes may change at large ΔT.

4.5 Suppose the interface between the β-phase precipitate and α-phase matrix in Fig. 4.4a has a surface energy, σ, which is less than half of the grain boundary energy,

σ_{gb}. What happens to the shape of a grain boundary precipitate as shown in Fig. 4.4a, and how do its size and shape evolve over time?

4.6 Prove that the surface area of the double spherical cap of Fig. 4.4 is $4\pi R^2(1 - \cos\theta)$, where R is the radius of curvature.

(*Hint*: One approach is to place a sphere of radius R at the origin and integrate its surface in rings from $z = 0$ to $z = h$. The radius of the ring is $R\cos\theta$, but its surface increases with tilt as $1/\cos\theta$.)

4.7 Consider a sequence of nucleations and precipitations from a supersaturated α-phase with $c = 0.22$ in Fig. 4.6.

(**a**) If the γ-phase is to form instead of the β-phase, how much could a nucleation barrier elevate the γ-phase free energy curve in Fig. 4.6b?

(**b**) What happens to the supersaturation of the α-phase during a precipitation sequence where the γ-phase goes to completion, and the β-phase forms later with its equilibrium free energy?

(**c**) Does the formation of the γ-phase increase or decrease the free energy barrier to nucleate the β-phase from the α-phase? Why?

(**d**) After forming the γ-phase, so the system is on the solid common tangent line of Fig. 4.6b, explain why the β-phase cannot form directly from the γ-phase, and why the γ-phase decomposes simultaneously into the α-phase and β-phase.

4.8 Richard's rule states that the entropy of melting $\Delta S_m = 1\,k_B$/atom.

(**a**) Neglecting pressure, show that the ratio $\Delta H_m/T_m = 1\,k_B$/atom (which is another statement of Richard's rule).

(**b**) Using Eq. 1.6 and Richard's rule, calculate $F_l - F_s$ for the metal nickel as a function of undercooling below T_m, where the subscripts denote liquid and solid.

(**c**) The solid/liquid surface of Ni has a surface energy of 0.255 J/m². Using your result from part b, and an undercooling of 10 K, graph Eq. 4.3 for Ni.

(**d**) Calculate the critical size of the nucleus, R^*, and the critical free energy of formation ΔG^* for Ni using the parameters of part c.

Effects of Diffusion and Nucleation on Phase Transformations

A phase diagram is a construction for thermodynamic equilibrium, a static state, and therefore contains no information about how much time is needed before the phases appear with their equilibrium fractions and compositions. It might be assumed that the phases found after practical times of minutes or hours will be consistent with the phase diagram, since most phase diagrams were deduced from experimental measurements on such timescales.[1] However, a number of nonequilibrium phenomena such as those described in this chapter are well known, and were likely taken into account when a $T–c$ phase diagram was prepared.

For rapid heating or cooling, the kinetic processes of atom rearrangements often cause deviations from equilibrium, and some of these nonequilibrium effects are described in this chapter. In general, the slowest processes are the first to cause deviations from equilibrium. For faster heating or cooling, however, sometimes the slowest processes are fully suppressed, and the next-slowest processes become important. Approximately, Chapter 5 follows a course from slower to faster kinetic processes. The last section, however, discusses why kinetic processes should bring materials to thermodynamic equilibrium.

5.1 Nonequilibrium Processing of Materials

5.1.1 Diffusion Lengths

The diffusional motion of atoms is thermally activated, and is more rapid at higher temperatures. An important consideration is the characteristic time τ for diffusion over a characteristic length x (see Eq. 3.45)

$$\tau = \frac{x^2}{D(T)}, \tag{5.1}$$

where $D(T)$ is the diffusion coefficient.

Minimizing the time when $D(T)$ is large serves to minimize x, so one way to classify either kinetic phenomena or methods of materials processing is by effective cooling rate. Table 5.1 lists some kinetic phenomena that are suppressed by cooling at increasingly rapid rates. Insights into the phenomena listed in Table 5.1 are obtained with a rule of thumb

---

[1] On the other hand, this does not necessarily mean that all phases on phase diagrams are in fact equilibrium phases. Exceptions are found, especially at temperatures below about half the liquidus temperature.

Table 5.1 Cooling rates, methods, typical kinetic phenomena

Cooling rate		Method	Phenomenon
Infinitesimal	10^{-10} K/s	Geological cooling	Equilibrium (sometimes)
Slow	10^0 K/s		Suppressed diffusion in solid
	10^1 K/s	Casting	Dendrites
	10^3 K/s	Iced brine quench	Suppressed precipitation
Medium	10^4 K/s		Suppressed diffusion in liquid
	10^5 K/s	Melt spinning	Extended solid solubility
Fast	10^6 K/s	Piston-anvil quench	Simple metallic glasses
	10^9 K/s	Laser surface melting	
Ultrafast	10^{10} K/s		Amorphous elements
	10^{11} K/s	Physical vapor deposition	
	–	Shock wave	Melting
	–	High-energy ball milling	Nanocrystallinity, glass formation
	–	Heavy ion irradiation	Chemical mixing, glass formation

that diffusion coefficients near the melting temperature of many (metallic) materials are $D(T_m) \sim 10^{-8}$ cm^2/s.[2] For an interatomic distance of typically 2×10^{-8} cm, Eq. 5.1 gives a characteristic time of 4×10^{-8} s. For ultrafast cooling, it seems possible to suppress all atom diffusion in solids below the melting temperature, suppressing crystal nucleation and growth. This allows the formation of amorphous metallic elements, but these are highly unstable and may persist only at cryogenic temperatures. At 2/3 of the melting temperature, the timescale for suppressing atom motion at the atomic scale is increased by a factor of 10^4 or so compared to the timescale at T_m, owing to a decrease in $D(T)$.

Box 5.1   **Thermodynamics and Kinetics**

Two necessary conditions for a phase transformation to occur in a material are:

- A driving force from a reduction of free energy. An equilibrium phase diagram gives important hints about differences in free energies.
- A mechanism for atoms to move towards their equilibrium positions. Diffusion is usually the mechanism, but nucleation can be involved, too.

5.1.2 Cooling Techniques

Some practical methods to achieve high cooling rates (called "quenching") are listed in Table 5.1. The cooling rates are approximate, since these depend on the thickness of the sample and its thermal conductivity, and these vary with the material and the method. Like the diffusion of atoms, the diffusion of heat has a quadratic relationship between the

---

[2] Sometimes an estimate of $D \sim 10^{-5}$ cm^2/s for the liquid proves useful, too.

characteristic cooling time, τ, and the sample thickness, x, as in Eq. 5.1, where $D(T)$ is now a thermal diffusivity. Approximately, the techniques listed in Table 5.1 make thicknesses of materials that are proportional to the square root of the inverse cooling rate. Samples from melt spinning are tens of micrometers thick, and samples from laser surface melting are often less than 0.1 μm, for example.

Iced brine quenching is an older technique, where a sample at elevated temperature is quickly immersed in a solution of rocksalt in water, cooled with ice. The salt serves to elevate the boiling temperature and improve the thermal conductivity. The rate of cooling depends on the thickness of the sample. The quench rate also depends on the formation of bubbles of water vapor on the sample surface, which suppress thermal contact to the water bath. Stirring the mixture can improve the cooling rate.

In melt spinning, a steady stream of liquid metal is injected onto the outer surface of a spinning wheel of cold copper, for example. The liquid cools quickly when in contact with the wheel, making a solid ribbon that is thrown off the wheel and spooled. This method is suitable for high-volume production. The liquid metal should have modest wetting of the spinning wheel, and optimizing the parameters of the system can be challenging.

Piston-anvil quenching, sometimes known as "splat quenching," uses a pair of copper plates that impact a liquid droplet from two sides. The alloy is typically melted by levitation melting in an induction coil. When the radiofrequency heating current is stopped, the liquid droplet falls under gravity past an optical sensor that triggers the pistons. Like melt spinning, the sample is thin, perhaps 20 to 30 μm, but wetting properties are less of a concern.

Laser surface melting can be performed with either a pulsed or continuous laser. The sample may be moved in a raster pattern under laser illumination so a significant area can be treated. Once melted, the surface is cooled by the underlying solid material, and the thinner the melted region, the faster the cooling.

Physical vapor deposition may use a high-temperature heater to evaporate a material under vacuum. The evaporated atoms move ballistically towards the cold surface of a substrate. When these atoms are deposited on the cold substrate, their thermal energy is removed quickly, in perhaps a hundred atom vibrations (approximately 10^{-11} s), leading to very high cooling rates when the deposition rate is not too rapid and the substrate is isolated thermally from the hot evaporator. Sputtering guns, in which a plasma of inert gas ions knock atoms off a surface, can also be used instead of a thermal evaporator.

Some of the other methods in Table 5.1 have more ambiguous temperatures, or their cooling rates are harder to define. Shock waves can be used to consolidate powders into bulk materials. A high-power shock, delivered by a detonation or a high-velocity impact, can melt the surfaces of the powder particles as it passes through them. The powder surfaces are cooled by thermal conduction to the powder interiors. Nonequilibrium materials such as metallic glasses can be prepared in bulk form by such methods, although these methods are not routine.

High-energy ball milling is typically performed by sealing the starting powder(s) in a steel vial containing hardened steel balls. Shaking the vial violently for several hours causes severe mechanical deformation, fracture, and rewelding of the metallic particles. The sizes of the powder particles are difficult to understand, but the process does produce chemical mixing on the atomic scale. How this occurs is not known in detail. The effective temperature cannot be particularly high, however, since this method is also useful for producing nanocrystalline materials.

Ion beam bombardment with Xe^+ ions accelerated to a few MeV is a way of knocking individual atoms off their lattice sites, sometimes moving them tens of nanometers inside the material. The ion or the primary knock-on atom may dislodge other atoms in a "collision cascade" that involves many atoms. A temperature is not appropriate for such ballistic processes. The effective "cooling rate" is very high.

Box 5.2	Historical Anecdotes

The liquid quenching of steels has a storied history, perhaps because early metalworkers did not have a good understanding of the relevant time-dependent phenomena. The cooling medium does affect the cooling rate, and hence the properties of steel. Especially prized were good mechanical properties of steel swords, which could make the difference between life and death. Different civilizations took different approaches to optimizing the quenching medium and quenching procedures. Quenching methods passed on from earlier generations of metallurgists include the following:

- The steel blade, heated to the color orange to bring it into the austenite phase, was quenched into the urine of a red-haired boy.
- Another practice was to use urine from a three-year-old goat, fed nothing but ferns for three days. Salt quenching does have merit, although a large quantity of liquid is usually required.
- A procedure chronicled in a temple in Asia Minor was "Heat until it glows like the rising sun in the desert, then cool it to the color of royal purple, plunging it into the body of a muscular slave. The strength of the slave, transferring to the metal, imparts to it hardness."

5.2 Alloy Solidification with Suppressed Diffusion in the Solid

5.2.1 Alloy Solidification with Solute Partitioning

Summary of Nonequilibrium Phenomena

Here we consider the kinetics of alloy solidification. Several issues in solidification are common to other first-order phase transformations, including:

- nucleation of the new phase(s);
- effects of temperature gradients on the movement of the reaction front;
- effect of the latent heat on the temperature profile; and
- restricted partitioning of chemical elements between two phases when diffusion is slow.

All these effects are considered in this chapter, but we begin with the partitioning of solutes between the solid and liquid phases. Slow diffusion in the solid suppresses equilibration and the solidified alloy has a heterogeneous chemical composition, differing from what is predicted by its equilibrium phase diagram. A good point of reference, however, is how solidification occurs under the equilibrium conditions given by a T–c phase diagram.

Alloy Solidification: Case of Equilibrium

We start with equilibrium predictions for phase fractions and compositions for an alloy of composition c_0, cooled very slowly from the liquid and allowed to freeze and equilibrate at all stages (Fig. 5.1), even though at lower temperatures equilibration may take geological times. The relevant graphs to examine are on the right of Fig. 5.1 (rotated by 90° for correspondence to the temperature scale of the diagram), specifically the thick solid curves. The six graphs are the fractions and compositions for each phase, l, α, and β, over a range of temperatures. At high temperatures the alloy is entirely liquid of composition c_0, but below the liquidus temperature the fraction of liquid decreases as crystals of the β-phase grow in the liquid. Using the lever rule (Eq. 2.3) for the c_0 marked on the phase diagram, the fractions of β-phase and liquid phase are approximately equal at the eutectic temperature of Fig. 5.1. Their chemical compositions are simply read from the solidus and liquidus lines on the phase diagram. Note that the α-phase does not exist above the eutectic temperature for this alloy of composition c_0, but the lever rule shows that immediately below the eutectic temperature the α-phase should jump to a phase fraction, f_α, of approximately 0.3.

5.2.2 Alloy Solidification: Suppressed Diffusion in the Solid Phase

Mixing in Liquid, Suppressed Diffusion in Solid (General Features)

Kinetic deviations from the equilibrium analysis of Section 5.2.1 can be predicted approximately. With faster changes in temperature, kinetic processes with the longest timescales will be the first to alter phase formation. Atom diffusion in the solid phases is a kinetic process that is typically slow, and is appropriate to consider first.

Solid-state diffusion modifies the equilibrium cooling curves of Fig. 5.1 as follows. At the highest temperature where the β-phase first forms during cooling (just below the

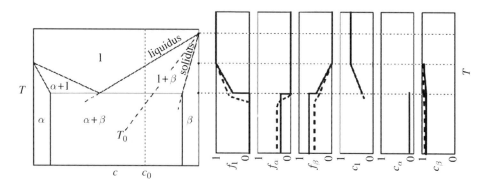

Figure 5.1 Analysis of phase fractions, f, and compositions, c, upon cooling under equilibrium conditions (thick curves at right), and when there is suppressed diffusion in the solid phases (thick dashed).

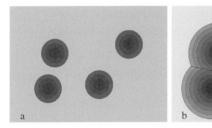

Figure 5.2 (**a**) Nucleation of four solute-rich precipitates, with increasing volume added in shells at lower temperatures. (**b**) Later stage of the phase transformation, with less solute in the liquid phase and less concentration of solute in the outer shells, following the composition of the β-phase solidus at T.

liquidus for the composition c_0), the β-phase is rich in B-atoms. With further cooling, the precipitates grow in shells of different solute concentrations, as shown in Fig. 5.2a. At lower temperatures, the phase diagram shows that this initial β-phase is too rich in B-atoms, and some B-atoms in the solid should exchange with A-atoms in the liquid. This usually does not occur because it requires atoms in the interior of the crystals of β-phase to diffuse to the outside of the crystal to reach the liquid. Diffusion in the solid is usually too slow to ensure equilibrium during cooling, so a heterogeneous solute concentration is expected in the solid, as shown in Fig. 5.2b. Although the average composition of the β-phase is too rich in B-atoms, at the surface of β-phase crystals, new layers of solid are forming with their equilibrium composition (for the appropriate temperature). One way to summarize this behavior is that the extra B-atoms in the β-phase are immobilized and unable to participate in the equilibrium process, and the effective alloy composition can be considered to decrease below c_0.

With the immobilization of an excess of B-atoms inside the β-phase, the effective composition of the alloy is enriched in A-atoms. For predicting the fractions of solid and liquid, the effective composition of the alloy is reduced below c_0 in Fig. 5.1. By the lever rule, less β-phase, and more liquid, will occur just above the eutectic temperature. The composition of the new β-phase, and the composition of the remaining liquid,[3] are still set by the compositions on the solidus and liquidus curves, however. Over a wide range of alloy compositions around c_0, the last liquid to freeze will have the eutectic composition (assuming that final solidification occurs at the eutectic temperature).

Incidentally, the areal fractions of solid and liquid in Fig. 5.2b depict the phase ratios at the eutectic temperature in Fig. 5.1. The "pro-eutectic" β-phase in Fig. 5.2b has formed above the eutectic temperature, but at lower temperatures both the α- and β-phases form from the remaining liquid in a "eutectic transformation," depicted in Fig. 14.9 and described in Section 14.3. Owing to the shift of the effective alloy composition below c_0, the amount of β-phase formed in the eutectic transformation, compared to the amount of pro-eutectic β-phase, is greater than expected from the phase diagram when there is suppressed diffusion in the β-phase. Finally, the eutectic transformation usually does not

---

[3] Here the composition of all remaining liquid is assumed homogeneous, since diffusion in the liquid is relatively fast. Hence the liquid in contact with the β-phase has the same composition as the rest of the remaining liquid.

begin until there is some undercooling below the eutectic temperature because the α-phase does not nucleate immediately. It may be useful to extend the liquidus curve below the eutectic temperature while pro-eutectic β-phase is still forming.

Mixing in Liquid, Suppressed Diffusion in Solid (Semiquantitative)

An approximate analysis of nonequilibrium solidification with no diffusion in the solid phase is performed with the geometrical construction of Fig. 5.3. The equilibrium solidus and liquidus curves are drawn with black lines that are straight, so for every temperature they give the same "partitioning ratio," k,

$$k \equiv \frac{c_s}{c_l}, \tag{5.2}$$

where the compositions of the solid and liquid are c_s and c_l. This assumption of straight lines and a constant k is often reasonable, and it simplifies enormously the arguments that follow. Figure 5.3 shows that as a liquid of composition c_0 is cooled, the first solid forms at temperature T_3, and has the composition $k c_0$. With the equilibrium solidus curve, by the temperature T_2 an equilibrium system would be completely solid, and the solid would have a uniform composition c_0. For nonequilibrium cooling with no diffusion in the solid, at this temperature T_2 the solidification is incomplete because there is an excess of B-atoms in the remaining liquid. We make the simple approximation that the last solid to form will have a composition Δc greater than c_0, because Δc is the deficiency in B-concentration of the first solid to form, and the final solid must have an average concentration of c_0. This is shown in Fig. 5.3, where the gray triangles are congruent. The phase fractions and their compositions are then drawn with the same rules as for the equilibrium phase diagram, except that the gray line at left is used for the solidus, not the equilibrium black line. This analysis is qualitatively correct, and highlights the importance of the partitioning ratio in Eq. 5.2 (often called the "partitioning coefficient," or "distribution coefficient"). The triangle construction is qualitatively reasonable, but not rigorous.

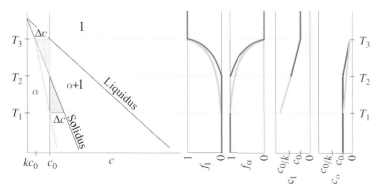

Figure 5.3 Liquid–solid phase boundaries for the A-rich region of a T–c phase diagram. Black lines are the equilibrium phase diagram. Using the gray triangles, the gray solidus curve is constructed as described in the text, giving the altered compositions and phase fractions in gray at right.

5.2.3 Scheil Analysis of Mixing in Liquid, Suppressed Diffusion in Solid

If we keep the assumption that the liquidus and solidus curves are straight lines, it is possible to perform a more rigorous analysis of the composition profile in the solid phase after solidification, again assuming no diffusion in the solid. Start with the lever rule, Eq. 2.3, which is an equation for the conservation of solute between solid and liquid phases

$$f_s[c_0 - c_s] = f_l[c_1 - c_0], \tag{5.3}$$

$$f_s[c_0 - c_s] = f_l\left[\frac{c_s}{k} - c_0\right], \tag{5.4}$$

where the partitioning ratio or Eq. 5.2 was used to relate $c_1 = c_s/k$. We consider how small reductions in temperature cause the transformation of small amounts of liquid into solid, and how small amounts of solute are rejected into the remaining liquid. This requires a differential form of the lever rule, obtained by differentiating Eq. 5.4

$$\frac{df_s}{dT}[c_0 - c_s] - f_s\frac{dc_s}{dT} = \frac{df_l}{dT}\left[\frac{c_s}{k} - c_0\right] + \frac{1}{k}f_l\frac{dc_s}{dT}. \tag{5.5}$$

The transformation of a small fraction of liquid, df_l, into a small amount of solid, df_s, requires $df_l = -df_s$, since only two phases are present (and $f_l = 1 - f_s$). After a convenient cancellation of all terms containing c_0, terms with df_s/dT are separated from those with dc_s/dT

$$\frac{df_s}{dT}\left[\frac{c_s}{k} - c_s\right] = \frac{dc_s}{dT}\left[f_s + \frac{1-f_s}{k}\right]. \tag{5.6}$$

At this point we make a simplification and an approximation. We simplify by ignoring the details of the temperature dependence, because we have equality for each temperature interval in Eq. 5.6. Our result below therefore gives c_s vs. f_s, but no dependence on T. The approximation is that the term $(dc_s/dT)f_s$ is small, and we neglect it. This is reasonable until f_s is large in the later stages of solidification. The approximate Eq. 5.6 is

$$\frac{df_s}{1 - f_s} = \frac{1}{1-k}\frac{dc_s}{c_s}. \tag{5.7}$$

For the limits of integration, start with $c_s = k c_0$ at $f_s = 0$ (because the earliest solid to form has concentration $k c_0$), and consider the different compositions $c_s = k c_1$ of the solid phase at the fractions f_s

$$\int_0^{f_s} \frac{1}{1-f_s'}\, df_s' = \frac{1}{1-k}\int_{kc_0}^{c_s} \frac{1}{c_s'}\, dc_s'. \tag{5.8}$$

The integrand on the LHS can be simplified by the change of variables from solid to liquid fractions, $f_l = 1 - f_s$, but the limits start at 1 (fully liquid) and decrease to $1 - f_s$.

$$\int_1^{1-f_s} \frac{1}{f_l'}(-1)df_l' = \frac{1}{1-k}\int_{kc_0}^{c_s} \frac{1}{c_s'}\, dc_s'. \tag{5.9}$$

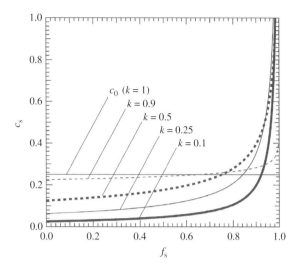

Figure 5.4 Solid composition versus fraction of solid from the Scheil equation 5.12. The overall alloy composition was $c_0 = 0.25$.

Both integrals are elementary, giving logarithm functions

$$-\ln f_1' \Big|_1^{1-f_s} = \frac{1}{1-k} \ln c_s' \Big|_{kc_0}^{c_s}, \tag{5.10}$$

$$(k-1)\ln(1-f_s) = \ln\left(\frac{c_s}{kc_0}\right). \tag{5.11}$$

Exponentiating

$$c_s = kc_0 \left(1-f_s\right)^{k-1}, \tag{5.12}$$

$$c_l = c_0 \left(1-f_s\right)^{k-1}. \tag{5.13}$$

Equation 5.12 is the "Scheil equation," and has been in use for a century [40, 41]. It gives the composition of the new solid that freezes when the fraction of solid is f_s. Equation 5.12 is especially useful in one-dimensional freezing, where the f_s is proportional to the distance across the solid.

When $k = 1$, so the solid and liquid have the same composition (the phase diagram has a liquidus and solidus that coincide), the composition of the solid is uniform, and Eq. 5.12 gives $c_s = c_0$ for all fractions of solid. As k decreases from 1, however, the composition profile of the solid becomes increasingly nonlinear. Figure 5.4 shows some examples.[4] For this alloy of composition $c_0 = 0.25$, the first solid forms when the temperature cools to intersect the liquidus at $c = 0.25$. This first solid has the composition $c_s = k c_0$, and this is the y-intercept at left of Fig. 5.4. With decreasing temperature, the new solid forms with increasingly more solute (crossing the composition c_0 when f_s is at least 0.5, depending on k). When k is small, the final solid has to incorporate a large amount of solute that was

---

[4] Figure 5.3, for which $k \simeq 0.35$, is generally consistent with Fig. 5.4.

rejected into the liquid during the first half of freezing, and there is a spike in composition for the final solid that freezes. Unfortunately, the Scheil equation is not quantitatively correct for the very last solid that forms, owing to the neglect of a term in Eq. 5.6.[5] Nevertheless, the accumulation of solute in the final solid is correct semiquantitatively, and an extremely heterogeneous composition is predicted when the partitioning ratio is small ($k \ll 1$).

Temperature was not considered in this analysis, but we can understand it qualitatively from the first analysis with Fig. 5.3 (which we now see is most reliable when k is close to 1). Most of the solid forms at higher temperatures that lie above the solidus for the composition of c_0. Nevertheless, a strong solute enrichment of the remaining liquid can delay the solidification of a small remainder of liquid until much lower temperatures.

5.3 Alloy Solidification with Suppressed Diffusion in Both Solid and Liquid

5.3.1 Concentration Gradients in the Liquid

Mixing in the liquid is usually incomplete, so the liquid is not homogeneous throughout. Typically there is a concentration gradient in the liquid near the solid–liquid interface, and this gradient depends on the rate of freezing. Consider these two different cases:

- If the interface moves very slowly, diffusion could occur over a long distance comparable to the size of the liquid region. This ensures homogeneity of the liquid composition, and was the case considered in Section 5.2.2. (Homogeneity will also be promoted by convective currents, or by stirring.)
- When freezing occurs rapidly, the solid–liquid interface can outrun the diffusion front in the liquid. The solid must accommodate all atoms, both A and B, so the composition of the new solid is the same as the alloy composition, c_0.

When the very first solid forms, the liquid has a concentration c_0 at the solid–liquid interface, but this concentration increases as solute is rejected into the liquid. The liquid with higher solute concentration has a lower freezing temperature, so solidification is pushed to lower temperatures. If liquid diffusion is extremely slow, the temperature may decrease to T_2 in Fig. 5.3. At this temperature T_2, the solid of composition c_0 is in equilibrium with a liquid of composition c_0/k, and there is no need to reject more solute at the growing interface. After a transient period to form the concentration gradient, this case of slow diffusion in the liquid leads to the growth of a solid with concentration c_0 at its surface.

If the solid–liquid interface is moving with velocity v, there is, however, a diffusion profile in the liquid near the interface. With a characteristic length $x = v\tau$, and a characteristic time $\tau = x^2/D$ (from Eq. 5.1), there is a concentration profile over the length $x = D/v$ in the liquid in front of the interface (more details are shown in Fig. 13.7). After an

---

[5] Neglecting the term $(dc_s/dT)f_s$ predicts $c_s > 1$ as $f_s \rightarrow 1$.

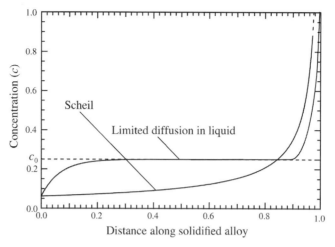

Figure 5.5 Concentration profile along solidified bar for the two cases of complete mixing in the liquid (Scheil, Eq. 5.12) and mixing by a diffusion front in the liquid that is smaller than the length of the liquid (limited diffusion in liquid). In both cases $c_0 = 0.25$ and $k = 0.25$, and no diffusion occurs in the solid after freezing.

early transient, this diffusion profile moves in front of the solid–liquid interface as the solid grows into the liquid.

This concentration profile with solute enrichment in the liquid ahead of the moving interface moves from left to right until it comes within the characteristic distance D/v to the end of the liquid. Here the excess solute in the liquid must freeze into an excess of solute in the final solid. Likewise, there must be a depletion of solute where the solidification started, and the solid had the lower concentration kc_0. These two features (a solute depletion to the left and an excess to the right) are seen in the concentration profiles of Fig. 5.4, but the shapes of the concentration profiles differ significantly – the two are compared in Fig. 5.5.

5.3.2 Practical Issues in Alloy Solidification, Dendrite Instability

At practical solidification rates, it is typical to ignore diffusion in the solid. Nevertheless, we made two different assumptions about the solute mixing in the liquid, which gave different concentration profiles in the solid, as shown in Fig. 5.5. In reality, the actual situation is usually somewhere between these two cases, with partial mixing in the liquid. This gives a concentration gradient in the growing solid, somewhat as described by the Scheil equation 5.12, but the moving diffusion front tends to homogenize the solid at the average composition c_0.

A more serious issue is that our one-dimensional analysis of a flat interface usually proves too simple. A flat interface can become unstable in three dimensions, as discussed in Section 13.1.2. The timescales for heat transport and solute transport are often incompatible. Diffusion coefficients for solutes in liquid metals are typically in the range of 10^{-5} to 10^{-4} cm^2/s. A diffusion time for a length of 1 cm is therefore $\tau = 10\,000$ s, or a few hours. The thermal diffusivities are a factor of 10^4 larger, so heat can be removed across a 1 cm distance in 1 s. This rapid removal of heat can freeze the liquid before solute can equilibrate across a casting, for example.

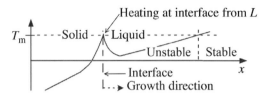

Figure 5.6 Temperature profile near solid–liquid interface showing effect of temperature rise from the release of latent heat at interface.

Nevertheless, chemical segregation on a small spatial scale, "microsegregation," can better minimize the free energy during solidification of an alloy. In the short cooling time of 1 s, solute segregation may occur over a distance of 0.1 mm, because $x \simeq \sqrt{D\tau}$. Such microsegregation is possible by the formation of tree-like "dendrites," shown in Figs. 13.3–13.6. Dendrites are small rods with tips that grow into the liquid. They are understood as an instability of a flat solid–liquid interface, caused by a temperature inversion in the adjacent liquid.

A real temperature inversion can occur if the latent heat of solidification raises the local temperature, as in Fig. 5.6. There is no superheating of the solid, so the interface temperature is T_m. If the interface is a locally hot spot owing to the latent heat, the temperature must be lower in the adjacent liquid. This temperature inversion from the latent heat at the interface occurs in both pure elements and in alloys, and facilitates the growth of solid protuberances into undercooled liquid. Another contribution to this dendrite instability occurs in alloys when solute is not incorporated into the new solid phase, and the extra solute accumulates in the adjacent liquid. If the solute lowers the melting temperature of the alloy, there is an effective temperature inversion at the interface from "constitutional supercooling."

Working against the formation of dendrites is the surface energy of the solid–liquid interface.[6] The surface energy, which prefers a flat surface, is important for setting the spatial scale of the dendritic microstructure. The tip of the solid dendrite is where most of the growth occurs, so the radius of the tip is a key feature. In short, the high surface-to-volume ratio at the dendrite tip provides a critical radius, much as for nucleation (Section 4.2). For dendrites, however, the undercooling is relatively small and the critical radius is relatively large, compared to other nucleation processes in materials. More features of dendrites and their growth are described in Section 13.1.3.

5.4 Time, Temperature, and Transformation

The time to complete a phase transformation varies with temperature and with the initial state of the material. This section explains the time required to form a new phase after its parent phase is quickly taken across the temperature of a phase boundary. The main

---

[6] It is also possible to suppress dendrite formation by moving the solid with respect to the liquid, such as rotating a crystal during growth.

example is an elemental liquid cooled quickly below its melting temperature, T_m, so a solid phase nucleates and grows.

Especially at temperatures near a phase boundary, nucleation slows the formation of a new phase. For solidification, undercooling below T_m is necessary before any solid phase forms.[7] Section 4.2 explained the energy penalty associated with the solid–liquid interface, so undercooling is needed to boost the thermodynamic driving force for creating a new volume of solid from the liquid. In practice, the required undercooling is not easy to predict because it depends on factors such as the contact between the liquid and the walls of its container. Nevertheless, at T_m the free energies of the solid and liquid are equal, so it takes infinite time to nucleate the solid phase. Undercooling is also required for the nucleation of solid $\rightarrow$ solid phase transformations.

The number of growing nuclei of a new solid phase increases rapidly with the undercooling, ΔT, below the melting temperature, T_m (Section 4.5.5). The nucleation rate has an exponential form $\sim \exp\left[- \mathcal{T}^2/(\Delta T)^2 \right]$. Here $\mathcal{T}^2$ is a gradual function of $1/T$ (and depends on surface energy and volume free energies near T_m). This exponential function is strictly zero at T_m when the undercooling ΔT is zero. At small ΔT, this exponential is nearly zero. The potent exponential suppression of nucleation requires a substantial undercooling ΔT to start the phase transformation, pushing it to temperatures observably below T_m.[8]

On the other hand, the diffusional processes needed for growth are slower at lower temperatures. The general kinetics of a nucleation-and-growth phase transformation is indicated at the left of Fig. 5.7. The rate of the phase transformation is zero at the critical temperature (owing to nucleation), and is small again at low temperatures owing to diffusion.

On the lower right of Fig. 5.7 is a kinetics map called a TTT diagram (for time–temperature–transformation). These diagrams cover a wide range of times; the x-axis uses a logarithmic scale. Each TTT diagram pertains to a particular phase transformation at a particular chemical composition. The assumption in using a TTT diagram is that the high-temperature phase is cooled instantaneously to the temperature on the y-axis of the plot, and held at this temperature for increasing time. The extent of completion of the phase transformation is often drawn on the TTT diagram as curves for 10% and 90% completion. A typical nucleation-and-growth transformation has a "nose" at a temperature where the transformation occurs in the shortest time. The nose is the leftmost part of the curves in the TTT diagram of Fig. 5.7. At temperatures above the nose, the phase transformation is suppressed because nucleation is rare. Below the nose, the rate of the phase transformation follows the kinetics of thermally activated processes such as diffusion.

The top of Fig. 5.7 shows microstructures for two nucleation-and-growth transforma-tions that came to completion in the same time, consistent with the TTT diagram. For the microstructure at the left, which formed at the lower temperature, the number of nuclei was

---

[7] On the other hand, superheating tends to be much less of an effect – a material usually begins to melt close to the temperatures predicted from the phase diagram, even for fast heating rates.

[8] With heterogeneous nucleation, the undercooling is not so large as would be predicted by the development of Section 4.5.5, which was based on homogeneous nucleation.

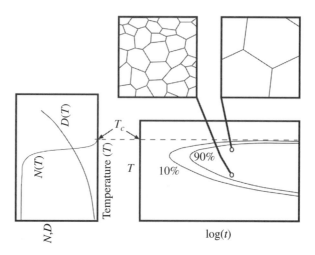

Figure 5.7 At left are shown the number of nuclei, *N*, and diffusivity, *D*, at temperatures near the critical temperature, T_c, of a nucleation-and-growth type of phase transformation. The TTT diagram is at the lower right, and typical microstructures are shown above it.

high, but their growth rates were low. The opposite case was true for the microstructure at the right. In general, a phase transformation that occurs at a lower temperature is finer grained than one occurring at a higher temperature.

The TTT diagram is an important tool for designing microstructures of materials. Section 14.4.3 presents TTT diagrams for the transformation upon cooling of the fcc "austenite" phase in different steels. Two added complexities of Fig. 14.11 are: (1) the fcc austenite transforms into two phases, and (2) these two phases can be configured in more than one microstructure.[9] The different microstructures grow by different mechanisms, and compete with each other. Figure 14.11d shows that, at low temperatures, a "bainite" microstructure grows more efficiently than a "pearlite" microstructure, for example. Both microstructures have the same two phases, bcc iron plus a carbide, but these two phases have different sizes and shapes in bainite and pearlite.

5.5 Glasses and Liquids

5.5.1 Glass Formation

Liquid diffusion is much faster than diffusion in a solid, but liquid-phase diffusion can be suppressed at high cooling rates. The partitioning of B-atoms between the solid and liquid is suppressed by high cooling rates, and with the phase diagram of Fig. 5.1 the β-phase forms with a concentration of B-atoms that is lower than predicted by the phase diagram.

---

[9] Generally the coarser microstructures are found at higher temperatures.

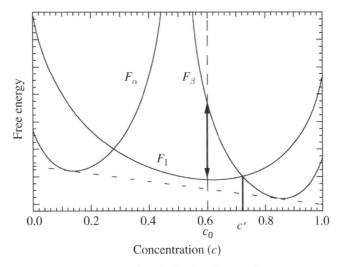

Concentration (c)

Figure 5.8 Free energy versus composition curves, similar to those for the phase diagram of Fig. 2.10 at a temperature below the eutectic temperature. If solute partitioning between phases is suppressed by kinetics, at this temperature the liquid phase may be expected for compositions $0.28 < c < 0.72$.

An extreme case, amenable to convenient analysis, can occur for cooling rates in excess of 10^6 K/s. At these rates, solute partitioning between the solid and liquid is impossible, and all regions of the alloy maintain a composition of c_0 at all temperatures.

The state of constrained equilibrium is then found in a simple way by fixing the compositions of all phases to be equal to the alloy composition c_0. A set of free energy versus composition curves is shown in Fig. 5.8 (consistent with Fig. 2.10). The common tangent construction for thermodynamic equilibrium predicts the coexistence of two solid phases (α and β) at this temperature, but without solute partitioning the liquid phase is preferable over a wide range of compositions. Without solute partitioning between the α- and β-phases, the formation of a crystal of β-phase of composition c_0 would cause an increase in free energy, as indicated by the line with the arrowheads in Fig. 5.8. Such crystallization does not occur. The liquid phase is especially stable for compositions in the middle of Fig. 5.8. Near the eutectic composition, deep undercoolings of the liquid are possible at high quench rates. At low temperatures, however, the liquid becomes increasingly more viscous. As the viscosity increases past 10^{12} Pa s, the alloy cannot crystallize in measurable times. This frozen liquid is a "metallic glass" [42]. Many alloys having phase diagrams with low eutectic temperatures can be quenched into a metallic glass by rapid cooling.

Figure 5.9 extends the previous eutectic free energy curves and phase diagram of Fig. 2.10. The new features in Fig. 5.9a, shown by gray dots and vertical gray lines, are the crossings of the free energy curves of the liquid and β-phase. If no diffusion were possible, the liquid would be most favorable at compositions to the left of these points, and the β-phase would be favored at compositions to the right. The compositions and temperatures of these points define the gray line labeled "T_0" in Fig. 5.9b.[10]

---

[10] Another such point for one temperature is indicated in Fig. 5.8 as the composition labeled c'.

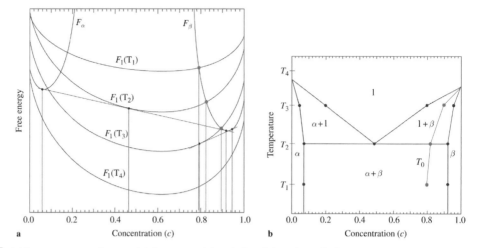

Figure 5.9 (**a**) Free energy curves for two solid phases, α and β, and a liquid phase drawn for four temperatures as in Fig. 2.10. The vertical gray lines mark the crossings of the curves for the liquid and the β-phase. These define the T_0-line in panel b. (**b**) T_0-line constructed from points in panel a. It can be considered a liquid–solid phase boundary if chemical segregation is suppressed.

The T_0 line denotes the limit of solubility of A-atoms in the β-phase competing with a liquid of the same composition, or equivalently the limiting composition where a metallic glass could form upon rapid cooling. It is an important construction from the free energy versus composition curves, but it is not directly measurable and does not appear on equilibrium $T–c$ phase diagrams. It is often approximated as being located halfway between the liquidus and β-solidus lines on an equilibrium phase diagram, but it probably is closer to the solidus line because the free energy of the solid phase usually increases more rapidly with composition.

5.5.2 The Glass Transition

Figure 5.10 shows the heat capacity of a metallic glass, CuZr, as it is heated. The alloy was prepared by rapid cooling from the liquid state ($>1000\,°C$) to room temperature by injection into a cold copper mold. The as-cast alloy was largely amorphous (a metallic glass). Upon heating in the calorimeter, it remains in the amorphous state at low temperatures up to the crystallization temperature around 445 °C. Before the large exothermic peak of crystallization (usually called "devitrification"), however, the amorphous material softens, becoming a viscous liquid that is deeply undercooled below the usual melting temperature of 935 °C. The heat capacity of this undercooled liquid rises about 2 k_B/atom above that of the amorphous alloy. This increase in heat capacity originates primarily from the additional degrees of freedom that accompany atom mobility. The onset of this change, labeled T_g in Fig. 5.10, is called the "glass transition." It is usually not considered a phase transition, although it can be reversible and it involves a change in free energy.[11]

[11] Please do not confuse the glass transition with devitrification; these are completely different processes.

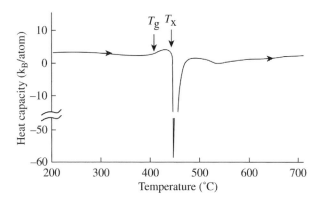

Figure 5.10 Calorimetry measurements on CuZr metallic glass heated at a rate of 20 °C/s. Heat absorption (endothermic feature) occurs at the glass transition temperature T_g, and a large heat emission (exothermic peak) occurs when the alloy crystallizes above T_x. (The broad feature around 540 °C is another crystal relaxation.)

The prominent devitrification (crystallization) peak in Fig. 5.10, starting at $T_x = 445\,°C$ with a maximum at $447\,°C$, is not part of the glass transition. Nevertheless, it is of practical importance because it is an irreversible transition from the amorphous state.[12] The nucleation and growth of crystals requires some diffusion of atoms. As such, the temperature T_x increases if the heating is faster, giving a larger temperature range for the undercooled liquid, and the opportunity to have a lower viscosity liquid for shaping or forming. The glass transition involves more localized motions of atoms, however, and does not involve significant diffusion distances. Typically T_g is less sensitive to heating rate than T_x. Shape-forming of metallic glasses by heating is considerably more robust when the heating and processing times are short.

5.6 Kinetics near Equilibrium

The diffusion equation was derived in Chapter 3 from the underlying assumption of independent atom jumps in random directions. Diffusion was used in the present chapter to understand the rates of phase transformations, but the diffusion equation itself offers no information about any tendency to form a new phase. Instead, thermodynamic tendencies during diffusion were inferred from phase diagrams. As atoms are diffusing, the average of their individual jumps moves the material towards thermodynamic equilibrium, but how? What follows below are two approaches to this question. The first shows why activated state rate theory for individual atom jumps can be consistent with thermodynamic equilibrium. The second concerns small deviations from equilibrium, and how we expect the system to return to the equilibrium state. These concepts are developed further in Chapter 23 (online).

---

[12] The kinetics of devitrification may follow the behavior shown in the TTT diagram of Fig. 5.7, since crystallization occurs by nucleation and growth.

5.6.1 Kinetic Master Equation and Equilibrium

The kinetic master equation offers a general description of how state variables of a system change with time. It is readily understood for atoms on an Ising lattice, where the ensemble-averaged concentrations of atoms on different sites are monitored in time as the atoms move from one site to another. We first consider what the master equation predicts about equilibrium, i.e., when none of the state variables are changing with time. We ask the question, "What are the conditions for which this steady state is actually the state of thermodynamic equilibrium?"

The master equation is readily formulated for one site i surrounded by z neighboring sites with labels j. The concentration on site i changes with time

$$\frac{dc_i}{dt} = +\sum_j^z W_{ij}c_j - \sum_j^z W_{ji}c_i, \tag{5.14}$$

where the matrix $\underset{\approx}{W}$ is a set of transition rates. The first term in Eq. 5.14 is the gain in concentration on site i as atoms move to it from the neighboring j, and the second term is the loss of concentration on site i from the reverse processes.

We now assume thermodynamic equilibrium, and then find what conditions are imposed on $\underset{\approx}{W}$. In thermodynamic equilibrium:

$$c_i = \frac{e^{-\varepsilon_i/kT}}{Z}, \tag{5.15}$$

$$c_j = \frac{e^{-\varepsilon_j/kT}}{Z}, \tag{5.16}$$

$$Z \equiv e^{-\varepsilon_i/kT} + \sum_j^z e^{-\varepsilon_j/kT}, \tag{5.17}$$

where ε_j is the energy associated with the atom on site j. Substituting into the kinetic master equation Eq. 5.14

$$\frac{dc_i}{dt} = +\sum_j^z W_{ij}\frac{e^{-\varepsilon_j/kT}}{Z} - \sum_j^z W_{ji}\frac{e^{-\varepsilon_i/kT}}{Z}. \tag{5.18}$$

The condition for kinetic equilibrium is that there are no changes in atom concentrations on the lattice sites, i.e., $dc_i/dt = 0$

$$0 = +\sum_j^z W_{ij}e^{-\varepsilon_j/kT} - \sum_j^z W_{ji}e^{-\varepsilon_i/kT}, \tag{5.19}$$

$$\sum_j^z W_{ij}e^{-\varepsilon_j/kT} = \sum_j^z W_{ji}e^{-\varepsilon_i/kT}. \tag{5.20}$$

By inspection, we can find two conditions on $\underset{\approx}{W}$ that satisfy Eq. 5.20. First, $\underset{\approx}{W} = 0$ will work. This solution eliminates all kinetics, so atoms stay on their sites with their thermodynamic probabilities. The second solution is more interesting. For the individual W_{ij} we pick exponentials with positive arguments that cancel the Boltzmann factors

$$W'_{ij} = e^{+\varepsilon_j/kT}, \tag{5.21}$$

$$W'_{ji} = e^{+\varepsilon_i/kT}. \tag{5.22}$$

Substituting into Eq. 5.20 gives equality, so the rates of Eqs. 5.21 and 5.22 are consistent with reaching thermodynamic equilibrium

$$\sum_{j}^{z} e^{+\varepsilon_j/kT} e^{-\varepsilon_j/kT} = \sum_{j}^{z} e^{+\varepsilon_i/kT} e^{-\varepsilon_i/kT}, \tag{5.23}$$

$$z = z. \tag{5.24}$$

If the rates of Eqs. 5.21 and 5.22 are both multiplied by the same constant, steady state is still achieved ($dc_i/dt = 0$). By choosing the constant factor $e^{-E^*/kT}$, Eqs. 5.21 and 5.22 become

$$W_{ij} = e^{-(E^*-\varepsilon_j)/kT}, \tag{5.25}$$

$$W_{ji} = e^{-(E^*-\varepsilon_i)/kT}. \tag{5.26}$$

Equations 5.25 and 5.26 can be understood with reference to Fig. 5.11. Consider equilibrium concentrations on the three states shown in the figure, all proportional to Boltzmann factors. The ratio of the equilibrium concentrations between the activated state and the initial state is

$$\frac{c_*}{c_i} = \frac{e^{-E^*/kT}}{e^{-\varepsilon_i/kT}} = e^{-(E^*-\varepsilon_i)/kT}, \tag{5.27}$$

which equals W_{ji} of Eq. 5.26. Finally, we assume that once an atom gets into the activated state, it will continue to the other side.[13] This is the basis for activated state rate theory, or at least this is the reason why it leads to thermodynamic equilibrium. When the two states i and j are in thermodynamic equilibrium with an activated state, they are in thermodynamic equilibrium with each other.

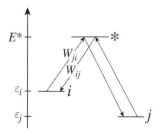

Figure 5.11   Energies and transitions between states i and j through an activated state of energy E^*.

[13] If it returns to state i with some probability, but has the same return probability for jumps out of the state j, there is no net effect on the equilibrium concentrations.

5.6.2 Linear Kinetic Response

Here is a simple and intuitive model for how a system moves towards equilibrium. This treatment is a handy one that can be applied to a number of nonequilibrium behaviors, but the system cannot be too far from equilibrium. The idea is to first obtain a "restoring force" when the system is displaced away from equilibrium, and this force is proportional to the displacement. A second assumption is that the system moves to equilibrium at a rate proportional to the restoring force. We begin with the free energy, without any kinetic considerations.

The free energy is approximated with a few terms in a Taylor series near the equilibrium value of a state variable. To be specific, choose an order parameter with an equilibrium value of L_0

$$F(L_0 + \Delta L) = F(L_0) + \left(\frac{dF}{dL}\right)_{L_0} \Delta L + \frac{1}{2}\left(\frac{d^2 F}{dL^2}\right)_{L_0} (\Delta L)^2 + \cdots . \tag{5.28}$$

Because F is a minimum at equilibrium, $(dF/dL)_{L_0} = 0$, and

$$F(L_0 + \Delta L) - F(L_0) \equiv \Delta F \simeq \frac{1}{2}\left(\frac{d^2 F}{dL^2}\right)_{L_0} (\Delta L)^2, \tag{5.29}$$

$$\Delta F \simeq \frac{1}{2}A(\Delta L)^2. \tag{5.30}$$

The second derivative evaluated at L_0 is a positive constant, which was relabeled as A. This ΔF acts as a potential energy, so it can be used to obtain a "force" that pushes the system towards equilibrium. The force, $\mathcal{F}$, is obtained in the usual way

$$\mathcal{F} = -\frac{d\Delta F}{d\Delta L} = -\frac{d}{dL}\frac{1}{2}A(\Delta L)^2, \tag{5.31}$$

$$\mathcal{F} = -A\,\Delta L. \tag{5.32}$$

The next assumption sets the kinetics. Assume that the rate of change of an order parameter is proportional to this force, $\mathcal{F}$

$$\frac{d\Delta L}{dt} = M\mathcal{F} = -MA\,\Delta L, \tag{5.33}$$

where M is the "mobility." Equation 5.33 has a simple solution,

$$\Delta L(t) = \Delta L(t = 0)\,e^{-MAt}. \tag{5.34}$$

where $A = d^2 F/dL^2 > 0$ for stable systems. Equation 5.34 predicts that both positive and negative deviations from the equilibrium L_0 will be restored with the same characteristic time, and the rate will also scale with the size of the initial deviation from equilibrium. Equation 5.34 is expected to be useful if the system is not far from equilibrium, and when the kinetic mechanism allows for a constant mobility, M. In reality, the mobility usually changes with temperature and state of order, but probably not much for small changes in T and L.

Similar analyses can be performed with other state variables. Consider chemical composition, c, for the unmixing system of Fig. 2.18. This problem is discussed in more detail in Section 16.1, but here are some essential points. Suppose that unmixing occurs by forming two regions with compositions that deviate by $+\delta c$ and $-\delta c$ from the overall composition c'. The total free energy has no linear term in δc because the region with $c' + \delta c$ and the region with $c' - \delta c$ would give opposing changes of free energy. The curvature, however, $\mathrm{d}^2 F/\mathrm{d}c^2$, causes a deviation in δc to decay when the curvature is positive in Fig. 2.18a. For compositions in the middle range of Fig. 2.18a where $\mathrm{d}^2 F/\mathrm{d}c^2 < 0$, however, an initial δc will not decay, but will grow. Unmixing alloys of these intermediate compositions are unstable at this low temperature. They are also far from equilibrium, and a different type of analysis is needed for their kinetics (as in Chapter 16).

Problems

5.1 The heat equation is the same as the diffusion equation (but temperature T replaces composition c).

  (**a**) Suppose you have a plate of steel of 1 cm in thickness, and you want to cool it to room temperature from a temperature of 1300 K. (The thermal diffusivity of steel is approximately $D_{\mathrm{th}} = 10^{-5}$ m^2 s.) You want to cool the interior of the plate as rapidly as possible, but you can remove heat only from the surfaces. Show that compared to casting there is a benefit to iced brine quenching, but there is no point in using a method such as piston-anvil quenching.

  (**b**) Suppose a binary alloy undergoes chemical unmixing by diffusion at low temperatures, and the spatial scale of the unmixing is the characteristic length $x = \sqrt{D\tau}$. Assuming the chemical and thermal diffusivities are constants, what is the relationship between the spatial scale of the unmixing and the cooling rate?

5.2 This problem uses the eutectic phase diagram of Fig. 2.23 with phase boundaries made from straight lines. The phases α, β, and liquid have the same density. An alloy of 70% B is cooled from the melt.

  (**a**) Plot the volume fraction of each phase versus temperature, assuming thermodynamic equilibrium.

  (**b**) Show how each plot of part a changes when there is no diffusion in the solid phase that has formed.

  (**c**) Using free energy versus composition curves for the phases α, β, and liquid, justify why an amorphous phase could form if there were no diffusion in the liquid during rapid cooling.

5.3 For the solidification profile "Limited diffusion in liquid" of Fig. 5.5, why does the final transient in the composition occur more abruptly than the initial transient?

5.4 The Scheil equation that relates c_s and f_s does not include temperature. Draw a diagram that estimates the freezing temperatures corresponding to f_s for the case of $k = 0.5$ in Fig. 5.4.

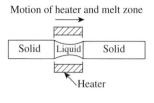

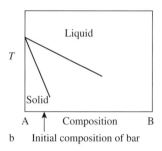

Figure 5.12 (a) Conceptual design of apparatus for zone refining. (b) A-rich region of phase diagram used for purifying A-rich solid by zone refining.

5.5 "Zone-refining" is a technique to make high-purity crystals of a pure element. Without zone-refining of silicon, it is unclear whether there would be a semiconductor industry. In this technique, a heater melts a small zone of material, and the heater and zone move along a bar of material (see Fig. 5.12a). The solutes segregate to the far end of the bar. The technique works for removing B-solutes from A-metal when the A-rich part of the A–B phase diagram is as is shown in Fig. 5.12b.

    (a) Why does zone-refining work?

    (b) Would the material become more pure after a second pass of the heater?

    (c) For what combination of slopes of the liquidus and solidus would zone-refining work best?

5.6 Use activated state rate theory for transition rates between states (using different activated states E^* for different transitions), and use the kinetic master equation to follow the probabilities of states to show:

    If state A is in equilibrium with state B, and state B is in equilibrium with state C, then state A is in equilibrium with state C.

5.7 In Section 5.6.2 the system was assumed to be near equilibrium, so the free energy was near a minimum. Consider some variations of this situation, all near the point where $dF/dL = 0$. Analyze them with an appropriate form of Eq. 5.28 and small values of ΔL, both positive and negative.

    (a) Suppose $d^2F/dL^2 < 0$. Is this system in equilibrium? How will ΔL change with time?

    (b) Suppose $d^2F/dL^2 = 0$ and $d^3F/dL^3 > 0$. Is this system in equilibrium? How will ΔL change with time?

    (c) Suppose $d^2F/dL^2 = 0$, $d^3F/dL^3 = 0$, and $d^4F/dL^4 > 0$. Is this system in equilibrium? How will ΔL change with time?

    (d) Suppose $d^2F/dL^2 = 0$, $d^3F/dL^3 = 0$, and $d^4F/dL^4 < 0$. Is this system in equilibrium? Explain how ΔL will change with time, and compare this to the case of part a.

PART II

THE ATOMIC ORIGINS OF THERMODYNAMICS AND KINETICS

Free energy is a central topic of this book because a phase transition occurs in a material when its free energy, or a derivative of its free energy, has a singularity. Chapter 2 showed how to use the dependence of free energy on composition or order parameter to obtain thermodynamic phase diagrams. Chapters 3 and 4 discussed the kinetics of diffusion and nucleation, using activated state rate theory with a free energy of activation. Chapter 5 showed how the free energies of equilibrium phases and the free energies of activation (such as ΔG^*) bring competition between the thermodynamics and kinetics of phase transformations.

The Gibbs free energy is

$$G = E - TS + PV ,$$

and the coverage of E, S, P, and ΔG^* is the core of Part II of this book. Chapter 6 discusses two sources of energy, electronic and elastic, that are important for phase transitions in materials. Chapter 7 explains configurational and vibrational entropy, and Chapter 8 discusses effects of pressure. The last two chapters of Part II are on kinetic phenomena. Chapter 9 describes how microstructural processes can interact and constrain each other. Chapter 10 explains how vacancy diffusion is altered by interactions between atoms.

6   Energy

The internal energy E is the first term in the free energy, $F = E - TS$. This chapter describes features of E that are important for phase transitions in materials. The internal energy originates from the quantum mechanics of electrons at atoms, specifically the chemical bonds between atoms. The simplest bond is between two atoms in a diatomic molecule, which is developed first to illustrate concepts of bonding, antibonding, electronegativity, covalency, and ionicity.

Crystals have long-range translational symmetry that brings a new quantum number $\vec{k}$ for states of delocalized electrons. The concept of energy bands is presented first by extending the concepts of molecular bonding and antibonding to electron states spread over many atoms. An even simpler model of a gas of free electrons is also developed for electrons in metals.

The strength of bonding depends on the distance between atoms. The interatomic potential of a chemical bond gives rise to elastic constants that characterize how a bulk material responds to small deformations. The chapter ends with a discussion of the elastic energy generated when a particle of a new phase forms inside a parent phase, and the two phases differ in specific volume.

6.1 Atomic Schrödinger Equations and Formalism

Start with two atoms, A and B, initially far apart. The properties of their electrons are calculated independently with two single-atom Schrödinger equations

$$-\frac{\hbar^2}{2m}\nabla^2 \psi_A(\vec{r}) + V_A(\vec{r})\,\psi_A(\vec{r}) = \epsilon_A \psi_A(\vec{r}), \tag{6.1}$$

$$-\frac{\hbar^2}{2m}\nabla^2 \psi_B(\vec{r}) + V_B(\vec{r})\,\psi_B(\vec{r}) = \epsilon_B \psi_B(\vec{r}). \tag{6.2}$$

At isolated atoms, the single-electron states (denoted $|A\rangle$ and $|B\rangle$ in Dirac notation) have wavefunctions ψ_A and ψ_B, and energies ϵ_A and ϵ_B. The potentials $V_A(\vec{r})$ and $V_B(\vec{r})$ are centered at the two separated nuclei. Each Laplacian, ∇^2, acts on the coordinates of one electron, wherever it is. To be more explicit, sometimes the Laplacian is given a subscript identifying the particular electron, e.g., the kinetic energy of electron number 1 is written $-(\hbar^2/2m)\,\nabla_1^2$, and electron number 2 has the kinetic energy $-(\hbar^2/2m)\,\nabla_2^2$.

We are interested in how the states for single electrons become something new when the two nuclei are brought together close enough so the wavefunctions of their atomic states overlap slightly. In a short time, an individual electron will have amplitude at both atoms, and a molecule is formed. A single-electron wavefunction for the diatomic molecule is called a "molecular orbital." A state for one electron now needs to include the effects of both nuclei and their core electrons, since the electron spends time about both. The potential is a real challenge for several reasons, primarily because the potential for one electron depends on the presence of other electrons.[1] We begin, however, by ignoring possible changes in potential, and we assume that the total potential for the molecule is simply the sum of potentials of the isolated atoms

$$V_\mathrm{M}(\vec{r}) = V_\mathrm{A}(\vec{r}) + V_\mathrm{B}(\vec{r}). \tag{6.3}$$

Another assumption, a guess really, is that the molecular orbital can be constructed as a weighted sum of the atomic orbitals as

$$|\mathrm{M}\rangle = c_\mathrm{A}|\mathrm{A}\rangle + c_\mathrm{B}|\mathrm{B}\rangle, \tag{6.4}$$

$$\psi_\mathrm{M}(\vec{r}) = c_\mathrm{A}\psi_\mathrm{A}(\vec{r}) + c_\mathrm{B}\psi_\mathrm{B}(\vec{r}), \tag{6.5}$$

written in Dirac notation (Eq. 6.4) and with explicit wavefunctions (Eq. 6.5). It is important to remember that the molecular state $|\mathrm{M}\rangle$ accommodates one electron, so the coefficients c_A and c_B are less than 1 (the atomic states $|\mathrm{A}\rangle$ and $|\mathrm{B}\rangle$ accommodate one electron each). The wavefunctions $\psi_\mathrm{A}(\vec{r})$ and $\psi_\mathrm{B}(\vec{r})$ are centered at different nuclei, but an electron can spend time in both of them (in states $|\mathrm{A}\rangle$ and $|\mathrm{B}\rangle$) when the two nuclei are brought near each other.

Box 6.1　　　　　　　　　　　　　　**Dirac Notation**

Many calculations in quantum mechanics involve integrations over regions containing electron wavefunctions. Working with conventional integral formalism and its coordinates leads to bulky equations where it can be hard to see the forest through the trees. Paul Dirac dispensed with the coordinates altogether with a clever and compact notation. The quantum state of an electron 1 at atom A, having the wavefunction $\psi_\mathrm{A}(\vec{r}_1)$, is denoted $|A\rangle$ in Dirac notation. Without coordinates, this state function is not the wavefunction, but the two are related by projecting the quantum state onto spatial coordinates (i.e., finding the presence of the state $|A\rangle$ at $\vec{r}_1$). This projection and its complex conjugate are written as

$$\psi_\mathrm{A}(\vec{r}_1) = \langle \vec{r}_1|A\rangle, \qquad \psi_\mathrm{A}^*(\vec{r}_1) = \langle A|\vec{r}_1\rangle. \tag{6.6}$$

A typical calculation of orthogonality or normalization takes the form

$$\int_{-\infty}^{\infty} \psi_\mathrm{A}^*(\vec{r}_1)\,\psi_\mathrm{A}(\vec{r}_1)\,\mathrm{d}^3\vec{r}_1 = \int_{-\infty}^{\infty} \langle A|\vec{r}_1\rangle \langle \vec{r}_1|A\rangle\,\mathrm{d}^3\vec{r}_1 = \langle A|\int_{-\infty}^{\infty} |\vec{r}_1\rangle \langle \vec{r}_1|A\rangle\,\mathrm{d}^3\vec{r}_1. \tag{6.7}$$

---

[1] This is, however, a problem already faced when calculating energies of electrons in a multi-electron atom. The effect of a second electron is to push around the first electron, but this also alters the potential and wavefunction of the second electron. Iterative methods are useful for these types of problems.

One interpretation of the integral on the RHS is that it means projecting $|A\rangle$ onto spatial coordinates for electron 1 as $\langle \vec{r}_1 | A \rangle$, using these projections to weight $|\vec{r}_1\rangle$, and then summing (integrating) over all $\vec{r}_1$. This reconstructs the state $|A\rangle$ from the $|\vec{r}_1\rangle$, so Eq. 6.7 takes the compact form[a]

$$\int_{-\infty}^{\infty} \psi_A^*(\vec{r}_1) \psi_A(\vec{r}_1)\, d^3 \vec{r}_1 = \langle A | A \rangle. \tag{6.8}$$

The form $\langle \; | \; \rangle$ denotes a definite integral, typically with limits over all space. The same is true of expectation values of operators. For example

$$\int_{-\infty}^{\infty} \psi_A^*(\vec{r}_1) \frac{-e^2}{r_1} \psi_B(\vec{r}_1)\, d^3 \vec{r}_1 = \left\langle A \left| \frac{-e^2}{r_1} \right| B \right\rangle, \tag{6.9}$$

or for another example:

$$\int_{-\infty}^{\infty} \int_{-\infty}^{\infty} \psi_A^*(\vec{r}_1) \psi_B^*(\vec{r}_2) H'(\vec{r}_2) \psi_B(\vec{r}_1) \psi_A(\vec{r}_2)\, d^3 \vec{r}_1\, d^3 \vec{r}_2 = \langle A | B \rangle_1 \langle B | H' | A \rangle_2, \tag{6.10}$$

where the electron subscripts on the RHS of Eqs. 6.9 and 6.10 are usually ignored.

Dirac notation allows rapid and lucid mathematical manipulations. In the end, however, calculating integrals requires thought about the best coordinate system, and typically multiple integrations over three spatial variables.

[a] The coordinates form a complete set for expressing any function of position. For complete sets we can invoke the property of "closure," defined with an integral or a sum as, for example, $\int_{-\infty}^{\infty} |\vec{r}\rangle \langle \vec{r}|\, d^3\vec{r} = 1$ or $\sum_i |i\rangle \langle i| = 1$.

6.2 Molecular Orbital Theory of Diatomic Molecules

6.2.1 Molecular Schrödinger Equation

The Schrödinger equation for a single electron in the molecule is

$$H|M\rangle = \epsilon |M\rangle, \tag{6.11}$$

where the Hamiltonian includes a kinetic energy operator for one electron, and the potential for this electron, the $V_M(\vec{r})$ of Eq. 6.3. Two key operations are now done to convert the RHS of the Schrödinger equation 6.11 into a real energy ϵ without the wavefunction:

1. multiply Eq. 6.11 by $\langle A|$, and make a second equation by multiplying Eq. 6.11 by $\langle B|$;
2. for both equations, integrate over the coordinates of the electron.

Doing both steps simultaneously with Dirac notation

$$\langle A | H | M \rangle = \epsilon \langle A | M \rangle, \tag{6.12}$$

$$\langle B | H | M \rangle = \epsilon \langle B | M \rangle. \tag{6.13}$$

Using Eq. 6.4 for $|M\rangle$

$$\langle A|H|\big(c_A|A\rangle + c_B|B\rangle\big) = \epsilon\langle A|\big(c_A|A\rangle + c_B|B\rangle\big), \tag{6.14}$$

$$\langle B|H|\big(c_A|A\rangle + c_B|B\rangle\big) = \epsilon\langle B|\big(c_A|A\rangle + c_B|B\rangle\big), \tag{6.15}$$

Moving the constants c_A and c_B through the integrals

$$c_A\,\langle A|H|A\rangle + c_B\,\langle A|H|B\rangle = \epsilon\, c_A\,\langle A|A\rangle + \epsilon\, c_B\,\langle A|B\rangle, \tag{6.16}$$

$$c_A\,\langle B|H|A\rangle + c_B\,\langle B|H|B\rangle = \epsilon\, c_A\,\langle B|A\rangle + \epsilon\, c_B\,\langle B|B\rangle. \tag{6.17}$$

Definitions and Approximations

The atomic states $|A\rangle$ and $|B\rangle$ are normalized to accommodate one electron, so

$$\langle A|A\rangle = \langle B|B\rangle = 1. \tag{6.18}$$

Although coordinates are not included in Dirac notation, the wavefunctions $\psi_A(\vec{r})$ and $\psi_B(\vec{r})$ are physically centered at different nuclei. The tails of the wavefunction for $|A\rangle$ overlap the tails of the wavefunction for $|B\rangle$. We define the overlap integral S

$$S \equiv \langle A|B\rangle, \tag{6.19}$$

$$S \equiv \int_{-\infty}^{+\infty} \psi_A^*(\vec{r})\,\psi_B(\vec{r})\,\mathrm{d}^3\vec{r}. \tag{6.20}$$

For now, for simplicity we set $S = 0$ (expecting that S is small). A closely related approximation is that the tails of the wavefunction for $|A\rangle$ overlap only weakly with the potential energy from the second nucleus at B, so we neglect energies like $\langle A|V_B|A\rangle$. This allows the approximations

$$\langle A|H|A\rangle \simeq \epsilon_A, \tag{6.21}$$

$$\langle B|H|B\rangle \simeq \epsilon_B, \tag{6.22}$$

where ϵ_A and ϵ_B are the energies of electrons in states at the isolated atoms A and B. With these simplifications, Eqs. 6.16 and 6.17 become

$$c_A\,\epsilon_A + c_B\,\langle A|H|B\rangle = \epsilon\, c_A, \tag{6.23}$$

$$c_A\,\langle B|H|A\rangle + c_B\,\epsilon_B = \epsilon\, c_B. \tag{6.24}$$

The important physical quantity that remains after our manipulations and approximations is the "bond integral," h

$$h \equiv \langle A|H|B\rangle, \tag{6.25}$$

$$h \equiv \int_{-\infty}^{+\infty} \psi_A^*(\vec{r})\,H\,\psi_B(\vec{r})\,\mathrm{d}^3\vec{r}. \tag{6.26}$$

Energies of Molecular States

Some consequences of the bond integral h are explored below, but first we complete the formal effort to obtain the electron energies in the molecule, ϵ. With Eq. 6.25 for h, the pair of Eqs. 6.23 and 6.24 is written as one compact matrix equation

$$\begin{bmatrix} \epsilon_A - \epsilon & h \\ h & \epsilon_B - \epsilon \end{bmatrix} \begin{bmatrix} c_A \\ c_B \end{bmatrix} = \begin{bmatrix} 0 \\ 0 \end{bmatrix}. \tag{6.27}$$

It proves convenient to replace ϵ_A and ϵ_B with their average $\bar{\epsilon} = (\epsilon_A + \epsilon_B)/2$ and the difference $\Delta = (\epsilon_B - \epsilon_A)/2$ so

$$\epsilon_A = \bar{\epsilon} - \Delta \quad , \quad \epsilon_B = \bar{\epsilon} + \Delta, \tag{6.28}$$

$$\begin{bmatrix} \bar{\epsilon} - \Delta - \epsilon & h \\ h & \bar{\epsilon} + \Delta - \epsilon \end{bmatrix} \begin{bmatrix} c_A \\ c_B \end{bmatrix} = \begin{bmatrix} 0 \\ 0 \end{bmatrix}. \tag{6.29}$$

The matrix equation 6.29 has a useful solution when the determinant is zero, giving the secular equation

$$(\bar{\epsilon} - \Delta - \epsilon)(\bar{\epsilon} + \Delta - \epsilon) - h^2 = 0, \tag{6.30}$$

$$(\epsilon - \bar{\epsilon})^2 = \Delta^2 + h^2, \tag{6.31}$$

$$\epsilon^{\pm} = \bar{\epsilon} \mp \sqrt{\Delta^2 + h^2}, \tag{6.32}$$

where the two values of ϵ are labeled as ϵ^+ and ϵ^-. Compared to the average of the atomic energies, $\bar{\epsilon}$, one molecular state has an energy higher by the amount $\sqrt{\Delta^2 + h^2}$, and the other has an energy lower by the amount $\sqrt{\Delta^2 + h^2}$. These states are naturally called "antibonding" (with higher energy ϵ^-) and "bonding" (with lower energy ϵ^+), respectively.

6.2.2 Molecular Wavefunctions of Bonding and Antibonding States

Homonuclear Molecular Orbitals

We now solve for the c_A and c_B in Eq. 6.4, from which we can obtain the electron density in the bonding and antibonding molecular states, or "molecular orbitals." For homonuclear molecular orbitals involving identical atoms, for which $\epsilon_B = \epsilon_A$ so $\Delta = 0$, Eq. 6.32 gives $\epsilon^{\pm} = \bar{\epsilon} \mp |h|$. Substituting $\epsilon^{\pm}$ for ϵ in Eq. 6.29

$$\begin{bmatrix} \pm|h| & h \\ h & \pm|h| \end{bmatrix} \begin{bmatrix} c_A^{\pm} \\ c_B^{\pm} \end{bmatrix} = \begin{bmatrix} 0 \\ 0 \end{bmatrix}. \tag{6.33}$$

The first line of Eq. 6.33 gives

$$\pm|h|\, c_A^{\pm} + h\, c_B^{\pm} = 0, \tag{6.34}$$

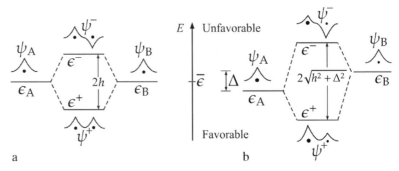

Figure 6.1 Energies and wavefunctions of molecular orbital formation in diatomic molecules with atom A on left and B on right in both panels a and b. Energy scale and definitions are in the middle. (**a**) Homonuclear molecule, such as H_2. With identical atomic energies and wavefunctions, the one-electron molecular states are symmetric. (**b**) Heteronuclear molecule, such as HF. With $\epsilon_B > \epsilon_A$, the bonding level ψ^+ is lower because it puts more of the electron in the favorable atomic wavefunction about atom A. The antibonding orbital ψ^- puts more electron in the unfavorable atomic wavefunction about atom B.

and since h is negative (bond energies are favorable)

$$c_A^\pm = \pm c_B^\pm. \tag{6.35}$$

For the bonding state with $\{\epsilon^+, c_A^+, c_B^+\}$, we obtain $c_A^+ = c_B^+$.
For the antibonding state with $\{\epsilon^-, c_A^-, c_B^-\}$, we obtain $c_A^- = -c_B^-$.

The two molecular states for single electrons, $|M^+\rangle$ and $|M^-\rangle$, are depicted in Fig. 6.1 as molecular orbital wavefunctions $\psi^+(\vec{r})$ and $\psi^-(\vec{r})$. The two panels a and b show bonding of two identical atoms (homonuclear), and bonding of two different atoms (heteronuclear). When identical atoms are brought together ($\Delta = (\epsilon_B - \epsilon_A)/2 = 0$), the bonding energy ϵ^+ and the antibonding energy ϵ^- are favored or disfavored over the isolated atoms by $\pm h$, where h is the bond integral of Eq. 6.25 (set $\Delta = 0$ in Eq. 6.32).

The $c_A^\pm$ and $c_B^\pm$ need normalization to 1 because there is one electron in either the bonding or antibonding molecular orbital

$$1 = \langle M^+|M^+\rangle = \left(c_A^+\langle A| + c_B^+\langle B|\right)\left(c_A^+|A\rangle + c_B^+|B\rangle\right), \tag{6.36}$$

$$1 = \langle M^-|M^-\rangle = \left(c_A^-\langle A| + c_B^-\langle B|\right)\left(c_A^-|A\rangle + c_B^-|B\rangle\right), \tag{6.37}$$

and with $c_A^+ = c_B^+$ and $c_A^- = -c_B^-$

$$1 = \langle M^+|M^+\rangle = c_A^{+2}\langle A|A\rangle + c_A^{+2}\langle A|B\rangle + c_A^{+2}\langle B|A\rangle + c_A^{+2}\langle B|B\rangle, \tag{6.38}$$

$$1 = \langle M^-|M^-\rangle = c_A^{-2}\langle A|A\rangle - c_A^{-2}\langle A|B\rangle - c_A^{-2}\langle B|A\rangle + c_A^{-2}\langle B|B\rangle. \tag{6.39}$$

Using the overlap integral S of Eq. 6.19 for $\langle A|B\rangle$ and $\langle B|A\rangle$

$$c_A^+ = +c_B^+ = \frac{1}{\sqrt{2}\,\sqrt{1+S}}, \tag{6.40}$$

$$c_A^- = -c_B^- = \frac{1}{\sqrt{2}\,\sqrt{1-S}}. \tag{6.41}$$

Comparing Eqs. 6.40 and 6.41, it is seen that with increasing overlap S, the bonding wavefunction needs some reduction in amplitude, and the antibonding some increase in amplitude. This can be understood from the electron density in the regions of overlap, as indicated schematically for ψ^+ and ψ^- in Fig. 6.1a. Midway between the two atoms, the wavefunction amplitude is doubled for the bonding wavefunction ψ^+, compared with the wavefunction in the tail of an atom wavefunction. It is zero in the middle of the antibonding wavefunction ψ^-. Obtaining the electron density as $\psi^*\psi$, the density for the bonding ψ^+ is increased by a factor of 2 over that for two independent atoms, whereas the density for the antibonding ψ^- is zero. Normalization of the whole wavefunction to 1 therefore decreases the total amplitude of the bonding wavefunction and increases it for the antibonding.

Heteronuclear Molecular Orbitals

The bonding energy increases with Δ, as given in Eq. 6.32 and shown in Fig. 6.1b. This Δ is a difference of atom properties, ϵ_A and ϵ_B. This difference is specified as the difference in "electronegativity," χ. The larger is χ, the more favorable are the single electron states at the atom. Our results of Eq. 6.32 show that the larger the difference in electronegativity $|\chi_A - \chi_B|$, the stronger the bonding energy. On the other hand, with increasing $|\chi_A - \chi_B|$ the antibonding state becomes even more unfavorable compared to the homonuclear molecule. As discussed further in Section 6.2.4, Fig. 6.1b illustrates what happens – the bonding level puts more electron density on the atom with the larger χ (and lower ϵ_A). The antibonding molecular orbital puts more electron density on the less favorable atom with the smaller χ. The difference between bonding and antibonding energies grows with Δ, or with increasing $|\chi_A - \chi_B|$.

6.2.3 Origin of the Bond Energy

A common argument is that the higher density of electrons in the region between the atoms, where the electron can see both nuclei with their attractive potentials, is why the bonding level is more favorable energetically. Unfortunately, this argument has a fundamental flaw. The electron density is normalized, so putting more electron density between the nuclei must remove electron density from other parts of the molecule. Indeed, Eqs. 6.40 and 6.41 show that $c_A^+ < c_A^-$, for example. Normalization removes electron density from nearest the nuclei where the potential energy is most favorable, and places it where the potential is modestly favorable. The consequence is that the antibonding level sees an increase in electron density at the nuclei, and the bonding level a decrease. The potential energy actually favors the antibonding level over the bonding level for molecular orbital theory with atomic wavefunctions.

The first step to repair this embarrassing situation is to include the electron kinetic energy. Here the disadvantage of the antibonding wavefunction is more obvious. Its node between the nuclei requires that the electron wavefunction undergoes a more rapid oscillation, increasing the kinetic energy because of a larger $\nabla^2 \psi$. In contrast, the bonding wavefunction is smoother in this region between the nuclei, so it has the lower kinetic energy. This explanation of bonding is now more correct.

Molecular wavefunctions of the H_2^+ molecular ion have been calculated accurately, since there is only one electron, and the complicated electron–electron interactions are absent. Both the electron potential energy (which decreases when the electron is bunched near the nucleus) and the electron kinetic energy (which increases when the electron is confined) need to be considered in detail, for both the bonding and antibonding states. One should also consider the neglected overlap energies such as $\langle A|V_B|A\rangle$, which destabilize the antibonding state and favor the bonding state. For atomic s-electron wavefunctions it is found, perhaps not surprisingly, that atomic wavefunctions are inadequate even for a qualitative explanation of the chemical bond. The antibonding state is still favored slightly over the bonding state for most interatomic separations. What needs to be done is alter the shapes of $\psi_A(\vec{r})$ and $\psi_B(\vec{r})$ to minimize the energies of the bonding and antibonding states. This can be done fairly well by changing the characteristic length for the exponential decay of the atomic s-electron wavefunctions. By making a more rapid decay with distance of $\psi_A(\vec{r})$ and $\psi_B(\vec{r})$ for the bonding state, it is possible to place more electron density at the nuclei. The electron kinetic energy is increased a bit because the wavefunction is more confined, but the potential energy becomes much more favorable, and the bonding state does become bonding. The antibonding wavefunctions tend to lower their energy if the atomic s-states are expanded a bit, reducing the kinetic energy somewhat.[2] The real situation points out the risk of relying too heavily on analytical models, even for calculating bonding and antibonding states of the simplest molecule, H_2^+.

6.2.4 Ionicity and Electronegativity

For a heteronuclear molecule the two atoms differ in electronegativity, so $\Delta = (\epsilon_B - \epsilon_A)/2 \neq 0$. We expect unequal electron densities on each atom, so $|c_A| \neq |c_B|$. Expressions analogous to Eqs. 6.40 and 6.41 are obtained directly from Eqs. 6.33. Approximately[3]

$$c_A^\pm = \frac{1}{\sqrt{2}}\sqrt{1 \pm \frac{\delta - S}{\sqrt{1 + \delta^2}}}, \qquad (6.42)$$

$$c_B^\pm = \pm\frac{1}{\sqrt{2}}\sqrt{1 \mp \frac{\delta + S}{\sqrt{1 + \delta^2}}}, \qquad (6.43)$$

where the new variable

$$\delta \equiv \frac{\Delta}{|h|} \qquad (6.44)$$

[2] A further improvement can be obtained by optimizing the atomic wavefunctions at each interatomic separation, but here the result is not changed qualitatively.
[3] When $\delta = 0$, there is a small difference between Eqs. 6.40 and 6.41 and Eqs. 6.42 and 6.43, since in Eqs. 6.42 and 6.43 some terms of order S^2 were neglected, and normalization was not done directly.

scales with the electronegativity difference of the atoms. Assuming the A-atom has the higher electronegativity as in Fig. 6.1b, electron density will transfer from B to A. We seek to write the molecular charge density $\rho = \psi^*\psi$ in the form

$$\rho(\vec{r}) = (1 + \alpha_i)\rho_A(\vec{r}) + (1 - \alpha_i)\rho_B(\vec{r}) + \alpha_c \rho_{bond}(\vec{r}), \qquad (6.45)$$

where α_i is a fractional charge transfer caused by the ionic character of the bond (the difference in electronegativity), and α_c weights the charge density of covalent bonding that was considered in the previous section for a homonuclear molecule such as H_2. The charge redistribution upon bonding, $\rho_{bond}(\vec{r})$, is defined with respect to the charge already present around two overlapping but independent atoms

$$\rho_{bond}(\vec{r}) = 2\psi_A(\vec{r})\psi_B(\vec{r}) - S\left[\rho_A(\vec{r}) + \rho_B(\vec{r})\right], \qquad (6.46)$$

where $\rho_A = \psi_A^*(\vec{r})\psi_A(\vec{r})$ and $\rho_B = \psi_B^*(\vec{r})\psi_B(\vec{r})$. Both ψ_A and ψ_B were assumed real in Eq. 6.46.

It turns out that

$$\alpha_i = \frac{\delta}{\sqrt{1 + \delta^2}}, \qquad (6.47)$$

$$\alpha_c = \frac{1}{\sqrt{1 + \delta^2}}, \qquad (6.48)$$

and $\alpha_i^2 + \alpha_c^2 = 1$. When $\delta = 0$, so the atoms have the same electronegativity, $\alpha_i = 0$ and $\alpha_c = 1$. This is a purely covalent bond.

When $\delta > 1$ and the effects of electronegativity are dominant, $\alpha_i > \alpha_c$, and the bond is mostly ionic. Assuming that these bonding tendencies are characteristics of atoms, or differences between atoms, it is appropriate to assign an electronegativity to each chemical element. Measures of electronegativity have been defined by different approaches, but the general trends of different electronegativity scales tend to be in agreement. It is a challenge, however, to account for bonding between different types of outer electrons of different atoms. Furthermore, there is a fundamental problem with all electronegativity scales – they predict that all pairs of chemical elements will form compounds. This is not so, and proper explanations of compound formation require concepts and physical properties that cannot be assigned to individual atoms alone.

6.3 Electronic Bands and the Tight-Binding Model

6.3.1 Translational Symmetry and Phase Factors

Translational symmetry is the fundamental difference between a molecule and a crystal, and translational symmetry has a profound effect on the spectrum of electronic energies in crystals. Neglecting surfaces, the electron density must be the same if position is shifted by a repeat distance of the crystal, i.e., a translation vector of the lattice. This does not apply, of course, to an isolated diatomic molecule.

With translational symmetry, the distinction between bonding and antibonding states needs to be modified. The atomic wavefunction, which switched from +1 to −1 for an antibonding state in a diatomic molecule, has a similar form in a periodic solid if it switches phase between atoms in a chain as $\ldots +1, -1, +1, -1, +1, -1, \ldots$ This alternate switching of phase gives a valid antibonding state. Unfortunately, it gives only one state for one electron. With many electrons in a solid, we need many more states. How can we construct enough states when atomic wavefunctions $\psi(x)$ are placed along a linear chain at positions $x = na$?

The trick is to multiply the ψ by phase factors like

$$\ldots \; e^{i\vec{k}\cdot 1\vec{a}}, e^{i\vec{k}\cdot 2\vec{a}}, e^{i\vec{k}\cdot 3\vec{a}}, e^{i\vec{k}\cdot 4\vec{a}} \ldots,$$

where $\vec{a}$ is a primitive translation vector of the lattice. The state and its energy depend on the new vector $\vec{k}$, which can take many values. As one example, if $\vec{k} = \pi/a\,\hat{a}$, we have the sequence of phase factors $\ldots \; e^{i\pi}, e^{i2\pi}, e^{i3\pi}, e^{i4\pi}, \ldots$, which is $\ldots -1, +1, -1, +1, \ldots$, because $e^{i\pi \; \text{even integer}} = +1$ and $e^{i\pi \; \text{odd integer}} = -1$. In this case when $\vec{k} = \pi/a\,\hat{a}$, we recover our antibonding state. An even easier case is $k = 0$. For $k = 0$, the phase factors are $e^{i\vec{k}\cdot n\vec{r}} = e^{i0} = +1$ at every atom, and all atomic wavefunctions have the same phase. When $\vec{k} = 0$, we obtain a fully bonding wavefunction, as for the diatomic molecule.

Box 6.2	Bloch's Theorem

This theorem justifies our use of phase factors. Assume the Hamiltonian of the Schrödinger equation is translationally symmetric, so

$$H_0 = H_{n\vec{a}}, \tag{6.49}$$

and a shift of the potential energy operator (and kinetic energy operator, but this is easy) by $n\vec{a}$ in an infinite crystal gives the same Hamiltonian. The wavefunctions, $|0\rangle$ and $|n\vec{a}\rangle$, of the unshifted and shifted Hamiltonians satisfy

$$H_0 \, |0\rangle = \epsilon_0 \, |0\rangle, \tag{6.50}$$

$$H_{n\vec{a}} \, |n\vec{a}\rangle = \epsilon_{n\vec{a}} \, |n\vec{a}\rangle, \tag{6.51}$$

but by Eq. 6.49

$$H_0 \, |n\vec{a}\rangle = \epsilon_{n\vec{a}} \, |n\vec{a}\rangle. \tag{6.52}$$

Do the wavefunctions change with this translation $n\vec{a}$ that leaves the Hamiltonian invariant? A naïve guess is that they are identical, but this is incomplete. It turns out we can have a multiplicative factor, λ_n, for the translation $n\vec{a}$ in the form

$$|n\vec{a}\rangle = \lambda_n \, |0\rangle. \tag{6.53}$$

The wavefunction must remain normalized, however, so

$$1 = \langle n\vec{a} | n\vec{a} \rangle = \lambda_n^* \lambda_n \langle 0 | 0 \rangle, \tag{6.54}$$

$$1 = \lambda_n^* \lambda_n, \tag{6.55}$$

$$e^{i\vec{k}\cdot n\vec{r}} = \lambda_n. \tag{6.56}$$

Equation 6.56 shows that λ_n must be a phase factor of modulus 1, and Bloch's theorem is

A wavefunction of a periodic Hamiltonian is a product of a function with the periodicity of the lattice times a phase factor of modulus unity, $e^{i\vec{k}\cdot\vec{r}}$.

What is new is that by changing the value of $\vec{k}$, we can now obtain a wide range of different wavefunctions, varying between these extremes of fully bonding and fully antibonding. This $\vec{k}$ sets the energy levels of the electrons, and is a new quantum number[4] to describe the electron states in a periodic potential. These phase factors are complex numbers, but this is not a problem when calculating electron density or energy, which involve taking products such as $e^{-i\vec{k}\cdot n\vec{r}}\,e^{+i\vec{k}\cdot n\vec{r}} = +1$.

Figure 6.2 shows the construction of three periodic wavefunctions on a linear solid of atom spacing a, using three different values of k. The repeat length of the wavefunction at the bottom is a, and it has the characteristics of a low-energy bonding orbital. At the top is the wavefunction of highest energy, an antibonding state. It has a repeat distance of $2a$. The middle construction has a repeat distance of $8a$, and is closer in energy to the bonding wavefunction than the antibonding. For the neighboring atoms in the middle construction, the wavefunctions generally overlap with the same sign, for both the real and imaginary parts. (Again, there is no problem with an imaginary phase factor because the electron density is $\psi^*(x)\,\psi(x)$, which is always real.)

The middle construction of Fig. 6.2 with $k = 1/4\ \pi/a$ is only one possibility. A wavevector $k = 1/8\ \pi/a$ would give a wavefunction with twice the repeat distance, and an energy that is closer to that of the fully bonding state. For a long chain there is a nearly continuous range of k between 0 and π/a, giving a "band" of electron energies from fully bonding to fully bonding.

6.3.2 Tight Binding on a Linear Chain

So if $\vec{k} = 0$ gives the most bonding state, and $\vec{k} = \pm\pi/a\ \hat{a}$ gives the most antibonding, how do we calculate energies for the intermediate values of k? This section explains a straightforward approach, sometimes called "tight-binding" theory. It is the extension of molecular orbital theory for a diatomic molecule to an infinite crystal of atoms.

For simplicity assume that only bonds between an atom and its first-nearest neighbors have significant energy, and continue with the example of a one-dimensional chain of atoms. An atom in the chain (at position $x = 0$) has a single electron in the state $|0\rangle$ that makes a bond to a neighboring state at an atom to its left (at position $\vec{x} = -\vec{a}$) and a neighbor to its right (at position $\vec{x} = +\vec{a}$). These interactions give the energy per atom

$$\epsilon = \langle -a|\mathcal{H}|0\rangle + \langle +a|\mathcal{H}|0\rangle. \tag{6.57}$$

[4] The number refers to an interval in k-space. The interval has the approximate width $1/L$, where L is a large length. One approach is to use the dimension of a large box as L. Another imposes periodic boundary conditions over the length L, which force the phase to be the same when the same atom is reached after the translation L.

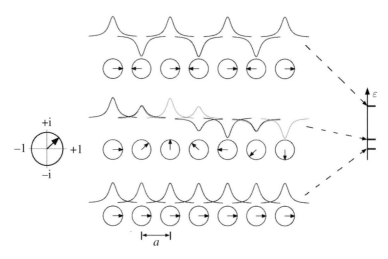

Figure 6.2 At left is the unit circle on the complex plane, intersecting the real axis at ±1 and the imaginary axis at ±i. The constructions of three electron wavefunctions are shown at right. The top is for the wavevector $k = \pi/a$, showing phase factors of $\ldots + 1, -1, +1, -1, \ldots$ The bottom is for the wavevector $k = 0$, showing phase factors of $\ldots + 1, +1, +1, +1, \ldots$ In the middle, at intermediate energy, is for the wavevector $k = \pi/4a$, showing phase factors of $\ldots +1, +0.7+0.7i, +i, -0.7+0.7i, -i, -0.7-0.7i, -i, \ldots$ The imaginary parts of the wavefunction are shown in gray.

The form of the bonding interaction $\mathcal{H}$ is not of concern at the moment. We ignore its covalency, for example, but we assume this is the same for all atoms along the chain. Imposing Bloch's theorem for the phase factors for the two neighbor positions at $\vec{x} = \pm\vec{a}$

$$\epsilon = e^{-i\vec{k}\cdot\vec{a}}\langle a|\mathcal{H}|0\rangle + e^{+i\vec{k}\cdot\vec{a}}\langle a|\mathcal{H}|0\rangle, \tag{6.58}$$

$$\epsilon = 2\cos(\vec{k}\cdot\vec{a})\,\langle a|\mathcal{H}|0\rangle. \tag{6.59}$$

The result for ϵ depends on the bond energy between a neighboring pair of atoms, $\langle a|\mathcal{H}|0\rangle$, a negative quantity that could be the same as the bond integral h of Eq. 6.25. Unlike the case of combining adjacent atomic wavefunctions in bonding or antibonding states with a simple ±1, here the phase of the wavefunctions gives a $\vec{k}$-dependence of $2\cos(\vec{k}\cdot\vec{a})$. Figure 6.3a shows $\epsilon(k)$ for our one-dimensional chain, shifting the lowest-energy (most bonding) state to $\epsilon = 0$ (and dividing by 2 to avoid overcounting of bonds).

Section 6.4.1 will show how equal intervals in k have equal numbers of electron states. This allows us to construct the "density of electron states," $\rho(\epsilon)$, a function of energy, using Fig. 6.3b. Although the energy intervals in Fig. 6.3b are the same, the intervals in k near $k = 0$ and $k = \pi/a$ are wider, and there are more electron states near the maximum and minimum energies. A histogram density of states (DOS) is shown in Fig. 6.3c. More rigorously, the density of states is proportional to the inverse of the slope of the dispersion $\epsilon(k)$, i.e., $1/(d\varepsilon/dk)$, with peaks where the dispersion is flat. For the cosine function of Fig. 6.3a, the inverse of the slope is $1/\sin(ka)$, so $\rho(\epsilon) \propto \csc(ka)$, and the DOS diverges at $\epsilon = 0$ and $\epsilon = 2\langle a|\mathcal{H}|0\rangle$.

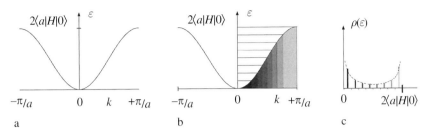

Figure 6.3 (**a**) Band structure of linear chain, as in Eq. 6.59. (**b**) In each interval of equal width in ϵ, there are more values of k near $k = 0$ and $k = \pi/a$ (giving peaks in the electronic density of states, $\rho(\epsilon)$). (**c**) Plot of widths in k-space for the energy intervals in part b, showing construction of the density of states $\rho(\epsilon)$.

Figure 6.3 illustrates key concepts of a band of energies (energy band) for electrons in solids:

- There is a bonding integral for atomic wavefunctions at neighboring sites.
- Bloch's theorem gives a phase relationship between wavefunctions at atoms in a periodic solid.
- With increasing $\vec{k}$, the increasing phase, $\vec{k} \cdot \vec{a}$, between the first neighbors decreases the constructive interference between the atomic wavefunctions, raising their bonding energy.

These assumptions, and the approach taken in this section, are the essence of the "tight-binding model" for electron energy bands in solids. The tight-binding model extends the concept of molecular orbitals to include the translational periodicity of a crystal. Additional features of tight binding in 2D and 3D materials are presented in Section 20.4.2.

6.4 Free and Nearly-Free Electrons

6.4.1 A Gas of Free Electrons in a Big Box

This section presents a completely different view of bonding in metallic solids, with no connection to the previous model of "tight binding." It is called the "nearly-free electron model," and begins with free electrons in a box. (It later adds in perturbations from the periodic crystal potential, but we will not take this step here.) The electrons stay in the box owing to a background of positive charges, but within this box they behave as free electrons in a vacuum. They do obey the Pauli principle, however, and no more than one electron can occupy each quantum state. The electrons behave as a gas, so there are no interactions between electrons, and no interactions with the atomic structure.

This free electron model sounds too simple to be true, and often it is. Nevertheless, it can account for the elastic stiffness of metals such as sodium. It can also explain electrical conductivity and electron screening of charges at impurities, which are basic properties of metals. Beware, though, the free electron model is unreliable for transition metals and main group elements, for which the tight-binding model is a better approximation.

Particle in a Box

The "particle in a box" is a classic problem in quantum mechanics because it is one of three for which the Schrödinger equation has an analytical solution in three dimensions.[5] The box is defined by the potential energy $V(x, y, z)$. Assume the potential box is cubic with edge lengths L along each Cartesian axis, with infinite walls at the edges and beyond. The wavefunction does not extend outside the box, and is zero at the edges of the box. The potential, V, is assumed a constant inside the box, which we set as 0 for convenience. The Schrödinger equation is

$$-\frac{\hbar^2}{2m}\left(\frac{\partial^2}{\partial x^2} + \frac{\partial^2}{\partial y^2} + \frac{\partial^2}{\partial z^2}\right)\Psi(x, y, z) + V(x, y, z)\Psi(x, y, z) = \epsilon\,\Psi(x, y, z). \tag{6.60}$$

This partial differential equation is solved by separation of variables, where the solution is the product of independent functions in the coordinates x, y, z, explicitly $\Psi(x, y, z) = \psi_x(x)\psi_y(y)\psi_z(z)$. The separations[6] lead to ordinary differential equations in one variable, such as $\psi_x(x)$

$$-\frac{\hbar^2}{2m}\frac{\partial^2}{\partial x^2}\psi_x(x) = \epsilon_x\psi_x(x). \tag{6.61}$$

The electron wavefunction has a harmonic form, and all the energy is kinetic

$$\epsilon_x = \frac{(\hbar k_x)^2}{2m}. \tag{6.62}$$

Within the box from 0 to L, the ψ_x, ψ_y, ψ_z must be sine functions that go to zero at 0 and L, so

$$\Psi(x, y, z) = \left(\frac{2}{L}\right)^{\frac{3}{2}} \sin(k_x x)\,\sin(k_y y)\,\sin(k_z z), \tag{6.63}$$

where the prefactor is a product of three normalization factors for the three dimensions.

Density of States in k-Space

The boundary conditions set the allowed values of k_x because a half-integer number of wavelengths must fit across the box. The result is

$$k_x = n_x \pi/L, \quad k_y = n_y \pi/L, \quad k_z = n_z \pi/L, \tag{6.64}$$

where n_x, n_y, n_z are positive integers. The quantum levels are separated by an amount $\Delta k_x = \pi/L$, and likewise for intervals in the y- and z-directions. We divide the three-dimensional k-space into small cubes of edge length π/L, and assign one quantum state to each cube. In

---

[5] The other two being the harmonic oscillator and the hydrogen atom.
[6] The separations are done much as in Section 3.4 by substituting the product $\psi_x(x)\psi_y(y)\psi_z(z)$ into Eq. 6.60, and isolating functions of an independent variable on one side of the equation.

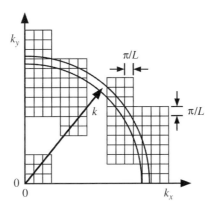

Figure 6.4 Planar cut showing quantum volumes in k-space for a particle in a box of edge length L. There are more of them in a shell at larger k.

three dimensions the number of states in the positive octant of a sphere (here n_x, n_y, n_z are positive numbers) is the volume divided by the volume of each element

$$N(k) = \frac{1}{8}\frac{4\pi}{3}k^3 \frac{1}{\left(\frac{\pi}{L}\right)^3}. \tag{6.65}$$

To find the number of states at a given $|\vec{k}|$, or the "density of states" (DOS), first find the volume in a spherical shell of radius k, as illustrated in Fig. 6.4.

$$\frac{dN}{dk} = \frac{1}{2}\pi k^2 \left(\frac{L}{\pi}\right)^3. \tag{6.66}$$

This DOS from the spherical shell increases as k^2, so in three dimensions there are few states at small k, but many at large k.

Density of States in Energy

It is often more useful to know the density of states as a function of energy, rather than k. An electron of wavevector $\vec{k}$, moving in a constant potential, has the momentum (de Broglie hypothesis) $\vec{p} = \hbar\vec{k}$ and energy

$$\epsilon = \frac{p^2}{2m} = \frac{\hbar^2}{2m}k^2, \tag{6.67}$$

so

$$k = \frac{\sqrt{2m\,\epsilon}}{\hbar}, \quad \text{and} \quad \frac{dk}{d\epsilon} = \frac{\sqrt{m}}{\sqrt{2}\,\hbar}\frac{1}{\sqrt{\epsilon}}. \tag{6.68}$$

With these simple results:

$$k^2 \propto \epsilon, \quad \text{and} \quad \frac{dk}{d\epsilon} \propto \frac{1}{\sqrt{\epsilon}}, \tag{6.69}$$

we convert Eq. 6.66 from a dependence on k to a dependence on ϵ

$$\frac{dN}{d\epsilon} = \left(\frac{dN}{dk}\right)\left(\frac{dk}{d\epsilon}\right) \propto (\epsilon)\left(\frac{1}{\sqrt{\epsilon}}\right) \propto \sqrt{\epsilon}. \tag{6.70}$$

If we retained the constants in the expressions above, we would have found

$$\frac{dN}{d\epsilon} = \frac{(2m)^{\frac{3}{2}} L^3}{4\pi^2 \hbar^3} \sqrt{\epsilon},$$ (6.71)

but the proportionality to $\sqrt{\epsilon}$ of the density of states of the free electron gas is easy to remember and simple to derive. Two electrons can be placed into each state if they have opposite spins, so the electronic DOS, $\rho(\epsilon) \equiv dN/d\epsilon$, is

$$\rho(\epsilon) = \frac{(2m)^{\frac{3}{2}} L^3}{2\pi^2 \hbar^3} \sqrt{\epsilon}.$$ (6.72)

Fermi Energy

What is the maximum energy of an electron in the box? The answer depends on the density of electrons, $n \equiv N/V$. The electrons first fill the states of lowest energy, which are at low values of k. With increasing electron density, electrons occupy states of increasingly higher energy. The highest energy electrons will be at the highest value of k, which we call k_F in honor of Enrico Fermi. At low temperatures, there are no electrons in states with k exceeding k_F, but all states with $k < k_F$ are filled. The total number of electrons in the box is (cf. Eq. 6.65 with two spin states per cube)

$$nL^3 = \frac{2}{8} \frac{4\pi}{3} k_F^3 \left(\frac{L}{\pi}\right)^3 = \frac{\pi}{3} k_F^3 \left(\frac{L}{\pi}\right)^3,$$ (6.73)

$$k_F = \left(3\pi^2 n\right)^{\frac{1}{3}}.$$ (6.74)

This gives k_F from the density $n = N/V$, which then gives the corresponding Fermi energy ϵ_F

$$\epsilon_F = \frac{(\hbar k_F)^2}{2m},$$ (6.75)

$$\epsilon_F = \frac{\hbar^2}{2m} \left(3\pi^2 n\right)^{\frac{2}{3}}.$$ (6.76)

6.4.2 Electronic Heat Capacity

The breadths of energy bands are often of order 1 to 10 eV – far wider than the spread of thermal energies, such as 0.025 eV at room temperature. To a first approximation, temperature does not alter the occupancy of electronic states – the states below the Fermi energy are fully occupied, and those above ϵ_F are unoccupied. (This is strictly true at $T = 0$.) To a second approximation, however, thermal energy can excite electrons to energies approximately $k_B T$ above ϵ_F, leaving holes within $k_B T$ below ϵ_F. Heat is absorbed by these electronic excitations, so the free electrons contribute to the heat capacity of a metal, and hence to its entropy at finite temperature.

The total energy of thermal electronic excitations is proportional to the fraction of electrons that are excited (approximately $k_B T/\epsilon_F$) and the average energy of the excitations (proportional to $k_B T$)

$$U_{el}(T) \simeq \frac{k_B T}{\epsilon_F} k_B T. \qquad (6.77)$$

The heat capacity is $C_{el} = \partial U_{el}(T)/\partial T$

$$C_{el}(T) \simeq 2 \frac{k_B^2}{\epsilon_F} T , \qquad (6.78)$$

which is found to be linear in T. A more accurate calculation [43] gives the electronic heat capacity

$$C_{el}(T) = \frac{\pi^2}{3} \rho(\epsilon_F) k_B^2 T, \qquad (6.79)$$

where $\rho(\epsilon_F)$ is the electronic DOS at the Fermi level. This $\rho(\epsilon_F)$ is assumed constant within a few $k_B T$ around ϵ_F. We also assumed that the electronic DOS is unchanged with temperature.[7]

The electronic heat capacity is predicted to increase linearly with temperature by Eq. 6.79, and this is approximately true for most metals. The electronic entropy can therefore be calculated easily as

$$S_{el}(T) = \int_0^T \frac{C_{el}(T')}{T'} \, dT'. \qquad (6.80)$$

$$S_{el}(T) = \frac{\pi^2}{3} \rho(\epsilon_F) k_B^2 T, \qquad (6.81)$$

and $S_{el}(T)$ also increases linearly with T.

Vibrational entropy and configurational entropy dominate the thermodynamics of metals over a wide range of temperatures, but the electronic entropy is important at low and high temperatures. At low temperatures, it is a well-known result that the phonon heat capacity increases at T^3, so at very low temperatures (under a few K), the electronic entropy is more important than the vibrational entropy. At temperatures of 1000 K, the electronic entropy of metals is typically tenths to one k_B/atom (or more if $\rho(\epsilon_F)$ is large). Although this is much less than the vibrational entropy, it is the difference in entropy between phases that is important for phase stability, so sometimes electronic entropy can play a role in the phase stability of metals at high temperatures.

6.5 Some Electronic Structures of Materials

6.5.1 From Molecules to Metals

Figure 6.5 presents electronic DOS curves for the systems considered so far – a diatomic molecule, a tight-binding crystal, and a free electron gas. The phases of the atomic wave-functions, constructively interfering for the lower-energy bonding state but destructively

---

[7] The reliability of these assumptions varies widely with the type of metal or alloy.

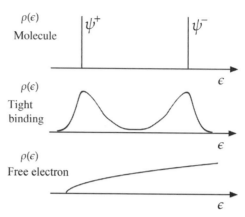

Figure 6.5   Electronic DOS for a diatomic molecule, with bonding and antibonding levels; a periodic solid in the tight-binding approximation (the shape depends on the nature of the wavefunctions and the dimensionality); and a free electron gas in three dimensions, with a parabolic shape.

interfering for the higher-energy bonding state, give the two states in the energy spectrum of the diatomic molecule. The tight-binding model of a periodic solid gives a distribution of phase interferences between the atomic wavefunctions in a crystal, and gives a distribution of energies between the fully bonding and the fully antibonding levels. The free electron gas is conceptually different.

The electron DOS curves of real materials show all features of Fig. 6.5 in varying degrees. Sharp spikes in the density of states are characteristic of electrons that do not move easily between atoms. Molecular crystals tend to have electronic states like the molecules of which they are composed, for example, and many oxides have bonding with discrete, and local, states for electrons. These sharp peaks become broad with increased overlap of the wavefunctions at neighboring atoms, for example when atoms are forced together by pressure. Pressure can drive crystals into a metallic state. It was mentioned earlier that alkali metals have a free-electron-like electronic DOS, but transition metals also have s-electrons that make a broad DOS. Nevertheless, transition metals have d-electrons that contribute a spiky electronic DOS function on top of the broad s-electron background. Such details of the electronic DOS often give specific materials their special character.

6.5.2 Fermi Surfaces

The total electronic DOS determines the bonding energy of a solid, and often its thermodynamic stability. If, compared with an alternative structure, a crystal structure has an electronic DOS with more states at low energy, it will be more stable energetically. Phase diagrams depend on such energy differences.

For electronic properties of materials, however, more important than the total electronic DOS is often the DOS near the Fermi energy. In essence, we want the electrons to do something interesting in response to a stimulus, and this means changing their states. Suppose, for example, that the electrons are all in states like the low-energy bonding state

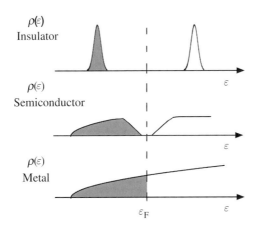

Figure 6.6   Electronic DOS for an insulator, semiconductor, and metal, all aligned with the same Fermi energy, ϵ_F. The gray shading indicates states that are filled, white indicates empty states. The ϵ_F lies in the band gaps of insulators and intrinsic semiconductors, but the gap of an intrinsic semiconductor is narrow enough so that temperature can excite some electrons from the lower valence band to the upper conduction band. For a metal, there are always states that are accessible by thermal excitation, even at very low temperature.

of a molecule in Fig. 6.5. Now apply a small electric field to a collection of such molecules, perhaps arranged as a molecular crystal. The electrons remain in their states because it costs too much energy for them to change. For a metal, however, there are electronic states just above the Fermi level, and it takes only a little energy for the electrons to change into these states. This, approximately, is the origin of electrical conductivity.

Figure 6.6 gives a general picture of insulators, semiconductors, and metals. The point to notice is the energy required to excite an electron from an occupied state to an unoccupied state.[8] Such excitations are required for moving electrons and electrical conductivity. Temperature can assist these excitations if the band gap energy is less than several $k_B T$, or impurity (dopant) atoms can be added to create states in the gap and reduce the energy required to change the states of electrons. For a metal, electrons with energies near the Fermi level can change their state in response to internal disturbances such as impurities or atom vibrations.

It is useful to display the wavevectors of all electron states with energy ϵ_F. In 1D this is nothing for the insulator and semiconductor in Fig. 6.6, and a mere point at k_F for the metal. In 3D the shapes of real Fermi surfaces are quite a bit more interesting and complicated, but many are well known. Usually Fermi surfaces are anisotropic because chemical bonding in many crystals is directional. The Fermi surfaces of metallic elements are shown in Fig. 6.7 and at the website www.phys.ufl.edu/fermisurface/.[9] These Fermi surfaces are worthy of study if the electronic properties of materials are of interest to you. For example, the Fermi surfaces of the column Cr, Mo, W look rather similar, and consist of pockets of electrons in the middle (a spoked object) and hole pockets on the outside faces and corners. When Cr or

[8] Of course the Pauli principle forbids electrons from entering a state that is already occupied.
[9] Please take at least a quick look at the University of Florida website to see the structures in three dimensions.

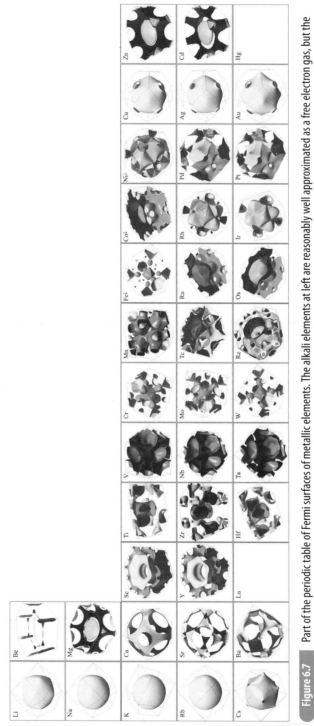

Figure 6.7 Part of the periodic table of Fermi surfaces of metallic elements. The alkali elements at left are reasonably well approximated as a free electron gas, but the transition metals are considerably more complicated. Note that isoelectronic elements in the same column tend to have similar Fermi surfaces. The hexagonal crystal structures of Zn and Cd at right are evident in the anisotropy of their Fermi surfaces. The different shaded "sheets" of the Fermi surfaces may correspond to bands made from different types of electrons, such as p- or d-electrons. The occupied states may be inside the surfaces, or outside in the case of "hole pockets" [45, 46]. Copyright (2000) by the American Physical Society.

Mo are alloyed with Re, which lies to the right on the periodic table, electrons are donated to the Cr band structure. The spoked object grows and the hole pockets shrink [44]. At a concentration of about 6% Re, the spoked object extends across the entire Brillouin zone, making continuous network in k-space. This is an example of an "electronic topological transition," and it is accompanied by a change in mechanical properties of Cr–Re alloys, for example.

Box 6.3   **Density Functional Theory (DFT)**

Since about 1990, DFT has made calculations of electronic structure possible and convenient for materials, molecules, and condensed matter. The thermodynamic internal energy, E, comes from the energies of electrons, which respond to the positions of the nuclei. The energy of an electron also depends on the positions of all other electrons, and this is where the quantum mechanics gets complicated. The Pauli principle of forbidden double occupancy of electron states is straightforward when the states are known, but in an arbitrary configuration of nuclei the electronic states need to be calculated consistently with the electron–electron "exchange interaction" (Section 21.2) that is responsible for the Pauli principle. The rigorous approach is to construct a total electron wavefunction that includes all single-electron states, but this is impractical for most computational materials science. Density functional theory dispenses with the complexity of the pure quantum mechanics by adding a simpler term to the potential energy in the Schrödinger equation to account for electron exchange interactions and Coulomb correlations between the instantaneous positions of electrons [47]. This term is a functional (a function of a function) of the electron density, and does not use the electron wavefunction itself. In its simplest form, the exchange and correlation energy of an electron depend on the local density of the other electrons in the material, in what is called the "local density approximation," or LDA.

Density functional theory is most reliable for a uniform electron gas, and its reliability for the complicated electron densities in real materials was in doubt in the early years of DFT. Nevertheless, there are theorems by Hohenberg, Kohn, and Sham proving that the ground state of the electron system has an energy that depends on the electron density alone (without details of the phases of the wavefunctions). Finding good density functionals was, and still is, a major effort in electronic theory, but today there are already excellent density functionals available for many materials calculations. DFT does not work for all materials, however. Weaknesses today include many magnetic materials, Van der Waals and dispersion interactions, correlated electron behavior, and electrons in excited states. For materials that avoid these issues, DFT can be surprisingly accurate for predicting the thermodynamic energy, E.

6.5.3 Alloy Band Structures

Free Electron Metals

Extending the concept of electronic energy bands from pure elements to alloys is sometimes possible in an easy way. The free electron gas model already ignores the chemistry of the elements, and all properties depend on the electron density, $\rho(\epsilon)$. The solute atoms in an alloy are assumed to donate (or remove) electrons to the gas, with the change in $\rho(\epsilon)$ proportional to the solute concentration. (Some thought may be needed about which valence electrons are available for donation – s-electrons may be more available for donation than d-electrons, for example.) In such cases, the $\rho(\epsilon)$ depends on the overall electron-to-atom ratio of the alloy, and this "e/a" ratio is commonly used as an axis when plotting properties of alloys. The e/a ratio serves a role like that of chemical composition, but it can accommodate multiple chemical elements at a time. Although it is often not quantitative, parameterization with the e/a ratio is easy and it can sometimes provide qualitative understanding.

Transition Metals

For more detailed band structures such as obtained with the tight-binding model or with modern electronic structure calculations, a central question is how much the electronic bands change with alloying. Specifically, does alloying cause the band structure to simply fill with electrons, or is the band structure also distorted? Distortions of the band structure of an A–B alloy are expected when there is a big difference in the bonding between A–B pairs and the A–A or B–B pairs of neighbors. Large differences in electronegativity or atomic size usually cause large distortions in the alloy band structure, for example. On the other hand, when the chemical elements are similar, the distortions can be small, and we may assume a "rigid band structure" that is filled with electrons during alloying. Even if the band structure is distorted by alloying, sometimes the distortions are not essential for understanding general trends, as for understanding the cohesive energy of d-electron metals (see Problem 12.5). In this case, the e/a ratio is a valuable parameter for organizing the properties of elements and alloys.

A generic electronic DOS for $3d$ transition metals is shown in Fig. 6.8. The big peaks are from the $3d$ electrons, but a free-electron-like contribution from $4s$ electrons runs through the DOS. The four peaks (two are overlapped) are from the splitting of the bonding (+, lower energy) and antibonding (–, higher energy) levels of the different $3d$ electrons.[10]

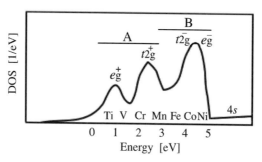

Figure 6.8   Approximate electronic DOS for $3d$ transition metals, with reference energy set at the bottom of the d-band.

[10] It is not so simple as bonding between the five atomic $3d$ orbitals because some orbitals change from bonding to antibonding for different k-vectors in the band. The designation is rather in terms of doubly (e) and triply (t) degenerate orbitals appropriate for the bcc structure. (With two spin states, this designation accounts for all 10 d-electrons.)

The approximate locations of the Fermi level in this rigid band approximation follow the positions of the chemical symbols below the graph (approximate because these elements do not all have the bcc crystal structure). The three main peaks are labeled as two groups, A and B, above and below the Fermi level of Mn. There is an important rule of thumb for alloying of elements in these two groups. When both elements are in group A, or both in group B, there is a tendency to form solid solutions upon alloying. This is as expected from a rigid band model. On the other hand, alloys of elements from the different groups A and B tend to form ordered compounds. The $3d$ electron states from the elements of groups A and B are rather different (changing from bonding to antibonding d-states), and different crystal structures can better accommodate these differences. Manganese itself seems a bit confused in choosing a crystal structure.

6.6 Elastic Constants and the Interatomic Potential

6.6.1 Metallic Bonding

If we calculate the average energy of the free electron gas, and how this energy depends on volume (it goes as $V^{-2/3}$), we can calculate pressures needed to reduce the volume of the electron gas. This is a mechanical response. This section calculates the bulk modulus of a "free electron metal," which has the electronic DOS at the bottom of Fig. 6.6 (examples are the alkali metals in Fig. 6.7). It is remarkable that a macroscopic property of a real metal can be calculated with reasonable accuracy with the simple assumption of an electron gas, and this success gives hope that some electrons in metals do behave as a gas in a box. The energy of the electrons increases as they are confined to a box of smaller volume, giving rise to a mechanical pressure.

The kinetic energy of an electron in the gas varies with wavevector. Electrons with states near $k = 0$ have the lowest energy, and those near the Fermi level the highest. The average electron energy $\overline{\epsilon}$ is calculated using the DOS of Eq. 6.72 as the probability distribution for computing an average

$$\overline{\epsilon} = \int_0^{\epsilon_F} \epsilon \, \rho(\epsilon) \, \mathrm{d}\epsilon, \tag{6.82}$$

$$\overline{\epsilon} = \frac{(2m)^{3/2}}{2\pi^2\hbar^3} L^3 \frac{2}{5} \epsilon_F^{5/2}, \tag{6.83}$$

and using the expression for ϵ_F from Eq. 6.76, the essence of $\epsilon_F^{-3/2}$ is identified in the first factors of Eq. 6.83

$$\overline{\epsilon} = \frac{3}{5} N \epsilon_F. \tag{6.84}$$

Pressure, P, is a response in the energy of the free electron gas when its volume is changed

$$P = -\frac{\mathrm{d}\overline{\epsilon}}{\mathrm{d}V}. \tag{6.85}$$

From the dependence of ϵ_F on $n = N/V$ in Eq. 6.76, it is straightforward to show

$$\frac{d\epsilon_F}{dV} = -\frac{2}{3}\frac{\epsilon_F}{V},\tag{6.86}$$

so with Eqs. 6.84 and 6.85

$$P = \frac{2}{5}\frac{N}{V}\epsilon_F.\tag{6.87}$$

The bulk modulus is defined as the pressure from a material with a fractional change in volume

$$B \equiv -V\frac{dP}{dV},\tag{6.88}$$

and with Eq. 6.87

$$B = -\frac{2}{5}NV\frac{d}{dV}\left(\frac{\epsilon_F}{V}\right),\tag{6.89}$$

$$B = -\frac{2}{5}NV\left(-\frac{\epsilon_F}{V^2} + \frac{1}{V}\frac{d\epsilon_F}{dV}\right).\tag{6.90}$$

Using the volume dependence of ϵ_F from Eq. 6.86

$$B = -\frac{2}{5}N\frac{\epsilon_F}{V}\left(-1 - \frac{2}{3}\right),\tag{6.91}$$

$$B = \frac{2}{3}\frac{N}{V}\epsilon_F.\tag{6.92}$$

From the electron density, $n \equiv N/V$, we can calculate ϵ_F. Equation 6.92 can then be used to obtain the bulk modulus, which has been measured for elemental metals and many alloys. For alkali metals, such as sodium or potassium, the success of this free electron gas model in predicting B is impressive (see Problem 6.3).

For other metals, such as transition metals, the electrons form chemical bonds that are not well described by the free electron gas model. Usually this more direct covalent bonding is stiffer, increasing B. In general, we need to know how the electronic energy depends on volume. To calculate B at low temperatures, for example, Eqs. 6.85 and 6.88 are appropriate. These equations are straightforward, but obtaining an accurate $\bar{\epsilon}(V)$ requires detailed calculation.

6.6.2 General Interatomic Potential

A simple but important relationship between an interatomic potential and bulk modulus is given here. Interatomic potentials, $\Phi(r)$, go to zero as the atoms are separated by long distances, where they no longer see each other. Typically the potential decreases as the atoms come together, owing to the attraction of a chemical bond. At very small separations, however, the energy increases rapidly because the electronic kinetic energy rises as the electrons are squeezed into smaller volumes. Interatomic potentials are typically represented as the sum of two functions, loosely called a "hard core repulsion," and an

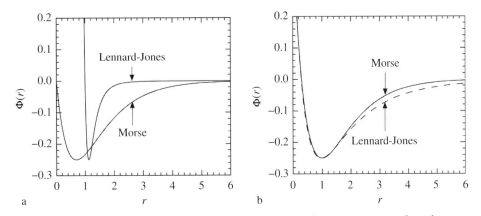

Figure 6.9 Functional forms of Morse and Lennard-Jones potentials. (**a**) Here $\Phi_M = e^{-2r} - e^{-r}$, and $\Phi_{LJ} = \frac{1}{r^{12}} - \frac{1}{r^{6}}$. (**b**) By choosing the appropriate parameters and rescaling the two axes, these two potentials can be made to look similar. Here the Morse potential was shifted right, and the Lennard-Jones potential was shifted in r and rescaled in r.

attractive contribution that has a more gradual variation with distance. A Morse potential, comprising two exponential functions, is popular

$$\Phi_M = C\left(e^{-2\alpha r} - e^{-\alpha r}\right), \tag{6.93}$$

as is a Lennard-Jones "6–12" potential

$$\Phi_{L\text{-}J} = \frac{A}{r^{12}} - \frac{B}{r^{6}}. \tag{6.94}$$

In both cases the first term is the hard core repulsion, and the second is the attractive, bonding contribution. Although there is no rigorous justification for either Φ_M or $\Phi_{L\text{-}J}$, in many applications it is possible to use either of them by suitable choice of parameters. They are shown in Fig. 6.9.

Here a bulk modulus is calculated from a generic interatomic potential. For clarity and convenience we first express the potential as a function of volume, $\Phi(V)$ (and transform to $\Phi(r)$ later)

$$\Phi(V_0 + \Delta V) = \Phi(V_0) + \left.\frac{d\Phi}{dV}\right|_{V_0} \Delta V + \frac{1}{2}\left.\frac{d^2\Phi}{dV^2}\right|_{V_0} \Delta V^2 + \cdots, \tag{6.95}$$

but when V_0 is the equilibrium volume (corresponding to r_0, the minimum of $\Phi(r)$ in Fig. 6.9), the second term is zero and the change in potential energy with ΔV is

$$\Delta\Phi(\Delta V) = \frac{1}{2}\left.\frac{d^2\Phi}{dV^2}\right|_{V_0} \Delta V^2 + \cdots, \tag{6.96}$$

The pressure and bulk modulus are (cf. Eqs. 1.24, 6.88)

$$P \equiv -\frac{d\Delta\Phi}{dV}, \tag{6.97}$$

$$B \equiv -V_0 \left.\frac{dP}{dV}\right|_{V_0}, \tag{6.98}$$

so

$$B = +V_0 \frac{d^2\Phi}{dV^2}\bigg|_{V_0}. \tag{6.99}$$

Since the curvature of $\Phi(V)$, like that of $\Phi(r)$ in Fig. 6.9, is positive at V_0, we find $B > 0$, as it must be if the crystal is stable. Equation 6.96, which depends on the volume change from equilibrium, is recognized as the elastic energy. With Eq. 6.99, Eq. 6.96 gives the expected proportionality between pressure and volume change

$$E_{el} \equiv \Delta\Phi(\Delta V) = \frac{1}{2}\frac{B}{V_0}\Delta V^2, \tag{6.100}$$

$$P \equiv -\frac{dE_{el}}{d\Delta V} = -B\frac{\Delta V}{V_0}. \tag{6.101}$$

It is helpful to transform the bulk modulus of Eq. 6.99 from a volume dependence to a radial dependence, and then relate B to a Morse or a Lennard-Jones potential of Fig. 6.9, for example. The function $\Phi(r)$ has no angular dependence, or anisotropy in direction, so it is too simple to describe some chemical bonds. Nevertheless, assuming it is valid, r and V are related as

$$V = g\,r^3, \tag{6.102}$$

where g is a constant geometrical factor, and $g = 1$ for a simple cubic crystal. A differential relationship

$$\frac{dV}{dr} = 3g\,r^2 = 3\frac{V}{r}, \tag{6.103}$$

as appropriate for three dimensions, is useful for relating the volume and radial dependences of Φ

$$\frac{d\Phi}{dV}\bigg|_{V_0} = \frac{d\Phi}{dr}\bigg|_{r_0}\frac{dr}{dV}\bigg|_{r_0} = \frac{1}{3}\frac{r_0}{V_0}\frac{d\Phi}{dr}\bigg|_{r_0}. \tag{6.104}$$

Repeating the derivative with respect to volume

$$\frac{d^2\Phi}{dV^2}\bigg|_{V_0} = \frac{1}{9}\left(\frac{r_0}{V_0}\right)^2\frac{d^2\Phi}{dr^2}\bigg|_{r_0}, \tag{6.105}$$

and substituting into Eq. 6.99, gives B in terms of r

$$B = \frac{1}{9}\frac{r_0^2}{V_0}\frac{d^2\Phi}{dr^2}\bigg|_{r_0} = \frac{1}{9g\,r_0}\frac{d^2\Phi}{dr^2}\bigg|_{r_0}. \tag{6.106}$$

Now B can be obtained directly from $\Phi(r)$ such as in Fig. 6.9. The result for B from Eq. 6.106 is especially simple for a sc crystal with $g = 1$. Crystals with higher coordination number tend to be more close packed, and have a smaller g. For the same $\Phi(r)$ for first neighbors, close-packed crystals tend to have a larger B, as expected when there is a higher density of bonds.[11] Problem 6.8 addresses this relationship for a bcc crystal.

---

[11] In general, however, $\Phi(r)$ will change with the crystal structure.

6.7 Linear Elasticity

6.7.1 Strains

An elastic field has an energy per atom that can be comparable to the thermal energy $k_B T$, and can be important for the thermodynamics of solids. The mechanics of elasticity is based on strains (fractional changes in length [dimensionless]), and stresses (forces per unit area [pressure]). The elastic energy distribution in a material can be calculated from the strain field, but the converse is not generally true. We begin by defining the strains with the help of Fig. 6.10.

Suppose a large elastic body is deformed significantly, so some of its internal volume elements are translated with respect to the laboratory frame by the vector $\vec{u}$. This does not necessarily create strains, as illustrated in Fig. 6.10a, which shows a displaced but undeformed square. For the square to move without deformation, all parts of the square must undergo the same translation $\vec{u} = u_x\hat{x} + u_y\hat{y}$, where u_x and u_y are simple constants.

Strains occur when u_x and u_y vary with position, as shown in Fig. 6.10b. When u_x varies with x and u_y with y, as in Fig. 6.10d, the strain is dilatational, but when u_x varies with y and u_y with x, as in Fig. 6.10c, the strains are shear strains. In three dimensions the displacement vector is

$$\vec{u}(x, y, z) = u_x\hat{x} + u_y\hat{y} + u_z\hat{z}. \tag{6.107}$$

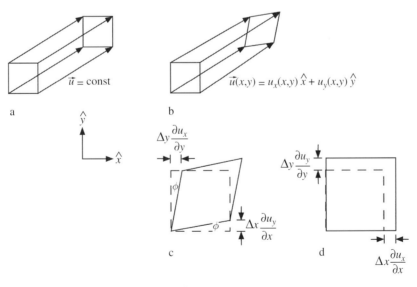

Figure 6.10 (**a**) Uniform displacement of a square with constant $\vec{u}$. The square is undistorted. (**b**) Nonuniform displacement of a square with varying $\vec{u}(x, y)$. Notice the distortion of the square. (**c**) After translating the distorted square of part b back to overlap the initial square. The effects of the variations of the components of $\vec{u}$ are labeled. This is an example of a shear strain. (Incidentally, the angle ϕ is equal to $\partial u_x/\partial y$.) (**d**) Example of a uniform dilatation, showing the effects of the variations of the components of $\vec{u}$ as labeled.

The three tensile (if positive) or compressive (if negative) strains are defined as

$$\epsilon_{xx} = \frac{\partial u_x}{\partial x}, \ \epsilon_{yy} = \frac{\partial u_y}{\partial y}, \ \epsilon_{zz} = \frac{\partial u_z}{\partial z}. \tag{6.108}$$

Notice the consistency of all indices for tensile or compressive strains, ϵ_{ii}, and how the indices are mixed for the three shear strains, γ_{ij}, defined as

$$\gamma_{xy} = \frac{1}{2}\left(\frac{\partial u_x}{\partial y} + \frac{\partial u_y}{\partial x}\right), \ \gamma_{xz} = \frac{1}{2}\left(\frac{\partial u_x}{\partial z} + \frac{\partial u_z}{\partial x}\right), \ \gamma_{yz} = \frac{1}{2}\left(\frac{\partial u_y}{\partial z} + \frac{\partial u_z}{\partial y}\right). \tag{6.109}$$

6.7.2 Stresses

Stresses are applied to a material by making contact with a surface and applying a force to it. The notation for a stress uses two indices, the first for the face to which the force is applied, and the second for the direction of the force, $\sigma_{\text{face,direction}}$. The notation is shown in Fig. 6.11, where the numbers $\{1, 2, 3\}$ correspond to $\{x, y, z\}$. A normal stress such as σ_{xx} or σ_{11} describes a force along $\hat{x}$ applied to the face with normal along $\hat{x}$. It is also necessary to apply a counteracting force (along $-\hat{x}$) to the opposite face of the sample to prevent movement. The stress σ_{xy} is a shear stress with the force along $\hat{y}$ that pulls on a surface with normal along $\hat{x}$. A shear stress has its force perpendicular to the normal of the face. Figure 6.11 depicts six shear stresses, but counteracting stresses are needed to avoid rotational acceleration. The six shear stresses therefore obey relationships such as $\sigma_{xy} = \sigma_{yx}$. There are only three independent shear stresses, and three normal stresses.

6.7.3 Elastic Response

In linear elasticity theory, the relevant material properties are the elastic constants C_{ij}, which give a linear relationship between the stresses and the strains through Hooke's law

$$\sigma_i = \sum_j^6 C_{ij}\,\epsilon_j. \tag{6.110}$$

Equation 6.110 is more complicated than it looks because the subscripts are not single Cartesian indices, but rather pairs of them. The elastic constants C_{ij} actually have four

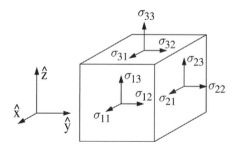

Figure 6.11 Cartesian coordinate systems, with nine stress vectors labeled in the notation $\sigma_{\text{face,direction}}$. Opposing vectors are assumed on the three opposite faces.

Cartesian indices. The contracted indices i and j in Eq. 6.110 are ordered pairs of Cartesian indices, with the following convention

$$\{\; 1, \quad 2, \quad 3, \quad 4, \quad 5\,, \quad 6\,\}$$

$$\{xx, \quad yy, \quad zz, \quad yz, \quad zx, \quad xy\} \tag{6.111}$$

There are 36 constants C_{ij} in Eq. 6.110, but symmetry across the diagonal reduces this number to 21, and crystal symmetry reduces them further. For a cubic crystal, the number of independent elastic constants is only three, the set $\{C_{11}, C_{12}, C_{44}\} \equiv \{C_{xxxx}, C_{xxyy}, C_{yzyz}\}$.[12] For a cubic system, Eq. 6.110 becomes

$$\sigma_{11} = C_{11}\epsilon_{11} + C_{12}\epsilon_{22} + C_{12}\epsilon_{33}, \tag{6.112}$$

$$\sigma_{22} = C_{12}\epsilon_{11} + C_{11}\epsilon_{22} + C_{12}\epsilon_{33}, \tag{6.113}$$

$$\sigma_{33} = C_{12}\epsilon_{11} + C_{12}\epsilon_{22} + C_{11}\epsilon_{33}, \tag{6.114}$$

$$\sigma_{23} = C_{44}\gamma_{23}, \tag{6.115}$$

$$\sigma_{31} = C_{44}\gamma_{31}, \tag{6.116}$$

$$\sigma_{12} = C_{44}\gamma_{12}. \tag{6.117}$$

For low-symmetry crystals, the elastic behavior is more complicated, sometimes requiring more parameters than can be measured.

For cubic crystals, an important relationship for the bulk modulus is

$$B = \frac{1}{3}\left(C_{11} + 2C_{12}\right), \tag{6.118}$$

and the shear modulus is

$$\mu = C_{44}. \tag{6.119}$$

A cubic crystal generally does not have isotropic elasticity, however. For cubic materials the anisotropy is parameterized by the Zener ratio, A, involving the three elastic constants

$$A \equiv \frac{2C_{44}}{C_{11} - C_{12}}. \tag{6.120}$$

In essence, A is a measure of the elastic stiffness along the body-diagonal direction, relative to the stiffness along the cube edge. The elastic anisotropy for three metallic elements is illustrated in Fig. 6.12, showing how the Young's modulus varies with crystallographic orientation. Incidentally, if $A = 1$ there are only two independent elastic constants since $C_{44} = (C_{11} - C_{12})/2$.

Linear elasticity theory considers only small strains, and a subtle but important consequence is that elastic energies from internal and external loadings are additive. For example, if a material with internal elastic energy from a misfitting precipitate is

---

[12] Symmetry requirements include (1) the fact that the three Cartesian axes are equivalent, but also (2) the fact that the sign of ϵ_{ij} cannot alter σ_{ii} (positive or negative shear forces perpendicular to $\hat{i}$ should cause the same effect), so $C_{iiij} = 0$, for example.

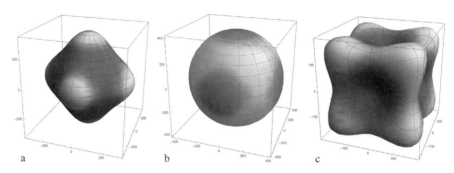

Figure 6.12 Orientation dependence of elastic stiffness for three bcc crystals. (**a**) Cr, $A = 0.7$. (**b**) W, $A = 1.0$. (**c**) Fe, $A = 2.5$. Cube edges are along $\hat{x}$, $\hat{y}$, and $\hat{z}$. [48]

simultaneously subjected to an external applied load, the elastic energies of both the precipitate and the external load can be calculated independently. There is no energy cross-term in linear elasticity theory.

6.7.4 Elastic Energy

Without the complexity of stress and strain tensors (of rank 2) and elastic constants (of rank 4), the elastic energy of a 1D solid is simple to calculate. Suppose we have a chain of N springs and masses, labeled by i, and suppose we move this entire chain by the distance u along $\pm\hat{x}$. If u is constant, meaning a uniform displacement, there will be no change in lengths of the springs. If there is a variation of $u(x)$ with x, however, the springs in regions of largest variation will have the largest stretching or compression. The energy of this elastic chain is

$$E_{\text{el}} = \frac{1}{2}k_x \sum_i^N \left(u(x_i) - u(x_{i-1}) \right)^2, \tag{6.121}$$

$$E_{\text{el}} = \frac{1}{2}C_x \int_0^L \left(\frac{\mathrm{d}u}{\mathrm{d}x} \right)^2 \mathrm{d}x, \tag{6.122}$$

where the transformation to a continuum was done by assuming the solid of N atoms (masses) has the length $L = Na$ (a is the interatomic separation), and by assuming that the spatial scale for variations in $u(x)$ is long compared to a. We recognize $\mathrm{d}u/\mathrm{d}x$ in Eq. 6.122 as the strain.

A most important aspect of elastic energy is that it depends on the square of the distortion in an element of volume. For example, if we stretch the ends of the chain by the length δx, we could put all the elongation in one spring, giving the energy $k_x/2\,(\delta x)^2$. On the other hand, spreading this distortion over n springs gives more springs that are distorted, but each has the energy $k_x/2\,(\delta x/n)^2$. The total elastic energy is reduced by a factor of n by spreading out the distortion. For this reason, stretching our chain of elastic springs will produce a uniform distortion of all springs. Within its elastic range, a bar of uniform thickness will elongate uniformly when pulled. Another important consequence is that strain fields tend to spread over long distances in materials.

For real 3D solids, the general expression for the elastic energy per unit volume is

$$E_{\text{el}} = \frac{1}{2} \sum_{i=1}^{6} \sum_{j=1}^{6} C_{ij}\, \epsilon_i \epsilon_j, \tag{6.123}$$

using the contracted notation of Eq. 6.111. Equation 6.123 is obtained by integrating force times distance (strain) in a small volume, i.e., $[\int \sigma_{ij} \mathrm{d}\epsilon_{ij}]\, \mathrm{d}v$, using Eq. 6.110 for σ_{ij}, and considering all six strains ϵ_j. In general, to calculate the elastic energy we need to calculate the strain fields inside a material. This often requires special methods of partial differential equations that can account for the shape of the body and the boundary conditions of free or constrained surfaces. Especially when the problem includes two phases with different elastic constants, full calculations may have to be done numerically.

6.8 Misfitting Particle

For phase transformations in solids, it is important to know the elastic energy when a volume of material changes its shape, but remains within a surrounding matrix of untransformed material. The new misfitting particle generates long-range strain fields in the surrounding matrix, in addition to the strains within itself. We want to know the elastic energy from these strains. This problem is unique to phase transformations in solids – in a liquid or gas, the surrounding material can flow out of the way, so there are no strains and no elastic energy. The problem of expanding a small sphere inside a large spherical body is one that can be solved analytically, and gives insight into the strain energy of solid–solid phase transformations. Figure 6.13 is a diagram with definitions of variables.

6.8.1 Eshelby Cycle and Elastic Energy

First consider a hypothetical series of steps to account for the elastic energy, sometimes called an "Eshelby cycle" [49, 50].

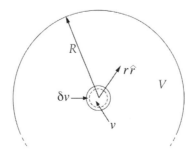

Figure 6.13   Expanded sphere inside a large spherical body.

- Start with a uniform body of material, and remove a small sphere from inside it. This step creates some surface energy of the small sphere and the hole, but surface energies are ignored. Neither the body nor the extracted sphere has strains, so the energy cost in this step is zero.
- Expand the small sphere to its new volume, which is the natural volume of the new phase. There is no internal stress if the phase transformation occurs in free space. The elastic energy for this "stress-free strain" is zero.
- The expanded sphere needs to be fitted back in the old hole, but it is too big. In this third step, forces are applied over the surface of the new sphere to compress it back to its original size. The energy cost is positive, and is not hard to calculate. This is the largest energy of the entire sequence.
- The compressed sphere is replaced into its original hole. It fits. The sphere bonds chemically to the surrounding material. This is a change in surface energy that we ignore, but it sets up a "coherent interface" for what follows.
- The surface forces that compressed the sphere are now removed. The sphere expands a bit, reducing its internal elastic energy, but the surrounding material is strained, especially near the sphere, giving a positive elastic energy. Overall, though, this step must give a reduction of energy, or it would not occur. The energy cost is negative, but less in magnitude than for the previous step of compressing the sphere. (Typical values are about a third of the compression step, depending on the elastic moduli.) This energy change is the most difficult to calculate, and is the subject of Chapter 24 (online).

6.8.2 Strain Fields for Misfitting Sphere in Infinite Matrix

For a misfitting sphere, the displacement field is naturally described in spherical coordinates $\vec{u}(r, \theta, \phi)$, and for isotropic elastic constants all displacements are radial, i.e., $\vec{u}(r, \theta, \phi) = u_r \hat{r}$. Near the sphere, but away from the outer surface of the body[13]

$$u_r = \frac{\delta v}{4\pi r^2}.$$ (6.124)

To calculate the strains, convert to Cartesian coordinates, recognizing that $r^2 = x^2 + y^2 + z^2$

$$\vec{u}(x, y, z) = \frac{1}{x^2 + y^2 + z^2}\left(\frac{x}{r}\hat{x} + \frac{y}{r}\hat{y} + \frac{z}{r}\hat{z}\right).$$ (6.125)

The denominator is $(x^2 + y^2 + z^2)^{3/2}$, and differentiating the component u_x with respect to x, gives

$$\epsilon_{xx} = \frac{\partial u_x}{\partial x} = \frac{1}{r^3} - \frac{3x^2}{r^5},$$ (6.126)

and in general

$$\epsilon_{ij} = \frac{\delta_{ij}}{r^3} - \frac{3x_i x_j}{r^5},$$ (6.127)

[13] There is a boundary condition of a free surface, however. An additional term proportional to rR^{-3} is added to Eq. 6.124 to ensure no radial stress at R, the outer radius of the body.

where i and j denote Cartesian coordinates x, y, z, and $x_1 \equiv x$, $x_2 \equiv y$, $x_3 \equiv z$. Here $\delta_{ij} = 1$ when $i = j$, and 0 otherwise.

An important result from Eq. 6.127 is that the total dilatational strain is zero

$$\epsilon_{xx} + \epsilon_{yy} + \epsilon_{zz} = 3\frac{1}{r^3} - \frac{3x^2 + 3y^2 + 3z^2}{r^5} = 0, \tag{6.128}$$

meaning that the volume elements in the matrix do not change their volumes, even though they change shape after the transformation. The strain in the matrix around the misfitting sphere is pure shear. This result (Eq. 6.128) is not intuitive, but we can understand it by considering shells of material around the sphere. These shells are compressed radially when the oversize sphere is inserted. By shifting outwards, however, they expand their surface area, and it turns out that the net volume of a shell remains the same. Each volume element in the shell keeps the same volume, but is compressed along $\hat{r}$ and expanded perpendicular to $\hat{r}$. (This can be accomplished by the shears shown in Fig. 19.15c, for example.) In spherical coordinates, obtaining the stress from this displacement involves another differentiation with respect to r, and the stress near the misfitting sphere turns out to be

$$\sigma_{rr} = -\frac{\mu \, \delta v}{\pi \, r^3}. \tag{6.129}$$

Note that this result depends only on the shear modulus μ, and not on the bulk modulus B. When the misfitting sphere has a larger volume than the original matrix, Eq. 6.129 shows that the stress on the surface of a spherical shell of radius r pushes it back towards the origin, much like the membrane of an inflated balloon.[14]

The transformed sphere is constrained to have a smaller volume than is natural for this new phase, so there are compressive strains inside the sphere. Actually, inside the sphere the strains are pure compressive, without shear contributions. This contribution to the strain energy scales with the bulk modulus, B. Around the interface of the misfitting sphere, the outwards elastic forces from the sphere are counteracted by the forces from the surrounding matrix. We can gain insight into this problem by considering two springs in series.

6.8.3 Elastic Springs in Series

Figure 6.14 illustrates how, when a force is shared by two springs in series, the stiffer spring (with large spring constant k_2) has the smaller displacement ($x_2 < x_1$). This is intuitive, but perhaps less intuitive is the fact that it also has the lower elastic energy. In Fig. 6.14b, the springs share the force F, and their displacement ratio is obtained from $F = k_1 x_1 = k_2 x_2$, so

$$\frac{x_2}{x_1} = \frac{k_1}{k_2}. \tag{6.130}$$

The energies in the two springs are in the ratio

$$\frac{E_2}{E_1} = \frac{\frac{1}{2} k_2 x_2^2}{\frac{1}{2} k_1 x_1^2} = \frac{k_2}{k_1} \left(\frac{x_2}{x_1}\right)^2, \tag{6.131}$$

---

[14] The strain has a component ϵ_{rr}, but also $\epsilon_{\theta\theta}$ and $\epsilon_{\phi\phi}$ because a differential volume is sheared when a spherical shell expands.

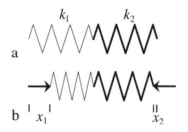

Figure 6.14 (**a**) Two springs in series – a soft spring at left and a stiff spring at right. (**b**) Force applied to the series springs. The displacement of the soft spring is much larger than that of the stiff spring.

and using Eq. 6.130 for the ratio x_2/x_1

$$\frac{E_2}{E_1} = \frac{k_1}{k_2}, \tag{6.132}$$

which is the same ratio as for the displacements in Eq. 6.130. The fractions of energy partitioned between the two springs, C_1 and C_2, are

$$C_1 = \frac{k_2}{k_1 + k_2} \quad \text{and} \quad C_2 = \frac{k_1}{k_1 + k_2}. \tag{6.133}$$

The total energy in the system is the sum of the energies in the two springs

$$E_{\text{tot}} = \frac{1}{2} k_1 x_1^2 + \frac{1}{2} k_2 x_2^2. \tag{6.134}$$

Rewriting E_{tot} in terms of the displacement of only the stiff spring x_2, using Eq. 6.130

$$E_{\text{tot}} = \frac{1}{2} k_1 x_2^2 \left(\frac{k_2}{k_1}\right)^2 + \frac{1}{2} k_2 x_2^2, \tag{6.135}$$

and multiplying through by the fraction $1 = (C_2/C_2)^2$

$$E_{\text{tot}} = \frac{1}{2} \left(\frac{k_1 + k_2}{k_1}\right)^2 \left[k_1 (C_1 x_2)^2 + k_2 (C_2 x_2)^2\right]. \tag{6.136}$$

6.8.4 Energy of Misfitting Sphere in Infinite Matrix

We return to the more complicated problem of the 3D misfitting sphere in an infinite matrix of Section 6.8.2, where elastic energies are calculated to be $1/2B\epsilon^2$ for compressive (or tensile) strains inside the sphere, and $1/2\mu\gamma^2$ for shear strains in the surrounding matrix. (Here B is the bulk modulus and μ is the shear modulus.) Comparing to the preceding description of the energy partitioning between springs in series (Eq. 6.132), it is not surprising that the ratio of the elastic energy in the spherical particle to that in the matrix proves to be inversely proportional to their relevant elastic constants

$$\frac{E_{\text{pt}}}{E_{\text{mx}}} = \frac{4\mu}{3B}, \tag{6.137}$$

where B is a property of the sphere with its compressive strain, and μ is a property of the infinite matrix with its shear strain. The constant 4/3 is not simple to calculate, and it is different for particle shapes other than spheres.

Equation 6.137 shows that the partitioning of elastic energy between the particle and matrix depends on the relative values of B and μ, which have comparable importance. The partitioning coefficients for the spherical particle and the infinite matrix are, respectively (cf. Eqs. 6.132, 6.133),

$$C_{\text{pt}} = \frac{4\mu}{3B + 4\mu}, \tag{6.138}$$

$$C_{\text{mx}} = \frac{3B}{3B + 4\mu}, \tag{6.139}$$

so that $C_{\text{pt}} + C_{\text{mx}} = 1$. As for Eq. 6.136, these energy partitioning fractions can be used to calculate the total elastic energy

$$E_{\text{tot}} = \frac{B}{2v}(C_{\text{pt}}\,\delta v)^2 + \frac{2\mu}{3v}(C_{\text{mx}}\,\delta v)^2, \tag{6.140}$$

where the first term gives the elastic energy in the spherical particle and the second the elastic energy in the infinite matrix. (Recall that the strain is $\delta v/(3v)$.) Although the pre-factors differ, the basic form is that of Eq. 6.136. The total elastic energy in the misfitting sphere of volume v and its surrounding infinite matrix from Eqs. 6.138 to 6.140 is

$$E_{\text{tot}} = \frac{2\mu B}{3B + 4\mu}\frac{(\delta v)^2}{v}, \tag{6.141}$$

where again the value of B is for the precipitate material and μ is for the matrix.

6.8.5 Energies for Other Shapes of Misfitting Particles in an Infinite Matrix

Eshelby solved the more difficult problem of a misfitting precipitate of ellipsoidal shape, with the assumption that the matrix and particle had the same elastic constants [51]. This is a classic result, but it is well beyond the scope of this text.

Nabarro considered ellipsoids of various shapes, with the further assumption that all elastic energy is in the strain field of the matrix (i.e., $B \gg \mu$, so the particle is incompressible and contains no elastic energy) [52]. With this simplifying assumption, $C_{\text{mx}} = 1$ and $C_{\text{pt}} = 0$, and Eq. 6.140 simplifies to

$$E_{\text{tot}} = \frac{2\mu}{3}\frac{(\delta v)^2}{v} \tag{6.142}$$

for a spherical particle. Nabarro described the ellipsoid of revolution by the ratio y/R, where R is the maximum radius of revolution, and y is half the length along the axis of revolution. For different shapes, the strain energy of Eq. 6.142 could still be used, if it is multiplied by a factor $E(y/R)$. When $y = R$, the particle is a sphere, so $E(1) = 1$. For $y < R$ the particle approaches a flat disk, and $E(y/R) = 3\pi y/(4R)$, which goes to zero for a very flat disk. When $y \gg R$, the particle shape approaches a long cylinder, and $E(\infty) = 3/4$. Summarizing for this case of a stiff particle in an isotropic matrix

$$E_{\text{disk}} \ll E_{\text{needle}} < E_{\text{sphere}}.$$

It is understandable that the strain energy is minimized when the particle is a flat plate. In the limit where the particle is infinitely flat and wide, it becomes a sheet that separates two regions of matrix. Since the two interfaces are free to slip, these three layers impose no stresses on each other and can relax their strains entirely. In the more typical case where a small, thin particle is embedded entirely within a matrix, it often finds an orientation where the elastic distortions are confined primarily to regions near its edge (see also Sections 15.4 and 24.3.2).

Now consider the effects of surface energy. A thin plate has the maximum surface area per atom in the particle, so thin plates are disfavored if interfacial energy is significant. In reality, both elastic energy and surface energy can be important, leading to complex shapes of precipitates. To make matters more complicated, the relative importances of surface energy and strain energy can change as a particle grows in volume and its surface-to-volume ratio is reduced.

Problems

6.1 Starting with the two Schrödinger equations 6.1 and 6.2 for isolated atoms A and B, write out the details of the derivations of molecular states and their energies. (Following the text is expected.)

(**a**) Obtain the matrix equation 6.29 for the molecular states of homonuclear diatomic molecules.

(**b**) From your result from part a, obtain Eq. 6.32.

(**c**) Obtain Eqs. 6.40 and 6.41.

(**d**) Explain the physical meanings of S, h, and Δ.

6.2 Consider the stability of a molecule of three identical atoms, where the atoms interact by this Lennard-Jones central-force potential

$$\phi(r) = \frac{A}{r^{12}} - \frac{A}{r^6}. \tag{6.143}$$

(**a**) Which structure of three atoms has the lower cohesive energy: an equilateral triangle or a linear molecule?

(**b**) By what percentage is the energy lower for the preferred structure?

6.3 The actual bulk modulus of potassium metal at room temperature is 3.1 GPa. Using the density for potassium (0.82 g/cm^3) and a molecular weight of 39, calculate its electron density, N/V, assuming each potassium atom contributes one electron to the electron gas. Calculate the bulk modulus of potassium and compare to its actual value. The agreement is not perfect, but it is rather good considering the simplicity of the analysis.

6.4 The 4*f* electrons of the rare earth elements tend to reside close to the nucleus, and are screened by the outer 6*s* (and 5*d*) electrons. With this information only, draw an approximate electron density of states for the elements from Ce to Lu.

6.5 A "rule of thumb" is that the stability of a crystal structure goes as the inverse of the density of states at the Fermi surface. Why might this be true, but then why is it no better than a rule of thumb?

6.6 The equilibrium separation of the K^+ and Cl^- ions in KCl is 2.79 Å.
  (**a**) Assuming the two ions to be point charges of $+e$ and $-e$, calculate the potential energy of attraction of the two ions.
  (**b**) The ionization potential of potassium is 4.34 eV, and the electron affinity of Cl is 3.8 eV. Find the dissociation energy, neglecting the energy of repulsion.
  (**c**) The measured dissociation energy is 4.42 eV. What is the energy due to repulsion of the ions?

6.7 Use the following 1*s* electron wavefunction of a hydrogen atom for this problem

$$\psi(r) = \frac{1}{\sqrt{\pi}} \left(\frac{1}{a_0}\right)^{3/2} \exp\left(-\frac{r}{a_0}\right). \qquad (6.144)$$

  (**a**) What is the electron density (electrons per $Å^3$) at a point 0.37 Å from the nucleus of an isolated hydrogen atom?
    Now suppose that two hydrogen atoms are brought together to form an attractive bond with 1*s* wavefunctions. Their internuclear separation is 0.74 Å.
  (**b**) What is the electron density midway between the two nuclei? Do not go to the trouble to renormalize the molecular wavefunction.
  (**c**) Explain in words why there is a difference between the results of parts a and b.
  (**d**) Make a semiquantitative estimate of how a proper normalization, including overlap effects, will alter the result of part b.

6.8 What is g of Eq. 6.102 for a bcc crystal? For the same $\Phi(r)$, is the value of B larger for a bcc or sc crystal? Why?

6.9 Complete the algebra to prove Eq. 6.128. This demonstrates that the elastic field in the matrix around the misfitting sphere has no tensile or compressive strains, and must be pure shear.

6.10 Eshelby cycle in one dimension. Consider two springs in series, such as in Fig. 6.14, but assume they have identical spring constants $k = k_1 = k_2$. Initially they both have the same length l. Their ends are confined by two rigid walls of a box of length $2l$, but both are initially relaxed and there is no strain energy. Now suppose the following steps are taken. Calculate the energy change at each step, and the total elastic energy after step v.

   (i) One of the springs is removed from the box.

  (ii) This spring expands to a new natural length by the small amount Δ, and it does so in free space.

 (iii) This spring is squeezed to its original length of l.

 (iv) It is held at this length l as it is inserted back into the rigid box in contact with the other spring.

  (v) The force holding the spring is removed, and both springs are allowed to reach their new equilibrium lengths.

 (vi) Confirm that your answer for the total elastic energy after step e is consistent with Eq. 6.136.

Entropy

Entropy allows temperature to control phase transitions in materials. The field of thermo-dynamics itself is based on how entropy, with temperature, balances the effects of internal energy. This chapter describes important types of entropy in materials, and how they can be calculated or estimated with the epitaph on Boltzmann's monument shown in Fig. 7.1

$$S = k_B \ln \Omega, \qquad (7.1)$$

modernized slightly, with k_B as the Boltzmann constant.

The nub of the problem is the number Ω, which counts the ways of finding the internal coordinates of a system for thermodynamically equivalent macroscopic states. Physical questions are, "What do we count for Ω, and how do we count them?" Sources of entropy are listed in Table 7.1, and some were discussed in Chapter 1. Configurational entropy in the point approximation was used throughout Chapter 2, and Section 21.3 accounts for magnetic entropy in essentially the same way. This chapter shows how configurational entropy can be calculated more accurately with cluster expansion methods. Atom vibrations are usually the largest source of entropy in materials, however, and the origin of vibrational entropy is explained in Section 7.4. Vibrational entropy is used to reassess the critical temperatures of ordering and unmixing, which were calculated in Chapter 2 with configurational entropy alone.

For metals there is a heat capacity and entropy from thermal excitations of electrons near the Fermi surface (Section 6.4.2). At lower temperatures the electronic entropy has modest thermodynamic effects because not many electrons are excited, but the electronic entropy of a metal grows with temperature. Also at high temperatures, electron excitations may alter the vibrational modes. The phonons in these vibrational modes can interact with other phonons, too, but the thermodynamics of these interactions is deferred to Chapter 26 (online).

7.1 Counting and Entropy

7.1.1 Static and Dynamical Sources of Entropy

It is usually reasonable to separate the internal coordinates of a material into configurational ones and dynamical ones. For example, when the number Ω is the number of ways to arrange atoms on the sites of a crystal, the method to count the configurations does not depend on temperature. We used this idea extensively in Chapter 2. On the other hand,

Table 7.1 Sources of entropy in materials	
Matter	Structural configurations
Nuclei	Sites for the nuclei (the electrons will adapt)
Electrons	Sites for electrons in mixed-valent compounds
Electron spins	Orientational disorder (magnetic disorder)
Energy	Dynamics
Nuclei	Vibrations (phonons)
Electrons	Excitations across Fermi level (electronic heat capacity)
Electron spins	Spin waves (magnons)
Interactions of above	Phonon–phonon, electron–phonon, electron–magnon …

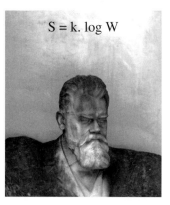

Figure 7.1   The Boltzmann monument in Vienna. Our notation uses Ω instead of W. Photograph courtesy of Clemens Pfeiffer cc-by 3.0.

when the number Ω counts the intervals of volume explored as atoms vibrate dynamically, this Ω increases with temperature, as does the vibrational entropy.

The configurational entropy of an alloy is changed when atoms reconfigure themselves in unmixing or ordering transitions. Configurational entropy was largely understood by Gibbs, who presented some of the combinatoric calculations of entropy that are used today [31]. The calculation of Ω is more difficult when there are partial correlations over short distances, but cluster approximation methods have proved powerful and accurate [53]–[55], and are presented in Section 7.2. In essence, new short-range order parameters are added to the description of atom configurations on lattice sites, giving a more precise description of the atom positions in a material. Although these configurational variables have different equilibrium values at different temperatures, temperature does not alter the combinatorial procedure for calculating Ω from the configurational variables.

Dynamical entropy increases with temperature as degrees of freedom, such as modes of vibration, are excited more strongly by thermal energy. In the quantum theory, thermal energy adds more quanta of vibrational energy to the vibrational modes (these quanta are called "phonons"). Either way, classical or quantum, temperature causes the vibrational excursions of atomic nuclei to be larger. Fundamentally, the material explores a larger volume in the hyperspace[1] spanned by position and momentum coordinates (Section 7.4). This volume, normalized by a quantum volume if necessary, is the Ω in Eq. 7.1 for vibrational entropy.

Box 7.1   **Heat Capacity and S**

Heat capacity has been the most important source of information about the entropy of solids through the relationship

$$S_P(T_2) - S_P(T_1) = \int_{T_1}^{T_2} \frac{C_P(T)}{T} \, dT,$$ (7.2)

or the equivalent relationship between S_V and C_V. For temperatures very close to T_c

$$S_P(T_c + \delta T) - S_P(T_c) = \frac{1}{T_c} \int_{T_c}^{T_c+\delta T} C_P(T) \, dT,$$ (7.3)

which emphasizes how the heat capacity is a derivative quantity of the entropy. The heat capacity is therefore highly sensitive to changes in the entropy near T_c. In fact, the heat capacity has a mathematical singularity at T_c.

- For a first-order phase transition with a latent heat (and a jump in entropy of Eq. 1.5), the heat capacity has a singularity that exists over an infinitesimal range of temperature, but has an infinite height and finite area. It is proportional to a Dirac δ-function, and integrating it in Eq. 7.3 gives the latent heat at T_c. Predicting the magnitude of a latent heat is challenging, since it requires an accurate difference in entropy of two phases near T_c.
- For a second-order phase transition without a latent heat, the singularity is milder. It can have infinite height, but must be rather narrow in temperature so it does not integrate to a finite value over an infinitesimal range of temperature. A function such as $-\ln |1 - T/T_c|$ has the appropriate properties. In fact, Onsager's exact solution for ordering on the 2D Ising lattice with 1nn interactions [1] gives the heat capacity near T_c

$$C_V(T) = -Nk_B \, 0.4945 \ln \left| 1 - \frac{T}{T_c} \right| + \text{constant}.$$ (7.4)

Box 7.2   **Heat Capacity and F**

The internal energy is related to the heat capacity as

$$E(T) = E_0 + \int_0^T C_P(T') \, dT',$$ (7.5)

---

[1] This is frequently called a "phase space," not to be confused with geometric properties of crystallographic phases.

where E_0 can include contributions ranging from gravitational potential to frozen-in defects. What is usually important are differences in $E(T)$ between different phases, so gravitation may cancel but defects may not. It would be easier to obtain $E(T)$ from heat capacity if thermodynamic equilibrium were assured at all temperatures. Equilibrium may sometimes be elusive at low temperatures, so handling E_0 can be tricky. Such concerns aside, $C_P(T)$ can be used to obtain both $S(T)$ and $E(T)$ from Eqs. 7.2 and 7.5, so calorimetric measurements can provide the free energy as $F(T) = E(T) - TS(T)$. Both historically and today, calorimetric measurements have been essential for understanding the free energies of materials.

7.1.2 Counting Atom Configurations in a Random Solid Solution

The process of placing atoms on crystal sites obeys "fermion statistics," meaning that only one atom is allowed per crystal site in the alloy.[2] Figure 7.2a begins to enumerate the random distribution of n atoms over N sites, when $n = 2$ and $N = 4$. Procedurally, the first atom can occupy any one of the N sites, leaving $N - 1$ for the second atom. The top three configurations show the first atom at left, leaving three positions for the second atom. The next step could put the first atom in the second site, as shown in the bottom two configurations of Fig. 7.2a, again allowing three sites for the second atom. This procedure of placing atoms $1, 2, 3, \ldots$ generates the number of configurations

$$\Omega' = N(N - 1)(N - 2) \ldots (N - (n - 1)) = \frac{N!}{(N - n)!},$$ (7.6)

where dividing by $(N - n)!$ cancels all factors in $N!$ that are $N - n$ and smaller. The fourth configuration of Fig. 7.2a illustrates a problem with this counting procedure. This fourth configuration is physically identical to the first. The procedure distinguishes between atoms by the sequence of their placement on n sites, but this is an unphysical distinction. It causes an overcounting of configurations. The overcounting is the number of ways of distributing n atoms over n sites, which is $n!$. The Eq. 7.6 needs to be reduced by this factor of $n!$ to give the correct Ω for Eq. 7.1

$$\Omega = \frac{N!}{(N - n)! \, n!}.$$ (7.7)

The configurational entropy for a solid solution, $S_{\text{conf}} = k_B \ln \Omega$, was previously evaluated with this same Ω in Eqs. 2.41–2.44, giving

$$\frac{S_{\text{conf}}}{N} = -k_B \Big[(1 - c) \ln(1 - c) + c \ln c \Big].$$ (7.8)

7.1.3 Phonon Statistics

Phonons, introduced in Section 7.5.1, can be counted like particles (sometimes we call them "quasiparticles"). They obey "bosons statistics," where there is no limit to how many

---

[2] For an A–B alloy, all sites without an A-atom are occupied by a B-atom, but this filling by B-atoms does not change the number of configurations.

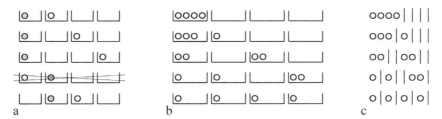

Figure 7.2 (**a**) Distribution of two (n) atoms over four (N) crystal sites. The gray atom was placed first, but here it is physically identical to the white. (**b**) Five configurations of placing four (m) phonons (circles) into four (N) oscillators (containers). (**c**) Abstraction of part b, with vertical lines demarking oscillators; circles denoting phonons.

phonons can be present in each vibrational mode. Assume each mode is an independent oscillator, and consider a set of N such oscillators, all with the same energy $\varepsilon = \hbar\omega$. Suppose there is a number m of phonons that can be placed in these different oscillators (the energy is therefore $m\varepsilon$). Figure 7.2b illustrates different configurations of four phonons among four equivalent oscillators. The top configuration in Fig. 7.2b has all four phonons in the left oscillator. The bottom configuration places one phonon in each oscillator. The middle three are other possibilities, such as three phonons in the first oscillator, one in the second, and none in the other two. Figure 7.2b inspires a trick for calculating Ω combinatorially, illustrated with Fig. 7.2c. The trick is to consider the three vertical bars and the four circles as distributed at random over the seven positions in each row. The number Ω is now the number of ways to distribute m indistinguishable phonons over $N+m$ possible sites.

The first phonon can be placed in $N + m$ possible sites, the second has one less, i.e. $N + m - 1$ possibilities, and the mth phonon can be placed in $N + 1$ ways. The product of these independent probabilities is $(N + m)! / N!$. There is, however, an overcounting because we put the phonons down individually, labeling the first, second, etc., but phonons are indistinguishable. The overcounting is the number of ways of distributing m phonons over m sites. This is $m!$, and is needed in the denominator of Eq. 7.9

$$\Omega = \frac{(N + m)!}{N!\ m!}. \tag{7.9}$$

Substituting Eq. 7.9 into 7.1, and using the Stirling approximation of $\ln(x!) \simeq x \ln x - x$

$$S_{\text{osc}} = k_{\text{B}}\Big[(N + m)\ln(N + m) - N \ln N - m \ln m\Big]. \tag{7.10}$$

Now consider the entropy per oscillator, S_{osc}/N, and define the occupancy of each oscillator, $n \equiv m/N$

$$\frac{S_{\text{osc}}}{N} = k_{\text{B}}\Big[(1 + n)\ln(N(1 + n)) - \ln N - n \ln(nN)\Big], \tag{7.11}$$

$$\frac{S_{\text{osc}}}{N} = k_{\text{B}}\Big[(1 + n)\ln(1 + n) + (1 + n)\ln N - \ln N - n \ln n - n \ln N\Big], \tag{7.12}$$

$$\frac{S_{\text{osc}}}{N} = +k_{\text{B}}\Big[(1 + n)\ln(1 + n) - n \ln n\Big]. \tag{7.13}$$

It is interesting to compare Eq. 7.13 with Eq. 7.8 for randomly distributing atoms over sites of a crystal. Equations 7.13 and 7.8 have two differences – the signs are reversed, and n can exceed 1, whereas $0 \le c \le 1$.

7.2 Short-Range Order and the Pair Approximation

Section 2.9 presented a thermodynamic analysis of the order–disorder transition in the "point approximation," which assumed that all atoms on a sublattice were distributed randomly. This assumption is best in situations when (1) the temperature is very high, so the atoms are indeed randomly distributed over the sublattice, (2) the temperature is very low, and only a few antisite atoms are present, or (3) a hypothetical case when the coordination number of the lattice goes to infinity. For more interesting temperatures around the critical temperature, for example, it is possible to improve on this assumption of sublattice randomness by systematically allowing for short-range correlations between the positions of atoms.

For example, a deficiency of the point approximation for ordering is illustrated with Fig. 7.3. Here the numbers of A-atoms and B-atoms on each sublattice are equal, so the long-range order (LRO) parameter, L, of Eq. 2.50 is zero, $L = (R - W)/(N/2) = 0$. Although $L = 0$, there is obviously a high degree of order within each of the two domains. It might be tempting to redefine the sublattices within each domain, allowing for a large value of L, but this gets messy when the domains are small. The standard approach to this problem is to use short-range order (SRO) parameters, such as the total number of A–B pairs. Since most pairs in Fig. 7.3 are of the A–B or B–A type, the structure of Fig. 7.3 has a high degree of short-range order. For an alloy that develops two sublattices, the SRO can be parameterized with the pair variables $\{N_{AA}^{\alpha\beta}, N_{AB}^{\alpha\beta}, N_{BA}^{\alpha\beta}, N_{BB}^{\alpha\beta}\}$, where an atom in the subscript is on the sublattice in its superscript. The point variables $\{N_A^\alpha, N_B^\alpha, N_A^\beta, N_B^\beta\}$ are still needed, however.[3]

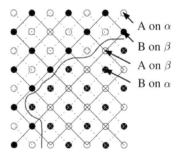

A on α
B on β
A on β
B on α

Figure 7.3 Alloy with two domains of chessboard order. The domain boundary is marked by the curved line. Equal numbers of A- and B-atoms are in each domain. The domains have opposite sublattice occupancies. Incidentally, the curved line is an "antiphase domain boundary" (APDB).

---

[3] Not all of these variables are independent, as explained in Section 23.3.1 (online).

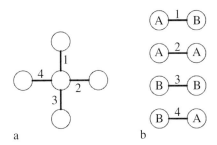

Figure 7.4 **(a)** Four numbered bonds to a central atom. **(b)** Random selection of the types of atom pairs across each bond.

In the pair approximation, we start with an arbitrary first-nearest-neighbor pair in the lattice (as opposed to starting with an arbitrary lattice site in the point approximation), and consider its probability of being an A–A, A–B, B–A, or B–B pair. For a disordered solid solution of equiatomic composition, these four pairs are expected to occur at random, so there are four equal possibilities for each bond. This is true, but when counting pairs an inconsistency arises, as illustrated with the help of Fig. 7.4. Figure 7.4a shows four bonds, each labeled with a number. Figure 7.4b shows a typical random selection of atom pairs for the four bonds. The problem is that the first two bonds put an A-atom at the center, whereas the second two put a B-atom at the center. This is obviously inconsistent. It is a problem that did not arise in the point approximation, but for the pair approximation (and for all higher-order clusters), it is not appropriate to treat the pairs (or clusters) as randomly selected. We need to ensure that the point variables (such as the central atom in Fig. 7.4a) are accounted for consistently, and atoms are conserved.

When calculating a configurational entropy of pairs, we need a correction factor that gives the correct configurational entropy for a fully disordered alloy [56]. For reference, start with the assumption that the pairs really are random. The entropy would be obtained from the combinatorics of placing one of four pairs on the locations of all the bonds in the crystal (a total of $zN/2$ bonds as shown in Fig. 2.14b). This gives (see Problem 7.2)

$$\Omega = K \frac{\frac{zN}{2}!}{N_{AA}^{\alpha\beta}! \; N_{AB}^{\alpha\beta}! \; N_{BA}^{\alpha\beta}! \; N_{BB}^{\alpha\beta}!} \tag{7.14}$$

for use in the equation for the configurational entropy

$$S_{conf}^{pair} = k_B \ln \Omega. \tag{7.15}$$

The correction factor, K, which prevents inconsistency when evaluating the point variables, turns out to be

$$K \equiv \left[\frac{N_A^\alpha! \; N_B^\alpha! \; N_A^\beta! \; N_B^\beta!}{\frac{N}{2}! \; \frac{N}{2}!} \right]^{z-1}. \tag{7.16}$$

It is not too difficult to derive K, but it should suffice to show how it works for a random, equiatomic alloy. For this alloy, there are $zN/8$ pairs of each type A–A, A–B, B–A, and B–B. The combinatorial factor of Eq. 7.14 is evaluated with the Stirling approximation

$$\ln \Omega = \frac{zN}{2} \ln \frac{zN}{2} - \frac{zN}{2} - 4\left(\frac{zN}{8} \ln \frac{zN}{8} - \frac{zN}{8}\right) + \ln K, \tag{7.17}$$

$$\ln \Omega = \frac{zN}{2} \ln \frac{8}{2} + \ln K = \frac{zN}{2} \ln 4 + \ln K. \tag{7.18}$$

The correction factor $\ln K$ is easy to calculate with Eq. 7.16 for a disordered equiatomic alloy, where all four point variables equal $N/4$

$$\ln K = \ln \left[\frac{\left[\left(\frac{N}{4}!\right)^4\right]^{z-1}}{\left(\frac{N}{2}!\right)^2} \right], \tag{7.19}$$

$$\ln K = (z-1)\left[4\left(\frac{N}{4} \ln \frac{N}{4} - \frac{N}{4}\right) - 2\left(\frac{N}{2} \ln \frac{N}{2} - \frac{N}{2}\right)\right], \tag{7.20}$$

$$\ln K = -(z-1)N \ln 2. \tag{7.21}$$

The $\ln \Omega$ of Eq. 7.15 can now be obtained by using Eq. 7.21 in Eq. 7.18, giving

$$S_{\text{conf}}^{\text{pair}} = k_{\text{B}} \ln \Omega, \tag{7.22}$$

$$S_{\text{conf}}^{\text{pair}} = k_{\text{B}} \left(zN \ln \sqrt{4} - (z-1)N \ln 2\right), \tag{7.23}$$

$$S_{\text{conf}}^{\text{pair}} = k_{\text{B}}N \ln 2, \tag{7.24}$$

which is the correct configurational entropy for an equiatomic random solid solution.[4] Notice that the correction factor K is negative, reducing the combinatorial result of Eq. 7.18 from the pair variables alone. This is expected, since the process of eliminating the inconsistency discussed with reference to Fig. 7.4 forces some correlations between the pair variables, reducing the apparent randomness of the alloy.

In the pair approximation we can still define a LRO parameter, L, as before. Using the pair variables, the number of "right" A-atoms on the α-sublattice, for example, is

$$R = \frac{N_{\text{AA}}^{\alpha\beta} + N_{\text{AB}}^{\alpha\beta}}{z}, \tag{7.25}$$

and with a similar expression for the "wrong" B-atoms on the α-sublattice, the LRO parameter is $L = 2(R - W)/N$. This would still yield $L = 0$ for the alloy depicted in Fig. 7.3, however. A better description of the order in this alloy is obtained with a short-range order (SRO) parameter, α

$$\alpha \equiv 1 - \frac{1}{c} \frac{N_{\text{AB}}^{\alpha\beta} + N_{\text{BA}}^{\alpha\beta}}{zN/2}, \tag{7.26}$$

which counts all A–B pairs, without regard to sublattice. For an equiatomic binary alloy, this SRO parameter[5] is -1 for perfect order, 0 for a fully random alloy, and $+1$ for a fully unmixed alloy. For the configuration of Fig. 7.3, $\alpha = -0.74$, using Eq. 7.26 and neglecting atoms without neighbors along two edges.

An important feature of SRO is that it does not vanish at high temperatures. Even though LRO is lost above the critical temperature, T_{c}, the SRO persists. The equilibrium SRO at

---

[4] It is the entropy of the point approximation for a fully disordered solid solution, for which the site occupancies are, in fact, independent.

[5] It is a "Warren–Cowley" SRO parameter, defined for its convenience in X-ray diffraction experiments.

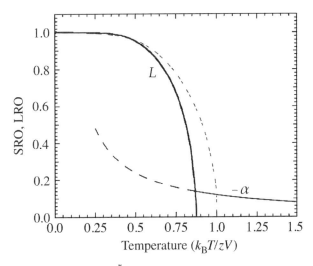

Figure 7.5 Long-range order parameter, L, and short-range order parameter, s, versus temperature for B2 ordering on an equiatomic bcc alloy in the pair approximation. The light dashed line is the LRO parameter in the point approximation.

high T can be calculated by considering the interchange of an A–B pair and a B–A pair into an A–A and B–B pair. By the method of Section 2.9.1, the equilibrium ratio of the numbers of pairs is

$$\frac{N_{AB}^{\alpha\beta}\, N_{BA}^{\alpha\beta}}{N_{AA}^{\alpha\beta}\, N_{BB}^{\alpha\beta}} = \exp\left(+\frac{4V}{k_B T}\right). \tag{7.27}$$

This result is valid at temperatures above T_c, showing that the SRO decreases monotonically with temperature. It can be shown with Eq. 7.27 that s decreases as $1/T$ at high temperatures. At temperatures below T_c, however, the SRO tracks the change of the LRO parameter with temperature.

Figure 7.5 shows results from the pair approximation for the SRO and LRO versus temperature for B2 ordering on a bcc lattice. The SRO is nonzero above the critical temperature of approximately 0.869. Below the critical temperature, the SRO parameter follows the LRO parameter, which rises a bit more steeply than for the point approximation. Also shown is the SRO below the critical temperature from Eq. 7.27. This dashed curve corresponds to a situation when sublattice formation is suppressed, perhaps owing to kinetics, as discussed in Chapter 23 (online).

7.3 Materials Structures and Properties Described by Clusters

7.3.1 Correlation Functions, ξ

A "cluster expansion" is a method to calculate a physical property that depends on local atom arrangements on a lattice. To account for atom configurations on lattice sites, it uses

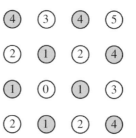

Chessboard ordering with atoms on numbered nearest-neighbor sites.

the concept of correlation functions. The range and complexity of the correlation functions increase with the level of the approximation. For example, the simplest "point correlation function" (used in Sections 2.7 and 2.9) is a type of average of the atom concentrations on one site. More detail is provided by pairs (Section 7.2), triangles, squares, etc. The best hierarchy of these clusters depends on the lattice type.

The first step for a binary alloy is to assign the "spin number," σ, to each atom, with $\sigma = +1$ for an A-atom and $\sigma = -1$ for a B-atom. Correlation functions are constructed as products of these spins, and their values depend on the atomic structure. For example, we now calculate pair correlation functions for the ordered structure in Fig. 7.6, where the numbers correspond to nearest-neighbor positions with respect to the site labeled "0." If the A-atoms are white, the pair correlation functions (with the first subscript of 2 denoting pairs, and the second subscript denoting neighbor distances) are[6]

$$\xi_{2,n} = (\sigma_0)(\sigma_n) = (+1)(\sigma_n),$$

$$\xi_{2,0} = (+1)(+1) = +1, \qquad \xi_{2,1} = (+1)(-1) = -1,$$

$$\xi_{2,2} = (+1)(+1) = +1, \qquad \xi_{2,3} = (+1)(+1) = +1,$$

$$\xi_{2,4} = (+1)(-1) = -1, \qquad \xi_{2,5} = (+1)(+1) = +1.$$

All correlation functions have the maximum value of ± 1 for this structure with perfect order, and the $\{\xi_{2,n}\}$ do a good job of describing the atoms at all sites of the structure, well past the fifth neighbor.

At the other extreme, consider a random solid solution. The point correlation function ξ_1 is constructed as an average of the $+1$ for all A-atoms and -1 for all B-atoms

$$\xi_1 = \sigma_A c_A + \sigma_B c_B, \tag{7.28}$$

$$\xi_1 = (+1)(1 - c) + (-1)(c) = 1 - 2c. \tag{7.29}$$

The pair correlation function is obtained similarly.[7] Adding the fractions of A–A, B–B, A–B, and B–A pairs

---

[6] The same result is obtained if the B-atoms are white because both signs are switched in each product.
[7] In a fully random alloy, the same analysis is valid for all pairs, no matter their separation from the atom at the origin.

$$\xi_{2,n} = \sigma_A c_A\, \sigma_A c_A + \sigma_B c_B\, \sigma_B c_B + \sigma_A c_A\, \sigma_B c_B + \sigma_B c_B\, \sigma_A c_A$$

$$= (+1)(1-c)(+1)(1-c) + (-1)(c)(-1)(c) + 2(+1)(1-c)(-1)(c)$$

$$= (1-2c)^2. \tag{7.30}$$

Compare Eqs. 7.29 and 7.30. For a random alloy, for each additional site added to the cluster comes an additional factor of $\xi_1 = 1 - 2c$, so for a cluster of m atoms of any separation n (using n to denote more than one distance)

$$\xi_{m,n} = (1-2c)^m. \tag{7.31}$$

These $\{\xi_{m,n}\}$ are obviously quite different from the previous case of the chessboard ordered structure in Fig. 7.6, where all $\{\xi_j\}$ are ± 1. An important concept is that any configuration of atoms on a crystal lattice can be described by the $\{\xi_j\}$ (where j denotes the m,n).

7.3.2 Interaction Functions, f

Since all properties of an alloy originate with its atom configurations, and since the $\{\xi_j\}$ can describe all atom positions in the alloy, the $\{\xi_j\}$ can account for alloy properties, at least those at low or moderate temperatures.[8] Alloys with similar correlation functions $\{\xi_j\}$ are expected to have similar properties, so we might expect to evaluate a property with contributions from the different correlation functions. For the free energy, F, or any ground state property of the alloy

$$F = \sum_{j}^{\infty} f_j\, \xi_j, \tag{7.32}$$

assuming we know the new set of all "interaction free energies," $\{f_j\}$. These interaction free energies are useful – when they are known, free energies for other atomic structures can be calculated, once we have their atom correlation functions. Obtaining the correlation functions $\{\xi_j\}$ is relatively straightforward, as demonstrated in Section 7.3.1.

Unfortunately, these interaction free energies $\{f_j\}$ are usually not well-defined physical quantities because their values depend on the selection of cluster variables. For example, suppose a chemical bond energy depends only on the electrons of first-nearest-neighbor pairs of atoms, but the strength of an A–B bond depends on the surroundings of the B-atom in the pair.[9] This effect of neighbors around the bond could be accounted for by the average chemical composition, i.e., a point variable like N_B. Alternatively, we could use triplet variables with a B-atom neighbor added to the A–B bond, so three-body clusters such as N_{ABB} might be used. We can reasonably account for the neighbors to our A–B bond by two sets of variables, either $\{N_B, N_{AB}\}$ or $\{N_{AB}, N_{ABB}\}$. Not surprisingly, the interaction free energy for first-neighbor pairs, $f_{2,1}$, is different for these two analyses. Which set of cluster variables is more physical? If the bond neighbor effect originates from free electrons, perhaps the point variable would be a better choice, since it accounts for the average

---

[8] The full proof is challenging [57], so we accept it on physical grounds.
[9] Suppose some atomic orbitals at a B-atom are filled by other neighbors, for example.

composition. On the other hand, if the bonding is molecular, the three-body term might have better justification, and it offers more local detail. Without a proper understanding of the electronic structure, the best set of cluster variables is not obvious. In the present case it is likely that the set $\{N_B, N_{AB}\}$ would be tried first, since the point variables are needed for defining the composition anyhow. Adding triplets, i.e., $\{N_B, N_{AB}, N_{ABB}\}$, would likely be tested, but the interaction free energies $f_{2,1}$ would be different. The additional triplet variables would undoubtedly allow the free energy functions to be calculated more accurately, but the uniqueness of the interaction functions could be questioned.

It is often expected that longer-range correlations of atom positions are less important for alloy thermodynamics, so it is hoped that correlation functions involving short distances (small j or small clusters) are sufficient. The cluster expansion method has proved most useful when a short-range description of the atomic structure accounts for the property of interest, and Eq. 7.32 can be truncated after only a few terms. If so, we can obtain the interaction free energies f_j by "calibrating" Eq. 7.32, using known values of F from alloy structures with known correlation functions $\{\xi_j\}$. In essence, a set of n interaction free energies can be obtained from a set of n different alloys with known F. The inversion method is not too difficult, but is beyond the present scope [58].

7.3.3 Thermodynamics with Cluster Approximations

For the same interchange energy V, increasing the order of the approximation causes T_c to decrease. The explanation is not obvious, since both the energy and the entropy are different in the different approximations. Table 7.2 presents some results and shows some trends. Notice that the T_c is less sensitive to the order of the approximation as the coordination number, z, is larger. In fact, in the limit as $z \to \infty$, the point approximation predicts the correct T_c.

For further accuracy, it is natural to extend the pair approximation in a hierarchy of approximations that goes from point to pair to larger clusters such as squares on the square lattice, for example. A systematic "cluster variation method" to do this was developed by Kikuchi [59], and has been in widespread use since the late 1970s [55, 60]. Variables for the populations of larger clusters, such as first-nearest-neighbor square variables, are

Table 7.2 Critical temperatures $[zV/k_B]$				
Lattice	z	Exact	Pair	Point
Hexagonal	3	0.5062		1
Square	4	0.5673	0.7212	1
Triangle	6	0.6062	0.8222	1
Simple cubic	6	0.7522	0.8222	1
bcc	8	0.7944	0.8690	1
fcc	12	0.8163	0.9142	1

placed randomly on the square lattice, for example, consistent with their concentrations. In this case the pair variables are overcounted, and the handling of this problem is the key to the method. In essence, it is necessary to correct for the overcounting of all subclusters within the largest cluster, using weights that depend on how many subclusters are present in the large, independent cluster (e.g., Eq. 7.16). The systematics to do this are elegant and straightforward [59, 61]. One measure of the success of these higher-order approximations is their convergence on the actual critical temperature for ordering, as determined, for example, by Monte Carlo methods. Discrepancies of 1% or so are possible with modest cluster sizes, but the analysis of accuracy is itself a specialty.

An important conceptual point about chemical unmixing by spinodal decomposition can be understood with cluster approximations. The phase diagram for unmixing, Fig. 2.17, is based on free energy versus composition curves of Fig. 7.11 (or 2.7). At intermediate compositions, these curves from the point approximation show a characteristic "hump" at temperatures below the critical temperature for unmixing. In the pair approximation, the magnitude of this hump is decreased considerably, and it continues to decrease as the cluster approximation extends to higher order [62]. In the limit of very large clusters, the hump in the free energy versus composition curve vanishes, and a straight line connects two solid solutions of different compositions. What happens is that the populations of higher-order clusters are themselves biased towards unmixed structures, where all A-atoms are on one side of the cluster and B-atoms on the other side, for example. With these internal degrees of freedom accounting for much of the unmixing enthalpy, there is less effect of composition variations on the free energy. Composition alone is therefore insufficient for characterizing the state of the alloy, and the $F(c)$ curve in the point approximation is naïve. This viewpoint helps to overcome a conceptual problem with the $F(c)$ curve of Fig. 7.11, "How reliable is an unstable free energy function for calculating equilibrium phase diagrams?" This question is avoided in part by treating spinodal decomposition as a kinetic process with a modified diffusion equation, as in Section 16.4.

Box 7.3   **Calorimeters**

A calorimeter measures small transfers of heat to samples after small changes in temperature. These data can give the heat capacity, $C_P(T)$, or heats of phase transitions (e.g., L). One approach to measuring $C_P(T)$ is adiabatic calorimetry, which has been in use for a century [63–66]. In adiabatic calorimetry, the sample, its holder, and its heater start at temperature T_0, and are thermally isolated from their surroundings. Measured electrical energy, ε_{elec}, from power times time is provided to the sample through the heater. After equilibration, the sample temperature is $T_0 + \delta T$, and the total heat capacity of the sample, holder and heater equals $\varepsilon_{elec}/\delta T$.

Today, with digital data acquisition and computerized data fitting, heat pulse (or relaxation) calorimetery has become popular because it is automated, accurate, fast, and works with small samples (typically ten milligrams or so). The sample is mounted in thermal contact with a platform that has thermal links (typically thin wires) to a thermal anchor of controlled temperature T_0. A pulse of electrical energy raises the temperature of the platform to $T_0 + \delta T(t)$, where $\delta T(t)$ rises quickly to a maximum, then decays to zero. The decay of $\delta T(t)$ typically shows a quick relaxation as the platform transfers heat to the specimen, and a second, longer relaxation as heat is transferred through the thin wires to the thermal anchor. Both relaxations are approximated as exponential functions. The time constant of the second relaxation of $\delta T(t)$ is proportional to the total heat capacity of the sample plus platform. The instrument then increments T_0 and repeats the process at the new temperature [67]–[70]. With care and with calibration, these heat capacity data can be accurate to 1% over the temperature range of 5 to 400 K.

For higher temperatures, a differential scanning calorimeter (DSC) may be useful. A DSC heats two identical sample pans. The electrical heating is performed with feedback control so the temperatures of both pans increase at the same rate. Differences in heat capacity are measured as the differences in electrical power required to maintain the same heating rates. This method of differential scanning calorimetry has the advantage of nulling the background from the sample holder, at least in principle. A simpler differential measurement technique, differential thermal analysis (DTA), places two samples in a common thermal environment, such as side-by-side in a furnace. The furnace temperature is increased uniformly, and the temperature difference between the two samples is measured to obtain their differential heat capacity. Differential measurements can be accurate to 2% of the total heat capacity, but like all methods of calorimetry, it is a challenge to obtain high-quality data over a wide range of temperatures.

7.4 Concept of Vibrational Entropy

We can understand the essence of vibrational entropy for phase transitions with Eq. 7.1, i.e., $S = k_B \ln \Omega$, while making reference to Fig. 7.7. Figure 7.7 depicts a multidimensional "hyperspace" spanned by the momentum and position coordinates of a three-dimensional

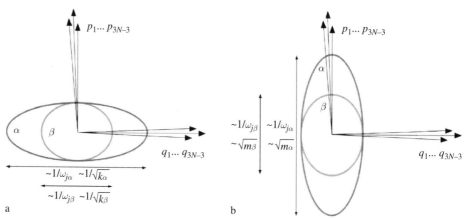

Figure 7.7 Multidimensional space of all momentum and coordinate variables for all atoms of a material. At any instant in time, the material is at a point in this space. (**a**) Phases α and β have the same atom masses, but different interatomic force constants. (**b**) Phases α and β have the same interatomic force constants, but different atom masses.

material with N atoms. At any instant in time, the material is at a point in this space that gives all positions, $\{q_i\}$, and momenta, $\{p_i\}$, of its atoms. The point moves around as the atoms vibrate. The larger the volume explored during vibration, the greater the number of ways, Ω, of finding the system, and the larger is $S = k_B \ln \Omega$. We obtain differences in vibrational entropy by finding the change in the volume explored in the hyperspace of Fig. 7.7 when there are changes in either the springs or the masses.

7.4.1 Changes in Springs

Suppose two phases α and β have the same atoms, but these phases have different structure and bonding. The changes in chemical bonds alter the forces between atoms, so they vibrate with different frequencies, $\{\omega_i\}$. Assume each vibrational mode can be treated independently of all the other modes of the material. Compare the maximum excursions of the coordinate q_j for the mode j in the α- and β-phases, assuming the same potential energy $P.E.$

$$P.E. = \frac{1}{2} k_{s\alpha}\, q_{\alpha\,\text{max}}^2 = \frac{1}{2} k_{s\beta}\, q_{\beta\,\text{max}}^2, \tag{7.33}$$

$$\frac{q_{\alpha\,\text{max}}}{q_{\beta\,\text{max}}} = \sqrt{\frac{k_{s\beta}}{k_{s\alpha}}} = \frac{\omega_\beta}{\omega_\alpha}, \tag{7.34}$$

where the relationship $\omega = \sqrt{k/m}$ for harmonic oscillators was used, and the masses are equal if the same atoms are in both phases. For a fixed amount of thermal energy (and hence the same maximum potential energy), a mode j explores a range of q_j that is inversely proportional to its frequency ω_j.

This is illustrated in Fig. 7.7a, which shows a lower frequency for the mode $\omega_{j\alpha}$ in the α-phase than $\omega_{j\beta}$ in the β-phase, and therefore a larger spatial excursion in the α-phase. From this one mode, the difference in vibrational entropy is[10]

$$S_j^\beta - S_j^\alpha = k_B \ln\left(\frac{\omega_0}{\omega_{j\beta}}\right) - k_B \ln\left(\frac{\omega_0}{\omega_{j\alpha}}\right), \tag{7.35}$$

$$S_j^\beta - S_j^\alpha = k_B \ln\left(\frac{\omega_{j\alpha}}{\omega_{j\beta}}\right). \tag{7.36}$$

Equation 7.36 gives the contribution to entropy from one vibrational mode. In a three-dimensional solid with N atoms there are $3N - 3 \simeq 3N$ vibrational modes. The volume in hyperspace explored by thermal vibrations is the product of all the ranges of coordinates explored

$$S^\beta - S^\alpha = k_B \ln\left(\frac{\prod_j^{3N} \omega_{j\alpha}}{\prod_j^{3N} \omega_{j\beta}}\right). \tag{7.37}$$

---

[10] For differences in entropy, the value of ω_0 can be neglected.

What we learn from Fig. 7.7a and Eq. 7.37 is that the vibrational entropy is larger if the springs between atoms are weaker, and the vibrational frequencies are smaller. Changes in the force constants between atoms are expected if the atoms are rearranged to change their chemical bonding. For example, suppose $0 < 4V = e_{AA} + e_{BB} - 2e_{AB}$, so A–B bonds are energetically favorable, compared with A–A and B–B. If the A–B pairs have a deeper interatomic potential well (e.g., Fig. 6.9), the curvatures at the bottoms of these wells are expected to be larger, so the interatomic springs are stiffer. The atom vibrational frequencies are higher, and the vibrational entropy decreases. In other words, we expect the vibrational entropy to decrease upon chemical ordering.

7.4.2 Changes in Masses

Now consider the alternative case where the masses of the atoms are altered, but the interatomic springs are kept the same. The maximum excursions of the vibrating atoms are unchanged because the energy at the turning points of a vibration is all potential energy. The potential energy depends only on the compressions of the springs, which are unchanged. Instead, the velocities of the atoms through their midpoint position are altered by the change in mass. Figure 7.7b shows how, for the same temperature, the excursions along the momentum coordinate, $p_{\max}$, depend on mass. For the same temperature, we expect equality of the kinetic energies in the phases α and β. For simple harmonic oscillators it is not difficult to show

$$S^\beta - S^\alpha = \frac{3}{2} k_B \ln\left(\frac{m_\beta}{m_\alpha}\right) \tag{7.38}$$

for the case of unchanged spring constants. Equation 7.38 is sometimes useful for intuition, but if an atom is substituted for another chemical species of different mass, the interatomic forces are usually changed, too. There is a case where Eq. 7.38 is quantitative, and that is the case of isotopic substitution. This is especially important for the isotopes of hydrogen.

7.5 Phonon Thermodynamics

7.5.1 Oscillators and Normal Modes

In 1907, Einstein [71] identified an essential point about atom vibrations – they are quantized in units of $\hbar\omega$, where $\omega/2\pi$ is a vibrational frequency and $2\pi\hbar$ is Planck's constant. This quantum behavior holds true for more realistic models of normal modes of vibrations in crystals. Like standing waves, normal modes extend across the entire crystal. Importantly, in a harmonic crystal the normal modes do not share energy. Harmonic normal modes behave as independent oscillators, each with its own frequency. Normal coordinates offer a sweeping simplification of vibrational dynamics. With increasing temperature, each normal mode of vibration takes in quanta of excitations; each quantum is called a "phonon." In a harmonic solid, a normal mode cannot transfer energy to the other modes, and without damping, a normal mode persists forever. This remains true in quantum mechanics, although integral numbers of excitations, phonons, exist in each normal mode. The energy cost for each additional phonon is always $\varepsilon_i = \hbar \omega_i$, and is unchanged as more phonons are added to the normal mode. Phonon modes behave as independent harmonic oscillators. This section explains how vibrational entropy originates with thermal phonons.[11]

7.5.2 Partition Function

The partition function for a quantum harmonic oscillator of energy $\varepsilon_i = \hbar \omega_i$ is

$$Z_i = \sum_n^{\infty} e^{-\beta(n+\frac{1}{2})\varepsilon_i}, \tag{7.39}$$

$$Z_i = \frac{e^{-\beta\varepsilon_i/2}}{1 - e^{-\beta\varepsilon_i}}, \tag{7.40}$$

where Eq. 7.40 was obtained by identifying Eq. 7.39 as a geometric series times the constant "zero-point" factor $\exp(-\beta\varepsilon_i/2)$, and $\beta \equiv (k_B T)^{-1}$. The partition function for a harmonic solid with N atoms and $3N$ independent oscillators is the product of these individual oscillator partition functions

$$Z_N = \prod_i^{3N} \frac{e^{-\beta\varepsilon_i/2}}{1 - e^{-\beta\varepsilon_i}}, \tag{7.41}$$

from which the phonon free energy is calculated by the prescription of Eq. 1.19, $F = -k_B T \ln Z$

$$F_{\text{vib}} = \frac{1}{2} \sum_i^{3N} \varepsilon_i + k_B T \sum_i^{3N} \ln\left(1 - e^{-\beta\varepsilon_i}\right), \tag{7.42}$$

and the phonon entropy (vibrational entropy) is found by differentiating with respect to T as in Eq. 1.24

$$S_{\text{vib}} = k_B \sum_i^{3N} \left[-\ln\left(1 - e^{-\beta\varepsilon_i}\right) + \frac{\beta\varepsilon_i}{e^{\beta\varepsilon_i} - 1} \right]. \tag{7.43}$$

Using the Planck distribution[12] of Eq. 7.44 for phonon occupancy versus temperature, $n(\varepsilon_i, T)$, and the following relationships

$$n(\varepsilon_i, T) = \frac{1}{e^{+\beta\varepsilon_i} - 1}, \tag{7.44}$$

$$-\ln\left(1 - e^{-\beta\varepsilon_i}\right) = \ln(1 + n), \tag{7.45}$$

$$\beta\varepsilon_i = \ln\left(\frac{1+n}{n}\right), \tag{7.46}$$

---

[11] The normal modes that accommodate the phonons are calculated with the Born–von Kármán model in Section 26.1 (online) and [72–74]. Here we assume that the normal modes are known.

[12] The Planck distribution is the same as the Bose–Einstein distribution when the chemical potential is zero, as is appropriate for harmonic phonons.

it is straightforward to show that Eq. 7.43 reduces to Eq. 7.13 for the vibrational entropy per oscillator, S_{osc}.

7.5.3 Vibrational Entropy of a Harmonic Crystal

A crystal has a dense spectrum of vibrations extending from long wavelength sound waves to vibrations between adjacent atoms. It is useful to work with a continuous spectrum of phonon frequencies called a phonon density of states (DOS), $g(\varepsilon)$. For a three-dimensional solid with N atoms, $3Ng(\varepsilon)\mathrm{d}\varepsilon$ phonon modes exist in an energy interval $\mathrm{d}\varepsilon$. For a DOS in digital form with m intervals of width $\Delta\varepsilon$ (so $\varepsilon_j = j\Delta\varepsilon$), the partition function can be computed numerically from Eq. 7.41

$$Z_N = \prod_{j=1}^{m} \left(\frac{e^{-\beta\varepsilon_j/2}}{1 - e^{-\beta\varepsilon_j}} \right)^{3Ng(\varepsilon_j)\Delta\varepsilon}, \tag{7.47}$$

which becomes at high temperatures

$$Z_N = \prod_{j=1}^{m} \left(\frac{k_{\mathrm{B}}T}{\varepsilon_j} \right)^{3Ng(\varepsilon_j)\Delta\varepsilon}. \tag{7.48}$$

Again, using Eqs. 7.44, 7.45, 7.46, a useful expression for the phonon entropy of a harmonic material at any temperature can be obtained from Eq. 7.47 (cf. Eq. 7.13)

$$S_{\text{vib}}(T) = 3k_{\mathrm{B}} \int_{0}^{\infty} g(\varepsilon) \left[\left(n(\varepsilon) + 1 \right) \ln\left(n(\varepsilon) + 1 \right) - n(\varepsilon) \ln\left(n(\varepsilon) \right) \right] \mathrm{d}\varepsilon, \tag{7.49}$$

where $g(\varepsilon)$ is normalized to 1 and $n(\varepsilon)$ is the Planck distribution (Eq. 7.44) for the temperature of interest.

For thermodynamic calculations, series expansions of Eqs. 7.42 and 7.43 can be useful at high temperatures. For $S_{\text{ph}}(T)$, Fig. 7.8 compares the exact expression of Eq. 7.13

$$S_{\text{vib},\omega}(T) = k_{\mathrm{B}} \left[\left(n(\varepsilon) + 1 \right) \ln\left(n(\varepsilon) + 1 \right) - n(\varepsilon) \ln\left(n(\varepsilon) \right) \right], \tag{7.50}$$

to approximations for up to four terms of a standard high-temperature expansion (terms are presented in the figure). At high temperatures, the following simple approximation works well for the entropy of one oscillator mode with energy $\varepsilon = \hbar\omega$

$$S_{\text{vib},\omega}(T) \simeq k_{\mathrm{B}} \left[\ln\left(\frac{k_{\mathrm{B}}T}{\hbar\omega} \right) + 1 \right]. \tag{7.51}$$

The vibrational entropy can be estimated with Fig. 7.8. A typical "Einstein temperature" for atom vibrations is 300 K, so a temperature of 1000 K corresponds to 3.3 on the x-axis of Fig. 7.8. The entropy of 2.2 k_{B}/mode is multiplied by 3 for vibrations in three dimensions, giving a vibrational entropy of 6.6 k_{B}/atom at 1000 K. This is an order-of-magnitude larger than the configurational entropy of a binary alloy (from Eq. 7.24, $k_{\mathrm{B}} \ln 2 = 0.69\, k_{\mathrm{B}}$/atom).

For comparing the thermodynamic entropies of two phases α and β, the important quantity is the difference in their vibrational entropies, $\Delta S_{\text{vib}}^{\beta-\alpha}$. A handy expression for

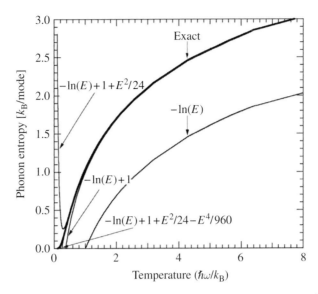

Figure 7.8 Temperature dependence of entropy calculated for one mode of energy $\hbar\omega$, with $E = \hbar\omega(k_B T)^{-1}$.

the high-temperature limit of the difference in vibrational entropy between two harmonic phases, α and β, can be obtained readily from Eq. 7.49

$$\Delta S_{\text{vib}}^{\beta-\alpha} = 3k_B \int_0^\infty \left[g^\alpha(\varepsilon) - g^\beta(\varepsilon) \right] \ln(\varepsilon) \, d\varepsilon. \tag{7.52}$$

The important point about Eq. 7.47 (or Eqs. 7.49 and 7.52) is that the only material parameter relevant to the thermodynamics is the phonon DOS, $g(\varepsilon)$.[13] Measuring or calculating the phonon DOS is central to understanding the vibrational entropy of materials.

7.6 Bond Proportion Model

For a phase transition, what is important is not necessarily the total vibrational entropy, but the difference in vibrational entropies of the two phases.[14] If two phases differ in their

---

[13] It turns out that, for most problems, $g(\varepsilon)$ must be known with high accuracy, and this has been a challenge until relatively recently. For example, if Eq. 7.52 is used for a case where the phonon density of states curves of the α- and β-phases have the same shape, but differ in energy scaling by 10%, the change in vibrational entropy is $\Delta S_{\text{ph}} = 3k_B \ln(1.1) \simeq 0.3 \, k_B/\text{atom}$. This change in vibrational entropy is almost half of the maximum configurational entropy of a binary alloy.

[14] The "Kopp–Neumann rule" from the nineteenth century states that the heat capacity of a compound is the sum of atomic contributions from its elements. By this rule, the vibrational entropy of a solid phase depends only on its chemical composition, and not on its structure. This rule is simply wrong for predicting the thermodynamics of phase transitions. Furthermore, the Kopp–Neumann rule has inconsistencies when picking an atomic heat capacity for carbon, for example.

energies by ΔE and entropies by ΔS, these two phases have the same free energy at the critical temperature T_c

$$\Delta F = \Delta E - T_c \Delta S = 0, \tag{7.53}$$

giving the condition $\Delta E = T_c \Delta S$. Suppose that T_c has been evaluated for the difference in configurational entropy alone. Call it T_c^{conf}. Now if, for the same $\Delta E'$, a difference in vibrational entropy ΔS_{vib} is added, the new prediction of the critical temperature, $T_c^{\mathrm{conf+vib}}$, compares to T_c^{conf} as

$$\Delta E' = T_c^{\mathrm{conf}} \Delta S_{\mathrm{conf}}, \tag{7.54}$$

$$\Delta E' = T_c^{\mathrm{conf+vib}} \left(\Delta S_{\mathrm{conf}} + \Delta S_{\mathrm{vib}} \right), \quad \text{so} \tag{7.55}$$

$$\frac{T_c^{\mathrm{conf+vib}}}{T_c^{\mathrm{conf}}} = \frac{\Delta S_{\mathrm{conf}}}{\Delta S_{\mathrm{conf}} + \Delta S_{\mathrm{vib}}}. \tag{7.56}$$

In what follows, ΔS_{vib} is calculated in a point approximation, added to ΔS_{conf}, and used to calculate T_c for chemical unmixing and ordering phase transitions. This extends the results of Sections 2.7–2.9.

7.6.1 Local Approximations for Bonds and Springs

Keeping the same atoms in the material, the vibrational entropy of a phase transition depends on changes in bond stiffnesses (force constants). A natural first approach to estimate the vibrational entropy of a phase transition is to count the different types of bonds between neighboring atoms, and how they are changed by the phase transition.

• With some risk, the accounting can be confined to first-nearest-neighbor (1nn) pairs.
• With more risk, stiffer springs can be assigned to the pairs with the stronger 1nn bonds.

These assumptions are enough to explain qualitatively the effects of vibrational entropy on the critical temperature, where a disordered solid solution at high temperatures transforms upon cooling to either an unmixed or ordered structure. The low-temperature phases are expected to have stronger 1nn bonds. By Fig. 7.7a, a phase with stiffer interatomic spring constants, k_s, has the smaller vibrational entropy (as does β in the figure), decreasing its relative stability at finite temperature. (In contrast, the solid solution has more bonds that are energetically unfavorable, weaker springs between atoms, and higher vibrational entropy, as does α in the figure.) The critical temperature for disordering is therefore expected to decrease when vibrational entropy is added to the configurational entropy. This is the usual case for real alloys. Assigning a fixed stiffness to a particular type of chemical bond, and counting these bonds, is called the "bond proportion model." It has a long history [75–80].[15]

Consider first the "Einstein model," in which each atom behaves as an independent harmonic oscillator. Figure 7.9a conveys the nature of this approximation. The Einstein

---

[15] Unfortunately, for quantitative predictions of the vibrational entropies of alloy phases, the bond proportion model is rarely adequate for reasons given in Section 7.7 and Chapter 26 (online).

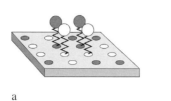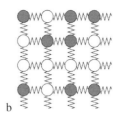

a   b

Figure 7.9 (**a**) Einstein model, with each atom attached to a heavy mass. (**b**) More realistic model of atoms and springs, which can be used to develop the Born–von Kármán model of lattice dynamics, for example.

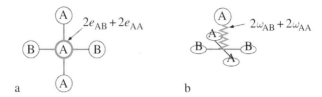

a   b

Figure 7.10 (**a**) Point approximation for the bond strength of the A-atom in the center. (**b**) Analogous approximation, adapting the Einstein model for vibrational frequencies, where the spring stiffness depends on the average first-nearest neighbors.

model assumes each atom is attached by a spring to a rigid point. It is of course more realistic to assume each atom is attached to its first-nearest neighbors (1nn) by springs, as in Fig. 7.9b. This complicates the problem because the tension in each spring then depends on the positions of two atoms instead of one. In the usual Einstein model, the oscillator coordinate of any atom is independent of the vibrations of all other atoms, even those in its 1nn shell.

Here we adapt the Einstein model so that the stiffness of each spring in Fig. 7.9a depends on the neighboring atoms. Our adapted Einstein model is designed to partner with the point approximation that was used for configurational entropy in Sections 2.7 and 2.9. A comparison is shown in Fig. 7.10 of the bond strength and bond stiffness of a central A-atom – both are influenced by its four 1nn atoms. Two neighbors are A, and two are B, and both the bond strength in Fig. 7.10a and the bond stiffness in 7.10b are proportional to these numbers through the parameters e_{AB} and ω_{AB}.

7.6.2 Bond Proportion Model and Unmixing on the Ising Lattice

Returning to Eq. 2.21, the partition function for a single site, Z_{1site}, is

$$Z_{1site} = e^{-\beta e_A} + e^{-\beta e_B}, \tag{7.57}$$

where $\beta \equiv 1/(k_B T)$. Assuming all site occupancies are independent, the total partition function of the alloy with N sites is

$$Z_N = Z_{1site}^N. \tag{7.58}$$

This Z_N can be evaluated by the binomial expansion of Eq. 7.58 plus Eq. 7.48 for the high-temperature limit of the vibrational partition function

$$Z_N = \frac{N!}{(N-n)!\,n!} \left(e^{-\beta e_A}\right)^{N-n} \left(e^{-\beta e_B}\right)^n \prod_{j=1}^{m} \left(\frac{k_B T}{\hbar \omega_j}\right)^{3N\,g(\hbar\omega_j)\,\hbar\Delta\omega}. \qquad (7.59)$$

As in Eq. 2.25, for a fixed $n = cN$, only one term of the binomial expansion is used.

To evaluate the last factor in Eq. 7.59, the partition function for the vibrational frequencies, we invoke a physical correspondence between nearest-neighbor pairs and interatomic force constants.[16] This correspondence works well for the simple cubic lattice – in three dimensions each atom vibrates independently in three coordinates, and there are three bonds to each atom. For a simple cubic lattice with $z = 6$, each atom has $z/2$ neighbors (bonds are not overcounted), and each atom contributes $z/2 = 3$ vibrational modes (modes are not overcounted).[17] In general, however, there is a risk of overcounting or undercounting the vibrational modes. When the coordination number z is not the same as $2\mathcal{D}$, where $\mathcal{D}$ is the number of dimensions, it is necessary to correct the mode counting with the ratio $\eta = 2\mathcal{D}/z$, which is $1/2$ for an fcc crystal with $z = 12$, for example. Each atom has the same three degrees of freedom in a sc or fcc crystal, but twice as many neighbors in fcc.

For our model pictured in Fig. 7.10b, the vibrational factors in Eq. 7.59 include only the frequencies $\{\omega_{AA}, \omega_{BB}, \omega_{AB}\}$, and the numbers of frequencies are ηN_{AB}, ηN_{AA}, ηN_{BB}, reducing the product factor in Eq. 7.59 to three factors

$$Z_N(c) \simeq \left(\frac{N!}{(N-n)!\,n!}\right)\left(e^{-\beta e_A}\right)^{N-n}\left(e^{-\beta e_B}\right)^n$$

$$\times \left(\frac{k_B T}{\hbar \omega_{AB}}\right)^{\eta N_{AB}} \left(\frac{k_B T}{\hbar \omega_{AA}}\right)^{\eta N_{AA}} \left(\frac{k_B T}{\hbar \omega_{BB}}\right)^{\eta N_{BB}}. \qquad (7.60)$$

To obtain N_{AA} for a random solid solution in the point approximation, the average number of A-atom neighbors of an A-atom is $(1-c)z$, so

$$N_{AA} = (1-c)^2 \frac{zN}{2}, \qquad (7.61)$$

where dividing by 2 avoids double-counting, as illustrated with Fig. 2.14. Likewise, $N_{BB} = c^2 zN/2$ and $N_{AB} = c(1-c)zN$, where A–B and B–A bonds are added together because they are equivalent in a solid solution.

Obtaining the Helmholtz free energy from the expression $F(c) = -k_B T \ln Z(c)$, and using the Stirling approximation $\ln(x!) \simeq x \ln x - x$, Eq. 7.60 gives

$$F(c) = E_{conf}(c) - T\left[S_{conf}(c) + S_{vib}(c)\right], \qquad (7.62)$$

---

[16] Equation 7.59 needs products of all the frequencies, not the individual frequencies. Mahanty and Sachdev [81] and Bakker [82, 83] addressed this problem, and showed for some crystals how the product of the $\{\omega_j^2\}$ equals the product of all interatomic force constants, $\{\gamma_{ij}\}$, between atom pairs ij, divided by the product of all atom masses, which is conserved in a closed system. This is consistent with our approach of adapting the Einstein model for vibrational dynamics to the thermodynamic point approximation.

[17] The correspondence between bonds and vibrational modes also works naturally for a linear chain and a square lattice, for which each atom is associated with one or two vibrational modes in one or two dimensions.

where

$$E_{conf}(c) = \Big[e_{AA} + 2c(e_{AB} - e_{AA}) + c^2(4V)\Big]\frac{zN}{2}, \tag{7.63}$$

$$S_{conf}(c) = -k_B\Big[c \ln c + (1 - c) \ln(1 - c)\Big]N, \tag{7.64}$$

$$S_{vib}(c) = k_B\Big[\mathcal{L}_{AA} + 2c(\mathcal{L}_{AB} - \mathcal{L}_{AA}) + c^2(\mathcal{L}_{AA} + \mathcal{L}_{BB} - 2\mathcal{L}_{AB})\Big]\frac{z\eta N}{2}, \tag{7.65}$$

$$S_{vib}(c) = k_B\left[\ln\frac{k_B T}{\hbar\omega_{AA}} + 2c \ln\frac{\omega_{AA}}{\omega_{AB}} + c^2 \ln\frac{\omega_{AB}^2}{\omega_{AA}\omega_{BB}}\right]\mathcal{D}N, \tag{7.66}$$

where $\mathcal{D} = 3$ for three-dimensional solids. Here $E_{conf}(c)$, $S_{conf}(c)$, and $S_{vib}(c)$ are the configurational energy, configurational entropy, and vibrational entropy of the random alloy. In Eq. 7.65 factors were defined such as $\mathcal{L}_{AB} \equiv \ln(k_B T/(\hbar\omega_{AB}))$.[18] This $k_B\mathcal{L}_{AB}$ is the vibrational entropy of an oscillator of frequency ω_{AB} in the classical limit (e.g., Eq. 7.51). Equations 7.63 and 7.65 show that $E_{conf}(c)$ and $S_{vib}(c)$ have identical quadratic forms in c because their energies or interatomic force constants depend in the same way on the numbers of nearest-neighbor bonds in this point approximation.

For chemical unmixing, $V < 0$ (from Eq. 2.32 this means that the average of A–A and B–B pairs is more favorable, i.e., their energies are more negative, than A–B pairs). The energy and entropy terms of Eqs. 7.63–7.66 are graphed in Fig. 7.11a. It was assumed

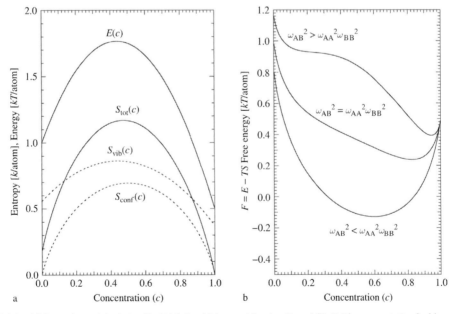

Figure 7.11   (a) $S_{conf}(c)$ for random solid solution (Eq. 7.64), $E_{conf}(c)$ for unmixing, i.e., $V < 0$ (Eq. 7.63), representative $S_{vib}(c)$ when $\omega_{AB}^2 < \omega_{AA}\omega_{BB}$ (Eq. 7.66), and $S_{tot}(c) = S_{conf}(c) + S_{vib}(c)$. (b) Free energy for $zV/(kT) = 1$ (critical condition at $c = 0.5$ without vibrational entropy), showing effects of different ratios of ω_{AB}^2 to $\omega_{AA}\omega_{BB}$.

---

[18] In the classical limit of the Planck distribution, $n(\varepsilon_{AB}, T)$ of Eq. 7.44, this $\mathcal{L}_{AB}$ is the logarithm of the number of phonons of frequency ω_{AB}.

that $\omega_{AB} = 0.77 \sqrt{\omega_{AA}\omega_{BB}}$, and $\omega_{AA} < \omega_{BB}$, as expected if $e_{AA} > e_{BB}$. Figure 7.11b shows curves of the total free energy versus composition for different ratios of vibrational frequencies. The logarithmic singularity of the derivative of $S_{conf}(c)$ always forces a downwards slope away from $c = 0$ and $c = 1$. The energy term, positive when $V < 0$, promotes unmixing at intermediate concentrations at low temperatures.

The critical temperature for chemical unmixing, T_c, is found by setting $\partial^2 F/\partial c^2 = 0$. It is straightforward to do this with Eqs. 7.63 through 7.66, with the result

$$k_B T_c = \frac{4\,|V|\,z}{\frac{1}{c(1-c)} + 2\mathcal{D}\,\ln\left(\frac{\omega_{AA}\omega_{BB}}{\omega_{AB}^2}\right)}. \tag{7.67}$$

For unmixing, Eq. 7.67 and its derivation predict:

- The T_c is reduced by vibrational entropy when $\omega_{AA}\omega_{BB} > \omega_{AB}^2$, and increased when $\omega_{AA}\omega_{BB} < \omega_{AB}^2$.
- When $\omega_{AB}^2 = \omega_{AA}\omega_{BB}$, Eq. 7.67 provides $T_c = z|V|/k_B$ for $c = 0.5$, the previous result of Eq. 2.48 without vibrational entropy. The middle curve in Fig. 7.11b shows $F(c)$ for this condition. The free energy without vibrational entropy, $F_{conf}(c)$, has a negative curvature in c below $T_c = z|V|/k_B$, characteristic of spinodal decomposition in the point approximation.
- When $\omega_{AB}^2 \neq \omega_{AA}\omega_{BB}$, $S_{vib}(c)$ is a quadratic function of c (Eq. 7.66), much like $E_{conf}(c)$.[19]
- The disordered solid solution has more A–B pairs, which are unfavorable for an unmixing system. The solid solution is therefore expected to have softer springs so $\omega_{AB}^2 < \omega_{AA}\omega_{BB}$. At the same temperature, the atoms have larger vibrational excursions in the solid solution. The vibrational entropy term $-TS_{vib}(c)$ is larger for the solid solution, helping to stabilize it against unmixing. This reduces T_c.
- If $\omega_{AB}^2 > \omega_{AA}\omega_{BB}$, T_c is raised and the whole unmixing phase boundary moves to higher temperatures. If $\omega_{AB}^2 \exp(-2/3) > \omega_{AA}\omega_{BB}$ (approximately $0.513\,\omega_{AB}^2 > \omega_{AA}\omega_{BB}$), disordering cannot occur at any temperature at $c = 0.5$. For yet larger ω_{AB}^2 with respect to $\omega_{AA}\omega_{BB}$, the entropy of mixing quickly becomes negative for intermediate compositions, suppressing disordering.[20]

7.6.3 Bond Proportion Model and Ordering on the Ising Lattice

Consider the chessboard ordering of an equiatomic A–B alloy on the square lattice shown in Fig. 7.12. Figure 7.12 identifies the α- and β-sublattices, each having $N/2$ sites for a crystal with N sites. Chemical ordering occurs when the interchange energy $V > 0$ (Eq. 2.32), so the energetics tends to maximize the number of A–B pairs at low temperatures. Finite temperatures favor some disorder in the structure – placing a few atoms on the wrong sublattice gives a big increase in the configurational entropy,[21]

---

[19] A simple approximation is that the strengths of the bonds and spring constants are proportional, giving $S_{vib}(c)$ the same shape as $E_{conf}(c)$. This is conceptually convenient, but it has limited reliability.

[20] A negative entropy of mixing has been observed in alloys of V–Pt at surprisingly low concentrations of Pt [84].

[21] The entropy has a logarithmic singularity in the derivatives of its sublattice concentrations analogous to the concentration dependence of Eq. 7.64 near $c = 0$.

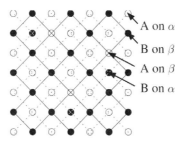

A on α

B on β

A on β

B on α

Figure 7.12 Ordered square lattice with two species {A, B}. Dashed and solid lines represent two interpenetrating square sublattices $\{\alpha, \beta\}$. A few misplaced A and B atoms are shown.

driving some mixing of A-atoms onto the β-sublattice, in exchange for B-atoms on the α-sublattice. At high temperatures, the alloy is expected to become a random solid solution, at least if the configurational entropy is dominant.

The labels in Fig. 7.12 help to identify the four types of sublattice concentration variables $\{N_A^\alpha, N_A^\beta, N_B^\alpha, N_B^\beta\}$, where for example N_A^α is the number of A-atoms on the α-sublattice. For an equiatomic alloy, by conservation of atoms and conservation of sublattice sites, $N_A^\alpha = N_B^\beta \equiv R$, where R denotes atoms that are on the "right" sublattice, and $N_A^\beta = N_B^\alpha \equiv W$, denoting "wrong" atoms. As in Section 2.9, the sublattice concentrations are used to define a "long-range order" (LRO) parameter, L, for the alloy [12, 15, 85]

$$L \equiv \frac{R - W}{N/2}. \tag{7.68}$$

This L can range from -1 to 1. The case of $L = 1$ corresponds to perfect order where $R = N/2$ and $W = 0$. The disordered alloy has as many right as wrong atoms on each sublattice, i.e., $R = W$, so $L = 0$ for a disordered alloy.

Assume that the atoms are placed randomly on each sublattice, e.g., the N_A^α A-atoms have no preference for any site on the α-sublattice, irrespective of the presence of A-atoms or B-atoms on neighboring sites. This is the "point" or "Gorsky–Bragg–Williams" approximation. This approximation makes it easy to count the bonds between the different species of atoms. To determine the numbers of pairs N_{AA}, N_{BB}, N_{AB}, an important point shown in Fig. 7.12 is that the first neighbors of each atom on the α-sublattice are on the β-sublattice, and vice versa (Section 2.9.2). To evaluate N_{AA}, pick one of the R A-atoms on the α-sublattice, and multiply by the probability of finding a wrong A-atom in each of its z neighboring sites on the β-sublattice (which has $N/2$ sites)[22]

$$N_{AA} = N_{BB} = RW\frac{z}{N/2}, \tag{7.69}$$

$$N_{AB} = (R^2 + W^2)\frac{z}{N/2}. \tag{7.70}$$

Using Eqs. 7.68 through 7.70, the numbers of pairs are expressed in terms of the LRO parameter

$$N_{AA} = N_{BB} = (1 - L^2)\frac{zN}{8}, \tag{7.71}$$

---

[22] Fig. 2.20 can be helpful for verifying these relations.

$$N_{\mathrm{AB}} = (1 + L^2)\frac{zN}{4}. \tag{7.72}$$

The partition function has a combinatoric factor from the random occupancies on two sublattices (giving two binomial probabilities) and three factors that depend on the numbers of bonds

$$Z = \left(\frac{(N/2)!}{R!\ W!}\right)^2 \left[\mathrm{e}^{-\beta\,e_{\mathrm{AA}}}\left(\frac{k_{\mathrm{B}}T}{\hbar\,\omega_{\mathrm{AA}}}\right)^\eta\right]^{N_{\mathrm{AA}}} \left[\mathrm{e}^{-\beta\,e_{\mathrm{BB}}}\left(\frac{k_{\mathrm{B}}T}{\hbar\,\omega_{\mathrm{BB}}}\right)^\eta\right]^{N_{\mathrm{BB}}}$$

$$\times \left[\mathrm{e}^{-\beta\,e_{\mathrm{AB}}}\left(\frac{k_{\mathrm{B}}T}{\hbar\,\omega_{\mathrm{AB}}}\right)^\eta\right]^{N_{\mathrm{AB}}}, \tag{7.73}$$

where the high-temperature limit is assumed for the phonons. As for Eq. 7.60, the number of vibrational modes must be consistent with the number of pairs. For the square lattice Eq. 7.73 works naturally with $\eta = 1$, but for other lattices it is necessary to avoid incorrect counting of the $3N$ vibrational modes by setting $\eta = 2\mathcal{D}/z$, where $\mathcal{D}$ is the dimensionality.

We want the partition function to depend on the LRO parameter, L, so the exponents in Eq. 7.73 are replaced with Eqs. 7.71 and 7.72, and Eq. 7.68 is used with the conservation of sites ($R + W = N/2$) to remove all instances of R, W, N_{AA}, N_{BB}, N_{AB} in Eq. 7.73, giving $Z(L)$. The free energy is then $F(L) = -k_{\mathrm{B}}T \ln Z(L)$. Subtracting the free energy of the disordered solid solution, the free energy of ordering, $\Delta F(L) \equiv F(L) - F(L = 0)$, is

$$\Delta F(L) = -L^2 \frac{zN}{8} 4V$$

$$- T\frac{N}{2}k_{\mathrm{B}}\left[-(1 + L)\ln(1 + L) - (1 - L)\ln(1 - L)\right]$$

$$- TL^2 \frac{\mathcal{D}N}{4}\ k_{\mathrm{B}}\ \ln\left(\frac{\omega_{\mathrm{AA}}\omega_{\mathrm{BB}}}{\omega_{\mathrm{AB}}^2}\right), \tag{7.74}$$

where the three lines of Eq. 7.74 are identified as the changes upon ordering of E_{conf}, $-TS_{\mathrm{conf}}$, and $-TS_{\mathrm{vib}}$, respectively. In equilibrium $\mathrm{d}\Delta F(L)/\mathrm{d}L = 0$, giving a transcendental equation for $L(T)$

$$2LzV = k_{\mathrm{B}}T \ln\left(\frac{1 + L}{1 - L}\right) + L\mathcal{D}\ k_{\mathrm{B}}T \ln\left(\frac{\omega_{\mathrm{AB}}^2}{\omega_{\mathrm{AA}}\omega_{\mathrm{BB}}}\right). \tag{7.75}$$

This differs from the usual Gorsky–Bragg–Williams solution of Eq. 2.55 by the extra term from vibrational entropy. This term is linear in L, so it can be considered as a temperature-dependent rescaling of the interatomic exchange energy, V.

There is no analytical solution to Eq. 7.75 for all T, but it is possible to obtain an analytical solution for the critical temperature, T_{c}. The trick is to assume L is infinitesimally small, so Eq. 7.75 can be linearized.[23] The result is

$$k_{\mathrm{B}}T_{\mathrm{c}} = \frac{zV}{1 + \frac{\mathcal{D}}{2}\ln\left(\frac{\omega_{\mathrm{AB}}^2}{\omega_{\mathrm{AA}}\omega_{\mathrm{BB}}}\right)}. \tag{7.76}$$

---

[23] This assumes that the ordering transition remains second order, which is true for Eq. 7.75. The curve of $L(T)$ from Gorsky–Bragg–Williams theory, which decreases continuously from $L = 1$ to $L = 0$ at T_{c}, is rescaled monotonically with S_{vib} – the effective V can be considered as altered by a temperature-dependent term.

The configurational entropy is always important in a chemical order–disorder transition. In the absence of vibrational entropy, i.e., $\omega_{AB}^2 = \omega_{AA}\omega_{BB}$, Eq. 7.76 reduces to $T_c = zV/k_B$, the well-known result of Gorsky–Bragg–Williams theory (Eq. 2.58). When both vibrational and configurational entropies are important, from Eq. 7.76 we find:

- For chemical ordering we expect A–B bonds to be stronger and stiffer than A–A and B–B, so $\omega_{AB}^2 > \omega_{AA}\omega_{BB}$. Vibrational excursions in the ordered phase are smaller, so the entropy is smaller and T_c is reduced, perhaps significantly. In the case where $\omega_{AB}^2 = 1.1\,\omega_{AA}\omega_{BB}$, for a simple cubic lattice T_c is reduced to 87% of its value without vibrational entropy. This suppression of T_c seems to be the usual effect of vibrational entropy for order–disorder transitions.
- The opposite effect is possible if $\omega_{AB}^2 < \omega_{AA}\omega_{BB}$. Although unexpected from physical intuition, if $\omega_{AB}^2/\omega_{AA}\omega_{BB} < \exp(-2/3)$, vibrational entropy suppresses disordering, and $T_c = \infty$.

7.7 Bond-Stiffness-versus-Bond-Length Model

7.7.1 Phonon Frequencies and Bond Lengths

The bond proportion model just described, which counts the different types of atom pairs, has a deficiency that is easy to understand. The bond lengths between pairs of atoms are different in different structures, and the stiffness of a bond is sensitive to its length.[24] The effects of bond stiffness versus bond length can have major effects on the phonon entropy and thermodynamics of phase transitions.

Consider a disordered solid solution with a distribution of interatomic distances, shorter and longer than optimal, owing to the different atomic sizes. The calculation of these distances is not simple because the relaxations in position for each atom depend in turn on the relaxations of its neighbors, but we can make an estimate. The Grüneisen parameter γ is defined as

$$\gamma \equiv -\frac{V}{\omega}\frac{\partial\omega}{\partial V}, \qquad (7.77)$$

where ω is the phonon frequency and V is volume. Recognizing that $V \propto r^3$ and $\omega \propto \sqrt{k_s}$, where k_s is an interatomic force constant

$$6\gamma = -\frac{r}{k_s}\frac{\partial k_s}{\partial r}. \qquad (7.78)$$

There is no universal value for a Grüneisen parameter, but sometimes $\gamma \simeq 2$. In this case, Eq. 7.78 predicts that if the interatomic distance changes by +1%, the interatomic force changes by –12%. This is a major effect. A change in interatomic distance of +10% could cause the interatomic force to be lost entirely.

---

[24] This same characteristic of the interatomic potential is used later in quasiharmonic theory, where interatomic forces weaken with thermal expansion (Sections 8.2, 12.3, 26.3).

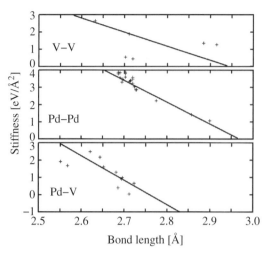

Figure 7.13   Stretching terms of the nearest-neighbor force constants as a function of bond length of Pd–V alloys. Each point corresponds to one type of bond in one of a set of fcc crystal structures of different volumes. Reproduced, with permission, from [87]. Copyright (2000) by the American Physical Society.

   Quantitative analyses of the effects of bond length on bond stiffnesses were performed by Sluiter *et al.* on Al–Li [86] and by van de Walle and Ceder on Pd–V [75, 87, 88]. A convenient analysis involves a bond "stretching" force between neighboring atoms, plus a bond "bending" force. The stretching forces (central forces) between 1nn atoms were most sensitive to bond length; the bond bending forces were smaller and less affected by bond length. Results for stretching stiffnesses for 1nn bonds in a set of fcc-based Pd–V structures with different interatomic distances are presented in Fig. 7.13. The points tend to decrease rapidly with bond length, as expected if the interatomic forces soften with expansion. Approximately, there is a range of about 10% in the interatomic separation where the force constants vary from their maximum value to zero, and the stretching stiffness has an approximately linear relationship with bond length. The effects of bond length on bond stiffness of Fig. 7.13 are large, so a Pd–Pd bond cannot be expected to have the same stiffness in any local environment of a random solid solution, for example.

7.7.2 Extending the Bond Proportion Model

For a chemical ordering transition, we use the bond-stiffness-versus-bond-length relationship to improve the free energy (Eq. 7.74) and the critical temperature (Eq. 7.76) of the bond proportion model. The free energy difference $\Delta F(L)$ of Eq. 7.74 is with reference to the disordered solid solution, and is zero for $L = 0$ (i.e., $\Delta F(0) = 0$). Consider again the case of B1 ordering (NaCl prototype) of an AB alloy on a simple cubic lattice, appropriate for Eq. 7.76. In this structure, all 1nn bonds are between A–B pairs, and there is only one bond length. It is the disordered solid solution where the distortions of bond lengths occur. This is a complex problem owing to the different relaxations in the different local environments. We can get to the essence of the problem by using the same Grüneisen parameter for all bonds (Eq. 7.77), and by simplifying the three curves of Fig. 7.13 by assuming the bond stiffnesses are zero at the same bond lengths for A–A, B–B, and A–B pairs. In this case, all three frequencies $\{\omega_{AA}, \omega_{BB}, \omega_{AB}\}$ change with volume as

$$\omega_{ij} = \omega_{ij}^0 \left(1 - \gamma \frac{\delta V}{V^0} \right), \tag{7.79}$$

where the superscript 0 denotes a reference value for the ordered structure. For small $\delta V/V^0$

$$\ln(\omega_{ij}) \simeq \ln(\omega_{ij}^0) - \gamma \frac{\delta V}{V^0}. \tag{7.80}$$

To the free energy of Eq. 7.74 we add the term

$$-TS_{\text{Gruen}} = k_B T \left(1 - L^2 \right) \mathcal{D} N \gamma \frac{\delta V}{V^0}, \tag{7.81}$$

which vanishes when $L = 1$, but grows with disorder as the number of R–W pairs increases as $1 - L^2$. After some analysis along the lines of Section 7.6.3, the critical temperature of Eq. 7.76 becomes

$$k_B T_c = \frac{zV}{1 + 2\mathcal{D}\gamma \frac{\delta V}{V^0} + \frac{\mathcal{D}}{2} \ln\left(\frac{\omega_{AB}^2}{\omega_{AA}\omega_{BB}} \right)}. \tag{7.82}$$

Consider a typical case where the disordered phase has a lattice parameter 1% larger than the ordered phase (so $\delta V/V^0 \simeq 0.03$). Assuming $\gamma = 2$, the second term in the denominator of Eq. 7.82 is +0.36. The critical temperature is suppressed substantially when the average bond length in the disordered solid solution is 1% longer than for the ordered phase.[25] Because we assumed the same Grüneisen parameter for all bonds, this effect of bond stiffness versus bond length is independent of the effect on T_c from stiffness differences of the different pairs of atoms. In a more complete treatment these effects are interrelated.

7.7.3 Other Effects on Bond Stiffness

A closer examination of Fig. 7.13 shows that the bond stiffnesses for V–V and Pd–V pairs have considerable scatter about the linear fit. Some of this can be accounted for by a dependence of bond stiffness on chemical composition. For example, the addition of B-atoms to a host crystal of element A may fill or deplete the electronic band structure of A, causing changes in the density of electron states at the Fermi level. The electron screening of the A-ion cores is thereby altered, changing the bond stiffnesses of A–A pairs (Section 26.5.1 [online]). Such effects may scale with the electronegativity difference between A- and B-atoms, and with the density of electron states at the Fermi level.

The harmonic model explains many physical phenomena, especially at low temperatures when the phonons are approximately independent of each other, and independent of other sources of dynamical entropy. It is often extended to higher temperatures as the "quasiharmonic model," which assumes independent phonons with altered frequencies.

---

[25] Contraction upon ordering is typical. In the less common case where the lattice parameter of the disordered alloy is smaller than the ordered, T_c would be raised.

The quasiharmonic model, discussed in Sections 8.2, 12.3, 26.3, is essentially the same as the bond-stiffness-versus-bond-length model when the bond lengths are determined by thermal expansion. At high temperatures, however, the different sources of dynamical entropy can interact in the important ways discussed in Section 12.3 and Chapter 26 (online). In short, the existence of phonons alters the energies required to create additional phonons (phonon–phonon interaction, PPI), and the phonon excitations are altered by electron excitations (electron–phonon interaction, EPI).

Problems

7.1 The combinatoric entropy becomes more tricky to evaluate for larger clusters of atoms. Consider all eight possible nearest-neighbor triplets of atoms on a linear chain for a binary alloy: {AAA, BAA, ABA, BBA, AAB, BAB, ABB, BBB}. At each site we might expect to have eight possibilities for a triplet, but the triplets overlap between sites. The result is that the random solid solution has a configurational entropy of $k_B \ln 2$ per atom, which is the same result obtained in the point approximation.

Prove this.

(*Hint*: It may be convenient to consider how many ways a new atom can be added to the end of a finite chain.)

7.2 Derive the combinatorial factor in Eq. 7.14.

(*Hint*: Here is an approach to counting the ways of assigning pairs to each bond, of which there are $zN/2$ in the crystal. Start with an empty set of bonds, and assign the first A–A pair to one of them. There are $zN/2$ ways to put down this first pair. The second pair cannot go into the first bond, so there are $zN/2 - 1$ ways to put it down. The last A–A pair can go down $zN/2 - (N_{AA}^{\alpha\beta} - 1)$ ways. Each pair goes down independently of the others, so show that the product of all these factors equals $(zN/2)! / (zN/2 - N_{AA}^{\alpha\beta})!$. Of course we could have filled these bonds in any order with A–A pairs, so we overcounted by the factor of $N_{AA}^{\alpha\beta}!$, and we need to divide by this factor. The A–B pairs go next, but they start with only a number of bonds $(zN/2) - N_{AA}^{\alpha\beta}$ to occupy. Be brave and continue through the B–A and B–B pairs. A marvelous simplification at the end occurs because $zN/2 - N_{AA}^{\alpha\beta} - N_{AB}^{\alpha\beta} - N_{BA}^{\alpha\beta} - N_{BB}^{\alpha\beta} = 0$.)

7.3 (**a**) Derive Eq. 7.27 for the short-range order in the absence of long-range order.

(**b**) Assuming $L = 0$, what value of α is predicted for B2-type short-range order at $T = T_c$? (See Fig. 7.5.)

7.4 The following questions ask you to use Eqs. 7.13 and 7.8 to compare the vibrational entropy and configurational entropy of an alloy with a composition c and a characteristic phonon frequency ω. For questions about vibrational entropy, don't forget that there are three times as many vibrational modes as the number of atoms.

(a) Show that the mathematical forms of Eqs. 7.13 and 7.8 are the same for small n and small c.

(b) What composition has the largest value of configurational entropy, and what is this value?

(c) Show that the Planck occupancy factor becomes $n(T) = k_B T/\hbar\omega$ at high temperatures.

(d) Using the simple approximation for the Planck occupancy at high temperature, $n(T) = k_B T/\hbar\omega$, at what temperature is the vibrational entropy equal to the highest value of the configurational entropy from part b when $\hbar\omega/k_B = 400\,K$?

(e) The temperature of part d is a bit low for the high-temperature expression for $n(T)$. What is the effect (up, down, no change) on your result of using the full expression for the Planck occupancy?

(f) For $\hbar\omega = 400\,K$, at a temperature of 800 K, calculate the vibrational entropy, and compare it to the maximum value of the configurational entropy of a binary alloy. (You can use the high-temperature approximation of part c.)

(g) Two materials have different characteristic phonon frequencies – one has $\hbar\omega = 400\,K\ k_B$, and the other has $\hbar\omega = 440\,K\ k_B$. At what temperature will their difference in vibrational entropy exceed $0.0693\ k_B$/atom? Same question for $0.693\ k_B$/atom?

7.5 This problem uses the cluster expansion method to evaluate the entropy of a disordered solid solution in one dimension. In this problem, we first use three alloys to calibrate the interaction entropies of first-neighbor pairs, from which we can calculate the entropy of disordered alloys.

The three samples with known atom arrangements and known entropies are: pure element A, with $S = 0.75\ k_B$/atom; an ordered equiatomic A–B alloy (having structure . . . ABAB. . .) with $S = 0.95\ k_B$/atom; pure element B, with $S = 0.40\ k_B$/atom.

(a) Using $\sigma = +1$ for A-atoms and $\sigma = -1$ for B, calculate the correlation function $\xi_0 = \langle \sigma_n^0 \rangle = 1$ for the empty lattice, $\xi_1 = \langle \sigma_n \rangle$ for the point variables, $\xi_2 = \langle \sigma_n \sigma_{n+1} \rangle$ for first-neighbor pair variables, and $\xi_3 = \langle \sigma_n \sigma_{n+1} \sigma_{n+2} \rangle$ for triplets of neighboring atoms, for all three alloys.

(b) Set up the 3×3 matrix equation

$$S_j = \xi_{j,i} s_i, \qquad\qquad (7.83)$$

where the three S_j were given to you above, but the s_i are unknown. You have already calculated the required nine elements of $\xi_{j,i}$ in part a – now put them in the correct places in the matrix. (Use only ξ_0, ξ_1, and ξ_2, not ξ_3.)

(c) Invert the matrix of part b, and obtain the three s_j.

(d) What is the mathematical form for the correlation functions $\xi_0 = \langle 1 \rangle$, $\xi_1 = \langle \sigma_n \rangle$, $\xi_2 = \langle \sigma_n \sigma_{n+1} \rangle$, and $\xi_3 = \langle \sigma_n \sigma_{n+1} \sigma_{n+2} \rangle$ for a disordered solid solution of B-atom concentration x, where $0 \le x \le 1$? Using the three s_j obtained from part c, what are the entropies of disordered solid solutions of concentrations $x = 0.25$, $x = 0.50$, and $x = 0.75$?

(*Hint*: Much of this is worked out for you, including the correlation functions of part a, by Matthew S. Lucas in his 2008 Caltech Ph.D. thesis, available here: https://thesis.library.caltech.edu/3032/.)

7.6 The vibrational entropy decreases when a Pt atom is added to bcc V by the amount $0.35k_B$/(Pt atom) (comparing the vibrational entropy of the pure elements and the Pt–V alloy at 300 K). The configurational entropy of chemical mixing increases with the addition of Pt to V (at least for Pt concentrations $c < 0.5$).
  (a) At what concentration of Pt in V will the addition of a Pt atom cause zero change to the total entropy (the sum of configurational plus vibrational)?
  (b) How does this concentration change if the amount $0.35\,k_B$/(Pt atom) is doubled?

7.7 Isotopic fractionation occurs in materials over geologic times, and can be useful as a method for estimating the age of minerals. In this problem, assume that atoms of different isotopes differ only by their nuclear mass. Consider equal amounts of two phases in equilibrium at temperature T. You may assume two isotopes that are equally abundant on average. Consider the total free energy or energy to answer these two questions.
  (a) Using high-temperature expressions for vibrational entropy, show that there is no preference of the heavy or light isotopes for the phase with stiffer interatomic force constants.
  (b) Show that the zero-point energy of these two phases can be lowered by isotopic segregation to one phase or the other. Do heavy isotopes prefer the phase with the stiffer or softer phonon spectrum?

7.8 **(a)** Explain the role of η in Eq. 7.60, and why it appears only in the factors involving temperature and vibrational frequency.
  (b) Starting with the partition function of Eq. 7.60, derive Eqs. 7.63, 7.64, and 7.66.
  (c) Defining $R \equiv \sqrt{\omega_{AB}^2/(\omega_{AA}\omega_{BB})}$ as the ratio of vibrational frequencies, make a graph of T_c for unmixing versus R for an equiatomic alloy. Be sure that your range of R extends from 0.7 to 1.3, and preferably further.
  (d) Usually the R defined in part c is in the range from 0.9 to 1.1. What is the fractional change of T_c over this range? Give a physical explanation for why T_c changes as the A–B bonds become stiffer than the average of the A–A and B–B bonds.

7.9 The arguments in this problem are typical of the "quasiharmonic approximation" for nonharmonic behavior at elevated temperatures. A typical Grüneisen parameter is 2.0, and a typical coefficient of linear (not volume) thermal expansion is 2.0×10^{-5}/K. Suppose the ratio of vibrational frequencies R is $R \equiv \sqrt{\omega_{AB}^2/(\omega_{AA}\omega_{BB})} = 1.05$.
  (a) For a temperature change of 500 K, what is the effect of thermal expansion on the critical temperature for ordering in an equiatomic alloy, compared with what would be expected from the alloy properties at low temperature?

(b) Suppose two disordered equiatomic alloys differ in their thermal expansion and Grüneisen parameter, even though they have the same $R = 1.05$. Suppose the first alloy has the properties given above, but the second has a larger Grüneisen parameter of 2.5 and a larger linear (not volume) thermal expansion of 2.5×10^{-5}/K. How much more vibrational entropy is expected of the second alloy compared with the first for a temperature increase of 500 K?

8

Pressure

Historically there has been comparatively little work on how phase transitions in materials depend on pressure, as opposed to temperature. For experimental research, it is difficult to generate and control pressures of thermodynamic importance, whereas high temperatures are easily achieved. More recently, the situation has become reversed for computational work. The thermodynamic variable complementary to pressure is volume, whereas temperature is complemented by entropy. It is comparatively easier to calculate the free energy of materials with different volumes, as opposed to calculating the free energy with all different sources of entropy.

There are now rapid advances in high-pressure experimental techniques, often driven by interest in the geophysics of the Earth. New materials are formed under extreme conditions of pressure and temperature, and some such as diamond can be recovered at ambient pressures. The use of pressure to tune the electronic structure of materials can be a useful research tool for furthering our understanding of materials properties. Sometimes the changes in interatomic distances caused by pressure can be induced by chemical modifications of materials, so experiments at high pressures can point directions for materials synthesis at ambient pressure.

Chapter 8 begins with basic considerations of the thermodynamics of materials under pressure, and how phase diagrams are altered by temperature and pressure together. Volume changes can also be induced by temperature, and the concept of "thermal pressure" from nonharmonic phonons is explained. Electronic energy is responsible for big contributions of PV to the free energy, and there is a brief description of how electron energies are altered by pressure. The chapter ends with a discussion on kinetic processes under pressure.

8.1 Materials under Pressure at Low Temperatures

8.1.1 Gases (for Comparison)

The behavior of solids under pressure, at least high pressures that induce substantial changes in volume, is more complicated than the behavior of gases. Nevertheless, it is useful to compare gases with solids to see how the thermodynamic extensive variable, V, depends on the thermodynamic intensive variables, T and P. Recall the equation of state for an ideal gas of noninteracting atoms

$$PV = Nk_BT. \tag{8.1}$$

Nonideal gases are often treated with two Van der Waals corrections:

- The volume for the gas is a bit less than the physical volume it occupies because the molecules themselves take up space. The quantity V in Eq. 8.1 is replaced by $V - Nb$, where b is an atomic parameter with units of volume.
- The interaction between gas molecules is usually attractive, and tends to increase the pressure a bit. This can be considered as a surface tension that pulls inwards on a group of gas molecules. The quantity P in Eq. 8.1 is replaced by $P + a(N/V)^2$. The quadratic dependence of $1/V^2$ is expected because the number of atoms affected goes as $1/V$, and the force between them may also go as $1/V$. (Also, if the correction went simply as $1/V$, it would prove uninteresting in Eq. 8.2.)

The Van der Waals equation of state (EOS) is

$$\left[P + a\frac{N^2}{V^2} \right]\left[V - Nb \right] = Nk_\mathrm{B}T. \tag{8.2}$$

Equation 8.2 works surprisingly well for the gas phase when the parameters a and b are small and the gas is "gas-like." Equation 8.2 can be converted to a dimensionless form

$$\mathcal{P} = \frac{\mathcal{T}}{\mathcal{V} - 1} - \frac{1}{\mathcal{V}^2}, \tag{8.3}$$

with the definitions

$$\mathcal{P} \equiv P\frac{b^2}{a}, \tag{8.4}$$

$$\mathcal{V} \equiv \frac{V}{Nb}, \tag{8.5}$$

$$\mathcal{T} \equiv k_\mathrm{B}T\frac{b}{a}. \tag{8.6}$$

Figure 8.1 shows the Van der Waals EOS of Eq. 8.2 for a fixed a and b, but with varying temperature. At high temperatures the behavior approaches that of an ideal gas, with $P \propto T/V$ (Eq. 8.1). More interesting behavior occurs at low temperatures. It can be shown that a two-phase coexistence between a high-density and a low-density phase appears below the critical pressure P_c and critical temperature T_c, where

$$P_\mathrm{c} = \frac{1}{27}\frac{a}{b^2}, \tag{8.7}$$

$$k_\mathrm{B}T_\mathrm{c} = \frac{8}{27}\frac{a}{b}. \tag{8.8}$$

This critical condition is at point "C" in Fig. 8.1a, for which the volume is

$$V_\mathrm{c} = 3\,b\,N. \tag{8.9}$$

At lower temperatures, such as $k_\mathrm{B}T = 0.26a/b$ shown in Fig. 8.1b, the same pressure corresponds to three different volumes V_1, V_2, and V_3. We can ignore V_2 because it is unphysical – at V_2 an increase in pressure causes an expansion (and likewise, the material shrinks if pressure is reduced). Importantly, the volumes V_1 and V_3 can be interpreted

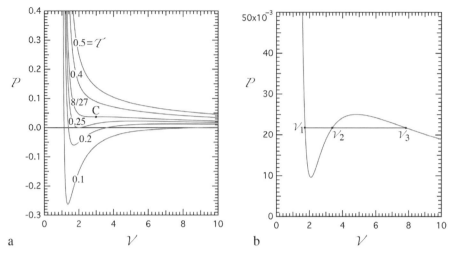

Figure 8.1 (**a**) Isothermals of the Van der Waals equation of state, Eq. 8.3 or 8.2, plotted with rescaled variables of Eqs. 8.4–8.6. (**b**) Maxwell construction for $T = 0.26a/(bk_B)$.

as the specific volumes of a liquid and as a gas, respectively. We find the pressure that defines V_1 and V_3 from the condition that the chemical potentials of the gas and liquid are equal in equilibrium, i.e., $\mu_3 = \mu_1$. Along a $P(V)$ curve, the change in chemical potential is $1/N \int P\,dV$. Starting at a chemical potential of μ_1 at the point V_1 in Fig. 8.1b

$$\mu_3 = \mu_1 + \frac{1}{N} \int_{V_1}^{V_3} P(V)\,dV. \tag{8.10}$$

The integral must be zero if $\mu_3 = \mu_1$. The areas above and below the horizontal line in Fig. 8.1b must therefore be equal, and this "Maxwell construction" defines the pressure of the horizontal line.

A dimensionless ratio can be formed from Eqs. 8.7, 8.8, and 8.9

$$\frac{P_c V_c}{Nk_B T_c} = \frac{3}{8}. \tag{8.11}$$

Rescaled appropriately, the Van der Waals equations of state for all gases are the same. This is approximately true in practice, although the dimensionless ratio is lower than 3/8, often around 0.25 to 0.3, and varies for different gases. Some characteristics of a generic gas are presented in Table 8.1, for comparison with the characteristics of a solid.

This instability of the Van der Waals EOS below a critical temperature can be used to model a pressure-induced liquefaction, for example, or the liquid–gas phase boundary at constant pressure. The approach has problems with quantitative details, but it predicts the essential behavior. More generally, the scaling relationships of Eqs. 8.4–8.6 can be used to identify "corresponding" temperatures and pressures for different gases, and their thermodynamic functions are often similar at corresponding conditions.

Table 8.1 Pressures and temperatures of gases and solids		
	Gas	Solid
Pressure	$P > 0$	$P > -P_{coh}$
Temperature	$T > 0$	$T \geq 0$
Stresses	Isotropic	Anisotropic
Typical pressure	1 atm = 0.1 MPa	10 GPa = 10^5 atm

8.1.2 Solids (for Comparison)

The ideal gas behavior shown at the top of Fig. 8.1a, i.e., $P \propto T/V$ for large T, V, is never appropriate for a solid. At $P = 0$, for example, the solid has a finite volume. Table 8.1 shows that on the scale of familiar pressures in gases, a solid is essentially incompressible.

More familiar are small compressions of solids and elastic behavior, where materials deform as springs. The bulk modulus of a solid, B,

$$B \equiv -V \frac{\mathrm{d}P}{\mathrm{d}V},$$ (8.12)

is typically a few times 100 GPa, and the elastic energy per unit volume under uniform dilation is

$$E_{el} = \frac{1}{2} B \, \delta^2,$$ (8.13)

where δ is the fractional change in volume. Equation 8.13 is handy if the elastic energy is from uniform compression, so the full analysis of elasticity in Section 6.7 is not required. The elastic constants originate from second derivatives of the interatomic potentials, which give "springs" between atoms, as explained in Section 6.6. These springs are loaded in different directions when different stresses are applied to a material, but all their strains are linear with stresses, so the macroscopic response to a small stress is still that of a spring. It is possible to relate the interatomic force constants to the macroscopic elastic constants, but this requires delicacy in accounting for the symmetry of the crystal and the geometry of the problem.

Like a spring, the elastic energy E_{el} increases as the square of δ. For a rod-shaped material stretched along its long dimension, the range of elastic behavior is typically below $\delta \sim 10^{-2}$. This gives an elastic energy density of order 10^7 J/m^3, or of order 10^{-3} eV/atom. This is small on the energy scale of most chemical bonds. Beyond this small elastic limit, materials deform plastically, and do not return to their original shapes. On the other hand, suppose a material is loaded on all surfaces simultaneously, such as by surrounding it with a liquid under high pressure (sometimes called "hydrostatic loading"). In this case it is possible to have compressions in all dimensions, and it is possible to achieve $\delta \sim 0.1$ or more. The increase in elastic energy is of order 0.1 eV/atom, comparable to differences between chemical bond energies and differences in energies of different crystal structures.

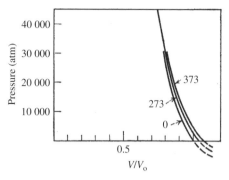

Figure 8.2 "Isothermals" for metallic sodium, giving pressure versus volume at three different temperatures, as labeled in K. Reproduced, with permission after [89].

The challenge is that such compressions require pressures in excess of 20 GPa (2×10^5 atm) for most inorganic materials.

Figure 8.2 is a plot of pressure versus volume, in this case for metallic sodium. Familiar pressures lie low on the y-axis, making the curves seem like vertical lines at familiar pressures of hundreds of atmospheres. In more detail, the slopes of the curves are seen to be steeper with pressure, since the bulk modulus increases with compression. Effects of temperature are also seen – as expected, the solid expands with temperature, although less so when the pressure is higher. Finally, the three curves are extrapolated to negative pressures, which is possible until the cohesion of the solid is lost.

Box 8.1	A Diamond-Anvil Cell

This cell can put a small sample under a pressure of a million atmospheres or higher ($>$100 GPa). The cell uses two diamonds with the "brilliant cut," common in diamond jewelry. The sharp point of each diamond is removed, making small, flat "culets" of perhaps 200 μm across, and the small sample is mounted between the culets of two diamonds. A modest force between the diamonds gives an enormous pressure on the small area of the sample. These methods demand high mechanical precision, and special attention to properties of the pressure medium (e.g., solid neon), gasket (to seal the cell), and methods to measure the pressure (e.g., pressure-dependent shifts in wavelength of the fluorescence line of ruby).

New techniques for focusing synchrotron x-rays on samples of 10 μm or so are companion developments. Synchrotron diffraction experiments with samples at high pressures are nearly routine today, and every material undergoes a significant change by 100 GPa. In-situ measurements with simultaneous and independent control of pressure, temperature, and magnetic field are becoming more common.

Typical curves of E versus V for sodium metal are shown in Fig. 8.3. The curves in Fig. 8.3a include the elastic energy, but also the phonon energy in the solid, causing a vertical shift with temperature. These curves are approximately parabolic owing to the elastic energy, and have some reduction in curvature as the elastic constants soften with temperature.

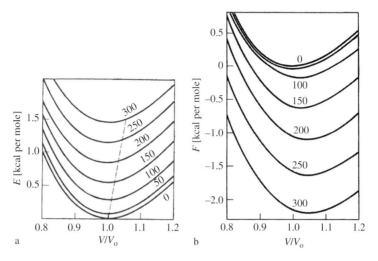

Figure 8.3 (**a**) Energy, E, versus volume for metallic sodium and different temperatures. The dashed line shows the equilibrium volume at the different temperatures in K. (**b**) Free energy, F, versus volume for metallic sodium and different temperatures in K. Curves are offset and shifted by effects of entropy. Reprinted, with permission, after [89].

At elevated temperature, the equilibrium volume of the crystal is not at the minimum of the energy. The equilibrium volume is instead at the minimum of the free energy, $F = E - TS$, which includes the entropy of the solid at elevated temperatures (Fig. 8.3b). Somehow the entropy contributes a "pressure" to expand the solid against its elastic constants, called the "thermal pressure," P_{th}. The dominant source of the entropy is from the phonons. Harmonic phonons do not change their frequency with the volume of a solid, however, so we need to go beyond the harmonic model to understand P_{th}.

8.2 Thermal Pressure, a Step beyond the Harmonic Model

At absolute zero temperature, we expect pressure from only the volume dependence of the internal energy, $E(V)$. At finite temperature, the pressure comes from the volume dependence of the free energy, $F(V)$. From Eq. 1.24

$$P = -\left(\frac{\partial F}{\partial V}\right)_T,$$ (8.14)

$$P = -\left(\frac{\partial E}{\partial V}\right)_T + T\left(\frac{\partial S}{\partial V}\right)_T.$$ (8.15)

The first term in Eq. 8.15 was discussed in Chapter 6 as the elastic energy. It accounts for how chemical bonds in a solid are distorted with changes in volume, increasing the energy, E_{el}. The second term in Eq. 8.15 is the "thermal pressure," P_{th}

$$P_{\text{th}} \equiv T \left(\frac{\partial S}{\partial V} \right)_T. \tag{8.16}$$

The entropy usually depends on volume, so thermal expansion alters the free energy. For a crystalline solid without magnetism or electronic peculiarities, most of this change is in the vibrational entropy. Confining the analysis to high temperatures, the volume dependence of the entropy can be found through Eq. 7.51, where the ith phonon mode contributes the entropy $S_{\omega_i}(T)$

$$S_{\omega_i}(T) = k_{\text{B}} \left[\ln \left(\frac{k_{\text{B}} T}{\hbar \omega_i} \right) + 1 \right]. \tag{8.17}$$

The only parameter in this equation that could depend on V is the phonon frequency. Differentiating the logarithm

$$\left(\frac{\partial S_{\omega_i}(T)}{\partial V} \right)_T = -k_{\text{B}} \frac{1}{\omega_i} \left(\frac{\partial \omega_i}{\partial V} \right)_T, \tag{8.18}$$

$$\left(\frac{\partial S_{\omega_i}(T)}{\partial V} \right)_T = \frac{k_{\text{B}}}{V} \left[-\frac{V}{\omega_i} \left(\frac{\partial \omega_i}{\partial V} \right)_T \right], \tag{8.19}$$

$$\left(\frac{\partial S_{\omega_i}(T)}{\partial V} \right)_T = \frac{k_{\text{B}}}{V} \gamma_i. \tag{8.20}$$

The (dimensionless) mode Grüneisen parameter γ_i was defined as

$$\gamma_i \equiv -\frac{V}{\omega_i} \left(\frac{\partial \omega_i}{\partial V} \right)_T = -\left(\frac{\partial \ln \omega_i}{\partial \ln V} \right)_T. \tag{8.21}$$

The Grüneisen parameter is the fractional change in phonon frequency per fractional change in volume.[1] When multiplied by T, Eq. 8.20 is the thermal pressure contributed by one phonon mode

$$P_{\text{th}_i} = \frac{k_{\text{B}} T}{V} \gamma_i. \tag{8.22}$$

For a real solid with $3N$ phonon modes, we sum the contributions from Eq. 8.22

$$P_{\text{th}} = T \left(\frac{\partial S}{\partial V} \right)_T = \frac{k_{\text{B}} T}{V} \sum_i^{3N} \gamma_i, \tag{8.23}$$

$$P_{\text{th}} V = k_{\text{B}} T \sum_i^{3N} \gamma_i. \tag{8.24}$$

It is common practice to replace the sum in Eq. 8.24 with $3N \overline{\gamma}$, using an average Grüneisen parameter, $\overline{\gamma}$. Doing so gives the simple expression $P_{\text{th}} V = 3 \overline{\gamma} N k_{\text{B}} T$, which has the same form as the ideal gas law of Eq. 8.1. It is tempting to attribute P_{th} to "the pressure of the phonon gas." This is an effective mnemonic, although dishonest as a physical

---

[1] The Grüneisen parameter γ is not a fundamental constant. It differs for different phonons (cf., Fig. 26.7), but often $\gamma \sim 2$.

explanation. The development of Eq. 8.24 is better understood as the "quasiharmonic model" of the free energy, described in Sections 12.3 and 26.3.

8.3 Free Energies and Phase Boundaries under Pressure

8.3.1 Clausius–Clapeyron Equation

At the phase boundary, the Gibbs free energies of two phases α and β are equal

$$G_\alpha(T, P) = G_\beta(T, P). \tag{8.25}$$

Starting at an equilibrium phase boundary, a change in temperature alone will cause the phases to have unequal free energies. However, by adjusting the pressure, we can bring back the equality of Eq. 8.25. We seek the relationship dP/dT that gives the change in pressure required to maintain equilibrium for a change in temperature

$$0 = d\Big[G_\alpha(T, P) - G_\beta(T, P)\Big], \tag{8.26}$$

$$0 = \left(\frac{\partial\Big[G_\alpha(T, P) - G_\beta(T, P)\Big]}{\partial T}\right)_P dT + \left(\frac{\partial\Big[G_\alpha(T, P) - G_\beta(T, P)\Big]}{\partial P}\right)_T dP. \tag{8.27}$$

The partial derivatives[2] of G equal $-S$ (Eq. 1.24), and V

$$0 = -\Big[S_\alpha - S_\beta\Big] dT + \Big[V_\alpha - V_\beta\Big] dP, \tag{8.28}$$

$$0 = -\frac{L}{T} dT + \Big[V_\alpha - V_\beta\Big] dP, \tag{8.29}$$

$$\frac{dP}{dT} = \frac{L}{\Big[V_\alpha - V_\beta\Big] T}, \tag{8.30}$$

where $L = T\Delta S$ is the latent heat of the phase change. Equation 8.30 is the "Clausius–Clapeyron equation," and is handy in high-pressure experiments.[3] It is usually impractical to obtain calorimetric data on heats of phase transformations at high pressures, but the latent heat L can be obtained from phase boundaries and volume differences between the phases. Measuring a phase boundary in a P–T diagram at high pressure is no small effort, however.

8.3.2 Characteristic Pressure of a Solid

Now consider the states of a single phase that have equal free energies at different combinations of T and P. The subscripts α and β are useful to denote two different states of equal free energies, following the previous analysis of the Clausius–Clapeyron equation.

---

[2] Because $dG = \mu dN - S dT + V dP$, for constant N and T, $\partial G/\partial P = +V$.
[3] For second-order phase transitions with $L = 0$, Ehrenfest's equations of Problem 8.7 give dP/dT.

Importantly, we can use our knowledge of phonon entropy to predict dP/dT if we rearrange Eq. 8.28

$$\frac{dP}{dT} = \frac{S_\alpha - S_\beta}{V_\alpha - V_\beta},$$
(8.31)

$$\left(\frac{dP}{dT}\right)_V = \left(\frac{\partial S}{\partial V}\right)_T$$
(8.32)

(which is a Maxwell relationship that we could have invoked immediately). Equation 8.20 gives $(\partial S/\partial V)_T$ for one oscillator mode, and for a solid of N atoms and $3N$ modes

$$\left(\frac{dP}{dT}\right)_V = \frac{k_B}{V} \sum_i^{3N} \gamma_i.$$
(8.33)

It can be useful to approximate $\gamma \simeq 2$, and V is of order $10\,\mathrm{cm}^3$/mole, giving

$$\left(\frac{dP}{dT}\right)_V = \left(\frac{\partial S}{\partial V}\right)_T \sim \frac{5\,\mathrm{MPa}}{\mathrm{K}}.$$
(8.34)

This type of scaling is also typical of how phase boundaries shift with pressure. Approximately, phase diagrams show phase changes over temperature ranges of $1000\,\mathrm{K}$, so pressure ranges of $5\,\mathrm{GPa}$ are typically required to observe similar effects. Hence the magnitude of the value listed in Table 8.1. This also suggests that at pressures of $100\,\mathrm{GPa}$ and above, materials behavior moves into unfamiliar territory. Such extreme energies in materials are not possible with temperature alone because boiling temperatures would be exceeded.

At high pressures, changes in crystal structure are probably the rule, rather than the exception. Sometimes the explanation seems simple. For materials such as iron that have multiple crystal structures of similar energies of formation, pressure favors structures that minimize the volume per atom. For example, bcc iron is driven to hcp iron under a pressure of $10\,\mathrm{GPa}$. Calculating this pressure might be done with an approach explained after Eq. 8.13. Ultimately, however, the bulk modulus B originates at the level of the electrons in the solid. Electronic energies are not always possible to understand with a simple isotropic model.

8.4 Chemical Bonding and Antibonding under Pressure

To delve deeper into why pressure causes phase transitions in materials, consider how pressure alters the energies of electrons and their chemical bonds. We can understand a general feature of chemical bonding at high pressures with the bond integral $h \equiv \langle a|\mathcal{H}|0\rangle$ of Eq. 6.25. The bond integral becomes larger as the atoms are pushed together, so the tails of the wavefunctions on atoms centered at sites 0 and a overlap more strongly in the region of interaction. The energy splitting between the bonding and antibonding levels therefore grows larger with pressure (in addition to shifting upwards in energy). The entire band

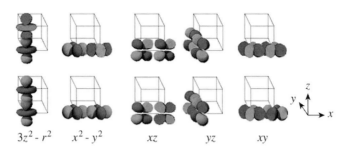

$3z^2 - r^2$ $x^2 - y^2$ xz yz xy

Figure 8.4 The five atomic d-electrons, showing their characteristic symmetries and geometries when centered on a simple cubic lattice. Bonding levels are on the bottom, antibonding on the top. Notice changes of gray for antibonding, as expected when the wavefunction is multiplied by a factor of -1.

of electron states widens in energy with increasing pressure. A common consequence is that insulators become metals at high pressures. This even occurs with some molecular solids, where molecular identity is lost as the atoms of separate molecules approach each other closely. For example, oxygen becomes a crystalline metal near 100 GPa, where it can no longer be considered as O_2 molecules [90, 91], and hydrogen becomes metallic near 500 GPa [92].

The individual electron energy levels respond differently to pressure. For an example, consider d-electrons, which have five characteristic shapes in an isolated atom as shown in Fig. 8.4 (with labels at the bottom). The atomic orbitals are placed at neighboring sites on a simple cubic lattice. Figure 8.4 shows how the lobes between atomic d-wavefunctions overlap each other to form bonding or antibonding orbitals. Some items to note in Fig. 8.4 are:

- The cube axes lie along the directions $\hat{x}$, $\hat{y}$, $\hat{z}$ that are used to define the orientation dependence shown at the bottom of the figure. For example, the wavefunction $x^2 - y^2$ has its maxima and lobes along the x-axis and y-axis, whereas the wavefunction xz has lobes along the diagonal of the x–z-plane.
- The two cases at left show strong overlaps of the lobes of the wavefunctions, which point directly at each other. The three cases at right have lobes along diagonal directions, so the overlap is not so strong. The orbitals shown in the two cases at left are called e_g, and t_{2g} orbitals are in the three cases at right.
- About the axis of the bond, the e_g states at left have a continuous density of overlap, whereas the t_{2g} states at right have one nodal plane that contains the bond axis. The former is a σ-bond, and the latter is a π-bond.
- The bonding cases are shown at the bottom of Fig. 8.4. The light gray overlaps other light gray especially strongly for the two cases at left (the σ-bonds of the e_g orbitals). This boosts the electron density between the atoms.
- The antibonding cases are shown at the top. The light gray overlaps the neighboring dark gray, meaning that a negative lobe of one wavefunction overlaps a positive lobe of its neighbor. This causes a node in the wavefunction perpendicular to the bond axis, between the atom centers, a characteristic of antibonding states.

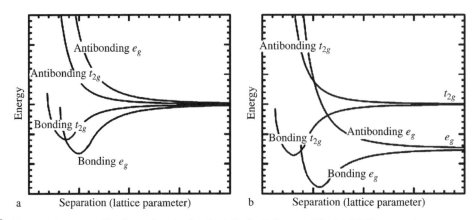

Figure 8.5 (**a**) Expected energies of bonding and antibonding levels for the d-electrons of Fig. 8.4. (**b**) More general case, with unequal energies for atomic levels.

The effect of pressure on the bonding can now be understood intuitively. If we compress the cubes in Fig. 8.4, the two e_g cases at left will overlap much more strongly than will the three t_{2g} at right, whose lobes are less direct in their overlap. For a simple cubic lattice, the e_g bonding is more sensitive to the decrease in volume, which causes a more rapid increase in the energy difference between the bonding and antibonding states. The energy change with volume is smaller for the t_{2g} states at right. The energies of the different levels are shown in Fig. 8.5a. When the atoms are far apart (large lattice parameter), there is no overlap of the lobes of the wavefunctions on different sites. As volume decreases, the bonding levels split away from the antibonding, and the effect is stronger for the e_g σ-bonds than the t_{2g} π-bonds.

Although we have justified why we might expect different electron levels in a solid to respond differently to pressure, our picture is in need of several refinements. First, the different levels need not start at the same energies at large separations. (This is more obvious when different electron states, e.g., $3p$ and $3d$, are involved.) This can lead to crossings of levels as shown in Fig. 8.5b. A level crossing may induce a change of crystal structure, since bonds have an orientational dependence that favors some near-neighbor atom configurations over others.

When we consider the spread of electron levels into bands, the widths of the different energy bands and their center shifts will change with pressure, usually differently for the different energy bands. Perhaps the simplest way to extend the argument of this section is to consider the curves in Fig. 8.5 to have some vertical spread in energy, spanning from the bonding curve to the antibonding curve. This energy spread increases with pressure. The details of the level crossings in Fig. 8.5b will be altered, of course, and detailed calculations of electron energies are needed to assess the effects. Fortunately, this can be done fairly directly with modern density functional theory (DFT) codes. Changing the lattice parameter and recalculating the electron energy levels is not a particularly difficult thing to do, once the DFT machinery is up and running.

Nevertheless, the geometrical approach of the present section often proves valuable for understanding changes in bonding under pressure. The example worked out with Fig. 8.4 placed d-orbitals on a simple cubic lattice. The different problem of d-orbitals on a face-centered cubic (fcc) lattice can be worked out the same way. In the fcc structure, with neighbors on the face centers, the diagonal lobes of the t_{2g} orbitals make σ-bonds, and these bonds are more sensitive to pressure than are the e_g orbitals on the fcc crystal. Examples with other crystal structures, or with other orbitals (such as p-electrons) can be understood similarly. It may be necessary to prepare the atomic orbitals by hybridization, such as forming sp^3 hybrids from the s- and p-orbitals for the tetrahedral coordinations of diamond cubic silicon or carbon.

8.5 Pressure-Driven Phase Transitions

8.5.1 Effects of Pressure on Free Energy

Section 8.4 showed how electronic states formed with atomic orbitals change energy with changes in volume. The idea can be extended beyond the use of atomic orbitals. The key assumption is that the electronic states $\{j\}$ change with volume, but with individual volume sensitivities $\{\Upsilon_j\}$

$$\Upsilon_j \equiv -\frac{V}{\epsilon_j}\frac{\partial \epsilon_j}{\partial V}, \tag{8.35}$$

which has the form of the Grüneisen parameter in Eq. 8.21, but is for electronic energies $\{\epsilon_j\}$, not phonon energies $\{\varepsilon_i\}$.

The free energy changes with V as

$$F(V + \Delta V) = E - TS + \left(\frac{\partial F}{\partial V}\right)_T \Delta V. \tag{8.36}$$

The partial derivative on the RHS is recognized as $-P$ from Eq. 1.24, but P originates from the energy and entropy (cf., Eq. 8.15)

$$F(V + \Delta V) = E - TS + \left(\frac{\partial E}{\partial V}\right)_T \Delta V - T\left(\frac{\partial S}{\partial V}\right)_T \Delta V. \tag{8.37}$$

The change in volume alters the electronic energy $E(V)$ through the electronic energies $\{\epsilon_j\}$, and the entropy $S(V)$ through the phonon energies $\{\varepsilon_i\}$. From Eqs. 8.23 and 8.35

$$\left(\frac{\partial E}{\partial V}\right)_T = -\frac{1}{V}\sum_j^n \epsilon_j \Upsilon_j, \tag{8.38}$$

$$\left(\frac{\partial S}{\partial V}\right)_T = \frac{k_B}{V}\sum_i^{3N} \gamma_i, \tag{8.39}$$

where the entropy is in the classical limit. We have assumed n relevant, occupied electron states, and $3N$ normal modes in the material

$$F(V + \Delta V) = E(V) - TS(V) - \frac{\Delta V}{V}\left(\sum_j^n \epsilon_j \Upsilon_j + k_{\mathrm{B}}T\sum_i^{3N} \gamma_i\right). \tag{8.40}$$

In general, the average $\overline{\Upsilon}$ and $\overline{\gamma}$ are positive, consistent with the expected increase in free energy with a decrease in volume. Many electron states and phonon modes are altered differently by volume, however. Realistic calculations require a high level of detail about the material.

8.5.2 Two-Level System under Pressure

For a workable example, we consider equilibrium in a two-level system subjected to increasing pressure at a fixed temperature. Assume the two electron energy levels (denoted 1 and 2) change linearly with volume, although with different rates

$$\epsilon_1 = \epsilon_1^0 - \Upsilon_1 \epsilon_1^0 \frac{\Delta V}{V}, \qquad \epsilon_2 = \epsilon_2^0 - \Upsilon_2 \epsilon_2^0 \frac{\Delta V}{V}, \tag{8.41}$$

$$S_1 = S_1^0 + 3k_{\mathrm{B}}\,\gamma_1 \frac{\Delta V}{V}, \qquad S_2 = S_2^0 + 3k_{\mathrm{B}}\,\gamma_2 \frac{\Delta V}{V}, \tag{8.42}$$

and the factor of 3 accommodates three phonon modes for each state of the system. Making definitions to simplify appearances

$$\Delta\epsilon \equiv \epsilon_1 - \epsilon_2, \qquad\qquad \Delta S \equiv S_1 - S_2, \tag{8.43}$$

$$\Delta\epsilon^0 \equiv \epsilon_1^0 - \epsilon_2^0, \qquad\qquad \Delta S^0 \equiv \Delta S_1^0 - \Delta S_2^0, \tag{8.44}$$

$$\delta_{\mathrm{el}} \equiv \frac{1}{B}\left(\Upsilon_1 \epsilon_1^0 - \Upsilon_2 \epsilon_2^0\right), \qquad \delta_{\mathrm{ph}} \equiv \frac{3k_{\mathrm{B}}}{B}\left(\gamma_1 - \gamma_2\right). \tag{8.45}$$

Recognizing that $-P/B = \Delta V/V$, the differences of Eqs. 8.43 are

$$\Delta\epsilon = \Delta\epsilon^0 + P\delta_{\mathrm{el}}, \qquad \Delta S = \Delta S^0 + P\delta_{\mathrm{ph}}. \tag{8.46}$$

At a fixed temperature, pressure sets the fractions f_1 and f_2 of the two levels as the ratio of their Boltzmann factors

$$\frac{f_1}{f_2} = \frac{\mathrm{e}^{-F_1/kT}}{\mathrm{e}^{-F_2/kT}} = \mathrm{e}^{-\Delta\epsilon/kT}\,\mathrm{e}^{\Delta S/k}, \tag{8.47}$$

$$\frac{f_1}{f_2} = \mathrm{e}^{-\Delta\epsilon^0/kT}\,\mathrm{e}^{\Delta S^0/k}\,\mathrm{e}^{-P\left(\delta_{\mathrm{ph}} + \delta_{\mathrm{el}}/k_{\mathrm{B}}T\right)}. \tag{8.48}$$

Pressure favors the level with the smaller Grüneisen parameter. The vibrational modes for the atom with smaller γ do not stiffen so strongly under pressure, so its vibrational entropy remains larger than the other.

Effects of pressure on both phonon frequencies and electronic levels are shown in Fig. 8.6, which uses one parameter, δ, for the exponential in Eq. 8.48 as $P\delta = P\left(\delta_{\mathrm{ph}} + \delta_{\mathrm{el}}/k_{\mathrm{B}}T\right)$. For these examples, a constant $k_{\mathrm{B}}T$ was assumed to be relatively small

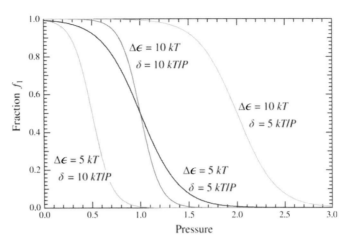

Figure 8.6   Fraction of the low-pressure level for the two-level model at different temperatures and pressures. Parameters $\Delta\epsilon^0$ and $\delta = \delta_{ph} + \delta_{el}/k_B T$ are as labeled on the curves.

compared with the other energies (although not overwhelmingly so, or the examples would be near $T = 0$). The parameters were set so $\Upsilon_1 > \Upsilon_2$, and the low-pressure level becomes energetically unfavorable with increasing pressure.[4] If there were no effect of temperature, the transition would be abrupt, and $f_1(P)$ would be a step function falling from 1 to 0 at the single pressure where the levels cross.

For the same differences in energy of the two levels at $P = 0$ (same $\Delta\epsilon^0 = \epsilon_1^0 - \epsilon_2^0$), Fig. 8.6 shows that a higher temperature broadens the pressure range of the transition. With a larger $\Delta\epsilon^0$, a larger δ is needed to get a transition at the same pressure. In addition, for large $\Delta\epsilon^0$ and large δ, the transition occurs over a narrower range of pressure (compare the two middle curves).

Now consider an average property of this two-level system, such as an equation of state $V(P)$, taken as the average of properties of the two levels. Here the two levels cause an atom to have two different specific volumes

$$V(P) = f_1 V_1(P) + f_2 V_2(P). \tag{8.49}$$

The fractions of the two levels vary with pressure according to Eq. 8.48. With the normalization $f_1 + f_2 = 1$, and a little manipulation

$$V(P) = \frac{1}{1 + e^{\Delta F/kT}} \left[V_1(P) + e^{\Delta F/kT} V_2(P) \right]. \tag{8.50}$$

This two-level model can be applied directly in the Weiss theory of Invar, a material with zero thermal expansion. In the Weiss model, the loss of magnetic moment with temperature causes the magnetic atoms to have a smaller volume. Through the magnetic transition, this contraction counteracts the normal thermal expansion. Although the Weiss theory is phenomenological, Eq. 8.50 can be used to model Invar behavior that is driven by either

---

[4] The alternative, where the low-pressure level becomes more favorable with pressure, has no phase transition under pressure.

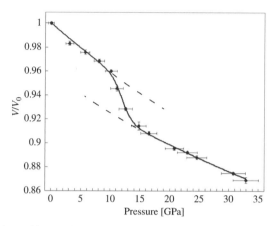

Figure 8.7   Volume–pressure data obtained from synchrotron X-ray diffraction measurements on Pd_3Fe under pressure (symbols), and the fit of the data to the Weiss-like equation of state of Eq. 8.50 with different $V_1(P)$ and $V_2(P)$ (line). Reproduced, with permission, from [93]. Copyright (2009) by the American Physical Society.

temperature or pressure. One way to interpret this result is to graph two equations of state $V_1(P)$ and $V_2(P)$. Start on the low-pressure curve $V_1(P)$, and make a smooth transition to $V_2(P)$ as predicted by the fractional change shown in Fig. 8.6. An example is shown in Fig. 8.7.

8.5.3 Simultaneous High P and T

In extreme environments, such as in the Earth at depths of the mantle or below, both temperature and pressure are high enough to be thermodynamically important. We have already shown how we expect contributions to the free energy that scale with ΔP, and others that scale with ΔT. There is another issue that is not evident from Eq. 8.37 for $F(V, T)$, however. A high T can alter the dependence of F on V. Likewise, a high P can alter the dependence of F on T. So, in addition to a contribution to F proportional to ΔV and another proportional to ΔT, we expect a new contribution proportional to the product $\Delta V \times \Delta T$.

1. Equation 8.39 took $\partial S/\partial V$ at fixed T, but over a range of temperatures the Grüneisen parameters $\gamma \propto \partial \omega/\partial V$ are likely to change. This temperature dependence is proportional to

$$\frac{\partial}{\partial T}\gamma \propto \frac{\partial}{\partial T}\left(\frac{\partial \omega}{\partial V}\right) = \frac{\partial^2 \omega}{\partial T\, \partial V}. \tag{8.51}$$

2. For the T-dependence of F, we could include a T-dependence of E and S in Eq. 8.37. Not all of this comes from thermal expansion. As discussed in Chapter 26 (online), phonon anharmonicity causes S to have a pure T-dependence that exists even without thermal expansion. This anharmonicity, $A(T) \propto \partial \omega/\partial T$, is expected to change with V, giving a volume dependence proportional to

$$\frac{\partial}{\partial V}A \propto \frac{\partial}{\partial V}\left(\frac{\partial \omega}{\partial T}\right) = \frac{\partial^2 \omega}{\partial V\,\partial T}. \qquad (8.52)$$

These effects 1 and 2 must be the same because $\partial^2 \omega/(\partial T\,\partial V) = \partial^2 \omega/(\partial V\,\partial T)$. The result is a contribution to the free energy that depends on the product of $\Delta V \times \Delta T$. There are cases where the term $\Delta F \propto \Delta P\,\Delta T\,\partial^2 \omega/(\partial V\,\partial T)$ is large when ΔP and ΔT are large, but the scale of this is not well understood today [94]. The coupling between T and P cannot be predicted from the variations of either one of them alone.

8.6 Activation Volume

In activated state rate theory, the jump frequency depends on temperature as

$$\Gamma(T) = \Gamma_0\,e^{-\Delta E/k_B T}, \qquad (8.53)$$

where ΔE is an activation energy. Section 5.6.1 explained the origin of this expression (cf., Eq. 5.27). The basic idea is that there is an equilibrium between the intitial state and the "activated state," or transition state, so the probabilities of these states are in the ratio of their Boltzmann factors (giving Eq. 8.53). This condition of equilibrium between the initial and activated states can be generalized for equilibrium in a system under a constant pressure and temperature. The probability ratio is set by a ratio of Gibbs factors, and the transition rate is

$$\Gamma(P, T) = \Gamma_0\,e^{-\Delta G/k_B T}, \qquad (8.54)$$

where ΔG is an activation free energy that includes the ΔE, and now includes a pressure term with ΔV as an "activation volume"

$$\Gamma(P, T) = \Gamma_0\,\exp\left(-\frac{\Delta E + P\Delta V}{k_B T}\right). \qquad (8.55)$$

An activation volume ΔV is analogous to an activation energy:

• The activation energy is the change in energy of the system that must be produced by a thermal fluctuation if the jump process is to occur.
• An activation volume is the change in volume of the system that must be produced by a thermal fluctuation if the jump process is to occur.

An activation volume gives insight into how atoms are rearranged during a kinetic process. For atomic diffusion, for example, activation volumes are positive because surrounding atoms must open a passage for the diffusing atom. The ΔV is comparable to an atomic volume. An applied pressure suppresses diffusion by the factor $\exp(-P\,\Delta V/k_B T)$.

Performing measurements of jump frequencies under pressure is a way to obtain information about how atoms are rearranged during the transient state of a jump. It is typical to interpret experimental data on activation energies with Eq. 8.53 rearranged as

$$\ln\left(\Gamma(T)\right) = \ln\left(\Gamma_0\right) - \frac{\Delta E}{k_B T}. \tag{8.56}$$

A graph of $\ln\left(\Gamma(T)\right)$ versus $1/k_B T$ gives a slope of $-\Delta E$, giving the activation energy. A similar rearrangement of Eq. 8.55 gives, for a fixed T,

$$\ln\left(\Gamma(P)\right) = \ln\left(\Gamma_0\right) - \frac{\Delta E}{k_B T} - \frac{P\Delta V}{k_B T}. \tag{8.57}$$

For a fixed T, the second term on the RHS is a constant when $\ln\left(\Gamma(P)\right)$ is plotted against $P/k_B T$. This graph gives a slope of $-\Delta V$, giving the activation volume.

In cases where ΔV is small, an improved analysis should consider how the prefactor depends on pressure. This is expected if the prefactor is proportional to a phonon frequency, and the phonon frequencies increase under pressure. An analysis gives

$$\left(\frac{\partial \Gamma(P,T)}{\partial P}\right)_T = -\left(\frac{\gamma}{B} + \frac{\Delta V}{k_B T}\right)\Gamma(P,T), \tag{8.58}$$

where γ is a Grüneisen parameter (Eq. 8.21), and B is the bulk modulus. It is often true that the effective γ is not known. In this case, measurements at one temperature cannot determine separately the two terms in parentheses in Eq. 8.58. Measurements at multiple temperatures and pressures can help sort out the different contributions.

Problems

8.1 The melting temperature of ice depends on pressure. The heat of fusion of ice at atmospheric pressure is 330 kJ/kg. The heat capacity of ice is 2.05 kJ/(kg K). The specific volume of ice is greater than that of liquid water by 9×10^{-5} m^3/kg. Thermal expansion and compressibility constants for ice are $V^{-1}\left(\partial V/\partial T\right)_P = 1.6 \times 10^{-4}$ /K, and $-V^{-1}\left(\partial V/\partial P\right)_T = 1.2 \times 10^{-10}$ m^2/N.

(a) Ice is compressed isothermally at $-2\,^\circ$C, starting at atmospheric pressure. At what pressure will the ice melt?

(b) Ice at $-2\,^\circ$C, starting at atmospheric pressure, is heated in a container that maintains constant volume. At what temperature will the ice melt?

(c) Ice at $-2\,^\circ$C, starting at atmospheric pressure, is compressed adiabatically in a container with good thermal insulation. At what pressure will the ice melt?

8.2 A tall, vertical column of a material is kept at constant temperature at all heights. When the temperature is T_0, the material below a height l in the column is solid, but the material above l is liquid. When the temperature is increased to $T_0 + 0.2$ K, the solid–liquid interface shifts upwards by 0.5 m. The heat of fusion is 10 kJ/mole, and the density of the liquid is 1000 kg/m^3.

What is the density of the solid near the interface?

8.3 Under pressure, fcc γ-cerium undergoes a volume collapse to fcc α-cerium at room temperature (see Fig. 1.3). One explanation of this effect includes an electronic transition, where one $4f$ electron at each cerium atom transfers to a free electron gas (in more detail, the gas may be a nearly free electron band composed of $6s$ electrons, but please assume a simple electron gas for this problem). Suppose that the bonding between the $4f$ electrons has an interatomic potential of a Lennard-Jones potential.

(**a**) Explain why this electronic transition might occur with decreasing volume.

(**b**) At an intermediate volume, there are two coexisting fcc phases of cerium with distinct lattice parameters. Why would there be a two-phase mixture of α- and γ-cerium instead of just one of them of intermediate lattice parameter?

8.4 Following the approach of Section 8.4, graph how the bonding and antibonding orbitals for the five types of d-electrons change with volume in the

(**a**) fcc structure;

(**b**) bcc structure. (Hybridize the three t_{2g} to obtain amplitude along the $\langle 111 \rangle$ directions. The two e_g already have lobes along the $\langle 100 \rangle$ directions, but they could be hybridized, too.)

You may assume that all levels have equal energy at large volumes.

8.5 (**a**) Explain why the parameter for the shift of electron energy with volume, Υ, alters thermal expansion. How does the effect depend on the sign of the applied pressure?

(**b**) Calculate the changes with thermal expansion of the energy of an electronic level and the elastic energy to obtain the increment to the linear coefficient of thermal expansion, $\Delta \alpha = \Upsilon/(3BVT)$.

(*Hint*: See the approach of Eqs. 26.34–26.37.)

8.6 Differentiate the fundamental relationship $E(S, V, N)$ to obtain dE, and using the definitions

$$\left(\frac{\partial E}{\partial S}\right)_{N,V} = T, \quad \left(\frac{\partial E}{\partial N}\right)_{S,V} = \mu, \quad \left(\frac{\partial E}{\partial V}\right)_{S,N} = -P, \tag{8.59}$$

make use of the equality of the mixed partial derivatives

$$\frac{\partial^2 E}{\partial V\, \partial S} = \frac{\partial^2 E}{\partial S\, \partial V}, \tag{8.60}$$

to derive the Maxwell relationship

$$\left(\frac{\partial T}{\partial V}\right)_{S,N} = -\left(\frac{\partial P}{\partial S}\right)_{V,N}. \tag{8.61}$$

8.7 The Clausius–Clapeyron Eq. 8.30 is appropriate for finding phase boundaries of first-order transitions that have a latent heat, L, but second-order phase transitions are continuous and have no latent heat. The slopes of P–T phase boundaries for second-order transitions are given by "Ehrenfest's equations"

$$\frac{\mathrm{d}P}{\mathrm{d}T} = \frac{C_{P\beta} - C_{P\alpha}}{VT\,(\beta_\beta - \beta_\alpha)}, \tag{8.62}$$

$$\frac{\mathrm{d}P}{\mathrm{d}T} = \frac{\beta_\beta - \beta_\alpha}{\kappa_\beta - \kappa_\alpha}, \tag{8.63}$$

where β is the coefficient of thermal expansion, κ is the isothermal compressibility (inverse of B_T), and C_P is the heat capacity at constant pressure.

Derive these two equations by seeking equality of *slopes* of the free energies of the α and β phases.

(*Hint*: You can save a few steps by taking partial derivatives of Eq. 8.28, first with respect to T and then with respect to P.)

9

Interactions in Microstructures and Constrained Equilibrium

Physical phenomena that occur at different scales of space and time can often be separated, simplifying the analysis into independent calculations. For example, mixing the orbital coordinates of the Moon around the Earth with the heating of water through tidal forces usually results in two separate problems. An analysis of how tidal flows alter the temperature of the oceans can be developed without many details of celestial mechanics.

Interactions between different physical processes often make rich contributions to phase transformations in materials, and are central to many discussions in Part III. The slow kinetics of one physical process can alter the thermodynamics of another process, confining it to a "constrained equilibrium." The facile state variables are constrained by sluggish variables, and the phase transformation does not take the most direct path to minimizing the free energy. The material can also become trapped in a local minimum of free energy, called a "metastable" (pronounced "meta-stable") state. Diffusion and nucleation can be sluggish in solids, so constrained equilibrium is commonplace.

A first example is the formation of a nonequilibrium phase, a glass, which we approach with the simplest assumption that some state variables remain constant, while others relax towards equilibrium (this approach was used several times in Chapter 5). This separability of physical processes, with its "divide and conquer" strategy, does not always give the full story. The work done on the oceans by tidal forces dissipates some of the kinetic energy of lunar rotation. This causes the orbit of the Moon to increase by more than 0.1 mm/day, thus reducing the tidal forces in geologic time. Section 9.6 builds an example of hydride phases in metallic alloys, where interstitial hydrogen atoms move much faster than substitutional metal atoms. Over longer times, the interactions between metal atoms and hydrogen atoms change the equilibrium for both. A general form for the free energy of a chemical unmixing transformation is developed for these coupled state variables. Coherency stresses in a microstructure are another source of coupling between state variables.

Sometimes "self-trapping" occurs, when the slowing of a facile variable enables the relaxation of a slower second variable coupled to it, and this relaxation impedes changes of the facile variable. Examples of self-trapping include the diffusion of atoms to locations where they relieve elastic energy, trapping the elastic distortion. The "small polaron" is described, in which an electron becomes trapped at an ion after the neighboring atoms nestle into relaxed positions around it.

There has been much recent interest in multiferroic materials, which have internal degrees of freedom for changing their structure, magnetization, or electric polarization. This chapter develops thermodynamic relationships between these different degrees of freedom, with the goal of moving the problem away from an analysis of complex material responses to intensive thermodynamic variables, and towards the extensive variables that are closer to the atoms and electrons in the material. The chapter concludes by addressing more deeply the meaning of "separability," showing some formal thermodynamic consequences.

9.1 Solid-State Amorphization

Figure 9.1 shows free energy versus composition curves for three solid phases and one liquid phase. Two crystalline phases α and β are rich in A- and B-atoms, respectively. The third crystalline phase, the γ-phase, is typically an ordered compound of A- and B-atoms. Many such compounds have large unit cells, requiring some time for nucleation and growth, so their formation can be suppressed by rapid cooling. The simplest way to impose a constraint on the state of equilibrium for rapid cooling is by assuming that the γ-phase cannot form, and erasing its free energy versus composition curve from Fig. 9.1a. Section 5.5.1 discussed how this constraint on the thermodynamics can make an amorphous phase favorable when F_L lies below the common tangent of F_α and F_β in Fig. 9.1a.

There is a more surprising effect if the intermetallic phase is suppressed upon heating, because this can cause crystalline materials to become amorphous. Suppose that crystalline phases of A-rich α-phase and B-rich β-phase are brought into contact at low temperatures, and atoms of A and B are allowed to interdiffuse across the interface. At low temperatures, an important consideration is that there are often large differences in the mobilities of A- and B-atoms. As an extreme case, suppose that only the A-atoms are diffusively mobile in the solid phases. The formation of the γ-phase, with its large unit cell, requires

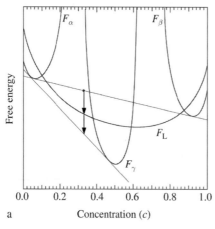
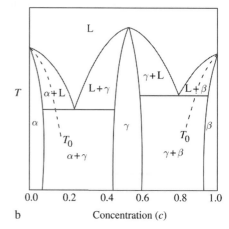

a   Concentration (c)   b   Concentration (c)

Figure 9.1   (**a**) Free energy versus composition curves, similar to those of Fig. 5.8, but with a curve for an additional γ-phase. Arrows indicate reductions in free energy that are possible from a starting mixture of α- and β-phases for overall composition $c_0 = 0.33$. (**b**) Equilibrium phase diagram for the four curves of part a. Two T_0 lines are sketched.

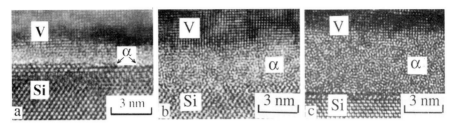

Figure 9.2 High-resolution TEM images of cross-sections of interfaces between V on a (001) Si surface after annealing at 440 °C for (**a**) 0 min, (**b**) 15 min, (**c**) 45 min, showing growth of the amorphous (α) layer. From L.J. Chen *et al.* [97].

the coordinated movements of B-atoms. If these movements are suppressed kinetically, nucleation of the γ-phase is suppressed. With no γ-phase, the free energy may favor the next-best phase, the liquid phase, as indicated by the top arrow in Fig. 9.1a.[1] This process of "solid-state amorphization" produces an amorphous phase at the interface between the crystalline α- and β-phases. Solid-state amorphization of metals was first reported in 1983 [95], but many examples have since been found [96]. In the time required for the amorphous layer to grow to macroscopic dimensions, however, the γ-phase usually nucleates and then grows quickly, consuming the amorphous layer. Early stages of solid-state amorphization between a layer of V deposited on Si are shown in Fig. 9.2.

9.2 Self-Trapping

9.2.1 Examples of Self-Trapping

A cooperative interaction occurs when a localized defect alters its immediate environment, and, in turn, the environment changes the energy of the defect. This interaction lowers the free energy of the system, stabilizing the defect in its relaxed configuration. By lowering the free energy, this local interaction then changes the defect mobility, causing the defect to become "self-trapped." An analogy is a very soft mattress that compresses slowly. It is easy to get on and off the mattress quickly, but by lingering you sink deeply, and exit is more difficult. Two timescales are involved. It is typical to consider first the effects of the slow relaxation that reduce the energy of the defect. If the defect is highly mobile, as it may be at high temperatures, the relaxation is avoided, and the defect remains mobile. With decreasing temperature, the defect slows, and relaxation can be large enough to trap it. Some "self-trapping" phenomena are:

• Gorsky realized that large solute atoms should diffuse away from regions of material that are in compression, moving to regions in tension [98]. This "Gorsky effect" also occurs for interstitial atoms such as hydrogen, which expand the unit cell and can diffuse

---

[1] A tricky issue in this analysis is the definition of the initial composition c_0, because the reaction layer can extend to various distances in the α- and β-phases below their interface.

across a bar of metal when it is bent elastically. The hydrogen atoms move into regions of tension, reducing the elastic stress over the timescale for diffusion.

- In a unit cell of bcc iron, an interstitial carbon atom causes a tetragonal distortion along one of the Cartesian axes. A single jump of a carbon atom can switch the unit cell between tetragonal variants along $\hat{x}$, $\hat{y}$, or $\hat{z}$. Changing the direction of an applied stress induces carbon atoms to move into new variants. Near room temperature, the jump frequency of carbon atoms is comparable to the mechanical resonant frequency of a steel tuning fork. Its vibrations are damped by the dissipation of energy into the activation of carbon atom jumps ($\Delta G_{m, i}$ in Eq. 3.1), giving a peak in damping versus temperature. This is the "Snoek effect" [99].

- In steels, carbon atoms can be attracted to the strain fields of dislocations. A "Cottrell atmosphere" of carbon atoms can form around the core of a dislocation [100, 101], much like the solute segregation to dislocations shown in Fig. 14.6. The stress–strain curves of bcc carbon steels often have an "upper yield point" in stress that drops abruptly (to a "lower yield point") when dislocations break free from their Cottrell atmospheres, and plastic deformation begins.

- Section 15.3.2 explains the slowing of grain growth by "solute drag." When solute atoms are favorably situated in grain boundaries, grain boundaries become more difficult to move, if they move slowly enough for the solutes to keep up.

- Not all interactions between solutes, defects, and elastic fields are understood as self-trapping phenomena. The pinning of dislocations or magnetic flux lines by hetero-geneities is expected to occur without changes to the pinning sites that make the whole system more favorable; at least this is a common assumption today.

9.2.2 Polarons

Polaron Formation

A classic example of a self-trapped defect is a "small polaron" in an ionic crystal. An electron can often move quickly from atom to atom, with residence times shorter than the periods of atom vibrations. Trapping occurs when this residence time is longer, and the energy of the electron on an atom is lowered because the neighboring atoms have time to reconfigure a bit. For example, consider a set of crystallographically similar sites for iron ions in an oxide crystal. Suppose most of the iron atoms are Fe^{3+}, with a few that are Fe^{2+}. The extra electron on the Fe^{2+} cation alters its interactions with neighboring O^{2-} anions, with an equilibrium bond length between Fe^{2+} and O^{2-} that is typically a small percentage longer than for Fe^{3+} and O^{2-}, which has a stronger electrostatic attraction. If the electron were to jump instantaneously from an Fe^{2+} to a nearby Fe^{3+}, the local ion distance would be in error by a small percentage. This would cause an extremely large strain and a large penalty in elastic energy, so the electron cannot move this way. Instead, the motion of the electron is accompanied by O^{2-} anion displacements. The composite entity of electron plus distortion field is called an "electron polaron."

A simple model of a polaron, using ideas of Holstein [102, 103], considers a crystal made of local molecules that have variable interatomic spacings, x. A molecule has an elastic energy, E_{elas}, and an electronic energy, E_{elec}, that both depend on x

$$E_{elas} = \frac{1}{2}\kappa x^2, \tag{9.1}$$

$$E_{elec} = -Ax, \tag{9.2}$$

where κ is a spring constant and A is a force. The parameter A couples the electronic energy to the local distortion.[2] Minimizing the total energy, E_{tot}, gives the equilibrium value of x_0 and a favorable formation energy of the polaron, $E_f = E_{tot}(x_0)$

$$E_{tot} = E_{elas} + E_{elec} = \frac{1}{2}\kappa x^2 - Ax, \tag{9.3}$$

$$\frac{dE_{tot}}{dx} = 0 = \kappa x - A, \tag{9.4}$$

$$x_0 = \frac{A}{\kappa}, \tag{9.5}$$

$$E_f = -\frac{A^2}{2\kappa}. \tag{9.6}$$

Polaron Hopping

Polarons move "adiabatically," with thermal activation, by hopping between neighboring atoms. When a large but improbable thermal fluctuation moves the neighboring O^{2-} anions further from a Fe^{3+} and closer to a nearby Fe^{2+}, the electron can move to a matched energy level at the neighboring site of the Fe^{3+} ion. The thermal jump rate is expected to be [104, 105]

$$\Gamma(T) = \nu_0\, e^{-2\alpha R_n}\, e^{-G/k_B T}, \tag{9.7}$$

where ν_0 is an attempt frequency that is set by local atom vibrations, R_n is the distance between initial and final sites of the polaron, α is an inverse localization length, and G is the activation free energy. The inverse localization length α accounts for overlaps of the tails of the electron wavefunctions at the two sites. With greater overlap, there is a greater probability for the electron to jump between the sites.

The activation free energy, G, is not well understood today, because the transition state involves rearrangements of the local ions and delocalization of the electron, as depicted in Fig. 9.3b. Nevertheless, activation energies measured experimentally are large, typically a few hundred meV, and are a good fraction of the formation energy, E_f. The Boltzmann factor, $\exp(-G/k_B T)$, therefore causes the electrical conductivity by polarons to increase rapidly with temperature. This is quite the opposite of conductivity by band electrons in crystalline metals, where thermal atom displacements scatter electrons from their states, decreasing the electrical conductivity with temperature.

---

[2] It describes an "electron–phonon interaction," which is developed more thoroughly in Sections 26.5, 28.4.1, and 28.4.2.

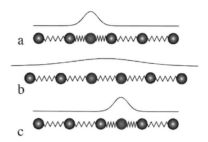

Figure 9.3 Hopping of a polaron from configuration in a, to configuration in c. To minimize the energy of the transition state in b, there is typically some delocalization of the electron (or hole) and delocalization of the elastic distortions.

9.3 Thermodynamics of Complex Materials

9.3.1 Multiferroic and Complex Materials

Many materials show "complex" responses to applied fields such as magnetic, electric, or stress. What is complex is how one applied field alters properties usually associated with a different field. For example, an applied magnetic field can alter the volume of a material, even though changes in volume are usually associated with an applied stress. Sometimes these cross-effects are large. The name "multiferroic," has been coined for a material that exhibits more than one of these properties: "ferroelectricity," "ferromagnetism," and "ferroelasticity" [106]. The prefix "ferro" means iron, but here it is used in reference to the magnetism of iron, specifically how iron develops a permanent dipole moment at ambient conditions (see Section 21.3). "Ferroelectric" means that the material sustains an internal electric polarization analogous to the aligned magnetic moments in ferromagnetism (Section 21.8). Ferroelastic materials transform to different variants of their crystal structure in response to an applied stress (Section 19.2 covers martensitic transformations). Some ferroelastic materials can recover an original shape by undergoing a phase transition with temperature, and some of these are "shape-memory" materials.[3]

9.3.2 Thermodynamic Relationships

Heckmann's Diagram

Figure 9.4 is Heckmann's diagram from a century ago [107, 108], slightly extended. Along the thick lines are four primary properties of materials: heat capacity, elasticity, electric permittivity, and magnetic permeability. These primary properties are links between factors in an extended Gibbs free energy

---

[3] An example of a multiferroic material could be a material that changes its volume in response to an applied magnetic field, and could be useful in an actuator or sensor. Additional properties such as electrical or thermal conductivity do not fit so easily into this scheme, but these may also be altered by multiple external fields. In these cases, the material may even be called "smart." Our focus is not so much on the interrelated engineering properties of materials and their applications, but how interacting thermodynamic variables alter the free energy and phase stability.

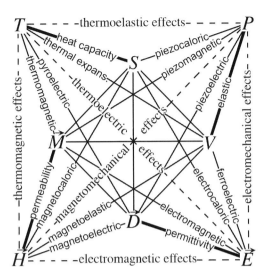

Figure 9.4 Relationships between thermal, mechanical, electrical, and magnetic properties of materials. Intensive thermodynamic variables are on the outside, and extensive on the inside. The primary relationships, T–S, P–V, $\vec{E}$–$\vec{D}$, $\vec{H}$–$\vec{M}$, are on dark lines, secondaries are on thin lines. (Dashed lines on outer edges and central diagonals are not fundamental, but suggest general phenomena.)

$$G = E - TS + PV + \vec{E} \cdot \vec{D} + \vec{H} \cdot \vec{M}, \qquad (9.8)$$

where we have added new terms associated with electric and magnetic degrees of freedom (assuming the material has them, and they are interesting). Here $\vec{E}$ is applied electric field, $\vec{D}$ is polarization, $\vec{H}$ is applied magnetic field, and $\vec{M}$ is magnetization.[4]

Numerous secondary relationships are indicated with thin lines in Fig. 9.4. For the $n = 4$ primary relationships, there are $n^2 - n = 12$ secondary relationships between intensive and extensive variables. In complex materials, volume (strain), magnetization, and electric polarization may all interact, so these 12 secondary relationships can be important. It is tricky to take all 16 relationships into account simultaneously, however, so the following strategy is suggested.

First, consider the most relevant primary relationships, T–S, P–V, $\vec{E}$–$\vec{D}$, $\vec{H}$–$\vec{M}$. These are always important thermodynamically. Second, focus on the extensive variables S, V, $\vec{D}$, $\vec{M}$ (not the intensive variables). The extensive variables scale with the amount of material and originate with the atoms and electrons inside the material.[5] These are located on the inner diamond of Fig. 9.4. For example, suppose the interest is in the entropy S, which is directly altered by temperature (a thick line connects T and S in Fig. 9.4). The strategy next considers the entropy as dependent on energy, volume, magnetization, and electric

---

[4] These are vector quantities. For crystalline solids it is usually important to consider PV as a product of tensor quantities, as in Eq. 6.123.

[5] It is common to consider $S(T, P, \vec{E}, \vec{H})$, since T, P, $\vec{E}$, $\vec{H}$ are the independent (intensive) variables that define the state of the material. This is the classical approach that views the material as responding to external forces, as opposed to viewing the properties as originating with the electrons and atomic structure.

polarization, of which all but the internal energy, E (not electric field $\vec{E}$), are on the inner diamond:

$$\hat{S} = S(E, V, \vec{D}, \vec{M}).\tag{9.9}$$

Assuming a fixed number of atoms and a constant composition, this expression for $\hat{S}$ has full thermodynamic information. Defining partial derivatives as (Section 1.6.2)

$$\left(\frac{\partial S}{\partial E}\right)_{V,\vec{D},\vec{M}} \equiv \frac{1}{T}, \quad \left(\frac{\partial S}{\partial V}\right)_{E,\vec{D},\vec{M}} \equiv \frac{P}{T}, \quad \left(\frac{\partial S}{\partial \vec{D}}\right)_{E,V,\vec{M}} \equiv \frac{\vec{E}}{T}, \quad \left(\frac{\partial S}{\partial \vec{M}}\right)_{E,V,\vec{D}} \equiv \frac{\vec{H}}{T},\tag{9.10}$$

gives the thermodynamic identity

$$T\mathrm{d}S = \mathrm{d}E + P\mathrm{d}V + \vec{E} \cdot \mathrm{d}\vec{D} + \vec{H} \cdot \mathrm{d}\vec{M}.\tag{9.11}$$

Equation 9.11 can be obtained systematically from Fig. 9.4. After adding dE, locate the entropy on the figure. Changes in the entropy, dS, can be constructed as changes in the other extensive variables dV, d$\vec{D}$, d$\vec{M}$ on the inner diamond of Fig. 9.4 by multiplying by their conjugate intensive variables (P, $\vec{E}$, $\vec{H}$ on the outer square of Fig. 9.4), and dividing by T (the conjugate variable of S). Other examples are developed in Problem 9.4.

Maxwell Relationships

The extensive variables in the inner diamond can be obtained as first partial derivatives of G with respect to an intensive variable on the corners of the outer square of Fig. 9.4 (some are given in Eq. 1.24). A second derivative of G relates some of the thin lines in the figure (secondary relationships). The resulting Maxwell relations, as in Section 1.6.2, come about because second derivatives can be taken in either order

$$\frac{\partial^2 G}{\partial P \partial T} = \frac{\partial^2 G}{\partial T \partial P},\tag{9.12}$$

$$\frac{\partial S}{\partial P} = -\frac{\partial V}{\partial T},\tag{9.13}$$

$$\text{piezocaloric} \leftrightarrow \text{thermal expansion},\tag{9.14}$$

where the sign was obtained from the rules of Fig. 1.8. Likewise

$$\frac{\partial^2 G}{\partial P \partial \vec{H}} = \frac{\partial^2 G}{\partial \vec{H} \partial P},\tag{9.15}$$

$$\frac{\partial \vec{M}}{\partial P} = -\frac{\partial V}{\partial \vec{H}},\tag{9.16}$$

$$\text{piezomagnetic} \leftrightarrow \text{magnetoelastic}.\tag{9.17}$$

Two subtleties in Eq. 9.16 are signs and vector quantities. If the volume increases with an applied magnetic field, then pressure, which would reduce volume, should reduce the magnetization. Hence the negative sign. The vector derivatives could be taken as one Cartesian component at time, such as $\partial M_x/\partial P = \partial V/\partial H_x$, and this would work for an isotropic response of $\vec{M}$ to $\vec{H}$. In general we expect anisotropy, so the two sides of Eq. 9.16 should be related by a second-rank tensor. Following Eqs. 9.12–9.17, four other relationships can be found with the aid of Fig. 9.4:

$$\text{electrocaloric} \leftrightarrow \text{pyroelectric}, \quad \text{magnetocaloric} \leftrightarrow \text{thermomagnetic},$$

$$\text{piezoelectric} \leftrightarrow \text{ferroelectric}, \quad \text{magnetoelectric} \leftrightarrow \text{electromagnetic}.$$

Thermal Expansion

Figure 9.4 can help organize secondary relationships between the intensive variables themselves, $\{\ S,\ V,\ \vec{D},\ \vec{M}\ \}$. Consider the volume coefficient of thermal expansion, $\beta \equiv 1/V\ dV/dT$, marked on a secondary line of Fig. 9.4. Expansion causes an increase in elastic energy E. If the entropy S increases with V, the free energy at finite T is minimized by an increase in V (see Section 12.3.1 for more detail). Nevertheless, thermal expansion involves more than the relationship $S(V)$:

- Figure 9.4 shows how the relationship $S \leftrightarrow V$ is similar to the relationship with magnetization $\vec{M} \leftrightarrow V$. When V changes with $\vec{M}$, which depends on T (i.e., $\vec{M}(T)$), thermal expansion is altered from predictions with $S(V)$ alone. This is the case for Invar, a magnetic alloy of Ni–Fe. Invar has cancelling contributions to its thermal expansion, which is nearly zero at room temperature. The thermal expansion of Invar is also altered by applied magnetic fields, $\vec{H}$, owing to the primary relationship $\vec{H} \leftrightarrow \vec{M}$.
- For most materials we expect the secondary relationship $S(V)$ to depend on T, owing to the primary relationship $S \leftrightarrow T$. In other words, the coefficient of thermal expansion, β, depends on the pure temperature dependence of S, the anharmonicity, in addition to the volume effects on entropy that are described in Section 12.3.1.

Today the extensive variables $\{\ S,\ V,\ \vec{D},\ \vec{M}\ \}$ are much better understood at the level of the atoms and electrons than in Heckmann's time. Knowing $\{\ S,\ V,\ \vec{D},\ \vec{M}\ \}$ is tantamount to knowing the state of the material, and therefore its thermodynamics and related properties. This is more direct than a traditional engineering approach that parameterizes the material responses to intensive variables $\{\ T,\ P,\ \vec{H},\ \vec{E}\ \}$ because a large number of secondary relationships are needed in the traditional approach.

9.4 Partitioning of Energy in Polycrystals and Single Crystals

9.4.1 Magnetic and Elastic Energy

It is usually not possible to assume isotropy of fields in a material because single crystals typically have "easy axes" for magnetization, polarization, or elastic distortions (cf. Fig. 6.12). The energy is lowest when the fields are aligned along these axes. On the other hand, if an external field $\vec{H}$ is applied, the magnetization $\vec{M}$ will tend to align along a direction that minimizes the product $\vec{M} \cdot \vec{H}$. When this direction for $\vec{M}$ is not along an easy axis, some compromise orientation is expected.

Orientations of field quantities along favorable crystallographic directions are not expected to be precise in polycrystalline materials, which typically have adjoining crystals with differing orientations. Although magnetic polarizations may be biased towards particular directions in two adjacent crystals, the magnetoelastic distortions create stresses as the two crystals push on each other. The elastic energy then interferes with the magnetic alignment. Recall that the magnetic and elastic energies increase as the squares of the magnetization and strain, respectively. An unproven ansatz is that for magnetomechanical effects in polycrystalline materials, the energy will be shared equally between the magnetic and mechanical coordinates. This minimizes the overall energy penalty of misorientations, although we expect local regions where one type of energy dominates over the other.

This ansatz is not appropriate for single crystals, or for carefully engineered microstructures. An isolated single crystal expands freely without impinging on a neighboring crystal. Without unfavorable elastic stresses from poor accommodation, a minimum magnetic energy can be achieved. It is also possible to configure crystals to minimize elastic energy, so the energy can remain in the magnetization, and need not be transferred to elastic stresses. An added complication to this simple story is the presence of domains in a material, where even a single crystal may contain internal regions that differ in the alignments of their magnetizations or strains. The magnetic domains shown in Fig. 21.10 are such an example.

9.4.2 Thermal Stresses in Anisotropic Microstructures

Consider internal stresses that arise from thermal expansion alone. If the thermal expansion of a crystal is isotropic, all microstructural features expand by the same fractional length. Even an irregular packing of crystals in a polycrystalline microstructure expands uniformly, and does not develop internal stresses with changes in temperature. On the other hand, when thermal expansion is crystallographically anisotropic, an irregular polycrystalline microstructure develops internal stresses as adjacent crystals, differing in crystallographic orientation, try to expand by different amounts along different spatial directions. Suppose a stress-free polycrystalline microstructure is prepared by annealing at high temperature, but anisotropies of thermal contraction create internal stresses upon cooling.[6] Assuming the strains are elastic, the internal stresses are relieved as the material is heated. This stored elastic energy is converted to heat. It therefore takes less heat to raise the temperature of this microstructure, so its heat capacity is suppressed by release of this stored elastic energy.

Such a suppression of heat capacity was found when comparing two types of samples of Ni_3V with (1) an isotropic cubic phase, prepared by quenching the sample from a high temperature in an fcc region of the phase diagram, and (2) an anisotropic tetragonal phase, prepared by annealing in the $D0_{22}$ region of the phase diagram [109, 110]. Figure 9.5 shows differences in the heat capacity of these two samples – the partially disordered cubic sample had the larger heat capacity. The shape of the measured differential heat

---

[6] Some polycrystalline ceramic materials with unit cells of low symmetry may fracture under these internal stresses upon cooling.

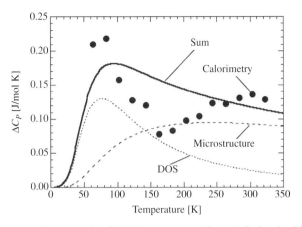

Figure 9.5 Dots: differential heat capacity of two samples of Ni$_3$V. Sign convention is for partially disordered fcc minus D0$_{22}$-ordered. After [109, 111].

capacity curve is quite different from the thin curve labeled "DOS" in Fig. 9.5, which is the usual differential heat capacity from vibrational entropy (cf. Fig. 26.2b). The curve labeled "Microstructure" in Fig. 9.5 is the contribution obtained by assuming that the D0$_{22}$-ordered sample retains internal stresses at low temperature that are relieved as the solid expands. The sum of the two contributions better accounts for the measured differential heat capacity than either contribution alone. The real situation is probably more complicated. Over a temperature range of a few hundred K, local plastic deformation can be induced by anisotropic thermal strains, especially at locations of stress concentrations. Plastic deformation requires mechanical work from thermal expansion, counteracting the decrease in measured heat capacity from relief of internal stress.

In summary, the energy considerations of elasticity, magnetization, and polarization, combined with energy penalties for external fields, make it difficult to know the internal structures of multiferroic and complex materials, and predict their properties. Computer simulations are appropriate for these problems [112, 113], as is "homogenization theory," a branch of mathematics that develops equations for averages [114].

9.5 Coherency Strains in Chemical Unmixing

The previous section gave examples of how polycrystalline materials have internal stresses and elastic energies that can be avoided in single crystals. Arguably, these energetic contributions can be discounted when calculating a thermodynamic free energy on the grounds that the polycrystalline material is not in equilibrium. Now consider an unmixing transformation within a single crystal. The alloy becomes chemically inhomogeneous, and these chemical heterogeneities generate internal stresses.

9.5.1 Chemical Effects on Elastic Fields

Consider unmixing on an Ising lattice as in Section 2.8, where each atom is associated with one crystallographic site. The free energy of Sections 2.7– 2.9 was obtained by summing energies of the A–A, A–B, and B–B bonds, and the entropy by the combinatorics of arranging A- and B-atoms on the lattice. Different atoms have different sizes, however, and the local lattice parameter depends on the local chemical composition, parameterized by η

$$\eta = \frac{1}{a}\frac{da}{dc}. \tag{9.18}$$

The optimum lattice parameter is different in a region of pure A than in a region of pure B. There is no elastic energy in the lattice when the overall composition is $c = 0$ or $c = 1$ (assume the crystal of pure A or pure B is in free space). A function of the form $c(1 - c)$ has the right qualitative features to describe the concentration dependence of the elastic energy. It has similarities to the chemical energy $-c(1 - c)\,4V\,zN/2$ in Eq. 2.30, and both have the same sign (for unmixing, $4V < 0$).

During and after chemical unmixing, the chemical composition varies throughout the material. This generates elastic energy by the effects of Eq. 9.18. Assuming the alloy remains coherent, the lattice planes bend to accommodate the regions of different composition, causing "coherency stresses" and a local elastic energy that changes the free energy by ΔF in the volume dV

$$\Delta F = +\frac{Y}{1 - v}(\eta\,\delta c)^2\,dV, \tag{9.19}$$

where Y is the Young's modulus along the direction of composition variation, and v is the Poisson ratio of lateral contraction to elongation.[7] In the early stages of unmixing, δc is the composition deviation from the average, and varies with location. The elastic energy is proportional to the square of δc. Elastic anisotropy, which causes variations of Y along different crystallographic directions (e.g., Fig. 6.12), can favor some directions for chemical unmixing [115].

9.5.2 Free Energies with Coherency Stresses

A complication for the thermodynamics of solids is that interface between the precipitate and matrix can change from "coherent" to "incoherent" as in Fig. 4.2. For a coherent precipitate, each site in the precipitate and matrix has a one-to-one correspondence to a lattice site in the original undeformed crystal. This structural constraint from coherency requires bending of lattice planes. A loss of coherency therefore allows a reduction in the energy of elastic fields, but at the expense of more energy for the interface between precipitate and matrix.

In deriving the common tangent construction in Section 2.3, the basic assumption of Eq. 2.6 was that the free energy depended only on the phase (α- or β-phase) and its

---

[7] Suppressing the lateral contraction raises the elastic energy by the factor $1/(1 - v)$. This expression is revisited in Section 16.5 after adding the "square gradient energy" for spinodal decomposition on small length scales.

Figure 9.6 Transmission electron micrograph of coherent θ''-precipitates in Al–Cu–Mg alloy oriented along [001] axis. The precipitates lie on {100} planes. They tend to have 0.03 μm separations from interactions through coherency stresses and solute diffusion lengths. Figure courtesy of J.M. Howe.

chemical composition (c_α or c_β). The phase fractions f_α or f_β weighted the free energies of each phase, but the free energy contribution from the α-phase had the form $F_\alpha(c_\alpha)$, not $F_\alpha(c_\alpha, f_\alpha)$. This essential assumption behind the common tangent rule is violated when coherency stresses evolve during a phase transformation. If the same phase fraction f_β is obtained by replacing one large β-phase particle with n particles of $1/n$ the size, the elastic energy is not the same.[8] Evidently F_β and F_α depend on the shape and distribution of the precipitates. This expands considerably the parameter space for free energy minimization, and a common tangent construction with composition and phase fraction may need modification to predict states of equilibrium.

Another feature of n small particles is that they tend to have regular separations. Two particles keep their distance in part because they compete for the solute atoms in the matrix. In addition, coherency stresses also tend to cause repulsion between precipitates, promoting preferential orientations and spacings between them. Elastic anisotropy and the effects of precipitate shape on elastic fields can favor arrangements as in Fig. 9.6. The effects of elastic interactions between particles can, in turn, alter the shape of a precipitate as it grows.

We are left in a quandary about how to draw phase boundaries in T–c phase diagrams when the two phases generate coherency stresses. The chemical considerations of Section 2.3 can be used alone for constructing phase boundaries. Coherency stresses are present in the material, however, and contribute to the free energy. Today the best practice is to use both criteria. Separate phase boundaries are drawn on T–c phase diagrams, labeled as "chemical equilibrium" and "coherent equilibrium," depending on the absence or presence of coherency stresses.[9] The shifts of boundaries with chemical compositions tend to be small, of order 1 at.% for coherent and incoherent precipitates in Ti–Al, for example [116]. Not much experimental data are available on compositional differences between coherent and incoherent equilibria, so the default assumption when interpreting a T–c phase diagram

---

[8] Under the very special circumstances of the Bitter–Crum theorem of Section 15.5.4, however, these two configurations could have the same energy.

[9] The partial relaxation of coherency stresses allows phase boundaries between these limits, too.

is that it shows incoherent (chemical) equilibrium. Section 15.5 continues this discussion, and describes effects in metal hydrides.

9.6 Coupling between Unmixing Processes

9.6.1 Crystal Lattice and Interstitial Lattice

This section develops a model of two processes of chemical unmixing, with a coupling between them that can be switched on or off. It is shown how the free energy can be characterized by finding eigenvectors of the two independent composition variables. Their eigenvalues describe the stability of the alloy against chemical unmixing. The concepts pertain generally to phase transformations involving multiple independent variables such as compositions or order parameters. One example is the "two-sublattice model," where there are two types of anions and two types of cations on the anion and cation sublattices of a crystal structure [117, 118].

Here we model the free energy of a hydrogen–metal system, where hydrogen atoms occupy interstitial sites in the metal crystal as in Fig. 9.7a for a square lattice. Assume the metal is an A–B alloy that tends to unmix at low temperatures. Likewise, the hydrogen–vacancy (H–V) system tends to unmix on the interstitial lattice, forming hydrogen-rich zones called "hydrides." If the hydrogen does not prefer A or B neighbors, it is appropriate to separate the two unmixing processes into two independent problems, one for the A–B alloy on the crystal lattice, and the other for the H–V system on the interstitial lattice. With separated problems, formulations for unmixing follow the approach of Section 2.7, starting with Eq. 2.28, obtaining $F(c)$ curves as in Fig. 2.16b, and identifying critical temperatures as in Eq. 2.48. An $F(c_B, c_H)$ at low temperature for the separated unmixing problems is shown in Fig. 9.8a. It may not be obvious in a first glance at this contour plot, but for any fixed value of c_B' (horizontal line on the figure), the shape of the curve $F_T(c_B', c_H)$ is the same, and likewise the shape is the same along c_B for any fixed c_H'. For these two uncoupled problems, the unmixing of either metal atoms or H atoms is not affected by the other.

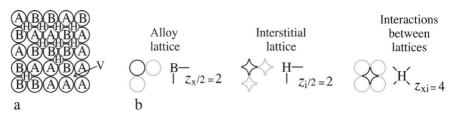

Figure 9.7 (**a**) Two-dimensional square lattice with A–B components and H on interstitial lattice, showing preference for B–H neighbors. (**b**) Bonds between atoms on crystal lattice, interstitial lattice, and between crystal and interstitial. Numbers at right are for counting pairs from sites on lattices or between them.

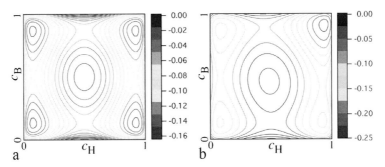

Free energy surfaces, $F_T(c_B, c_H)$ of Eq. 9.20, for unmixing of both A–B and H–V systems. Minima are near corners, maximum is at center. (**a**) Without H–metal interactions, $z_x V_{AB} = -1.28\,k_B T$ (n.b., V defined in Eq. 2.32, so e.g., $e_{AB} > 0$), $z_i V_{HV} = -1.45\,k_B T$ (n.b., e.g., $e_{HV} > 0$). (**b**) With H–B interactions, $z_x V_{AB} = -1.28\,k_B T$, $z_i V_{HV} = -1.45\,k_B T$ as in panel a, plus $z_{xi}^{(12)} e_{BH}^{(12)} = -0.06\,k_B T$, $z_{xi}^{(22)} e_{BH}^{(22)} = -0.06\,k_B T$.

9.6.2 Energies of Interactions between Substitutional and Interstitial Solutes

A more interesting and realistic total free energy has the form

$$F_T(c_B, c_H) = E_{AB}(c_B) + E_{HV}(c_H) + E_{BH}(c_B, c_H) + E_{AH}(c_A, c_H)$$
$$- T[S_{AB}(c_B) + S_{HV}(c_H)], \tag{9.20}$$

where E_{BH} and E_{AH} are interaction energies between the metal lattice and the interstitial lattice (see third diagram of Fig. 9.7b).[10] The $E_{BH}(c_B, c_H)$ includes both concentration variables, c_B and c_H, and could have various terms,

$$E_{BH} = N\Big[z_{xi}^{(11)} e_{BH}^{(11)} c_B c_H + z_{xi}^{(21)} e_{BH}^{(21)} c_B^2 c_H + z_{xi}^{(12)} e_{BH}^{(12)} c_B c_H^2 + z_{xi}^{(22)} e_{BH}^{(22)} c_B^2 c_H^2 + \cdots\Big], \tag{9.21}$$

and similarly for $E_{AH}(c_A, c_H)$. Each interaction term in Eq. 9.21 has a constant energy parameter e, and a dependence on the two chemical compositions c_B and c_H. The coordination number z_{xi} counts bonds between the interstitial and substitutional lattices, as shown in Fig. 9.7b.[11]

With the substitutions $c_A = 1 - c_B$ (and $c_V = 1 - c_H$ if necessary), $E_{AH}(c_A, c_H)$ becomes a function of the two variables c_B and c_H. As the simplest example, keep only the first term in Eq. 9.21 for E_{BH} (and E_{AH})[12]

$$E_{BH} + E_{AH} \simeq N z_{xi}^{(11)} c_B c_H V^{(11)}, \tag{9.22}$$

$$\text{where} \quad V^{(11)} \equiv \Big(e_{BH}^{(11)} - e_{BV}^{(11)}\Big) - \Big(e_{AH}^{(11)} - e_{AV}^{(11)}\Big), \tag{9.23}$$

$$\text{or} \quad V^{(11)} \equiv \Big(e_{BH}^{(11)} - e_{AH}^{(11)}\Big) - \Big(e_{BV}^{(11)} - e_{AV}^{(11)}\Big). \tag{9.24}$$

---

[10] With separate interstitial and metal lattices, the configurational entropy in the point approximation depends only on the concentration on each lattice.

[11] Here $z_{xi}^{(11)}$ is equal to the number of substitutional sites around each interstitial site. The converse z_{ix} for the number of interstitials around each substitutional equals $z_{xi} = 4$ for a square lattice. For a hexagonal lattice, however, $z_{ix} = 6$ and $z_{xi} = 3$, but there are twice as many interstitial sites as substitutional sites. These substitutional–interstitial bonds are not double-counted when summing over only interstitial sites, so there are no factors of 1/2 in Eq. 9.21.

[12] We also neglect terms linear or constant in composition because they will not alter the common tangent construction between different parts of the free energy surface.

Equation 9.23 gives the preference for forming hydride in B-rich regions (and rearranging it as Eq. 9.24 gives the preference for B-atoms to reside in the hydride). The full interaction energy Eq. 9.22 scales with $c_B c_H$. The second term in Eq. 9.21, which scales with $c_B^2 c_H$, can account for a change in energy of a B–B bond in the presence of a H-atom, and this term allows potent effects of hydrogen on the unmixing of the metal lattice.

9.6.3 Free Energies with Interactions between Substitutional and Interstitial Solutes

Including interactions between the interstitial atoms and substitutional atoms demands a more general formalism for finding the conditions for unmixing. For readability in this section, we make the change in notation of $c_B \rightarrow x$ and $c_H \rightarrow y$. We seek the free energy near the equilibrium concentrations (x_0, y_0) as a Taylor series in two coordinates

$$F(x_0 + \Delta x, y_0 + \Delta y) = F(x_0, y_0) + \overrightarrow{\nabla F}(x_0, y_0) \begin{bmatrix} \Delta x \\ \Delta y \end{bmatrix}$$

$$+ \frac{1}{2} \begin{bmatrix} \Delta x & \Delta y \end{bmatrix} \mathbf{H}_F(x_0, y_0) \begin{bmatrix} \Delta x \\ \Delta y \end{bmatrix} + \cdots, \tag{9.25}$$

where the gradient is written in vector form

$$\overrightarrow{\nabla F}(x_0, y_0) \equiv \begin{bmatrix} \frac{\partial F}{\partial x} & \frac{\partial F}{\partial y} \end{bmatrix}_{x_0, y_0}, \tag{9.26}$$

and the "Hessian matrix," which contains all second derivatives of F evaluated at (x_0, y_0), is

$$\mathbf{H}_F(x_0, y_0) \equiv \begin{bmatrix} \frac{\partial^2 F}{\partial x^2} & \frac{\partial^2 F}{\partial x \partial y} \\ \frac{\partial^2 F}{\partial y \partial x} & \frac{\partial^2 F}{\partial y^2} \end{bmatrix}_{x_0, y_0}. \tag{9.27}$$

If we null all terms in Eq. 9.27 involving y, the last term in Eq. 9.25 reduces to the familiar $\frac{1}{2} \frac{\partial^2 F}{\partial x^2} \Delta x^2$ for one dimension (see $\frac{d^2 F}{dc^2}$ in Fig. 2.18a). We also recover this one-dimensional result when $\Delta y = 0$ in Eq. 9.25, i.e., when one of the compositions is fixed.

Box 9.1   **Laplacian Limitations**

The Laplacian has limitations when we need second derivatives in more than one dimension. The Laplacian does contain multiple second derivatives as $\nabla^2 F = \frac{\partial^2 F}{\partial x^2} + \frac{\partial^2 F}{\partial y^2}$. If the Laplacian of F at a point $\{x_0, y_0\}$ is negative, $F(x_0, y_0)$ is larger than its average value in a small ring around it. We therefore expect unmixing by the common tangent construction in two dimensions (cf. Fig. 2.18a). This is indeed helpful, but as a scalar quantity, the Laplacian cannot tell us what direction in composition space is most favorable for unmixing. Furthermore, a negative Laplacian is too conservative a criterion for unmixing. Consider a saddle point where $\frac{\partial^2 F}{\partial x^2} = -\frac{\partial^2 F}{\partial y^2}$. At this special saddle point the Laplacian is zero, but the system is still unstable in the direction with negative second derivative. Finally, the Laplacian is often celebrated because it is separable into terms of single coordinates, but this very feature precludes its use when we need physical interactions between different coordinates.

9.6.4 Stability near Equilibrium

Since the x_0 and y_0 are equilibrium concentrations, the first derivative of F is zero, or in two dimensions its gradient is zero, i.e., $\overrightarrow{\nabla F}(x_0, y_0) = 0$. By seeking the change in free energy ΔF for small departures from $F(x_0, y_0)$, Eq. 9.25 becomes[13]

$$\Delta F(x_0 + \Delta x, y_0 + \Delta y) = \frac{1}{2} \begin{bmatrix} \Delta x & \Delta y \end{bmatrix} \underline{\mathbf{H}}_F(x_0, y_0) \begin{bmatrix} \Delta x \\ \Delta y \end{bmatrix}. \tag{9.28}$$

Since the space of c_B–c_H is two dimensional, the problem of finding the critical temperature for unmixing now takes on a new twist – we need to know the principal directions of curvature of the free energy surface at x_0, y_0. Consider a saddle point, for example, when one principal direction has a positive second derivative, and the other principal direction has a negative second derivative. In the composition space of c_B and c_H, unmixing is expected along the second principal direction, but not the first. Directions of principal curvature (generally not along c_B or c_H alone) are eigenvectors of the Hessian matrix of F. The eigenvalue equation is[14]

$$\begin{bmatrix} \frac{\partial^2 F}{\partial x^2} & \frac{\partial^2 F}{\partial x \partial y} \\ \frac{\partial^2 F}{\partial y \partial x} & \frac{\partial^2 F}{\partial y^2} \end{bmatrix}_{x_0, y_0} \begin{bmatrix} \Delta x_i \\ \Delta y_i \end{bmatrix}_i = \lambda_i \begin{bmatrix} \Delta x_i \\ \Delta y_i \end{bmatrix}_i. \tag{9.29}$$

The procedure to find the two {eigenvalue, eigenvector} pairs is a standard one where a determinant of a matrix is first set equal to zero and a quadratic secular equation is solved for two values of λ.

The secular equation is

$$F_{xx} F_{yy} - \lambda (F_{xx} + F_{yy}) + \lambda^2 - F_{xy}^2 = 0, \tag{9.30}$$

with the notation $F_{xy} \equiv \partial^2 F/(\partial x \partial y)$, and $F_{xy} = F_{yx}$. This quadratic equation for λ gives two roots

$$\lambda = \frac{F_{xx} + F_{yy}}{2} \pm \frac{1}{2} \sqrt{(F_{xx} - F_{yy})^2 + (2F_{xy})^2}. \tag{9.31}$$

When F_{xy} is small, the two eigenvalues of Eq. 9.30 are simpler

$$\lambda_1 \simeq F_{xx} + F_{xy}^2/(F_{xx} - F_{yy}), \quad \text{and} \quad \lambda_2 \simeq F_{yy} - F_{xy}^2/(F_{xx} - F_{yy}), \tag{9.32}$$

---

[13] The gradient $\overrightarrow{\nabla F}$ in Eq. 9.25 does not play a role in the unmixing thermodynamics because the gradient gives a linear tilt of the $F(x, y)$, which does not affect the ΔF between $F(x, y)$ and a common tangent to the minima of $F(x, y)$.

[14] The Hessian matrix of Eq. 9.28 transforms the vector to its right into a new vector (times curvatures of free energy). This new vector in composition space is generally in a different orientation than the original. For two special vectors, the "eigenvectors" of $\mathbf{H}$ in Eq. 9.29, the transformation gives a vector along the same direction. The vector to the left of $\mathbf{H}$ in Eq. 9.28 is parallel, so Eq. 9.28 becomes a scalar product of the magnitudes of two vectors. Eigenvectors of $\mathbf{H}$ behave just like the simpler variable c in the one-dimensional problem. The Hessian adds a sign and magnitude with a scalar "eigenvalue," λ_i, made of curvatures of the free energy. The eigenvalues of $\mathbf{H}$ act like $\partial^2 F/\partial c^2$ in the one-dimensional Eq. 2.46.

and substitution into Eq. 9.29 gives the directions of the two eigenvectors

$$[\Delta x_1, \Delta y_1] \quad \text{for} \quad \lambda_1, \text{ where} \quad \Delta x_1/\Delta y_1 \simeq (F_{xx} - F_{yy})/F_{xy}, \text{ and} \qquad (9.33)$$

$$[\Delta x_2, \Delta y_2] \quad \text{for} \quad \lambda_2, \text{ where} \quad \Delta x_2/\Delta y_2 \simeq -F_{xy}/(F_{xx} - F_{yy}). \qquad (9.34)$$

In Eqs. 9.33 and 9.34, the directions of the eigenvectors (in the composition space of c_B and c_H) are more important than their magnitudes.

These results reduce to two uncoupled behaviors when $F_{xy} = 0$, so the Hessian of Eq. 9.27 is diagonal. We recover the two separate problems for (1) $\Delta x = \Delta c_B$, and (2) $\Delta y = \Delta c_H$, with the expected results for no coupling

$$\lambda_1 = F_{xx}, \; [\Delta x_1, \Delta y_1] = [1, 0] \quad \text{and} \quad \lambda_2 = F_{yy}, \; [\Delta x_2, \Delta y_2] = [0, 1], \qquad (9.35)$$

where the eigenvectors were normalized to 1.

The signs of the eigenvalues λ_1 and λ_2 give key information about the stability of the alloy against compositional unmixing.

- When both λ_1 and λ_2 are positive, ΔF increases for any deviation from equilibrium at x_0, y_0. It is a minimum (perhaps a local minimum).
- If λ_1 is negative and λ_2 is positive, there is a saddle point at $F(x_0, y_0)$. The alloy is unstable against compositional unmixing in the direction of the eigenvector $[\Delta x_1, \Delta y_1]$ (which may not align along Δc_B or Δc_H).
- When both λ_1 and λ_2 are negative, $F(x_0, y_0)$ is a maximum, and the alloy is unstable against unmixing for any combination of Δc_B and Δc_H.

The eigenvalues λ_1 and λ_2 are sometimes used to deduce characteristic rates of compositional unmixing – the more negative, the faster the unmixing. The idea is based on Eq. 5.33, assuming that

$$\partial/\partial t \, [x, y]^T \propto \mathbf{H}_F [x, y]^T. \qquad (9.36)$$

The resulting first-order differential equations with constant coefficients give exponential decays when both λ_1 and λ_2 are positive, but give runaway unmixing when either λ_1 or λ_2 is negative. Chapter 23 (online) gives a better approach that allows degrees of freedom in the kinetic mechanisms, too.

9.6.5 Critical Temperature in a Coupled System

Consider a narrow range of temperature where λ_1 changes from positive to negative, while λ_2 remains positive. This corresponds to cooling a solid solution from a high temperature through an unmixing phase boundary. When there is no coupling term and $F_{xy} = 0$, Eq. 9.35 shows that $\lambda_1 = F_{xx} = \partial^2 F/\partial c_B^2$ (recall $x \equiv c_B$). The critical temperature T_c is obtained by setting this second derivative equal to zero, giving the result of Eq. 2.48 for one composition variable, i.e., $T_c = -zV/k_B$.

Now couple the composition variables so $F_{xy} \neq 0$. At the former T_c where $F_{xx} = 0$, λ_1 (Eq. 9.31 or 9.32) is no longer F_{xx}, but is now

$$\lambda_1 = -\frac{F_{xy}^2}{F_{yy}}. \tag{9.37}$$

We are considering the case $F_{yy} > 0$, so with a positive-definite numerator, Eq. 9.37 shows that λ_1 is always negative at the critical temperature of the uncoupled alloy. This means that the critical temperature for unmixing is always increased when the composition variables interact. The unmixing line on the phase diagram moves to a higher temperature.

This result may seem surprising because there were no assumptions about the signs of the terms in Eq. 9.21, which could be positive or negative. Equation 9.28 offers physical insight. After multiplying it out,

$$\Delta F(\Delta x, \Delta y) = F_{xx}\Delta x^2 + 2F_{xy}\Delta x\Delta y + F_{yy}\Delta y^2. \tag{9.38}$$

The coupling term goes as the product $\Delta x \, \Delta y$, which can be positive or negative, allowing the coupling term to be negative for either sign of F_{xy}. Consider the case when F_{xy} is positive. The coupling term is negative in a region of alloy where Δx is negative and Δy is positive. By conservation of solute, on average there must be another region in the alloy with opposite composition variations, i.e., Δx is positive and Δy is negative. In this other region the coupling term is negative, too. (The alternative case with negative F_{xy} works similarly, but Δx and Δy have the same sign.) With coupling, the unmixed alloy is stabilized by appropriate choice of Δx and Δy, compared to the uncoupled case with Δx alone. Compared to a homogeneous alloy, the many possibilities for unmixed compositions allow an alloy to better optimize its free energy in the presence of new interactions, stabilizing an unmixed state to higher temperatures.

9.6.6 Examples of Unmixing with Interactions between Substitutional and Interstitial Solutes

The previous analysis gave useful results for small deviations from equilibrium, but it is helpful to consider examples of full free energy surfaces across the space spanned by c_B and c_H. Some specific cases are shown in Fig. 9.9. Figure 9.9a is a reference system, with a symmetrical pair of independent unmixing reactions for A–B and H–V, and no B–H interactions. For the central composition of $c_B = 0.5$ and $c_H = 0.5$, there is ambiguity about which unmixing compositions would be favorable – the H-atoms can reside in either B-rich regions or A-rich regions. Likely both will occur at different locations in the material. With the addition of a modestly attractive B–H interaction, however, there is a preference for the unmixed alloy to place the H-atoms in B-rich regions, as shown in Fig. 9.9b. Without the B–H interactions, the second derivatives of F are equal along the c_B and c_H coordinates. This gives zeros in the denominators of Eq. 9.32, and large effects of B–H interactions on the direction for chemical unmixing. Below a composition near $c_B = 0.5$ and $c_H = 0.5$, the lowest common tangent touches the two minima at left and right in Fig. 9.9b.

The two surfaces of Fig. 9.9c,d show how unmixing of metal atoms can be induced by H–B interactions. In the reference case of Fig. 9.9c, the A–B substitutional system does

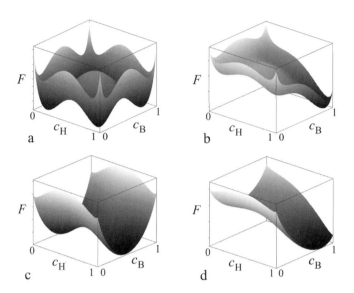

Figure 9.9 Free energy surfaces $F_T(c_B, c_H)$ of Eq. 9.20. Surfaces a and c are for independent A–B and H–V systems; b and d include interaction terms. (**a**) Without H–metal interactions, $z_x V_{AB} = -1.3\, k_B T$ (n.b., e.g., $e_{AB} > 0$) and $z_i V_{HV} = -1.3\, k_B T$. (**b**) Same as a, but with attractive B–H interactions, $z_{xi} e_{BH}^{(11)} = -0.3\, k_B T$. (**c, d**) Free energy surfaces below the critical temperatures for unmixing H–V system, but A–B alloy stable against unmixing. (**c**) Without H–metal interactions, $z_x V_{AB} = 0.15\, k_B T$ and $z_i V_{HV} = -1.85\, k_B T$ (n.b., e.g., $e_{HV} > 0$). (**d**) Same as c, but with attractive B–H interactions, $z_{xi} e_{BH}^{(11)} = -1.5\, k_B T$.

not unmix, since V_{AB} has the wrong sign. An equiatomic A–B alloy remains stable when the H–V unmixing occurs. Figure 9.9b shows, however, that with the addition of a simple H–B attraction ($E_{BH} = z_{xi} N e_{BH}^{(11)} c_B c_H$ of Eq. 9.21), there is a tendency for the B-atoms to prefer the H-rich regions in the unmixed alloy, shifting the minimum in Fig. 9.9d towards higher c_B (with higher c_H). It is interesting that this B–H interaction energy can overcome the stability of the A–B alloy, since it does not alter the strengths of bonds between A- or B-atoms. One interpretation is that there is substantial unmixing of the H–V system, and the attraction of B-atoms to H-rich regions leads to some chemical separation of the B-atoms. Even larger effects occur when atoms on one sublattice alter the bonding on the other. Such changes are caused by the two terms used for constructing Fig. 9.8b, $z_{xi} N e_{BH}^{(12)} c_B c_H^2$ and $z_{xi} N e_{BH}^{(22)} c_B^2 c_H^2$. Both strengthen the H–H bond in the presence of a B-atom, and the latter strengthens the B–B bond in the presence of an H-atom.

9.6.7 Metal Hydride Alloys

In general, there is an enormous difference in the diffusion coefficients of interstitial H atoms and substitutional B atoms in an A–B–H alloy. The H atom is small and fast, and many interstitial sites are vacant. The B atoms are much more sluggish, and need a neighboring vacancy for movement. For hydride formation in an alloy solid solution at low temperatures, a reasonable first approximation is that the B atoms are immobilized, and Eq. 2.48 can be used directly (with a constant V_{HV} appropriate for the metal atom configuration). It is also useful to consider the total free energy in Fig. 9.8b. For simplicity, suppose the alloy composition is $c'_B = 1/2$ and $c'_H = 1/2$ (at the center of the diagram). If the B-atoms are immobile, the H-atoms can unmix along a horizontal line with $c'_B = 1/2$, giving a modest reduction in free energy. A hydride phase can form even if the metal atoms are not configured to give the state of thermodynamic equilibrium.

Nevertheless, a larger reduction in free energy can occur if the A–B alloy were to unmix into zones that are rich in B and rich in H, where there is the deepest minimum in Fig. 9.8b. This minimum is 50% deeper than without the B–H interactions in Fig. 9.8a, even though the B–H interaction parameters are an order-of-magnitude smaller than the A–B or H–V interactions. Interactions between the crystal and interstitial lattices can be especially efficient because the atom configurations on one lattice do not impose conservation constraints on the other.[15] The unmixing of the metal atoms is expected to take a long time compared to hydride formation, but interaction terms in the free energy can promote this "disproportionation." Disproportionation reactions are very sensitive to temperature, usually for reasons of kinetics. Sometimes they can be reversed by removing the hydrogen and heating the metal to a temperature where the metal atoms are mobile. Disproportionation is well studied for the hexagonal Haucke phase of $LaNi_5$, which can absorb hydrogen to a composition of $LaNi_5H_6$. This phase disproportionates at temperatures above 525 K – it tends to separate into LaH_2 and Ni.[16] The stability of $LaNi_5$ against disproportionation can be improved by alloying, which may alter either the kinetics of atom movements or the thermodynamic tendency for unmixing [121].

A solid solution of fcc $Zr_{82}Pd_{18}$ transforms to an amorphous phase when it absorbs hydrogen at temperatures below 500 K [120]. The amorphous phase begins to grow at the grain boundaries of polycrystalline material, and can fully amorphize the alloy. One interpretation is that the H atoms strongly prefer a tetrahedral site surrounded by Zr atoms, but the tetrahedral sites in the fcc structure do not have the right volume. Without adequate diffusion of the metal atoms, but with rapid diffusion of hydrogen, the free energy can be most efficiently reduced by the collapse of the fcc structure into an amorphous metal hydride with tetrahedral sites favorable for hydrogen. Removing the hydrogen at low temperatures leaves behind a bulk metallic glass [120, 122].

9.7 Factoring the Partition Function

When two systems (denoted 1 and 2) are independent, they have separate free energies that are minimized in equilibrium. The total free energy is therefore also minimized, since it is the sum of two minimized parts: $F_{tot} = F_1 + F_2$. Recalling that $F = -k_B T \ln Z$

---

[15] The condition for equilibrium is not the same as the common tangent plane for a ternary alloy, because there are two binary solute conservations, one for the crystal lattice and one for the interstitial lattice. See Problem 9.2.

[16] Another example of disproportionation is fcc $Zr_{80}Rh_{20}$, which decomposes into fcc $Zr_2 + Zr_2RhH$ in the presence of hydrogen [119, 120].

$$F_{\text{tot}} = F_1 + F_2, \tag{9.39}$$

$$-k_{\text{B}} T \ln Z_{\text{tot}} = -k_{\text{B}} T \ln Z_1 - k_{\text{B}} T \ln Z_2, \tag{9.40}$$

$$Z_{\text{tot}} = Z_1 Z_2. \tag{9.41}$$

The additivity of subsystem free energies is equivalent to factoring the partition function. This factorization is a common feature of many useful models, and discovering factorizations can be a goal of theoretical work.

9.7.1 Kinetic Energy Is Separable

For a system of atoms, the energy, E, is a sum of kinetic and potential energies. In three dimensions, each atom has three momentum coordinates $\{p_i\}$ and three position coordinates $\{q_i\}$. With full generality, the energy of N atoms is

$$E(\{q_i\}, \{p_i\}) = \sum_{i=1}^{3N} \frac{p_i^2}{2m_i} + V(\{q_i\}). \tag{9.42}$$

The partition function Z (cf. Eq. 1.7) for this system is

$$Z = \frac{1}{h^{3N}} \int_{-\infty}^{\infty} \int_{-\infty}^{\infty} \cdots \int_{-\infty}^{\infty} \int_{-\infty}^{\infty} e^{-\frac{E(\{q_i\},\{p_i\})}{k_{\text{B}}T}} \, dp_1 dq_1 \ldots dp_{3N} dq_{3N}, \tag{9.43}$$

which is dimensionless because each product $dp_i dq_i$ is normalized by h.[17] The E in the exponential is a sum of the terms in Eq. 9.42, so it can be written as a product of individual exponentials that contain one term each (e.g., $e^{-(d+e+f)} = e^{-d}e^{-e}e^{-f}$). The exponentials with the momentum coordinates (from the kinetic energy) are easiest to integrate. All $3N$ integrals are independent, and each is a Gaussian integral that is evaluated with Eq. 9.52 below

$$Z_{\text{KE}} \equiv \int_{-\infty}^{\infty} \cdots \int_{-\infty}^{\infty} \exp\left(-\frac{\sum_i^{3N} p_i^2}{2m_i k_{\text{B}} T}\right) dp_1 \ldots dp_{3N}, \tag{9.44}$$

$$Z_{\text{KE}} = \prod_{i=1}^{3N} \int_{-\infty}^{\infty} e^{-\frac{p_i^2}{2m_i k_{\text{B}}T}} \, dp_i = \prod_{i=1}^{3N} \sqrt{\pi 2 m_i k_{\text{B}} T}. \tag{9.45}$$

This factor from the total kinetic energy, Z_{KE}, simplifies Eq. 9.43

$$Z = \frac{1}{h^{3N}} Z_{\text{KE}} \int_{-\infty}^{\infty} \cdots \int_{-\infty}^{\infty} e^{-V/k_{\text{B}}T} \, dq_1 \ldots dq_{3N}. \tag{9.46}$$

Reconfiguring the atoms does not alter Z_{KE}. The momentum coordinates and the kinetic energy are therefore not interesting for the thermodynamics of phase transitions that rearrange atoms in materials.[18]

---

[17] We do not need a factor $\Omega(E)$ as in Eq. 1.8 because the allowable states are distributed evenly over the phase space spanned by $\{q_i\}$ and $\{p_i\}$.

[18] Effects of kinetic energy can occur with changes in the atom masses $\{m_i\}$ by chemical or isotopic substitution.

Box 9.2	**Gaussian Integrals**

To evaluate

$$I = \int_{-\infty}^{+\infty} e^{-\beta x^2} \, dx, \tag{9.47}$$

first take its square

$$I^2 = \int_{-\infty}^{+\infty} \int_{-\infty}^{+\infty} e^{-\beta(x^2 + y^2)} \, dx \, dy, \tag{9.48}$$

allowing the integration to be done in polar coordinates with $r^2 = x^2 + y^2$

$$I^2 = \int_{0}^{+\infty} \int_{0}^{2\pi} e^{-\beta r^2} r \, d\theta \, dr. \tag{9.49}$$

The angular integral is trivial

$$I^2 = \int_{0}^{+\infty} e^{-\beta r^2} 2\pi r \, dr. \tag{9.50}$$

Perform the substitution $\xi \equiv \pi r^2$, so $d\xi/dr = 2\pi r$ and $d\xi = 2\pi r \, dr$

$$I^2 = \int_{0}^{+\infty} e^{-\beta/\pi \, \xi} \, d\xi = \frac{\pi}{\beta}, \tag{9.51}$$

$$I = \sqrt{\frac{\pi}{\beta}} = \int_{-\infty}^{+\infty} e^{-\beta x^2} \, dx. \tag{9.52}$$

9.7.2 Potential Energy and Separability

The personality of a system of atoms is determined by the potential energy $V(\{q_i\})$. There is a vast breadth of interesting choices for V. For example, there may be coordinates of V that follow a transformation from an fcc crystal to a bcc crystal, or coordinates that accommodate chemical segregation between a parent phase and a new precipitate. For now, assume the positional coordinates, $\{q_i\}$, are confined to small displacements of atoms around their equilibrium positions[19]

$$V(\{q_i\}) = V_0 + \sum_{i=1}^{3N} V_i \, q_i^2. \tag{9.53}$$

The factors V_i are constants, and a V_0 is assigned to the phase. Substituting Eq. 9.53 into 9.46

$$Z = \frac{1}{h^{3N}} Z_{\text{KE}} \, e^{-V_0/k_B T} \int_{-\infty}^{\infty} \cdots \int_{-\infty}^{\infty} \exp\left(-\sum_{i=1}^{3N} \frac{V_i}{k_B T} \, q_i^2\right) dq_1 \ldots dq_{3N}. \tag{9.54}$$

[19] These $\{q_i\}$ would be zero if the atoms were stationary on their crystal sites.

The sum of terms in the exponential of Eq. 9.54 gives a product of $3N$ integral evaluations with Eq. 9.52. It looks much like Eq. 9.45 for the factor Z_{KE}, written to its left as

$$Z = \frac{1}{h^{3N}} \left(\prod_{i=1}^{3N} \sqrt{2\pi m_i k_{\mathrm{B}} T} \right) \mathrm{e}^{-V_0/k_{\mathrm{B}} T} \left(\prod_{i=1}^{3N} \sqrt{\frac{\pi k_{\mathrm{B}} T}{V_i}} \right). \tag{9.55}$$

The indices are the same in the two product functions, making one of them redundant [20]

$$Z = \mathrm{e}^{-V_0/k_{\mathrm{B}} T} \prod_{i=1}^{3N} \sqrt{\frac{2m_i}{V_i}} \frac{\pi k_{\mathrm{B}} T}{h}. \tag{9.56}$$

The partition function Z is now a product of factors, where each factor depends on only one coordinate. The free energy, $F = -k_{\mathrm{B}} T \ln Z$, becomes a sum of logarithms, one for each coordinate

$$F = V_0 - k_{\mathrm{B}} T \sum_{i=1}^{3N} \ln\left(\sqrt{\frac{2m_i}{V_i}} \frac{\pi k_{\mathrm{B}} T}{h} \right). \tag{9.57}$$

The important point is that with the quadratic potential energy of Eq. 9.53, the partition function can be factored, and the free energy of Eq. 9.57 has each coordinate i in an independent, additive term.

Consider now a potential energy that differs from Eq. 9.53 by cubic terms

$$V(\{q_i\}) = V_0 + \sum_{i=1}^{3N} \left(V_i q_i^2 + V_i^{(3)} q_i^3 \right). \tag{9.58}$$

Sadly, the integrations over q_i as in Eqs. 9.54 and 9.55 are no longer elementary. We could do numerical integrations for each temperature, T. After this extra work, the results would separate into individual factors, which would be individual terms in the free energy.

This separability is lost, however, if the potential energy contains a term with products of different coordinates, such as

$$V(\{q_i\}) = V_0 + \sum_{i=1}^{3N} V_i q_i^2 + \sum_{i=1}^{3N} \sum_{j=1}^{3N} \sum_{k=1}^{3N} V_{ijk}^{(3)} q_i q_j q_k. \tag{9.59}$$

The free energy has terms that couple the position coordinates, much like the coupling between c_{H} and c_{B} in Eq. 9.21. Sometimes a perturbation theory can be developed if $V_{ijk}^{(3)}$ is small.

Entropy

Somewhat hidden in Eq. 9.57 is the entropy of harmonic oscillators. The potential energy of a mass m_i that oscillates with angular frequency ω_i and displacement q_i is $(1/2)m_i \omega_i^2 q_i^2$, so for each coordinate

$$\frac{1}{2} m_i \omega_i^2 q_i^2 = V_i q_i^2, \tag{9.60}$$

$$\sqrt{\frac{2m_i}{V_i}} = \frac{2}{\omega_i}. \tag{9.61}$$

---

[20] This product of two products equals a product of pairs: $\prod_{i=1}^{3N} a_i \prod_{i=1}^{3N} b_i = \prod_{i=1}^{3N} a_i b_i$.

The vibrational entropy term of Eq. 9.57 can be cast in a more familiar form, with $2\pi/h = 1/\hbar$. To be rigorous, there is also a change in i from denoting atom coordinates to denoting normal modes of vibration (details are in Section 26.1 [online])

$$F = V_0 - T\,k_B \sum_{i=1}^{3N} \ln\left(\frac{k_B T}{\hbar\omega_i}\right). \tag{9.62}$$

We did not do a proper quantum mechanical partition function with Planck factors and zero-point contributions, but Eq. 9.62 is consistent with Eq. 7.42 for the vibrational free energy in the high-temperature limit. In our harmonic approximation, the vibrational modes of frequency $\{\omega_i\}$ are independent of each other, and each contributes its own term to the entropy.

9.7.3 Natural or Convenient Coordinates

Squared Coordinates in the Potential Energy

We just saw that quadratic terms in the energies of Eqs. 9.42 and 9.53 give simple Gaussian integrals, a clean factoring of the partition function, and a clean separation of terms in the free energy. This motivates transformations of coordinates to obtain quadratic terms in the potential energy.

For example, the potential energy of Eq. 9.53 with local coordinates of atoms is usually too simple for physical predictions. A better physical picture is that the potential energy depends on the distance between one atom i and its neighbor j, much as if a spring connects them. A crystal of masses on springs has normal modes of vibration, constructed from a Fourier transform of real-space displacements across a large, periodic crystal (Section 26.1 [online]). The amplitudes of the normal modes make quadratic contributions to the energy, allowing factoring of the partition function and Gaussian integrations. This separability gives noninteracting phonon modes. Crystals with linear springs between atoms (harmonic crystals) have an elegance that has spawned an enormous number of mathematical results.[21] Electric and magnetic fields often have energies that go as squares of coordinates, and can also contribute independent terms to free energies.

Quasiparticles

In quantum mechanics, excitations of the independent coordinate k_j are quantized with the excitation energy $\hbar\omega_{k_j}$. An individual quantum of excitation of the coordinate is called a "quasiparticle," and given a name that ends with "on." Examples include the phonon (a vibrational excitation), magnon (a spin wave), polaron (an electron coupled with phonons), plasmon (a collective oscillation of an electron gas), exciton (an excited electron coupled to a hole), and of course a photon (an excitation of the electromagnetic vector potential). Quasiparticles are treated as independent particles, in analogy to the real particles: electron, proton and neutron. However, quasiparticles are created or annihilated in materials with thermal energies, unlike the high energies needed to create or annihilate real particles.

---

[21] Anharmonicity causes interactions between phonons, making the analysis ponderous (Section 26.4 [online]).

Problems

9.1 Figure 5.8 shows the sequence of free energy reductions that pertain to the formation of a glass in a binary alloy system with a composition $c_0 = 0.6$.

(**a**) How do you expect the glass-forming tendency to change for compositions c_0 greater than or less than 0.6?

(**b**) Explain why phase diagrams with "deep eutectics" are promising for glass formation by quenching from the liquid. In your explanation, draw and discuss the T_0 lines on a eutectic diagram.

9.2 Consider a free energy surface much like the one shown in 9.8b, but for simplicity assume that the four local minima are all at compositions 10% away from the corner of the diagram (e.g., the minimum at upper left is at $c_B = 0.9$ and $c_H = 0.1$). Suppose the alloy composition is $(c_B, c_H) = (0.5, 0.5)$. Obtain the compositions and phase fractions in equilibrium, following these steps:

(**a**) Draw lines connecting the two pairs of minima at constant c_H. Why do you expect the equilibrium phases to have these values of c_H, and what are their volume fractions?

(**b**) Now allow some mobility of B-atoms. Show that for this special case, movement up the line to the deep minimum on the upper right requires an equal movement down the line at left towards the shallower minimum in the lower left corner. Also show that total free energy is minimized when the alloy has equal fractions of B-rich H-rich phase, and B-poor H-poor phase.

(**c**) Suppose the overall composition of the alloy is now $(c_B, c_H) = (0.75, 0.75)$. What alteration is required for the step b? What is the final state of equilibrium in this case?

(**d**) Describe an algorithm to obtain the state of equilibrium for an arbitrary alloy composition with c_B, c_H between 0.1 and 0.9.

9.3 Suppose the mixed derivatives F_{xy} are small (in a sense determined in part c). The eigenvectors should be similar to those of Eqs. 9.35, with some deviation that can be expressed as with the parameter δ_i as follows

$$\begin{bmatrix} \frac{\partial^2 F}{\partial x^2} & \frac{\partial^2 F}{\partial x \partial y} \\ \frac{\partial^2 F}{\partial y \partial x} & \frac{\partial^2 F}{\partial y^2} \end{bmatrix}_{x_0, y_0} \begin{bmatrix} 1 \\ \delta_1 \end{bmatrix}_1 = \lambda_1 \begin{bmatrix} 1 \\ \delta_1 \end{bmatrix}_1, \tag{9.63}$$

$$\begin{bmatrix} \frac{\partial^2 F}{\partial x^2} & \frac{\partial^2 F}{\partial x \partial y} \\ \frac{\partial^2 F}{\partial y \partial x} & \frac{\partial^2 F}{\partial y^2} \end{bmatrix}_{x_0, y_0} \begin{bmatrix} \delta_2 \\ 1 \end{bmatrix}_2 = \lambda_2 \begin{bmatrix} \delta_2 \\ 1 \end{bmatrix}_2. \tag{9.64}$$

(**a**) It is straightforward to work with Eq. 9.63 when δ_1 is small and the term $\delta_1 \frac{\partial^2 F}{\partial x \partial y}$ can be neglected. Show that

$$\delta_1 = \left[\frac{\partial^2 F}{\partial x \partial y} \right] \left[\frac{\partial^2 F}{\partial x^2} - \frac{\partial^2 F}{\partial y^2} \right]^{-1}. \tag{9.65}$$

(b) For the eigenvector λ_2 of Eq. 9.64, show that δ_2 has the opposite sign as δ_1.

(c) When the difference $F_{xx} - F_{yy}$ is small, F_{xy} becomes more important in determining the direction of unmixing in the space of c_B and c_H. Considering the shape of the free energy surface near equilibrium, why should the coupling term F_{xy} become more important when $F_{xx} - F_{yy}$ is small?

9.4 **(a)** Using Fig. 9.4 and the prescription described after Eq. 9.11, obtain expressions for the change PdV of a multiferroic material.

(b) Do the same for the change in magnetization $\vec{H} \cdot d\vec{M}$ of a multiferroic material. What is $\vec{H} \cdot d\vec{M}$ when volume is constrained and $dV = 0$? Why is there a different effect on $\vec{H} \cdot d\vec{M}$ when pressure is constant (i.e., what are the other sources of dV)?

9.5 **(a)** Assuming elastic strains only, why does the curve labeled "Microstructure" in Fig. 9.5 have a maximum, and then decrease to zero at the temperature where it was originally equilibrated?

(b) How would the measured $\Delta C_V(T)$ of Fig. 9.5 be changed if plastic deformation were induced by thermal expansion at a temperature of, say, 500 K?

9.6 **(a)** The result of Eq. 9.52 can be extended to integrals of the form

$$\mathcal{I}_{sq} = \int_{-\infty}^{\infty} \exp\left[-(\alpha q + \beta q^2)\right] dq. \tag{9.66}$$

Start by defining a new variable η, and use the trick of completing the square

$$\eta \equiv \frac{\alpha}{2\sqrt{\beta}} + \sqrt{\beta}q, \tag{9.67}$$

$$\exp\left[-\left(\alpha q + \beta q^2\right)\right] = \exp\left[-\eta^2 + \frac{\alpha^2}{4\beta}\right]. \tag{9.68}$$

Show that Eq. 9.66 becomes

$$\mathcal{I}_{sq} = \exp\left(\frac{\alpha^2}{4\beta}\right) \sqrt{\frac{\pi}{\beta}}, \tag{9.69}$$

which is the same as Eq. 9.52 when $\alpha = 0$. (*Hint:* The differential $dq = d\eta/\sqrt{\beta}$.)

(b) Show that for a potential V of this form in the position coordinates $\{q_i\}$

$$V(\{q_i\}) = V_0 + \sum_{i=1}^{3N} \left[q_i V_i^{(1)} + q_i^2 V_i^{(2)}\right], \tag{9.70}$$

Eq. 9.57 becomes

$$F = V_0 - \sum_{i=1}^{3N} \frac{V_i^{(1)\,2}}{4V_i^{(2)}} - k_B T \sum_{i=1}^{3N} \ln\left(\sqrt{\frac{2m_i}{V_i^{(2)}}} \frac{\pi k_B T}{h}\right). \tag{9.71}$$

What are the dimensions of the new term in Eq. 9.71, not present in Eq. 9.57? Is the entropy affected by the new terms $q_i V_i^{(1)}$ in the potential energy?

10   Atom Movements with the Vacancy Mechanism

Chapter 3 derived the diffusion equation with the assumption of random atom jumps. Solutions to the diffusion equation were presented, but the reader was warned that these solutions require that the diffusion constant D remains constant, and this is rarely true as an alloy evolves. As the state of the material evolves during the phase transformation, changes in both chemical interactions and jump frequencies can alter D by orders of magnitude, especially at low temperatures.

Even if the state of the material is unchanged over a reasonable time, physical details of atom movements require modifications to the random walk model of diffusion. When atom movements occur by the vacancy mechanism, where a mobile vacancy rearranges atoms in its wake, the motions of the atoms are not random – the direction of the next jump is correlated to the direction of the previous jump. This chapter explains why atom jumps with a vacancy mechanism are not random, even when the vacancy itself moves by random walk. In an alloy with chemical interactions strong enough to cause a phase transformation, however, the vacancy frequently resides in energetically favorable locations, so any assumption of random walk can be seriously in error.

When materials with different diffusivities are brought into contact, their interface is displaced with time because the fluxes of atoms across the interface are not equal in both directions. Even the meaning of the interface, or at least its position, requires some thought. Stresses and voids may develop during interdiffusion. Cross-effects from gradients in temperature and composition are discussed.

An applied field can bias the diffusion process towards a particular direction, and such a bias can also be created by chemical interactions between atoms. Chapter 10 ends with two other topics of diffusion – one is atom diffusion that occurs in parallel with atom jumps forced without thermal activation, and the second is a venerable statistical mechanics model of diffusion that has components used today in many computer simulations of diffusion. The statistical mechanics model also gives a deeper understanding of the activation free energy of a diffusive jump.

10.1 Random Walk and Correlations

10.1.1 Correlation Factor (General Features)

Section 3.2 (and 23.1) derived the diffusion equation. The derivation assumed random jumps of individual atoms, but used a coarse average over many atoms. Here we focus on the motions of an individual diffusing atom, and average these motions later. Figure 10.1 is a picture of the process in two dimensions. It shows where the atom is located, with respect to its start at the origin, after six jumps. Six jump vectors are added to obtain this distance, $\vec{R}_6$. In general, after n jumps

$$\vec{R}_n = \sum_{i=1}^{n} \vec{r}_i. \tag{10.1}$$

We calculate the square of $\vec{R}_n$, the "mean-squared displacement," which is a scalar: $R_n^2 = \vec{R}_n \cdot \vec{R}_n$. Writing out in detail the product of two forms of Eq. 10.1

$$\begin{aligned} R_n^2 = \vec{r}_1 \cdot \vec{r}_1 + \vec{r}_1 \cdot \vec{r}_2 + \vec{r}_1 \cdot \vec{r}_3 + \cdots \vec{r}_1 \cdot \vec{r}_n \\ + \vec{r}_2 \cdot \vec{r}_1 + \vec{r}_2 \cdot \vec{r}_2 + \vec{r}_2 \cdot \vec{r}_3 + \cdots \vec{r}_2 \cdot \vec{r}_n \\ + \vec{r}_3 \cdot \vec{r}_1 + \vec{r}_3 \cdot \vec{r}_2 + \vec{r}_3 \cdot \vec{r}_3 + \cdots \vec{r}_3 \cdot \vec{r}_n \\ + \cdots \\ + \vec{r}_n \cdot \vec{r}_1 + \vec{r}_n \cdot \vec{r}_2 + \vec{r}_n \cdot \vec{r}_3 + \cdots \vec{r}_n \cdot \vec{r}_n. \end{aligned} \tag{10.2}$$

Writing the terms in this square form makes it easier to identify terms with equal differences in their indices. Regroup the big sum of Eq. 10.2 into sums of terms along diagonals from upper left to lower right:

- The single sum along the diagonal has n terms with indices that differ by 0.
- The second sum (two of them) from one off the diagonal has $n - 1$ terms with indices that differ by 1 (note that $\vec{r}_i \cdot \vec{r}_{i+j} = \vec{r}_{i+j} \cdot \vec{r}_i$, so these two sums are equal).
- The third sum (two of them) from two off the diagonal has $n - 2$ terms with indices that differ by 2.
- …
- The nth sum (two of them) from $n - 1$ off the diagonal has one term with indices that differ by $n - 1$ (these are the single elements in the upper right and lower left corners).

Writing these sums explicitly

$$R_n^2 = \sum_{i=1}^{n} \vec{r}_i \cdot \vec{r}_{i+0} + 2\sum_{i=1}^{n-1} \vec{r}_i \cdot \vec{r}_{i+1} + 2\sum_{i=1}^{n-2} \vec{r}_i \cdot \vec{r}_{i+2} + \cdots + 2\sum_{i=1}^{1} \vec{r}_i \cdot \vec{r}_{i+n-1}. \tag{10.3}$$

Keep the first sum separate from the others, but sum the others by making a double sum

$$R_n^2 = \sum_{i=1}^{n} r_i^2 + 2\sum_{j=1}^{n-1}\sum_{i=1}^{n-j} \vec{r}_i \cdot \vec{r}_{i+j}, \tag{10.4}$$

where the first sum contains the n diagonal elements of the square form of Eq. 10.2, and all off-diagonal elements are in the double sum. The variable j in Eq. 10.4 is the difference in indices between two vectors $\vec{r}_i$ and $\vec{r}_{i+j}$. The variable j also accounts for the reduced number of terms in the sums that are further off the diagonal, and j ranges from 1 to $n - 1$, which is the number of (off-diagonal) elements across the square structure of Eq. 10.2. The dot products of vectors in Eq. 10.4 are now written with the cosine of the angles, θ, between the vectors and their lengths

$$R_n^2 = \sum_{i=1}^{n} r_i^2 + 2 \sum_{j=1}^{n-1} \sum_{i=1}^{n-j} r_i\, r_{i+j} \cos \theta_{i,i+j}. \tag{10.5}$$

Box 10.1   **Random Walk Result**

The random walk result is easily obtained from Eq. 10.4 if the directions of successive jumps are completely uncorrelated, i.e., a subsequent jump has no bias in orientation with respect to the previous jump. In this case, either the time average or the ensemble average $\langle \vec{r}_i \cdot \vec{r}_{i+j} \rangle$ is zero for all $j > 0$, giving the result for a random walk

$$R_n^2 = n\, r_i^2, \tag{10.6}$$

assuming all n jumps have the same length $r_i = a$. Using the relationship between jump rate, jump distance, and diffusion coefficient in three dimensions, $D = \Gamma\, a^2/6$ (Eq. 3.18)

$$R_n^2 = n\, \frac{6D}{\Gamma}. \tag{10.7}$$

Finally, recognizing $1/\Gamma$ as a time and n/Γ as the total time for diffusion, τ

$$R_n^2 = 6D\, \tau, \tag{10.8}$$

adequately consistent with Eq. 3.45.

Now assume that all jumps have the same length, r, as is usually expected when diffusion occurs on a crystal lattice. The first sum is simply a sum of n identical terms

$$R_n^2 = n\, r^2 + 2r^2 \sum_{j=1}^{n-1} \sum_{i=1}^{n-j} \cos \theta_{i,i+j}. \tag{10.9}$$

We have been considering a single chain of jumps as depicted in Fig. 10.1. Our interest, however, is in macroscopic behavior, which averages many jumps of many atoms. There are subtle issues in how to perform the averaging. A thermodynamic average is an ensemble average, which averages quantities from a large number of identical systems.

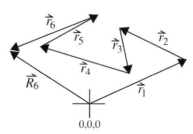

Figure 10.1   Six random jump vectors, summed to give $\vec{R}_6$.

An ensemble average would average the double sums in Eq. 10.9. An alternative average is a time average of many jumps of a single species. We assume the equivalence of the ensemble average and the time average.

We work with the time average of a system that is in steady state (so there is no phase transformation underway, for example). If the system is in steady state, there is no special meaning of $t = 0$. The time-averaged angle between a pair of jumps cannot depend on whether the first jump of the pair is the very first jump at $t = 0$ or the nth jump at a later time. What is important in averaging the jump angles is the time between jumps, not absolute time. The system loses memory of the direction of a previous jump, but the time for losing memory is the same for jumps that happen earlier or later in the jump sequence. We therefore replace $\cos \theta_{i,i+j}$ with $\cos \theta_j$ and average it

$$\overline{R_n^2} = n r^2 + 2r^2 \sum_{j=1}^{n-1} \sum_{i=1}^{n-j} \overline{\cos \theta_j}, \tag{10.10}$$

where the overlines indicate time averages. The sum over i in Eq. 10.10 is no longer needed (i is the number of the jump in the sequence), because all $n-j$ terms in this sum are the same

$$\overline{R_n^2} = n r^2 + 2r^2 \sum_{j=1}^{n-1} (n - j) \overline{\cos \theta_j}. \tag{10.11}$$

We rewrite Eq. 10.11 as

$$\overline{R_n^2} = f n r^2, \tag{10.12}$$

where the "correlation factor," f, is defined as

$$f = 1 + \lim_{n \to \infty} \frac{2}{n} \sum_{j=1}^{n-1} (n - j) \overline{\cos \theta_j}. \tag{10.13}$$

After a large number of jumps, the jumping species is expected to lose memory of its initial jump, so $\overline{\cos \theta_j} \to 0$ for large j. As $n \to \infty$, however, $j \ll n$, so

$$f = 1 + 2 \sum_{j=1}^{\infty} \overline{\cos \theta_j}. \tag{10.14}$$

We next make use of the relationship $\overline{\cos \theta_j} = (\overline{\cos \theta_1})^j$, which is expected because each subsequent jump has the average correlation $\overline{\cos \theta_1}$ along the direction of the previous jump.

$$f = 1 + 2 \sum_{j=1}^{\infty} \left(\overline{\cos \theta_1} \right)^j. \tag{10.15}$$

Recognizing a geometric series in the sum (and $x^0 = 1$)

$$\sum_{j=1}^{\infty} x^j = \sum_{j=0}^{\infty} x^j - 1 = \frac{1}{1 - x} - 1, \tag{10.16}$$

Eq. 10.15 reduces to

$$f = \frac{1 + \overline{\cos \theta_1}}{1 - \overline{\cos \theta_1}}. \tag{10.17}$$

Equation 10.17 has a very simple form. In the case of a truly random walk, where the $\cos \theta_1$ averages to zero (which it will if the jumps are equally likely left, right, up, down, in, out), Eq. 10.17 gives $f = 1$ and Eq. 10.12 becomes the result for the mean-squared displacement of a random walk, $R_n^2 = nr^2$. This amounts to adding all elements along the long diagonal in Eq. 10.2, and ignoring the off-diagonal elements that give the $\cos \theta_j$ terms. It turns out, however, that atom motions are usually correlated, so they usually depend on the direction of previous jumps. In these cases $f < 1$, so R_n^2 is smaller than expected from random jumps as shown by Eq. 10.12. When chemical interactions are included, as in Sections 10.2.1 and 10.2.2, it is possible that $f \ll 1$.

10.1.2 Correlation Factor for Atoms with Uncorrelated Vacancy Jumps

From Eq. 10.17 we are left with the task of calculating $\overline{\cos \theta_1}$. This depends on the diffusion mechanism, lattice, and composition. Here we work the example of the motions of a tracer atom[1] shown as a gray circle in Fig. 10.2. Assume there has just been an exchange of sites between the vacancy and the tracer, so the vacancy must have jumped to the right, and the tracer to the left. What is the average cosine of the angle between the next jump of the tracer with respect to its original jump to the left?

In working this calculation for the tracer, we watch the vacancy because the vacancy jump sequence sets the probabilities. Assume the vacancy executes a random walk on the lattice, so its probability of exchange with any one of its six neighbors is 1/6. The probability is $(1/6)^m$ for any specific sequence of m jumps of the vacancy. For the tracer atom to move, it must jump into one of the six sites of its first neighbor shell. Figure 10.2 shows all possible ways that the tracer can move into a neighboring site within three jumps of the vacancy. The average $\overline{\cos \theta_1}$ is calculated as the sum of the values of $\cos \theta_{sq}$ for the sequences (in the right column of Fig. 10.2), times the number of sequences N_{sq}, times the probability of the vacancy jump sequence, $(1/6)^m$

$$\overline{\cos \theta_1} = \sum_{n=m}^{m_{max}} \frac{N_{sq}}{6^m} \cos \theta_{sq}, \tag{10.18}$$

$$\overline{\cos \theta_1} = \frac{1}{6}(-1) + \frac{2}{6^2}\frac{-1}{2} + \frac{2}{6^3}\frac{1}{2} + \frac{2}{6^3}\frac{-1}{2} + \frac{5}{6^3}(-1), \tag{10.19}$$

$$\overline{\cos \theta_1} = -\frac{47}{216}. \tag{10.20}$$

Using Eq. 10.17, the correlation factor is evaluated as $f = 0.64$. This differs from the more accurate result of $f = 0.56$ because we stopped our analysis at only three jumps of the vacancy. Table 10.1 presents correlation factors for tracer atoms on a number of

---

[1] This means that an atom is "tagged" in some way to follow its motion amid numerous similar atoms. Radioactive atoms were the first robust tracer atoms.

Table 10.1 Correlation factors for tracer atoms with vacancy diffusion	
Lattice	f
fcc	0.78145
hcp	0.78146 parallel to c
hcp	0.78121 normal to c
bcc	0.72719
sc	0.6531
dc	0.5
Honeycomb	0.5
Square	0. 46694
Hexagonal	0.56006

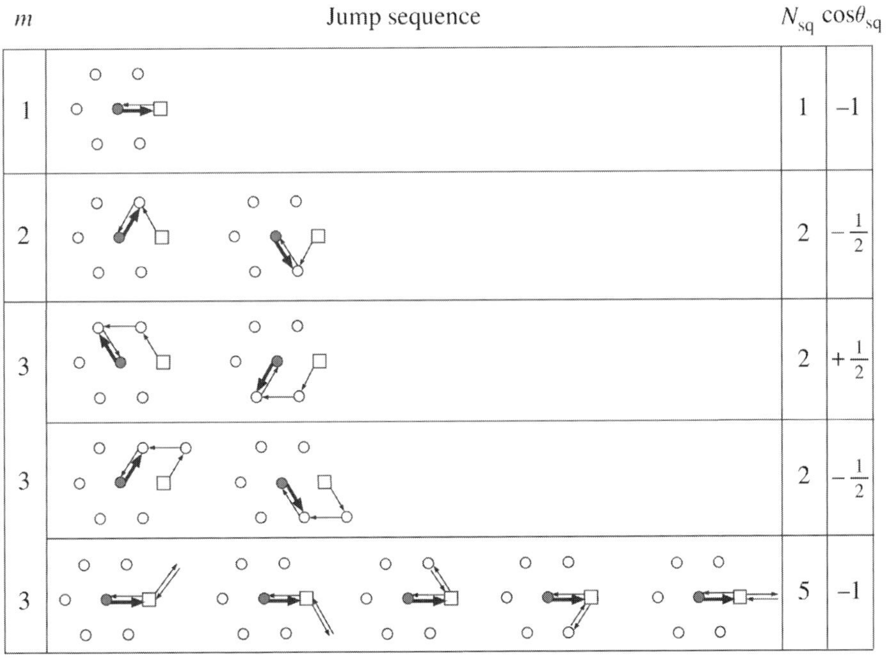

Figure 10.2 First jump of tracer atom (gray, at center) after specific sequences of m jumps of a vacancy (square in initial position). Light arrows are vacancy jumps, heavy arrows are for the tracer. It is assumed that the vacancy and tracer have just interchanged positions, so the previous tracer jump was to the left.

lattices, assuming the vacancy executes a random walk. These are all less than 1, but not dramatically so. Notice that lattices with higher coordination numbers z tend to have larger values of f. Sometimes this approximation is used: $f \simeq 1 - 2/z$.

When enumerating sequences of vacancy jumps, it is useful to draw arrows for both the vacancy and the tracer atom jumps, as in Fig. 10.2. This practice will avoid the error of

Figure 10.3   Three jumps of a vacancy (light arrows), with the first involving the jump of the tracer (heavy arrow). The third jump of the vacancy involves the second jump of the tracer, which contributes to the average of $\cos\theta_2$, not $\cos\theta_1$.

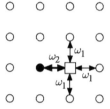

Figure 10.4   Vacancy interchanges with first neighbors on a square lattice, with a different interchange frequency ω_2 for the solute atom (black) than the majority species (white).

including sequences such as shown in Fig. 10.3, which shows two arrows for the tracer. This means that the sequence pertains to the second jump of the tracer, i.e., $\cos\theta_2$, which is not to be included in the average of $\cos\theta_1$.

10.2 Correlation Factors for Atoms and Vacancies in Alloys

10.2.1 Correlation Factor for a Vacancy in a Heterogeneous Alloy

We just showed how atomic diffusion with a vacancy mechanism is correlated, even though the vacancy executes a random walk. In an alloy the vacancy motion becomes correlated too, and its correlation factor can be much less than 1. Consider the black atom in Fig. 10.4 as a different species from the white, and with a much higher probability of exchanging sites with the vacancy. The frequencies of exchange are ω_1 and ω_2, and $\omega_2 \gg \omega_1$.[2]

There is a competition between the four neighbors for jumping into the vacant site in Fig. 10.4, and all four jumps should be considered for calculating the average $\overline{\cos\theta_1}$ of the vacancy. Since $\omega_2 \gg \omega_1$, the solute at left is most likely to jump with probability p_2, but the other three jumps have the probability p_1

$$p_2 = \frac{\omega_2}{\omega_2 + 3\omega_1}, \tag{10.21}$$

$$p_1 = \frac{\omega_1}{\omega_2 + 3\omega_1}. \tag{10.22}$$

---

[2] In practice, different chemical species will have different atom–vacancy interchange frequencies, and for thermally activated jumps these frequencies become increasingly different at lower temperatures.

The average of $\cos\theta_1$ is the sum of these probabilities times the cosines of the angles between the original vacancy jump (to the right) and the next jump of the vacancy

$$\overline{\cos\theta_1} = p_2(-1) + 2p_1(0) + p_1(+1), \tag{10.23}$$

$$\overline{\cos\theta_1} = \frac{-\omega_2 + \omega_1}{\omega_2 + 3\omega_1}, \tag{10.24}$$

$$\overline{\cos\theta_1} \simeq \frac{-\omega_2 + \omega_1}{\omega_2}\left(1 - \frac{3\omega_1}{\omega_2}\right), \tag{10.25}$$

$$\overline{\cos\theta_1} \simeq \frac{1}{\omega_2}\left(\omega_1 - \omega_2 + 3\omega_1 - \frac{3\omega_1^2}{\omega_2}\right), \tag{10.26}$$

$$\overline{\cos\theta_1} \simeq -1 + 4\frac{\omega_1}{\omega_2}, \tag{10.27}$$

which is substituted into Eq. 10.17 for f to obtain

$$f = \frac{1 - 1 + 4\frac{\omega_1}{\omega_2}}{1 + 1 - 4\frac{\omega_1}{\omega_2}}, \tag{10.28}$$

$$f \simeq 2\frac{\omega_1}{\omega_2}. \tag{10.29}$$

The correlation factor for the vacancy, and its mean-squared displacement after n jumps, becomes small as $\omega_2 \gg \omega_1$. Successive jumps tend to reverse each other – the solute atom tends to jump back and forth into the vacancy. Notice that f is proportional to ω_1, the frequency of vacancy exchanges with the slower-moving majority species. This is not unusual for diffusion problems in an alloy. Usually both species must move for macroscopic diffusion to occur. If one species remains unmoved on its lattice sites, rapid motions of the other species around it cannot lead to chemical rearrangements [123, 124].

In the limit where $\omega_1/\omega_2 = 0$, the vacancy and solute atom exchange are the only events that occur. Alternate jumps of the vacancy are in opposite directions, so $\cos\theta_1 = -1$ and $f = 0$, and similarly for the solute atom. After numerous jumps of the solute atom, Eq. 10.12 gives $\overline{R_n^2} = 0$, and neither the solute atom nor the vacancy has moved a macroscopic distance.

10.2.2 Correlation Factors for Ordered Alloys

The previous section on correlated vacancy motion used a simple parameterization of jump frequencies, without considering the reasons for the different frequencies. Correlations of vacancy motion can be caused by the same interatomic interactions that are responsible for phase transitions, and we can assess these effects on the temperature scale of the critical temperature.

Homogeneous Order

The correlation factor f becomes small in ordered alloys at low temperatures. Consider the B2 ordered structure or the chessboard structure of Fig. 2.19, where each first-nearest neighbor (1nn) is on the other sublattice. Assuming A-atoms prefer the α-sublattice (and B the β), the probability to reverse an atom–vacancy exchange depends on the starting sublattice of the A-atom. An A-atom that started on the α-sublattice and moved to the β is more likely to reverse its interchange with the vacancy than an A-atom that started on the unfavorable β-sublattice. An increased probability for reversal leads to a large, negative contribution to $\cos \theta_1$ in Eq. 10.17, and a lower correlation factor f for the A-atom (although this needs to be considered in conjunction with those A-atoms that start on the β-sublattice).

Because local chemical preferences play a more important role at low temperatures, the A jumps from $\alpha \rightarrow \beta$ sublattices become very slow at low temperatures, and the reversal of A-atom jumps (from $\beta \rightarrow \alpha$) becomes even more probable at low temperatures. Finally, the higher degree of chemical order at low temperatures gives fewer unfavorably placed atoms to exchange sites with the vacancy. H. Bakker adapted the random walk problem to account for different sublattices of an ordered alloy [125], using some of the methods of Chapter 23 (online) and a pair approximation for chemical interactions. Some aspects of this analysis are presented in Problem 10.7. Bakker found that at half the critical temperature for ordering it is reasonable to expect correlation factors of 0.01 or less. Correlations of vacancy jumps are much more important in ordered alloys than in crystals of pure elements.

Vacancy Traps

Figure 10.5 shows an A-atom that is likely to exchange sites with a vacancy in an ordered alloy, in which the unlike A–B pairs are preferred. All neighbors of the vacancy are A-atoms, but three of them are bonded to two or three B-atoms, and are less likely to move than the central A-atom (it is bonded to three A-atoms and zero B-atoms). Especially at low temperatures, the vacancy is "trapped," and will exchange sites numerous times with this A-atom. A closer examination of the local atomic configuration in Fig. 10.5 shows an antiphase domain boundary in the middle, running from top to bottom. An antiphase domain boundary (see Fig. 7.3) separates two ordered regions where the A- and B-atoms switch lattice sites. At low temperatures, vacancies are expected to become trapped at antiphase domain boundaries, or in regions with local disorder.

Figure 10.5 (a), (b) Transitions between these two configurations are expected to be most probable because in both configurations the central A-atom is unfavorably bonded to its neighbors (this alloy favors A–B pairs). Unfavorable pairs are indicated with lines.

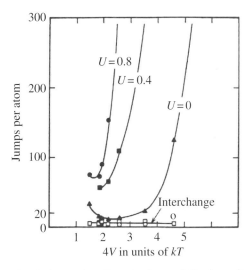

Figure 10.6 Number of vacancy jumps per atom to change a two-dimensional square lattice from a disordered solid solution to a partially ordered alloy (with a Warren–Cowley short-range order parameter of 0.25) versus $4V$. Reprinted from [127] with the permission of AIP Publishing.

The temperature dependence of trapping is shown in Fig. 10.6 as an overall effect on the kinetics of a disorder → order transformation. The thermodynamics of ordering depends only on the parameter $4V$ (cf. Eq. 2.49)

$$4V \equiv e_{AA} + e_{BB} - 2e_{AB}, \tag{10.30}$$

which averages the chemical preferences of the A-atoms and B-atoms. Kinetic processes depend on the individual atom preferences, however, for which the parameter U is defined [126] as

$$U \equiv (e_{AA} - e_{BB})/4V. \tag{10.31}$$

Figure 10.6 shows that the effects of vacancy trapping occur at much higher temperatures when $|U| > 0$, and the chemical preferences are dominated by one atom over the other. From a control experiment, there is no vacancy trapping for the (physically unrealistic) interchange mechanism, where adjacent pairs of atoms interchange positions without the use of a vacancy, and the efficiency of ordering by atom interchange does not depend strongly on temperature.

At lower temperatures vacancies tend to spend most of their time in the strongest traps, owing to the exponentially decreasing probability of exiting the trap. The strengths of traps for vacancies at antiphase domain boundaries increase as $(z - 1)(e_{AA} - e_{AB})$ because there are more unfavorable bonds to a trapped atom in lattices with larger coordination number z. This is approximately the same scaling as the critical temperature for ordering. Nevertheless, as a fraction of the critical temperature, lattices of low coordination number are more susceptible to vacancy trapping at antiphase domain boundaries (e.g., Fig. 10.5) because for low z a strong trap can be constructed with fewer atoms.

Box 10.2   **Positron Annihilation Spectroscopy**

This spectroscopy uses antiparticles to measure vacancy concentrations in materials. A common source of positrons is ^{22}Na, which decays to ^{22}Ne by emitting a positron from its nucleus. The ^{22}Ne nucleus is initially in an excited state, and quickly emits a 1.27 MeV γ-ray that can be used to mark the birth of the positron. The positrons have energies of hundreds of keV, but after they enter a material they lose energy in sequence to core electron ionizations, valence electron ionizations, plasmon creation, electron–hole pair generation, and phonons. Conveniently, a positron reaches thermal energy in only 10^{-12} to 10^{-11} s. This is short compared to the positron lifetime, so positrons are usually "thermalized," and diffuse through the material until they become "trapped" in favorable locations. Positron traps have relatively low electron density (as opposed to cores of atoms), and the positrons have longer lifetimes at locations of low electron density. After times from fractions to several ns, the positron annihilates with an electron, emitting two photons of energy 0.511 MeV (the rest mass of the electron or positron).

Two types of data are measured by positron annihilation spectroscopy [128, 129]. Using the 1.27 MeV γ-ray to start a clock and the annihilation photon to stop a clock, the lifetime of each positron can be measured. The lifetime is greater when annihilation occurs in a location of low electron density such as a vacancy, so spectra of positron lifetimes can determine vacancy concentrations. A second approach uses Doppler energy shifts of annihilation photons to distinguish between positrons that annihilated with fast-moving core electrons, or with slower electrons in regions of low density. The energy spectrum of annihilation photons is interpreted as a central peak that originates from positrons in vacancies, and a broadened component with "wings" that extend to positive and negative energies from annihilations with core electrons. The "peak-to-wings ratio" is a measure of the vacancy concentration. Positron annihilation spectroscopy is sensitive to vacancy concentrations in the range from 10^{-7} to 10^{-3}, but empirical calibrations are required and artifacts are possible [130]. Calculations of the rates of positron annihilations have been approached by extending density functional theory [131]. Assuming the positron has a low energy, new terms in the Hamiltonian for a crystal with a positron include the electron–positron correlation energy, and how the electron–electron interactions are altered in the presence of a positron. Accurate calculations are a challenge [132].

10.3 Phenomena in Alloy Diffusion

The previous section showed that the prediction of a diffusion coefficient from atomic jump frequencies, interatomic interactions, and the geometry of the lattice is more complicated than expected from the simple expression $D = \Gamma a^2/6$ (Eq. 3.18). Nevertheless, by the substitution of $\Gamma \rightarrow f\Gamma$, where f is the correlation factor, we can, in principle, calculate a diffusion coefficient that works well for a homogeneous material in steady state. Heterogeneities like the one of Fig. 10.5 are a concern for alloys, however.

Now we consider diffusion in a material with an inhomogeneous chemical composition. Some issues for diffusion in a chemically inhomogeneous material are:

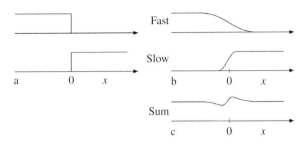

Figure 10.7 Composition profiles for diffusional intermixing of two ideal gases. (**a**) Initial condition at $t = 0$ – a sharp interface between the two gases. (**b**) Independent diffusion of fast, light gas (top) and slow, heavy gas (bottom). (**c**) Sum of concentration profiles, showing increased and decreased total atom density near interface.

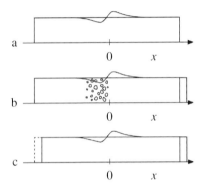

Figure 10.8 Resolution of atom concentration problems for a bar of crystalline material. (**a**) Original composition profile from Fig. 10.7c. (**b**) Excess matter forms more crystal on the right, deficit of matter causes Kirkendall voids on the left. (**c**) Excess matter forms more crystal on the right, deficit of matter causes a loss of crystal on the left.

- mobility differences between different chemical species;
- density conservation (or not);
- chemical interactions.

The first two issues are illustrated with Fig. 10.7, which shows two ideal gases that mix by diffusion. An ideal gas has no chemical interactions between point-like atoms as they move past each other, and independent diffusion equations can be used for both chemical species. The fast, light gas and the heavy, slow gas are brought into contact at $t = 0$. The initial conditions are sharp concentration cutoffs at $x = 0$ for both gases, and appropriate solutions to both diffusion equations are the error functions of Section 3.3.2. These two error function solutions are illustrated in Fig. 10.7b. If we sum the concentrations from the two error function solutions, we find the total concentration of Fig. 10.7c. Note the peculiar density variation near the original interface.

For diffusion of two chemical species in a solid, however, the profile of Fig. 10.7c will not do. To the right of the interface, the increased atomic density is not possible, especially on a crystal lattice that has a fixed number of sites per unit volume. Instead, as shown in Figs. 10.8b and 10.8c, the extra atoms that have moved to the right must cause the bar to grow to the right. Most likely, there will be additional crystal sites created to the right of the original interface. The atom deficit on the left of the original interface can be accommodated two ways, and both occur in nature. For the case of Fig. 10.8b, vacant crystal sites are formed on the crystal lattice, and these coalesce into voids, or small holes to the left of the original interface. These are called "Kirkendall voids." Another possibility is the annihilation of crystal sites as in Fig. 10.8c. In this case, the left end of the bar moves to the right, displacing an amount of volume needed to account for the missing atoms (minus any voids that have formed).

10.3.1 Marker Velocity

Suppose the difference in atom mobility is accommodated as shown in Fig. 10.8c, where the entire bar of material crawls a bit to the right. It is rarely practical to measure such tiny motions in the laboratory reference frame, especially when the sample is quite hot, but we can design an experiment to keep track of the original interface. The idea is to put inert "markers" at the original interface between the two different materials at $x = 0$ and $t = 0$. Suitable markers should be materials with negligible diffusivity at the temperature of interest, and no reactivity with either of the chemical species. (Tungsten, with its very high melting temperature, is often a suitable marker for metallic systems.) For the case of Fig. 10.8c, the markers move to the left, owing to the greater mass transport to the right. This is also true for Fig. 10.8b, although to a lesser extent. Measuring the "marker velocity" is an important way to understand differences in the diffusivity of two chemical species.

The classic marker velocity experiment was performed by Smigelskas and Kirkendall with a diffusion couple, as shown in Fig. 10.9 [133]. A brass rod was wrapped with molybdenum wire, and electroplated with a thick layer of copper around it. The distance d between the Mo wires was measured, and after the diffusion couple was heated it was found that d decreased. The interpretation is that Zn diffuses faster than Cu, so more atoms moved outside the markers than moved in.

The markers define a plane through which there is a diffusive flux. The marker plane has a positive velocity (along $\hat{x}$ to the right) if there is a greater flux of atoms moving through it to the left than to the right – the marker velocity is in the direction opposite to that of the dominant flux. Any extra atoms moving to the left will add volume in proportion to $1/c$, which has units of cm^3/atom. The greater the flux to the left, the greater the volume added to the left, and the higher the marker velocity to the right. The ratio J/c has units of [atoms/(s cm^2)] [cm^3/atoms] = [cm/s], which is a velocity. The marker velocity is

$$v = -\frac{J_A + J_B}{c},\qquad(10.32)$$

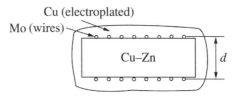

Figure 10.9 Cross-section cut through a diffusion couple with Mo wire markers used for the Kirkendall experiment.

where c is the total number of atoms/cm^3. These relations between composition variables prove useful below

$$c = c_A + c_B, \tag{10.33}$$

$$c_A = c - c_B, \tag{10.34}$$

$$\frac{\partial c_A}{\partial x} = -\frac{\partial c_B}{\partial x}. \tag{10.35}$$

When the fluxes J_A and J_B in Eq. 10.32 are equal and opposite, there is no marker velocity, as expected. Fick's first law gives, for the two chemical species in Fig. 10.7a,b

$$J_A = -D_A \frac{\partial c_A}{\partial x}, \tag{10.36}$$

$$J_B = -D_B \frac{\partial c_B}{\partial x}, \tag{10.37}$$

which upon substitution into Eq. 10.32 gives

$$v = \frac{D_A \frac{\partial c_A}{\partial x} + D_B \frac{\partial c_B}{\partial x}}{c}, \tag{10.38}$$

$$v = \frac{\frac{\partial c_B}{\partial x}(D_B - D_A)}{c}. \tag{10.39}$$

If there is a difference in the diffusivity of two species in a diffusion couple, there will be a marker velocity. As a check, note that the marker velocity is to the right if $\partial c_B/\partial x$ is positive and the B-atoms are the faster-moving species. The net flux is dominated by B-atoms, which move to the left by Eq. 10.37.

10.3.2 Interdiffusion Coefficient

For a diffusion couple of elements A and B, the interdiffusion dynamics can be parameterized with two diffusion coefficients, D_A and D_B. It turns out that this problem can also be parameterized with another pair of variables, the marker velocity v, and an "interdiffusion coefficient" $\tilde{D}$. In this case we solve one diffusion equation instead of two, and this may serve our needs. For a full picture, however, we also need the marker velocity.[3] The interdiffusion coefficient is a composition-weighted average of D_A and D_B, but curiously, Eq. 10.46 shows that the compositional weights are for the opposite species.

Suppose the marker moves with velocity $-v$ with respect to the far ends of the diffusion couple. It sweeps past B-atoms that are stationary. In the moving coordinate system, the flux of B-atoms must include the flux vc_B of B-atoms that do not move. If the solute concentration is in [atoms/cm^3], and v is the velocity of the coordinate system in [cm/s], the product vc has units of [atoms/(s cm^2)], as appropriate for a flux. The total flux of B-atoms across the moving marker (here $+v$ is to the right along $+x$) is

$$J_B = -D_B \frac{\partial c_B}{\partial x} + vc_B. \tag{10.40}$$

---

[3] This is much like the moving solid–liquid boundary considered in Section 5.3.

Substituting the expression Eq. 10.38 for v

$$J_B = -D_B \frac{\partial c_B}{\partial x} + c_B \left(\frac{D_A \frac{\partial c_A}{\partial x} + D_B \frac{\partial c_B}{\partial x}}{c} \right), \tag{10.41}$$

$$J_B = -D_B \frac{\partial c_B}{\partial x} \frac{c_A + c_B}{c} + D_A \frac{\partial c_A}{\partial x} \frac{c_B}{c} + D_B \frac{\partial c_B}{\partial x} \frac{c_B}{c}. \tag{10.42}$$

From Eq. 10.35, $\partial c_A/\partial x = -\partial c_B/\partial x$

$$J_B = -D_B \frac{\partial c_B}{\partial x} \frac{c_A + c_B}{c} - D_A \frac{\partial c_B}{\partial x} \frac{c_B}{c} + D_B \frac{\partial c_B}{\partial x} \frac{c_B}{c}, \tag{10.43}$$

and with the cancellation of two terms

$$J_B = -\left(\frac{D_B c_A + D_A c_B}{c} \right) \frac{\partial c_B}{\partial x}, \tag{10.44}$$

$$J_B = -\tilde{D} \frac{\partial c_B}{\partial x}, \tag{10.45}$$

where the definition of the "interdiffusion coefficient" is

$$\tilde{D} \equiv \frac{D_B c_A + D_A c_B}{c}. \tag{10.46}$$

We now have two equivalent formulations of interdiffusion for a diffusion couple. The first uses the crystal lattice as a frame of reference, and two diffusion equations (Eqs. 10.36 and 10.37) with two diffusion coefficients $\{D_A, D_B\}$. The second uses the ends of the sample for reference. It uses one diffusion equation (Eq. 10.45). The variables in this second approach are the interdiffusion coefficient and the marker velocity $\{\tilde{D}, v\}$ (Eqs. 10.46 and 10.39). The second formulation can be more convenient in practice.

10.3.3 Variable $\tilde{D}(c)$

Unfortunately, there is no general analytical approach to solving a diffusion equation with variable $\tilde{D}(c)$. A nonconstant $\tilde{D}(c)$ violates our basic assumption of Section 3.2 that all atoms jump independently without regard for their local environment. The alternative approach of using two diffusion coefficients $\{D_A, D_B\}$ (instead of $\tilde{D}(c)$ and marker velocity v) is based on the difficult practice of monitoring the diffusion process in a laboratory coordinate system. A further challenge comes from how chemical interactions give a concentration dependence to $D_A(c)$ and $D_B(c)$.

In the present section we do not solve the diffusion equation for a concentrated alloy. Instead, for a one-dimensional diffusion couple we accomplish a related goal of extracting the concentration dependence of $\tilde{D}(c)$. We do not need to measure the velocity v of physical markers, but in the course of the analysis of the diffusion profile, we extract a position for a "Boltzmann–Matano" interface by a condition of conservation of solute.

The mathematical trick for what follows is the "Boltzmann substitution"

$$\eta \equiv \frac{x}{\sqrt{t}}. \tag{10.47}$$

The x and t are independent variables, and they are the only variables upon which η depends (see discussion of Eq. 3.45). For this reason the partial derivatives of η are equal to the total derivatives. The deeper observation is that for interdiffusion across a moving interface, this is also true for c

$$\frac{\partial \eta}{\partial x} = \frac{\mathrm{d}\eta}{\mathrm{d}x}, \qquad \frac{\partial \eta}{\partial t} = \frac{\mathrm{d}\eta}{\mathrm{d}t}, \qquad \frac{\partial c}{\partial \eta} = \frac{\mathrm{d}c}{\mathrm{d}\eta}. \tag{10.48}$$

In what follows, some derivatives of Eq. 10.47 are needed

$$\left.\frac{\partial \eta}{\partial x}\right|_t = \frac{1}{t^{1/2}}, \tag{10.49}$$

$$\left.\frac{\partial \eta}{\partial t}\right|_x = -\frac{1}{2}\frac{x}{t^{3/2}}, \tag{10.50}$$

$$\left.\frac{\partial c}{\partial x}\right|_t = \left.\frac{\partial c}{\partial \eta}\frac{\mathrm{d}\eta}{\mathrm{d}x}\right|_t = \frac{1}{t^{1/2}}\frac{\mathrm{d}c}{\mathrm{d}\eta}, \tag{10.51}$$

$$\left.\frac{\partial c}{\partial t}\right|_x = \left.\frac{\partial c}{\partial \eta}\frac{\mathrm{d}\eta}{\mathrm{d}t}\right|_x = -\frac{1}{2}\frac{x}{t^{3/2}}\frac{\mathrm{d}c}{\mathrm{d}\eta}. \tag{10.52}$$

The one-dimensional diffusion equation is now written in a way that D could be a function of composition, and therefore may vary along x when there is a concentration gradient

$$\frac{\partial c}{\partial t} = \frac{\partial}{\partial x}\left(D\frac{\partial c}{\partial x}\right). \tag{10.53}$$

Substituting the partial derivatives from Eqs. 10.52 and 10.51

$$-\frac{1}{2}\frac{x}{t^{3/2}}\frac{\mathrm{d}c}{\mathrm{d}\eta} = \frac{\partial}{\partial x}\left(D\frac{1}{t^{1/2}}\frac{\mathrm{d}c}{\mathrm{d}\eta}\right), \tag{10.54}$$

$$-\frac{1}{2}\frac{x}{t^{3/2}}\frac{\mathrm{d}c}{\mathrm{d}\eta} = \frac{1}{t^{1/2}}\frac{\partial D}{\partial x}\frac{\mathrm{d}c}{\mathrm{d}\eta} + \frac{1}{t^{1/2}}D\frac{\partial}{\partial x}\left(\frac{\mathrm{d}c}{\mathrm{d}\eta}\right). \tag{10.55}$$

Since the concentration profile depends on η (and not x and t independently), the x-dependence of D also originates from its dependence on η. The two derivatives with respect to x in Eq. 10.55 are more fundamentally written as derivatives with respect to η

$$-\frac{1}{2}\frac{x}{t^{3/2}}\frac{\mathrm{d}c}{\mathrm{d}\eta} = \frac{1}{t^{1/2}}\left(\frac{\mathrm{d}D}{\mathrm{d}\eta}\frac{\mathrm{d}\eta}{\mathrm{d}x}\right)\frac{\mathrm{d}c}{\mathrm{d}\eta} + \frac{1}{t^{1/2}}D\left(\frac{\mathrm{d}^2c}{\partial\eta^2}\frac{\mathrm{d}\eta}{\mathrm{d}x}\right). \tag{10.56}$$

The two derivatives $\mathrm{d}\eta/\mathrm{d}x$ are simply $t^{-1/2}$, by Eq. 10.49. Multiplying through by t helps to simplify

$$-\frac{1}{2}\frac{x}{t^{1/2}}\frac{\mathrm{d}c}{\mathrm{d}\eta} = \frac{\mathrm{d}D}{\mathrm{d}\eta}\frac{\mathrm{d}c}{\mathrm{d}\eta} + D\left(\frac{\mathrm{d}^2c}{\partial\eta^2}\right). \tag{10.57}$$

The definition of η (Eq. 10.47) is found on the left, and the right can be simplified with the chain rule

$$-\frac{\eta}{2}\frac{\mathrm{d}c}{\mathrm{d}\eta} = \frac{\mathrm{d}}{\mathrm{d}\eta}\left(D\frac{\partial c}{\partial\eta}\right). \tag{10.58}$$

The units of this Eq. 10.58 are now concentration [atoms/cm^3], making it appropriate for further analysis of concentration profiles $c(x)$. Setting equal increments $d\eta$ on the two sides of Eq. 10.58, and integrating from $c = 0$ to a $c = c'$ that is less than the maximum c_0 for the diffusion couple

$$-\frac{\eta}{2}\,dc = d\left(D\frac{dc}{\partial\eta}\right),$$

(10.59)

$$-\frac{1}{2}\int_0^{c'}\eta\,dc = D(c')\frac{dc}{\partial\eta}\bigg|_{c'} - D(0)\frac{dc}{\partial\eta}\bigg|_0.$$

(10.60)

Finally, we return from the variable η to the coordinates x and t that are more appropriate for analyzing a measured concentration profile in a diffusion couple. We do so by fixing a time, $t = t'$, which corresponds to the amount of time that diffusion has occurred in the diffusion couple. Derivatives with respect to η are transformed to derivatives with respect to x at a fixed time t'. Using Eq. 10.47 and noting that $dc/d\eta = dc/dx\,\partial x/\partial\eta = dc/dx\,t^{1/2}$

$$-\frac{1}{2}\int_0^{c'}\frac{x}{t'^{1/2}}\,dc = D(c')\frac{dc}{dx}\bigg|_{c'}t'^{1/2} - D(0)\frac{dc}{dx}\bigg|_0 t'^{1/2}.$$

(10.61)

The last term in Eq. 10.61 is zero, however, because there is no variation in concentration in the diffusion couple at $c = 0$, which corresponds to distances far from the interface

$$-\frac{1}{2}\int_0^{c'}x\,dc = t'\,D(c')\frac{dc}{dx}\bigg|_{c'}.$$

(10.62)

Figure 10.10 shows a representative concentration profile from a diffusion couple. The first step in extracting a $\tilde{D}(c)$ from a profile $c(x)$ is to find the origin $x = 0$, called the "Matano interface." (It corresponds to the marker position if there is no change in atomic density.) The condition $dc/dx = 0$ is true at locations far from the interface where the diffusion profile does not yet reach. In particular, far to the left the concentration is c_0 and $dc/dx = 0$. From Eq. 10.62 we obtain, when integrating to $c' = c_0$

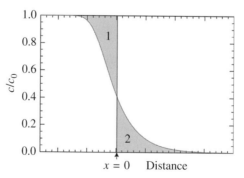

Figure 10.10   The integration of Eq. 10.62 over the full range of c gives a value of 0 (Eq. 10.63), implying that the areas of regions 1 and 2 are equal if $x = 0$ is chosen properly.

$$-\frac{1}{2}\int_0^{c_0} x\,dc = 0. \tag{10.63}$$

Equation 10.63 tells us that, integrated along the vertical "c-direction" in Fig. 10.10, the net integral is zero. This implies that the areas "1" and "2," marked in the figure, are equal when we have chosen the correct Matano interface, i.e., the correct $x = 0$. This can be done iteratively with experimental data.

After finding the Matano interface, $\tilde{D}(c)$ can be evaluated with a slight rearrangement of Eq. 10.62

$$\tilde{D}(c') = -\frac{1}{2t'}\frac{dx}{dc}\bigg|_{c'}\int_0^{c'} x\,dc. \tag{10.64}$$

The evaluation of a particular $\tilde{D}(c')$ is shown in Fig. 10.11a, with the range of integration shown, along with the tangent curve $\frac{dc}{dx}|_{c'}$ needed for the analysis. Since the range of integration is vertical, it can be more convenient to redraw the diffusion profile as shown in Fig. 10.11b. Here the tangent line is used without inversion, and integration is along the horizontal axis. The time t' is assumed known from the experimental record, so it is possible to use Eq. 10.64 to evaluate $\tilde{D}(c)$ at different compositions c' between 0 and c_0.

Knowledge of $\tilde{D}(c)$ is valuable for estimating diffusion behavior. Even without marker velocity information, $\tilde{D}(c)$ can be used to provide the spatial extent of the interdiffusion zone, and can give some information about the diffusion profile.[4] The concentration dependence of $\tilde{D}(c)$ can originate with mobility differences between the chemical species, such as differences in their jump frequencies or correlation effects. As discussed in Section 10.2.2,

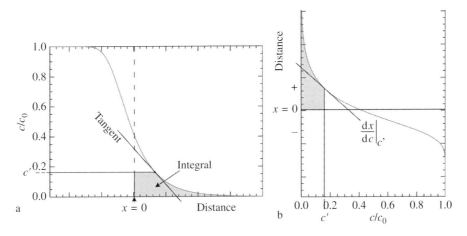

Figure 10.11 (a) Use of the Matano analysis of interdiffusion, showing the integral of Eq. 10.64 and the (inverse) slope needed for evaluation of $\tilde{D}(c')$. (b) Reorientation of the functions in part a, for convenience.

---

[4] Comparing Fig. 10.10 with Fig. 10.7b, we see that in both cases the B-atoms diffuse faster than A-atoms. In Fig. 10.10, the largest displacement of solute along x is at small c, indicating that $\tilde{D}(c)$ is largest for small c. This is consistent with Eq. 10.46 when D_B is large because $D_B\,c_A$ is large when c_B is small.

big correlation effects (small f) can originate with chemical interactions between the species. Sometimes these are the same interactions that lead to phase transformations, and it is quite common for new phases to form in the interdiffusion zone of diffusion couples (Section 11.6). The next section addresses such chemical issues in more detail.

10.4 Diffusion in a Potential Gradient

In inhomogeneous materials, a moving atom will find some regions that are more favorable energetically, with a lower chemical potential. The (negative) gradient of a potential is a force, which tends to move the atom into these regions. A force on a mass causes acceleration, and an increase in the velocity of the atom. Counteracting this acceleration are the interactions of the moving atom with all its surrounding neighbors, since it bumps into them continuously. Before addressing the trajectory of the atom in a chemical potential gradient, we first consider the simpler problem of the effect of collisions on a particle moving under a general force. The physical picture is shown in Fig. 10.12a. Under a steady force the particle accelerates, with linearly increasing velocity. A collision occurs at time t_c. We do not expect to know the details of each collision process. We assume it takes a brief time, and our particle may come away from a collision with positive or negative velocity. The details depend on the temperature and of course the motions of the other atoms during the interaction.

We therefore make approximations, illustrated in Fig. 10.12b, which prove to work well when averaged over numerous collisions. Specifically, we assume that the velocity of the particle is reset to exactly zero after a collision. Collisions are assumed to occur after an average time t_c, from which we obtain the average velocity of the particle, $\bar{v}$. Figure 10.12b reminds us that this average velocity is half of the maximum velocity, which is the acceleration a during t_c

$$\bar{v} = \frac{1}{2} a t_c. \tag{10.65}$$

The acceleration is $\mathcal{F}/m$, where m is the mass of the particle, and the force $\mathcal{F}$ is obtained from the potential U as

$$\mathcal{F} = -\frac{dU}{dx}, \tag{10.66}$$

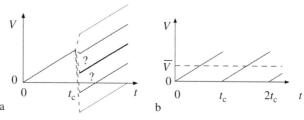

Figure 10.12   (**a**) A collision occurs at time t_c, but the velocity of the particle is unknown. After the collision, the acceleration of the particle is as before. (**b**) Approximation of the process in part a.

so

$$\bar{v} = -\frac{1}{2m} \frac{\mathrm{d}U}{\mathrm{d}x} t_{\mathrm{c}}. \tag{10.67}$$

So far, these motions of atoms in a potential gradient are not diffusive behavior. The motion of an atom in a potential gradient with collisions is parameterized by a "mobility," M

$$M \equiv \frac{\bar{v}}{\mathcal{F}}, \tag{10.68}$$

whereas diffusive motion is parameterized by a diffusion coefficient, D. The two types of dynamics can be related through the flux, J, and Fick's first law. Again, a flux of B-atoms is a concentration times a velocity with units of [atoms/cm^3] [cm/s] = [atoms/(cm^2 s)]

$$J_{\mathrm{B}} = c_{\mathrm{B}} \bar{v}_{\mathrm{B}}, \tag{10.69}$$

and with Eq. 10.68

$$J_{\mathrm{B}} = c_{\mathrm{B}} M_{\mathrm{B}} \mathcal{F}_{\mathrm{B}}, \tag{10.70}$$

$$J_{\mathrm{B}} = -c_{\mathrm{B}} M_{\mathrm{B}} \frac{\mathrm{d}\mu_{\mathrm{B}}}{\mathrm{d}x}, \tag{10.71}$$

where μ_{B} is the chemical potential that generates the force on the B-atom. The chemical potential depends inherently on composition, not distance, so it is more useful to write

$$J_{\mathrm{B}} = -c_{\mathrm{B}} M_{\mathrm{B}} \frac{\partial \mu_{\mathrm{B}}}{\partial c_{\mathrm{B}}} \frac{\mathrm{d}c_{\mathrm{B}}}{\mathrm{d}x}. \tag{10.72}$$

Box 10.3   **Ohm's Law**

Ohm's law for electrical conductivity is now obtained easily. The flux of current, J, is the product of the electron charge e, the electron density n, and the average electron velocity $\bar{v}$

$$J = en\bar{v}. \tag{10.73}$$

The force on the electron is $-eE$, so with Eq. 10.67

$$J = -en\left(\frac{1}{2m} eE t_{\mathrm{c}}\right), \tag{10.74}$$

$$J = -\frac{ne^2 t_{\mathrm{c}}}{2m} E. \tag{10.75}$$

Defining the electrical conductivity σ as

$$\sigma \equiv \frac{ne^2 t_{\mathrm{c}}}{2m}, \tag{10.76}$$

Ohm's law takes the familiar form

$$J = -\sigma E. \tag{10.77}$$

It is easy to work with the chemical potential of an ideal gas (where $c_B V_q = e^{\mu_B/k_B T}$), or an ideal solution

$$\mu_B = k_B T \left(\ln c_B + \ln V_q \right),$$ (10.78)

$$\frac{\partial \mu_B}{\partial c_B} = \frac{k_B T}{c_B},$$ (10.79)

and

$$J_B = -M_B k_B T \frac{dc_B}{dx}.$$ (10.80)

Comparing to Fick's first law, Eq. 3.16

$$J_B = -D_B \frac{\partial c_B}{\partial x},$$ (10.81)

it is found that

$$D_B = M_B k_B T.$$ (10.82)

It is traditional to use this expression for D_B, appropriate for an ideal gas or dilute solution, and add corrections later for more realistic expressions for chemical potentials.

Box 10.4   **The Isotope Effect**

The isotope effect on diffusion can be understood by combining Eqs. 10.66–10.68 and 10.82 to obtain

$$D = \frac{t_c}{2m} k_B T.$$ (10.83)

The collision time is a tricky quantity to obtain, since it originates with details of interatomic interactions. Likewise the factor of $k_B T$ is not what dominates D in a thermally activated process. Nevertheless, the mass denominator in Eq. 10.83 is useful for relating diffusion coefficients of different isotopes, assuming that an isotope has identical chemical interactions with surrounding atoms. For example, the diffusion coefficient of deuterium, ^2H, is half that of protium, ^1H. A characteristic diffusion distance $x = \sqrt{Dt}$, so protium diffuses about 41% further than deuterium under the same conditions. This isotope effect can sometimes be used for isotope separation.

We can obtain a more realistic chemical potential from Eq. 2.27 to describe diffusion in a solid solution with chemical interactions, recalling that the chemical potential μ_B is the change in free energy per change in number of B-atoms

$$\mu_B = \frac{\partial G}{\partial N_B},$$ (10.84)

$$\mu_B = \frac{1}{N} \frac{\partial G}{\partial c_B}.$$ (10.85)

Using this Eq. 10.85 with Eq. 2.27

$$\mu_B = z \left(e_{AB} - e_{AA} + c_B 4V \right) + k_B T \ln \left(\frac{c_B}{1 - c_B} \right),$$ (10.86)

and likewise for the A-atoms ($c_A = 1 - c_B$)

$$\mu_A = z\left(e_{AB} - e_{BB} + c_A\,4V\right) + k_B T\,\ln\left(\frac{c_A}{1 - c_A}\right). \tag{10.87}$$

The concentration dependences of the chemical potentials are calculated

$$\frac{\partial \mu_B}{\partial c_B} = \frac{\partial \mu_A}{\partial c_A} = \left[\frac{4Vz\,c_A c_B + k_B T}{c_A c_B}\right], \tag{10.88}$$

and substituted into Eq. 10.72 and its equivalent for A-atoms to give

$$J_B = -D_B\left[\frac{4Vz\,c_A c_B}{k_B T} + 1\right]\frac{dc_B}{dx}, \tag{10.89}$$

$$J_A = -D_A\left[\frac{4Vz\,c_A c_B}{k_B T} + 1\right]\frac{dc_A}{dx}, \tag{10.90}$$

where

$$D_B = \frac{M_B}{c_A}k_B T, \tag{10.91}$$

$$D_A = \frac{M_A}{c_B}k_B T. \tag{10.92}$$

The terms in the square brackets in Eqs. 10.89 and 10.90 are called "chemical factors." They can effectively increase or decrease the diffusion coefficient.

For the flux of B-atoms, Eqs. 10.89 (and 10.91) transform to Eqs. 10.80 (and 10.82) when $c_B \to 0$ (so $c_A \to 1$). In this limit, all B-atoms are in the same chemical environment, each surrounded by all A-atoms. The other limit, where $c_B \to 1$, might seem to give a peculiar result with Eq. 10.91, since $c_A \to 0$ and D_B apparently diverges in an unphysical way. The resolution to the dilemma is in the mobility, M_B, which is proportional to c_A. In calculating a mobility for use in a diffusive flux, there is no contribution from B-atoms that change sites with other B-atoms because this has no effect on the concentration profile. It is necessary for a B-atom to exchange sites with an A-atom if it is to contribute to J_B. As the concentration of A-atoms goes to zero in regions rich in B-atoms, there is no contribution to J_B. We could add a correction such as $M_B \to c_A M_B$ in Eq. 10.91, but other considerations are important, too.[5] It is perhaps best to ignore the explicit temperature and concentration dependences of D_B and D_A in Eq. 10.91 and 10.92, and work with constant diffusivities in Eqs. 10.89 and 10.90.

Consider the effects of chemical preferences on interdiffusion. First suppose we have an unmixing alloy, so $4V < 0$ (from Eq. 2.32, $4V = e_{AA} + e_{BB} - 2e_{AB}$). The chemical factor in the square brackets of Eq. 10.89 and 10.90 is less than 1, suppressing the effective value of D_B. For an equiatomic alloy at the special temperature $T' = -Vz/k_B$, the chemical factor is $[-1 + 1] = 0$. There is no diffusive flux in any concentration gradient, $J_B(T') = 0$, and no leveling of heterogeneities in chemical composition. This temperature is the critical

---

[5] The mobility usually depends exponentially on temperature as $e^{-Q/k_B T}$, and this overwhelms the proportionality to $k_B T$ in Eq. 10.91. There are also important factors in M_B from a vibrational attempt frequency and lattice dynamics.

temperature of chemical unmixing (Eq. 2.48), where chemical preferences are lost to thermal disorder.

Now consider an alloy with ordering tendencies so $4V > 0$. The chemical factor in Eqs. 10.89 and 10.90 is greater than 1, so intermixing is enhanced, as expected if unlike atom pairs are preferred, and heterogeneities in composition are leveled more rapidly. For an equiatomic alloy at a temperature $T' = Vz/k_B$, the chemical factor is $[+1 + 1] = 2$. This particular T' is the critical temperature of ordering (Eq. 2.58), so the atom movements are not those of a disordered solid solution. We saw in Section 10.2.2 that diffusion in an ordered alloy at lower temperatures can have a small correlation factor f, which can overwhelm the effects of the chemical factor in Eqs. 10.89 and 10.90. Actually, even for temperatures above T_c there will be significant short-range order in the alloy, and atom movements will be altered. Chapter 23 (online) shows a framework for analyzing such problems.

It is risky to use Eqs. 10.91 and 10.92 at low temperatures – the alloy may be undergoing a phase transformation so that its diffusion coefficients are changing with time. Nevertheless, over short time intervals it may be useful to consider atom fluxes with a diffusion equation, provided there is not too much change in the local atomic configurations.

10.5 Diffusion in a Temperature Gradient

10.5.1 Jump Frequency and Temperature

In general, jump frequencies increase with temperature, so along a temperature gradient there will be a change in jump rate. Even if the jumps are in random directions, there can be a migration of atoms or vacancies down the temperature gradient from hot regions to cold. It is possible for a vacancy or interstitial flux to be driven by a temperature gradient,[6] as explained here.

Consider a flux of species as in Section 3.2. An expression more general than Eq. 3.11 allows the jump frequency Γ to vary with position

$$j = -\frac{a}{2}\left[\Gamma(x + a, t)\, c(x + a, t) - \Gamma(x, t)\, c(x, t)\right], \tag{10.93}$$

so the net flux through the dashed central plane in Fig. 3.2 can be driven by a difference in Γ for the two atomic planes. This can be caused by a difference in temperature, as in a temperature gradient along x. Instead of assuming Γ is constant with a gradient in c (as in Section 3.2.1), now assume that c is constant, but there is a gradient in Γ

$$j = -\frac{a}{2}c\left[\Gamma(x + a, t) - \Gamma(x, t)\right], \tag{10.94}$$

$$j = -\frac{a^2}{2}c\,\frac{\partial \Gamma}{\partial x}. \tag{10.95}$$

---

[6] This migration is suppressed by a conservation condition such as single atom occupancy on crystal sites, of course.

Here the positional variation of the jump frequency Γ is caused by a temperature gradient

$$\frac{\partial \Gamma}{\partial x} = \frac{\partial T}{\partial x} \frac{d\Gamma}{dT}. \tag{10.96}$$

For the variation of Γ with temperature, use Eq. 3.9, rewritten here as

$$\Gamma(T) = \Gamma_0 \, e^{-Q/k_B T}, \tag{10.97}$$

$$\frac{d\Gamma}{dT} = \Gamma_0 \frac{Q}{k_B T^2} \, e^{-Q/k_B T}. \tag{10.98}$$

With Eq. 10.95, 10.96, and 10.98

$$j = -\frac{\Gamma_0 \, a^2}{2} c \frac{Q}{k_B T^2} e^{-Q/k_B T} \frac{\partial T}{\partial x}. \tag{10.99}$$

The flux is in the direction opposite to the temperature gradient, i.e., the vacancies or interstitials move from the hot location to the cold location. Physically, this is expected because the jump rates have a gradient from hot to cold, and jumps from hot to cold are more probable than cold to hot.

10.5.2 Vacancies, Solutes, and Interstitials in Temperature Gradients

Equation 10.99 predicts a diffusive flux of vacancies or interstitials down a temperature gradient. For vacancies, the effect is enhanced by the temperature dependence of the vacancy concentration c. The vacancy concentration increases with temperature, as in Eq. 3.5 for example. By a similar analysis for Eq. 10.99 (see Problem 10.9), with Eq. 3.5 the vacancy flux is

$$j_v = -\frac{\Gamma_0 \, a^2}{2} \frac{Q + \Delta G_{f,v}}{k_B T^2} e^{-(Q+\Delta G_{f,v})/k_B T} \frac{\partial T}{\partial x}. \tag{10.100}$$

There is a flux of matrix atoms in the direction opposite to that of the vacancy flux, but the fraction of moving atoms is relatively small. On the other hand, solute atoms that interact with vacancies as in Section 10.2.1 can be "dragged" by vacancies down the temperature gradient. In this "solute drag model," the solute and vacancy remain bound, but the vacancy still tends to move down the concentration gradient, dragging the solute with it. Solutes that do not interact so strongly with the vacancy will tend to behave as matrix atoms, and will move in the opposite direction as the vacancy, but slowly.

The migration of interstitials from a hot region to a cold region generates a concentration gradient in interstitial composition. This concentration gradient drives interstitial diffusion that counteracts the flux from the temperature gradient in the opposite direction. The balance of these two fluxes leads to a steady state concentration gradient,

$$\frac{\partial c}{\partial x} = -\left[c \frac{Q}{k_B T^2} e^{-Q/k_B T} \right] \frac{\partial T}{\partial x}, \tag{10.101}$$

which is proportional to the temperature gradient (and of opposite sign). For typical values of Q, the exponential factor in Eq. 10.101 is small for most solids, and the concentration gradient of interstitial solutes caused by temperature gradients is small.

10.5.3 Coupled Fluxes and Gradients

There are four equations relating fluxes of vacancies and heat to gradients of concentration and temperature:

- Fick's first law relating the flux of vacancies j_v to the gradient of vacancy concentration $\vec{\nabla}c_v$, as in Eq. 3.17.
- A relationship between the flux of vacancies j_v and the temperature gradient $\vec{\nabla}T$, as in Eq. 10.100.
- A relationship between the flux of heat j_T and the gradient of vacancy concentration $\vec{\nabla}c_v$.
- A relationship between the flux of heat j_T and the temperature gradient $\vec{\nabla}T$ (the usual Fourier law of thermal conductivity, with the same form as Fick's first law, Eq. 3.17).

In the relationships above we have two gradients, $\vec{\nabla}c_v$ and $\vec{\nabla}T$, and two fluxes j_v and j_T. The first and fourth relationships have the same form but for vacancies and heat (or temperature times heat capacity). The second and third relationships are cross-relationships between a flux of one quantity and a gradient of the other. We obtained Eq. 10.100 for the second relationship, but we did not derive the third. However, the Onsager reciprocity relationships from nonequilibrium thermodynamics require that the coefficient be the same as in Eq. 10.100. For completeness we write a standard matrix form for the coupled fluxes and gradients

$$\begin{bmatrix} j_v \\ j_T \end{bmatrix} = - \begin{bmatrix} L_{v,v} & L_{v,T} \\ L_{T,v}, & L_{T,T} \end{bmatrix} \begin{bmatrix} \frac{\partial c_v}{\partial x} \\ \frac{\partial T}{\partial x} \end{bmatrix}, \tag{10.102}$$

where we know that $L_{v,v} = D_v$, $L_{T,T} = D_T$ (thermal diffusivity), $L_{v,T} = L_{T,v}$ by reciprocity, and both are given by Eq. 10.100. To this set of coupled equations we could add the solute concentration, c_B (giving a 3×3 matrix). The new cross-terms relating fluxes of solutes to gradients of vacancy concentration, and vice versa, must be equal because the underlying microscopic mechanism is the same for both – it is the interchange of vacancies and solute atoms.

Cross-terms relating fluxes of vacancies to gradients of temperature, and vice versa, are also equal because the same physical process underlies both. At the microscopic level, the vacancy jump takes in thermal (vibrational) energy to initiate the jump over a barrier, and releases thermal energy after the barrier is surmounted. A flux of vacancies can move heat.[7]

10.6 Nonthermodynamic Equilibrium in Driven Systems

In a model for driven systems developed by Georges Martin [134, 135], some atom movements occur by thermally activated processes that move an alloy towards thermodynamic

[7] These cross-terms describe a heat engine to transport vacancies. A temperature gradient provides the thermal energy to run the engine.

equilibrium. The new feature of Martin's model is a second type of "ballistic" atom movement that occurs without any thermodynamic bias. These ballistic atom jumps occur at random, without influence of the local chemical environment of the moving atom. Examples of ballistic jumps are atom motions when an alloy is bombarded by energetic radiation such as fast neutrons, or perhaps during the severe mechanical damage inflicted by high-energy ball milling. Under such conditions, over time the alloy reaches a state of equilibrium for the conditions of temperature and rate of forced ballistic jumps, although this is not a state of thermodynamic equilibrium for the temperature T. Since it is a steady state, however, the approach of Section 5.6.2 for calculating the response of a system to small departures from the steady state can be used to obtain a modified critical temperature for unmixing.

This section uses the model of Martin to treat the stability of concentration heterogeneities in an alloy driven towards chemical unmixing by thermodynamics, but driven towards mixing by ballistic atom movements. It is assumed that a fraction f of atom movements occur ballistically, and a fraction $1-f$ occur thermodynamically. Conservation of solute requires any concentration fluctuation of positive sign to be compensated by a fluctuation of negative sign.[8] In essence, regions depleted in solute must account for the same amount of solute as the regions enriched in solute. This conservation condition can be obtained by matching pairs of small subvolumes, one enriched and the other depleted in solute, so that

$$\delta V_1 \, \delta c_1 + \delta V_2 \, \delta c_2 = 0, \tag{10.103}$$

where the subvolumes δV_1 and δV_2 have compositions that deviate from the mean composition of the alloy, c_0, by the amounts δc_1 and δc_2. When the alloy is homogeneous (i.e., $\delta c_1 = \delta c_2 = 0$), the free energy of the two subvolumes is $(\delta V_1 + \delta V_2)F(c_0)$, where $F(c)$ is the free energy per unit volume. With the composition fluctuation, the change in free energy, δF, is

$$\delta F = \delta V_1 \, F(c_0 + \delta c_1) + \delta V_2 \, F(c_0 + \delta c_2) - (\delta V_1 + \delta V_2)F(c_0), \tag{10.104}$$

$$\delta F = \delta V_1 \, \delta c_1 \left(\frac{dF}{dc}\right)_{c_0 + \delta c_1} + \delta V_2 \, \delta c_2 \left(\frac{dF}{dc}\right)_{c_0 + \delta c_2}. \tag{10.105}$$

Using the conservation of solute of Eq. 10.103

$$\delta F = \delta V_1 \, \delta c_1 \left[\left(\frac{dF}{dc}\right)_{c_0 + \delta c_1} - \left(\frac{dF}{dc}\right)_{c_0 + \delta c_2}\right]. \tag{10.106}$$

For small δc_1 and δc_2, as expected near the critical temperature for unmixing, the higher derivatives of $F(c)$ can be ignored. If the subvolumes are equal, $\delta V_1 = \delta V_2 \equiv \delta V$, $\delta c_1 = -\delta c_2 \equiv \delta c$ by Eq. 10.103, so

$$\frac{\delta F}{\delta c} = \delta V \, 2\left(\frac{d^2 F}{dc^2}\right)_{c_0} \delta c. \tag{10.107}$$

---

[8] As shown in Section 16.3.2, conservation of solute can be included with the method of Lagrange multipliers for the volume integral of the free energy density. Alternatively, the concentration profile can be expressed in a Fourier series. For each sine wave in the series, we can match positive and negative subregions so that Eq. 10.103 is true.

The change of free energy $F(c)$ caused by the composition fluctuation δc can be either positive or negative, depending on the sign of d^2F/dc^2, as discussed in Section 2.7.

The sign of Eq. 10.107 determines the thermodynamic tendency for the composition fluctuation to grow or shrink. In our present treatment, however, there is no parameter such as surface energy or gradient energy to set the spatial scale of the fluctuations, so absolute rates of growth or shrinkage of a composition fluctuation are not found. Nevertheless, the kinetic tendencies can be interpreted with the results of Section 5.6.2. Following Eq. 5.29, the effective potential energy for a composition fluctuation (as opposed to an order parameter in Section 5.6.2) is

$$\delta F = \delta V\, 2 \left(\frac{d^2 F}{dc^2} \right)_{c_0} (\delta c)^2. \tag{10.108}$$

The effective restoring force against a composition fluctuation is

$$\mathcal{F} = -\frac{d\delta F}{d\delta c} = -\delta V\, 2 \frac{d}{d\delta c} \left(\frac{d^2 F}{dc^2} \right)_{c_0} (\delta c)^2, \tag{10.109}$$

$$\mathcal{F} = -4\, \delta V \left(\frac{d^2 F}{dc^2} \right)_{c_0} \delta c, \tag{10.110}$$

recognizing that the second derivative of F evaluated at c_0 is a constant.[9]

We make the same type of assumption as we did after Eq. 5.32 for the ordering kinetics, i.e., the rate of change of δc is proportional to $\mathcal{F}$

$$\frac{d\delta c}{dt} = M\mathcal{F}, \tag{10.111}$$

$$\left(\frac{d\delta c}{dt} \right)_{\text{thermo}} = -4M\delta V \left(\frac{d^2 F}{dc^2} \right)_{c_0} \delta c. \tag{10.112}$$

The subscript "thermo" indicates that the rate of change equals a mobility times a thermodynamic driving force.

The general form of d^2F/dc^2 was given in Eq. 2.46. Substituting this into Eq. 10.112 gives[10]

$$\left(\frac{d\delta c}{dt} \right)_{\text{thermo}} = -4M\delta V\delta c \left(z\, 4V + \frac{k_B T}{c(1-c)} \right). \tag{10.113}$$

We now add the ballistic atom movements to the thermodynamic ones. These occur without consideration of the chemical environment, so their contribution to the rate of change of the concentration can be obtained by making T very large in Eq. 10.113

$$\left(\frac{d\delta c}{dt} \right)_{\text{ball}} = -\frac{4M'\, \delta V\, \delta c\, k_B T}{c(1-c)}, \tag{10.114}$$

where a different mobility M' was assumed for ballistic jumps.

---

[9] In the case where the alloy of composition c_0 is stable against unmixing, d^2F/dc^2 is a positive constant. The negative sign of Eq. 10.110 shows that $\mathcal{F}$ is then a restoring force that will suppress a concentration fluctuation δc.

[10] A second way to obtain the critical temperature is to find the condition where $(d\delta c/dt)_{\text{thermo}} = 0$ for small δc. This is proportional to d^2F/dc^2.

To get the total rate of change of concentration, we add the rates for ballistic and thermal jumps (recall that f is the fraction of ballistic jumps)

$$\left(\frac{\mathrm{d}\delta c}{\mathrm{d}t}\right)_{\text{total}} = (1-f)\left(\frac{\mathrm{d}\delta c}{\mathrm{d}t}\right)_{\text{thermo}} + f\left(\frac{\mathrm{d}\delta c}{\mathrm{d}t}\right)_{\text{ball}}. \tag{10.115}$$

Substituting from Eqs. 10.113 and 10.114

$$\left(\frac{\mathrm{d}\delta c}{\mathrm{d}t}\right)_{\text{total}} = -\left((1-f)M4Vz + \frac{M+f(M'-M)}{c(1-c)}k_BT\right)4\,\delta V\,\delta c. \tag{10.116}$$

The critical temperature is obtained by setting $(\mathrm{d}\delta c/\mathrm{d}t)_{\text{total}} = 0$ for small δc

$$k_BT_c = -z\,4V\,c(1-c)\left[\frac{(1-f)M}{M+f(M'-M)}\right]. \tag{10.117}$$

Equation 10.117 is equivalent to Eqs. 21 and 22 in Martin's paper [134]. When $f = 0$, Eq. 10.117 reduces to the thermodynamic result of Eq. 2.47. Likewise, when the ballistic atom movements have no effect on the concentration fluctuation, i.e., $M' = 0$, Eq. 2.47 is again recovered. The other extreme of $f = 1$ or $M = 0$ gives a $T_c = 0$, so unmixing does not occur. (This is intuitive – it means that there are only random atom motions, which ought to mix the alloy.) In intermediate cases, the critical temperature for unmixing is reduced by the effects of ballistic jumps. When $M = M'$, Eq. 10.117 reduces to the simple result

$$k_BT_c = -z\,4V\,c(1-c)(1-f), \tag{10.118}$$

showing that the critical temperature is directly proportional to the fraction of thermodynamic atom movements. (Recall $4V < 0$ for unmixing, so $T_c > 0$.)

10.7 Vineyard's Theory of Diffusion

George Vineyard developed an elegant and general theory of diffusion [136]. The theory is based on the statistical mechanics of atom vibrations and velocities. It predicts the familiar activation energy for an atom movement in a potential energy landscape. What it does that is new is address the vibrational degrees of freedom, from which it shows how vibrational entropy gives a prefactor for diffusive jumps.

10.7.1 Degrees of Freedom for Atoms in a Crystal

Assume we have a crystal with N atoms. To describe uniquely all atom positions in the crystal, some set of $3N$ coordinates is required. We transform to normal coordinates $y_i = \sqrt{m_i}\,x_i$, where m_i is an effective mass associated with the displacement x_i of the ith normal coordinate. Normal coordinates are independent of each other, and we make the big supposition that the normal coordinates $\{y_i\}$ are all known.

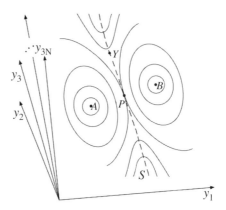

Figure 10.13 Potential energy hypersurface $\Phi(\{y_i\})$, showing saddle point P and a general point Y on the ridge S. Initial and final points of the crystal are A and B.

At any instant in time, the crystal is at a single point in a hyperspace spanned by its $\{y_i\}$ coordinates, as depicted in Fig. 10.13. The potential energy is a hypersurface in this space, and the crystal explores different regions on this hypersurface as all atoms vibrate, and as one atom makes a diffusive jump. A general shape of the potential energy hypersurface $\Phi(\{y_i\})$ with some labeled points is shown in Fig. 10.13. A key feature of this hypersurface is that it has local minima at points A and B, which correspond to the crystal before and after the diffusive jump of one atom. Our goal is to calculate the average rate at which systems make the transition from A to B. These two points are separated by a ridge in the potential marked by S, and crossing the ridge at any point Y will suffice to get the diffusive jump. There is a saddle point on the ridge at P.

Box 10.5   **Gaussian Integrals Again**

Starting with the result of Eq. 9.52

$$I_0 = \int_{-\infty}^{+\infty} e^{-\beta x^2}\, dx = \sqrt{\frac{\pi}{\beta}}, \tag{10.119}$$

we can calculate the Gaussian-weighted average of x^2, or the second moment of a Gaussian function

$$I_2 = \int_{-\infty}^{+\infty} x^2\, e^{-\beta x^2}\, dx. \tag{10.120}$$

The relationship between this I_2 and I_0 is found by differentiating with respect to β

$$\frac{d}{d\beta} I_0 = \int_{-\infty}^{+\infty} \frac{d}{d\beta} e^{-\beta x^2}\, dx, \tag{10.121}$$

$$\frac{d}{d\beta} I_0 = -\int_{-\infty}^{+\infty} x^2\, e^{-\beta x^2}\, dx = -I_2, \tag{10.122}$$

$$I_2 = -\frac{d}{d\beta}\sqrt{\frac{\pi}{\beta}} = \frac{1}{2}\sqrt{\frac{\pi}{\beta^3}} = \int_{-\infty}^{+\infty} x^2\, e^{-\beta x^2}\, dx. \tag{10.123}$$

It can be shown that the even moments of a Gaussian function are

$$\mathcal{I}_{2n} = \int_{-\infty}^{+\infty} x^{2n} \, e^{-\beta x^2} \, dx = \frac{1 \cdot 3 \cdot 5 \cdots (2n-1)}{2^{n+1}} \sqrt{\frac{\pi}{\beta^{2n+1}}}. \tag{10.124}$$

The first moment, or the Gaussian-weighted average of positive x, is evaluated easily by making the substitution of $\eta = x^2$, so $d\eta/dx = 2x$ and $x \, dx = 1/2 \, d\eta$

$$\mathcal{I}_1 = \int_0^{+\infty} x \, e^{-\beta x^2} \, dx = \frac{1}{2} \int_0^{+\infty} e^{-\beta \eta} \, d\eta, \tag{10.125}$$

$$\int_0^{+\infty} x \, e^{-\beta x^2} \, dx = \frac{1}{2\beta}. \tag{10.126}$$

It can be shown that the odd moments of a Gaussian function are

$$\mathcal{I}_{2n+1} = \int_0^{+\infty} x^{2n+1} \, e^{-\beta x^2} dx = \frac{n!}{2 \, \beta^{n+1}}. \tag{10.127}$$

10.7.2 Ensemble-Averaged Jump Rate

To get an average rate with statistical mechanics, we need to make an ensemble of many similar crystals, and define Γ as the ensemble average for the rate of transitions from A to B. It is

$$\Gamma = \frac{I}{Q_A}, \tag{10.128}$$

where

- I is the fraction of systems that reach the surface S per second, and are moving to the right in Fig. 10.13, and
- Q_A is the fraction of systems around point A, to the left of S.

Obtaining Q_A is by a straightforward thermal average of systems in the potential well around A

$$Q_A = \rho_0 \int_A e^{-\Phi(\{y_i\})/k_B T} \, dV, \tag{10.129}$$

where ρ_0 is a normalization factor, and the integration is over the volume of the hyperspace around A. The potential well is expected to be deep, so most systems are concentrated near the point A and the integration is not sensitive to its upper limit.

10.7.3 Transition States

Obtaining I requires the rates of change of the coordinates $\{y_i\}$, written as $\{\dot{y}_i\}$. The velocity of the independent normal coordinates is defined as

$$\vec{v} = (\dot{y}_1, \dot{y}_2, \ldots, \dot{y}_{3N}). \tag{10.130}$$

Its square is twice the kinetic energy of all atoms in the crystal

$$v^2 = \dot{y}_1^2 + \dot{y}_2^2 + \cdots + \dot{y}_{3N}^2, \tag{10.131}$$

$$v^2 = m_1 \dot{x}_1^2 + m_2 \dot{x}_2^2 + \cdots + m_{3N} \dot{x}_{3N}^2. \tag{10.132}$$

It is assumed that v^2 is independent of location on the hypersurface of Fig. 10.13.[11] The density at Y of crystals having velocity v is

$$\rho(Y, v)\, dv = \rho_1\, e^{-\Phi(Y)/k_B T}\, e^{-v^2/2k_B T}\, d\dot{y}_1 d\dot{y}_2 \ldots d\dot{y}_{3N}. \tag{10.133}$$

The normalization factor ρ_1 is calculated with the condition that all atoms in the crystal each have some velocity, and the probability of the system at Y is $\rho_0\, e^{-\Phi(Y)/k_B T}$

$$\rho_0\, e^{-\Phi(Y)/k_B T} = \int_{-\infty}^{\infty} \rho(Y, v)\, dv, \tag{10.134}$$

$$\rho_0\, e^{-\Phi(Y)/k_B T} = \rho_1\, e^{-\Phi(Y)/k_B T} \int_{-\infty}^{\infty} e^{-\sum_{i=1}^{3N} \dot{y}_i^2/2k_B T}\, d\dot{y}_1 d\dot{y}_2 \ldots d\dot{y}_{3N}. \tag{10.135}$$

The independent normal coordinates $\{y_i\}$ are separated

$$\rho_0 = \rho_1 \prod_{i=1}^{3N} \int_{-\infty}^{\infty} e^{-\dot{y}_i^2/2k_B T}\, d\dot{y}_i, \tag{10.136}$$

and from Eq. 10.119

$$\rho_0 = \rho_1 \prod_{i=1}^{3N} \sqrt{2\pi k_B T}, \tag{10.137}$$

$$\rho_1 = \frac{\rho_0}{(2\pi k_B T)^{3N/2}}. \tag{10.138}$$

10.7.4 Transition Rate

Next consider the contribution dI to the flow of systems across the surface S at point Y. First, assume that $\vec{\dot{y}}_1$ is aligned along the direction of $d\vec{S}$, i.e., normal to the surface S at Y. The transition through the point Y occurs by varying $\vec{\dot{y}}_1$, as shown in Fig. 10.13

$$d\vec{I} = d\vec{S} \cdot \hat{\dot{y}}_1 \int_{\dot{y}_1=0}^{\infty} \int_{\{\dot{y}_i\}=-\infty, i \ne 1}^{\infty} \dot{y}_1\, \rho(Y, \vec{v})\, d\vec{v}' d\dot{y}_1, \tag{10.139}$$

where the integration over $\dot{y}_1$ is separated from the others. Using Eqs. 10.133 and 10.138

$$dI = dS \frac{\rho_0}{(2\pi k_B T)^{3N/2}} e^{-\Phi(Y)/k_B T} \int_0^{\infty} \dot{y}_1\, e^{-\dot{y}_1^2/2k_B T}\, d\dot{y}_1$$

$$\times \prod_{i=2}^{3N} \int_{-\infty}^{\infty} e^{-\dot{y}_i^2/2k_B T}\, d\dot{y}_i. \tag{10.140}$$

---

[11] This is reliable if the system is in equilibrium. A conceptual difficulty arises if the diffusing atom moves ballistically over the barrier, and does not equilibrate as it moves.

The two integrals are evaluated with the help of Eqs. 10.126 and 10.119 as k_BT and $\sqrt{2\pi k_BT}$, respectively, giving

$$dI = dS \frac{\rho_0}{(\sqrt{2\pi k_BT})^{3N}} e^{-\Phi(Y)/k_BT} k_BT (\sqrt{2\pi k_BT})^{3N-1}, \tag{10.141}$$

$$dI = \rho_0 e^{-\Phi(Y)/k_BT} \sqrt{\frac{k_BT}{2\pi}} dS, \tag{10.142}$$

$$I = \rho_0 \sqrt{\frac{k_BT}{2\pi}} \int_S e^{-\Phi(Y)/k_BT} dS. \tag{10.143}$$

With Eqs. 10.129 and 10.143, the transition rate of Eq. 10.128 is

$$\Gamma = \frac{I}{Q_A} = \sqrt{\frac{k_BT}{2\pi}} \frac{\int_S e^{-\Phi(Y)/k_BT} dS}{\int_A e^{-\Phi(\{y_i\})/k_BT} dV}. \tag{10.144}$$

Equation 10.144 is a ratio of two configurational partition functions. It is a many-body expression that includes coordinates of all the atoms in the crystal.

Something we do not know is the contour of the ridge along S. We expect that the ridge is high, so there is a low rate for a crystal in the ensemble to attain this potential energy. We therefore approximate the transitions over the ridge as all occurring at the saddle point P, which has the lowest potential energy along S. In what follows we replace the energy at a general point Y with the energy at point P.

10.7.5 Harmonic Vibrations

Further progress is made by considering harmonic vibrations of the atoms in the crystal, with all atoms vibrating about their crystal sites. Small displacements of harmonic vibrations occur as the crystal coordinates are near both points A and P, but there is a difference in the number of normal coordinates at these points. All $3N$ coordinates are available for vibrations around point A, but there is one less coordinate for point P – one of the coordinates, y_1, is used to constrain the system to move across the surface S. Harmonic vibrations modify the potential energies[12] at points A and P as

$$\Phi(\{q_i\}) = \Phi(A) + \sum_{j=1}^{3N} \frac{1}{2}\omega_i^2 q_i^2, \tag{10.145}$$

$$\Phi(\{q_i'\}) = \Phi(P) + \sum_{j=2}^{3N} \frac{1}{2}\omega_i'^2 q_i'^2, \tag{10.146}$$

where the small displacement coordinates around A are $\{q_i\}$, and $\{q_i'\}$ around P. The q_1 from displacements along y_1 is missing from the $\{q_i'\}$. Substituting Eqs. 10.145 and 10.146 into Eq. 10.144

---

[12] The potential energy of a harmonic oscillator of angular frequency ω is $\frac{1}{2}\omega_i^2 q_i^2$. (Recall that the energy of a spring is $1/2kx^2$, and the oscillation frequency of a mass m on this spring is $\omega = \sqrt{k/m}$.)

$$\Gamma(T) = \sqrt{\frac{k_B T}{2\pi}} \frac{\int_S e^{-\Phi(P)/k_B T} e^{-\sum_{j=2}^{3N} \frac{1}{2}\omega_i'^2 q_i'^2/k_B T} dA}{\int_A e^{-\Phi(A)/k_B T} e^{-\sum_{i=1}^{3N} \frac{1}{2}\omega_i^2 q_i^2/k_B T} dV} . \tag{10.147}$$

The integrations in Eq. 10.147 are Gaussian integrals of Eq. 10.119 which evaluate to the form $\sqrt{2\pi k_B T}/\omega$

$$\Gamma(T) = \sqrt{\frac{k_B T}{2\pi}} \frac{e^{-\Phi(P)/k_B T} \prod_{i=2}^{3N} \frac{\sqrt{2\pi k_B T}}{\omega_i'}}{e^{-\Phi(A)/k_B T} \prod_{i=1}^{3N} \frac{\sqrt{2\pi k_B T}}{\omega_i}} , \tag{10.148}$$

$$\Gamma(T) = \frac{1}{2\pi} \frac{\prod_{i=1}^{3N} \omega_i}{\prod_{i=2}^{3N} \omega_i'} e^{-[\Phi(P)-\Phi(A)]/k_B T} . \tag{10.149}$$

10.7.6 Diffusion Coefficient $D(T)$

The expression 10.149 for $\Gamma(T)$ allows us to write the diffusion coefficient of Eq. 3.18 as

$$D(T) = D_0\, e^{-\Delta E/k_B T} , \tag{10.150}$$

where (adding the correlation factor f of Section 10.1.1)

$$D_0 = \frac{f\,a^2}{12\,\pi} \frac{\prod_{i=1}^{3N} \omega_i}{\prod_{i=2}^{3N} \omega_i'} , \tag{10.151}$$

$$\Delta E = \Phi(P) - \Phi(A). \tag{10.152}$$

Using Eq. 7.8 for the vibrational entropy

$$\Delta S_{\mathrm{vib}} \equiv S_P - S_A = k_B \left[\sum_{i=2}^{3N} \ln\left(\frac{k_B T}{\hbar \omega_i'}\right) - \sum_{i=2}^{3N} \ln\left(\frac{k_B T}{\hbar \omega_i}\right) \right], \tag{10.153}$$

$$\Delta S_{\mathrm{vib}} = \sum_{i=2}^{3N} \ln\left(\frac{\omega_i}{\omega_i'}\right), \tag{10.154}$$

$$e^{+\Delta S_{\mathrm{vib}}/k_B} = \frac{\prod_{i=2}^{3N} \omega_i}{\prod_{i=2}^{3N} \omega_i'} . \tag{10.155}$$

so the prefactor D_0 is

$$D_0 = \frac{f\,a^2}{12\,\pi} \omega_1\, e^{+\Delta S_{\mathrm{vib}}/k_B}. \tag{10.156}$$

Recognizing the change in free energy between points A and P as $\Delta F = \Delta E - T\Delta S_{\mathrm{vib}}$

$$D(T) = \frac{f\,a^2}{6} \nu_1\, e^{-\Delta F/k_B T}, \tag{10.157}$$

where the attempt frequency is $\nu_1 \equiv \omega_1/2\pi$. This is the frequency of the vibrational mode that moves the atom at A towards the saddle point at P. The ΔF includes both the activation energy and the activation entropy of the diffusive jump.

The Vineyard theory advances our understanding of diffusion. It is already a multibody treatment of diffusion, but a more detailed theory would consider other atom movements around a vacant site. Instead of one coordinate y_1 that reaches a critical condition, one might add coordinates for the neighboring atoms that need to be displaced to allow the diffusing atom to enter the vacancy [137]. The Vineyard theory and its extensions assume a quasistatic process for the atom to surmount the activation barrier. This is necessary so that a temperature can be defined and the tools of statistical mechanics are applicable. Molecular dynamics computer simulations are not subject to the same constraints, and can allow for ballistic motions of a vacancy across more than one neighbor in a quick sequence, for example. Such phenomena may be important at high temperatures.

Problems

10.1 (**a**) Explain in words (with diagrams and/or equations if helpful) why solid-state interdiffusion involving two species of atoms, one fast and the other slow, differs from interdiffusion of two ideal gases. Assume that the atom density [atoms/cm^3] remains constant during interdiffusion.

   (**b**) Explain the origin of the Kirkendall effect and the origin of Kirkendall voids.

   (**c**) Explain the origin of a marker velocity in a diffusion couple.

10.2 The diffusion coefficient in one dimension with uncorrelated atom jumps between adjacent planes is $D = 1/2\,\Gamma\,a^2$. Calculate in an analogous way the diffusion coefficient in one dimension when there are both jumps between neighboring planes (separated by a) of frequency Γ_1, and jumps between second-neighbor planes (separated by $2a$) of frequency Γ_2.

10.3 One hundred jumping beans are placed along the center line of a large floor at 6 cm intervals. Twelve hours later the distance of each bean from the center line is measured, and the sum of the squares of the distances divided by 100 is 36 cm^2. (This is a two-dimensional problem, but the distance from the line is along one dimension.)

   (**a**) Calculate the diffusion coefficient of the jumping beans.

   (**b**) If the mean jump distance of a bean is 0.1 cm, estimate the mean jump frequency of a bean.

10.4 In a one-dimensional random walk, there is the probability p of a jump to the right, and the probability q of a jump to the left. The probabilities of a set of multiple jumps are products of p and q; for example the probability of two jumps to the right is the product pp, and the probability of one jump to the right and one to the left is pq.

   (**a**) Show that the probability for moving a total of n steps to the right out of a total of N steps is the binomial probability

$$P(N, n) = \frac{N!}{(N - n)!\, n!}\, p^n\, q^{N-n}. \tag{10.158}$$

(b) Show that the average value of n can be expressed as

$$\langle n \rangle = p \frac{\partial}{\partial p} (p + q)^N. \qquad (10.159)$$

(c) Calculate $\langle n \rangle$, and use the same technique to calculate $\langle n^2 \rangle$.
(d) What is the mean displacement after N jumps?
(e) What is the mean squared displacement after N jumps?
(*Hint*: The function $(p + q)^N$ is the "generating function" for the various $P(N, n)$ because of the binomial expansion

$$(p + q)^N = \sum_{n=0}^{N} \frac{N!}{(N - n)! \, n!} p^n q^{N-n}. \qquad (10.160)$$

It is sometimes more convenient to manipulate $(p + q)^N$ than the individual $P(N, n)$.)

10.5 **(a)** Calculate the correlation factor, f, for diffusion on a simple cubic lattice. In obtaining $\cos \theta_1$, consider only the first three jumps of the vacancy.
    (b) How would your result change if you considered the first four jumps? (Reasonable answers get credit.)

10.6 For equiatomic ordering alloys with the same T/T_{crit} and $|U|$, show that more vacancy jumps tend to be strongly trapped when the lattice has a lower coordination number, z (at least for $z \geq 3$). You will need to consider the relationship between $e_{\text{AA}} - e_{\text{AB}}$ and both the normalized critical temperature and the probability of exiting a strong trap, but you will also need the probability of finding a trap. The combinatorics of finding a trap in a largely disordered solid solution can be approximated combinatorially as $2^{-\zeta}$, where ζ is the number of atoms needed to construct a strong trap like that in Fig. 10.5.

10.7 Consider a random walk of an A-atom on an ordered structure such as a B2 or chessboard structure, where each jump is to a nearest neighbor on the opposite sublattice (α and β denote the sublattices). We use the definitions of Fig. 10.14 from Bakker [125], where the notation P means the probability of a jump from the α to β, and Q from β to α. The variables $+$ and $-$ mean that the jump is in the same direction or reversed from the previous jump (examples are in Fig. 10.14). Define also

$$q \equiv Q_- - Q_+, \qquad (10.161)$$

$$p \equiv P_- - P_+. \qquad (10.162)$$

The correlation factor is analogous to that of Eq. 10.14

$$f = 1 + 2 \sum_{j=1}^{\infty} \overline{\cos \theta_j}. \qquad (10.163)$$

To calculate $\cos \theta_1$, for example, note that

$$\overline{\cos \theta_1} = Q_+(+1) + Q_-(-1) = -q. \qquad (10.164)$$

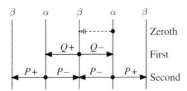

Figure 10.14 Definition of jump probabilities for random walk on an ordered structure with sublattices α and β. Solid circles are starting points for the nth jump, where n is labeled on the right.

Show that the correlation factor is

$$f = \frac{1 - p - q + pq}{1 - pq}.$$ (10.165)

(*Hint*: Perform the mechanics of long division on Eq. 10.165, and work more $\cos\theta_j$ terms with Fig. 10.14.)

10.8 For convenience, assume an interdiffusion profile for a fast and a slow species with the form

$$c(x) = c_0 \tanh(e^{-x}).$$ (10.166)

(**a**) Use an expression of this form to find the position of a Matano interface.
(**b**) Calculate $\tilde{D}(c)$ for a few positions at equal distances to the left and right of the Matano interface.
(**c**) How and why would you expect the results of part b to change if the $c(x)$ were

$$c(x) = c_0 \frac{1 + \tanh(-x)}{2} \quad ?$$ (10.167)

(*Hints*: Use a mathematical software package for parts a and b. Consider the inversion symmetry in x for part c.)

10.9 The vacancy flux down a temperature gradient in Eq. 10.100 accounted for the effect of more rapid vacancy jumps in the hotter region (Eq. 10.99), and also accounted for the gradient in vacancy concentration from more vacancy formation in hotter regions.

(**a**) Explain why both effects serve to move vacancies from hot to cold.
(**b**) Assuming the two effects are small and additive, complete the derivation of Eq. 10.100.

PART III

TYPES OF PHASE TRANSFORMATIONS

Part III describes the important and established families of phase transitions in materials. Chapters 11–20 describe structural and chemical phase transformations of materials that occur by movements of atoms. These include heterogeneous first-order transitions such as melting and precipitation, and spinodal decomposition or ordering that may occur homogeneously as second-order phase transitions. Interfaces between phases are given special attention, since interfaces are where much of the dynamical change occurs, and intrinsic properties of interfaces are discussed in Chapter 11. Martensite and other displacive phase transformations are the subject of Chapter 19, and microstructural and nanostructural aspects of phase transformations are covered in Chapter 20. All these phase transitions involving atom rearrangements are historical figures in the field of materials science, and new phenomena are often explained with reference to them. Chapter 21 describes some of the major phase transitions involving electrons and spins. Electronic and magnetic phase transitions in materials can sometimes be understood with similar approaches as phase transformations involving atom rearrangements, although some aspects of electronic or magnetic excitations are not classical.

11

Thermodynamics and Phase Transitions at Surfaces

Up to this point in the book, we have considered only bulk materials, nominally infinitely large, so there has been no role for a surface surrounding the material. Such an approach can explain much about phase transitions between two different crystal structures, for example, even when the materials are of micrometer dimensions. At nanometer dimensions, the situation is altered considerably by the high surface-to-volume ratio of the material, and this is a topic of Chapter 20. Even when materials are larger, however, they still have surfaces, and the structures and dynamics of surfaces affect their chemical reactivity and their growth characteristics, for example. This chapter describes atomistic structures of surfaces of crystalline materials, and describes how a crystal may grow by adding atoms to its surface.

Most inorganic materials are polycrystalline aggregates, and their crystals of different orientation abut each other at "grain boundaries." Some features of atom arrangements at grain boundaries are explained, as are some aspects of the energetics and thermodynamics of grain boundaries. In general, grain boundaries are not features of a material in thermodynamic equilibrium. Since they alter the internal energy and entropy of materials, however, some properties of grain boundaries can be understood by the concept of a material in a constrained equilibrium (Chapter 9). Today this understanding is limited primarily to the energetics of grain boundaries.

Later sections describe how surface energy varies with crystallographic orientation, and how this affects the equilibrium shape of a crystal. The interaction of gas atoms with a surface, specifically the topic of gas physisorption, is presented at the end of this chapter.

11.1 Surface Structure

11.1.1 Surface Reconstruction

Figure 11.1 shows how arrangements of atoms at free surfaces differ from atom arrangements in the bulk. At the top of the crystal, the figure depicts "relaxations" vertically, and "reconstructions" horizontally. Compared with the atomic spacing in the bulk, relaxation usually causes the top layer of atoms to be a bit closer to the layer below by the amount δ, but this effect diminishes rapidly below the first layer. For the first layer, δ is typically several percent of the interatomic distance, a, although it depends strongly on the crystallographic orientation of the surface.

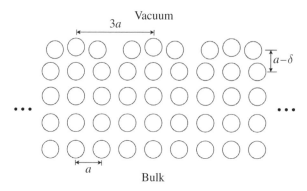

Figure 11.1   Generic surface relaxation (vertical shrinkage) and surface reconstruction (new arrangements of atoms in the plane of the surface).

The explanation of this contraction during relaxation is sometimes based on the transfer of electrons from the missing bonds above the surface to bonds between the first and second atomic layers. A higher electron density does not necessarily imply a closer interatomic distance, however, so this is not the full explanation. Over the surfaces of metals, the electron density tends to smooth out, reducing the electronic kinetic energy. This altered electron density tends to reduce the attractive electrostatic potential for the positive ion cores, so the nuclei in the top layer move towards the atoms beneath. This is a subtle effect, so the relaxations depend on the crystallographic structure and are generally small. For ionic solids, as a rule the surfaces are composed of structural units with charge neutrality, such as pairs of ions having opposite charges.

Surface reconstructions involve horizontal (and vertical) repositioning of atoms in the plane of the surface. Figure 11.1 shows what would be described as a [Po(100)-(3 × 3)] reconstruction (assuming a similar rearrangement out of the plane of the paper, and assuming that Po has a simple cubic structure). Like relaxations, surface reconstructions are also driven by the altered electronic bonding characteristics of atoms at a free surface. For semiconductors, sometimes the geometries of the surface reconstructions can be understood by a tendency to eliminate dangling bonds by displacing atoms across atomic-scale valleys to make new bonds that bridge across the valley. These bonds break the translational periodicity of the bulk. Some surface reconstructions are quite intricate, such as the [Si(111)-(7 × 7)] reconstruction shown in Fig. 11.2. Today it seems that many complicated reconstructions like these can be predicted by electronic energy calculations that assume zero temperature, so the energy contribution to the free energy is large, and the entropic contributions seem to be small. An issue, however, is that many surface structures are metastable, and may require an activation to achieve equilibrium. Such is the case for the Si (111) reconstruction. Although it is stable at low temperatures, it may require a high-temperature annealing before it forms.

11.1.2 Terraces, Ledges, Kinks, and Roughness

Figure 11.3a shows the surface of a simple cubic crystal, and defines pictorially a terrace, ledge, kink, vacancy, and adatom. Such a surface may be expected at a vacuum interface,

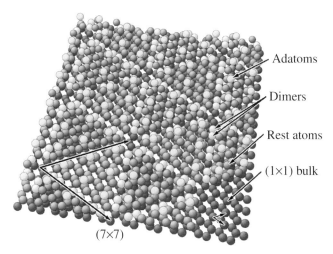

Figure 11.2 The (7×7) reconstruction of a (111) surface of Si, which occurs after annealing under vacuum. Atom arrangements in the unreconstructed surface are shown in the lower right. Structure taken from the NIST Surface Structure Database (SSD, by P.R. Watson, M.A. Van Hove, and K. Hermann). Visualization with Balsac (by K. Hermann, Fritz–Haber Institute Berlin).

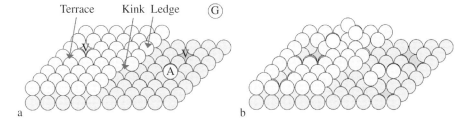

Figure 11.3 (**a**) A surface of a simple cubic crystal with two terraces, one kink, and a ledge as labeled. Two vacancies are labeled as "V," an adatom as "A," and an atom in the vapor as "G." (**b**) Surface with more defects and more roughness.

but the same concepts apply to an interface between a precipitate and a matrix crystal, for example. Consider the growth dynamics of this surface if atoms are slowly added by physical vapor deposition. At intermediate temperatures, surface diffusion of the adatoms is expected to be relatively rapid, and an adatom moves over the surface until it finds an energetically favorable site. Such sites are at ledges, where the adatom does not have just one neighbor as it does on a terrace, but two. A kink is even more favorable because the adatom has three neighbors. If adatoms move rapidly, the atoms deposited on the surface will attach to the kinks, causing the ledge to grow horizontally one atom at a time. When the kink reaches the edge of the crystal, a new ledge forms and it grows horizontally. Eventually the upper terrace moves from left to right across the surface in Fig. 11.3a, and the process begins anew with another terrace.

The process of surface growth need not be so orderly, however. Especially at lower temperatures there may be kinetic limitations for adatom migration. For vapor deposition at high rates, there may be much more disorder in the surface. Figure 11.3b, for example, shows a surface where terraces, ledges, and kinks are numerous, and surface growth may occur by other mechanisms. Although the surface structure may be dominated by kinetics, as suggested here, a thermodynamic analysis can make a prediction about surface roughness.

11.2 Thermodynamic Roughening Transition

Surface steps are defects with an energy cost, and there are ways to calculate their thermodynamic probability in equilibrium. Suppose each surface atom has z_s neighboring sites for possible lateral bonds. On a (001) surface of a simple cubic crystal, this would be $z_s = 4$, for example. An unsatisfied lateral bond between an atom and a vacant surface site has the energy $+\varepsilon$. There will be many such unsatisfied bonds on a rough surface, and few on a smooth surface. On the other hand, the rough surface has many more equivalent configurations than a smooth surface, and a higher configurational entropy. Smooth surfaces are therefore expected at low temperatures where energy dominates the thermodynamics, and rough surfaces are expected at high temperatures where entropy is dominant [138, 139].

11.2.1 High-Temperature Approach to T_c

Approximating a crystal surface as a solid solution of adatoms and vacancies, we can explain the onset of a thermodynamic "roughening transition" at an intermediate temperature. With decreasing temperature, the interactions between atoms on a rough surface lead to a cooperative transition, where reductions in roughness lead to increasingly favorable bonding. At lower temperatures, a smoothing of the surface will set in by lateral interactions between the adatoms. The atoms begin to come together into terraces, and the more ordered the terrace, the more favorable it is to add an atom to it.[1]

Assume there are N_B atoms on the N sites of a two-dimensional surface. At sufficiently high temperature we expect a random occupancy of these atoms on the surface sites, with the probability $c = N_B/N$ of a site being occupied. The other surface species is the vacancy, denoted here as an A-atom for consistency with notation of Section 2.7.2. There are a number $N_A = N - N_B$ vacancies on the surface. The calculation of the free energy is now identical to that of a two-dimensional solid solution as performed in Section 2.7.2 [140]. For a surface coverage with $c = 1/2$, the result is the critical temperature of Eq. 2.48

$$T_c = -z_s V/k_B. \tag{11.1}$$

Here $4V = e_{AA} + e_{BB} - 2e_{AB}$.[2]

---

[1] There are also vertical interactions to consider. These may not be the same strength because between any two adjacent positions on the surface there may be a vertical change of more than one atom in height. The "solid-on-solid" model assumes that the energy of a step between adjacent surface sites scales as their difference in height, squared.

[2] Of course, there are no lateral bonds between vacancies, so e_{AA} can be neglected on physical grounds, and perhaps e_{AB} can also be set to zero for reference.

The phase diagram is that of Fig. 2.17, where the α' phase, i.e., regions of empty sites, is in equilibrium with the α'' phase containing the clusters of adatoms. The phase diagram shows that at lower temperatures the empty regions contain fewer adatoms, and the terraces have fewer vacancies. This approach is with risk, especially at higher temperatures, because the surface structure is not confined to a plane. This approach works best when there is a substrate of immobile atoms, and a layer of different chemical species is deposited on top of it at modest temperature. We expect that at temperatures near T_c, however, the surface is close to being flat, and large regions of flat surface are predicted at temperatures modestly below T_c.

11.2.2 Low-Temperature Approach to T_c

Approaching the roughening transition temperature from below is a more reliable way of finding T_c. At a low temperature, suppose we place some atoms on a flat surface. To minimize the energy cost of vertical steps, these atoms will cluster together to form a mesa of one atom in height, with a perimeter length of L. The energy and entropy depend on L, and L is assumed large because there are few mesas on the surface. Assume the energy scales with L, being close to $z_s V$ per unit length. For a simple cubic lattice with $z_s = 4$, the energy is

$$E = -\frac{L\,4V}{a},\tag{11.2}$$

but this will be different for other lattices. [3]

For the configurational entropy, we follow the atoms around the perimeter of the mesa and allow a change in direction at each unit length step, giving a kink in the edge. The number of such locations is L/a, where a is the interatomic spacing. Of the z_s candidate sites to place the next atom on the edge, we disallow the site that goes backwards to our previously placed atom and another site is eliminated if we forbid overhangs. We make the simple assumption that the number of possible directions for the next atom on the edge of the mesa is $z_s - 2$. The total number of ways of arranging the edge of length L around a mesa is

$$\Omega = (z_s - 2)^{L/a},\tag{11.3}$$

so the configurational entropy is

$$S_{\rm conf} = k_B \ln \Omega = k_B \frac{L}{a} \ln(z_s - 2).\tag{11.4}$$

With Eq. 11.2, the free energy, $F = E - T S_{\rm conf}$, is

$$F = \frac{L}{a}\Big(-4V - T k_B \ln(z_s - 2)\Big).\tag{11.5}$$

[3] To understand the sign, assume zero energy for interaction with vacancies (here the A-atoms), so $e_{AB} = e_{AA} = 0$ and $4V = e_{BB}$. The B–B bonding is favorable, so $4V < 0$, but a B-atom along the perimeter has lost a B-neighbor.

Our comparison is to a free energy for a flat surface with no mesas, so the critical temperature occurs when F of Eq. 11.5 is zero. This gives

$$T_c = -\frac{4V}{k_B \ln(z_s - 2)}. \tag{11.6}$$

A comparison of Eqs. 11.1 and 11.6 shows differences in T_c, although the two equations are equal when $z_s \simeq 4.455$.

This cooperative transition from a smooth surface at low temperatures to a rough surface at high temperatures is generally expected over a narrow range of temperatures. If this transition occurs below the melting temperature, the roughening transition is expected to have important effects on the surfaces of crystals. The surface roughness will also affect surface properties such as gas adsorption.

11.2.3 Entropy and Dimensionality for Line Defects

Adatoms in Fig. 11.3 have a higher configurational entropy than atoms in a ledge. For this two-dimensional surface bounding a three-dimensional crystal, all atoms in a ledge are constrained to be neighbors along a line, whereas each adatom is independent and can assume any position on the surface. For a one-dimensional surface between two-dimensional regions, however, an adatom and a ledge are equivalent. The thermodynamics of ledges requires an interesting adaptation from one to two dimensions (explored in Problem 11.2 with Fig. 11.20). Each atom at a ledge in 1D or 2D has a missing bond or two, so in either 1D or 2D the energy E scales with the number of atoms at ledges. The configurational entropy, S_{conf}, scales with the number of ledges, and how they can be placed on the surface. There are as many locations for ledges in 1D of Fig. 11.20 as there are in 2D (but in 2D the ledges go in and out of the plane of the paper). In 2D, however, the atoms in a ledge are all connected, and span the width of the surface. The S_{conf} is from configurations of ledges, not atoms, so in 2D the S_{conf} needs to be normalized by the number of atoms along the ledge. The consequence is that a ledge in 2D has a configurational entropy that is vanishingly small per atom, making a ledge thermodynamically improbable at finite temperature.[4] This same argument applies to dislocations, which have even more energy per atomic length. Line defects such as ledges or dislocations are therefore considered to be non-thermodynamic defects in three-dimensional crystals (however, see Section 12.7.2 for two-dimensional crystals).

11.3 Surface Structure and Kinetics

11.3.1 Screw Dislocation Mechanism

A screw dislocation intersecting a surface, as shown at the top of Fig. 19.1b, offers a mechanism for continuous surface growth, as depicted in Fig. 11.4. The screw dislocation

[4] Ledges can have some orientational freedom in 2D that is not available in 1D. This is not a large effect, especially when ledges follow crystal orientations.

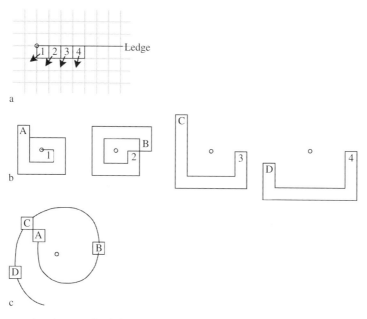

Adding 13 atoms to sites along a surface ledge created by a screw dislocation. Dislocation core intersects the surface at the small circle. (**a**) Initial position of the ledge, and four sites in front of the ledge. Arrows point perpendicular to the radial line to the dislocation core. (**b**) Pattern of adding 13 atoms to the 4 sites at different radius shown in part a. (**c**) Set of the four endpoints (A–D) of the patterns in part b, showing spiral pattern.

creates a ledge on the surface that extends from the dislocation core to the edge of the crystal surface. Figure 11.4a marks the ledge in a view from above the surface. Atoms can be added in front of the ledge, and four locations for atom addition are labeled 1–4. An important feature of this growth mechanism is that if atoms are added at the same rate to all locations along the ledge, the ledge evolves into a spiral.

Figure 11.4b shows why a spiral pattern is expected. An equal number of atoms (13 atoms, to be specific) have been attached to each point 1–4 in Fig. 11.4a. Because the starting points 1–4 lie at different radial distances from the dislocation core, the inner radii make more wraps around the dislocation core. Treating the problem as a continuum without crystallography, the rotation angle decreases with the initial radius

$$\theta = \frac{Na}{2\pi r}, \tag{11.7}$$

where N is the number of atoms added to each atom position at the initial radius r, and a is the interatomic separation.[5] Figure 11.4b suggests a physical picture of adding rows of atoms as strings of equal length from the initial position, and Fig. 11.4c shows how the terminations of these strings are consistent with a spiral.

A feature of a spiral structure is that, at small radii, the high curvature causes atoms at the ledge to have a smaller average number of neighboring atoms than the atoms along the

---

[5] We assume there is no large change of radius for the atoms added to the spiral around the angle θ.

ledge at larger radii. (Equivalently, there is a higher density of corner atoms for the structure evolving from point 1 on the left of Fig. 11.4b than from point 4 on the right.) This is more costly energetically, so this energy penalty should slow the growth rate of the inner parts of the growth spiral. At small radii, the ledges are expected to have larger separations.[6] Nevertheless, spiral structures are often found at surfaces, sometimes with ledges at regular radial separations of tens of micrometers. These structures were often formed by growth at high temperatures, and the material was cooled to low temperatures where atomic mobility was suppressed. They are usually not in thermodynamic equilibrium.

11.3.2 Layer or Island Growth

The fabrication of multilayered thin film devices by methods of atom deposition usually requires attention to the roughness of interfaces between layers of different materials. The thermal issues described in Section 11.2 can be relevant, but these considerations were for free surfaces of one chemical element. The problem is more complicated when B-atoms are deposited onto a substrate of A-atoms. Such is the case shown in Fig. 11.5 for a flat substrate surface indicated by the dashed line. The figure shows the three common types of thin-film growth:

- "layer-by-layer" or Frank–Van der Merwe (FV) growth in Fig. 11.5a;
- "island" or Volmer–Weber (VW) growth in Fig. 11.5b;

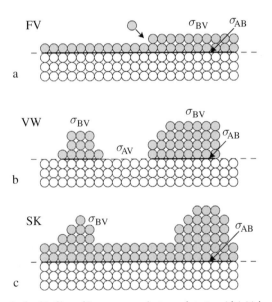

Figure 11.5 Three common growth modes for thin films of B-atoms on an A-atom substrate, with initial surface indicated by a dashed line. Specific surface energies, σ_{XY}, are indicated between A-atoms, B-atoms, and vacuum. (**a**) Layer-by-layer growth (FV), (**b**) island growth (VW), (**c**) Stranski–Krastanov growth (SK).

---

[6] If you like the string analogy, consider the structure to be formed with shorter strings at smaller radii.

• Stranski–Krastanov (SK) growth in Fig. 11.5c, which starts with layers and later develops islands on top of them.

Island growth is promoted by a high specific surface energy of the A/B interface, σ_{AB}, relative to the specific surface energies of the vacuum interfaces. A comparison of Figs. 11.5a and 11.5b shows that those regions of surface covered by islands in Fig. 11.5b have equal projected areas of A–B and B–V interfaces, as does the FV growth in Fig. 11.5a. The difference in surface energy for FV and VW growth comes from the fraction $1 - f$ of surface that is not covered by islands in Fig. 11.5b. The critical condition for island growth is

$$E_{FV} > E_{VW}, \tag{11.8}$$

$$\sigma_{AB} + \sigma_{BV} > \sigma_{AV}. \tag{11.9}$$

On the other hand, when $\sigma_{AB} + \sigma_{BV} < \sigma_{AV}$, layer-by-layer FV growth is expected. When $\sigma_{AB} + \sigma_{BV} \simeq \sigma_{AV}$, Stranski–Krastanov growth often occurs. The real situation is more complicated than this, since we considered only a two-dimensional structure. We ignored the vertical sides of the islands, which add energy cost. Furthermore, the sides may have a different surface energy owing to anisotropy.

Elastic strains usually accompany thin film growth, owing to lattice mismatch between the substrate surface and the B-atom layers. The strains can be large. In the early stages of growth, the strains of mismatch are taken up by the thin layer of B-atoms. As the deposited layer becomes thicker and more robust, however, the substrate may accommodate an increasing fraction of the strain, and the substrate may warp. These elastic strains can cause a thin film to break free of the substrate, and peeling of the film is common for film thicknesses of 1 µm or more. Another consequence of elastic strains may be to cause the deposited atoms to arrange as islands, often in a semiregular array on the surface. Some "quantum dot" structures have been fabricated with the help of elastic interactions through the substrate.

11.4 Energies of Grain Boundaries and Interfaces

11.4.1 Grain Boundary Structure

In polycrystalline materials, as in Fig. 11.6, interfaces between adjoining crystals are called "grain boundaries." They usually form an interconnected network, and are important features of polycrystalline microstructures. Grain boundaries make a major contribution to the free energy of consolidated nanocrystalline materials.

The local atomic configurations in grain boundaries accommodate the misorientations between the adjacent crystals. Important features of grain boundary structures are seen in "bubble raft models" from the 1950s, shown in Fig. 11.7. Some features can only be seen by orienting the plane of the paper at a shallow angle to the eye, and sighting along the rows of bubbles (please look approximately horizontally in the image). In particular,

Figure 11.6   Image made by electron backscatter diffraction showing individual crystals in polycrystalline microstructure of a slowly cooled carbon steel. Image courtesy of A. Khosravani.

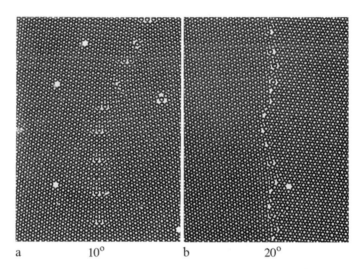

a   10°   b   20°

Figure 11.7   Physical models of grain boundaries between crystals, made by the coalescence of uniform soap bubbles on the surface of water. To see the structures, please sight horizontally along rows of bubbles that intersect the boundary, which is approximately vertical in the center of each image. Misorientations of the crystals are approximately (**a**) 10°, (**b**) 20°. There are a few point defects in the "crystal interiors." Republished with permission from [141]; permission conveyed through Copyright Clearance Center, Inc.

notice that these boundaries have structures that repeat along their length. In small zones of a bubble or two in width, the local environments in the grain boundary are distorted severely from their configurations in the crystal, and these zones are closer together when the tilt of the boundary is larger. Sometimes the atoms in grain boundaries are assumed to have the thermodynamic properties of a liquid. This may be adequate for rough estimates, but atom configurations in grain boundaries differ from those of liquids, and as described in Section 11.4.2 their energies may differ considerably.

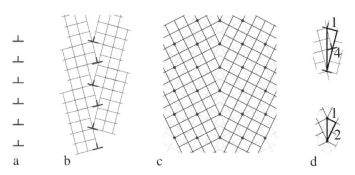

Figure 11.8 (**a**) Stable structure of edge dislocations. (**b**) Grain boundary constructed with edge dislocations for two cubic crystals tilted by 28.07°. (**c**) Coincidence grain boundary for symmetric 53.13° tilt. Atoms are located on all corners of the dark lattices. The gray dots are sites of the "coincidence lattice." (**d**) Selections from panels b and c, showing right triangles with tangents of 1/4 and 1/2.

The specification of an individual grain boundary requires a minimum of five coordinates – two for the orientation of each crystal, and one for the rotation in the plane of the boundary.[7] This complexity allows for numerous types of grain boundaries.

For a small misorientation between two crystals, Read and Shockley proposed a grain boundary structure composed of an array of dislocations [142]. The six edge dislocations in Fig. 11.8a are in a stable configuration because the compressive stress above each dislocation cancels partially the tensile stress below its neighboring dislocation. Figure 11.8b shows in more detail the extra half-planes of the six edge dislocations between two tilted simple cubic crystals. This dislocation array creates a low-angle tilt boundary when the atoms relax their positions. For a higher angle of tilt, more dislocations are necessary geometrically.[8] Other types of boundaries can be constructed from patterned dislocations, such as twist boundaries comprising arrays of screw dislocations.

When adjoining crystals have larger misorientations, their grain boundaries show other features of geometry and symmetry. For example, a rotation of 90° around a fourfold axis of a cubic crystal cannot produce a grain boundary because the crystals are matched perfectly by symmetry. Most tilt angles, however, cannot bring two crystals into crystallographically equivalent orientations. Nevertheless, special tilts can bring a high degree of crystallographic registry at the interface itself. A special tilt condition, easily seen in two-dimensional figures, occurs for boundaries of tilt angles θ' when $\tan(\theta'/2)$ is a ratio of small integers. Figures 11.8b and 11.8c show such conditions for the integer ratios 1/4 and 1/2, corresponding to tilt angles of 28.07° and 53.13°, respectively. (See panel d in Fig. 11.8.) More geometric detail is shown in Fig. 11.8c, with each crystal

---

[7] Displacement coordinates may also be required for a full specification, but it is usually assumed that once the orientations are fixed, chemical bonding will select the displacements by nudging the atoms into a favored configuration at the boundary.
[8] The terminating half-planes of edge dislocations are directly analogous to the ledges on vicinal surfaces shown in Fig. 11.11b.

extended into the other with thin gray lines. Notice the periodic matching of sites on the black and gray lattices, comprising a "coincidence lattice" for this case of $\tan(\theta'/2) = 1/2$. The coincidence lattice is always less dense than the crystal lattice. From Fig. 11.8c it is straightforward to confirm that there are five times more sites per area in the crystal lattice than in the coincidence lattice. We therefore call this particular boundary a $\Sigma 5$ boundary.

11.4.2 Grain Boundary Energy

Atoms in grain boundaries are forced into peculiar local coordinations, raising their bond energies. The energy penalty varies for different materials. Metals that prefer close-packed structures seem to have lower grain boundary energies than those with the bcc structure, for example. Materials with stronger bonding and higher melting temperatures are generally expected to have higher grain boundary energies. From work on a number of grain boundary structures in fcc metals, the grain boundary energy varies from approximately 0.4 to 1.2 J/m^2 in the sequence Au, Ag, Al, Pd, Cu, Co, Ni, Pt [143, 144]. For fcc Ni metal, however, grain boundaries of different orientations have energies that span much of this entire range [145]. The bcc metals generally have higher grain boundary energies, such as about 1.2 J/m^2 for iron and between 3 and 4 J/m^2 for tungsten (tungsten has a very high melting temperature) [146].

From computational work it is known that coincidence grain boundaries such as in Fig. 11.8c often have low specific surface energy. Results for grain boundaries in fcc Ni metal are shown in Fig. 11.9. Some $\Sigma 3$ boundaries have very low specific surface energies. (These $\Sigma 3$ boundaries are fcc twin boundaries.) A $\Sigma 11$ twist boundary also has relatively low energy, although a $\Sigma 5$ tilt boundary,[9] perhaps surprisingly, does not.

Accommodating small deviations from these special boundary orientations can be accomplished by adding some extra dislocations along the boundary, as suggested by Fig. 11.8a,b. Additional positive edge dislocations can accommodate higher tilt angles than that of Fig. 11.8b, for example, but these additional dislocations disrupt the geometrical matching. The energy of the grain boundary increases with the number of these additional dislocations – both are approximately proportional to the extra tilt angle if it is small and the extra dislocations are far from each other.

A standard picture of a dislocation assigns elastic energy to its surrounding strain field, and a smaller nonelastic energy from poorly bonded atoms in the dislocation core. This same reasoning has been applied to grain boundary energies, in part because of the Read–Shockley model of grain boundaries as arrays of dislocations as in Fig. 11.8b,c. The idea of an elastic contribution to grain boundary energy suggests correlations between elastic constants and grain boundary energies. From studies on cubic metals, it seems that the grain boundary energy correlates reasonably well with the shear constant times lattice parameter, $C_{44} a_0$, which has units of J/m^2. The energy of a dislocation is often approximated this way (cf. Eq. 19.2), so even if a dislocation model of a grain boundary is naïve, the shear

---

[9] This is the same $\Sigma 5$ coincidence boundary shown in Fig. 11.8c. An fcc crystal is invariant after a tilt of 90° along a $\langle 001 \rangle$ tilt direction, so a tilt of 53° along $\langle 001 \rangle$ is equivalent to a tilt of $(90-53)° = 37°$. The $\Sigma 5$ tilt boundary is therefore shown at 37° in Fig. 11.9.

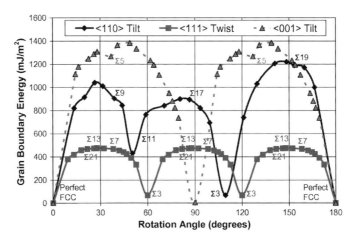

Figure 11.9 Calculated grain boundary energy shown as a function of the rotation angle for nickel in the $\langle 110 \rangle$ tilt, $\langle 111 \rangle$ twist, and $\langle 001 \rangle$ tilt directions. From embedded atom method calculations. Reprinted from [147], with permission from Elsevier.

distortions of atom bonds may be a key parameter for grain boundary energy [148]. For example, the temperature dependence of grain boundary energy tracks the temperature dependence of $C_{44}\, a_0$ [149].

Grain Boundary Free Energy

Much less is known about the entropy of grain boundaries, although there are calculations of configurational entropies of mixing when solute atoms segregate to grain boundaries [150–152]. There are degrees of freedom in the positions of the atoms in grain boundaries, but structural units as in Fig. 11.7 reduce the configurational entropy from positional disorder. For pure elements, most of the entropy of formation of grain boundaries is likely vibrational, although electronic entropy may play a role in metals. Today there is almost no experimental data on the vibrational entropies of grain boundaries, except that deduced for compacted nanocrystals as described in Section 20.5.2. From work on consolidated nanocrystals, the vibrational entropy of formation of a grain boundary appears to be positive, helping to stabilize the grain boundary at finite temperatures.

Especially with the emergence of modern transmission electron microscopes, there has been a renewed interest in the atomic structure within grain boundaries. For some time, however, it has been understood that different structural units can be periodically arranged along different grain boundaries, as seen by comparing Figs. 11.7a and b. Tendencies for chemical segregation to grain boundaries also depend on these detailed atomic structures. We call such structures and compositions "grain boundary complexions." Complexions are essentially two-dimensional phases. Transitions from one complexion to another are possible within the same grain boundary. These may be driven by chemical segregation or by temperature.

Transitions between different grain boundary complexions can be thermally reversible [153–155]. This implies a temperature where the free energy of one complexion equals that of another. We therefore expect an entropy difference between the two complexions. The thermodynamics of grain boundary complexions is not well understood. The kinetics of grain boundary transitions [156] should also be influenced by the difference in the free energies of the complexions.

In ceramic materials it is well known that an amorphous phase can form at grain boundaries between two crystals [157, 158]. This phase is three-dimensional, but thin. In spite of its higher free energy than crystalline material, it is favorable compared to a grain boundary formed by continuation of the structures of the two abutting crystals.

Some solutes tend to segregate to grain boundaries – the different local structure at the interface can favor a different chemical composition. We expect a grain boundary to be more accommodating of oversized solute atoms, for example. Arguments based on elastic misfit in the host crystal have been used to explain why solute atoms of mismatched size tend to segregate to grain boundaries [159]. A Hume-Rothery rule of Section 2.1.4 suggests that a solute atom with radius differing from the host by 15% may be better accommodated in a grain boundary, where the atoms have more freedom to change their positions. The segregation of solute atoms to grain boundaries may later promote the heterogeneous nucleation of a precipitate phase.

11.4.3 Chemical Energy of a Precipitate Interface

When a precipitate of different composition nucleates in a parent phase, there is necessarily an interface between the precipitate and the parent phase, with unfavorable atom pairs across the interface. The details of the bonding across the interface are generally complicated by differences in the lattice parameters and crystal structures. The example here considers only the chemical unmixing between the parent and precipitate phases, and assumes these α- and β-phases are solid solutions.

A chemical contribution to the interface energy can be estimated with a simple picture of two solid solutions in contact, as shown in Fig. 11.10. A bond-counting argument is used, with the assumption of a surface coordination number z_s for atom pairs across the plane of the interface. The plane of the interface has the average bond energy $E_{\alpha\beta}$, which differs from the average of the bond energy across equivalent planes in the α-phase and β-phase by

$$\gamma_{ch} = E_{\alpha\beta} - \frac{E_{\alpha\alpha} + E_{\beta\beta}}{2}, \tag{11.10}$$

where γ_{ch} is the chemical energy per atom at the interface.

For comparison, we need to count the bonds across equivalent planes in the α-phase and β-phase. This amounts to finding the number of like and unlike pairs of atoms across planes in solid solutions with concentrations of B-atoms c_α and c_β

$$E_{\alpha\alpha} = \frac{z_s}{2} \left[e_{AA}(1 - c_\alpha)^2 + e_{BB}c_\alpha^2 + 2e_{AB}c_\alpha(1 - c_\alpha) \right], \tag{11.11}$$

$$E_{\beta\beta} = \frac{z_s}{2} \left[e_{AA}(1 - c_\beta)^2 + e_{BB}c_\beta^2 + 2e_{AB}c_\beta(1 - c_\beta) \right], \tag{11.12}$$

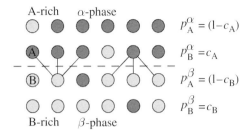

Figure 11.10 Interface (dashed line) between two phases of different chemical composition.

where all pairs of atoms in the bonds are in the same phase α or β. The bond counting is in fact the same as for solid solutions in Eq. 2.34, although z_s is different here.

Across the α/β interface shown as the dashed line in Fig. 11.10, each atom sees atoms in the other solid solution, which has a different composition

$$E_{\alpha\beta} = \frac{z_s}{2}\left[e_{AA}(1-c_\alpha)(1-c_\beta) + e_{BB}c_\alpha c_\beta + e_{AB}c_\alpha(1-c_\beta) + e_{AB}c_\beta(1-c_\alpha)\right], \quad (11.13)$$

where the counting is reminiscent of Fig. 2.20 for counting between two sublattices of an ordered alloy. (All bonds across the interface begin and end in different phases.)

Substituting Eqs. 11.11–11.13 into Eq. 11.10, and grouping terms in powers of c_α, c_β, and $c_\alpha c_\beta$

$$\gamma_{ch} = -\frac{z_s}{2}\left(c_\alpha - c_\beta\right)^2 4V, \quad (11.14)$$

where again

$$4V = e_{AA} + e_{BB} - 2e_{AB}. \quad (11.15)$$

For an unmixing alloy $4V < 0$, so $\gamma_{ch} > 0$. The chemical energy per atom at the interface makes a positive contribution to the interface energy, as expected.

This counting of bonds implies that the energy of chemical bonding is independent of, and additive to, the energy of the interface structure (e.g., Section 11.4.1). Reality is not so simple because different chemical environments prefer different local structures. A more subtle issue is that the chemical composition may not change abruptly at the interface. Section 16.2 discusses the "square gradient energy," which discourages an abrupt change in chemical composition. These additional chemical effects, plus structural effects on the interfacial energy, cause significant departures of the interface energy from the γ_{ch} that we obtained from a simple bond-counting model.

11.5 Anisotropic Surface Energy

11.5.1 Surface Structures

Free surfaces of crystals have energies that depend on crystallographic orientation. Close-packed crystal planes tend to have lower specific surface energies (i.e., energy per unit

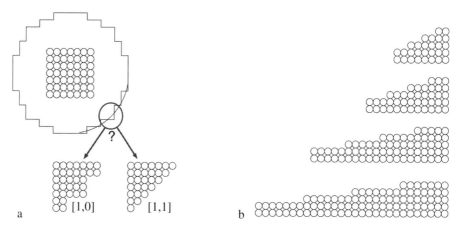

Figure 11.11 (**a**) Approximation of a circle by square elements. An alternative at lower right is to have a diagonal [1,1] surface, which will have a different surface energy than the [1,0] surface. (**b**) Construction of a series of "vicinal" surfaces with [1,0] planes and one-atom ledges of different separations.

area) than higher-index planes. The idea is that although atoms in close-packed planes have lost neighbors on one side, they are surrounded snugly by other atoms, giving favorable chemical bonding. Geometrically, it is possible to construct a macroscopic surface shape with zigs and zags of atom planes, as illustrated in Fig. 11.11a for surfaces of [1,0] planes of a square lattice.

A choice is presented in the lower part of Fig. 11.11a, and this choice illustrates key issues for the surface energies of crystals. Along the circled part of the surface, a diagonal [1,1] surface may be favored over the zig-zagged [1,0] surface because it has less length – for a 45° slope a [1,0] surface is $\sqrt{2}$ times longer than the [1,1]. However, the atoms at a [1,1] surface have two bonds to neighboring atoms, rather than the three bonds for atoms at a [1,0] surface. Assuming large areas of [1,0] surface compared with the atom dimensions (so corner atoms can be neglected), for this 45° angle a bond-counting model predicts that a [1,0] surface would have 2/3 the surface energy, but must be 1.414 times longer. The total energy for the [1,0] surface segments is then $2/3 \times 1.414 = 0.94$ times that for the [1,1] surface. This simple bond-counting model predicts that a zig-zagged [1,0] surface is favored over straight lengths of [1,1] surface.[10]

Figure 11.11b shows how other angled surfaces can be constructed from sections of [1,0] surfaces. The lower pictures in Fig. 11.11b are sometimes called "vicinal" [1,0] surfaces.

11.5.2 Wulff Construction

There is a clever geometrical construction, presented in Fig. 11.12, that predicts the optimal shape of a crystal when the shape is determined entirely by surface energy. This "Wulff construction" arranges the planes of low specific surface energy to optimize their coverage

[10] A more realistic electronic structure calculation of surface energy may well give a different result, especially since the bond counting approach gave a difference of merely 6%.

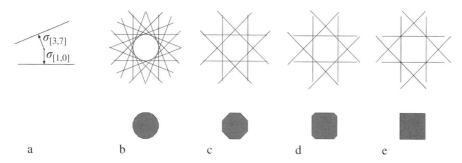

Figure 11.12 (**a**) Steps of Wulff construction. (**b**) Equilibrium crystal shape for isotropic surface energy, with equal specific surface energies for all planes. (**c**)–(**e**) Equilibrium crystal shapes with increasing specific surface energy of [1,1] planes.

of the crystal surface. It can sometimes be used backwards to find the relative surface energies for different crystallographic planes, given the equilibrium shape of a crystal. The Wulff construction is presented here without proof. To make a Wulff construction:

- Starting at a point, draw vectors radially outwards, each one pointing along a direction normal to a crystal plane.
- The length of the vector should be proportional to the surface energy of its corresponding plane.
- Draw planes perpendicular to the arrows, intersecting their tips (Fig. 11.12a).
- The crystal shape is bounded by the planes closest to the central point. These close planes are the ones that form the crystal surfaces.

The first three steps are illustrated in Fig. 11.12a for two planes. Complete sets of planes are shown (without the vectors) in Figs. 11.12b–e. Figure 11.12b shows the Wulff construction for a crystal with isotropic surface energy. All planes have the same specific surface energy, so all planes are equidistant from the central point. This construction naturally gives a circle, which has the lowest surface area.[11] In Fig. 11.12c we assume that the higher index (approximately [3,7]) planes have a high surface energy, and need not be considered. The [1,0] and [1,1] surfaces are still assumed to have equal energy, however. Figure 11.12d shows the case where the specific surface energy of a [1,1] plane is a bit higher than for a [1,0] plane. There is less surface covered by [1,1] planes, and the equilibrium shape is altered as shown at the bottom of Fig. 11.12d. Figure 11.12e shows the case where the specific surface energy of the [1,1] plane is $\sqrt{2}$ times larger than for [1,0] planes. In this case the crystal is bounded entirely by [1,0] planes.

11.5.3 Wulff Construction with Grain Boundaries

Figure 11.13 shows an interesting modification of the Wulff construction for predicting the shapes of precipitates that nucleate at grain boundaries. An important issue is that the

---

[11] For isotropic surface energy in 3D, a sphere, which minimizes surface-to-volume ratio, also minimizes the total surface energy.

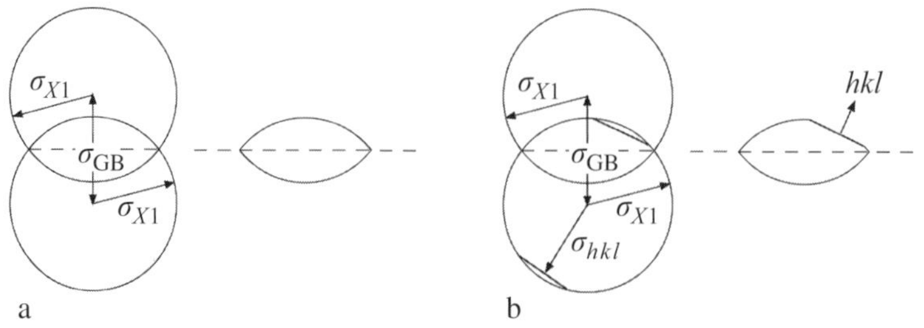

Figure 11.13 (**a**) Displacement of surface energy vectors σ_{X1} by the amount σ_{GB} in the direction normal to the grain boundary, and the resulting lenticular precipitate shape. (**b**) Crystallographic preference of the surface energy for *hkl* planes, showing effect on precipitate shape.

grain boundary energy, σ_{GB}, is unfavorable, so eliminating it effectively reduces the surface energy cost for forming the precipitate at a grain boundary. A precipitate shape with more surface area parallel to the grain boundary is more favorable because more grain boundary energy is eliminated. The Wulff construction for isotropic surface energy is modified as shown in Fig. 11.13a, with a displacement of the Wulff construction by σ_{GB} along the normal of the grain boundary. The resulting shape is the lens discussed in Section 4.3. The curved surfaces have the same energetics as for the homogeneous nucleation of the precipitate, but the width of the "lenticular" precipitate allows more elimination of grain boundary area.

The limits are informative. If σ_{GB} were zero, the precipitate would be a sphere, as expected for homogeneous nucleation where no grain boundary is present. If σ_{GB} were slightly less than $2\sigma_{X1}$, the spheres in Fig. 11.13a would overlap slightly, and the precipitate would be nearly flat and spread widely across the boundary. If $\sigma_{GB} > 2\sigma_{X1}$, the surface energy of forming a precipitate would be negative, and nucleation would occur slightly above the critical temperature for the bulk free energy. Other structural rearrangements at the grain boundary might be expected for this unusual case, however.

Figure 11.13b shows another modification that is expected in general. Here there is low surface energy of the {*hkl*} planes in contact with the matrix phase, so these {*hkl*} surfaces are included in the favored shape of the precipitate. These crystallographic features on grain boundary allotriomorphs are fairly common – Fig. 4.5 is a clear example. With images such as that of Fig. 4.5, it is possible to assess the relative energies of the different surfaces between the precipitate and matrix, and the grain boundary [160].

11.6　Reactions at Surfaces

11.6.1　Linear and Parabolic Oxidation

The kinetics of phase transformations involving reactions at surfaces are often analyzed in one dimension.[12] Oxidation of metal surfaces is a typical example, but the analysis here applies to many other surface reactions. A cross-section is shown in Fig. 11.14, with a

---

[12] Microscopically, or even macroscopically, the interface is rarely flat, however.

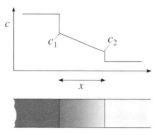

Figure 11.14 Top: concentration profile for oxygen atoms perpendicular to a metal surface. At the surfaces of the oxide layer, the concentrations of oxygen will differ from the oxygen concentration in either the gas on the left, or the metal on the right. Bottom: schematic diagram of layers.

metal on the right and a gas on the left. Between them is an oxide of thickness x. In this problem we assume that the oxygen atoms must move through the oxide layer to reach the metal surface and react. The opposite case is often seen, however, where the metal atoms diffuse through the oxide layer. Not surprisingly, the permeability of the oxide layer for gases, or the diffusivity of ions in the layer, plays a major role in determining the kinetics.

The simplest model for an interface reaction ignores rate limitations caused by diffusion of atoms through the reaction layer, and sets a constant rate of growth

$$\frac{dx}{dt} = K, \tag{11.16}$$

$$x(t) = Kt. \tag{11.17}$$

Equation 11.17 is often relevant when the reaction layer is thin, so the reaction time at the interface, rather than atom transport through the reaction layer, controls the rate of growth.

The second simple reaction model assumes that the rate of growth of the reaction layer is determined by the rate of arrival of atoms that diffuse across the reaction layer

$$\frac{dx}{dt} = \frac{J}{\rho}, \tag{11.18}$$

where ρ is the density of the oxide (atoms/cm^3). Fick's first law (Eq. 3.16) gives the flux of atoms across the reaction layer

$$J = -D \frac{dc}{dx'}. \tag{11.19}$$

The concentration gradient is obtained approximately from Fig. 11.14

$$\frac{dc}{dx'} = \frac{c_2 - c_1}{x}. \tag{11.20}$$

With this concentration gradient, and equating the two fluxes of Eqs. 11.18 and 11.19

$$\rho \frac{dx}{dt} = -D \frac{c_2 - c_1}{x}, \tag{11.21}$$

$$\rho \int_0^{x(t)} x \, dx = D(c_1 - c_2) \int_0^t dt', \tag{11.22}$$

$$x(t) = \left(\frac{2D(c_1 - c_2)}{\rho} \right)^{1/2} \sqrt{t}. \tag{11.23}$$

Thin-film reactions that follow the kinetics of Eq. 11.23 are called "parabolic," and those following Eq. 11.17 are called "linear." Both can occur in the same phase transformation – in the early stages the reaction may have linear kinetics because it is not limited by the diffusive flux, but with a thicker reaction layer the atom diffusion through the layer may be rate-limiting. Fitting experimental data to both types of curves is typical practice. Sometimes the rates of the two processes, set by K and D, differ enormously. Aluminum metal, when exposed to air, forms a reaction layer of Al_2O_3 almost instantly, but this layer allows very slow diffusion for either aluminum or oxygen ions. The reaction effectively stops after forming about 10 nm of Al_2O_3. We sometimes say that the reactive surface of aluminum metal has been "passivated."

11.6.2 Phases in Concentration Gradients

Consider a flat interface between two pure elemental materials, heated at temperature T so interdiffusion occurs. The top left plot in Fig. 11.15 shows a typical interdiffusion profile for a single phase. A single solid phase occurs for some systems, but more typically the material prefers different phases at different chemical compositions, as indicated in the (rotated) double eutectic phase diagram to the right of the profile. In this case the phases α and β remain pure A and pure B outside the diffusion zone, but in the zone they accommodate atoms of the other chemical element. The phase diagram indicates that a γ-phase of intermediate composition will form between the α- and β-phases. This requires two distinct interfaces between the phases, α/γ and β/γ. At these interfaces, the chemical compositions in the adjacent phases are expected to be those at phase boundaries in the phase diagram. The locations of the interfaces cannot be predicted from the phase diagram, of course, and they move as the reaction proceeds.[13]

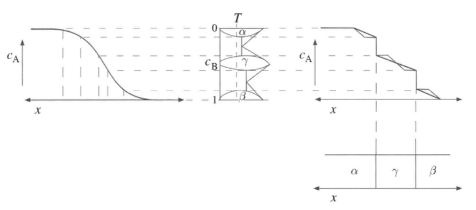

Figure 11.15 Left: a generic but naïve concentration profile for element A after some interdiffusion with element B (cf. Fig. 3.5b). To its right is a phase diagram giving compositions of phases at temperature T. Top right: the composition profile shows how the different phases can be accommodated in the diffusion zone. Discontinuities in the composition are of course expected at sharp interfaces, such as those at the bottom right.

---

[13] It is also common for the new phase to nucleate as three-dimensional particles, breaking the one-dimensional analysis that is so convenient for thin-film reactions.

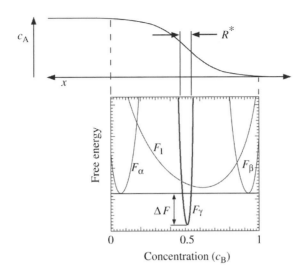

Figure 11.16 Top shows a generic interdiffusion concentration profile. The critical radius for nucleation has the width R^* as marked in the figure. Lower figure, with composition aligned approximately in correspondence with the top spatial profile, shows free energy versus composition curves. The curve for the γ-phase rises steeply for small changes in composition.

The γ-phase is expected from the equilibrium phase diagram of Fig. 11.15, but it is not necessarily present at all stages of the reaction. It is possible for the interdiffusing atoms to spread over a wide spatial range of the α- and β-phases, even if the composition variations in these phases are not large. This may be the case if nucleation of the γ-phase is suppressed. (Solid-state amorphization, Section 9.1, is one effect of suppressed nucleation.)

Nucleation can be suppressed by the large concentration gradients in the early stages of interdiffusion [161]. Consider the situation shown in Fig. 11.16, where the γ-phase has a rather narrow range of composition where its free energy is low compared with other phases. In a thin-film reaction with a steep composition gradient, there may be only a narrow range of x where the composition is appropriate for the γ-phase to have a low free energy. When this range of x is smaller than the size of the critical nucleus of the γ-phase, nucleation of the γ-phase cannot occur. If $F(c)$ is parabolic with curvature $\partial^2 F/\partial c^2$, nucleation will be suppressed when

$$\Delta F < \frac{1}{2} \frac{\partial^2 F}{\partial c^2} \left(\frac{dc}{dx} R^* \right)^2 , \qquad (11.24)$$

where ΔF is the free energy change favoring the formation of the new phase (see Fig. 11.16), R^* is the critical radius for nucleation, and dc/dx is the concentration gradient. Equation 11.24 states that nucleation is suppressed if the edges of the new phase are at compositions with specific free energies above the common tangent of Fig. 11.16. The larger the concentration gradient, the more difficult is nucleation. Nucleation in a diffusion zone is also suppressed if the new phase has a large critical nucleus.

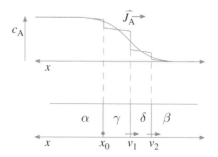

Top shows a typical average interdiffusion concentration profile, with typical concentration discontinuities at interfaces between phases. Lower figure shows two new phases, γ and δ, that have nucleated and started to grow.

11.6.3 Phase Growth in Thin Films

A phase diagram often contains many phases of different compositions. For thin-film reactions, for example at the interface between silicon and a transition metal, important questions are "Which phase forms first?" and "Do all the phases form?" These questions sound simple, but prove complicated, and require kinetic information well beyond what can be gleaned from the equilibrium phase diagram, especially information on nucleation and rates of diffusive growth.

The growth velocities of the phases are an important way to understand the kinetics of thin film reactions [162]. Consider the picture shown in Fig. 11.17, where for simplicity it is assumed that only A-atoms move (with net flux to the right), and the interface velocities depend on the arrival of A-atoms, and their attachment to the growing phase. Both processes cause the interfaces to move at different velocities, labeled as v_1 and v_2 in Fig. 11.17. Using the position of the α/γ interface as a reference at position x_0, suppose the interfaces move to the right.[14] If growth is to the right, and if $v_1 < v_2$, the widths of both the γ-phase and the δ-phase will grow with time. This need not be the case, and it is quite possible that $v_1 > v_2$, so the γ–δ interface at left will outrun the δ–β interface to its right. In this case the δ-phase will disappear. Perhaps the δ-phase will never form, since it would always be overtaken by the γ-phase. On the other hand, perhaps the δ-phase nucleates more easily than the γ-phase, so there is an early period of time where only the δ-phase is present, but once the γ-phase forms, it quickly consumes the original δ-phase. All these cases have been observed in different interfacial reactions.

Owing to their technological importance, some of the most thoroughly studied thin-film reactions involve metals and silicon, typically with metal layer thicknesses of 10 to 100 nm [163]. In these cases when the metal layer is thin, and silicon is the majority species, eventually the final phase is a compound with the largest fraction of silicon. This is the state of equilibrium expected for a silicon–metal alloy with low metal composition.

---

[14] Section 10.3 showed that the entire block of material moves a bit to the right under the conditions shown, so all interfaces can be argued to move to the left.

The binary phase diagrams of transition metals and silicon have numerous phases at low temperatures. What is interesting is that usually only one phase forms at a time. It often consumes all the metal layer, and remains metastable until a second phase forms and consumes it. For predicting the first phase that forms, an empirical correlation is the Walser–Bené rule, which states that the first silicide to grow is the highest-melting compound next to the lowest-melting eutectic in the phase diagram [164], but exceptions occur. The next phase to form is often a phase with high melting temperature that is rich in the unreacted element, but again this rule is not fundamental. Another independent, and perhaps contradictory, rule for silicide formation is that the phase formation sequence depends on whether the silicon atoms or the metal atoms are the more mobile. When the metal atoms are more mobile, the equilibrium phase tends to form first, but when the silicon is more mobile a series of silicide compounds form first. In spite of the technological importance of phase transformations in thin films of metals and silicon, the reasons for their kinetics are not well understood.

11.7 Gas Adsorption

Gases interact chemically with surface atoms. Sometimes these interactions lead to covalent chemical bonds, with large bond energies. In such cases it is difficult for these "chemisorbed" gas molecules to be released at convenient temperatures, pressures, and times. An alternative process of "physisorption" is of interest for reversible gas storage on surfaces of nanomaterials. In physisorption, the gas molecule such as H_2 remains intact, and does not dissociate or form a covalent bond with atoms of the sorbent. Bonding by "dispersion forces," or Van der Waals interactions, holds the molecule to the surface at low temperatures or high pressures. The gas molecules can be released by heating or evacuating the storage system.

Nanomaterials have high "specific surface areas," quantified as "volumetric" with units of $[cm^2/cm^3 = 1/cm]$, or "gravimetric" with units of $[cm^2/g]$. Molecules of gas can "adsorb" onto exposed surfaces of materials, and relatively large amounts of gas can do so on some nanomaterials. Adsorption of gases such as hydrogen or methane on nanomaterials is an active research topic for energy storage materials.

Here we derive the thermodynamic "Langmuir isotherm" for gas physisorption on a surface, obtaining the amount of sorption as a function of pressure, at a fixed temperature. Assume:

- There is only one energy in the problem, ε, the binding energy of one gas molecule to one surface site. Binding is favorable, so $\varepsilon < 0$.
- The sorbent material has a fixed number of surface sites, and the adsorbed molecules have a random distribution over these sites.
- There is no interaction between the molecules adsorbed on the surface, and there is zero vibrational entropy associated with the physisorbed molecules.
- The gas is assumed an ideal gas, simplifying the treatment of the chemical potential of molecules in the gas phase.

For a fixed temperature, the probability that a surface site is occupied is

$$f_{ad} = \frac{e^{(\mu-\varepsilon)/k_B T}}{1 + e^{(\mu-\varepsilon)/k_B T}}, \tag{11.25}$$

where the denominator normalizes the Gibbs factor by the two possibilities – an empty site (with Gibbs factor $e^0 = 1$) and a site with an adsorbed molecule (with Gibbs factor $e^{(\mu-\varepsilon)/k_B T}$).

In equilibrium, the chemical potential for an adsorbed molecule on a site equals the chemical potential of the molecule in the gas. Assume the chemical potential is that of the ideal gas, which is related to the quantum volume V_Q divided by the volume per atom[15]

$$e^{\mu/k_B T} = \frac{N V_Q}{V} = \frac{N}{V}\left(\frac{2\pi\hbar^2}{m k_B T}\right)^{3/2}. \tag{11.26}$$

Using Eq. 11.26 and the ideal gas law, $pV = N k_B T$, to rearrange Eq. 11.25

$$f_{ad} = \frac{p}{\frac{k_B T}{V_Q}e^{\varepsilon/k_B T} + p}. \tag{11.27}$$

The first term in the denominator has units of energy/volume, or pressure. Its value is fixed for a given temperature, so it is defined as $p_0(T)$

$$p_0(T) \equiv \frac{k_B T}{V_Q}e^{\varepsilon/k_B T}, \tag{11.28}$$

to give the compact expression

$$f_{ad} = \frac{p}{p_0(T) + p}. \tag{11.29}$$

Because the surface sites are noninteracting, all sites have the same probability of occupancy, f_{ad}. Therefore f_{ad} is the surface coverage, or the fraction of sites occupied by gas molecules. The surface coverage increases with pressure as shown in Fig. 11.18, reaching an asymptotic value of 1. The increase in coverage is faster at lower temperatures because p_0 is smaller, both from the prefactor $k_B T$ and from the exponential (which has a negative argument since $\varepsilon < 0$).

If each adsorption site has the same energy for physisorption, the number of adsorbed molecules should scale with the specific surface area of the material. This is often true. Some results for hydrogen physisorption on different carbon materials are shown in Fig. 11.19. The top dashed line in the figure is the capacity expected if hydrogen molecules, approximated as spheres, are packed hexagonally on a surface. The lower solid line is a rule of thumb that carbon materials at 77 K accommodate 1 wt.% hydrogen per 500 m^2/g of gravimetric specific surface area. Deviations from this rule-of-thumb behavior can be attributed to differences in site adsorption energy.

---

[15] The quantum volume is discussed with Eqs. 12.22–12.24. Here we also assume no internal degrees of freedom in the molecule of mass m.

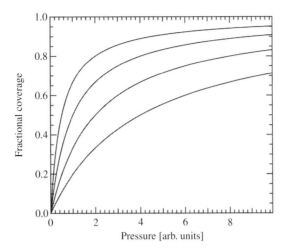

Figure 11.18   Series of Langmuir isotherms for four values of p_0 ranging from 0.5 (top) to 4 (bottom).

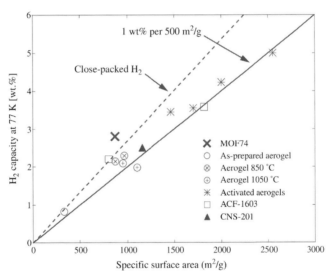

Figure 11.19   Maximum hydrogen adsorption capacities at 77 K of various carbon adsorbents plotted against their specific surface areas. Materials include a series of carbon aerogels, two activated carbon fibers (ACF-1603), one activated coconut carbon (CNS-201), and one metal-organic-framework (MOF-74). From [165].

Problems

11.1   Construct a coincidence lattice for the grain boundary orientation in Fig. 11.8b, and compare its density to the coincidence lattice shown in Fig. 11.8c.

11.2   A one-dimensional model of roughening is shown in Fig. 11.20. Each upward step produces an atom with an unsatisfied bond on its left, marked with an "X." If the

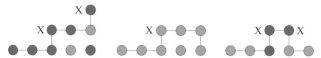

Three possibilities for a surface segment with ledges. Each upward step causes a loss of a bond to the left, and implies a downward step without a bond to its right.

average surface height is zero for the N sites across the surface, for every up step there must be a down step without a bond to its right. The excess bond energy can therefore be formulated in terms of $pN/2$, the number of up steps (p is the probability of a step at a site).

(a) Show that the energy cost of unsatisfied bonds is

$$E_s = z_s \, \varepsilon \, \frac{pN}{2}, \qquad (11.30)$$

where z_s is the coordination number ($z_s = 2$ for the 1D case of Fig. 11.20).

(b) Assume the up steps and down steps are distributed at random over the surface. Show that the number of ways of arranging pN steps over N surface sites times the number of ways of orienting each step (up or down) gives a configurational entropy

$$S_s = -k_B N \big(p \ln p + (1-p) \ln(1-p) \big) + k_B \, pN \ln 2, \qquad (11.31)$$

and compare your result to Eq. 2.31.

(c) For a given temperature, find the equilibrium value of p by minimizing $F_s = E - TS$, obtaining

$$p(T) = 2 \exp\left(-\frac{z_s \varepsilon}{2k_B T}\right), \qquad (11.32)$$

assuming $p \ll 1$.

(d) Explain how Eq. 11.32 shows that a surface becomes rougher with increased temperature (at least at low temperature).

11.3 (a) Calculate and compare the total surface energies for Frank–Van der Merwe and Volmer–Weber growth shown in Figs. 11.5a and 11.5b, ignoring the vertical edges. For Volmer–Weber growth, assume that the fraction of surface f is covered by islands, and the fraction $1 - f$ is not. Use your expression for total energy to derive Eq. 11.9.

(b) Explain how elastic energy from the crystallographic mismatch of the film on the substrate could lead to Stranski–Krastanov growth.

11.4 Arguing with a Wulff construction, what will be the equilibrium shapes of crystals with these surface energies:

(a) The specific surface energy of an (h, k, l) crystal plane is proportional to $\sqrt{h^2 + k^2 + l^2}$.

(b) The specific surface energies of $(1,1,1)$ planes and $(1,0,0)$ planes are equal, and all other specific surface energies are at least twice this value.

11.5 Wulff construction for a vicinal surface

Suppose that a flat surface has a surface energy of γ_0, but a vicinal surface as shown in Fig. 11.11b has an extra energy proportional to the number of steps per unit area. This problem calculates the surface energy for surfaces tilted by the angle θ from the flat surface.

(**a**) Show that the specific surface energy γ can be written

$$\gamma(\theta) = \gamma_0 \cos \theta + \frac{\gamma_1}{a} \sin \theta, \qquad (11.33)$$

where a is the lattice parameter of the simple cubic crystal. (*Hint*: The fraction of flat surface can be written as $\cos \theta$, and the density of steps on it is $\tan \theta / a$.)

(**b**) Use the trigonometric identity

$$\cos(\theta - \phi) = \cos \theta \cos \phi + \sin \theta \sin \phi \qquad (11.34)$$

to find a relation between γ_0 and γ_1 / a in terms of $\tan \phi$.

(**c**) Figure 11.21 shows a generalized Wulff plot for the surface energy of parts a and b. Show that it is a graph of

$$\gamma = 2\Gamma \cos(\theta - \phi). \qquad (11.35)$$

(**d**) Identify 2Γ in Eq. 11.35 from the terms in Eqs. 11.33 and 11.34.

(**e**) For the case shown in Fig. 11.21 (i.e., this ratio of γ_0 to γ_1 / a), will the equilibrium shape of a crystal have surfaces other than the {100} family?

11.6 Interfacial reactions

(**a**) Consider nucleation in a concentration gradient of 0.01/nm. If the $F(c)$ for a new phase is quadratic in composition, and rises from its minimum to 1.0 eV/atom over a composition range of 0.05, what is the largest critical radius that is possible if the change in free energy, ΔF, must be less than 0.1 eV/atom?

(**b**) Interfacial reactions are often performed at low temperatures, where there are substantial differences in rates of nucleation and growth of the different phases. Consider the diagram of Fig. 11.17, and suppose that the growth velocity v_1 is set by the diffusion of A-atoms through the α-phase, whereas the velocity v_2

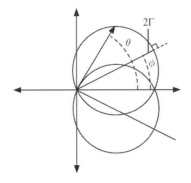

Figure 11.21   Figure for Problem 11.5, showing two rotated circles that intersect the origin.

is set by the diffusion of A-atoms through the γ-phase. Suppose the diffusion coefficients for the two phases are both of the form $D(T) = D_0 \exp(-Q/RT)$, where $D_0 = 10 \, \text{cm}^2/\text{s}$ for both phases, but $Q_\alpha = 50 \, \text{kcal/mole}$ and $Q_\gamma = 60 \, \text{kcal/mole}$. Approximate the velocity of growth as the characteristic diffusion length for a time interval τ.

At a temperature of 400 °C, what is the ratio of v_1 and v_2?

Approximately how long will it take for the δ-phase to disappear at 400 °C if it has an initial thickness of 10 nm?

11.7 Consider two dilute lattice gases (i.e., a system of lattice sites where one atom may or may not be located). Each lattice gas has no interactions between atoms, but the two differ in the energy of binding of an atom to a site. Allow diffusive contact between the two lattice gases. Please start with the fractional occupancy, f, of each lattice gas in the form

$$f = \lambda \, \frac{\exp(-\varepsilon/k_B T)}{Z}, \tag{11.36}$$

$$Z = 1 + \lambda \, \exp(-\varepsilon/k_B T), \tag{11.37}$$

where $\lambda = \exp(\mu/k_B T)$ is the activity.

(a) Explain the form of the partition function, Z. Explain its similarity to fermion statistics, such as electron occupancies.

(b) Calculate the ratio of atom occupancies on each lattice at equilibrium.

11.8 In physisorption, a question that arises naturally is how the gas molecules spread themselves out over the surface, i.e., what is the thickness and density of the surface layer of adsorbed gas molecules? This information is not easy to obtain by measurement. A "Gibbs excess" of gas adsorption ignores the physical structure of the adsorbed gas layer, and defines operationally the number of adsorbed gas molecules for an adsorbent material in a vessel:

1. The amount of empty volume available to free gas molecules is determined when no adsorption occurs (perhaps by placing a dummy adsorbent of equal volume in the vessel). The number of moles of gas molecules in the vessel is determined (typically from knowledge of the volume and pressure).

2. The number of moles of gas molecules in the vessel is determined when the adsorbent material is active, for a pressure equal to that of the first measurement.

The extra moles of gas in the vessel in step 2, compared to 1, are the Gibbs excess adsorption.

(a) When gas pressures are low and the adsorption is strong, argue that the Gibbs excess is the same as the total number of molecules in the physical adsorbed layer.

(b) When the gas pressure and density are high, show that the Gibbs excess adsorption is smaller than the total number of molecules in the physical adsorbed layer. Explain why the Gibbs excess adsorption can decrease at high gas pressures even as the adsorbed layer becomes thicker.

(*Hint for Parts a and b*: The space occupied by the adsorbed layer of gas molecules would have gas molecules in it even without attraction of the gas molecules to the surface. The number of gas molecules in this space is larger at higher pressures.)

(**c**) If you were deciding if an adsorbent material would have practical benefit for storing more gas molecules in a vessel at a given pressure, would you prefer information about the total number of molecules in the adsorbed layer, or information about the Gibbs excess? Is there also an issue with the volume occupied by the adsorbent material itself? Explain.

12

Melting

The free energy of melting was discussed in Section 1.3.1. Thermodynamic equilibrium between the solid phase and the liquid phase was the starting point for the discussion on alloy solidification in Chapter 5, and of course the equilibrium thermodynamics are the same for either melting or solidification. Chapter 12 presents additional considerations about the enthalpy and entropy of the solid and liquid phases near the melting temperature T_m, and highlights rules of thumb, such as the tendency for the entropy of melting to be similar for different materials. Correlations between T_m and the amplitude of thermal displacements of atoms ("Lindemann rule"), and between T_m and the bulk modulus B are justified with simple models, but these correlations are semiquantitative at best.

Melting and solidification are first-order phase transitions that take place at an interface between the solid and the liquid, introduced in Section 12.1. Sections 5.2.1–5.3 described solute partitioning at an interface, and how faster rates of solidification lead to a number of nonequilibrium states in the solid. Interface phenomena during solidification are covered in more detail in, for example, Sections 13.2 and 15.3, including how the velocity of the two-phase interface controls the chemical composition of the solid. (The approach uses bulk thermodynamics for modest interface velocities, and atomistic kinetics for faster rates.)

At a temperature well below the melting temperature T_m, a glass undergoes a type of melting called a "glass transition." The glass transition was described in Section 5.5.2, but is discussed in more detail in Section 12.6. Finally, this chapter describes some features of melting in two dimensions, which is quite different from melting in three dimensions.

12.1 Structure and Thermodynamics of Melting

12.1.1 Liquids and Solids

A liquid is typically the middle phase in the temperature sequence: solid, liquid, gas. It is common to consider the liquid as somehow intermediate between a solid and a gas, but its similarity to a solid or a gas depends on the state of the liquid and on the properties of interest. At the critical point, for example, a liquid is indistinguishable from a dense gas, as are its properties. A characteristic property of a liquid or gas is that they cannot support shear stress (or shear waves) because their atoms are relatively mobile. For other properties, however, a liquid behaves more like a solid than a gas. Atoms in a liquid adhere together and are a condensed phase of matter, unlike a diffuse gas. Chemical bonds between atoms or molecules in a liquid are more like those in a solid than in a gas of low density, where there is a loss of bonding.

The melting of a crystal brings a partial loss of bonding energy, and this depends on details of changes in atom positions. Any atomistic understanding of the free energy of a liquid requires atom positions, but details about atom configurations in a liquid, and their time evolution, are not easily expressed. It is therefore not surprising that calculations of the change in bonding energy upon melting are not yet precise. An even greater challenge is that the sources of entropy upon melting are not fully understood. Certainly there is an increase in the configurational entropy, since many more atom configurations are explored in the liquid than in the solid. Upon melting, the new atom configurations could also give a different vibrational entropy. It is also known that some materials change from insulators to metals upon melting, so electronic entropy can change, too.

12.1.2 Premelting

Atoms at the surface of a crystal have lower coordination than in the bulk, and the surface layer of a crystal is liquid and mobile at temperatures below the melting temperature of the bulk [166]. When the very top layer melts, it disrupts the coordination below it, and the liquid layer can extend down into the next layer and more. Within tens of degrees below T_m, this "premelted" surface layer may be several atomic layers thick.[1] Nucleation of the liquid phase is therefore not an initial step of a melting transformation – the liquid phase is already nucleated as a premelted surface, and the thickness of the premelted layer diverges at T_m. This explains the experimental asymmetry of freezing and melting – it is commonplace to undercool below T_m, but superheating a crystal above T_m is almost impossible when heat is applied to the surface of the crystal.

The depth of this premelted layer is expected to change if the crystal is in contact with a liquid, rather than a vacuum. A liquid of Pb over Al can promote surface melting [168]. Although the density in this liquid is changing dynamically, it is expected that the crystal periodicity influences the liquid in contact with the crystal. The tendency of the liquid density to follow the periodicity of the neighboring crystal may extend over a few atomic planes, as shown in Fig. 12.1. As internal surfaces in polycrystalline materials, grain boundaries are also expected to melt at temperatures below the bulk T_m [169, 170].

12.1.3 Superheating

A brief superheating of a crystal above its melting temperature is possible if the heating rate is so fast that the atoms or molecules in the solid are unable to reach thermodynamic equilibrium. This can occur for polymers heated at rates of 10^5 K/s [173]. Much higher heating rates are required for metals and inorganic compounds. It has been observed, however, that a silver particle, coated with gold, can be superheated up to 25 K above T_m for times of about a minute [174]. Presumably the gold coating suppresses the premelted layer

---

[1] Surface premelting of ice may explain its slippery behavior at low contact pressure [167].

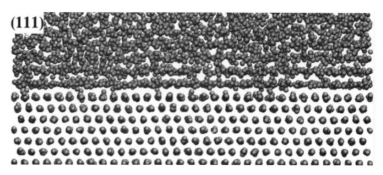

Snapshot from simulation of liquid Pb above Al (111) crystal surface at 625 K, showing layering tendency in the liquid. The Al surface is faceted, and the image is a projection of several overlapping atomic planes out of the plane of the paper. Reprinted from [171], with permission from Elsevier. Also see [172].

of the silver particle. Sometimes an oxide layer on the surface of a crystal can suppress surface melting.

Depositing the heat away from the premelted surface is another approach that has been successful – for example, an intense light source can deposit heat inside a transparent material such as ice, superheating it approximately 1 K [175]. When computer simulations of melting are performed without crystal surfaces, it is found that the melting temperature is higher than expected, maybe 20–30 %. For internal melting when heat is deposited inside a crystal, away from its surfaces, defects such as vacancies and interstitial atoms can be formed with a high density. A nucleation barrier is likely involved, since these defects tend to rejoin the surrounding crystal structure after a peak in the local thermal fluctuation [176]. This surrounding crystal also constrains the configurations of the defect-rich zones, helping to set a bound on the amount of superheating. The usual condition for melting (Eq. 12.1) is that the free energy of the liquid is equal to the free energy of the solid. For internal melting, however, the configurational entropy of the defect-rich region is suppressed compared to a liquid phase. The full entropy of the liquid, S_l, is not available, so $S_l - S_s$ is smaller and melting requires a greater energy in the zone. An upper bound for superheating can be approximated as $E_s(T_{ls}) = E_l(T_m)$, so internal melting now requires a temperature T_{ls} that is higher than the usual T_m [177]. There is also the implicit requirement the enthalpy of defects in the crystal rises rapidly with temperature, and at high temperatures the defect-rich zones have greater molar energy than a liquid. An entropy catastrophe, where the entropy of a superheated solid equals that of the liquid, has also been proposed as a limit of superheating that might extend to almost 40% above the melting temperature [178]. There are many opportunities to test these concepts by experiment or simulation.

12.1.4 Free Energy and Latent Heat

As described in Section 1.3.1, the criterion for melting, and for determining the melting temperature T_m, is the equality of the free energies of the solid and liquid phases

$$F_s(T_m) = F_l(T_m). \tag{12.1}$$

Schematic behavior is shown in Fig. 1.2, and discussed in Eq. 1.4 (repeated here)

$$S_l - S_s = \frac{E_l - E_s}{T_m} = \frac{L}{T_m}, \tag{12.2}$$

where the latent heat, L, is

$$L \equiv \left[S_l(T_m) - S_s(T_m) \right] T_m. \tag{12.3}$$

The sign of the latent heat is such that L is absorbed during melting, and released during freezing.

The release or absorption of latent heat has a significant effect on how a material is heated or cooled through its melting temperature. Without latent heat there is only a monotonic change in temperature with time. With latent heat comes an arrest or slowing of either heating or cooling at T_m. As discussed below, $S_l(T_m) - S_s(T_m)$ is similar for many materials, so from Eq. 12.3 the latent heat is expected to scale with the melting temperature. The duration of the arrest may also be expected to scale with T_m, after accounting for the rate of heat removal. A thermal arrest during cooling is usually a robust and reliable way to determine T_m.[2]

In many cases of rapid cooling, the liquid can be cooled below T_m before the solid phase has a chance to nucleate and grow. When the solid forms, the latent heat raises its temperature and the temperature of the adjacent undercooled liquid. This process is called "recalescence." Often the heat release during recalescence is enough to raise the local temperature to T_m, where the temperature remains constant for a time. The situation can be similar to that shown and discussed in Fig. 13.2e for one-dimensional solidification. A modest difference is that recalescence need not occur at the surface of a growing solidification front, but can occur when crystals nucleate inside the liquid phase. During recalescence, the local temperature of the new solid does not rise above T_m, however.

From Eq. 12.2 we see that a detailed calculation of T_m requires the functions $\{E_l, E_s, S_l, S_s\}$, or at least the differences $E_l - E_s$ and $S_l - S_s$. The energies and entropies of solids vary with the nature of their chemical bonds and crystal structures. Variations in the energy and entropy of liquids are similarly expected. This chapter presents some trends of melting temperatures of the elements in the periodic table, with general relationships to bonding energy and entropy. For realistic calculations, many details are needed for assessing chemical trends for the thermodynamic functions of solids and liquids that determine T_m, and this is a research topic today.

12.2 Chemical Trends of Melting

Melting temperatures of the elements are listed and depicted graphically in Fig. 12.2. There are several important trends in T_m across the periodic table that can be assessed with trends in chemical bonding:

- The three rows of transition metals (Sc–Ni, Y–Pd, Lu–Pt) have a rise and fall of T_m from left to right. The elements with five d-electrons (Cr, Mo, W) have the highest T_m.

---

[2] Arrests or kinks in cooling curves of $T(t)$ are common identifiers of other phase transitions, too, see Box 2.4.

- From left to right across the main group, the elements with four s- and p-electrons (C, Si, Ge) have the highest T_m.
- The rare earth metals from La to Tm have similar melting temperatures, but some tendency for the melting temperature to increase towards the end of the series.
- The actinide metals from Ac to Es have a peculiar trend of T_m.
- The alkali metals and noble gases have very low melting temperatures.

The first two or three trends can be understood with the bonding and antibonding arguments of Section 6.3. For a tight-binding electronic band structure, half the states in the band will be bonding and half will be antibonding. The maximum cohesive energy occurs when the band structure is half full, so all the bonding states are occupied and the antibonding states are not. The maximum T_m is therefore expected when the band structure is half full. This argument can be used to explain the chemical trends of T_m for the transition metals with their ten d-electrons, and for the main group elements with their eight s- and p-electrons. From this argument, the midpoints with maximum bond energy are the C column of the periodic table for the main group elements, and the Cr column for the transition metals (assuming one s-electron in a band of its own). Figure 12.2 shows this

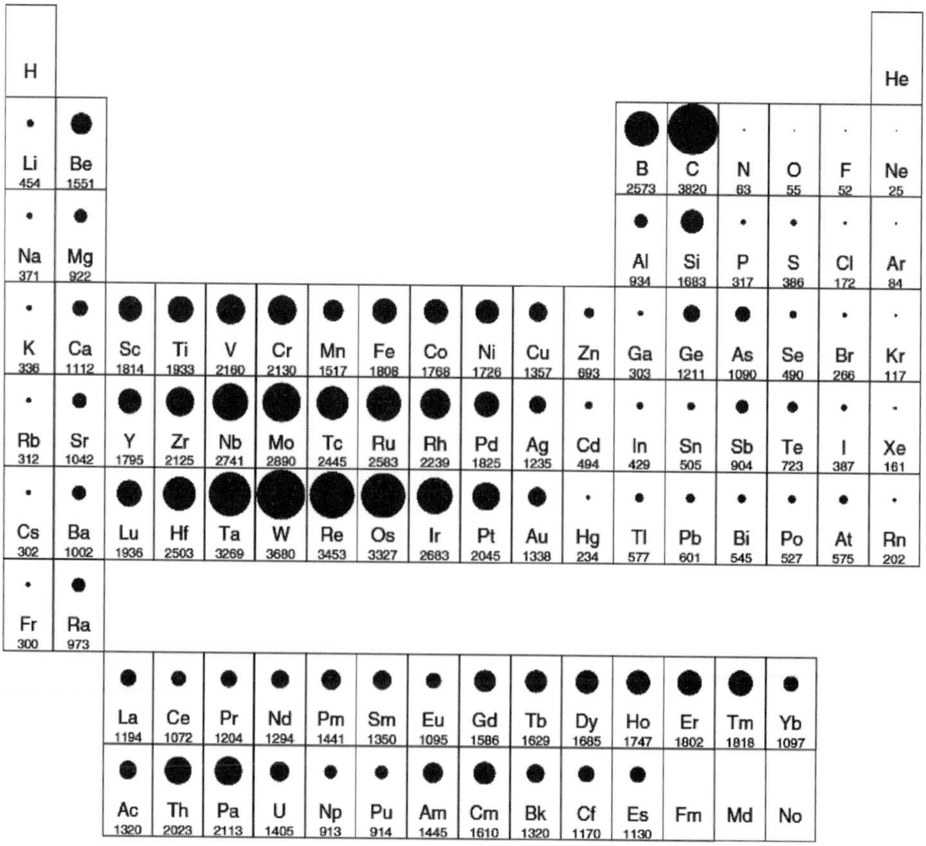

Melting temperatures of the elements (numbers in K):

Element	T_m	Element	T_m	Element	T_m
Li	454	Be	1551	B	2573
C	3820	N	63	O	55
F	52	Ne	25	Na	371
Mg	922	Al	934	Si	1683
P	317	S	386	Cl	172
Ar	84	K	336	Ca	1112
Sc	1814	Ti	1933	V	2160
Cr	2130	Mn	1517	Fe	1808
Co	1768	Ni	1726	Cu	1357
Zn	693	Ga	303	Ge	1211
As	1090	Se	490	Br	266
Kr	117	Rb	312	Sr	1042
Y	1795	Zr	2125	Nb	2741
Mo	2890	Tc	2445	Ru	2583
Rh	2239	Pd	1825	Ag	1235
Cd	494	In	429	Sn	505
Sb	904	Te	723	I	387
Xe	161	Cs	302	Ba	1002
Lu	1936	Hf	2503	Ta	3269
W	3680	Re	3453	Os	3327
Ir	2683	Pt	2045	Au	1338
Hg	234	Tl	577	Pb	601
Bi	545	Po	527	At	575
Rn	202	Fr	300	Ra	973

Lanthanides:

La	Ce	Pr	Nd	Pm	Sm	Eu	Gd	Tb	Dy	Ho	Er	Tm	Yb
1194	1072	1204	1294	1441	1350	1095	1586	1629	1685	1747	1802	1818	1097

Actinides:

Ac	Th	Pa	U	Np	Pu	Am	Cm	Bk	Cf	Es	Fm	Md	No
1320	2023	2113	1405	913	914	1445	1610	1320	1170	1130			

Figure 12.2 Melting temperatures of the elements, with numbers in K, and graphical depiction with diameters proportional to melting temperature. From Lawson [179].

to be approximately true. Problem 12.5 describes the simple but semiquantitative Friedel model which includes some of the essential features of chemical trends of band filling.

On the other hand, the $4f$-electrons of the rare earth metals do not give this trend because they are somewhat contracted, lying inside the outer $6s$-electrons. The outer radii of the rare earth metals contract across the series from La to Tm, however, allowing some f-electron bonding in the later part of the series and a gradual rise in T_m. The filling of the $5f$-states in the actinide elements from Ac to Es gives a peculiar trend in T_m. The electron energy levels in the actinides are strongly sensitive to the atom positions and coordinations.[3] One consequence is that a large number of crystal structures occur in actinide elements as functions of temperature or pressure. Predicting properties of the actinides, including melting, remains a research challenge.

12.3 Free Energy of a Solid

At the melting temperature T_m, the free energies of the solid and liquid are equal. Knowing the free energy of the solid is half of the problem of understanding the thermodynamics of melting.

12.3.1 Energy, Thermal Expansion, Phonon Softening

All parabolas look alike. That is, all functions $y_j = y_{0j} + a_j(x - x_{0j})^2$ can be superimposed into the same curve if we rescale them with the three constants

(1) y_{0j}, (2) a_j, (3) x_{0j}.

These are the

(1) vertical offset,   (2) curvature,   (3) horizontal offset.

This obvious fact about parabolas motivates the rescaling of more realistic interatomic potentials of different elements by their

1. binding energy (vertical offset),
2. bulk modulus (curvature at bottom of the potential; cf. Section 6.6.2),
3. interatomic distance (horizontal offset).

In an interesting analysis by Rose *et al.*, this rescaling of different interatomic potentials places them on one universal curve shown in Fig. 12.3 [180, 181]. In general, an interatomic potential $\Phi(r)$ departs from a parabola because $\Phi(r)$ is skewed to larger r. What we learn from Fig. 12.3 is that this ratio of skewness to curvature is similar for many metallic bonds.[4]

---

[3] This was once attributed to the strong angle and distance dependence of the f-electron wavefunctions, but this is oversimplified.

[4] When a cubic term is added to the quadratic potential to approximate a range near the bottom of the potential, Fig. 12.3 shows implicitly that the interatomic potentials for many elements have similar ratios of cubic to quadratic components.

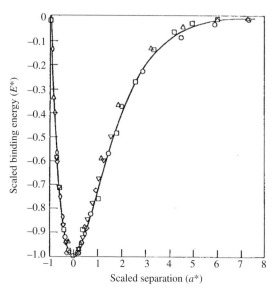

Figure 12.3 "Universal" interatomic potential of Rose et al. [180, 181] (solid curve). Points were obtained from Mo, K, Sm, Ba, Cu. Reprinted figure with permission from [181]. Copyright (1981) by the American Physical Society.

Two consequences of the skewed potential of Fig. 12.3 are obtained easily: (1) thermal expansion, and (2) a type of phonon softening (meaning a reduction in phonon frequency) caused by thermal expansion. Both can be calculated approximately if a small, negative cubic contribution is added to the interatomic potential to give it skewness

$$\Phi(x) = \frac{1}{2}kx^2 - bx^3, \tag{12.4}$$

where x is the displacement from the minimum of the potential. To calculate the thermal average $\bar{x}$, follow the usual prescription for probabilistic averages

$$\bar{x} = \frac{\int_{-\infty}^{+\infty} x\, e^{-\Phi(x)/k_B T}\, dx}{Z} = \frac{\int_{-\infty}^{+\infty} x\, e^{-\frac{1}{2}kx^2/k_B T}\, e^{bx^3/k_B T}\, dx}{Z}, \tag{12.5}$$

where the classical partition function for normalization is

$$Z = \int_{-\infty}^{+\infty} e^{-\Phi(x)/k_B T}\, dx = \int_{-\infty}^{+\infty} e^{-\frac{1}{2}kx^2/k_B T}\, e^{bx^3/k_B T}\, dx. \tag{12.6}$$

To evaluate Eq. 12.5, assume $bx^3/k_B T$ is small and

$$e^{bx^3/k_B T} \simeq 1 + bx^3/k_B T. \tag{12.7}$$

We therefore approximate Eq. 12.5 as

$$\bar{x} \simeq \frac{\int_{-\infty}^{+\infty} x\, e^{-kx^2/2k_B T}\left(1 + bx^3/k_B T\right) dx}{\int_{-\infty}^{+\infty} e^{-kx^2/2k_B T}\left(1 + bx^3/k_B T\right) dx}. \tag{12.8}$$

A motivation for the approximation of Eq. 12.7 is the convenience of evaluating moments of Gaussian functions. Recall that for integration over $\pm\infty$, all odd moments are zero, whereas even moments are given by Eq. 10.124. Discarding an odd moment in the numerator and one in the denominator of Eq. 12.8, and with $n = 2$ in Eq. 10.124 for the numerator

$$\bar{x} \simeq \frac{\frac{3}{4}\frac{b}{k_B T}\sqrt{\pi}\left(\frac{k_B T}{k/2}\right)^{5/2}}{\sqrt{\pi}\left(\frac{k_B T}{k/2}\right)^{1/2}} = \frac{3b\,k_B T}{k^2}, \tag{12.9}$$

or

$$\frac{\bar{x}}{x_0} = \left(\frac{3b\,k_B}{k^2\,x_0}\right)T = \alpha\,T, \tag{12.10}$$

where x_0 is the interatomic distance used for defining the potential, and α is a linear coefficient of thermal expansion, giving the fractional change in length per interval in temperature. In our convention $b > 0$, so thermal expansion is positive by the argument here, and this is usually true. The skewness of the interatomic potential of Fig. 12.3, approximated by adding a cubic term to a harmonic potential, causes the Boltzmann factor average of atom displacements to favor the positive $\bar{x}$ side of the potential.

Phonon frequencies change with temperature. One way to calculate this is to consider how the curvature of the interatomic potential changes with thermal expansion. The second derivative of Eq. 12.4 is

$$\frac{d^2\Phi}{dx^2} = k - 6bx, \tag{12.11}$$

and we are interested in the curvature at $\bar{x}$, so substituting for x from Eq. 12.9

$$\left.\frac{d^2\Phi}{dx^2}\right|_{\bar{x}} = k - 18\frac{b^2}{k^2}k_B T \equiv k'. \tag{12.12}$$

The "quasiharmonic model" assumes that phonons remain harmonic with increasing temperature, but are shifted in frequency because the atom vibrations explore a greater range of the interatomic potential. This frequency shift can be calculated with the effective force constant k' of Eq. 12.12. As usual for a harmonic oscillator, the angular frequency is $\bar{\omega} = \sqrt{k'/m}$. Assuming that the dimensionless number $18\,b^2\,k_B T/k^3$ is small, the square root is expanded to obtain

$$\bar{\omega}(T) \simeq \sqrt{\frac{k}{m}}\left(1 - \frac{9b^2\,k_B T}{k^3}\right). \tag{12.13}$$

All factors in the second term of Eq. 12.13 are positive, so $\bar{\omega}$ decreases linearly with T owing to thermal expansion.

This "quasiharmonic" effect on phonon frequencies is important, but it is usually only part of the reason why phonons shift their frequencies with temperature. For example, if volume were fixed so there is no thermal expansion, the quasiharmonic model predicts no changes in phonon frequencies. A pure temperature dependence of phonon frequencies usually occurs, however, and originates with phonon–phonon interactions and electron–phonon interactions, as described in Chapter 26 (online).

12.3.2 Separable Contributions to the Entropy

The entropy of a nonmagnetic, elemental solid at T_m comes from its vibrational and electronic degrees of freedom. Both these dynamical contributions to entropy grow with temperature. A temperature-dependent expression for the entropy of a solid includes these individual terms

$$S_s(T) = S_h(T) + S_{qh}(T) + S_{anh}(T) + S_{el}(T) + S_{epi}(T). \tag{12.14}$$

Here $S_h(T)$ is the harmonic contribution, which is the dominant term. The other terms are smaller corrections for physical effects described below. The harmonic entropy is calculated with a phonon density of states (DOS), $g(\varepsilon)$, characteristic of a low temperature, but $S_h(T)$ is evaluated with the Planck distribution for elevated temperatures. Near T_m, the simpler high-temperature expression of Eq. 7.51 may be appropriate. Here, for illustration, an average energy for the phonons $\hbar\overline{\omega}$ is used, with three modes per atom

$$S_h(T) = 3k_B \left[\ln\left(\frac{k_B T}{\hbar\overline{\omega}}\right) + 1 \right]. \tag{12.15}$$

The phonon DOS and $\overline{\omega}$ change with thermal expansion because the interatomic potential is not parabolic. The fractional shift of a phonon frequency with fractional change in volume is the Grüneisen parameter γ of Eq. 7.77

$$\gamma \equiv -\frac{V}{\omega}\frac{\partial\omega}{\partial V}, \tag{12.16}$$

predicting a phonon frequency at T that differs from the frequency $\overline{\omega}_0$ at $T = 0$

$$\overline{\omega}(T) = (1 - 3\gamma\alpha T)\,\overline{\omega}_0. \tag{12.17}$$

(Since α is the linear coefficient of thermal expansion, $3\alpha T$ is the fractional change in volume with temperature.) The sign is negative because phonon frequencies usually decrease with expansion, as shown by Eq. 12.13, for example. Using Eq. 12.17 in Eq. 12.15 gives, for small α,

$$S_h(T) + S_{qh}(T) = 3k_B \left[\ln\left(\frac{k_B T}{\hbar\overline{\omega}_0(1 - 3\gamma\alpha T)}\right) + 1 \right], \tag{12.18}$$

$$S_h(T) + S_{qh}(T) = 3k_B \left[\ln\left(\frac{k_B T}{\hbar\overline{\omega}_0}\right) + 1 \right] + 9k_B\gamma\alpha T. \tag{12.19}$$

We can obtain a Grüneisen parameter from the universal equation of state of Fig. 12.3 by comparing Eqs. 12.13 and 12.17, and using Eq. 12.10 for the thermal expansion

$$\gamma = \frac{b\,x_0}{k}. \tag{12.20}$$

We cannot know γ without the bond length x_0. For the scaled potential of Fig. 12.3, this x_0 may be 6 to 8 for most metals [181]. With $b/k \simeq 1/4$ in these units, γ is from 1.5 to 2. More details are given in the original reference [181]. It may be more convenient to obtain Grüneisen parameters from other sources. The best sources are phonon frequencies, and Grüneisen parameters from analyses of heat capacity data are also appropriate. Often we assume $\gamma = 2$, but it varies by a factor of 2 or so for different elements, and usually varies for different phonons in the same element (see Section 26.4.1 [online]).

The additional terms of Eq. 12.14, S_{anh}, S_{el}, and S_{epi}, are less well understood, so it is typical to hope they are small. The anharmonic entropy, S_{anh}, depends on temperature and not volume. It accounts for how phonons, initially described as harmonic normal modes that do not interact, do interact with each other to change their frequencies when there are cubic or quartic components of the potential. When the cubic and quartic components are small, one approach is to treat them as perturbations that couple harmonic phonons. The cubic perturbation causes a shortening of the lifetimes of these phonons, which cannot be predicted by the quasiharmonic model, and anharmonicity also causes additional shifts in phonon energies that may augment or detract from the quasiharmonic shift of Eq. 12.17. These effects are described in Section 26.4 (online).

The electronic entropy S_{el} depends on the electronic states near the Fermi level (Eq. 6.81). For insulators S_{el} is zero, and it is small for semiconductors and free electron metals. It does make an important contribution to the entropy of transition metals and their alloys, however. Finally, the adiabatic electron–phonon interaction, described in Section 26.5 (online), makes a contribution to the entropy, S_{epi}, which can be important in some metals at elevated temperatures. The electron–phonon interaction requires thermal excitations of both electrons and phonons simultaneously, and is not present with the thermal excitations of either electrons or phonons alone.

12.3.3 Contributions to the Entropy of fcc Aluminum

In summary, Section 12.3.2 started with the entropy of a harmonic crystal at low temperatures, and contributions were added to correct for various nonidealities that increase with temperature. This approach is appropriate because in practice $S_h(T)$ is the largest contribution, even at elevated temperatures. Consider the harmonic part of the entropy of aluminum shown in Fig. 12.4b. At 900 K the harmonic contribution to the free energy, $-TS_h$, is approximately –48 kJ/mol. This is very large. Contributions of a small percentage of this are still important when comparing differences in free energies of phases. Many of the other terms in Eq. 12.14 are a few kJ/mole at 900 K, but they vary considerably for different materials.

Figure 12.4b seems to show that the total entropy of fcc aluminum, assessed by the Scientific Group Thermodata Europe (SGTE) [184, 185], is nearly equal to the sum of the harmonic, quasiharmonic, and electronic contributions, $S_h(T) + S_{qh}(T) + S_{el}(T)$. (The contribution S_{epi} is negligible for aluminum [186, 187], and aluminum is nonmagnetic.) The details prove a bit more complicated, however. As shown in Fig. 12.4a, the phonon DOS at low temperatures is rather different from the phonon DOS at higher temperatures, primarily because of a large lifetime broadening of the phonon modes [182, 188]. The overall shift of the phonon frequencies is approximately as predicted from the quasiharmonic model, even though the large lifetime broadening implies a big cubic anharmonicity. It turns out that the phonon frequency shifts caused by the quartic part of the phonon anharmonicity largely cancel the frequency shifts from the cubic, so the large anharmonicity of aluminum has no net effect on the entropy. The phonon frequencies

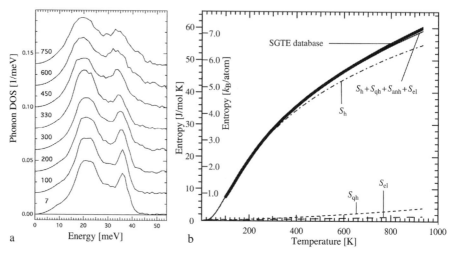

Figure 12.4 (**a**) Phonon DOS curves of fcc aluminum metal, obtained from inelastic neutron scattering measurements at the temperatures specified in the labels (units are K) [182]. (**b**) Entropy versus temperature for fcc aluminum from the Scientific Group Thermodata Europe "SGTE database" as top dark curve. The harmonic curve S_h was calculated from Eq. 7.49 with a phonon density of states (DOS) curve $g(\varepsilon)$ measured at low temperatures. The anharmonic vibrational entropy was obtained from experimental phonon DOS measured on Al metal at intermediate temperatures [182], and corrected for the lifetime broadening of the phonon spectrum. The electronic contribution S_{el} [183], shown as the bottom curve, was added, giving the thin curve $S_h(T) + S_{qh}(T) + S_{anh}(T) + S_{el}(T)$. For reference, the quasiharmonic contribution S_{qh}, i.e., $9Bv\alpha^2 T$ of Eq. 26.39, is shown, but this effect is already included in the measured DOS. Recall that 1 k_B/atom corresponds to 8.314 J/(mol K).

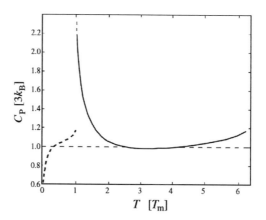

Figure 12.5 Heat capacity of liquid aluminum in units of $3k_B$ per atom, versus temperature in units of T_m. The heat capacity of solid aluminum is shown as a dashed curve below T_m. After [189].

change with temperature approximately as predicted by the quasiharmonic model. Using the anharmonic phonon spectra measured at elevated temperatures, with an approximate correction for the lifetime broadening, the phonon part plus the electron part of the entropy successfully give the entropy of fcc aluminum ($S_h(T) + S_{qh}(T) + S_{anh}(T) + S_{el}(T)$).

At very high temperatures there is another contribution from vacancy formation that causes a discrepancy in Fig. 12.4b, and this is seen more clearly in the heat capacity of the solid in Fig. 12.5. Aluminum is a bit unusual in that it is nearly a free-electron metal with a low DOS at the Fermi level, so its electronic entropy is several times lower than for some transition metals such as iron.[5] The quasiharmonic contribution, $S_{qh}(T)$, in aluminum is fairly typical of other materials, but its cancellation of entropic contributions from cubic and quartic anharmonicities is not usually true.

12.4 Entropy of a Liquid

The free energy of the liquid phase is the second essential piece of the thermodynamics of melting. Unfortunately, it is more difficult to understand than the free energy of the solid. For the entropy of a liquid we take the opposite approach than for a solid, and start with a liquid having complete disorder. This starting point requires a substantial correction, however, because the atoms in a liquid have chemical bonding and are not so randomly dispersed as in a gas. The challenge is to identify the correlations between atom positions in the liquid that reduce the initial overestimate of the entropy. This section explains how this can be done, but a quantitative analysis is out of reach. The next section, Section 12.5, invokes an experimental observation to obtain a useful thermodynamic criterion for melting.

Atoms are mobile in the liquid phase, but unlike a gas they stick together as they move around. Nevertheless, the concept of a quantum volume is useful for the positional degrees of freedom of a liquid, and we use it to write a trial partition function for one atom in the liquid

$$Z_{1t} = \frac{\Omega}{V_q}, \tag{12.21}$$

where Ω is the volume per atom, so the total volume of the liquid is $V = N\Omega$. Here V_q is the quantum volume

$$V_q = \lambda^3, \tag{12.22}$$

where λ is the thermal de Broglie wavelength, obtained approximately by equating the thermal energy to the quantum kinetic energy (with h/λ the momentum)[6]

$$k_B T \simeq \frac{(h/\lambda)^2}{2m}, \tag{12.23}$$

$$\lambda = \frac{h}{\sqrt{2\pi m k_B T}}. \tag{12.24}$$

---

[5] Iron has a substantial contribution from magnetic entropy, which dominates the heat capacity near the Curie temperature. Magnetism gives an important contribution to the entropy [190], and has effects on phonons [191].
[6] For both the gas and the liquid, a factor V/V_q can be obtained by integrating a Boltzmann factor over phase space as $\int \int \int \exp[-(p_x^2 + p_y^2 + p_z^2)/k_B T]h^{-3}dp_x dp_y dp_z dx dy dz$. This gives the factor of $\sqrt{\pi}$ in Eq. 12.24.

For an idealized liquid of N noninteracting atoms

$$Z_t = \left(\frac{\Omega}{V_q}\right)^N. \tag{12.25}$$

Equation 12.25 differs from that of the ideal gas where the large volume V is available to each atom, but indistinguishability of atoms gives a factor of $1/N!$ to the RHS.

Following the prescriptions $F = -k_B T \ln Z$ and $S = -\partial F/\partial T$, we obtain a trial entropy per atom for the liquid, S_{lt}

$$S_{lt} = k_B \ln\left(\frac{\Omega}{V_q}\right) + \frac{3}{2} k_B. \tag{12.26}$$

Equation 12.26 looks similar to the Sackur–Tetrode equation for the translational entropy of ideal gases, which has V instead of Ω, and a term $5/2\, k_B$ on the right-hand side instead of $3/2\, k_B$. The difference for the gas is that the volume explored by each indistinguishable atom is its fraction of the entire volume occupied by the gas, but the volume explored in a solid is mostly from the local excursions of a distinguishable atom.

Atoms in a liquid are excluded from overlapping with other atoms, and at the high densities of a liquid this reduces considerably the fraction of their accessible volume. For this reason we need to add a correlation correction to the entropy of the liquid, and this S_{corr} is of negative sign

$$S_l = k_B \ln\left(\frac{\Omega}{V_q}\right) + \frac{3}{2} k_B + S_{corr}. \tag{12.27}$$

The correlation correction, S_{corr}, is expected to be largest just above T_m where the liquid is most viscous, and its atoms are packed most tightly. Assessing the correlation correction can be done by analysis of the atomic structure of the liquid. The pair correlation function for liquids shows a peak at first-nearest-neighbor (1nn) distances. For temperatures above T_m, this 1nn peak decreases in amplitude, giving an approximate result found by Wallace [192]

$$S_{corr} = -k_B \left[2.6 - 1.7 \ln\left(\frac{T}{T_m}\right)\right]. \tag{12.28}$$

This is an average expression – the detailed behavior varies with chemical element.[7] Unfortunately, S_{corr} is fairly large, being around 26% of the entropy of the liquid at temperatures near T_m, impeding quantitative thermodynamic predictions. A typical S_{lt} is $12.5\, k_B$/atom near T_m, but chemical elements are as low as 8 (Li) and as high as 16 (Th) [192]. In what follows, we will use the corrected entropy of the liquid, S_l, which is approximately $9.9\, k_B$/atom near T_m. Sometimes an electronic contribution needs to be added, as discussed below.

Figure 12.5 shows the heat capacity of liquid aluminum from molecular dynamics simulations [189].[8] As a rough approximation, the liquid has a constant heat capacity of

---

[7] There are also higher-order correlations with three- and four-body terms, as suggested in Section 7.3, and these are poorly understood. Wallace [192] suggests they can be neglected.

[8] The results of Fig. 12.5 were obtained from simulations from the melting temperature to the boiling temperature of aluminum, calculated to be approximately 5000 K. The actual boiling temperature is half this value, so the scaling of the temperature axis needs to be taken with some caution.

$3\,k_B$ per atom, the same as the vibrational heat capacity of a harmonic solid. Evidently the atomic dynamics in the liquid absorb thermal energy in ways similar to vibrations of atoms in a crystal. There is a prominent peak in the heat capacity just above the melting temperature (just above $1\,T_m$ in the figure). This high value of C_P is associated with the loss of structure characteristic of a more ordered state, and a spike in C_P is expected when there is a discontinuity of entropy at T_m. This loss of structure occurs as shear stiffness is diminished over various length scales in the liquid. There is a minimum in C_P at approximately $3\,T_m$. At higher temperatures the heat capacity increases again, and this is associated with thermal expansion against a bulk modulus – the C_V does not show this increase at high temperatures.

12.5 Thermodynamic Condition for the Melting Temperature

12.5.1 Entropy Difference of Solid and Liquid

We use Eqs. 12.19 and 12.27 to obtain the entropy change at melting. Conveniently, at T_m the second term in Eq. 12.28 is zero (i.e., $1.7\ln(T_m/T_m) = 0$)

$$\Delta S = S_l - S_s, \tag{12.29}$$

$$\Delta S = -4.1\,k_B - 9k_B\gamma\alpha T_m + k_B\,\ln\left[\frac{\Omega}{V_q}\left(\frac{\hbar\overline{\omega}_0}{k_B T_m}\right)^3\right]. \tag{12.30}$$

The factors in the last term are from the physical volume explored by atoms in the liquid, Ω/V_q, and the number of phonons in an oscillator, $k_B T_m/\hbar\overline{\omega}_0$ of the crystal. As suggested by Fig. 7.7a, the number of phonons can be related to the physical volume explored by an oscillating atom. Nevertheless, the compatibility of these factors is approximate, and this is a risk. To calculate T_m with Eq. 12.2, we also need $E_l - E_s$. This is frustrating because the atomic structure of the liquid is not well known, so the chemical bond energy is not possible to calculate with confidence.

Our approach, which should seem surprising, is to set $E_l - E_s = 0.8\,k_B T_m$. Such a consistent difference of the energy of the solid and liquid cannot be justified rigorously, but often this approach is not too bad. For example, if a free electron metal does not change substantially in density upon melting, the free electron density and the energy of Eq. 6.83 will be similar for both the solid and liquid phases.[9] If $E_l - E_s = 0.8\,k_B T_m$, then Eq. 12.2 gives $\Delta S = 0.8\,k_B$/atom. We can use this ΔS for the LHS of Eq. 12.30, and then solve for T_m.

It turns out that $\Delta S = 0.8\,k_B$/atom for many elements, although it is larger for many others.[10] Interestingly, the elements for which ΔS is much larger than $0.8\,k_B$/atom are those that undergo a significant change in electronic structure upon melting. Silicon, for example, is a semiconductor as a solid, but a metal as a liquid. Germanium and tin also

---

[9] They are very similar for the alkali metals, which therefore have low melting temperatures.

[10] Another average over elements for ΔS is $1.1\,k_B$/atom, known as "Richard's rule." It has been noted that $\ln 3 = 1.1$, which could imply that an atom in a liquid has an extra degree of freedom with three possibilities. Perhaps the atom could have a displacement in one of three directions, but this is speculation.

become more metallic upon melting. For Si and Ge, $\Delta S \simeq 3.8\,k_B$/atom upon melting [192]. A large change in entropy is not surprising if there are large changes in electronic and atomic structure upon melting. Large changes in bonding and structure imply big changes in vibrational entropy, and changes in configurational and electronic entropy, too.

For most elements, though, if $\Delta S = 0.8\,k_B$/atom, and S_l is approximately $9.9\,k_B$/atom near T_m, then for the entropy of the solid near T_m, $S_s(T_m) \simeq 9.1\,k_B$/atom. Perhaps, then, we can estimate melting temperatures in terms of the vibrational properties of the crystal itself?

12.5.2 Lindemann Rule for T_m

All of the unknown parameters in Eq. 12.30, γ, α, $\overline{\omega}_0$, are properties of the solid phase. If we know them, and we assume $\Delta S = 0.8\,k_B$/atom, we can solve for T_m. The most important of these parameters is $\overline{\omega}_0$, an average phonon frequency defined so that

$$\ln(\hbar\overline{\omega}_0) = \int_0^\infty g(\varepsilon)\ln\varepsilon\,d\varepsilon. \tag{12.31}$$

This average frequency is designed to be representative of the phonons that contribute most to the vibrational entropy in Eq. 12.15. There is a temptation to use a Debye temperature in place of this average. This is reasonable if the Debye temperature is determined from the heat capacity of the solid, rather than sound wave velocities that sample only phonons of the longest wavelengths. The thermal expansion, α, is often known, but the Grüneisen parameter, γ, is more of a problem. Frequently the term in Eq. 12.30 from quasiharmonic-ity, $S_{qh}(T) = 9k_B\gamma\alpha T_m$, is neglected because it is assumed small (it is graphed on Fig. 12.4b for fcc Al). Also neglected is any true anharmonicity of the solid phase, $S_{anh}(T)$, and although this is acceptable for aluminum, $S_{anh}(T)$ is not small for many elements.

In short we take several risks, especially the risk that always $S_l - S_s = 0.8\,k_B$, but they allow Eq. 12.30 to be rearranged as

$$0.8\,k_B = -4.1\,k_B + k_B\,\ln\left[\Omega\left(\frac{\sqrt{2\pi m k_B T_m}}{h}\frac{\hbar\overline{\omega}_0}{k_B T_m}\right)^3\right], \tag{12.32}$$

$$\frac{3}{2}\ln T_m = -4.9 + \ln\left[\Omega\left(\sqrt{\frac{m}{2\pi k_B}}\,\overline{\omega}_0\right)^3\right], \tag{12.33}$$

$$k_B T_m = 0.038\,\Omega^{2/3}\frac{m}{2\pi}\,\overline{\omega}_0^2. \tag{12.34}$$

Emerging from Eq. 12.34 is the energy of a harmonic oscillator $1/2\,m\omega^2 x^2$, where x is its maximum displacement. When this energy is equal to $k_B T_m$, we have the vibrational energy at melting, and we can deduce the maximum atom displacement in the crystal at melting. To complete this connection, express the atomic volume, Ω, as a sphere around each atom with a radius called the "Wigner–Seitz" radius, r_{WS}

$$k_B T_m = 0.038\left(\frac{4}{3}\pi r_{WS}^3\right)^{2/3}\frac{m}{2\pi}\,\overline{\omega}_0^2, \tag{12.35}$$

$$k_B T_m = \frac{1}{2}\,m\overline{\omega}_0^2\left(0.177\,r_{WS}\right)^2. \tag{12.36}$$

The solid melts when its atom displacements are 17.7% of r_{WS}. Specifically, r_{WS} is the radius of the Wigner–Seitz volume of an atom, V_{WS}, defined so all the spherical V_{WS} add up to the volume of the crystal. For a simple cubic crystal, r_{WS} is 62% of the interatomic spacing, for a bcc crystal it is 57%, and for a fcc crystal, r_{WS} is 55% of the first-nearest-neighbor (1nn) distance. Using the fcc crystal, the "Lindemann rule for melting" is

A crystal melts when its atom vibrations reach 10% of the interatomic spacing.

Aspects of this criterion have been known for perhaps 100 years, and there have been a number of attempts to explain it [193, 194]. An often-mentioned but tortured idea is that the "crystal shakes itself apart." The obvious problem with this explanation is that it misses the thermodynamic condition of equal free energies of solid and liquid phases at T_m. The reader should be a bit annoyed that the thermodynamic properties of the liquid are not part of the Lindemann condition for melting. Liquid properties were neglected primarily because they are not well understood, but this is an excuse and not a reason. This neglect is partly responsible for the inaccuracies of the rule – it is reliable only to 20–25%, according to Lawson [179]. Improvements are also possible by adding corrections for how the electronic entropy S_{el} changes upon melting.[11] Nevertheless, the Lindemann rule survives to this day because Eq. 12.36 offers good value, giving a plausible result for little effort.

12.5.3 Correlation of T_m with Bulk Modulus

Suppose we make a natural assumption for Eq. 12.34 that the $m\overline{\omega}_0^2$ is a characteristic of the interatomic force, and this force is proportional to the bulk modulus B. We may therefore expect that the melting temperature T_m is proportional to B, although with a correction for the atomic volume $\Omega^{2/3}$ (or r_{WS}^2). A proportionality between T_m and B is a fairly common assumption in arguments about materials properties. Figure 12.6 shows the bulk moduli for the elements, and it is interesting to compare this figure to Fig. 12.2 for the melting temperatures. The agreement is qualitative, although some elemental trends are improved by adding the factor of r_{WS}^2. Nevertheless, it is evident that across most rows of the periodic table, the range in melting temperatures is not so large as the range in bulk modulus.

12.6 Glass Transition

12.6.1 The Kauzmann Paradox

In 1948 Walter Kauzmann noted that with decreasing temperature the heat capacity of a liquid decreased more rapidly than that of a solid [196]. Extrapolating these trends, Kauzmann found that the entropy of the liquid would be lower than that of the crystal phases at a finite temperature, sometimes called the "Kauzmann temperature." This would be odd because the liquid phase is the high-temperature phase, and has a higher entropy

[11] The anharmonicity of the crystal at T_m can be large, too.

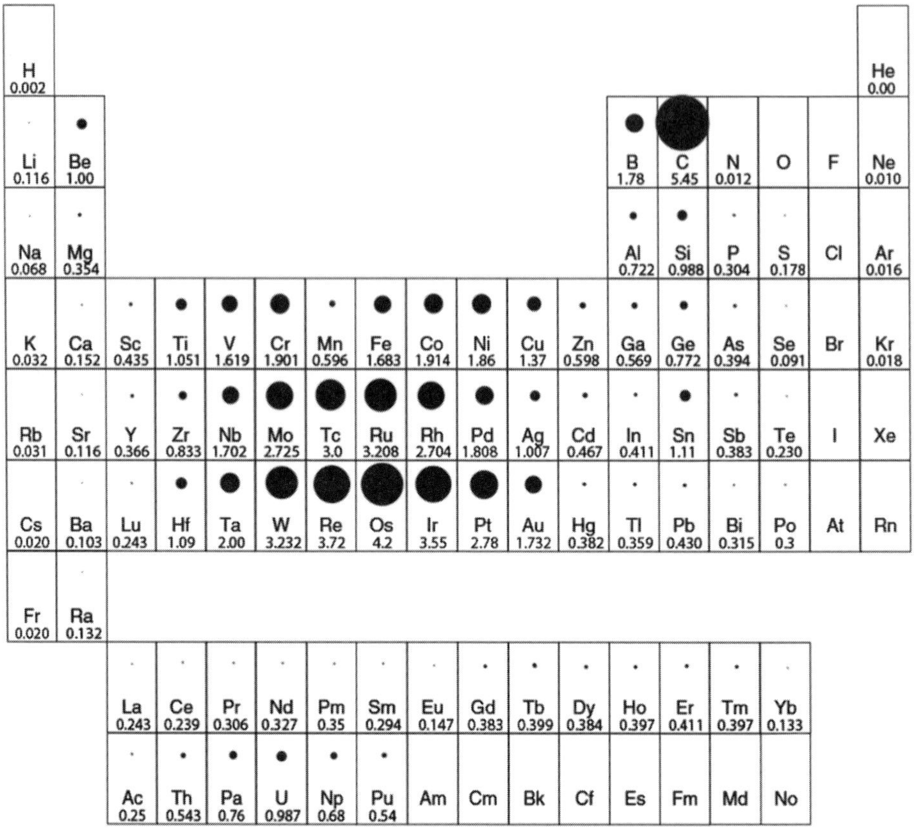

H 0.002																		He 0.00
Li 0.116	Be 1.00												B 1.78	C 5.45	N 0.012	O	F	Ne 0.010
Na 0.068	Mg 0.354												Al 0.722	Si 0.988	P 0.304	S 0.178	Cl	Ar 0.016
K 0.032	Ca 0.152	Sc 0.435	Ti 1.051	V 1.619	Cr 1.901	Mn 0.596	Fe 1.683	Co 1.914	Ni 1.86	Cu 1.37	Zn 0.598	Ga 0.569	Ge 0.772	As 0.394	Se 0.091	Br	Kr 0.018
Rb 0.031	Sr 0.116	Y 0.366	Zr 0.833	Nb 1.702	Mo 2.725	Tc 3.0	Ru 3.208	Rh 2.704	Pd 1.808	Ag 1.007	Cd 0.467	In 0.411	Sn 1.11	Sb 0.383	Te 0.230	I	Xe
Cs 0.020	Ba 0.103	Lu 0.243	Hf 1.09	Ta 2.00	W 3.232	Re 3.72	Os 4.2	Ir 3.55	Pt 2.78	Au 1.732	Hg 0.382	Tl 0.359	Pb 0.430	Bi 0.315	Po 0.3	At	Rn
Fr 0.020	Ra 0.132																	

La 0.243	Ce 0.239	Pr 0.306	Nd 0.327	Pm 0.35	Sm 0.294	Eu 0.147	Gd 0.383	Tb 0.399	Dy 0.384	Ho 0.397	Er 0.411	Tm 0.397	Yb 0.133
Ac 0.25	Th 0.543	Pa 0.76	U 0.987	Np 0.68	Pu 0.54	Am	Cm	Bk	Cf	Es	Fm	Md	No

Figure 12.6 Bulk moduli of the elements, with numbers in units of 10^{11} Pa, and graphical depiction with diameters proportional to bulk modulus. Temperatures of measurement were from 1 to 300 K. After [43, 195].

than the crystals. This puzzling hint of a very low entropy of the amorphous solid is known as the "Kauzmann paradox," and is part of the history of glass science. The resolution of the paradox is that the liquid loses atom mobilities at low temperature as it becomes a solid glass. This loss of degrees of freedom means that the entropy of the glass-forming liquid cannot decrease so rapidly at low temperature. Kauzmann's extrapolation is simply unreliable.

The loss of atomic degrees of freedom upon cooling originates from a chemical bonding that is stronger in some regions than others. There are numerous local configurations of atoms or molecules in a liquid, and some are more favorable than others for molecular or atomic motions. The increased viscosity of a liquid upon cooling originates from the locking-up of atom positions by chemical bonding. For different materials, this lock-up with temperature may occur more quickly or slowly. If the viscosity decreases rapidly above T_g, as is the case for nonpolar organic molecules such as toluene or metallic glasses, the glass is termed "fragile." A "strong" glass, such as SiO_2, undergoes a more gradual change in viscosity with temperature [197, 198].

12.6.2 Potential Energy Landscape (PEL)

A number of ways have been proposed to classify thermophysical properties of different glasses. One way is to assess the heat evolved when the glass is allowed to transform to an undercooled liquid, typically by heating through the glass transition at the temperature T_g. Less enthalpy is released from glasses formed at slower rates of cooling, or from glasses annealed at temperatures below T_g. Annealing causes small increases in density of the glass. Historically, the lower enthalpy in the glass caused by annealing or slow cooling was attributed to a decrease in the "free volume" in the glass, but a free volume coordinate can be too simplistic.

Even glasses with the same enthalpy have many equivalent atomic configurations, unlike a crystal which has only a few, but also unlike a gas which has far more [198–200]. The number of spatial coordinates for all atoms in a glass is of course $3N$, where N is the number of atoms. We hope, however, to describe the atomic structure with a smaller number of parameters, perhaps tens. These coordinates could represent relationships in positions of atoms, but for transitions between configurations it is more useful to transform the coordinates to a set of "reaction coordinates," $\{\chi_i\}$, along which changes proceed. These changes in configurations cause changes in the internal energy, $\Phi(\{\chi_i\})$, and the distributions of these energies and entropies set the thermodynamics of the glass. It is therefore interesting to consider an energy surface, termed the "potential energy landscape" (PEL) of a glass, as depicted in a one-dimensional cartoon in Fig. 12.7a. Six local minima, or "basins," are shown in Fig. 12.7a. We expect relatively few basins with the deepest energy minima (or few peaks with the highest maxima). The density of minima versus energy, $\rho_m(E)$, is depicted at the right of Fig. 12.7a. Finally, the small local minima in Fig. 12.7a represent atomic configurations about which atoms vibrate at finite temperatures.

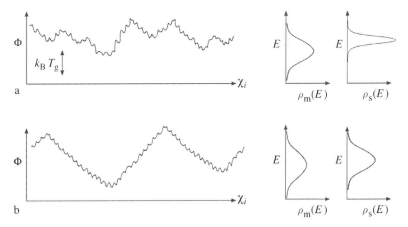

Figure 12.7 (**a**) Schematic diagram of energy landscape of a "fragile" glass, with label for the characteristic energy of a glass transition. The functions $\rho_m(E)$ and $\rho_s(E)$ are schematic densities of minima in $\Phi(\chi')$, and the density of saddle points in a multidimensional $\Phi(\{\chi_i\})$. (**b**) Energy landscape of a "strong" glass.

During annealing, thermally activated diffusion causes transitions between the six basins of Fig. 12.7. Because the annealing is at a low temperature, more regions leave states of higher energy than enter them, and there is an average movement down the energy landscape. The enthalpy of crystallization is reduced after such annealings, so we know that an annealed glass moves into basins at the lower energies of the PEL.

For local minima and local maxima, the complicated PEL function $\Phi(\{\chi_i\})$ should have many points where a coarse average of its derivatives is zero with respect to reaction coordinates (i.e., $\{\partial\Phi/\partial\chi_i = 0 \ \forall \ \chi_i\}$ for the six basins in Fig. 12.7a). From Fig. 12.7a it is easy to understand the zero slope at maxima where $\{\partial^2\Phi/\partial\chi_i^2 < 0 \ \forall \ \chi_i\}$, or minima with $\{\partial^2\Phi/\partial\chi_i^2 > 0 \ \forall \ \chi_i\}$. For a more realistic, multidimensional function $\Phi(\{\chi_i\})$ having more coordinates, however, we expect by statistics to find even more possibilities for mixed second derivatives, where some $\partial^2\Phi/\partial\chi_i^2 < 0$ and other $\partial^2\Phi/\partial\chi_i^2 > 0$.[12] These points are saddle points of the PEL $\Phi(\{\chi_i\})$. By increasing the energy in Fig. 12.7a above the lowest-energy states, we expect more saddle points, as shown with the distribution $\rho_S(E)$. Saddle points display instabilities in $\Phi(\{\chi_i\})$ in at least one reaction coordinate, χ_i, as described in Section 9.6.4. Qualitatively, this general behavior may help explain the origin of the glass transition – when more unstable modes are available by thermal activation, they can cooperate to give shear flow, and therefore give liquid-like behavior.

12.6.3 Fragile and Strong

The transitions across Figs. 12.7a and b can be understood with activated state rate theory [197, 201, 202]. Figure 12.7b, which depicts the PEL of a "strong" glass, is the simplest to understand because temperature gives access to additional states in a systematic way. The idea is that the activation energy for diffusional processes remains approximately constant with temperature. The "fragile" glass of Fig. 12.7a is more complicated because at temperatures above the glass transition temperature, T_g, more barriers are overcome, more basins are explored, and motion along the coordinates $\{\chi_i\}$ is much more rapid. The viscosity of a fragile glass decreases rapidly above T_g (in the liquid phase).

In contrast, the glassy state in a strong glass-forming material is much more robust, and the glass transition appears less abrupt with temperature. As a general rule, materials such as SiO_2, which form random networks of strongly bonded atoms, are strong glass formers, as are some entangled polymers. On the other hand, metallic alloys and small molecules tend to be moderately fragile glass formers.

12.6.4 Entropy and Heat Capacity

We expect the entropy of a glass to differ from the entropy of a liquid owing to differences in configurational entropy and perhaps vibrational entropy (ignoring magnetic

---

[12] If the signs of the $\partial^2\Phi/\partial\chi_i^2$ were arbitrary when $\{\partial\Phi/\partial\chi_i = 0\}$, the ratio, f, of the number of saddle points to the number of extrema in $\Phi(\{\chi_i\})$ could be obtained through binomial combinatorics of positive and negative signs. The analysis gives $f = \frac{1}{2}\sum_{j=0}^{3N}\frac{(3N)!}{(3N-j)!j!} - 1$, so for one coordinate (i.e., $3N = 1$) we obtain $f = 0$, for two coordinates $f = 1$, for three $f = 3$, and saddle points dominate rapidly with higher numbers of coordinates. For large N, f can be approximated with the maximum term in the sum, giving $f \simeq 2^{3N}$.

or electronic contributions to the entropy). We know that the atoms in a liquid explore more configurations than the atoms in a glass, which are essentially frozen in place. The configurational entropy of the liquid is therefore larger than that of the glass.[13]

Approximately, the heat capacity $C_P(T)$ of the glass at temperatures below T_g is comparable to that of a crystal – the atoms vibrate, but do not change their configurations. Above T_g, the undercooled liquid has a jump in C_P compared to the solid glass – the atoms explore many different configurations in the liquid state. Assuming the heat capacity of the glassy solid is comparable to that of a crystalline solid, between T_g and T_m the liquid must accumulate an extra entropy characteristic of the entropy of melting. By Richard's rule this is $1.1\,k_B/$atom, as marked on the right in Fig. 12.8b. Therefore, assuming $T_g = T_m/2$ and a constant heat capacity for the undercooled liquid

$$1.1\,k_B/\text{atom} = \int_{T_m/2}^{T_m} \frac{\Delta C_P}{T'}\, dT' = \Delta C_P \int_{T_m/2}^{T_m} \frac{1}{T'}\, dT' = \Delta C_P \ln 2, \tag{12.37}$$

$$\Delta C_P = \frac{1.1\,k_B/\text{atom}}{\ln 2} = 1.6\,k_B/\text{atom}. \tag{12.38}$$

In spite of these rough approximations, such as constant heat capacities and use of Richard's rule, this result generally accounts for the increase that occurs in ΔC_P just above T_g (cf. Fig. 5.10).

The types of entropy of the glass and the undercooled liquid are becoming clearer. The translational degrees of freedom are larger in the liquid than the glass, so a large contribution to ΔC_P from configurational entropy seems plausible. Recent experimental measurements of phonon spectra in a CuZr metallic glass, however, found a negligible difference in the vibrational entropy of the glass and the undercooled liquid [203]. This suggests the picture of Fig. 12.8a, which shows little change in vibrational heat capacity through the glass transition, but a jump in the configurational heat capacity and a growing configurational entropy in Fig. 12.8b. On the other hand, analyses of heat capacity data

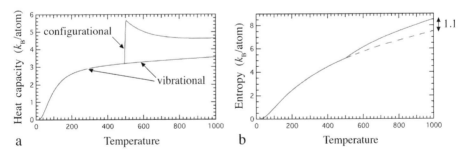

Figure 12.8 (**a**) The heat capacity of a typical solid of fixed configuration has a vibrational component as labeled. At a glass transition temperature of 500 K, the increase in configurational entropy is shown. (**b**) Entropy, obtained by integrating the heat capacity as $\int_0^T C_P/T'\,dT'$. The dashed curve is the entropy expected from vibrations in a solid glass, and the melting temperature is assumed to be 1000 K.

---

[13] Nevertheless, it proves useful to consider most glasses as keeping some residual configurational entropy even when cooled to $T = 0$.

on network glasses indicated that contributions from vibrational entropy are comparable to contributions from configurational entropy, but these were indirect assessments of vibrational entropy [204–206].

Consider again the glass transition in the context of the potential energy landscape depicted in Fig. 12.7. When the glass is heated above T_g, the undercooled liquid makes transitions from basin to basin. If the average vibrational characteristics in each basin are similar, the excess entropy must come from the greater number of basins available above T_g. A fragile glass, which has access to many basins with a small increase of temperature above T_g, has a rapid increase in its configurational entropy. A strong glass, which requires higher temperatures before it has access to as many basins, has a smaller jump in heat capacity above T_g. Owing to anharmonicities of vibrations, a strong glass could have a greater change in its vibrational entropy over this larger temperature range, but this is not yet clear.

12.7 Two Dimensions

There are major differences between melting in two dimensions and in three. We have better intuition about what "melting" means in three dimensions compared with two, having dealt with liquid water and ice in daily life. Curiously, however, the theory of melting in two dimensions is better developed than melting in three dimensions. It is unfortunate that the results from two-dimensional melting do not transfer well to three dimensions.

12.7.1 Two-Dimensional Models

We need to be careful about formulating a two-dimensional model because we have a choice of:

- Three-dimensional vector interactions between atoms or spins arranged on a plane (such as a crystal surface). This includes a Heisenberg model between spins, where the pairwise interaction strength goes as $\vec{s}_1 \cdot \vec{s}_2$.
- Two-dimensional vector interactions where the spin is confined to a plane. This includes the x–y model, which is a Heisenberg model with $s_z = 0$, so spins are confined to the x–y plane.
- Two-dimensional Potts model, where there is a finite set of orientations of spins, q of them (where q is perhaps 3 or 4). It therefore has a finite set of pairwise interaction energies that do not require a computation of $\vec{s}_1 \cdot \vec{s}_2$ for every pair. While less convenient for some analytical approaches than the Ising model, the Potts model can be handy for computer simulations.
- Two-dimensional Ising model, with the spins parallel or antiparallel to a z-axis. The convenience is that the site variables are $+1$ or -1, corresponding to some models of an A–B alloy.

We know that the Ising model has a phase transition in two dimensions, but the story is more complicated for the models with vectorial interactions. The proofs are beyond the present scope, but it is useful to know that the Heisenberg model in two dimensions does not have a transition to an ordered phase at finite temperature. With cooling, the x–y model has a transition to a phase with order over a mesoscopic distance, but it does not have a transition to a phase with true long-range order. When $q = 2$, the Potts model is identical to the Ising model, and it exhibits a continuous phase transition upon cooling, such as ordering or unmixing. When $q > 4$, a phase transition still occurs in the Potts model, but it becomes first order.

12.7.2 Kosterlitz–Thouless Melting Transition

In three dimensions, dislocations in a crystal have a high energy and low entropy. The dislocation energy increases with the volume of the crystal. The configurational entropy, however, depends on the density of dislocations, which is in units of dislocations per unit area. Being line defects, once a dislocation is known to pass through a plane, it is also known that the line extends continuously above and below the plane. Dislocation arrangements do not provide enough configurational entropy for dislocations to be thermo-dynamic defects in three-dimensional crystals, similar to the case for ledges discussed in Section 11.2.3. This argument does not hold in two dimensions, however.

In the simplest approach to the Kosterlitz–Thouless transition, we seek the critical temperature for having a dislocation in a two-dimensional crystal. The idea is that, once a dislocation is expected in equilibrium, the crystal can deform easily by shear. The vanishing of the macroscopic shear modulus is recognized as the onset of melting. The energy per unit length of an edge dislocation is

$$E = \frac{Ga^2}{4\pi \, (1 - \nu)} \ln \left(\frac{R}{r_c} \right), \tag{12.39}$$

where a is the Burgers vector, assumed to be the same as the lattice parameter and interatomic spacing, G is the shear modulus, ν is the Poisson ratio, R is the radius of the crystal, and r_c is the "core radius" of the dislocation, perhaps $5a$. The configurational entropy of centering the dislocation on any crystal site of the 2D crystal is

$$S = k_B \ln \left(\frac{R^2}{a^2} \right), \tag{12.40}$$

and the free energy of adding a dislocation is

$$F = \frac{Ga^2}{4\pi \, (1 - \nu)} \ln \left(\frac{R}{r_c} \right) - T \, k_B \ln \left(\frac{R^2}{a^2} \right). \tag{12.41}$$

At the temperature T_c there is no cost in free energy to add one dislocation

$$\frac{Ga^2}{4\pi \, (1 - \nu)} \ln \left(\frac{R}{r_c} \right) = T_c \, k_B \ln \left(\frac{25 \, R^2}{r_c^2} \right), \tag{12.42}$$

$$\frac{Ga^2}{4\pi \, (1 - \nu)} \ln \left(\frac{R}{r_c} \right) = T_c \, k_B \left[2 \ln \left(\frac{R}{r_c} \right) + \ln(25) \right], \tag{12.43}$$

$$\frac{Ga^2}{k_B \, 8\pi \, (1 - \nu)} = T_c, \tag{12.44}$$

where $r_c = 5a$ was assumed, and the constant $\ln(25) \simeq 3.2$ was neglected for $R \gg r_c$. The cancellation of the logarithm factors for energy and entropy was the key to this analysis.

A more detailed study of the Kosterlitz–Thouless transition [207] has shown that Eq. 12.44 gives an overestimate of the melting temperature. Consider a system of twice the size that contains two dislocations. The present approach would give the same critical temperature if the dislocations were independent. This is not the case, however, because dislocations interact strongly, and they interact over large distances. Two edge dislocations can configure themselves to reduce the total elastic energy, for example by having the compression field from one dislocation overlap the tension field from the other. This reduction in elastic energy lowers the melting temperature of the Kosterlitz–Thouless transition.

Problems

12.1 Use the quasiharmonic approximation of Section 12.3 to calculate α, the linear coefficient of thermal expansion of a crystal, assuming $k = 50\,\text{N/m}$, $\gamma = 2$, and an interatomic distance is $2 \times 10^{-10}\,\text{m}$. How does α change with the Grüneisen parameter γ?

(*Hint*: Note the common dependence of γ and α on b, the weight of the cubic term in Eq. 12.4.)

12.2 (**a**) Estimate a thermal velocity for an iron atom at its melting temperature. Suppose an iron atom in the liquid must travel a distance of $1\,\text{Å}$ to reach a suitable site on a growing crystal surface. If the growth rate of the solid into the liquid is $1\,\text{cm/min}$, what fraction of iron atoms attach to the surface sites?

(**b**) Why is this fraction so low? (*Hint*: You may challenge the assumption of a $1\,\text{Å}$ travel.) Why is this fraction even lower if the undercooling is smaller, and the growth rate is slower?

12.3 For an anharmonic oscillator with a potential of the form

$$U(x) = kx^2 + cx^3 + dx^4 \tag{12.45}$$

use harmonic expressions in the classical limit to calculate the vibrational entropy for small c and d.

12.4 Assume you have found a stationary point in a function of all $3N$ coordinates of a system of 100 atoms. Assuming random signs for the partial derivatives with respect to these individual coordinates, what is the relative probability of the stationary point being a minimum, compared with the probability of it being a saddle point?

12.5 Compare the melting temperatures of the $4d$ transition metals with the cohesive energy predicted by the following simple model by J. Friedel.

In this model the d-electrons form a rectangular band, with a width E_w and density of states $10/E_w$ (for 10 d-electrons). The state at the middle of this band has the same energy as a collection of isolated atoms, which we take as $E = 0$. The antibonding and bonding states are symmetrically disposed above and below this state.

(a) Assuming the band is filled from bottom (most bonding at $-E_w/2$) to top (most antibonding), show that the cohesive energy for an element with n d-electrons is

$$E_c(n) = -\frac{E_w}{20} n(10 - n). \tag{12.46}$$

(*Hint*: See Eq. 6.82, but the integral starts at $-E_w/2$, not 0.)

(b) Assuming one nonparticipating $5s$ electron, graph the cohesive energy for the elements from Sr to Ag, compared with the isolated atoms. Plot a suitably scaled set of melting temperatures from Fig. 12.2 on the same graph. Make an estimate of E_w.

12.6 Suppose the Kauzmann paradox really occurred, i.e, the entropy of the undercooled liquid decreased below the entropy of the optimal crystal structures.

(a) Why would this pose difficulties for thermodynamics, or would it?

(b) Suggest a resolution to the Kauzmann paradox, different from the one in the text.

(c) Is it possible in principle for an amorphous phase to be the thermodynamic ground state of an alloy? Why or why not?

13   Solidification

Chapter 12 discussed the thermodynamic criterion for melting, that is, the equality of free energies of the solid and liquid phases. The same thermodynamic criterion applies to solidification, of course, so solidification temperatures equal melting temperatures, T_m. There is a marked kinetic asymmetry between melting and solidification, however, so as phase transformations in materials, the two are quite different. In many ways melting is the simpler of the two, since melting begins at the surface and propagates inwards. Solidification can occur by different mechanisms that create very different solid microstructures. This chapter emphasizes processes at the solid–liquid interface during solidification, and the microstructure and solute distributions in the newly formed solid. The continuum approach used here has no intrinsic spatial scale, so the concepts and results apply to the large tonnage production of continuously cast basic metals or to the production of high-purity semiconductors. Many of the same concepts are pertinent to the processing of organic materials, too.

During solidification, a solid–liquid interface moves forward in a direction opposite to that of heat extraction. The velocity of the interface, v, increases with the rate of heat extraction. Instabilities set in even at relatively small v, however, and a flat interface evolves into finger-like columns or tree-like dendrites of growing solid. This instability is driven by the release of latent heat or the partitioning of solute atoms at the solid–liquid interface. Finger-like solids have more surface area, so countering the instability is surface energy. Some trends are predicted, but many predictions prove elusive with the classical analysis of this chapter.

Modern analyses of solidification, typically implemented as computational models, include several coupled fields that evolve with time. Heat extraction drives solidification, and the heat flux is a vector field. Scalar fields are used to describe the temperature distribution, solute distribution, and even the solid–liquid interface (as described in Chapter 17). Crystallographic orientation of the growing solid phase is important, since this affects growth rate and the surface energy. In short, solidification involves several competing and interrelated processes with coupled fields, and these issues are surveyed at the end of this chapter.

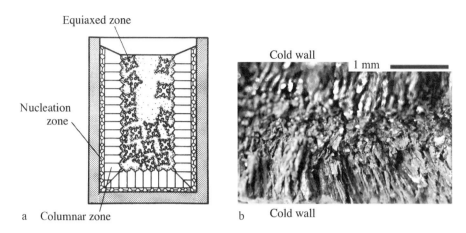

Equiaxed zone

Cold wall

Nucleation zone

a   Columnar zone

b   Cold wall

Figure 13.1 (**a**) Crystal growth of liquid metal freezing in a mold (shown as gray containment). From [208]. (**b**) Micrograph of a fracture surface of a metal casting. Mold walls were at top and bottom of image.

13.1 Solidification Microstructures

13.1.1 Casting Microstructure

Figure 13.1a shows the microstructure some time after hot liquid metal was cast into a cool mold. There is a region of dense nucleation of small crystals in immediate contact with the wall of the mold. Columnar crystals grew into the liquid, away from the direction of heat flow. As the rate of cooling slowed, crystals nucleated independently in the interior. These crystals tend to be isotropic in shape, called "equiaxed." Figure 13.1b shows the fracture surface of a metal that was cast in a mold. The cracking followed approximately the crystal boundaries in a process called "intergranular fracture," so this fracture surface reveals the microstructure of the cast metal. Columnar grains extend from the top and bottom of the image, with columnar spacings of approximately 0.2 mm. There are small equiaxed grains in the center.

Much attention has been given to solidification microstructures, since they affect the mechanical properties of cast metals (sometimes causing a tendency for intergranular fracture, as in Fig. 13.1b). The spacing between the growing columns or dendrites is an important quantity, but as discussed previously with Fig. 5.6, columns and dendrites form from an instability, and their details are challenging to calculate.

13.1.2 Heat Flow and Interface Motion

The rate of solidification depends on the rate of heat removal. The usual situation is to have the liquid hotter than the solid, where the solid phase nucleates at a cold surface and conducts heat out of the liquid as solidification proceeds. If there were no latent heat released at the solid–liquid interface, the diffusional flux of heat in the solid, j_s^h, would equal that of the liquid, j_l^h, where the subscripts denote solid and liquid, and the superscript denotes heat. The latent heat is always a consideration, however. Latent heat (L [energy/mol]) is released as liquid becomes solid, and the rate of heat release is greater when the interface moves faster. In a one-dimensional problem (as in the growth of a flat interface between solid and liquid), consider first the case where the solid must conduct the latent heat away. The extra flux of heat through the solid depends on the velocity of growth v as

$$j_s^h = j_l^h - \frac{v}{V_{mol}}L, \tag{13.1}$$

where V_{mol} is the volume per mole, so Lv/V_{mol} has the same units, [energy/(s cm^2)], as the heat flux terms. The heat flux follows a form of Fick's first law (Eq. 3.16). For later compatibility with solute diffusion, we transform from the diffusion of heat to the diffusion of temperature, assuming a constant molar heat capacity C_P [energy/(mol K)]

$$-D_{h,s}\frac{dT_s}{dz} = -D_{h,l}\frac{dT_l}{dz} - v\frac{L}{C_P}, \tag{13.2}$$

where $D_{h,l}$ and $D_{h,s}$ are thermal diffusivities of the solid and liquid [cm^2/s]. Rearranging,

$$v = \frac{C_P}{L}\left(D_{h,s}\frac{dT_s}{dz} - D_{h,l}\frac{dT_l}{dz}\right). \tag{13.3}$$

The velocity of the solid–liquid interface is determined by the difference in temperature gradients in the solid and liquid (assume, for now, $D_{h,s} = D_{h,l}$). Figure 13.2a shows a typical situation where heat is extracted to the left, and the solidification front moves to the right. Here the temperature gradient in the liquid is less than in the solid, so the difference in Eq. 13.3 is positive, and v is positive. With latent heat released at the moving interface, more heat needs to be transported through the solid, slowing the solidification. Figure 13.2b shows a smaller temperature gradient in the solid. This causes heat to flow from the hot liquid to the interface faster than the solid can conduct the heat away, so v is negative. There is melting of solid near the interface, and the interface moves to the left.

Figure 13.2c has a change of sign of the temperature gradient in the liquid, and this makes an important difference. The solid cannot be overheated above T_m. In essence,

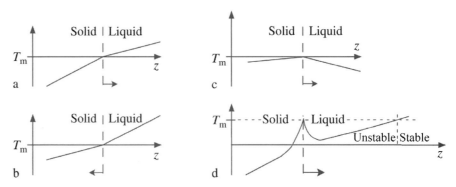

Figure 13.2 (a)–(c) Three possible temperature gradients near solid–liquid interface. (d) Modification of profile in part a with latent heat release at interface.

the solid melts instantaneously if heated above T_m, in contrast to a liquid, which can be undercooled below T_m for observable times. Since the temperature of the solid is always below T_m, the liquid near it is undercooled below T_m. This is an unstable situation. The solid–liquid interface will grow rapidly into the liquid as a protuberance, a "dendrite."

13.1.3 Dendrite Instability

"Dendritic" or tree-like microstructures (see Figs. 13.3 and 13.6) are common features of solidification microstructures. Dendrites are formed when the solid–liquid interface becomes unstable against asperities in the solid, which grow forward into the liquid [209, 210]. Dendrites grow along crystallographic directions, typically $\langle 1\,0\,0\rangle$ for fcc and bcc crystals (Fig. 13.4). The solid structure of Fig. 13.3 shows two crystal orientations along which the dendrites grew (on top and on bottom).

Section 5.3.2 pointed out with Fig. 5.6 that one important cause of this instability is the latent heat released during freezing, which makes the interface warmer than the solid and liquid immediately around it. The interface (as in Fig. 13.2d), is the warmest part of the growing solid – it is at the melting temperature, T_m. If the latent heat causes the interface

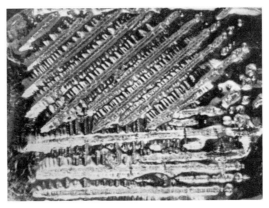

Figure 13.3 Dendritic structures in two adjacent crystals that grew into the liquid. From [210].

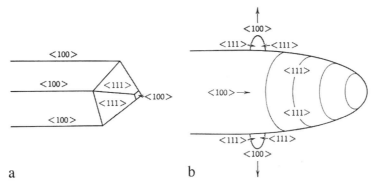

Figure 13.4 Crystallographic planes near a dendrite tip: (**a**) idealized, (**b**) more realistic. From [210].

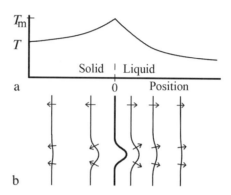

Figure 13.5 (**a**) Temperature profile near a solid–liquid interface. (**b**) The isothermals (lines of constant temperature) near the interface (dark line). Arrows, nearly perpendicular to the isothermals, show local directions of heat flow locally.

to be warmer than the nearby liquid, there is a temperature inversion in the liquid near the interface. A protuberance of solid that juts into this undercooled liquid can grow quickly into the undercooled liquid, and can grow to the right into the region labeled "unstable" in Fig. 13.2d.

A dendrite is a three-dimensional structure, so although the instability of a planar solidification front can be justified with a one-dimensional analysis, its detailed behavior cannot be predicted. For example, consider the temperature profile near a solid–liquid interface as shown in Fig. 13.5a (with the same form as Fig. 13.2d). Figure 13.5b shows the lines of constant temperature (isothermals) in two dimensions near the interface. There is an asperity in Fig. 13.5b, where the interface with a temperature of nearly T_m protrudes into the undercooled liquid. Heat flows down temperature gradients, in directions normal to the isothermals. In the liquid around the asperity, there is a heat flow away from the asperity (negative divergence of heat flux from the liquid near the tip), moving heat in the liquid near the tip to its surroundings, promoting growth of the solid asperity. Also, along the line of the asperity, the isothermals in the liquid have protrusions that push them tighter together along the centerline of the asperity, giving a steeper temperature gradient and more rapid heat flow ahead of the tip.

Chapter 17 describes phase field theory, which is well suited for computer simulations of solidification kinetics. This is an active field of research, but it has already given a number of important insights into dendrite growth [211–214], and some results from phase field simulations are presented in Fig. 13.6. A high density of crystals nucleates initially. The crystallographic direction for growth of the light gray nuclei (at top and bottom) is horizontal, along the direction of heat extraction, but the darker gray nuclei were misoriented at a 25° angle. The third and fourth frames show instances where secondary arms from the longer primary dendrites block the growth of other dendrites. After time, many of the misoriented dendrites become blocked, and tertiary arms take their place. A steady-state growth morphology is perhaps expected after this coarsening process is complete.

347

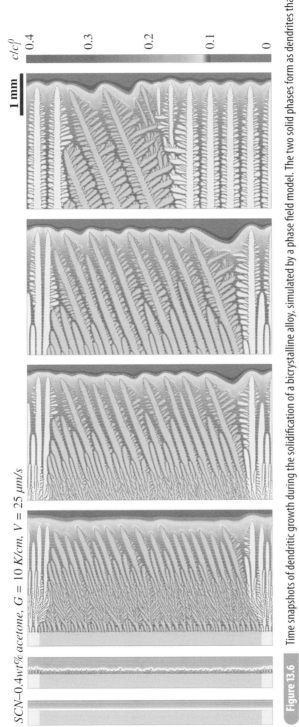

SCN−0.4wt% acetone, G = 10 K/cm, V = 25 μm/s

c/c_l^0

0.4

0.3

0.2

0.1

0

1 mm

Figure 13.6 Time snapshots of dendritic growth during the solidification of a bicrystalline alloy, simulated by a phase field model. The two solid phases form as dendrites that grow to the right, away from the direction of heat extraction. Please compare to Fig. 13.3. The contours in the liquid phase denote solute concentration, as indicated by the scale at far right. Parameters were for a succinonitrile−0.4 wt.% acetone alloy in a fixed temperature gradient of 10 K/cm, moving at 25 μm/s. (Courtesy of A. Karma and D. Tourret.)

13.2 Alloy Solidification with Suppressed Diffusion in the Liquid

The analysis of Section 5.2.2 assumed that during solute partitioning at the solid–liquid interface, the new solid retains its composition of formation, but solute diffusion in the liquid is sufficiently fast so the liquid maintains a uniform composition everywhere. Often, however, the physical length of the liquid is too large for homogenization by diffusion. Instead, we expect (or hope) that, after some time, solidification occurs as a steady-state process where the solid–liquid interface maintains the same shape as it moves with velocity v.

The assumption of a steady-state solidification process makes it convenient to define a new coordinate along the direction of growth

$$\tilde{z} = z - vt. \tag{13.4}$$

With time, the interface remains at $\tilde{z} = 0$, even though it moves to the right with velocity v in the laboratory coordinate frame of z. In steady-state, the interface carries with it the concentration profile. In the moving coordinate system, the diffusion equation (Eq. 23.13) gives the concentration of solute in the liquid

$$\frac{\partial c'}{\partial t} = D\frac{\partial^2 c'}{\partial \tilde{z}^2}. \tag{13.5}$$

For a moving coordinate system, Fig. 13.7 shows a second contribution to the change in composition with time. When the interface and its attached diffusion profile move to the right by the distance Δz per time Δt (with interface velocity $v = \Delta z/\Delta t$), the concentration at a point in the liquid increases by Δc per shift Δz, as shown in the figure. This $\Delta c/\Delta z$ gives a change with time as $(\Delta c/\Delta z)(\Delta z/\Delta t)$, or $(\Delta c/\Delta z)v$, giving a second term as

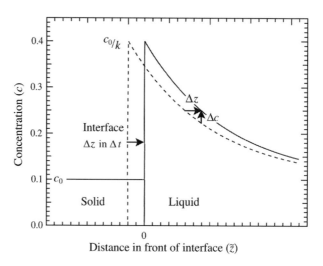

Distance in front of interface ($\tilde{z}$)

Concentration profile in front of a moving solid–liquid interface when the interface velocity $\Delta z/\Delta t$ is constant, and when the interface is far from the initial and final surfaces of the solid. Here $c_0 = 0.1$ and $k = 0.25$.

$$\frac{\partial c(\bar{z}, t)}{\partial t} = D \frac{\partial^2 c}{\partial \bar{z}^2} + v \frac{\partial c}{\partial \bar{z}}. \tag{13.6}$$

In steady state with an interface velocity v, the concentration profile in front of the moving interface remains unchanged, so $\partial c / \partial t = 0$

$$0 = D \frac{\partial^2 c}{\partial \bar{z}^2} + v \frac{\partial c}{\partial \bar{z}}. \tag{13.7}$$

The solution has the form $\exp(-\bar{z}v/D)$, as can be readily verified by substitution into Eq. 13.7. After scaling this function to match the boundary conditions of the problem, the following solution gives the correct features of Fig. 13.7

$$c(\bar{z}) = c_0 + \left(\frac{c_0}{k} - c_0 \right) e^{-\bar{z}v/D}. \tag{13.8}$$

The concentration profile $c(\bar{z})$ of Fig. 13.7 moves from left to right until it comes within the characteristic distance D/v to the end of the liquid. At the end, the excess solute in the liquid must freeze into solute-rich solid. Likewise, there must be a depletion of solute at the far left where the solidification started and the solid had the lower concentration kc_0, as expected from the phase diagram.

Following Section 5.2.2 and Fig. 5.3, we have assumed a phase diagram with solidus and liquidus curves that are straight lines, so at any temperature the local equilibrium between solid and liquid gives a solid composition

$$c_s^* = k\, c_l^*, \tag{13.9}$$

where the asterisk of c_l^* means that the liquid composition is the composition immediately at the interface (likewise for c_s^*). Here k is the partitioning ratio[1] of Eq. 5.2 (i.e., the ratio of the compositions of solidus to liquidus lines at the same temperature, T). The relevant distances for solute partitioning across an interface are quite short, probably a nm or less (see Fig. 12.1). With a diffusion constant in the liquid of 10^{-4} cm^2/s and in the solid of $D = 10^{-8}$ cm^2/s the characteristic times for 1 nm (x^2/D) are 10^{-10} s and 10^{-6} s, respectively. A velocity, v, of 1 mm/s moves an interface by 1 nm in 10^{-6} s, so there is enough time for solute partitioning at the interface. Equation 13.9 will be used throughout the rest of this chapter, but this constraint will be relaxed in Chapter 15.

13.3 Constitutional Supercooling

13.3.1 Concept of Constitutional Supercooling

Section 13.2 analyzed the steady-state composition profile around a moving solid–liquid interface. The important point was that solute is rejected into the liquid in front of the interface, increasing the chemical composition in the liquid to c_0/k. Nevertheless, solute diffusion in the liquid does not outrun the moving interface, so as the interface moves with

---

[1] This k is the same at any T because the solidus and liquidus curves are assumed to be straight lines.

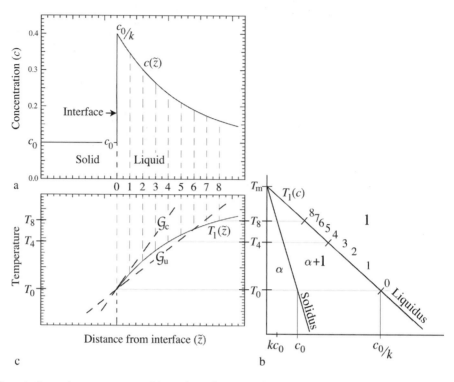

Figure 13.8 Concentrations and temperatures at positions in front of a moving solid–liquid interface in an alloy with composition c_0. The interface is at position $\bar{z} = 0$ with temperature T_0. At $\bar{z} = 0$ the solid has composition c_0 and the liquid has composition c_0/k. (**a**) From Fig. 13.7, showing a steady-state composition profile in front of the moving solid–liquid interface (no diffusion in solid, limited diffusion in liquid). (**b**) Phase diagram near A-rich α-phase. The unevenly spaced numbers on the liquidus are from the compositions in panel a. (**c**) Solid curve is the liquidus temperature in front of the moving solid–liquid interface. The tangent to the curve at the interface is the critical temperature gradient G_c.

a steady velocity v, it leaves behind a solid with the composition c_0 of the alloy overall (as shown in Fig. 13.7).

The solute-rich liquid in front of a moving solid–liquid interface tends to solidify at a lower temperature than the average liquid, so, ahead of the interface, some solute-rich liquid may be supercooled owing to its local composition (called "constitutional supercooling"). The liquidus temperature is the lowest temperature where the liquid phase is stable, so if the liquid temperature ever falls below T_1, solid can form. Figure 13.8 helps show how the liquidus temperature $T_1(\bar{z})$ varies with the solute composition ahead of the interface along $\bar{z}$. The composition profile in the liquid, $c(\bar{z})$, spans a range from c_0 far ahead of the interface, where the liquid has not yet seen diffusion from the solutes, to a larger composition c_0/k at the interface, where the solute was rejected from the newly formed solid.

For each local composition, it is easy to read off the liquidus temperature T_1 from a phase diagram. The exponential decay of the composition profile $c(\bar{z})$ in Fig. 13.8a gives a set of points on the liquidus line in Fig. 13.8b. (Please compare the locations of the numbers $0-8$ in the three panels of Fig. 13.8.) The liquidus temperatures on the $T_1(c)$ curve in Fig. 13.8b give the liquidus profile, $T_1(\tilde{z})$ in Fig. 13.8c. The lowest T_1 is at $\tilde{z} = 0$, immediately in front of the interface where the solute concentration in the liquid is highest, being c_0/k. Far ahead of the interface, $c(\tilde{z})$ changes little, so $T_1(\tilde{z})$ is nearly constant.

The rising liquidus temperature ahead of the interface can cause an instability if the temperature gradient in the liquid, $dT/d\tilde{z}$, is too small. Two temperature gradients, G_c (for "critical") and G_u (for "unstable"), are drawn in Fig. 13.8c. This figure shows that the steeper gradient G_c, tangent at the interface to the liquidus profile $T_1(\tilde{z})$, ensures that no location in front of the solid–liquid interface falls below T_1. On the other hand, the liquid is unstable for the smaller temperature gradient G_u from $\tilde{z} = 0$ to $\tilde{z} = 6$. For G_u there is "constitutional supercooling."

Constitutional supercooling can occur with much smaller undercoolings of the liquid below the liquidus temperature $T_1(c_0)$ than this conservative analysis with Fig. 13.8 may imply. Some solute enrichment in the liquid near the interface occurs for small undercoolings below $T_1(c_0)$, and this can be sufficient to drive the dendrite instability.[2]

Constitutional supercooling depends on solute diffusion in the liquid, whereas effects of latent heat depend on heat diffusion in the solid and liquid, which are much faster. If solute segregation is required during solidification, the constitutional supercooling can be more important for the kinetics than the thermal diffusion of the latent heat.

13.3.2 Critical Temperature Gradient

Figure 13.8c shows that the condition of stability against constitutional supercooling is a sufficient temperature gradient in the liquid

$$\frac{dT}{d\tilde{z}} \geq G_c, \tag{13.10}$$

where the critical gradient G_c is the variation of the liquidus profile T_1 in front of the moving interface

$$G_c = \left.\frac{dT_1}{d\tilde{z}}\right|_{\tilde{z}=0} = \left.\frac{dT_1}{dc_1}\frac{dc_1}{d\tilde{z}}\right|_{\tilde{z}=0}. \tag{13.11}$$

Figure 13.8a shows that the enhancement of solute in the liquid just in front of the interface is $c_0/k-c_0$, and the enhancement occurs over the characteristic distance of D/v in the liquid, so the concentration profile in the liquid in front of the interface is

$$\left.\frac{dc_1}{d\tilde{z}}\right|_{\tilde{z}=0} = -\left(\frac{c_0}{k} - c_0\right)\frac{v}{D}, \tag{13.12}$$

which also can be obtained readily by differentiating the $c(\tilde{z})$ of Eq. 13.8.

Define m as the slope of the liquidus line $T_1(c)$ shown in the phase diagram of Fig. 13.8b, so

$$T_1(c) = T_m + m\,c_1, \tag{13.13}$$

---

[2] Small undercoolings will be assumed in Section 13.4.2 when calculating the velocity of the dendrite tip.

and m can be evaluated by inspection of Fig. 13.8b

$$m = \frac{T_0 - T_m}{c_0/k} < 0. \tag{13.14}$$

The critical temperature gradient is found by substituting Eqs. 13.12 and 13.14 into Eq. 13.11. It can be expressed in different but equivalent ways

$$\mathcal{G}_c = -m \left(\frac{c_0}{k} - c_0 \right) \frac{v}{D}, \tag{13.15}$$

$$\mathcal{G}_c = (T_m - T_0)(1 - k) \frac{v}{D}, \tag{13.16}$$

$$\mathcal{G}_c = \left(T_l(c_0) - T_s(c_0) \right) \frac{v}{D}, \tag{13.17}$$

$$\mathcal{G}_c \equiv \Delta T_0(c_0) \frac{v}{D}. \tag{13.18}$$

Equation 13.18 is physically revealing. The critical temperature gradient $\mathcal{G}_c$ for constitutional supercooling is the difference between the liquidus and solidus temperatures, $\Delta T_0 \equiv T_l(c_0) - T_s(c_0)$, over the characteristic solute diffusion distance in front of the solid–liquid interface, D/v. Equation 13.18 is a stability criterion for the growth of a flat interface between solid and liquid. It predicts that a flat interface is stable at low velocity until $v_c = \mathcal{G}_c D/\Delta T_0(c_0)$. Considering that the analysis ignored surface energy, this criterion works surprisingly well.

Here is an interpretation of Eq. 13.18. At the interface, assume the liquid is in local equilibrium with the solid. The solid is forming with composition c_0 behind the moving interface, so the temperature at the interface is T_0 in Fig. 13.8b,c. The liquidus temperature is lowest for the liquid of composition c_0/k at the interface, and rises rapidly in the liquid ahead of the interface. If the actual temperature in the liquid does not rise at least as rapidly, some liquid in front of the interface will be constitutionally supercooled. Approximately, the diffusion distance D/v sets the spatial range where the composition of the liquid changes from c_0/k to c_0. An alloy of overall composition c_0 can have an unstable liquid over the temperature range $\Delta T_0(c_0) = T_l(c_0) - T_s(c_0)$, where the liquidus composition changes from c_0/k to c_0. If the temperature in the liquid changes by less than $\Delta T_0(c_0)$ over the spatial range D/v, there will be constitutional supercooling.

Numerical estimates show that constitutional supercooling is generally important for alloy solidification. Suppose the cooling rate of a 1 cm casting is 1–10 K/s. It may take 100 s for the solidification process to complete, corresponding to an interface velocity v of 10^{-2} cm/s. The diffusion coefficient for solute in a liquid is in the range of 10^{-5} to 10^{-4} cm^2/s. The characteristic solute diffusion distance in front of the moving interface is therefore $D/v = 10^{-3} - 10^{-2}$ cm. Using Eq. 13.18 and a ΔT_0 of 100 K, the critical gradient, $\mathcal{G}_c$, is therefore 10^4 K/cm. The diffusion of heat is much faster than the diffusion of solute, so we might expect a temperature gradient of 100 K/cm in our small casting. This is far below the critical gradient, so constitutional supercooling is expected in alloy solidification, and columnar growth or dendrites tend to be the rule rather than the exception. Although Fig. 13.8a was appropriate for evaluating $\mathcal{G}_c$, constitutional supercooling does not necessarily require such large undercoolings below the liquidus temperature, as for T^* in Fig. 13.10a for cellular solidification.

It is interesting to apply this analysis to solid–solid phase transformations. The diffusion coefficients are smaller by many orders of magnitude, but then the velocity of the interface is also set by this slow diffusion. Nevertheless, the zones of solute enrichment around precipitates are very small, typically less than a micrometer. The critical temperature gradients are therefore enormous, at least 10^7 K/cm. Such steep temperature gradients do not occur in bulk materials, so supercooling is always an issue in solid–solid phase transformation.

13.4 Cellular and Dendritic Microstructures

13.4.1 Formation of Cells (and Dendrites)

A flat solid–liquid interface can be stable at low growth velocity, v. With increasing v, however, the interface undergoes a "morphological instability," where the flat interface evolves into a columnar or dendritic structure having solid asperities that jut out into the undercooled liquid. Two important causes of this instability are (1) the latent heat released as the interface moves and liquid becomes solid, and (2) constitutional supercooling from the rejection of solute into the liquid near the interface. Counteracting this instability is (3) surface energy, which favors a flat interface.

Dendrite growth was introduced in Section 13.1.3. Cellular growth is similar, but it tends to occur at lower growth velocities and the growth direction is opposite to the direction of heat flow, not necessarily along a crystallographic direction, as for dendrites (Figs. 13.3 and 13.4). Figure 13.9 shows a growing instability leading to a cellular microstructure of the solid. In a three-dimensional cellular microstructure, elongated columns of A-rich phase are often packed side-by-side in a hexagonal pattern, separated by thin regions enriched in solute (B-atoms).

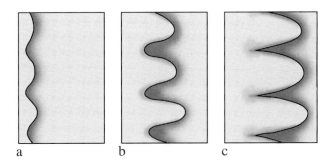

             a   b   c

Figure 13.9 Time sequence for evolution of a cellular microstructure. Solid line is the interface, with solid phase at left, liquid at right in each panel. Darker shading indicates higher solute concentration. (**a**) Early surface fluctuations of interface, with solute rejection into the liquid. (**b**) Accumulation of solute between the growing asperities. (**c**) Solidification between the columns at low temperature.

Figure 13.9a shows the rejection of solute into the liquid around a periodic modulation of the surface. The average solute flux away from the solid tends to be perpendicular to the solid–liquid interface (although partly along the growth direction, too). This causes an accumulation of solute in the valleys of the solid surface, which gain solute from both the bottom of the valley and from its sides. This accumulation of solute lowers the liquidus temperature, and delays solidification between the columns. As more heat is removed from the left of Fig. 13.9, the solute-rich regions between the columns solidify. Figure 13.9c depicts some of the solute distribution in the liquid and solid as the cellular structure grows to the right.

The issue of undercooling has been implied, since the solid would not grow into the liquid if the system were in thermodynamic equilibrium. We expect that the greater the undercooling, the faster the solidification, and the higher the velocity of the solid–liquid interface. The next section develops some relationships between undercooling, supersaturation, interface velocity, and spatial dimensions.

13.4.2 Growth of a Columnar Dendrite

Diffusion ahead of the Tip

Most models of solidification are organized to predict trends of microstructure as functions of the growth velocity v, which can be controlled by the rate of heat extraction. For a convenient microstructural model [208], consider the growth of a cylindrical column of solid of radius $\mathcal{R}$ with a hemispherical cap, also of radius $\mathcal{R}$. We analyze the solute profile around this spherical interface as the capped column grows into the liquid with velocity v.

In spherical coordinates, simplified to neglect compositional variations along θ or ϕ, the diffusion equation is

$$\frac{\partial c}{\partial t} = D\frac{1}{r^2}\frac{\partial}{\partial r}\left(r^2\frac{\partial^2 c}{\partial r^2}\right). \tag{13.19}$$

Make the key assumption of steady-state growth with $\partial c/\partial t = 0$, and multiply the RHS by r^2/D

$$0 = \frac{\partial}{\partial r}\left(r^2\frac{\partial^2 c}{\partial r^2}\right). \tag{13.20}$$

By substitution into Eq. 13.20, this steady-state solution for $c(r)$ can be confirmed

$$c(r) = B + \frac{A}{r}. \tag{13.21}$$

The boundary condition away from the tip at $r = \infty$ gives $B = c_0$, the overall composition of the alloy. At the solid–liquid interface at $\mathcal{R}$, the liquid has the composition c_1^*, so $A = \mathcal{R}(c_1^* - c_0)$ and

$$c(r) = c_0 + \left(c_1^* - c_0\right)\frac{\mathcal{R}}{r}. \tag{13.22}$$

The solute diffusion profile extends into the liquid by approximately the radius $\mathcal{R}$ of the spherical surface, and has a longer spatial tail than for a planar interface (Eq. 13.8).[3]

Solute Flux Balance and Growth Velocity in Steady State

The growth velocity can be obtained from a flux balance of solute atoms. There is a flux of solute atoms rejected at the interface,[4] j_1, equal to the interface velocity v times the difference in solute concentrations at the interface of the liquid, c_1^*, and solid, kc_1^*

$$j_1 = \frac{1}{2}v\left(c_1^* - kc_1^*\right). \tag{13.23}$$

There is also a diffusional flux j_2 as this rejected solute diffuses forward into the liquid, calculated from Fick's first law and Eq. 13.22 as $\overrightarrow{j_2} = -D\overrightarrow{\nabla}c(r)$

$$j_2 = -D\left(c_1^* - c_0\right)\frac{\mathcal{R}}{r^2}, \tag{13.24}$$

which we evaluate at the solid–liquid interface at $\mathcal{R}$

$$j_2 = -D\left(c_1^* - c_0\right)\frac{1}{\mathcal{R}}. \tag{13.25}$$

In steady state the two fluxes must be the same – the solute rejected by the growing solid must diffuse away into the liquid to avoid accumulation. Equating them[5]

$$j_1 = j_2, \tag{13.26}$$

$$\frac{1}{2}v\left(c_1^* - kc_1^*\right) = D\left(c_1^* - c_0\right)\frac{1}{\mathcal{R}}, \tag{13.27}$$

$$\frac{v\mathcal{R}}{2D} = \frac{c_1^* - c_0}{c_1^* - kc_1^*} \equiv \Omega, \tag{13.28}$$

$$v = \frac{2D\Omega}{\mathcal{R}}. \tag{13.29}$$

Figure 13.10 shows how Ω is a measure of supersaturation of the alloy during solidification, with larger values indicative of greater supersaturation. For an alloy of composition c_0, consider the growth of a solid tip at temperature T^*, which is below the liquidus temperature $T_1(c_0)$.[6] The composition c_1^* is slightly larger than c_0 in Fig. 13.10, and Ω is approximately $1/3$. The supersaturation Ω increases with undercooling, as does v. Equation 13.28 also implies that a higher v promotes a smaller tip radius $\mathcal{R}$, but surface energy has not yet been included.

---

[3] This may challenge the assumption of steady state.
[4] For a cylindrical column with diameter $\mathcal{R}$ and a hemispherical cap, the area swept by the column during forward growth, $\pi\mathcal{R}^2$, is half the surface area of the hemispherical cap, $2\pi\mathcal{R}^2$, that rejects the solute, hence the factor $1/2$ in Eq. 13.23.
[5] This is called the "Stefan boundary condition," $c_1^*(1 - k)v = -D\,\partial c_1^*/\partial\tilde{z}$.
[6] For the alloy composition c_0, note how the temperature T^* in Fig. 13.10 is much closer to the liquidus temperature than is T_0 in Fig. 13.8.

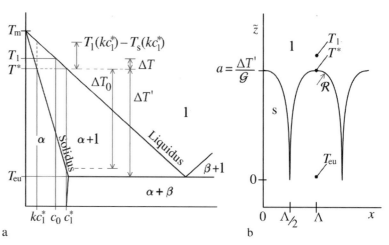

(a) Portion of phase diagram for cellular solidification of alloy with composition c_0. It is assumed that some undercooling of ΔT is required for growth of the solid, which gives the temperature T^* at the dendrite tip and enrichment of the adjacent liquid to c_1^*. (b) Spatial distribution of temperature around a solid–liquid interface that moves along $\hat{z}$ with velocity $\vec{v}$. Cell tips are separated by the distance Λ along $\hat{x}$, and extend along the $\hat{z}$ direction by the amount a from base to tip. In a constant temperature gradient $\mathcal{G}$, the cell tip has temperature T^*, and the bases of the solid columns merge at a temperature near T_{eu}.

Without undercooling at the dendrite tip, the dendrite will be in thermodynamic equilibrium with the adjacent liquid, giving a static situation with $v = 0$. Intuitively, we expect the tip velocity to be proportional to ΔT, which is the undercooling below the equilibrium liquidus temperature (labeled on Fig. 13.10a). The result for v in Eq. 13.29 is proportional to the supersaturation Ω, however. Using Fig. 13.10a, we now show how Ω and the tip undercooling ΔT are interchangeable. The numerator and denominator of Ω in Eq. 13.28 are readily identified as horizontal line segments in Fig. 13.10a, both at the temperature labeled T^*. Using each of the two line segments as the base of a triangle, and the liquidus line for the hypotenuse, the two triangles are closed by the vertical line segments of length ΔT (smaller, shaded triangle) and $T_l(kc_1^*) - T_s(kc_1^*)$ (larger triangle with dashed lines). Physically, the ΔT is the undercooling below the liquidus temperature. The vertical segment $T_l(kc_1^*) - T_s(kc_1^*)$ is itself a factor of k smaller than ΔT_0, which is defined as the vertical distance between the liquidus and solidus lines at composition c_1^*

$$\Delta T_0 \equiv T_l(c_1^*) - T_s(c_1^*), \tag{13.30}$$

$$T_l(kc_1^*) - T_s(kc_1^*) = k\left[T_l(c_1^*) - T_s(c_1^*)\right] \equiv k\,\Delta T_0. \tag{13.31}$$

With ratios of similar triangles in Fig. 13.10a, the supersaturation is converted to undercooling as

$$\Omega = \frac{c_1^* - c_0}{c_1^* - kc_1^*} = \frac{\Delta T}{k\,\Delta T_0}, \text{ or} \tag{13.32}$$

$$\Delta T = k\,\Delta T_0\,\Omega. \tag{13.33}$$

The growth velocity of Eq. 13.29 is rewritten in terms of the undercooling

$$v = \frac{2D}{\mathcal{R}} \frac{\Delta T}{k \Delta T_0}. \tag{13.34}$$

Estimating the gradient in temperature as $G \simeq \Delta T / \mathcal{R}$

$$v \simeq \frac{2DG}{k \Delta T_0}. \tag{13.35}$$

This allows an estimate of the growth velocity. Assuming a $G = 10^3$ K/cm, $D = 10^{-4}$ cm²/s, $\Delta T_0 = 200$ K, $k = 0.2$ gives $v \simeq 10^{-2}$ cm/s. Such a velocity is appropriate for some castings, but faster solidification is possible. In such cases we cannot ensure that chemical partitioning follows our expectations from the equilibrium phase diagram.

Heat Flux Balance and Growth Velocity in Steady State

For a pure element without solute segregation during solidification, the analysis of dendrite growth follows a parallel analysis, with $T(r)$ replacing $c(r)$. Here we adapt some equations from the previous derivation, such as Eq. 13.22

$$T(r) = T_0 + \left(T_1^* - T_0\right) \frac{\mathcal{R}}{r}, \tag{13.36}$$

where we consider fluxes of heat rather than fluxes of solute. Instead of the rejection of solute at the solid–liquid interface, there is a release of latent heat. The latent heat raises the temperature of the liquid at the tip T_1^* above that of the liquid some distance away, which is at temperature T_0.[7] The steady state growth velocity can be obtained from a flux balance of heat. The flux from the latent heat generated at the interface, j_1^h, is proportional to the interface velocity v as in Eq. 13.1 (rewritten for temperature in Eq. 13.2)

$$j_1^h = \frac{1}{2} \frac{vL}{C_P}. \tag{13.37}$$

There is also a heat flux j_2^h as this heat diffuses away from the interface. Like solute diffusion, there is some diffusion of heat into the undercooled liquid. Unlike solute diffusion, there is heat transport into the solid phase, too. Using Fick's first law for fluxes in both directions away from the solid–liquid interface

$$j_2^h = -D_{h,l}\left(T_1^* - T_0\right)\frac{1}{\mathcal{R}} - D_{h,s}\frac{dT_s}{d\tilde{z}}. \tag{13.38}$$

The latent heat generated by the growing solid must diffuse away into the liquid, so equating the fluxes $j_1 = j_2$

$$v = \frac{2C_P}{L}\left(D_{h,s}\frac{dT_s}{d\tilde{z}} + \frac{D_{h,l}}{\mathcal{R}}(T_1^* - T_0)\right), \tag{13.39}$$

a result that compares easily to that of Eq. 13.3, but now with heat flow in both directions away from the interface. Intuitively, v is slower if L is large, and more heat needs to be

---

[7] Mathematically, this distance should be ∞, but several times $\mathcal{R}$ should suffice. We implicitly assume that macroscopic temperature gradients in the liquid are smaller than that caused by the latent heat.

removed. Of course, a larger temperature gradient $dT_s/d\bar{z}$ drives more thermal diffusion, and a larger v. When the second term is important, a smaller tip radius, $\mathcal{R}$, gives a larger v. If the difference in temperature $T_1^* - T_0$ originates entirely from the latent heat, L, released at the interface, its proportionality to L cancels the L in the prefactor, giving a second term without dependence on the temperature increase at the interface. This result will be modified by surface energy, however, which disfavors small $\mathcal{R}$ (Section 13.6). Finally, we expect the radius at the tip of a growing column to be a fraction of the spacing between adjacent columns, so we expect higher growth rates to give smaller Λ of Fig. 13.10b.

13.5 Dendrite Growth with Solute Segregation

13.5.1 Scheil Equation in a Model of Dendrite Growth

The Scheil result of Fig. 5.4 finds a use for understanding solidification by cells or dendrites. Recall this result was obtained from the lever rule, assuming the solute rejected from the new solid is spread uniformly throughout the remaining liquid. The same approach can be used with dendrites, but now most of the solute finds itself trapped in the liquid between the growing dendrite columns. To use the analysis of Section 5.2.3, dendrites are modeled as flat plates [215, 216], and the Scheil analysis is performed for a direction perpendicular to that of tip growth. These dendrites grow forward at the leading edges of their plates,[8] but our first analysis is for the solidification of the liquid trapped between the plates.

In a short interval of time, the fraction of solute atoms transferred from the new solid to the liquid is $df_s\,(c_1^* - c_s^*)$ (the new bit of solid df_s pushes solute into the liquid at its interface), and this equals the solute transfer into the liquid of $dc_1 f_1$ (the remaining liquid gains a little solute concentration dc_1)

$$df_s\left(c_1^* - c_s^*\right) = dc_1 f_1. \tag{13.40}$$

This is just what we expect from Eq. 5.7 (including the approximation that $(dc_s/dT)f_s$ is negligible).

Some solute diffuses forward into the liquid ahead of the dendrite tips, but the tips constitute a rather small volume of material. Assuming steady state growth, however, the entire structure of dendrites and their attached concentration profiles moves forward with velocity $\vec{v}$. The simplest assumption is that solute diffusion along $\vec{v}$ occurs down a constant gradient in composition.[9] Assuming the liquid at any part of the solid–liquid interface is at the liquidus temperature, with Fig. 13.10a a constant gradient in composition is consistent with a constant gradient in temperature, $\mathcal{G} = dT_1/d\bar{z}$. A change in the liquidus temperature

---

[8] Imagine the dendrites are plates that extend above the plane of the paper for the images of Figs. 13.3 and 13.6. Plates avoid some complexities of three-dimensional columns.

[9] A constant gradient avoids solute accumulations that would distort a steady-state profile over time.

ΔT_1 corresponds to a change in composition as $m \, \Delta c_1^* = \Delta T_1$ (m is the slope of the liquidus line in Fig. 13.8b)

$$\frac{dc_1^*}{d\tilde{z}} = \frac{dc_1^*}{dT_1} \frac{dT_1}{d\tilde{z}} = \frac{\mathcal{G}}{m}. \tag{13.41}$$

This gradient generates a diffusive flux in the liquid by Fick's first law

$$j_1^{ch} = -D\frac{dc_1^*}{d\tilde{z}} = -D\frac{\mathcal{G}}{m}. \tag{13.42}$$

In steady state, j_1^{ch} must match the flux of solute across the solid–liquid interface. The growing dendrite tip sweeps up a solute flux

$$j_2^{ch} = v(c_{1,\,tip}^* - c_0), \tag{13.43}$$

where $c_{1,\,tip}^* - c_0$ is the difference between the liquid composition at the growing tip $c_{1,\,tip}^*$ and the average composition c_0. Equating j_1^{ch} and j_2^{ch}

$$c_{1,\,tip}^* - c_0 = -\frac{D\,\mathcal{G}}{m\,v}, \tag{13.44}$$

$$T_{tip} - T_0 = -\frac{D\,\mathcal{G}}{v}. \tag{13.45}$$

Equation 13.44 states that the interface moves slower if large solute segregations are required. Equation 13.45 states that the tip undercooling is set by how far the characteristic diffusion length D/v extends into the temperature gradient $\mathcal{G}$. If $c_{1,\,tip}^* = c_0$, there would be no undercooling of the tip, and the analysis would be that used to obtain the Scheil result of Section 5.13.

13.5.2 Forward Velocity of Solidification Profile

We return to Eq. 13.40, which was for solidification along $\hat{x}$, perpendicular to the $\hat{z}$ direction of tip growth. The flux of solute j_1^{ch} (Eq. 13.42) along $\hat{z}$ gives another term to Eq. 13.40 for the solute segregation during solidification. One key assumption is translational invariance of the steady-state solidification profile as it grows[10]

$$c(\tilde{z}, x, t) = c(\tilde{z}, x, t + \Delta t), \tag{13.46}$$

and likewise for the shapes of the columns as they advance with velocity $\vec{v} = v\hat{z}$. A second key assumption is that the temperature profiles have no dependence on x, and remain perpendicular to z as they move.[11] This second assumption makes it plausible that the same amount of solute must move forward from any part of the growth front, and the flux j_1^{ch} (Eq. 13.42) does not vary along $\hat{x}$. For any x on the moving solid–liquid interface, the solute diffusion profile in the liquid is then the same as at the tip. In essence, the

---

[10] This is not strictly true – a dendrite structure typically has secondary arms, so a forward translation puts the arms where there were gaps between the secondary arms. We hope, however, that this analysis will be true in an average sense.

[11] This is also risky, and inconsistent with Fig. 13.5b.

undercooling required to grow the solid gives an additional contribution of solute to the new solid by adding a new term from Eq. 13.44 to Eq. 13.40

$$df_s\left(c_1^* - c_s^* - \frac{D\mathcal{G}}{mv}\right) = dc_1 f_1, \tag{13.47}$$

which we rearrange for integration, with $b \equiv D\mathcal{G}/(mv)$ and $df_s = -df_1$

$$-df_1\left(c_1^*(1-k) - b\right) = dc_1 f_1, \tag{13.48}$$

$$\int_1^{f_1} \frac{df_1}{f_1} = -\int_{c_{1,\,tip}^*}^{c_1} \frac{dc_1}{\left(c_1^*(1-k) - b\right)}, \tag{13.49}$$

$$\ln(f_1) = -\frac{\ln\left((1-k)c_1^* - b\right)}{1-k}\Bigg|_{c_{1,\,tip}^*}^{c_1}, \tag{13.50}$$

$$f_1 = \left[\frac{(1-k)c_{tip}^* - b}{(1-k)c_1^* - b}\right]^{\frac{1}{1-k}}. \tag{13.51}$$

When there is no undercooling of the dendrite tip, $b = 0$, the RHS of Eq. 13.44 is zero, and $c_{1,\,tip}^* = c_0$. In this case $c_1^* = c_0 f_1^{k-1}$, consistent with the Scheil result of Eq. 5.13. With some algebra it is possible to rearrange Eq. 13.51 to obtain the expression for c_s^* given below in Eq. 13.75.

The new result for f_1 in Eq. 13.51, or the new result for $c_s^*(f_s)$ in Eq. 13.75, depends on the partitioning ratio k and the lever rule as does the Scheil result, but with additional features of undercooling and growth rate. Like the Scheil result of Fig. 5.4, however, the new analysis is one-dimensional. Compositions of the solid along $\hat{z}$ are averages through the whole material, and do not give details about the solute distribution in the interdendritic region along $\hat{x}$. Nevertheless, we may expect the interdendritic composition to look something like that of Fig. 5.4. For alloys with eutectic phase diagrams, this enriched liquid between the dendrites may solidify as a eutectic microstructure (Section 14.3) if the undercooling allows some liquid to reach a temperature below the eutectic temperature, T_{eu}. Dendritic solidification produces compositional inhomogeneities in the solidified alloy, and further processings such as forging may be required to eliminate them.

13.6 Surface Energy

13.6.1 Effects of Surface Energy on Undercooling and Growth Velocity

In the analysis so far, the radius $\mathcal{R}$ of the tip of a growing dendrite was considered for its effect on the solute diffusion kinetics. There is, however, an additional and essential effect of $\mathcal{R}$ that comes from thermodynamics. The thermal stability of the curved solid against a liquid is less than that of a flat interface. Near the curved solid–liquid interface at the tip of a growing dendrite, the chemical potential is increased by the interface energy in the same way as for atoms near a critical nucleus of a solid in a liquid. This increase of the chemical potential for atoms in local equilibrium near curved interfaces, called the Gibbs–Thomson effect, is discussed further in Section 20.2. In particular, the melting temperature is suppressed for solids of small radius as (Eq. 20.8)

$$T_{\mathrm{m}}(\mathcal{R}) = \left(1 - \frac{2\sigma}{\mathcal{R}L}\right) T_{\mathrm{m}}^{\infty},$$

(13.52)

where L is the latent heat per unit volume, σ is surface energy per unit area, and T_{m}^{∞} is the bulk melting temperature.

We ignore chemical effects on the surface energy, and with small $\mathcal{R}$ we assume the entire liquidus line $T_{\mathrm{l}}(c)$ is lowered uniformly. In particular, the liquidus temperature $T_{\mathrm{l}}^{*}(c_{\mathrm{l}}^{*})$ is lowered by this amount, so the tip undercooling is less than predicted by the liquidus line on the phase diagram. The undercooling obtained from Eq. 13.34, ΔT, is therefore reduced as in Eq. 13.52

$$\Delta T - \frac{2\sigma}{\mathcal{R}L} T_{\mathrm{l}}^{*}(c_{\mathrm{l}}^{*}) = \frac{v \mathcal{R} k}{2 D} \Delta T_{0},$$

(13.53)

where ΔT_{0} is from Eq. 13.30.

Equation 13.53 contains a relationship between v and ΔT and $\mathcal{R}$. The combination of these three variables can be rebalanced to satisfy Eq. 13.53, so it is not obvious how dendrites make their choice of $\mathcal{R}$ during solidification. For many years it was assumed that the choice was made for fastest growth, and we consider this possibility first. The expectation is that for the same thermal conditions (temperatures and gradients) the fastest growth occurs for $\mathcal{R}_{\mathrm{m}}$ that minimizes the undercooling ΔT. We therefore set $\partial \Delta T / \partial \mathcal{R} = 0$ as

$$0 = \frac{v k \Delta T_{0}}{2 D} - \frac{2\sigma}{\mathcal{R}_{\mathrm{m}}^{2} L} T_{\mathrm{l}}^{*}(c_{\mathrm{l}}^{*}),$$

(13.54)

$$\mathcal{R}_{\mathrm{m}} = 2 \sqrt{\frac{D \sigma T_{\mathrm{l}}^{*}(c_{\mathrm{l}}^{*})}{L k \Delta T_{0}}} \frac{1}{\sqrt{v}}.$$

(13.55)

Using this $\mathcal{R}_{\mathrm{m}}$ in Eq. 13.53, the critical undercooling ΔT_{c} is

$$\Delta T_{\mathrm{c}} = 2 \sqrt{v} \sqrt{\frac{L k \Delta T_{0}}{D \sigma T_{\mathrm{l}}^{*}(c_{\mathrm{l}}^{*})}}.$$

(13.56)

Figure 13.10b is an idealized picture of the solidification front, but such models have inspired the expectation that the spacing Λ between dendrites scales with the tip radius $\mathcal{R}$. From Eq. 13.55 we therefore expect the dendrite spacing to decrease with growth velocity as $v^{-1/2}$. This seems to be true for some systems, but it is not universal. Curiously, it is now known that this reasonable result has been obtained for the wrong reason – the choice of tip radius is not determined by a condition of fastest growth.

Box 13.1 **Péclet Number**

This entire chapter has avoided mention of atomic processes (except to set a constraint on the timescale for local equilibrium at a solid–liquid interface). The diffusion and growth equations have no intrinsic spatial scale. There is essentially no lower bound, and upper boundaries such as container walls can be scaled to larger sizes as necessary. Solidification with widely different spatial scales and times can have processes and microstructures that look similar when rescaled in space or time. The Péclet number, $\mathcal{P} \equiv v\mathcal{R}/2D$, is a dimensionless ratio of v, $\mathcal{R}$, and D, which can often be used to classify the different microstructures and instabilities of solidification.[a] Pairing these variables:

- $\mathcal{P}$ is the ratio of $\mathcal{R}$ to D/v, which means the structural length to the diffusion length,
- $\mathcal{P}$ is the ratio of v to $D/\mathcal{R}$, or the growth velocity to the diffusional velocity,
- $\mathcal{P}$ is the ratio of $v\mathcal{R}$ to D, or the ratio of a mean-squared growth scale to the mean-squared diffusion distance (in the same time).

For very small Péclet numbers, $\mathcal{P} \ll 1$, the dominance of diffusion gives uniform mixing of all solute in the liquid, following the assumptions of the Scheil analysis of Fig. 5.4, for example. (The solid–liquid interface can show finger-like instabilities as when a fluid is driven into another fluid of higher viscosity [217]. These unstable fluid–fluid interfaces look much like the one in Fig. 13.9b.) With chemical homogeneity, there are no localized diffusion fields, and the radii of the finger tips, $\mathcal{R}$, are comparable to the spacing between them, Λ. With larger $\mathcal{P}$, the tip radii resemble paraboloids.

At high Péclet numbers, diffusion is less effective in homogenizing the material during growth.[b] The situation may be that of Fig. 13.7, where a narrow diffusion front is pushed ahead of a rapidly moving interface. Chemical tendencies are less important, and thermodynamic equilibrium is less likely. High Péclet numbers are a way of classifying rapid cooling methods and their nonequilibrium processes.

[a] For transport of heat in gases and liquids, the Péclet number is a ratio of convective transport to diffusional transport. A low $\mathcal{P}$ indicates the dominance of thermal conduction.
[b] The Péclet number is often used in chemical engineering to characterize the efficiency of a process of chemical mixing.

13.6.2 Heat Transport

The LHS of Eq. 13.28 is the Péclet number, defined as the dimensionless ratio of the tip radius to the diffusion length D/v for solute in the liquid

$$\mathcal{P} \equiv \frac{v\mathcal{R}}{2D}. \tag{13.57}$$

When $\mathcal{P} < 1$, the growth velocity allows approximately enough time for solute diffusion over the tip of the growing column.

Comparing Eqs. 13.28, 13.32, and 13.57, we see that the steady-state flux balance gives the condition where the Péclet number equals the normalized chemical supersaturation or undercooling

$$\mathcal{P}_{\rm s} = \Omega \tag{13.58}$$

(where Ω, the supersaturation, or undercooling with our linearized phase diagram, is obtained from Eq. 13.32). This is a remarkably compact expression for describing steady-state solidification.

Now consider steady-state solidification without solute segregation (as for a pure element). The steady-state balance of a growth flux and a diffusional flux involves heat transport (rather than solute transport). The thermal Péclet number, $\mathcal{P}_h$, is

$$\mathcal{P}_h \equiv \frac{v\,\mathcal{R}}{2D_h}. \tag{13.59}$$

The normalized thermal undercooling is $\Delta T\, C_P/L$ (where L/C_P is the maximum temperature rise from the latent heat), so for the thermal flux balance analog of Eq. 13.58

$$\mathcal{P}_h = \Omega_h, \tag{13.60}$$

$$\mathcal{P}_h = \Delta T\, \frac{C_P}{L}, \tag{13.61}$$

Unlike solute diffusion, this expression in Eq. 13.60 is less correct because heat can diffuse into both the solid and the liquid (see comments after Eq. 13.37), but we use it for conceptual completeness.

13.6.3 Growth Velocity with Surface Energy

It is straightforward to rearrange the undercooling given in Eq. 13.53 as

$$\frac{v\,\mathcal{R}}{2D} = \frac{\Delta T}{k\,\Delta T_0} - \frac{2\sigma}{\mathcal{R}\,L}\,\frac{T_1^*}{k\,\Delta T_0}, \tag{13.62}$$

which is written compactly as

$$\mathcal{P}_s = \Omega_s - \frac{2\sigma}{\mathcal{R}\,L}\,\frac{T_1^*}{k\,\Delta T_0}. \tag{13.63}$$

Likewise for thermal diffusion for steady-state alloy solidification

$$\mathcal{P}_h = \Omega_h - \frac{2\sigma}{\mathcal{R}\,L}\,\frac{T_1^*}{L/C_P}. \tag{13.64}$$

A classical approach to finding the conditions of steady-state growth is to solve for the velocity v (contained within the Péclet number, $\mathcal{P} \equiv v\mathcal{R}/2D$)

$$v = \frac{2D}{\mathcal{R}}\left[\Omega - \frac{2\sigma}{\mathcal{R}\,L}\,\frac{T^*}{\Delta T}\right], \tag{13.65}$$

$$v = 2D\Omega\,\frac{1}{\mathcal{R}}\left[1 - \frac{\mathcal{R}_c}{\mathcal{R}}\right], \text{ where} \tag{13.66}$$

$$\mathcal{R}_c \equiv \frac{2\sigma}{L}\,\frac{T^*}{\Delta T}. \tag{13.67}$$

Using the relationship $\Delta G_V = L\,\Delta T/T_m$ from Eq. 1.6, this $\mathcal{R}_c$ is recognized as the critical radius R^* for nucleation (Eq. 4.4). A nucleus of the critical radius neither grows nor shrinks, so it is expected that $v = 0$ when the tip of the column has radius $\mathcal{R}_c$. Equation 13.66 is obtained from either Eqs. 13.63 and 13.64 with the appropriate form of undercooling

$$\Omega_h = \frac{\Delta T}{L/C_P} \text{ or } \Omega_s = \frac{\Delta T}{k\,\Delta T_0}, \tag{13.68}$$

where ΔT is the undercooling below the melting temperature or the liquidus temperature (Fig. 13.10a). At the critical radius, $\mathcal{R}_c$, the surface energy is too high, and the solid stops growing ($v = 0$). Equation 13.66 has the functional dependence on the dendrite tip radius, $\mathcal{R}$, of

$$v(\mathcal{R}) \propto \frac{1}{\mathcal{R}} \left(1 - \frac{\mathcal{R}_c}{\mathcal{R}}\right), \tag{13.69}$$

and this is graphed in Fig. 13.11.

13.6.4 Combined Undercoolings

We have not yet considered simultaneously the constitutional supercooling from the solute diffusion field, and effects of latent heat release at the solid–liquid interface. The three types of undercooling at a dendrite tip, denoted as (1) ΔT_s (solute), (2) ΔT_r (radius), (3) ΔT_h (heat), should be considered together, perhaps as

$$\Delta T = \Delta T_s + \Delta T_r + \Delta T_h. \tag{13.70}$$

There are too many independent parameters for a convenient analysis of all these effects, but this is a fine topic for a computational model.

For alloy solidification, the criterion for fastest dendrite growth involves a competition between the unfavorable effects of surface energy (sometimes called "capillarity") for small dendrites, against the shorter (and faster) diffusion lengths across small dendrites. From Fig. 13.11, or the maximum of Eq. 13.69, this optimal tip radius would be $\mathcal{R} = 2\mathcal{R}_c$. Perhaps it would be a bit larger if the total volume of transformed material were taken into

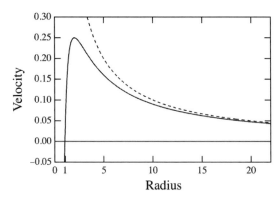

Figure 13.11   Dendrite growth velocity versus tip radius of dendrite (solid curve). Velocity is $(1 - \mathcal{R}_c/\mathcal{R})/\mathcal{R}$, so units of velocity are $[2D\,\Omega]$ (Eq. 13.66). Units of radius are $[\mathcal{R}_c]$ of Eq. 13.67. Dashed curve is $\mathcal{R}_c/\mathcal{R}$.

account. Unfortunately, experiment [218] and modern theory [219] find that the optimal tip radius is more like $\mathcal{R} = 50\mathcal{R}_c$, way off to the right of Fig. 13.11.

This discrepancy in tip radius between $2\mathcal{R}_c$ and $50\mathcal{R}_c$ brings an important insight. It shows that although the classical analysis of the previous sections can predict a number of useful relationships in terms of a Péclet number, $\mathcal{P} \equiv v\mathcal{R}/2D$, the analysis so far cannot predict the "operating point" on the v–$\mathcal{R}$ curve such as Fig. 13.11. Additional physics is required to determine the operating point. Also, the fact that the real operating point is at such a large value of $\mathcal{R}$ implies that the surface energy does not play quite the role suggested by the previous analysis.[12] (The relative difference between the solid and dashed lines in Fig. 13.11 is the relative effect of surface energy, and this becomes rather small at high $\mathcal{R}$.)

13.7 Developments in Solidification Science

13.7.1 Stability Analysis of Solid–Liquid Interface

The analysis so far has considered the movement of a flat solid–liquid interface, or a columnar one, in steady state. The process of evolving towards steady-state growth is also important, and this stage will leave a record preserved in the solid. A full picture of the dynamics is not practical without a computer simulation, which has therefore taken on an increasing role in solidification science. As an intermediate step, however, it is valuable to know if the growing interface is stable, and if steady-state growth is likely or even possible.

Stability of Flat Interface

Surface instabilities during solidification have a broad literature spanning from materials science to applied mathematics. For example, Fig. 13.5 shows how a bulge in the solid surface can release heat away from its forward direction, and how the bulge tends to create a steeper temperature gradient ahead of itself. Even an infinitesimal bulge will be promoted by these processes. There is, however, an increase in surface area associated with a wavelike variation in surface height, counteracting the tendency for growth.

The seminal work by Mullins and Sekerka [220, 221] analyzed the stability of a flat interface against wavelike variations in its height.[13] The Mullins–Sekerka analysis of interface stability considers three diffusion equations, two for heat (in solid and in liquid) and one for composition (in liquid). It begins with solutions for the flat interface such as Eq. 13.8, but adds a variation to $\tilde{z}$ at $\tilde{z} = 0$, so at the interface $\tilde{z} = \varepsilon \sin(2\pi x/\lambda)$ (which varies about 0 along the x-direction). The sharp jump in composition at the interface seen in Fig. 13.7 is located at different $\tilde{z}$ along the x-direction. At the interface, the exponential in Eq. 13.8 includes an argument with $\varepsilon \sin(2\pi x/\lambda)$. The analysis assumes that ε is small, so the exponential can be linearized, and hence the name "linear stability analysis."

---

[12] The critical gradient of Eq. 13.18 works reasonably well, even though it ignores surface energy.
[13] The approach has similarities to the Fourier transform analysis of spinodal decomposition, Section 16.4.2.

However, the problem is not simple. Even in steady state ahead of the interface, the composition variation with $\sin(2\pi x/\lambda)$ needs to include the diffusion length and the interface velocity through the Péclet number, for example. The next steps are difficult. The work involves matching temperature and compositions at the interface, which depend on the constitutional supercooling and surface energy. Heat and solute conservation at the interface must occur at the same growth rate, for example. The local velocity at the interface is $v + \partial\varepsilon/\partial t\, \sin(2\pi x/\lambda)$, and the goal is to obtain $\partial\varepsilon/\partial t$, or at least its sign. The sign of $\partial\varepsilon/\partial t$ determines whether the sinusoidal perturbation at the interface will grow or shrink with time.

Consider a wavelike variation of surface height in the x-direction, $h(x) = \varepsilon\sin(2\pi x/\lambda)$. The compression of the isothermals ahead of the bulge, as in Fig. 13.5, increases in proportion to the height variation, and for the sinusoidal modulation this is $h(x) = \lambda^0\,\varepsilon\sin(x/\lambda)$. Rejection of heat to the side of the bulge goes as the slope, or $dh/dx = -\lambda^{-1}\varepsilon\cos(2\pi x/\lambda)$. The surface energy increases as the curvature, which is $d^2h/dx^2 = -\lambda^{-2}\varepsilon\sin(2\pi x/\lambda)$. The relative effects of surface energy are therefore largest at small λ. Under appropriate conditions, three regions of negative, positive, negative $\partial\varepsilon/\partial t$ are found:

1. Since surface energy stabilizes a flat interface, a flat interface is therefore expected at small λ.
2. At intermediate λ, the $\partial\varepsilon/\partial t > 0$, so the interface is unstable against fluctuations of these λ.
3. It turns out that a flat surface can be stabilized again at very long wavelengths if λ greatly exceeds the diffusion length D/v, and the behavior is dominated by the temperature field.

Stability of Dendrite Growth

Steady state dendrite growth has also been analyzed by linear stability analysis. The model of a cylindrical column with a spherical cap offers the virtue of easy analysis, but it does not match well the dendrite shapes of Figs. 13.3 and 13.6. A parabola of revolution is a more reasonable shape for a growing dendrite, and the Laplacian is separable in parabolic coordinates. Ivantsov solved the diffusion problem around a paraboloidal dendrite tip, relating the tip curvature to the growth rate and supersaturation of the liquid [222]. The Ivantsov result involves an exponential integral that depends on the Péclet number, and it is often evaluated in various levels of approximation. The first approximation gives Eq. 13.58, for example [208], but in general

$$Iv(\mathcal{P}_s) = \Omega_s, \tag{13.71}$$

where Iv is the Ivantsov function.

The Ivantsov solution is beyond the scope of this text, as is the next step, a linear stability analysis of the Ivantsov solution that was performed by Langer and Müller-Krumbhaar [219, 223]. They analyzed the stability of a paraboloidal dendrite tip against surface ripples, which are essentially the incipient formation of secondary arms of a dendrite. The tendency for the ripples to grow depends on the Péclet number, undercooling, and critical radius, with instability behavior parameterized by the dimensionless number Σ

$$\Sigma = \frac{\mathcal{R}_c}{\mathcal{R}} \frac{\Omega}{\mathcal{P}}, \tag{13.72}$$

where the critical radius, normalized undercooling, and Péclet number are defined in Eqs. 13.67, 13.68, and 13.57. This parameter Σ can be seen as a normalized driving force for dendrite growth. It increases with v, undercooling, and $1/\mathcal{R}$.

Langer and Müller-Krumbhaar [223] discovered that at small Σ the surface instability on a paraboloidal dendrite becomes so severe that even the tip of the dendrite is altered. Low Σ corresponds to low surface energy and low driving force, so it is conducive to branching of the solid in many directions. This observation of tip bifurcation led to another hypothesis – the operating point of the dendrite is selected by a condition of "marginal stability," where the tip structure is just stable against branching. This corresponds to $\Sigma = 0.025$ [219].[14] The Σ for the dendrite operating point is an important metric for comparing theory to experiments on different systems.

Experimental Studies of Dendrite Growth

Some of the most influential experimental work on dendrite growth was performed by Glicksman and colleagues using succinonitrile (NC-(CH$_2$)$_2$-CN) because of its conveniently low melting temperature, transparency, and availability in high purity [218, 224]. Many of these experiments indicated a Σ of about 0.02 for the operating point of dendrite growth, which seemed consistent with the marginal stability hypothesis. Nevertheless, the agreement was questioned as fortuitous, since there were possibilities of convection currents in the experimental cells. In the early 1990s, a carefully planned dendrite growth experiment with succinonitrile was performed in the zero-gravity environment of the Space Shuttle, giving a $\Sigma = 0.0157$, which is a significant difference from the 0.025 of the marginal stability hypothesis [225].

Phase-contrast imaging with coherent X-rays offers a new capability for studying dendrite growth in metal alloys [226]. Some results from a recent study, which acquired hundreds of X-ray images of in-situ dendrite growth of an Al–Cu alloy, are shown in Fig. 13.12. With increasing growth velocity v, there is a clear transition from the planar growth front to cellular to dendritic solidification. However, the velocity of the planar-to-cellular transition was six times larger than predicted by Eq. 13.15 (determined by the critical velocity in the fixed experimental gradient), and the cellular-to-dendritic transition occurred at a factor of 4 higher velocity. The tip radius was found to scale with v as expected: $\mathcal{R} \propto v^{-1/2}$. There were, however, discrepancies between the measurements and the extensive phase field calculations of solidification performed by the same authors. We expect the spacing between cellular columns, Λ, to decrease with v, perhaps as

$$\Lambda \simeq v^{\alpha}, \tag{13.73}$$

---

[14] Recall that the hypothesis of maximum growth velocity predicted an operating point on Fig. 13.11 at $2\mathcal{R}_c$, so with $\Omega/\mathcal{P} \simeq 1$, this incorrect hypothesis predicts $\Sigma \simeq 1/2$; we already noted that the operating point is far larger than $2\mathcal{R}_c$.

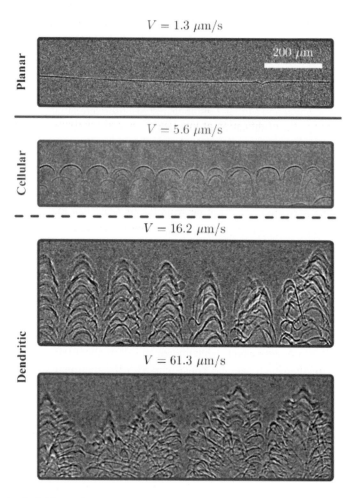

$$V = 1.3 \ \mu\text{m/s}$$

$$V = 5.6 \ \mu\text{m/s}$$

$$V = 16.2 \ \mu\text{m/s}$$

$$V = 61.3 \ \mu\text{m/s}$$

Figure 13.12   X-ray radiographs of solidification fronts in a Al–0.6 at.% Cu alloy in a temperature gradient of 50 K/cm, with different *v* as labeled. Reprinted from [226], with permission from Elsevier.

where $\alpha = -0.5$. The work by Clarke *et al.* [226] showed quite the opposite behavior, with the growth velocity $\alpha \simeq +0.5$. The authors were able to attribute most of these discrepancies to effects of convection mixing in the liquid, which tends to homogenize the chemical composition, much like a lower *v* allows better mixing. Convection is difficult to calculate precisely, however, and remains a challenge for solidification theory.

Current Challenges in Solidification Science

Convection currents are driven by buoyancy changes of a liquid with temperature, owing to thermal expansion. Mixing of solute or thermal fields by convection is important at typical velocities of solidification, and is of practical importance for most studies and practice of solidification. Convection currents are research topics in computational fluid mechanics (which is a field unto itself, spanning from nanofluidic devices to phenomena at the galactic scale). A proper, or at least adequate, approach to accounting for convection currents in solidification is emerging, but approximations should be appropriate when convective effects are small.

There are differences in surface energy of different crystallographic surfaces, as discussed in Section 11.5. The energy variations of this anisotropic surface energy are not large compared to the energy of rearranging atoms and creating the surface in the first place, but with solidification at low velocities and relatively large $\mathcal{R}$, it becomes important. For example, Figs. 13.3 and 13.4 show the crystalline symmetry reflected in the morphology of dendrites. This is expected to originate from effects of surface energy (the liquid is isotropic, as is thermal transport in cubic crystals). The Ivantsov solution to the diffusion equation near a paraboloid of revolution is appropriate for the relatively high curvature near the dendrite tip, but not far from the tip there is a four-fold set of ridges that evolve over time into secondary arms of the dendrite. This requires anisotropy of the surface energy. Even if the anisotropy is small, less than 1% for succinonitrile, a small bias towards specific growth directions can be magnified over time through growth exponents that favor some crystallographic directions and disfavor others. The analytical approach to this problem is forbidding, but analyses of the allowed solutions for the growing fields and interfaces have given "solvability theory," which allows statements to be made about the nature of steady-state solutions, assuming such solutions exist [227–229]. For example, anisotropy is a requirement for the existence of dendrites and their side branches, and even dendrites themselves.

Solidification structures have small features, although these are much larger than the atomic scale. Nevertheless, thermal fluctuations can have substantial effects on the structures of nanoparticles (Section 20.3.1). There could be a role for fluctuations in composition, density, or temperature on the early stages of dendrite growth, and such variations could be amplified over time during later growth.

13.7.2 Summary of Morphological Transitions with $\mathcal{G}, v$

There is some practical control over the rate of heat extraction and temperature gradients during solidification, so solidification processes can often be presented as functions of v and $\mathcal{G}$. Above $\mathcal{G}_c$ a planar interface is stable (Eq. 13.18).[15] For smaller $\mathcal{G}$ there are transitions in growth morphology with increasing range of v, as shown in Fig. 13.12 and summarized in Fig. 13.13.

- Planar solid–liquid interfaces are stable at low v. Planar interfaces can be difficult to achieve, and Fig. 13.13 may overstate their prevalence.
- At modest v, planar interfaces give way to a cellular solidification front, with an array of rounded columns that grow antiparallel to the direction of heat extraction. Approximately, the spacing Λ between these columns is proportional to the radius of their growing tips. Cellular structures often form within a narrow region of v and $\mathcal{G}$.

---

[15] Also, a higher $\mathcal{G}$ confines solidification processes to a narrow range of distance.

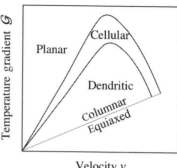

Figure 13.13 Map of solidification structures for different G, v. After [208, 226].

- At higher v, solidification becomes dendritic, where the smooth sides of the columns become unstable, and secondary branches are formed. The shape of the growing tip also tends to be more parabolic than in cellular columns. The distribution of solute in the final solid also differs from that of the cellular structure, and there are preferred crystallographic directions along the dendrites. Practical combinations of v and G often favor dendritic microstructures.
- At low G the cooling is more homogeneous through the material, and equiaxed microstructures occur instead of the columnar growth forms of cells and dendrites. Microscopically, it is possible for equiaxed microstructures to form as dendrites grow in multiple directions, as indicated in the "equiaxed zone" of Fig. 13.1a, somewhat like the shapes of snowflakes.
- At still higher v, solidification is expected to return to cellular growth, and then to a planar front at the highest velocities. There is uncertainty about the regime of high v and low G because high cooling rates usually imply high G. Cooling is fastest when the melted region is very thin, as in pulsed laser surface treatments, but then larger solidification morphologies are less relevant. The regime where the diffusion length falls to atomic distances is the topic of Section 15.3.3. An issue is how local chemical equilibrium is suppressed across a fast-moving interface.

Other solidification morphologies are possible [230], but these are the most studied. Under practical conditions where v and G are not independent, dendritic growth is perhaps the most common.

Problems

13.1 The formation of "pipes," or sometimes internal cavities, is a common problem in the centers of thick castings, where a liquid metal fills a cup-shaped mold and freezes.

   (a) Considering the sequence of how the liquid solidifies when poured into a cold mold, and how the solid phase contracts as temperature is reduced, explain the formation of cavities or pipes.

(b) If the linear thermal expansion of the solid is 2×10^{-5} K and the melting temperature is 1300 K, estimate a size of a pipe or cavity for a mold of thickness 10 cm.

(c) Water and bismuth expand when they solidify. What happens when they are frozen in a mold?

13.2 The portion of the phase diagram shown in Fig. 13.8b could have been part of a eutectic diagram. Consider now the A-rich side of a peritectic diagram, where the liquidus line is at a lower solute concentration than the solidus line.

    (a) With m as the slope of the liquidus line, show that $m(k-1)$ is positive for both the eutectic and peritectic case.

    (b) With the equivalent of Fig. 13.8c, show that constitutional supercooling is possible in a peritectic alloy, too.

13.3 Why are equiaxed crystals inherently unstable when they grow to a certain size? Does this instability occur more (or less) easily than for a flat interface under the same conditions?

13.4 Thin films of alloys can be prepared by high-vacuum evaporation, where an alloy is heated above its melting temperature. A substrate is placed some distance away, and a thin film of solid phase is deposited on this substrate with a chemical composition determined by the fluxes of atoms leaving the hot metal. This problem considers the challenge of controlling the chemical composition of the deposited film.

    (a) Assume the film composition is the same as the density of the elements in the vapor above the evaporating metal surface. For an ideal gas ($p_B V = N_B RT$) with pressure p_B, and similarly for the A-atoms. The pressure of element B above its liquid, p_B, is proportional to a vapor pressure $p_B^v(T)$ that depends strongly on temperature, but is also proportional to the concentration of B at the surface of the liquid, c_B^s. For evaporation of an alloy of two elements, show that the partial pressures at temperature T are in the ratio

$$\frac{p_B}{p_A} = \frac{c_B^s \, p_B^v(T)}{c_A^s \, p_A^v(T)}. \qquad (13.74)$$

    The ratio of vapor pressures $p_B^v(T)/p_A^v(T)$ can be orders of magnitude, but then the surface concentration ratio, c_B^s/c_A^s, can also vary by a large factor if the surface becomes enriched in the element that is less volatile.

    (b) There is a moving interface as the alloy evaporates, and the hot liquid shrinks during evaporation. When the B-atoms are the more volatile species and $p_B^v(T) > p_A^v(T)$, an analysis parallel to that of Fig. 13.7 can be adapted to the problem of alloy evaporation. If the evaporation rate is much faster than the characteristic diffusion time across the remaining liquid, the concentration of atoms in the vapor must be c_0 because the atoms cannot diffuse away from the surface before they evaporate. Show that this requires an inverse relationship between the surface concentration and the vapor pressure of the element.

(c) Using a typical $D = 10^{-4}\text{cm}^2/\text{s}$ and a metal pellet for evaporation of 0.3 cm, estimate a time for evaporation that will ensure that the thin film has the same chemical composition as the alloy that was evaporated.

13.5 Give physical reasons why (i) as k approaches 1 and $\Delta T(c_0)$ is smaller in Eq. 13.18, a smaller temperature gradient can avoid constitutional supercooling, and (ii) if the solute can diffuse far ahead of the interface (v is small or D is large), G_c is small.

13.6 Rearrange Eq. 13.51 to obtain the Bower *et al.* result [216]

$$c_s^* = k c_0 \left[\frac{a}{k-1} + \left(1 - \frac{ak}{k-1}\right)(1 - f_s)^{k-1} \right], \tag{13.75}$$

where

$$a \equiv \frac{DG}{m v c_0}. \tag{13.76}$$

14   Phase Transformations with Interfaces: 1. Microstructure

Most solid-to-solid phase transformations occur by nucleation and growth, where a small particle of the new phase nucleates in (or on) the parent phase, and then grows as atoms diffuse to its surface. The phase transformation is much more interesting, however, than just the enlargement of a small particle.

For reasons of both kinetics and thermodynamics, the small new particles differ from the large particles expected in thermodynamic equilibrium. Typically there are differences in crystal structure, chemical composition, interface structures, defects, elastic energies, and shapes. As the phase transformation proceeds and the particles grow larger, the "microstructure," comprising all these structural and chemical features, evolves with time as equilibrium is approached. As described in Box 1.1, understanding and controlling microstructure is central to obtaining desired properties of materials. In general, equilibrium is not necessary (or even desired) for a material to be useful.

This chapter gives an overview of processes that occur as microstructures evolve during nucleation and growth from a parent phase. It covers essential features of precipitation in a solid, with a few traditional examples from steels and aluminum alloys. Much of the content is central to physical metallurgy, but the presentation is at a more conceptual level and neglects many interesting details. An important model of the rates of nucleation and growth transformations is summarized, and the late-stage coarsening process is also discussed. Chapter 15 explains in greater depth the atom movements across interfaces, and the role of elastic energy.

14.1   Guinier–Preston Zones and Precipitation Sequences

14.1.1   Thin Plates

Guinier–Preston (G–P) zones are extremely small, nonequilibrium structures of atoms that form at low temperatures in unmixing alloy systems. They are precipitates of a sort, but have received special status owing to their nonequilibrium character and extremely small size. G–P zones form at a surprisingly rapid rate, considering the low temperatures where they form. They are often shaped as thin plates, suggesting an important role for elastic energy. Figure 14.1 is a high-resolution transmission electron microscopy image of a Cu-rich G–P zone in a crystal of fcc Al. A plate of Cu atoms, seen edge-on, gives a dark row in the center of the image. The details of the black and white contrast are not simple

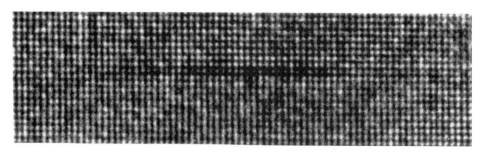

Figure 14.1 G–P(1) zone in Al–Cu crystal aligned along a ⟨100⟩ direction. The separation between neighboring white (or black) dots is 0.202 nm, half the lattice parameter of fcc Al. Image courtesy of Y.C. Chang and J.M. Howe.

to interpret as atoms or channels. Nevertheless, the G–P zone is seen to be about one atom in thickness, with a diameter of approximately 8 nm.

14.1.2 Quenched-in Vacancies

The growth of G–P zones usually involves an anomalously fast transport of atoms, such as the transport of Cu atoms to the plate-like G–P zones in fcc Al. The transport seems anomalous because the temperatures are only 150 °C or so. Also, the alloys have Cu concentrations of only a couple of percent, so the diffusion distances must be longer than expected from estimates of $x \simeq \sqrt{D(T)\,t}$, where $D(T)$ is the diffusion coefficient at the temperature T. G–P zones do not form in alloys cooled slowly from high temperature, since this results in larger precipitates of the equilibrium phase.

To form G–P zones, an alloy is quenched rapidly from a high temperature where the solid solution is stable. Besides preserving the high-temperature solid solution, the quench preserves a high concentration of vacancies, whose equilibrium concentration increases with temperature as $e^{-\Delta G_{f,v}/k_B T}$ (see Section 3.1.2). In a quenched solid solution, the excess vacancies provide a burst of diffusivity that allows rapid G–P zone formation. Diffusion is enhanced by a few orders of magnitude after the rapid quench, although this is difficult to assess quantitatively. With time the vacancies are annihilated by migrating out of the crystal to grain boundaries, or to vacancy sinks such as dislocations.

There is evidence, however, for a "vacancy pump" mechanism that makes the vacancies effective for longer times than expected by their random walk. The idea is that the vacancies are partly bound to the solute atoms, and help the solute atoms move to a growing G–P zone. After depositing the solute atom at the edge of the zone, the vacancy becomes free to diffuse in the matrix until it finds another solute atom. The process repeats.

14.1.3 Coherency and Strains

Figure 14.1 shows another important structural feature of G–P zones. The Cu atoms are located in registry with Al atoms, and the zones are coherent with the matrix (as in Fig. 4.2b). Copper is also an fcc metal, but with a considerably smaller lattice parameter (0.361 nm instead of 0.405 nm for Al). In the G–P zone the Cu atoms are not positioned

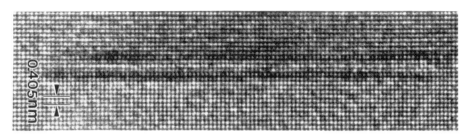

Figure 14.2   G–P(2) zone in Al–Cu crystal aligned along a ⟨100⟩ direction. Image courtesy of Y.C. Chang and J.M. Howe.

comfortably with respect to their neighbor Cu atoms owing to the constraint of the surrounding Al crystal. This is a source of positive electronic energy. There is also some strain in the surrounding Al-rich α-phase, but much less than would occur if the Cu atoms had their natural lattice parameter. Owing to the shorter length of the Cu–Al bond than the Al–Al, "coherency strains" distort the Al matrix by pulling the (200) matrix planes closer to the plane of the G–P zone.[1] This coherency strain energy is another positive (unfavorable) contribution to the thermodynamic free energy of zone formation. It is not so large in Al–Cu as in other alloys, and there is a general expectation that this elastic energy will increase with the square of the difference in atomic radii of the solute and solvent atoms.

With time, or at a slightly higher temperature, the G–P(1) zones of Fig. 14.1 evolve into G–P(2) zones that have two parallel plates of Cu atoms spaced by three (200) planes (see Fig. 14.2). Larger G–P(2) zones are also called θ'' precipitates, since they have a crystal structure with a tetragonal unit cell and a chemical composition of Al$_3$Cu. This structure is somewhat closer to the equilibrium θ-phase of Al$_2$Cu, but a significant rearrangement of atoms is still necessary for the transformation to the equilibrium phase.

14.1.4 Precipitation Sequence

These nonequilibrium precipitates are organized on Fig. 14.3, which shows part of the Al–Cu T–c phase diagram (expanded at low Cu concentrations). At temperatures below 550 °C, for most compositions below 33% Cu, the two phases in equilibrium are the α-phase (solid solution of fcc Al with some dissolved Cu), and the θ-phase, which has the composition Al$_2$Cu. We should expect equilibrium between the α-phase and the θ-phase after sufficient time, but this may be geological time at low temperatures. Instead, other structures such as the G–P(1) or G–P(2) precipitates can form before the θ-phase, and may effectively replace it in the material. When most Cu atoms are incorporated into these G–P zones, precipitation of the equilibrium θ-phase is suppressed.

At higher temperatures, however, specifically above the "solvus lines" indicated in Fig. 14.3, these transient precipitates are dissolved and a new phase forms. The sequence of precipitates is often summarized as

  Solution $\rightarrow$ G–P(1) $\rightarrow$ G–P(2) / θ'' $\rightarrow$ θ' $\rightarrow$ θ-phase.

---

[1] These "coherency strains" would vanish if there were a loss of registry between the fcc Al matrix and the Cu atoms in the zone as in Fig. 4.2c, but the bonding across the interface would be disrupted.

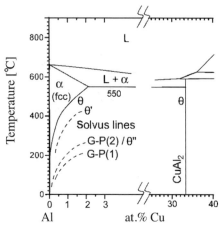

Figure 14.3 Al-rich side of Al–Cu phase diagram, showing metastable solvus lines as dashed curves.

However, the details of the sequence depend on temperature and time [231, 232], so the full sequence does not occur at one temperature. Not all details are understood, such as whether the G–P(2) zones evolve continuously into thicker θ' precipitates. These θ' precipitates are much larger than the G–P(2) zones, and have a composition of Al_2Cu instead of Al_3Cu. The θ' precipitates are found as flat plates along the {100} planes of the Al matrix, but these precipitates have lost coherency with the matrix around them. This helps to reduce the elastic energy that would otherwise impede their growth to larger sizes.

The precipitation sequence of Al–Cu is altered significantly by adding small amounts (a couple of at.%) of Mg or Sn, for example. New precipitate structures are observed, and some are quite effective in improving the mechanical strength of fcc Al alloys. The addition of Mg, for example, promotes the formation of Ω-phase precipitate plates on {111} planes of the fcc α-phase matrix, although these can coexist with θ'' precipitates on {100} planes [233–235]. The effect of Sn, a relatively large atom, seems to occur through interactions with vacancies. Vacancies tend to bind to Sn atoms, suppressing the availability of vacancies to form G–P(1) zones [236, 237]. Many structural Al alloys have been developed around families of compositions and their thermomechanical treatments.

14.2 Precipitation at Grain Boundaries and Defects

14.2.1 Collector Plate Mechanism

Precipitates often form at grain boundaries. A grain boundary "allotriomorph," rich in Pb, was shown in Fig. 4.5 (an allotriomorph is a crystal of abnormal shape). The same figure also shows a small spherical Pb precipitate within one of the Al grains. There are many

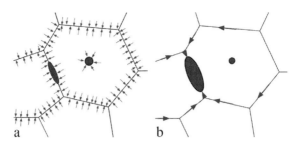

Figure 14.4 (**a**) Small arrows of approximately equal separation indicate fluxes into areas of grain boundary and into areas of two precipitates. (**b**) The rapid diffusion of atoms along grain boundaries gives a much greater growth rate of the grain boundary allotriomorph than the isolated precipitate.

hundreds of times more Pb atoms in the allotriomorph than in the spherical precipitate. The growth rate of the grain boundary allotriomorph is hundreds of times faster.

One mechanism for the diffusional growth of precipitates at grain boundaries is the "collector plate mechanism." It can be understood by considering fluxes of atoms through surfaces of different areas. To grow a precipitate that has nucleated homogeneously in the middle of a crystal, a solute flux must pass through a surface around the precipitate. This surface area is small when the precipitate is small.

To understand the collector plate mechanism, assume that after solute reaches the grain boundary it has infinite diffusivity, so it can move instantaneously to a precipitate situated on a grain boundary. The collector plate mechanism is illustrated with Fig. 14.4. The idea is that the total flux of solute atoms through a surface near a grain boundary is large because the grain boundary is large. The growth rate of a precipitate located at the grain boundary is therefore enhanced by a factor equal to the ratio of the net surface area of the two grains around it to area of the precipitate. In practice, this factor can be hundreds or thousands. For example, there are about 150 flux arrows into the grain boundary of Fig. 14.4a, versus only six into the homogeneous precipitate. The figure is two-dimensional – the enhancement factor is much larger in three-dimensional processes.

It is possible to improve on the optimistic assumption of an infinite diffusivity for atoms inside a grain boundary, and sometimes estimates of the ratio of grain boundary diffusivity to bulk diffusivity are possible from experimental measurements of the growth rate of grain boundary allotriomorphs. Nevertheless, the problem seems to be a bit more complex than such simple corrections allow. In particular, diffusion-controlled processes predict a growth rate of precipitate volume in proportion to $t^{1/2}$ (t is time), as is typical of diffusion over a fixed distance.[2] Perhaps curvature alters the growth rate of small particles, as discussed in Section 14.6, moving the time dependence towards $t^{1/3}$. However, the growth rate of grain boundary allotriomorphs seems to be more typically $t^{1/4}$. There are a number of processes that could influence these kinetics, such as a nonequilibrium concentration of vacancies or other transient phenomena. Nevertheless, the basic principle of the collector plate mechanism is well accepted.

[2] Typically a one-dimensional error function solution (Eq. 3.44) is used for the diffusional transport into the grain boundary when the grain boundary width is large compared with the diffusion length.

14.2.2 Precipitate-Free Zone

Figure 14.5 shows a microstructure formed by chemical unmixing with the collector plate mechanism. Notice the "precipitate-free zone" around the grain boundary. Far away from the grain boundary, homogeneous nucleation occurred efficiently, and the sizes and separations of the precipitates are not affected by the grain boundary. Near the grain boundary, however, the nucleation and growth of the allotriomorphs consumes considerable solute that is transported by the collector plate mechanism, and the solute atoms in these large precipitates come from the region near the grain boundaries. This region is depleted in solute, so precipitates are not found in this region. To some extent, this can be understood as a rescaling of distances, where the efficiency of the collector plate mechanism allows for more rapid growth, larger distances between the precipitates, and a large zone between the allotriomorphs and the precipitates in the interior of the crystals. However, there is more complexity to this problem because the precipitates in the interior of the crystals are interchanging atoms by diffusion, and undergoing a coarsening process, as described in Section 14.6. The chemical potential for solute atoms in the large allotriomorphs is probably lower than for solute atoms in the small precipitates, so the allotriomorphs may consume the homogeneous precipitates in their vicinity.

Incidentally, coarse precipitates on grain boundaries can have important consequences for the fracture toughness of polycrystalline materials. If the atoms are not well bonded across the surface between the allotriomorphs and the crystal interiors, a fracture crack may propagate along grain boundaries in a process called "intergranular fracture." (A process of "tearing along the perforated line," roughly speaking.)

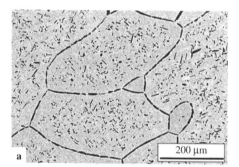 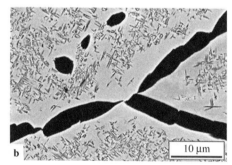

Figure 14.5   Precipitate-free zone near grain boundaries, with grain boundary allotriomorphs at the boundaries, and homogeneously nucleated precipitates within the grains. Images are scanning electron microscopy backscatter images of an annealed TiMoZrFe orthopedic alloy. (**a**) Overall microstructure showing grains of β-Ti; (**b**) higher magnification image showing the intragranular primary α-precipitates and fine-scale secondary α forming as a result of the ageing treatment. Note the precipitate-free zone next to the grain boundary. Reprinted from [238], with permission from Elsevier.

14.2.3 Other Sites for Heterogeneous Nucleation

Nucleation at Interfaces

The previous sections discussed the formation of precipitates at grain boundaries, a process that can be favored by kinetics, thermodynamics, or both. For example, grain boundary allotriomorphs can be favored by the kinetics of nucleation (heterogeneous nucleation is usually faster than homogeneous), the kinetics of diffusion (the grain boundary "collector plate" speeds solute atoms to the precipitate), and thermodynamics (the chemical potential is more favorable for atoms added to a large allotriomorph, and the elimination of grain boundary energy is favorable, too.) With so many mechanisms, it is not surprising that other internal interfaces such as antiphase domain boundaries or stacking faults can also serve as sites for heterogeneous nucleation.

Nucleation at Dislocations

The one-dimensional defect in crystals, the dislocation, has two salient features. One is the poor atomic coordination in its core, which is several atoms in diameter. The lower density of the core, like that of a grain boundary, can lead to rapid solute diffusion down the dislocation line in a process called "pipe diffusion." The thermodynamics of precipitation can also be favored at a dislocation because a precipitate crystal has a more favorable atom configuration than the dislocation core it replaces. The second important feature of a dislocation is its strain field. Edge dislocations generate both compressive and tensile stresses in the matrix around them. The strain fields can attract or repel solute atoms having specific volumes different from matrix atoms. For example, a small substitutional solute atom may be drawn into the compressive field of the dislocation, and may move to where the field is strongest near the dislocation line. Not surprisingly, dislocations are often found around precipitates, but the cause and effect may not be obvious. The size mismatch of a precipitate in the matrix may be responsible for generating the dislocation, or attracting it, or perhaps the dislocation was the nucleation site for the precipitate.

Figure 14.6 shows atom probe tomographic images from an Fe–6 at.% Cr– 0.04 at.% Si alloy, where the dark dots indicate Si atoms and the majority light dots are Cr atoms in the bcc Fe-rich matrix. A solid solution is evident in the image of Fig. 14.6a, which was annealed in the single-phase bcc region of the Fe–Cr phase diagram. Neutron irradiation creates vacancies and interstitial atoms, which tend to be mobile at the temperature of 560 K where the irradiation was performed. Some of these defects collapse into small dislocation loops, with local environments favorable to Cr and Si atoms, as shown in Fig. 14.6b. Grain boundaries are also favorable for solute segregation, as evident in Fig. 14.6c. Although these compositional heterogeneities were promoted by neutron irradiation, similar solute segregations are generally expected around dislocations and grain boundaries. Such chemically enriched zones offer locations for precipitate formation.

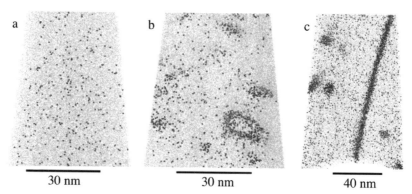

Figure 14.6 Two-dimensional projections of atom probe tomographic reconstructions of atoms in three conical samples of Fe–6 at.% Cr–0.04 at.% Si alloy. (**a**) Sample annealed to produce a solid solution of Cr and Si in bcc Fe matrix. (**b, c**) After irradiation at 560 K with a neutron dose that produced approximately 1.8 displacements per atom. (**b**) Cr (light) and Si (dark) atoms segregated around a small dislocation loop are visible near the lower right. (**c**) Cr (light) and Si (dark) atoms segregated to a grain boundary are seen as a dark band from upper right to bottom. Reprinted from [239], with permission from Elsevier.

Nucleation at Existing Precipitates

Precipitates can themselves serve as sites for the heterogeneous nucleation of new phases. When new phases form in a precipitation sequence, the new phases often form at interfaces between pre-existing precipitates and the matrix. Both the interface energies and the strain fields around precipitates can play a role in the kinetics or thermodynamics of forming the new phases.

The elastic strain field around a precipitate can alter the tendency to form similar precipitates in its vicinity. The effect can be positive or negative. A misfitting precipitate can make it less likely for another precipitate to form nearby because the elastic energy penalty will increase as the square of the distortion. On the other hand, a precipitate that creates shear stresses around it may make the local environment favorable for forming another precipitate with a counteracting shear stress. The formation of precipitates can therefore be "autocatalytic," in that the formation of one precipitate makes it energetically favorable to form another. (The burst formation of martensite plates in Figs. 19.6 and 19.7 is an example of this, but martensitic transformations are not precipitation processes.)

14.3 The Eutectoid Transformation and Pearlite

14.3.1 The Iron–Carbon Phase Diagram

Much of ferrous metallurgy is centered around the eutectoid point in the Fe–C phase diagram at 732 °C and 0.77 wt.% C, shown in Fig. 14.7. The full T–c phase diagram at ambient pressure has five phases: liquid, α (bcc), γ (fcc), δ (bcc), and cementite (Fe_3C, θ-carbide). The three of these at the eutectoid point, α, γ, and θ, are essential

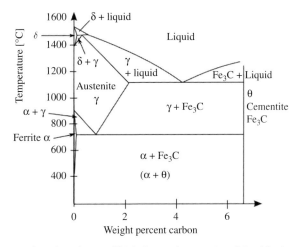

Figure 14.7 Iron-rich side of the iron–carbon phase diagram. Weight (or mass) percent is traditional for the composition axis.

for understanding steels. There is also a metastable martensite phase (α'), which is a supersaturation of up to 1% of C in α-like Fe. Over the years, there has been an accumulation of names for these phases:

ferrite = α = bcc
austenite = γ = fcc
martensite = α' = bct
cementite[3] = θ = θ-carbide = Fe_3C
at pressures above 12 GPa, hcp ϵ-Fe phase is formed from α-Fe
β-Fe was once defined as nonmagnetic bcc iron at temperatures above the Curie temperature.

Perhaps more baffling are the names for steel microstructures, such as pearlite, upper bainite, lower bainite, tempered martensite, spherodite, and perhaps ledeburite. These microstructures are not phases. They are mixes of α plus θ crystals, configured in different shapes or "morphologies," resulting from different processes of the phase transformations.[4] We consider first the pearlite transformation.

14.3.2 Pearlite: a Eutectoid Transformation

Lamellar Structure

This section describes pearlite, a microstructure shown in Fig. 14.8. Pearlite is named for its iridescent pearly appearance when its wave-like periodicity is the wavelength of optical light. The pearlite transformation is the classic eutectoid transformation, where one parent

---

[3] Cementite is metastable, with α-Fe and graphite being the stable phases. Since cementite is so robust, and graphite formation suppressed, it is appropriate to place cementite on the phase diagram at 25 at.% C or 6.7 wt.% C, without consideration of graphite.

[4] Different microstructures can contain the same crystalline phases in different arrangements. There is an apt correspondence with rocks. Different rocks can contain the same minerals in different arrangements, depending on their conditions of formation.

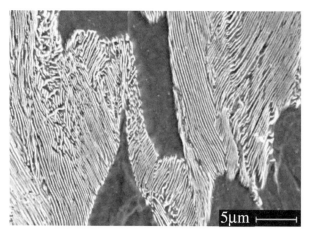

Figure 14.8   Pearlite colonies, surrounded by dark regions of α-Fe. Image is a scanning electron micrograph of an etched specimen of steel. Image by Michel Shock (public domain).

phase (γ) transforms into two new phases (α plus θ) that grow together in a diffusion-controlled process.

Groups of wave-like patterns in Fig. 14.8, called "colonies" or "cells," began to grow separately, but collided later. Within the colonies are thin lamellae of ferrite and cementite phases, seen as thin strips in cross-section in Fig. 14.8, and shown schematically in Fig. 14.9a. Figure 14.7 shows that austenite can dissolve up to 2 wt.% carbon, whereas the ferrite contains very little carbon. The carbon needs to redistribute spatially during the pearlite transformation $\gamma \rightarrow \alpha + \theta$, and the diffusion of carbon encourages the lamellae to maintain a small periodicity λ along the growth front shown in Fig. 14.9a. A smaller λ creates more surface energy $\sigma_{\alpha\theta}$, however, so there is an optimal λ for a given temperature of the pearlite transformation.

Lamellae and Surface Energy

We first find a minimum value of λ, balancing the surface and volume energies as in Section 4.2. As for Eq. 4.3, we seek the free energy, ΔG, of transforming a volume of austenite into ferrite plus cementite, now including interfaces between the ferrite and cementite. Working with a cubic centimeter of material, with $\Delta G_V = G_\gamma - G_{\alpha+\theta}$ as energy per cm^3, $\sigma_{\alpha\theta}$ as energy per cm^2, and λ in cm,

$$\Delta G = \Delta G_V + 2\frac{\sigma_{\alpha\theta}}{\lambda}, \tag{14.1}$$

where the factor of 2 accounts for two interfaces in the distance λ (see Fig. 14.9a). For this first-order phase transition, the bulk free energy difference, ΔG_V, is expected to scale linearly with undercooling, ΔT, as for Eq. 1.6,

$$\Delta G_V = \frac{-L\,\Delta T}{T_c}, \tag{14.2}$$

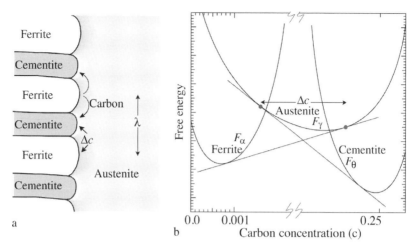

Figure 14.9 (**a**) Growth front of pearlite in steel, showing depletion of carbon in front of the cementite and excess carbon in front of the ferrite phase. (**b**) Free energy versus composition curves for steel undercooled below the eutectoid temperature. The parent phase, in this case the γ-phase (austenite), has a high free energy and is unstable.

where L is the latent heat of the phase transformation (and all factors are positive). Substituting into Eq. 14.1 and setting $\Delta G = 0$ gives the minimum value of $\lambda_{\min}$. For $\lambda > \lambda_{\min}$, the favorable ΔG_V overcomes the surface energy, favoring the pearlite transformation

$$\lambda_{\min} = \frac{2\sigma_{\alpha\theta} T_c}{L\,\Delta T}. \tag{14.3}$$

Figure 14.9a depicts the heterogeneity of the carbon in the austenite in front of the growing pearlite colony. The ferrite takes iron from the austenite but takes little carbon, whereas the cementite needs more carbon than the average composition of the austenite. As the pearlite colony grows to the right in Fig. 14.9a, carbon is rejected in front of the ferrite, and diffuses down a concentration gradient to the carbon deficit in front of the cementite.

Growth Rate by Diffusion

We estimate the concentration gradient with $\lambda/2$ for distance, and a Δc from a condition of local equilibrium at the interfaces of the austenite and the growing ferrite and cementite. The free energy versus composition curves (cf. Section 2.5) of Fig. 14.9b are for a generic eutectoid transformation below the eutectoid temperature. For local equilibrium at the interface between the α- and γ-phases, the common tangent line drawn in Fig. 14.9b shows a feature that is consistent with the picture in Fig. 14.9a – the austenite in front of the growing pearlite has an excess of carbon. Note how the opposite is true for the cementite. The larger the undercooling, the larger the value of Δc for the difference in carbon concentration in front of the ferrite and cementite (see the construction for Δc in Fig. 14.9b). This difference gives the concentration gradient $\Delta c\,2/\lambda$ that drives diffusion of carbon in Fig. 14.9a

$$J = -D\frac{2\,\Delta c}{\lambda}. \tag{14.4}$$

The velocity of growth, v, is proportional to the carbon flux, J (but in the opposite direction). It is also assumed that the velocity of growth is proportional to the free energy of the reaction ΔG, so

$$v = -\kappa' D \frac{2\,\Delta c}{\lambda} \Delta G, \tag{14.5}$$

$$v = -\kappa D \frac{\Delta c}{\lambda} \left(\Delta G_{\mathrm{V}} + \frac{2\sigma_{\alpha\theta}}{\lambda} \right), \tag{14.6}$$

$$v = -\kappa D \frac{\Delta c}{\lambda} \left(\frac{-L\Delta T}{T_{\mathrm{c}}} + \frac{2\sigma_{\alpha\theta}}{\lambda} \right), \tag{14.7}$$

making use of Eqs. 14.1 and 14.2, and absorbing the 2 and other geometrical factors in the constant κ. Equation 14.7 shows how the undercooling ΔT must be large enough to overcome the surface energy if growth is to occur (i.e., $v > 0$). It is essentially a recasting of Eq. 14.3.

To obtain the lamellar periodicity with the highest growth velocity, $\lambda_{\max}$, set the first derivative equal to zero

$$\frac{\mathrm{d}v}{\mathrm{d}\lambda} = 0 = \kappa D \Delta c \left(\frac{L\Delta T}{T_{\mathrm{c}} \lambda_{\max}^2} - \frac{4\sigma_{\alpha\theta}}{\lambda_{\max}^3} \right), \tag{14.8}$$

$$\lambda_{\max} = \frac{4\sigma_{\alpha\theta} T_{\mathrm{c}}}{L\Delta T}, \tag{14.9}$$

$$\lambda_{\max} = 2\lambda_{\min}. \tag{14.10}$$

Recall that $\lambda_{\min}$ of Eq. 14.3 is set by the thermodynamic condition of equality of surface energy and bulk free energy. The kinetics of diffusion favor a small value of $\lambda_{\max}$ to minimize diffusion distances. The optimal lamellar spacing turns out to be fairly close to $\lambda_{\min}$, indicating the dominant role of carbon diffusion in the austenite for kinetics of the pearlite transformation.

Substituting $\lambda_{\max}$ from Eq. 14.9 into Eq. 14.7 gives a maximum growth velocity

$$v_{\mathrm{m}} = -\kappa D \frac{\Delta c\, L\Delta T}{4\sigma_{\alpha\theta} T_{\mathrm{c}}} \left(\frac{-L\Delta T}{T_{\mathrm{c}}} + \frac{2\sigma_{\alpha\theta}\, L\Delta T}{4\sigma_{\alpha\theta} T_{\mathrm{c}}} \right), \tag{14.11}$$

$$v_{\mathrm{m}} = \frac{1}{8}\kappa D \frac{\Delta c}{\sigma_{\alpha\theta}} \left(\frac{L\Delta T}{T_{\mathrm{c}}} \right)^2. \tag{14.12}$$

The growth velocity depends strongly on the undercooling, ΔT. A larger undercooling gives a larger ΔG_{V}, which allows for more surface energy, closer lamellar spacing, and shorter diffusion distances.

There are a number of other issues that were not treated carefully in the present analysis. Although Eq. 14.12 is appropriate for a specific Δc, the specific Δc depends on λ. At a fixed temperature, a smaller lamellar spacing λ causes the free energy curves F_α and F_θ in Fig. 14.9b to be raised with respect to F_γ (owing to surface energy), so Δc is decreased. The concentration gradient is decreased proportionately, as is the growth velocity of Eq. 14.12. We also neglected details of the carbon concentration field and details of the curvatures of the interfaces between austenite and ferrite and between austenite and cementite. The carbon concentration in the austenite alters its lattice parameter, so the diffusion coefficient of carbon in austenite may have a spatial variation, too. Nevertheless, the present analysis does show the decrease in lamellar spacing with undercooling, and the increase in growth velocity with undercooling. The pearlite microstructure grows faster and finer with increased undercooling, at least until the diffusion coefficient D for carbon in austenite becomes small.

14.4 Heat Treatments of Steel

14.4.1 Austenitizing

To begin with a blank slate, so to speak, carbon steels are usually given an "austenitizing" heat treatment to start them as a single-phase solid solution of C in austenite (fcc Fe), and to help erase compositional inhomogeneities of other solutes. Temperatures well above 900 °C are usually required for good diffusional homogenization. At these temperatures, the steel is relatively easy to deform, at least compared with room temperature, and blocks of steel can be forged into useful shapes when in the austenite phase (see Fig. 14.10). Forging can reduce the grain size of austenite crystals, and this is also helped by the addition of one percent of Mn to the alloy. Austenite is a high-temperature phase, and will transform or "decompose" to other phases when a carbon steel is cooled to room temperature. Much of the personality of a steel comes from the phase transformations of austenite upon cooling, so a steel is defined as an Fe alloy with a C concentration that falls within the austenite single-phase field (and therefore can be fully austenitic when heated).

a   b

Figure 14.10 (**a**) Iconic image of a blacksmith forging a steel bar. Image reproduced under license from Aloysius Patrimonio, http://retroclipartz.com. (**b**) Large forging press at Japan Steel Works, Muroran, Japan (human in lower right). In both cases the steel is austenite, glowing with yellow blackbody radiation.

This limits steel compositions to no more than about 2 wt.% C, but carbon concentrations are usually much lower than this.[5]

14.4.2 Decomposition of Austenite

Most steels have carbon concentrations less than the eutectoid composition of 0.77 wt.% C (hypoeutectoid steel), but some have more (hypereutectoid steels). For hypoeutectoid steels, cooling from the austenite single-phase region passes through a two-phase region of ferrite plus austenite, so some ferrite forms during cooling. This "proeutectoid" ferrite usually nucleates and grows along the grain boundaries of the austenite. Few austenite grain boundaries have special orientations, so if the ferrite finds a preferential crystallographic orientation with one austenite crystal, it is probably less favorably oriented with respect to the grain on the other side of the boundary. Typically a ferrite crystal grows more quickly into the austenite crystal with which it has a poor orientation compatibility. When the steel is cooled below 732 °C, these ferrite crystals can serve as nuclei for the pearlite transformation.

On the other hand, cooling a hypereutectoid steel from the austenite region passes through a two-phase region of austenite plus cementite. In this case it is proeutectoid cementite that nucleates on the austenite grain boundaries, and tends to grow into the austenite grain with which the cementite does not have a good surface orientation. Below 732 °C, these cementite crystals can serve as nucleation sites for the pearlite microstructure.

The pearlite transformation speeds up with undercooling below 732 °C, reaching a maximum rate at a temperature of about 550 °C. At lower temperatures, the diffusion of carbon becomes slow (the D in Eq. 14.12 should be considered $D(T)$), slowing the pearlite transformation. Nevertheless, the free energy for the decomposition of austenite into ferrite plus carbide continues to increase with undercooling, and a consequence is a change in transformation mechanism and a different microstructure.

14.4.3 TTT Diagrams of Steels

Figure 14.11 shows TTT diagrams of four different steels. The simplest TTT diagram is shown in Fig. 14.11b. This steel has a carbon concentration close to the eutectoid composition, so the austenite is stable until the temperature decreases below the eutectoid temperature, T_{eu}, of approximately 723 °C. When cooled below about 700 °C, in a reasonable time the austenite transforms to ferrite plus cementite by the pearlite transformation.

The austenite is less stable when the carbon concentration is below the eutectoid composition. The phase diagram of Fig. 14.7 shows that we should expect some "proeutectoid" ferrite to form directly from the austenite above 723 °C, and it adds a new feature to the TTT diagram of Fig. 14.11a at short times and high temperatures (the "pro" in proeutectoid means before, or prior). Likewise, a carbon concentration above the eutectic composition

---

[5] Fe–C alloys with C concentrations above 2 wt.% C are called "cast irons." For example, ledeburite is a cast iron microstructure from a eutectic reaction below the eutectic point at 4.3 wt.% C and 1153 °C on the Fe–C phase diagram.

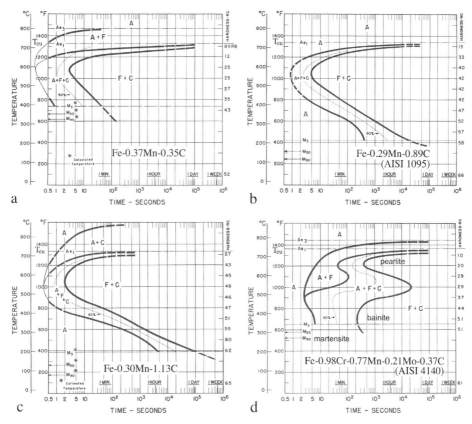

Figure 14.11 For these TTT diagrams, all four steels were held at approximately 890 °C in the fcc austenite phase, then cooled quickly to the temperature on the *y*-axis for times indicated on the *x*-axis. Compositions are listed in wt.%. Letters designate phases as: A: austenite (fcc), F: ferrite (bcc), C: cementite (Fe$_3$C). Eutectoid temperature is T_{eu}. (**a**) Fe–0.37Mn–0.35C below eutectoid composition (hypoeutectoid). Note formation of proeutectiod ferrite at temperatures above T_{eu} and somewhat below at short times. (**b**) Fe–0.29Mn–0.89C near eutectoid composition. (**c**) Fe–0.30Mn–1.13C above eutectoid composition (hypereutectoid). Note formation of proeutectiod cementite at temperatures above T_{eu} and somewhat below at short times. (**d**) Fe–0.98Cr–0.77Mn–0.21Mo–0.37C below eutectoid composition (hypoeutectoid). Note distinct transformations for pearlite (fastest at 650 °C) and bainite (fastest at 400 °C). Faster cooling gives martensite for all four steels. From [241].

such as 1.13 wt.% for the steel of Fig. 14.7c is expected to have proeutectoid cementite, and this feature is also seen in the TTT diagram at short times and higher temperatures, labeled "A+C."

At low temperatures, all four steels of Fig. 14.11 transform eventually into ferrite plus cementite (unless the cooling is fast enough to form martensite), and the transformation is the pearlite transformation at temperatures moderately below T_{eu}. The bainite transformation dominates at lower temperatures. The end product is still ferrite plus cementite, but the carbides are not continuous lamellae, but are islands surrounded by ferrite. The details of this transformation have been controversial [240], but it is a nucleation and growth type of phase transformation. Upper bainite, which forms from approximately 350 to 550 °C, has a more angular morphology than pearlite, but is still a mixture of ferrite and cementite. Lower bainite, which forms at lower temperatures, contains nonequilibrium carbide phases such as "ϵ-carbide" inside the ferrite crystals. Lower bainite is sometimes distinguished from upper bainite by its feathery appearance in an optical microscope.[6] The transition between the pearlite transformation and the bainite transformation is more obvious for the alloy steel of Fig. 14.11d.

14.4.4 Martensite

If a steel is cooled rapidly to a low temperature to suppress carbide formation, the austenite transforms into martensite in a diffusionless process involving shear. Martensitic transformations are the subject of Section 19.2. Martensite in steel is approximately described as a supersaturated solid solution of carbon in α-iron, but the carbon imparts a tetragonality to the bcc unit cell (so it is bct and designated α'). Martensite is extremely hard mechanically, and is a main reason why steel has been so important in the industrial age. Martensite can assume a large number of microstructures or morphologies, offering opportunities for balancing the mechanical properties of strength (resistance to deformation) against toughness (resistance to crack propagation), for example. The high cooling rate required to form martensite is a problem for thick sections of steel, since the cooling rate of the interior depends on diffusion of heat to the surface. One of the most important roles of alloying elements in steel is to lower the cooling rate required to form martensite. Figure 14.11d is a TTT diagram for an alloy steel (AISI 4140) that forms martensite more easily for two apparent reasons. The temperature where martensite is first detectable, denoted M_s for "martensite start," and lower temperatures where the amount of transformation is 50% and 90% (denoted M_{50} and M_{90}) are all three higher than for the two steels of Figs. 14.11b,c. The second, and more important, reason is that the pearlite transformation is slowed. This allows a slower cooling rate so martensite can form in the interior of thicker plates. The relative ease of forming martensite is called "hardenability," and is defined empirically by a type of quenching test.

14.4.5 Tempering, Carbides, and Retained Austenite

A polycrystalline martensite microstructure has high internal stresses and a mechanically hard α' phase. It can be brittle, and difficult to bend or cut. Heating at a low temperature can alleviate these difficulties, in a process called "tempering." There is stress relief during tempering, but more notably the supersaturation of carbon and the α' tetragonality are reduced by segregating carbon into carbide precipitates. Generally the tempering causes some loss of yield strength but an increase in fracture toughness. The selection of tempering temperatures and times varies with the type of steel and the properties desired, but for the

---

[6] Some of this appearance, and the change in type of carbide formed, is attributed to a greater tendency for transformations at low temperatures to involve cooperative atom movements by shear processes, rather than simple diffusion of carbon in fcc austenite.

high-strength steel AISI 4140 of Fig. 14.11d it might be 200 °C for high strength, or 600 °C for high toughness.[7]

It is possible to "spherodize" the carbides in most microstructures of steel by reheating to a temperature just below T_{eu}. At this elevated temperature, the carbides grow in size and reduce their surface-to-volume ratio as discussed in Section 14.6. For higher-carbon hypereutectoid steels, spheroidizing may start with a heating to a temperature of approximately 800 °C. This is in the γ+Fe$_3$C two-phase region of the Fe–C phase diagram of Fig. 14.7, i.e., where there is equilibrium between austenite and cementite.[8] There is a growth of carbides, but there is also some reversion of ferrite to austenite. This step is followed by slow cooling through T_{eu} to avoid the formation of fine-structured pearlite or martensite as the reverted austenite decomposes. The resulting "spherodite" microstructure is soft and ductile.

Besides martensite, ferrite, and carbides, steels can contain some austenite at room temperature.[9] Approximately one percent of austenite can be "retained" after the quench from the austenite region of the phase diagram if internal stresses between martensite plates favor the lower-volume austenite phase. Alternatively, austenite can be "reverted" from martensite by reheating into a region of the phase diagram where austenite is present, and in some alloy steels the reverted austenite crystals can be retained at low temperatures when they are small in size.

Steels contain ferrite, carbide, martensite, and austenite phases in different microstructures such as pearlite, bainite, and various martensites. These microstructures deliver a wide range of mechanical properties, sustaining the field of ferrous metallurgy over many years. In the broader field of materials science, steels helped make microstructure a core concept, and highlighted the difference between phase transformations and phase transitions.

14.5 The Kolmogorov–Johnson–Mehl–Avrami Growth Equation

14.5.1 Growth Rate before Particle Contact

Nucleation was considered in detail in Chapter 4, and a steady-state nucleation rate was obtained as Eq. 4.52. At longer times, this rate is no longer accurate for two reasons. For the formation of β-phase after the quench of Fig. 4.1, as more and more β-nuclei form and grow, the matrix α-phase is depleted of solute. The activation free energy for nucleation and the nucleation rate depend strongly on supersaturation, so the nucleation rate J decreases with time. The interplay of solute depletion, nucleation, and growth is complicated, especially in multicomponent alloys, and remains a topic of research [39, 242, 243].

---

[7] The temperature of 600 °C will increase the fraction of carbides, but not as predicted with Fig. 14.11d, which is for decomposition of austenite, not martensite.
[8] This is called an "intercritical annealing."
[9] There are fully austenitic steels, but these are highly alloyed with elements such as Ni.

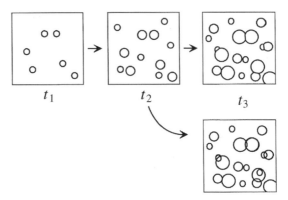

Evolution of the microstructure of a single-phase α transforming to the single-phase β at three times $t_1 < t_2 < t_3$. At time t_3, β-particles begin to impinge on one another; the extended volume concept allows the β-phase to grow and nucleate in previously transformed material, as shown in the bottom diagram. Reprinted from [244].

Even the nucleation of a new phase in a single-component system (which does not require long-range diffusion) is challenging at intermediate times when the new particles grow to impingement. The particles stop growing where they touch, altering the time dependence of the fraction of new β-phase. This overlap problem is handled by a clever method devised by Kolmogorov [245], Johnson and Mehl [246], and Avrami [247–249], and is explained here. Figure 14.12 depicts the microstructure of a nucleation and growth phase transformation at three times, where $t_1 < t_2 < t_3$. At t_1 the density of nuclei is predicted by the nucleation rate J, and each β-particle is approximately the critical size. At t_2, further nucleation gives an increased density of β-phase particles, and the nuclei that formed at earlier times have grown larger.

We assume that all spherical particles grow with the radial velocity v, which is constant at all times. At time t, a particle that nucleated at the earlier time t' has the radius $v(t - t')$ and the volume $4\pi/3 \, [v(t - t')]^3$. Without impingement of particles, the total volume of β-phase is

$$V_{\text{ex}}^{\beta}(t) = V \frac{4\pi}{3} \int_0^t J v^3 \, (t - t')^3 \, \mathrm{d}t', \tag{14.13}$$

where V is the total volume of the system. The integration over t' accounts for the nucleation times of all particles. For determining the volume of β-phase, Eq. 14.13 is valid only at early times before particle impingement. Nevertheless, it proves valuable for calculating the fraction of β-phase at later times too, and for this it is called the "extended volume" (and hence the subscript "ex"). Note, however, that this $V_{\text{ex}}^{\beta}(t)$ grows to arbitrary size at large times, whereas the actual $V^{\beta}(t)$ cannot exceed V, the volume of the alloy.

14.5.2 Growth Rate after Particle Contact

The essential trick to handle impingement is that, in an increment of time, only a part of the change in $V_{\text{ex}}^{\beta}(t)$ can contribute to the real $V^{\beta}(t)$. The fraction of $\mathrm{d}V_{\text{ex}}^{\beta}(t)$ that contributes to

$dV^\beta(t)$ is the fraction of untransformed phase that remains at time t. (All the rest of $V^\beta_{ex}(t)$ are in regions that have transformed already.) This remaining fraction of untransformed volume is

$$f_{un}(t) = \left(1 - \frac{V^\beta(t)}{V}\right), \qquad (14.14)$$

and the fraction of new extended volume that contributes to the new real volume of β-phase is

$$dV^\beta(t) = \left(1 - \frac{V^\beta}{V}\right) dV^\beta_{ex}(t). \qquad (14.15)$$

By integration

$$V^\beta_{ex}(t) = -V \, \ln\left(1 - \frac{V^\beta(t)}{V}\right). \qquad (14.16)$$

Using Eq. 14.13, the actual volume fraction of β-phase is

$$\frac{V^\beta(t)}{V} = 1 - \exp\left[-\frac{4\pi}{3} \int_0^t J v^3 (t - t')^3 \, dt'\right]. \qquad (14.17)$$

This is the Kolmogorov–Johnson–Mehl–Avrami (KJMA) equation.

14.5.3 Nucleation Processes

The KJMA Eq. 14.17 predicts that the volume fraction at $t = 0$ is zero, but has a sigmoidal shape with increasing slope that later gives way to an asymptote at long time with $V^\beta/V = 1$. The actual shape of V^β/V is typically calculated for two cases. The first case assumes that all nucleation occurs simultaneously, so $JV = N\delta(t' = 0)$, where N is the number of nucleation sites per unit volume. The Dirac delta function sets $t' = 0$ and selects the integrand as

$$\frac{V^\beta(t)}{V} = 1 - \exp\left[-\frac{4\pi}{3} v^3 N t^3\right]. \qquad (14.18)$$

The second case assumes the nucleation rate, J_{ss}, is a constant for all times. Integration of Eq. 14.17 gives a more rapid dependence on time

$$\frac{V^\beta(t)}{V} = 1 - \exp\left[-\frac{\pi}{3} v^3 J_{ss} t^4\right]. \qquad (14.19)$$

Notice the change in exponent from t^3 to t^4 between Eqs. 14.18 and 14.19. Since the above two examples are limiting cases of nucleation behavior, Avrami suggested that a plot of $\ln[\ln(1 - V^\beta/V)]$ vs. t will result in a straight line with a slope between 3 and 4, and the exponent of t will indicate the dominance of homogeneous or heterogeneous nucleation. Consistent with Eq. 14.18, Avrami assumed the heterogeneous sites were distributed randomly throughout the bulk material. This assumption is not good for polycrystalline solids when nucleation tends to occur at grain boundaries or perhaps dislocation lines. Equation 14.18 can be modified to accommodate these processes, however [250].

14.6 Coarsening

Suppose a diffusional phase transformation is complete, so all A- and B-atoms are approximately unmixed into their expected concentrations. Suppose the microstructure has B-rich β-phase precipitates distributed within an A-rich α-phase matrix. It is nearly in equilibrium. There remains only surface energy which can be reduced if the β-phase particles grow larger. This section estimates the time dependence of the average particle radius.

Conservation of solute requires that if some β-phase particles grow, others must shrink. The smallest particles have the highest surface energy per atom so these tend to disappear. This process is called "coarsening" of the microstructure.

14.6.1 Two Particles

Chemical Potential and Curvature

We formulate a simple problem with two β-phase particles in an α-phase matrix. The particles are assumed to be spherical, with radii r_1 and r_2, and separated by the distance R. Diffusion of B-atoms from particle 1 (with smaller radius, $r_1 < r_2$) to particle 2 will decrease continuously the total surface area of the particles. The essential picture and variables are shown in Fig. 14.13. Because the unmixing is complete, the total volume of particles, V_t, is a constant with time

$$V_t = \frac{4\pi}{3}(r_1^3 + r_2^3).$$ (14.20)

On the other hand, the total surface area, A_t, is not conserved, where

$$A_t = 4\pi(r_1^2 + r_2^2).$$ (14.21)

For example, it is straightforward to show that two equal-sized spheres, each of volume $V_t/2$, have a total surface area of

$$A_{t2} = (3V_t)^{2/3}(8\pi)^{1/3},$$ (14.22)

whereas a single sphere with volume V_t has a total surface area of

$$A_{t1} = (3V_t)^{2/3}(4\pi)^{1/3},$$ (14.23)

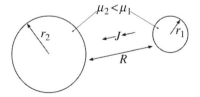

Figure 14.13   Two particles of β-phase coarsening in an α-phase matrix. The chemical potential for atoms in particle 1 is higher because of its larger surface-to-volume ratio, and this drives a diffusive flux to particle 2.

and A_{t1} is smaller. It can also be shown that A_{t2} is the maximum surface area for the two particles, and as one particle becomes larger at the expense of the other, A_t decreases continuously until only one particle remains with area A_{t1}. Nature will find a way to reduce the surface area by growing particle 2 at the expense of particle 1, but how do we calculate the rate?

The coarsening problem is essentially a diffusion problem, with a flux of atoms driven by the difference in chemical potential of B-atoms in the two particles. Since unmixing is complete, there is no reduction in the chemical free energy per unit volume, but there is an addition to the free energy in the microstructure from the surface area. For a particle of radius r, surface energy adds to the free energy a ΔG [energy per unit volume] [10]

$$\Delta G \simeq \frac{A\sigma}{V} = \frac{4\pi r^2 \sigma}{\frac{4\pi}{3}r^3}, \tag{14.24}$$

$$\Delta G \simeq +\frac{3\sigma}{r}, \tag{14.25}$$

where σ is the specific surface energy. For our two particles separated by the distance R, the gradient in chemical potential is approximately

$$\vec{\nabla}\mu = v\,\frac{\Delta G_1 - \Delta G_2}{R}, \tag{14.26}$$

$$\vec{\nabla}\mu = 3\sigma v \left(\frac{1}{r_1} - \frac{1}{r_2}\right)\frac{1}{R}, \tag{14.27}$$

where v is the volume per atom.

Chemical Potential Gradient and Diffusional Growth

This $\vec{\nabla}\mu$ drives a diffusion flux from the smaller particle 1 to the larger particle 2. Using a variant of Fick's first law for diffusion in a potential gradient (e.g., Eq. 10.71) with a "mobility" M having units [velocity per force], and solute concentration c

$$\vec{J} = -cM\,\vec{\nabla}\mu, \tag{14.28}$$

$$\vec{J} = -\frac{3cM\sigma v}{R}\left(\frac{1}{r_1} - \frac{1}{r_2}\right). \tag{14.29}$$

It is tempting to make an approximation

$$\left(\frac{1}{R}\right)\left(\frac{1}{r_1} - \frac{1}{r_2}\right) = \zeta = \text{constant.} \tag{14.30}$$

The idea is that as one particle grows at the expense of the other, the two factors in parentheses in Eq. 14.30 will evolve in opposite directions (which is true), and will compensate each other (which is not quite right). There is also a question of the effective area over which the diffusion flux occurs. These geometrical issues are challenging, but assuming a constant ζ of Eq. 14.30 enables the elementary integration in Eq. 14.34 below.

---

[10] Section 20.2, specifically Eq. 20.4, shows that a more accurate expression for the chemical potential should be $2\sigma v/r$, an equilibrium quantity. We ignore the constants in what follows, however.

The growth of particle 2 is proportional to the magnitude of the flux J, and particle 1 loses volume at the same rate

$$\frac{dV_2}{dt} = -JvA = -\frac{dV_1}{dt}, \tag{14.31}$$

$$4\pi r_2^2 \frac{dr_2}{dt} = -JvA, \tag{14.32}$$

where A is an effective area for the flux. Returning to Eq. 14.29 with Eq. 14.30

$$4\pi r_2^2 \frac{dr_2}{dt} \frac{1}{vA} = 3cM\sigma v\zeta, \tag{14.33}$$

$$\int_0^r r_2^2 \, dr_2 = \frac{3cM\sigma Av^2\zeta}{4\pi} \int_0^t dt', \tag{14.34}$$

$$r_2 = \left(\frac{9cM\sigma Av^2\zeta}{4\pi}\right)^{1/3} t^{1/3}. \tag{14.35}$$

For coarsening, the radius of the larger particle 2 grows as $t^{1/3}$.

Local Equilibrium

Perhaps the most subtle point in this analysis is the gradient of chemical potential that drives diffusion. The essential concept, covered later in Section 20.2, is the "Gibbs–Thomson effect." It considers changes in the free energy when atoms are added to a small particle, and relatively more surface area is created than for a large particle. This makes it less favorable to add B-atoms to small particles, increasing the chemical potential for the B-atoms in them (recall that $\mu = (\partial G/\partial N)_{T,P}$, from Eq. 1.24). The B-atoms are in local equilibrium between the small particle and the nearby matrix, so their chemical potential in the matrix near a small particle is higher than near a large particle. This gives the $\vec{\nabla}\mu$ that drives the flux of B-atoms through the matrix from the small particles to the large ones.

Several modifications could improve our analysis. Larger particles likely have more neighbors around them than do smaller particles. Fortunately, adding more neighbors as a particle grows serves to compensate some of the error in Eq. 14.30. A particle-by-particle analysis tends to lose sight of the forest for the trees, however. In practice there are numerous particles, and a spread of particle sizes that can be described by a smooth distribution function. This particle size distribution is bounded by maximum and minimum particle sizes. The number of particles of the smallest size goes to zero, because particles of infinitesimal size will shrink infinitely quickly. At the other extreme, we expect no particles above a maximum size, since very large particles take a very long time to appear.

14.6.2 Self-Similarity of the Microstructure during Coarsening

Assume the equations for diffusion and surface energy are valid over all spatial scales. In other words, the spatial and time coordinates of the coarsening process work in the same way when microstructures are rescaled in size. This motivates an approach to coarsening problems that concentrates on the relative sizes of the particles, not their absolute sizes.

A powerful assumption is that the distribution of particle sizes and shapes is "self-similar," meaning that the microstructures at different times look the same, although all dimensions are rescaled by a length ratio. In essence, the microstructure is magnified with time, without changing its characteristic appearance. For a self-similar microstructure, all features are proportional to an average radius $\bar{r}$, including those of Fig. 14.13

$$r_1 = k_1 \bar{r}, \quad r_2 = k_2 \bar{r}, \tag{14.36}$$

$$R = k_3 \bar{r}, \quad A = k_4 \bar{r}^2, \tag{14.37}$$

where $\{k_1, k_2, k_3, k_4\}$ are constants for all time. The ratio of distances $r_1/r_2 = k_1/k_2$ therefore remains constant during the coarsening of a self-similar microstructure because both r_1 and r_2 are proportional to $\bar{r}$. The flux to particle 1 from Eq. 14.29 is

$$\vec{J} = -\frac{3cM\sigma v}{k_3 \bar{r}} \left(\frac{1}{k_1 \bar{r}} - \frac{1}{k_2 \bar{r}} \right), \tag{14.38}$$

$$\vec{J} = -\frac{K_1}{\bar{r}^2}, \tag{14.39}$$

where we have collected all constants into K_1. Again, following Eq. 14.32, where the volume change is proportional to the flux, and the change in volume involves a geometrical constant K_2 (which was 4π for a sphere)

$$K_2 r_2^2 \frac{dr_2}{dt} = -JvA, \tag{14.40}$$

$$K_2 r_2^2 \frac{dr_2}{dt} = \frac{K_1}{\bar{r}^2} vk_4 \bar{r}^2, \tag{14.41}$$

$$\bar{r}^3 = K t. \tag{14.42}$$

All particle radii in a self-similar distribution grow as $t^{1/3}$.

A more difficult problem, beyond the present scope, is to calculate the mathematical form of the self-similar particle size distribution function. An important analytical treatment of coarsening was published in 1961 by Lifshitz and Slyozov [251] and independently by Wagner [252]. They confirmed the $t^{1/3}$ coarsening law, and found this nonintuitive functional form for the self-similar particle size distribution

$$f(\rho) = \frac{4}{9} \rho^2 \left(\frac{3}{3+\rho} \right)^{7/3} \left(\frac{1.5}{1.5-\rho} \right)^{11/3} \exp\left(\frac{\rho}{\rho - 1.5} \right), \tag{14.43}$$

where $\rho = r/\bar{r}$ is a dimensionless radius, and Eq. 14.43 is valid for $0 < \rho < 1.5$. With increasing time, the peak of the distribution (shown in Fig. 14.14) grows outwards to larger radius as $t^{1/3}$, and the full distribution follows by scaling the radial dimension (and decreasing the height of the distribution for normalization). This "LSW" distribution is strictly valid only for low volume fractions of particle, but it can be successful for volume fractions approaching 0.5.

A deeper analysis shows that the self-similar particle size distribution is stable once it has formed [253]. Its practical applicability requires that an initial distribution of particle sizes evolves quickly to the self-similar distribution. Also, this analysis does not account

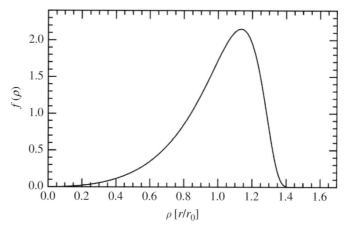

Figure 14.14 Grain size distribution function from LSW theory, presented with dimensionless radius. The function is negligible for $\rho > 1.4$ and strictly zero for $\rho \geq 1.5$.

for effects of elastic energy, and elastic energy does not scale simply with $\bar{r}$. Nevertheless, many experimental tests seem to confirm the self-similarity of the LSW distribution of Fig 14.14 during coarsening. Today, computer simulation is a more general method for calculating coarsening behavior [254], and X-ray tomographic methods promise more detailed experimental measurements in the near future.

Problems

14.1 Calculate the vacancy concentration in local equilibrium with a spherical void of radius r, specific surface energy γ, and with atomic volume v. The energy of formation of the vacancy in bulk material is E_f, and the temperature is T.

14.2 Suppose the grain size of an alloy is 1 μm, and precipitates, both homogeneous and heterogeneous, have sizes of 50 nm. Use the collector plate mechanism to answer the following.

(a) How much faster will a precipitate at the grain boundary grow, compared with a precipitate in the bulk, if the grain boundary diffusivity is infinite? (Assume one precipitate of each type per grain. You can also assume all shapes are spheres.)

(b) Suppose the grain boundary diffusivity is 100 times that of the bulk. What is the correction to your answer from part a?

14.3 As shown in Fig. 14.7, the austenite (γ) phase boundaries intersect at the eutectoid point at 732 °C and 0.77 wt.% C. It is a common practice to extrapolate these austenite phase boundaries to lower temperatures, and state that pearlite can form between these extrapolated lines. Justify this common practice.

(*Hint*: Assume that the austenite must be near local equilibrium with both the ferrite and cementite phases, as in Fig. 14.9b.)

14.4 **(a)** Using Fig. 14.9b, show that local equilibrium between the α-phase and the liquid gives more solute in the liquid for greater undercooling, and vice versa for the θ-phase. Why does this facilitate a smaller lamellar spacing with increased undercooling?

(b) With smaller lamellar spacing, there is an increase in the surface energy term of Eq. 14.1. Explain how this alters the relative positions of the free energy curves in Fig. 14.9b. Also explain how this could slow the growth rate of a pearlite colony.

14.5 Consider a lamellar eutectic microstructure where the β-phase is the minority phase. If the center-to-center separation between sheets of the β-phase is fixed at λ, the total surface area between β and α remains fixed even as the amount of β decreases and its lamellae are thinner. It may become favorable for the β-phase to form as rods when its fraction of volume is small. Assuming the rods grow in the arrangement of a square of edge length λ, what is the critical volume fraction of β in α at which the transition from lamellae to rods is favored by surface energy?

14.6 Consider a boundary between three grains in a pure material, where all three grain boundaries have specific surface energy σ. Orient the line of three-grain contact along $\hat{z}$. In the x–y-plane, the equilibrium configuration would have angles between the grain boundaries of $120°$, $120°$, $120°$.

(a) Suppose the boundaries between the pairs of grains make angles of $95°$, $95°$, $170°$. What is the magnitude and direction of the effective force on the line of three-grain contact, moving it towards equilibrium? Draw it.

Do the same analysis for a grain boundary junction closer to the equilibrium configuration, with angles of $115°$, $115°$, $130°$.

(*Hint*: See Fig. 4.4.)

(b) Suppose a polycrystalline material is prepared with a wide distribution of angles between boundaries. This material is heated moderately, or "annealed," so that the boundaries can move slowly. Assume a local relaxation towards equilibrium structures where each three-grain junction moves with a velocity $\vec{v} = M\vec{F}$, where M is a mobility (assumed constant), and $\vec{F}$ is a force.

Considering the results of part a, qualitatively describe the evolution during annealing of the distribution of grain boundary angles.

Calculate a time dependence for the relaxation of one grain boundary with angles of $(120 - \delta)°$, $(120 - \delta)°$, $(120 + 2\delta)°$ (i.e., calculate $\delta(t)$).

Phase Transformations with Interfaces: 2. Energetics and Kinetics

The previous chapter described the microstructural evolution in a material as it undergoes a phase transformation by nucleation and growth. Interface energy was used to explain, for example, features of lamellar pearlite colonies, and how their spatial scale is expected to be smaller at lower temperatures. Diffusion was not considered in detail, although it is expected to change as defects are removed, chemical segregations occur, and the phase transformation slows down. For general explanations, however, the diffusion processes were simplified to a single constant that depends on temperature. This chapter puts more emphasis on concepts that underlie precipitation phase transformations, specifically the diffusional motions of atoms from one phase to another across an interface, and the effects of elastic energy on an evolving microstructure.

Atoms move across an interface as one of the phases grows at the expense of the other. There is an energy penalty for having the interface, but the interface is an essential feature of having two adjacent phases. The interface has an atomic structure and chemical composition that are set by local thermodynamic equilibrium, but interface velocity constrains this equilibrium. Extending a thermodynamic treatment, we can understand fluxes through the interface by concentration gradients across it. When an interface moves at a high velocity, however, chemical equilibration by solute atoms does not occur in the short time when the interface passes by. These issues also pertain to rapid solidification, and extend the ideas of Chapter 13.

Solid–solid phase transformations also require consideration of elastic energy and how it evolves during the phase transformation. The balance between surface energy, elastic energy, and chemical free energy is altered as a precipitate grows larger. The optimal shape of the precipitate therefore changes as it grows. The chapter ends with some discussion of the elastic energy of interstitial solid solutions, with an application to metal hydrides.

15.1 Interface Thermodynamics and Kinetics

15.1.1 Interfaces in Phase Transformations

Here some features of a first-order phase transition in a binary alloy. Specifically, we consider how a new phase grows into its parent phase when the two phases differ in chemical composition [255–258]. This is a large topic with multiple degrees of freedom and multiple phenomena. For example, we know that the thermodynamic driving force for growth depends on the undercooling, but the rate of growth also depends on diffusion across the interface between the two phases. Diffusion alters the chemical profile, but the problem is more than a standard solution of the diffusion equation as in Sections 3.3 and 3.4. In the present case the interface is moving, perhaps rapidly. Concepts of incomplete diffusive mixing (Sections 5.2.2 and 5.3) are important. Some of these ideas can be extended to other relationships between the rate of growth of the new phase and the solute partitioning that accompanies it.

The focus of our analysis is on the interface between the parent phase β and the new phase α. Essential features are:

- There must be an interface whenever the α- and β-phases are together in the material.
- The growth rate of the new α-phase is set by the velocity of the interface. If the interface does not move, the new phase does not grow.
- The interface moves if A-atoms and B-atoms tend to go through it in opposite directions.
- As it moves, the interface keeps the same shape and chemical profile.

15.1.2 Diffusion across the Interface

Here we assume that the interface has a finite width.[1] We give the interface, or "I-phase," a $F_I(c)$ curve on the free energy versus composition diagrams of Fig. 15.1. Figure 15.1a shows the common tangent for the equilibrium between the α- and β-phases (presented in Section 2.3). The $F_I(c)$ curve for the interface is at a high free energy, so the interface would not be expected in equilibrium. The interface exists only because it is a geometrical requirement of having two adjacent phases.[2]

Figure 15.1a also shows a tangent to the $F_I(c)$ curve with same slope as the common tangent line for the α- and β-phases. Lines with parallel slopes in Fig. 15.1 have a special significance. Any point in the figure can be converted to free energy per atom, i.e., a chemical potential. As we follow a straight line from left to right, we replace A-atoms with B-atoms. At the extremes we have the chemical potential for pure A at $c = 0$, and the chemical potential for pure B at $c = 1$ (see discussion of Eqs. 2.13 and 2.14). This range of c is from 0 to 1, so the slope of the straight line, $d\mu/dc$, must equal the difference in chemical potentials of B-atoms and A-atoms

$$\frac{d\mu}{dc} = \mu_B - \mu_A \propto \frac{dF}{dc}, \tag{15.1}$$

where the constant of proportionality relates energy per mole to energy per atom, for example.

We now consider the diffusional fluxes of atoms through the I-phase that can have a concentration gradient across it. When we combine diffusional fluxes, J, with free energy curves, $F(c)$, there is an issue that we have been using dimensionless c for the latter, where $0 \leq c \leq 1$, but Fick's first law uses c with dimensions of [atoms/vol]. The tasteful solution

---

[1] There is, however, a large and successful body of work based on the concept of a "Gibbs excess," where the three-dimensional properties of the surface are ignored. Surface chemical segregation is expressed as moles per square meter, for example.

[2] The Gibbs phase rule allows a third phase in a T–c phase diagram of a binary alloy only at a special point such as the eutectic point, so the I-phase is not an equilibrium phase.

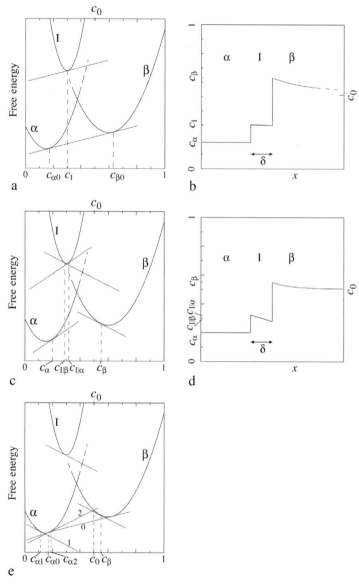

Figure 15.1 Left: free energy versus composition curves for two-phase alloy with interface. Right: steady-state profiles of composition versus position around the interface. Overall alloy composition is $c_0 = 0.5$; interface moves from left to right. (**a, b**) Equilibrium phase formation with slow interface velocity. Concentration gradient in interface is barely visible. (**c, d**) Rapid interface velocity. (**e**) Range of possible c_α with three tangent lines $\{1, 0, 2\}$ discussed in text.

is to modify the diffusion equations with density ρ with dimensions [atoms/vol]. Atom diffusion is driven by a gradient in chemical potential[3]

$$\vec{J} = -c\rho M \vec{\nabla}\mu , \qquad (15.2)$$

$$J(x) = -c\rho M \frac{d\mu}{dc}\frac{dc}{dx}, \qquad (15.3)$$

where a thin interface can exhibit one-dimensional behavior. The pair of parallel lines tangent to the I-curve and β-phase curve in Fig. 15.1a therefore have the same diffusive flux per concentration gradient

$$J(x) = -c\rho M \left(\mu_B - \mu_A\right)\frac{dc}{dx} = -c\rho M' \frac{dF}{dc}\frac{dc}{dx}. \qquad (15.4)$$

We cannot determine an equilibrium composition of the I-phase with the usual common tangent construction because it is not an equilibrium phase. We can, however, impose the condition of local equilibrium. In local equilibrium there is no thermodynamic drive for partitioning of B-atoms between the I-phase and the β-phase. This requires that the difference $\mu_B - \mu_A$ is the same in the two phases where they are in contact, which is the condition of parallel common tangents shown in Figs. 15.1a,c. The I-phase and the β-phase may have different chemical compositions, but neither prefers to change its composition more than the other, and further chemical segregation does not occur. Likewise, the same condition of local equilibrium is used for the I-curve and the α-phase curve on the other side of the interface. Note, however, in Fig. 15.1c, the common tangent curves on the two sides of the I-phase do not have the same slope. This will drive diffusion across the interface.

15.1.3 Net Flux across a Moving Interface

Now consider a steady diffusive flux through the interface, so the α-phase grows steadily into the β-phase. We assume the interface maintains the same composition profile as it moves, with local equilibrium where the I-phase contacts the α-phase and the β-phase. In general there is a concentration gradient within the I-phase (see Fig. 15.1d). To make the flux of B-atoms move to the right in Fig. 15.1d, the concentration c must be larger on the side of the I-phase in contact with the α-phase than on the side in contact with the β-phase. The compositions at the edges of the I-phase were obtained from Fig. 15.1c using the two tangent lines. Please confirm that the composition gradient in Fig. 15.1d is appropriate for an interface that moves to the right, so the α-phase grows at the expense of the β-phase. The new α-phase in Fig. 15.1d forms with a composition consistent with Fig. 15.1c. Although the parent β-phase has its favored composition near the interface, some time may be required for diffusion to homogenize the bulk of the β-phase.

---

[3] A simple chemical potential is $\mu = -k_B T \ln c$ (consistent with the Boltzmann factor relationship $c = \exp(-\mu/k_B T)$). By taking its derivative $d\mu/dc$, Eq. 15.3 becomes

$$J(x) = -Mk_B T \rho \frac{c}{c}\frac{dc}{dx} = -D\rho \frac{dc}{dx}, \qquad (15.5)$$

where we recovered Fick's first law of Eq. 3.16 by appropriate definition of the diffusion constant D (Eq. 10.82).

We cannot expect this thermodynamic analysis with Fig. 15.1 to account fully for the interface velocity. Nevertheless, we can develop a trend between the interface velocity and the solute partitioning. A key point involves conservation of atom density.[4] If there is a greater net flux out of one interface of the I-phase than its other interface, the center of the I-phase will move away from the first interface. (The alternative of changing the width of the I-phase is disallowed by the assumption of steady state.) The fluxes *from* the I-phase *to* the adjacent α and β-phases are

$$J^{\text{I}\|\alpha} = +c\rho M \left(\mu_B^{\text{I}\|\alpha} - \mu_A^{\text{I}\|\alpha} \right) \overrightarrow{\nabla c}, \qquad (15.6)$$

$$J^{\text{I}\|\beta} = -c\rho M \left(\mu_B^{\text{I}\|\beta} - \mu_A^{\text{I}\|\beta} \right) \overrightarrow{\nabla c}, \qquad (15.7)$$

where the flux from the I-phase to the α-phase at the I$\|\alpha$ interface is to the left (hence the sign change). The velocity of the interface to the right, v, is the net flux through the interface, normalized by density ρ [atoms/vol]

$$v = \frac{1}{\rho} \left(J^{\text{I}\|\alpha} + J^{\text{I}\|\beta} \right), \qquad (15.8)$$

$$v = cM \left[\mu_B^{\text{I}\|\alpha} - \mu_A^{\text{I}\|\alpha} - (\mu_B^{\text{I}\|\beta} - \mu_A^{\text{I}\|\beta}) \right] \overrightarrow{\nabla c}. \qquad (15.9)$$

These differences in chemical potential of A- and B-atoms are obtained from the tangent lines in Fig. 15.1 with Eq. 15.1. Finally, we obtain the gradient in c from the concentration difference across the interface $c_{\text{I}\|\alpha} - c_{\text{I}\|\beta}$, with an interface width of δ:

$$v = \frac{cM'}{\delta} \left(\left. \frac{dF_\alpha}{dc} \right|_{c_{\text{I}\|\alpha}} - \left. \frac{dF_\beta}{dc} \right|_{c_{\text{I}\|\beta}} \right) \left(c_{\text{I}\|\alpha} - c_{\text{I}\|\beta} \right). \qquad (15.10)$$

A faster v, and a more rapid growth of α-phase from the parent β-phase, could be driven by greater undercooling as follows. With faster v, we expect an increasingly large deviation of the compositions of the α-phase and β-phase from those of the common tangent of Fig. 15.1a and a larger difference in the slopes dF/dc in Eq. 15.10. In Fig. 15.1c,d there is a larger composition gradient across the interface, giving a larger diffusive flux.

15.1.4 Chemical Composition of the New Phase

The analysis to this point has related the interface velocity v to slopes of the tangent lines to the free energy curves, $F_\alpha(c)$ and $F_\beta(c)$. There are two pairs of tangent lines, however, and one mathematical relationship. The discussion so far has been silent about the mobility M, which was assumed a simple constant. The mobility likely differs for the different chemical species, and differs in the three phases. Quite possibly, one of the species, A or B, may be slow to move in one of the phases, and this slow process may dominate the interface velocity. A thermodynamic treatment cannot address this. Nevertheless, Fig. 15.1e allows us to address the allowable range of c_α. Tangent line 1 is parallel to the tangent line of the I-phase. If the composition $c_\alpha = c_{\alpha 1}$ (Fig. 15.1e), the side of the I-phase in contact with

---

[4] Actually the atoms are the conserved quantity, so if the volumes of A- and B-atoms differ significantly, corrections will be required.

the α-phase would have the same composition as the other side of the I-phase in contact with the β-phase. With no concentration gradient across the I-phase, there is no flux and no growth. Therefore $c_\alpha > c_{\alpha 1}$.

The tangent line 2 in Fig. 15.1e intersects the free energy curve of the β-phase $F_\beta(c)$ at c_0, the composition of the original parent β-phase. If this tangent line 2 had a greater slope, the free energy after the phase transformation would be larger than the initial material (recall that the total free energy for a mixture lies on the straight line between its end members). Therefore $c_\alpha < c_{\alpha 2}$. We find that the composition of the new α-phase lies between $c_{\alpha 1}$ and $c_{\alpha 2}$. Less solute partitioning is required if the α-phase has a composition closer to $c_{\alpha 2}$, however, and kinetics may favor less partitioning when the interface velocity is high.

A large concentration gradient across the interface drives more diffusive flux, and the interface moves faster. However, faster growth affects the compositions of the α- and β-phases. Comparing Fig. 15.1a,b to 15.1c,d, we see that as the composition difference across the interface becomes larger, the slopes of the tangents depart further from the common tangent at equilibrium (Fig. 15.1a), and the concentrations of the α- and β-phases become more similar.

The extreme case occurs when the compositions of the α-phase and β-phase are equal. The phase transformation is then "partitionless," with no solute partitioning between the α-phase and the β-phase. Partitionless phase transformations occur when the interface velocity exceeds a critical velocity

$$v_c \equiv \frac{M' k_B T}{\rho\, \delta} \left(\left.\frac{dF_\alpha}{dc}\right|_{c_c} - \left.\frac{dF_\beta}{dc}\right|_{c_c} \right) \left(c_{I|\alpha} - c_{I|\beta} \right), \tag{15.11}$$

where c_c is the composition of crossover of $F_\alpha(c)$ and $F_\beta(c)$ (approximately 0.38 in Fig. 15.1). This result for v_c is from a thermodynamic analysis, however, which assumes that all kinetic processes can be bundled into one constant M'.

15.2 Atomistic Model of Interface Motion

For a more detailed understanding of the kinetics of interface motion in a binary alloy, we recast the processes at the interface as forward and reverse atom jumps, using elementary

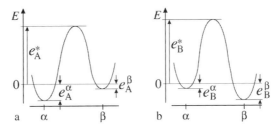

Figure 15.2 Activation barriers and phase preferences of A- and B-atoms. The minima are at atomic sites in the α- and β-phases, immediately adjacent to an interface. (**a**) Energies for A-atoms. (**b**) Energies for B-atoms.

jump processes based on the energetics shown in Fig. 15.2. The model is simplified by considering atom jumps directly between the α- and β-phases, without an I-phase for setting concentration gradients and diffusion. An activation energy e_A^* is required for an A-atom to reach the transition state between the two phases. The rate for an A-atom to jump from the α-phase to the β-phase is $\Gamma_A^{\alpha\beta}$. With analogous notations, the four elementary processes are

$$\Gamma_A^{\alpha\beta} = \Gamma_0 \, e^{-(e_A^* + e_A^\alpha)/k_B T}, \tag{15.12}$$

$$\Gamma_A^{\beta\alpha} = \Gamma_0 \, e^{-(e_A^* + e_A^\beta)/k_B T}, \tag{15.13}$$

$$\Gamma_B^{\alpha\beta} = \Gamma_0 \, e^{-(e_B^* + e_B^\alpha)/k_B T}, \tag{15.14}$$

$$\Gamma_B^{\beta\alpha} = \Gamma_0 \, e^{-(e_B^* + e_B^\beta)/k_B T}. \tag{15.15}$$

Four state variables are used with these four rates, $\{N_A^\alpha, N_A^\beta, N_B^\alpha, N_B^\beta\}$, which are atoms per unit area on the atomic planes at the interface. An important point is that there are no rules of conservation of sites because sites of the α-phase and the β-phase are being created or destroyed as atoms jump across the interface.[5] Conveniently, we can ignore differences in the initial chemical neighborhood of the moving atom because we assume a steady-state situation where the chemical compositions at the interface are constant.

The total flux of both species of atoms from the α-phase to the β-phase is

$$J_{tot}^{\alpha\beta} = +N_A^\alpha \, \Gamma_A^{\alpha\beta} - N_A^\beta \, \Gamma_A^{\beta\alpha} + N_B^\alpha \, \Gamma_B^{\alpha\beta} - N_B^\beta \, \Gamma_B^{\beta\alpha}. \tag{15.16}$$

We relate the (positive) velocity of the interface to the total flux of atoms through the interface, recognizing that the directions of flux and velocity are opposite

$$v = -\frac{1}{\rho} \, J_{tot}^{\alpha\beta}, \tag{15.17}$$

and in detail

$$v = -\frac{\Gamma_0}{\rho} \left[e^{-e_A^*/k_B T} \left(N_A^\alpha \, e^{-e_A^\alpha/k_B T} - N_A^\beta \, e^{-e_A^\beta/k_B T} \right) \right.$$
$$\left. + e^{-e_B^*/k_B T} \left(N_B^\alpha \, e^{-e_B^\alpha/k_B T} - N_B^\beta \, e^{-e_B^\beta/k_B T} \right) \right]. \tag{15.18}$$

Solute partitioning does not depend on the total flux through the interface but on the difference in fluxes of A-atoms and B-atoms, hence the opposite signs of the two terms in the square brackets in Eq. 15.19

$$J_{AB}^{\alpha\beta} = \Gamma_0 \left[e^{-e_A^*/k_B T} \left(N_A^\alpha \, e^{-e_A^\alpha/k_B T} - N_A^\beta \, e^{-e_A^\beta/k_B T} \right) \right.$$
$$\left. - e^{-e_B^*/k_B T} \left(N_B^\alpha \, e^{-e_B^\alpha/k_B T} - N_B^\beta \, e^{-e_B^\beta/k_B T} \right) \right]. \tag{15.19}$$

---

[5] This is unlike the situation of chemical ordering, for example, where the jumps of atoms from the α-sublattice to the β-sublattice must be compensated by jumps of atoms in the other direction. If the interface maintains a constant shape and composition as it moves, however, it is possible to impose additional constraints such as $N = N_A^\alpha + N_B^\alpha = N_A^\beta + N_B^\beta$.

It is not practical to explore all relationships between solute segregation and interface velocity with 10 free parameters in the full formulation of Eqs. 15.18 and 15.19. Fortunately, additional assumptions have been developed and tested [258, 259]. Sometimes the mobililty of the species differs considerably,[6] such as from differences of e_A^* and e_B^* in Fig. 15.2. Suppose that $e_A^* \gg e_B^*$, so the A-atoms are immobile and the kinetics are dominated by the motions of B-atoms across the interface. With $e^{-e_A^*/k_B T} = 0$, the first term in Eq. 15.19 vanishes and

$$J_{AB}^{\alpha\beta} = -\Gamma_0 \, e^{-e_B^*/k_B T} \left(N_B^\alpha \, e^{-e_B^\alpha/k_B T} - N_B^\beta \, e^{-e_B^\beta/k_B T} \right). \tag{15.20}$$

Combine three exponentials into two by making two definitions

$$G^* \equiv e_B^* + e_B^\alpha, \tag{15.21}$$

$$\Delta G \equiv e_B^\beta - e_B^\alpha. \tag{15.22}$$

Set the surface concentrations as equal to $N_0/2$ because we are interested in the dominant effect of the exponentials, but we could include the concentration dependence at one temperature by altering the exponentials

$$J_{AB}^{\alpha\beta} = -\frac{\Gamma_0 N_0}{2} \left(e^{-(e_B^*+e_B^\alpha)/k_B T} - e^{-(e_B^*+e_B^\beta)/k_B T} \right), \tag{15.23}$$

$$J_{AB}^{\alpha\beta} = -\frac{\Gamma_0 N_0}{2} \left(e^{-G^*/k_B T} - e^{-(G^*+\Delta G)/k_B T} \right). \tag{15.24}$$

With Eq. 15.24, our atomistic model considers (1) a B-atom jump from the α- to β-phase with a rate proportional to $e^{-\Delta G^*/k_B T}$, and (2) a B-atom jump from the β- to α-phase with a rate proportional to $e^{-(\Delta G^*+\Delta G)/k_B T}$. The activation barrier for the jump from β to α is larger because the B-atom is more favorably bound in the β-phase. The chemical flux from Eq. 15.24 is

$$J_{AB}^{\alpha\beta} = -\frac{\Gamma_0 N_0}{2} \left[e^{-G^*/k_B T} \right] \left[1 - e^{-\Delta G/k_B T} \right], \tag{15.25}$$

where the first factor in square brackets is the atom jump rate without chemical bias, and the second is a thermodynamic bias accounting for a reduced jump rate out of the β-phase, where the B-atom is more favorably bound.[7] The density is ρ, the atomic jump length is a, and their product is N_0, the number of atoms per unit area at the interface.

---

[6] Such an assumption is often used for carbon in iron, where carbon atoms have a much higher mobility than iron atoms.

[7] At small $\Delta G/k_B T$ we can linearize the second factor in Eq. 15.25

$$e^{-\Delta G/k_B T} \simeq 1 - \frac{\Delta G}{k_B T}, \tag{15.26}$$

and calculate the chemical flux assuming the β-phase is to larger x than the α-phase

$$J_{AB}^{\alpha\beta} = -\frac{\Gamma_0 N_0}{2} \left[e^{-\Delta G^*/k_B T} \right] [\Delta G/k_B T]. \tag{15.27}$$

The two square brackets are in the form of [flux][affinity], where the first factor is purely kinetic, and the second is a thermodynamic driving force.

The growth rate of the α-phase follows the chemical flux across the interface, so the interface velocity is $J_{AB}^{\alpha\beta}$ normalized by density

$$v = -\frac{J_{AB}^{\alpha\beta}}{\rho} = \frac{\Gamma_0\, a}{2}\left[e^{-G^*/k_B T}\right]\left[1 - e^{-\Delta G/k_B T}\right].$$

(15.28)

Using Eq. 3.18 for the diffusion coefficient, i.e., $D(T) = 1/6\,\Gamma\, a^2$

$$v = \frac{3\,D(T)}{a}\left[1 - e^{-\Delta G/k_B T}\right].$$

(15.29)

It can be shown (Problem 15.4) that in the limit of small $\Delta G/k_B T$, Eq. 15.29 is equivalent to Eq. 15.9, but we expect Eq. 15.29 to be more appropriate at rapid velocities where kinetic issues play a larger role. Consider two limits for Eq. 15.29

- If there is no chemical bias, then $\Delta G = 0$ and $v = 0$. There is no thermodynamic driving force to grow the α-phase, and it does not grow.
- If $\Delta G/k_B T$ is large, $v = 3D(T)/a$. This is the maximum possible velocity of the interface, since every B-atom jump contributes to interface motion. (Recall that A-atoms are assumed immobile.) Since ΔG does not appear in the expression for v, it involves no solute partitioning.

To estimate an interface velocity high enough to suppress solute partitioning, assume $D = 10^{-8}\ \mathrm{cm^2/s}$, typical of metals near their melting temperatures. For large ΔG we obtain $v = 3\ \mathrm{cm/s}$. (This is essentially v_D of Section 15.3.3.) Diffusion coefficients decrease exponentially with temperature, so solute partitioning is expected to be suppressed at lower velocities. This process has been called "solute trapping" [259]. Experimentally, however, it has been shown that solid–liquid interfaces can move even faster than this, sometimes up to 100 m/s. Evidently these high-velocity interfaces do not depend on atomic diffusion, but involve atom rearrangements by small displacements at the interface. These motions are not able to accommodate the equilibrium segregation of solutes across the interface, of course.

15.3 Local Nonequilibrium at Fast Interfaces

15.3.1 Assumption of Local Equilibrium

In Section 5.3, departures from macroscopic equilibrium were accommodated by suppressing the diffusion in the solid and liquid phases. Nevertheless, local equilibrium was still assumed at the solid–liquid interface. With the assumption of local equilibrium, the jump in chemical composition across a moving interface was obtained with the partitioning ratio, k, obtained from the compositions of the solidus and liquidus lines on the equilibrium phase diagram. Section 15.1 used free energy relationships to predict concentration gradients across interfaces and interface velocities. Activated state rate theory, as in the previous

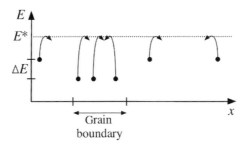

Figure 15.3 Potential energy of solute atoms in and near a grain boundary. Solutes in the boundary have potential energies lowered by ΔE, and higher activation energies for diffusive jumps.

Section 15.2, is also based on the assumption of local equilibrium between initial states and activated states (illustrated with Fig. 15.2).

This section treats the motion of a phase boundary that is controlled by solute atom diffusion until it moves so fast that local chemical equilibrium is not possible. More specifically, Section 15.3.2 explains the slowing of a moving interface by "solute drag," which was originally envisioned to apply to dilute solutes in a nearly pure material. Section 15.3.3 explains the transition from diffusion-limited to kinetic-limited motions of phase boundaries.

15.3.2 Description of Solute Drag

The motions of grain boundaries and grain growth in pure materials are often significantly faster than for grain boundaries in impure materials. J.W. Cahn made an early advance by considering an attractive potential for impurities at a grain boundary [260]. In essence, an attactive energy $\Delta E(x)$ is added to the chemical potential of a solute atom when it is positioned in or near the grain boundary. Figure 15.3 depicts more solutes in the more favorable potential of a grain boundary, and indicates that these solutes have higher activation barriers for diffusive jumps. If the solutes move along with the grain boundary, there is a retarding force called "solute drag."[8]

First consider how a quick motion of the grain boundary away from a solute atom would increase the chemical potential of the solute atom, so a solute atom exerts a force $-\Delta E/\Delta x$ that retards the motion of the grain boundary as it moves the distance Δx and no longer contains the solute atom. When the solute moves along with the grain boundary, it continues to need the extra energy ΔE to overcome a potential barrier for diffusion (as in Figs. 15.2 and 15.3).

When the solute atom stays with the moving grain boundary, the activation energy of solute diffusion is dissipated as heat. This energy cannot be re-used for another diffusive jump, so each jump requires the extra energy ΔE. As the grain boundary moves faster, more solute atoms per unit time jump along with the boundary, and energy dissipation is faster. The drag force therefore increases with velocity, at least for modest velocities of the

---

[8] This retardation drag differs from the "solute drag" of Section 10.5.2, where solute atoms were pulled by migrating vacancies.

grain boundary. Incidentally, the behavior is similar if the sign of ΔE is switched so that the solutes avoid the grain boundary – in this case the solutes jump out of the way from the advancing interface, and these extra jumps dissipate energy.

At high velocities of the interface, the solute atoms are unable to keep up. They get left behind if the interface moves faster than the solutes can jump along with the interface, and their favorable chemical potential no longer contributes to solute drag. What this means is that the solute drag increases with interface velocity up to a maximum velocity. Beyond this maximum, there are fewer solute atoms that move with the interface, and the solute drag force diminishes. As the grain boundary outruns the solute atoms, the condition of local equilibrium no longer applies to the solute concentration at the grain boundary. The solute drag effect goes away.

15.3.3 Analysis of Local Nonequilibrium and Solute Drag

Composition Profiles and Diffusion near Interfaces

The solute concentration near a rapidly moving interface is a topic of research today [261–264]. It is important for the nonequilibrium processing of materials by rapid solidification, for example, and brings new concepts beyond those of Chapter 13. For a fast-moving interface between two phases α and β, some important questions are

- What is the relationship between the velocity of the α/β interface and the chemical partitioning?
- How abruptly does the solute drag diminish with interface velocity?
- How does the interface velocity depend on undercooling?

For a starting point, consider Eq. 13.6 from Chapter 13

$$\frac{\partial c}{\partial t} = D\frac{\partial^2 c}{\partial x^2} + v\frac{\partial c}{\partial x}, \tag{15.30}$$

which gives the concentration c with coordinate x that is affixed to the interface moving with velocity v. The concentration profile around the interface had the steady-state solution in local equilibrium (shown in Fig. 13.7)

$$c(x) = c_0 + \left(\frac{c_0}{k} - c_0\right)e^{-xv/D}, \tag{15.31}$$

where the partitioning ratio k was obtained from the equilibrium phase diagram. Intuitively, we may expect that, at a high interface velocity v, the value of k increases from its equilibrium value, reaching 1 if the interface moves so fast that chemical partitioning does not occur. In this limit of $k = 1$, Eq. 15.31 gives $c(x) = c_0$, or no partitioning.

The change in the partitioning ratio at high interface velocities is caused by the finite time, τ, between diffusional jumps of solute atoms. If an interface moves very rapidly, τ may be too long for solute atoms to keep up. This section gives one approach to understanding the changes in diffusion local to a moving interface, and how it changes when the interface velocity reaches the "diffusion velocity," v_D

$$v_D = \frac{1}{6}\Gamma a, \tag{15.32}$$

which gives the characteristic time, τ

$$\tau = \frac{6}{\Gamma} = \frac{D}{v_D^2},$$

(15.33)

using the expression $D = \Gamma a^2/6$ (Eq. 3.18) to obtain Eq. 15.33. The diffusion velocity v_D is the highest velocity of an interface that solutes can follow as they jump the distance a in the time $1/\Gamma$. Equation 15.32 also points out that only 1/6 of the jumps of solute atoms can accommodate the direction of the moving interface.

Solute Response Time

When an interface moves slowly, the solute jumps are fast enough to maintain local equilibrium, and diffusion across the interface is not suppressed by the velocity of the interface. At the other extreme, an interface could move so fast that a solute atom cannot jump in the short time when the interface is in its vicinity. We modify Fick's first law to account for the time of the solute response

$$J + \tau \frac{\partial J}{\partial t} = -D \frac{\partial c}{\partial x}.$$

(15.34)

The new second term on the LHS is negligible when the flux changes over a time much greater than τ, but it becomes a large fraction of the flux when J undergoes a significant change in a time comparable to τ. (The flux from this new term goes down the concentration gradient, of course.) Preparing to obtain the diffusion equation, we take the spatial derivative of Eq. 15.34

$$\frac{\partial J}{\partial x} + \tau \frac{\partial}{\partial x} \frac{\partial J}{\partial t} = -D \frac{\partial^2 c}{\partial x^2},$$

(15.35)

$$\frac{\partial J}{\partial x} + \tau \frac{\partial}{\partial t} \frac{\partial J}{\partial x} = -D \frac{\partial^2 c}{\partial x^2},$$

(15.36)

noting the change in order of derivatives in Eq. 15.36. We now use the divergence theorem to replace two instances of $\partial J/\partial x$ with $-\partial c/\partial t$

$$\frac{\partial c}{\partial t} + \tau \frac{\partial^2 c}{\partial t^2} = D \frac{\partial^2 c}{\partial x^2}.$$

(15.37)

Equation 15.37 is the diffusion equation with a new term, a second time derivative of composition. This is an acceleration of the composition change, and τ is much like an inertial mass that diminishes the importance of this acceleration when τ is small. For typical diffusion problems with slow or stationary interfaces, we set $\tau = 0$ because jump dynamics are fast enough to ignore.

Coordinate Frame of Moving Interface

The case of interest now, however, is a moving interface that remains near a solute atom for only a short time. The situation is a transient one where the solute needs to respond to a quick change in its environment. Following the approach of Section 5.3 for an interface moving with a velocity v, it is useful to change to a coordinate frame affixed to the moving interface. Specifically, x is replaced with $\eta \equiv x - vt$. With this substitution for x, we see that position derivatives remain unchanged in form, but time derivatives gain the term

$$\frac{\partial c}{\partial x} = \frac{\partial c}{\partial \eta} \frac{d\eta}{dx} = \frac{\partial c}{\partial \eta}(+1),$$ (15.38)

$$\frac{\partial c}{\partial t} = \frac{\partial c}{\partial \eta} \frac{d\eta}{dt} = \frac{\partial c}{\partial \eta}(-v).$$ (15.39)

Comparing Eqs. 15.38 and 15.39 we find

$$\frac{\partial c}{\partial t} = -v \frac{\partial c}{\partial x},$$ (15.40)

$$\frac{\partial^2 c}{\partial t^2} = v^2 \frac{\partial^2 c}{\partial x^2}.$$ (15.41)

Suppression of Diffusion at the Moving Interface

A stationary solute atom would see a composition variation as the interface moves by. The composition profile moves past the solute with velocity v, much like a wave that is implied by the wave equation form of Eq. 15.41. The second term in the diffusion equation of Eq. 15.37 is the one that includes the solute transient response to the moving interface. Because we want to follow the solute profile in the frame of the moving interface, we transform this second term from a t-dependence in the laboratory frame to the x-coordinate of the moving interface.[9] Using the coordinate transformation of Eq. 15.41 for the second term in Eq. 15.37

$$\frac{\partial c}{\partial t} = -\tau v^2 \frac{\partial^2 c}{\partial x^2} + D \frac{\partial^2 c}{\partial x^2} + v \frac{\partial c}{\partial x}.$$ (15.42)

The last term $v \, \partial c / \partial x$ comes from the sweeping of the interface through a material with a variable composition, and does not depend on the solute response time τ. (It appears in Eq. 15.30, for example.) For simplicity in what follows, we assume the initial composition in the material is uniform and $\partial c / \partial x = 0$. Using Eq. 15.33 for τ

$$\frac{\partial c}{\partial t} = D \left(1 - \frac{v^2}{v_D^2}\right) \frac{\partial^2 c}{\partial x^2},$$ (15.43)

$$\frac{\partial c}{\partial t} = D^* \frac{\partial^2 c}{\partial x^2},$$ (15.44)

where D^* is an effective diffusion coefficient that accounts for the reduced number of atoms that can jump diffusively in the time when the interface moves past them. With Eq. 15.32, another form of Eq. 15.43 is

$$\frac{\partial c}{\partial t} = \left(D - 6 \frac{v^2}{\Gamma}\right) \frac{\partial^2 c}{\partial x^2},$$ (15.45)

---

[9] We do not transform the first term on the LHS of Eq. 15.37 from t to x because it is the composition evolution at a location within the moving interface.

which shows more clearly how v reduces the opportunity for solute atoms to make a diffusive jump at the interface. For even higher interface velocities when $v \geq v_D$, we set $D^* = 0$ (D cannot be negative because entropy increases with time). Without diffusion across the interface there is no chemical partitioning. Also at high velocities, the chemical potential of an atomic species is discontinuous across the interface because local equilibrium is not preserved.

Velocity Transition with Undercooling

Higher velocities of a two-phase interface are typically obtained by increasing ΔT, the difference between T and T_c, the critical temperature of the phase transition (e.g., increasing the undercooling below the melting temperature T_m). The thermodynamic driving force increases with undercooling, increasing the interface velocity. Interesting discontinuities have been observed in the interface velocity, however, as shown in Fig. 15.4. The figure shows how the solidification of pure Cu is relatively fast, compared to the Cu–Ag alloy, especially at low undercoolings (small ΔT). At a critical undercooling of 115 K, the interface velocity in Cu–Ag jumps by a factor of 20 or so. This velocity of approximately 0.1 m/s corresponds approximately to the condition $v = v_D$. Above this velocity, solute drag is no longer effective, and the solidification velocity becomes more like that of a pure metal. This sudden discontinuity in growth rate is suggestive of a cooperative effect, much like a phase transition.

There is a related phenomenon for the solidification of an alloy that is chemically ordered in equilibrium. "Disorder trapping" occurs with the transition to high interface velocity, where the solid is formed as a solid solution without the sublattice occupancies of chemical species as expected in equilibrium. This is again a consequence of solute jump rates being too low to maintain local equilibrium at a fast-moving interface. Finally, the grain structure of the solid also undergoes a change with the transition in interface velocity. Grain refinement, an abrupt reduction in crystal size, often occurs when the solidification front makes a transition to high velocity.

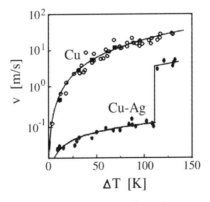

Figure 15.4   Interface velocities, deduced from recalescence times across a small sample, versus undercooling of pure Cu metal and Cu–Ag alloy. A critical value of $\Delta T = 115$ K is found for the alloy. Reprinted from [262], with permission from Elsevier.

15.4 Elastic Energy and Shape of Growing Plate-Like Precipitates

Figure 15.5 shows images of plate-like precipitates seen edge-on. They are thin in one dimension, and wide in two (one wide dimension extends upwards, out of these 2D images). It is obvious that these precipitates are not shaped at all as spheres, as would be expected if they were to minimize their surface energy in a cubic matrix. Here elastic energy plays a bigger role in determining shape than does surface energy. Approximately, the surface energy E_s promotes a thickening of the precipitate, and the elastic energy E_{el} tends to keep it thin in one dimension, effectively widening it. Some issues of preferred shapes for precipitates were mentioned in Section 6.8.5, but as the precipitate grows, there is a change in balance between surface energy and elastic energy. This section shows how E_{el}, which increases rapidly with the distortion caused by the precipitate, increasingly favors flat precipitates as precipitates grow larger. We assume no loss of coherency between the precipitate and matrix, but coherency loss often occurs to reduce the elastic energy around larger precipitates.

Although both E_{el} and E_s are positive, their relative importance changes with the volume of the precipitate particle. Consider a relatively simple case where the precipitate is

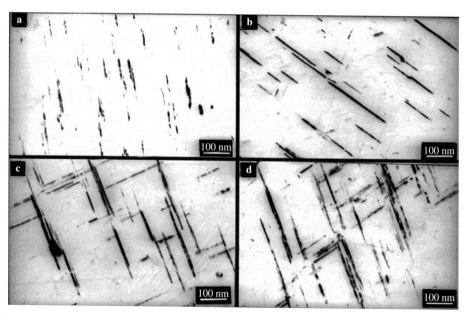

Figure 15.5 Transmission electron microscopy images showing plate-like precipitates in a Al−Cu−Mg alloy annealed for increasing times at the same temperature. The plates widen and thicken as they grow, while maintaining a crystallographic orientation relationship with the Al matrix. Reprinted from [265], with permission from Elsevier.

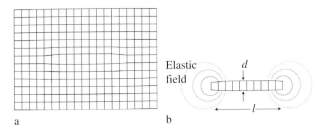

Figure 15.6 (**a**) Preferred shape of a tetragonal transformation product if good lattice matching is along horizontal and out-of-plane directions as described in text. (**b**) Redrawn precipitate from panel a with features labeled.

tetragonal and the parent phase is cubic. The stress-free transformation strain[10] for the z-variant of a tetragonal transformation is

$$\epsilon_{ij}^0 = \begin{bmatrix} \epsilon_1 & 0 & 0 \\ 0 & \epsilon_1 & 0 \\ 0 & 0 & \epsilon_3 \end{bmatrix}, \tag{15.46}$$

so the strain ϵ_1 is equal along $\hat{x}$ and $\hat{y}$, but the strain along $\hat{z}$ is different. (Incidentally, if $\epsilon_3 = \epsilon_1$, Eq. 15.46 would describe a cubic-to-cubic transformation.) For simplicity, assume that $\epsilon_1 \sim 0$, so the only significant transformation strain is the fractional change in length ϵ_3 along the z-axis. Without surface energy, a situation like Fig. 15.6 is expected, where the size of the precipitate d is minimized in the $\hat{z}$-direction of the large mismatch, thereby minimizing the elastic energy that goes as $(d\,\epsilon_3)^2$ (times geometrical factors).

An important point evident from the geometry of Fig. 15.6a is that the elastic distortion is largest around the edge of the precipitate. This region has a volume that scales as the circumference, πl, where l is the long direction in the plane of the plate. The elastic energy scales with the length of circumference, and is multiplied by $C\,(d\epsilon_3)^2$, where C includes elastic moduli and geometrical factors. An E_{tot} is the sum of this elastic energy and a surface energy from two flat circular surfaces with specific surface energy σ_s

$$E_{\text{tot}} = C\pi\,l\,d^2\,\epsilon_3^2 + 2\pi\sigma_s(l/2)^2. \tag{15.47}$$

We seek the thickness of the precipitate versus the volume of the precipitate, $V = d\,\pi\,(l/2)^2$, so l is replaced with $\sqrt{(4V)/(\pi d)}$

$$E_{\text{tot}} = C\pi\sqrt{\frac{4V}{\pi}}d^{3/2}\,\epsilon_3^2 + 2\sigma_s\frac{V}{\pi d}. \tag{15.48}$$

---

[10] The transformation strain describes the change in shape of a volume of parent material when transformed. A "stress-free" transformation occurs in free space, without the parent material around it to provide elastic constraint.

For a given volume, V, the total energy is a minimum for the optimal thickness, d

$$\frac{dE_{\text{tot}}}{dd} = 0 = 3C\sqrt{\pi V d}\,\epsilon_3^2 - \frac{2\sigma_s V}{\pi d^2}, \tag{15.49}$$

$$d^5 = \left(\frac{2\sigma_s}{3C}\right)^2 \frac{V}{\pi^3 \epsilon_3^4}, \tag{15.50}$$

so the thickness d increases as $V^{1/5}$. Consequently the length l must increase as $V^{2/5}$ because $V \propto d\, l^2$. The aspect ratio of the precipitate, d/l, therefore changes as $V^{-1/5}$, and the aspect ratio decreases as the precipitate grows. Notice that Eq. 15.50 predicts more thickening for larger surface energy, which appears with the elastic energy in the ratio of $\sigma_s/(C\,\epsilon_3^2)$. For plate-like precipitates, the relatively high elastic energy causes the precipitate to change its shape as it grows larger, thickening somewhat more slowly than it widens. It seems that the precipitates in Fig. 15.5 grow in width more than in thickness, for example.

A closely related concept is that the presence of a second precipitate in the matrix will alter the strain field around the first precipitate. Often the mutual arrangements of precipitates are determined in part by how they can configure themselves to minimize the total elastic energy. Of course, the diffusion process also sets a distance between precipitates because they need to collect solute from the matrix, and a second precipitate may not form in a location that has been depleted of solute during the formation of the first precipitate. The full competition between surface energy, elastic energy, and diffusion is complicated, and is appropriate for computer simulations.

15.5 Elastic Energy and Solute Atoms

15.5.1 Context

In the days before computational quantum mechanics, analytical solutions for continuum elasticity were available and celebrated. A feature of continuum mechanics is that it has no intrinsic spatial length such as a Bohr radius, so there is no reason why it cannot make predictions at the atomic scale. The question, of course, is how reliable are continuum calculations of atomic properties, since the true behavior is set by quantum mechanics and not classical mechanics. Nevertheless, the twentieth century saw a large number of atomistic studies based on continuum mechanics, such as the model of alloy solid solutions described next. The solute atoms are assumed to be small, misfitting spheres, and the surrounding crystal is an elastic continuum. This viewpoint is consistent with the second Hume-Rothery rule of Section 2.1.4 on metallic radius, for example. It does give some useful insights, and some semiquantitative results.

15.5.2 Components of the Elastic Model

When a solute atom or an interstitial atom fits perfectly on its site in a crystal, there are no displacements of neighboring atoms, and there is no elastic distortion in the matrix around the inserted atom. The elastic energy is zero in this very special case. This is not true in general, of course. When there is misfit of solute atoms or interstitial atoms on their sites in a crystal, elastic energy is expected. It can be separated into these contributions:

1. The "self-energy" of the misfitting atom is the energy cost to insert one atom into the crystal (replacing either an interstitial, vacancy or a matrix atom). This contribution scales with the number of misfitting atoms, but does not include their mutual interactions.
2. Elastic interaction between each misfitting atom and the surface of the material. This is called an "image force," in analogy to an electrostatic interaction near a conducting surface.[11]
3. The interactions between the misfitting atoms in the material. Such interactions could cause the misfitting atoms to cluster together, or spread apart. Surprisingly, this contribution can be zero owing to the "Bitter–Crum theorem" described below.
4. Higher-order interactions between misfitting atoms, or between misfitting atoms and atoms of the matrix. These are expected if distortions are large, and the elastic response differs from the linearity of Hooke's law. It is typical to assume these contributions to be zero, however, and we will do so owing to a lack of information about higher-order elasticity.

Item 1 on the list is often calculated with methods described in Section 6.8 – the elastic energy of each misfitting atom is treated as that of an elastic sphere inserted into a hole in a continuum. Item 2 on the list is ignored because the material is assumed infinite. Item 3 on the list is much more subtle, as discussed next.

15.5.3 Elastic Energy of a Solid Solution

There are two limiting concentrations for misfitting solute atoms where we expect minimal elastic energy. The first is obvious – when the concentration of misfitting atoms $c = 0$. The other limit, perhaps less obvious, is when $c = 1$, so all sites are filled with B-atoms.[12] At $c = 1$ there is a new unit cell that fills space without internal stresses.

There is an elastic energy penalty for intermediate concentrations of misfitting atoms. For example, if all misfitting atoms were placed into one spherical zone inside an elastic matrix, we expect the usual energy penalty of a misfitting particle (Section 6.8). The simplest functional form for the elastic energy of misfitting atoms goes as $c(1 - c)$, and it can be shown [266, 267] that

$$E_{\mathrm{el}3} = 2NV_0\mu\,\frac{1 + \nu}{1 - \nu}\,\eta^2\,c(1 - c), \tag{15.51}$$

---

[11] In electrostatics, a conductive surface is at a constant potential, so field lines are perpendicular to the surface. This imposes a symmetry on the field lines from a positive charge near a thin, conductive plate that is equivalent to placing a negative charge, an "image" of the positive charge, at the same distance on the other side of the plate. For elasticity, there is no net force normal to a free surface. When a center of dilatation is near a free surface, its elastic field in the material is equivalent to its field in a continuous material when a center of opposite dilatation is located symmetrically on the other side of the surface.
[12] We assume the macroscopic crystal expands or contracts freely as solute atoms are added.

where N is the number of misfitting solutes, V_0 is the volume per misfitting atom, v is the Poisson ratio (often about 1/3), η is the concentration dependence of the lattice strain ($\eta \equiv 1/a\, \mathrm{d}a/\mathrm{d}c$, where a is the lattice parameter), and μ is the shear modulus. There is an implied expectation that the lattice parameter depends linearly on the solute concentration.[13] The energy of Eq. 15.51 is a maximum at $c = 1/2$. It goes to zero at $c = 0$ and $c = 1$ because the crystals have no internal distortions.

Box 15.1   **Equivalence to Eshelby Formalism for a Misfitting Sphere**

The compact formulation of Eq. 15.51 by Khachaturyan [267] combines the compressive stress inside the sphere (that depends on bulk modulus B) with the shear stress in the surrounding matrix (that depends on the shear modulus μ) by using relationships such as

$$C_{11} = 2\mu \,\frac{1-v}{1-2v}, \quad C_{12} = 2\mu \,\frac{v}{1-2v}, \tag{15.52}$$

and Eqs. 6.118 and 6.119 for B and μ. For one atom in a big matrix where $c(1-c)N = 1$, some algebra will show the equivalence of Eqs. 6.141 and 15.51, recognizing that $3\eta = \delta V/V$. Note that the bulk modulus, and compression of the misfitting sphere (or atom), is hidden by use of the Poisson ratio v.

15.5.4 The Bitter–Crum Theorem

Suppose these conditions are satisfied:

1. The matrix and precipitates have the same elastic constants.
2. The misfit strain is isotropic (i.e. dilatational).
3. The elasticity tensor is isotropic.
4. The crystal is infinite.
5. The elasticity is linear.
6. Vegard's law is valid.

Under these six conditions a surprising result can be derived – the elastic energy does not depend on the configurations of the misfitting atoms. The solute atoms can be clustered into precipitate zones, or dispersed at random through the matrix, and the elastic energy remains the same. This is the "Bitter–Crum theorem" [268–270]. For example, the two configurations of interstitial atoms shown in Fig. 15.7a,b have the same elastic energy, given by Eq. 15.51.

Here we test this theorem for an interstitial alloy with two phases as shown in Fig. 15.7a having a concentration of interstitials c_α for the A-rich α-phase and c_β for the B-rich β-phase, with an overall concentration

$$\bar{c} = (1-f)c_\alpha + fc_\beta, \tag{15.53}$$

---

[13] A proportionality between solute concentration and lattice parameter is called "Vegard's law," which tends to be a successful approximation for fcc metal alloys. It is too inconsistent to be a real "law," however.

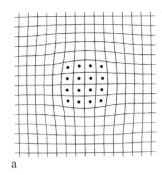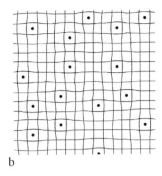

a   b

Figure 15.7   (**a**) Clustering of misfitting interstitial atoms into a precipitate. (**b**) Redistribution of the same misfitting atoms.

where f is the fraction of the β-phase. Each of these phases has an elastic energy of Eq. 15.51

$$E_{\text{el}\alpha} = 2N_\alpha V_0 \mu \frac{1+\nu}{1-\nu} \eta^2 c_\alpha (1 - c_\alpha), \tag{15.54}$$

$$E_{\text{el}\beta} = 2N_\beta V_0 \mu \frac{1+\nu}{1-\nu} \eta^2 c_\beta (1 - c_\beta). \tag{15.55}$$

There is also an elastic energy for placing the oversize β-phase precipitate in the α-matrix, and the elastic energy of this misfitting sphere also has the form of Eq. 15.51

$$E_{\text{el}\alpha/\beta} = 2N V_0 \mu \frac{1+\nu}{1-\nu} [\eta(c_\beta - c_\alpha)]^2 f(1 - f). \tag{15.56}$$

The total energy is the sum of the three contributions

$$E_{\text{tot}} = E_{\text{el}\alpha} + E_{\text{el}\beta} + E_{\text{el}\alpha/\beta}. \tag{15.57}$$

The number of interstitial sites within the α-phase is $N_\alpha = (1-f)N$, and likewise $N_\beta = fN$.

$$E_{\text{tot}} = 2N V_0 \mu \frac{1+\nu}{1-\nu} \eta^2 \left[(1-f)c_\alpha(1 - c_\alpha) + f c_\beta(1 - c_\beta) \right.$$

$$\left. + (c_\beta - c_\alpha)^2 f(1 - f) \right]. \tag{15.58}$$

With a little algebra it can be shown that the terms inside the large square brackets are equal to $\bar{c}(1 - \bar{c})$ (with $\bar{c}$ from Eq. 15.53)

$$E_{\text{tot}} = 2N V_0 \mu \frac{1+\nu}{1-\nu} \eta^2 \bar{c}(1 - \bar{c}). \tag{15.59}$$

This analysis of the misfitting precipitate of Fig. 15.7a gave the same total elastic energy as for the misfitting B-atoms in the solid solution of Fig. 15.7b (i.e., Eq. 15.51 is the same as Eq. 15.59). This Bitter–Crum theorem depends on the validity of the six assumptions at the beginning of this section [268–270]. This analysis predicts no role of elastic energy in unmixing transformations. Unmixing the solutes into two phases reduces the average size mismatch of the unit cells within each phase, but there will be a larger elastic energy between the two phases. If coherence at the interface is lost, however, unmixing will be favored because of the better homogeneity within each phase.

In real materials the assumptions behind the Bitter–Crum theorem are not generally realized, so elastic energy does affect the mutual interactions and configurations of misfitting atoms. We know that anisotropic transformation strains often cause precipitates to form as thin plates (Fig. 15.5), and with preferred mutual orientations (Figs. 9.6 and 19.8), owing to a violation of assumption 2 above.

15.5.5 Coherency Stresses

The misfit of an elastic inclusion generates elastic energy, as discussed in Sections 6.7.4 and 9.5. For an oversized ellipsoidal particle, compressed to fit back into its original matrix, there is a hydrostatic compression in the misfitting particle and a shear strain in the surrounding matrix (Fig. 15.7a). The same is true for a particle that transforms to a smaller size – it pulls the matrix inwards, perhaps as shown in Fig. 15.8a. Figure 15.8b illustrates how it is possible for the interface to break free, disrupting the continuity of the lattice planes. Now the particle is able to add or subtract lattice sites to relax the coherency stresses. There is an energy penalty, however, for creating the new surface between the particle and matrix. The surface-to-volume ratio is especially large for small particles, suggesting that very small particles and critical nuclei remain coherent with the matrix.

There is some ambiguity about definitions, but here we consider misfit stresses as originating from elastic particles in an elastic continuum, without consideration of the details of the interface between particle and matrix. We define "coherency strains" as strains that depend on the lattice continuity across the interface, as shown in Fig. 15.8a. There is elastic energy from the coherency stresses and strains that increases with the volume of the precipitate ($\sim r^3$)

$$E_{\text{coh}} = \frac{1}{2} \kappa r^3 (h \delta_0)^2, \tag{15.60}$$

where κ includes elastic constants and a geometric factor, δ_0 is the fractional lattice mismatch between the precipitate and the matrix, and h is the degree of coherence, defined as the fraction of the lattice mismatch that contributes to the elastic energy.

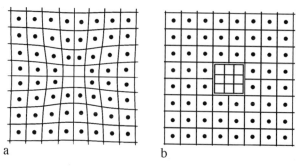

a   b

Figure 15.8 (a) Coherent interface around small precipitate showing strains. (b) Loss of coherence, showing reduced strains, more lattice sites, and a new interfacial region.

For an incoherent interface $h = 0$, but for a fully coherent interface $h = 1$. The full problem is quite beyond the present scope, and depends on details of how coherence is lost.

Consider an example where coherence is lost by forming dislocations at the interface, with the dislocation density being proportional to $1 - h$. (When $h = 1$ there are no dislocations at the interface, but when $h = 0$ there are enough dislocations to fully counteract the fractional lattice mismatch, δ_0.) The dislocations give an energy that scales with the amount of surface

$$E_{\text{surf}} = \sigma\, r^2\, (1 - h), \tag{15.61}$$

where σ contains a geometric factor and a surface energy. The total energy of the interface, E_{int}, is the sum of Eqs. 15.60 and 15.61

$$E_{\text{int}} = \frac{1}{2}\, \kappa\, r^3\, (h\, \delta_0)^2 + \sigma\, r^2\, (1 - h). \tag{15.62}$$

Minimizing E_{int} with respect to h (i.e., $dE_{\text{int}}/dh = 0$), gives

$$h_0 = \frac{\sigma}{\kappa}\, \frac{1}{r}\, \frac{1}{\delta_0^2}. \tag{15.63}$$

This optimum interface coherence h_0 changes with precipitate size – coherence is expected to be lost as r becomes larger. Likewise, the larger the lattice mismatch, δ_0, the lower the coherence. A surface energy in σ promotes interfacial coherence, whereas a larger shear modulus within κ works against coherence.

The coherency parameter need not be h_0, however. In nucleation and growth, precipitates often remain coherent longer than expected. Breaking coherence at the interface requires the formation of new dislocations, which requires some activation, so h may not decrease with r as expected from Eq. 15.63. On the other hand, an interface that moves into a microstructure containing dislocations may accumulate dislocations to reduce the coherency stresses and increase the surface energy (so h may be smaller than expected in this case).

Coherency energy involves an interaction between the precipitate and the matrix. If there is a change of coherence with growth, or a changing precipitate shape to optimize its energy, the free energy per unit volume of the precipitate is not a constant. In other words, even when two equilibrium phases are in contact, the chemical potential evolves during the phase transformation. This makes it inappropriate to use the common tangent construction for assessing the compositions of the two coherent phases in equilibrium, as discussed in Section 9.5. A consequence is altered chemical compositions of both the precipitate and the matrix, as discussed next.

15.5.6 Free Energy of Coherent Precipitation

Coherency Stresses and Energy

The Bitter–Crum theorem of Section 15.5.4 predicts that the elastic energy of misfitting atoms depends only on their number, and does not change if the solutes are configured into precipitates. Its underlying assumptions are problematic in practice, but it conveys the right

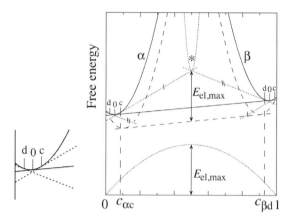

Figure 15.9 $F(c)$ curves for solid solution phase α with coherent precipitate phase β. Effects of elastic energy are described in the text. Enlargement at left shows tangents at the compositions $c_{\alpha d}$, c_0, $c_{\alpha c}$, relevant for forward and reverse transformations. Parallel dotted lines are marked by "|" and "||."

idea that there is a cost in elastic energy for adding misfitting solute atoms if the crystal lattice remains coherent.

Figure 15.9 shows free energy curves for an unmixing alloy system, with an elastic energy (Eq. 15.51 or 15.59) shown at the bottom. The free energy curves of the α- and β-phases, drawn with solid lines, include this elastic energy. (Dashed curves are hypothetical curves without elastic energy.) The solid common tangent line in Fig. 15.9 shows that when the concentration in the solid solution α-phase exceeds $c_{\alpha 0}$, it is thermodynamically favorable to form the β-phase. (The elastic energy is also low after full transformation to β-phase because its unit cells fill space with less stress.)

Assuming the β-phase particles are coherent with the α-matrix (e.g., Fig. 15.7a), so lattice planes are continuous through them, coherency stresses around the interface region alter the chemical potential per atom. From Eqs. 15.51 or 15.59 at $c = 1/2$, we expect an energy cost of the general form

$$E_{el, \, max} = N/2 \, V_0 \, \mu \, \frac{1 + \nu}{1 - \nu} \, \eta^2. \tag{15.64}$$

One way to estimate effects of elastic energy for forming β-phase is to include a free energy curve for an activated state (at "*" in Fig. 15.9), which occurs as the β-phase forms from α. The *-state is not known in detail, but it is associated with a coherent interface between the α- and β-phases. The discussions of Eq. 2.13 and Section 15.1.2 explain that the slopes of these common tangents are the differences in chemical potentials of a B-atom and an A-atom. There are two common tangents to the *-state in Fig. 15.9. The steep common tangent to the activated state allows the composition of the α-phase to be as high as $c_{\alpha c}$ as B-atoms move through the interface to the β-phase, but Fig. 15.1d shows that the slope can be smaller.

Assuming local equilibrium at the α/β interface, and therefore equal slopes of the tangent curves, the composition of the β-phase on Fig. 15.9 is shifted towards $c_{\beta c}$, which is also richer in solute than expected without coherency strains. The reverse transformation of the β-phase to the α-phase moves towards the tangent lines marked with "||," giving compositions smaller than the hypothetical concentrations without the interface (and as low as $c_{\beta d}$ and $c_{\alpha d}$).

Consider the steady-state situation where an interface already exists in a material, and the growth of one phase occurs as the interface moves into the other phase. Diffusion is required to move solute across the interface, and was illustrated previously with Figs. 15.1c and d. The common tangents to the *-phase in Fig. 15.9 (or I-phase) give compositions across the interface, and the concentration gradient is then set by an effective interface width, δ. The interface velocity, v, was given in Eq. 15.9. For hydride formation, discussed next, for clarity we set the chemical potential of vacancies equal to zero, giving a velocity

$$v = cM \left[\mu_H^{I|\alpha} - \mu_H^{I|\beta} \right] \overrightarrow{\nabla c}. \tag{15.65}$$

The interface velocity is zero when $\mu_H^{I|\alpha} = \mu_H^{I|\beta}$, even when $\overrightarrow{\nabla c}$ across the I-phase is nonzero, as occurs when the chemical potential is the same everywhere. The velocity can be reversed by controlling the chemical potential of hydrogen.

Coherency Stresses in Hydride Formation

Hydrogen diffuses quickly, so the velocity v can rapidly accommodate an increase or decrease of hydrogen concentration in the system. Differences in chemical potential of hydrogen set the direction of the phase transformation by changing the sign of $\mu_H^{I|\alpha} - \mu_H^{I|\beta}$ in Eq. 15.65. For hydriding ($\alpha \rightarrow \beta$), the chemical potential must exceed the equilibrium value, giving a more positive slope for the common tangents in Fig. 15.9, and the β-phase grows as in Figs. 15.1c and d. De-hydriding requires a lower chemical potential than the equilibrium value.

Because hydrogen atoms are highly mobile, they quickly enter or leave a material, and their chemical potential soon matches that of the surrounding hydrogen gas. For the chemical potential of the diatomic H_2 gas

$$\mu_g(P, T) = \mu_g^{P_0}(T) + \frac{1}{2} k_B T \, \ln \left(\frac{P}{P_0(T)} \right), \tag{15.66}$$

where we select the reference pressure P_0 for the equilibrium hydrogen concentration in the metal, using the solid common tangent line in Fig. 15.9.

A proposal by Schwarz and Khachaturyan is that the difference in the pressures for hydriding and de-hydriding is obtained by adding an elastic energy barrier to the chemical potential for hydriding, and subtracting it for de-hydriding. The result is [271, 272]

$$\ln \left(\frac{P_c}{P_d} \right) = \frac{2 V_0 \, \mu \, \frac{1+\nu}{1-\nu} \eta^2}{k_B T}, \tag{15.67}$$

where P_c is the gas pressure for charging (hydriding), and P_d for de-hydriding. This would be an overestimate if the new phases form as plates to reduce their elastic energy (so at least one of the conditions for the Bitter–Crum theorem is not met). Also, the V_0 is the atomic volume of the metal atom, and this is probably too large.

Figure 15.10 shows absorption and desorption isotherms of metal alloys with the hexagonal Haucke phase of $LaNi_5$, and some alloys where Sn or Ge atoms were substituted

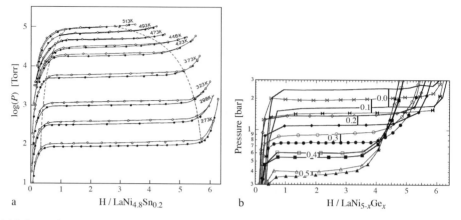

a   H / LaNi$_{4.8}$Sn$_{0.2}$ b   H / LaNi$_{5-x}$Ge$_x$

Figure 15.10 (a) Hydrogen absorption and desorption isotherms of LaNi$_{4.8}$Sn$_{0.2}$, measured at temperatures as labeled from 273 to 513 K. Reprinted from [273], with permission from Elsevier. (b) Hydrogen absorption and desorption isotherms at 300 K of LaNi$_{5-x}$Ge$_x$, with x as labeled from 0.0 to 0.5 [274].

on some sites for Ni atoms. The absorption isotherms are consistently higher in pressure than the desorption isotherms. From the logarithmic plots we can find a hysteresis as high as 0.17 in the log$_{10}$ of pressure, or approximately 0.4 for the LHS of Eq. 15.67. Using $T = 300$ K, this corresponds to an energy of approximately 10 meV/atom, or 1 kJ/mole. The reduction in hysteresis at high temperature may originate from a narrower composition range of the two-phase plateau, and smaller coherency strains. A risk with this analysis is that the hydride interface in La–Ni alloys does not seem to be fully coherent.

Perhaps the biggest cause for concern in using the hysteresis of Eq. 15.67 is that the pressure is measured only after achieving steady-state conditions. The free energy is not changed after the interface moves, since the interface probably remains the same as the $\alpha \rightarrow \beta$ transformation proceeds. This implies that there is a minimum velocity of the hydride interface, below which it stops moving. Perhaps there is an activation required to move the interface forward by each incremental step, and this activation requires a finite difference in chemical potential across the interface.

In metal hydride alloys, it is a rule of thumb that the larger the lattice parameter, the lower the pressure for hydriding. When some Ge atoms are substituted for Ni, the crystal lattice of LaNi$_5$ is expanded. With this increase in lattice parameter comes both a decrease in pressure for hydriding, a narrower two-phase plateau, and a smaller hysteresis, as shown in Fig. 15.10b for different Ge concentrations, x, in LaNi$_{5-x}$Ge$_x$.

Problems

15.1 (a) Explain why the different slopes of the tangent lines in Fig. 15.1c correspond to a departure from the equilibrium of Fig. 15.1a. (Are chemical potentials of A-atoms and B-atoms equal in the two phases?)

(b) Use the result of part a and a concentration gradient in the I-phase to explain why A-atoms will move from right to left across the interface.

15.2 Using the three free energy versus composition curves of Fig. 15.1, switch the roles of parent and new phase by assuming the parent phase is now the α-phase with chemical composition $c = 0.3$ (near c_1 in Fig. 15.1a).

(a) Draw common tangents for solidification at a moderately fast velocity (perhaps comparable to the velocity pertinent to the case in Fig. 15.1c).

(b) Draw the composition profile $c(x)$ near and through the interface as the β-phase grows from the parent α-phase. Explain how the interface compositions of the I-phase allow an appropriate diffusive flux through the interface for this growth process.

15.3 Suppose a liquid with composition c_{T_0} is frozen into the α-phase and β-phase of Fig. 15.1. Here this composition c_{T_0} is the composition near $c = 0.37$ where there is a crossing of the curves $F_\alpha(c)$ and $F_\beta(c)$.

(a) Draw common tangent curves when freezing occurs at a high velocity, and for freezing without solute partitioning between the α-phase and β-phase.

(b) Draw the composition profile $c(x)$ near and through the interface as solidification occurs without solute partitioning. Explain the average composition of the interface, and its composition gradient.

15.4 (a) Show the equivalence of Eqs. 15.9 and 15.29. You will need to linearize an exponential to do this (see Footnote 7). In Eq. 15.29, identify the concentration gradient across the interface.

(b) For Eq. 15.29, identify the differences in chemical potential using a diagram of energy versus reaction coordinate for an atom jump as in Fig. 5.11.

15.5 (a) Starting with Eq. 15.50, calculate the aspect ratio d/l of a plate-like precipitate as a function of precipitate volume, V.

(b) How does the aspect ratio scale with surface energy σ_s, elastic stiffness C, and transformation strain ϵ_3? Why?

15.6 Suppose the free energy of a metal–hydrogen system is that of the umixing alloy of Section 2.8.2, and regions on the phase diagram of Fig. 2.17 translate to the hydride as $\alpha' \rightarrow \alpha$ and $\alpha'' \rightarrow \beta$.

(a) If the elastic energy has the form of Eq. 15.51, why is there no change in the critical temperature (Eq. 2.48) or shape of the phase diagram when E_{el} is added to the energy of Eq. 2.40?

(b) Why might the hysteresis of Section 15.5.6 approach zero as T approaches T_c?

Spinodal Decomposition

Figures 1.5c,d and 1.7a,b illustrate the difference between chemical unmixing that occurs by nucleation and growth (the topic of the previous Chapters 14 and 15) and spinodal decomposition (the topic of Chapter 16). Nucleation creates a distinct surface between the new phase and the parent phase, and the two phases differ significantly in their chemical composition or structure.

Spinodal decomposition does not involve a surface in the usual sense because it begins with infinitesimally small changes in composition. Nevertheless, there is an energy cost for gradients in composition, specifically the square of the gradient, since a region with a large composition gradient begins to look like an interface. The "square gradient energy" is an important new concept presented in this chapter, but it is also essential to phase field theory (Chapter 17), and to the Ginzburg–Landau theory of superconductivity (Section 28.5.3 [online]).

At the end of Section 2.8.2 on unmixing phase diagrams, it was pointed out that there are conceptual problems with a free energy that is concave downwards because the alloy is unstable, but the free energy pertains to equilibrium states. An unstable free energy function may prove useful for short times, however. Taking a kinetic approach, we use the thermodynamic tendencies near equilibrium to obtain a chemical potential to drive a diffusion flux that causes unmixing. Following the classic approach of John Cahn [115, 275] we add a term to the free energy that includes the square of the composition gradient. This term affects the time and length scales of the unmixing transformation. There is also an elastic energy that increases with the extent of unmixing, and gives the "coherent spinodal" on the unmixing phase diagram.

16.1 Concentration Fluctuations and the Free Energy of Solution

First consider the free energy without a gradient energy. Consider the free energy to depend only on the local composition, without concern for the spatial size of the region, essentially as we have done since Chapter 2. This is easy for a homogeneous sample of concentration c_0. Taking the cross-sectional area of the sample to be unity, and the total length L, the total free energy is simply $Lf(c_0)$, where f is the Helmholtz free energy per unit volume. Figure 16.1 shows a simple one-dimensional concentration fluctuation in the material. Over the spatial range, x, from 0 to $+\zeta$, the composition of the sample is slightly greater than the average, i.e., $c(x) = c_0 + \delta c$ for $0 < x < +\zeta$. Similarly, in the range from $-\zeta$ to 0 the concentration is less than average, and by conservation of solute, $c(x) = c_0 - \delta c$ in this region.

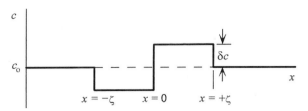

Figure 16.1 Idealized, one-dimensional concentration fluctuation in a material of average composition c_0. The positive and negative parts of the fluctuation account for the same amount of solute.

We seek the change in the total free energy, ΔF, when a homogeneous solid solution of composition c_0 forms the state depicted in Fig. 16.1. If the amplitude of the fluctuation is small, the free energy of the material can be found by Taylor-expanding about c_0. Therefore, ΔF for the region 0 to $+\zeta$ becomes

$$\Delta F^+ = \zeta f(c_0) + \zeta \left.\frac{\partial f}{\partial c}\right|_{c=c_0} \delta c + \zeta \frac{1}{2}\left.\frac{\partial^2 f}{\partial c^2}\right|_{c=c_0}(\delta c)^2 + \cdots - \zeta f(c_0), \tag{16.1}$$

where, for now, any energy contribution from the "surface" at $x = 0$ is neglected. Similarly, the free energy change over the region where $x < 0$ is

$$\Delta F^- = \zeta f(c_0) - \zeta \left.\frac{\partial f}{\partial c}\right|_{c=c_0} \delta c + \zeta \frac{1}{2}\left.\frac{\partial^2 f}{\partial c^2}\right|_{c=c_0}(\delta c)^2 + \cdots - \zeta f(c_0). \tag{16.2}$$

The change in total free energy after introducing the fluctuation is the sum of Eqs. 16.1 and 16.2

$$\Delta F = \zeta \left.\frac{\partial^2 f}{\partial c^2}\right|_{c=c_0}(\delta c)^2, \tag{16.3}$$

where the second derivative $\partial^2 f/\partial c^2\big|_{c=c_0}$ is evaluated at c_0.

In Eq. 16.3 both the terms ζ and $(\delta c)^2$ are positive, so the sign of the total free energy change is governed by the curvature $\partial^2 f/\partial c^2\big|_{c=c_0}$. Figure 16.2 shows the curvature of $f(c)$ for a binary system exhibiting a miscibility gap (cf. Fig. 2.16b). Over the ranges of composition labeled "stable" and "metastable," the second derivative of f is positive. In these ranges, small fluctuations in concentration cause an increase in the free energy of solution, so the fluctuations are unfavorable. Small fluctuations in composition will decay, and the alloy will return to a homogeneous state.[1] The two concentrations c_s in Fig. 16.2, inflection points where $\partial^2 f/\partial c^2 = 0$, are the limits of stability of the solution. They are called "spinodal compositions," and with temperature the locus of these points defines the "chemical spinodal." The chemical spinodal is shown in Fig. 16.5a. It lies inside the "miscibility line" where unmixing occurs across the common tangent of Fig. 2.18a. In this highly unstable range of compositions of Fig. 16.2 where $\partial^2 f/\partial c^2 < 0$, even infinitesimally small fluctuations will grow in amplitude. A curious feature of an infinitesimal change in composition is that there is no surface energy, and therefore no nucleation barrier to start

---

[1] In the metastable region, however, large composition fluctuations will grow if they create a critical nucleus of the stable α_2-phase.

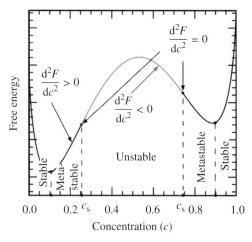

Free energy, F, as a function of composition for a system exhibiting a miscibility gap. Also shown are the ranges of concentration for which the solution is stable, metastable, and unstable. Compare with Fig. 2.18.

the unmixing transformation. What happens between the regions of different compositions as the unmixing transformation evolves, and the two phases become better defined? When does surface energy appear, or does it?

16.2 A Square Gradient Term in the Free Energy

16.2.1 Justification for the Square of the Composition Gradient

The composition fluctuation of Fig. 16.1 is quite abrupt, with its sharp interface between regions of different composition at $x = 0$. We expect to pay an energy penalty for such an abrupt modulation in composition since this is, in fact, a surface between regions of different compositions. Now suppose we broaden out the interface between regions of different composition, as shown in Fig. 16.3. Compare Figs. 16.3a and 16.3b. Both are composition profiles with the same volumes for each composition (although there are two triangles in Fig. 16.3b versus one in 16.3a, the net positive or negative areas are equal). If the free energy were a function of the composition only, the free energies of the profiles of Figs. 16.3a and 16.3b would be equal. We expect, however, that there will be a greater free energy for the composition profile of Fig. 16.3b, since it has steeper composition gradients that are more like surfaces. It also has more such interfaces per unit length, however, so this is not a proper comparison.

More subtly, when we compare the energies of the gradient regions of Figs. 16.3c and 16.3d, we notice that although the gradients are steeper in Fig. 16.3d, they occur over a proportionately smaller range in x. If we were to add a term to the free energy that is proportional to the magnitude of the composition gradient, when integrated over volume it would not matter if the gradient were steeper because a steeper gradient occurs

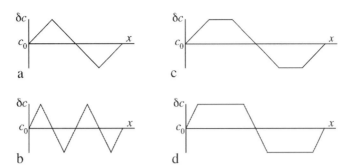

Figure 16.3 Four cases of composition fluctuations with composition gradients. The gradients, dc/dx, are the same for panels a and c, but are twice as large for panels b and d. The gradients in panels b and d extend over only half the range of x for panels a and c, however.

over a proportionally smaller volume. The free energy therefore does not depend on the magnitude of the gradient. It turns out that we need to add a positive term with the square of the gradient to the free energy to account for "surface-like" effects. The square of the gradient favors the profile of Fig. 16.3c over Fig. 16.3d, for example.

16.2.2 Derivation for the Square of the Composition Gradient

For a simple solid solution that has uniform composition everywhere, a simple expression for the free energy density such as $f(c)$ will suffice.[2] With inhomogeneities of composition, the free energy density is more complicated than a simple function of average composition. A natural next step in the complexity is to allow the free energy to depend in some way on the gradient of composition. (This additional dependence can be neglected when the composition is uniform since $\vec{\nabla}c = 0$.) When there is a small local fluctuation of Δc in an average composition of c_0, there will be gradients and curvatures of composition that can affect $f(c)$. Including all terms to second order

$$f(c_0+\Delta c, \vec{\nabla}c, \nabla^2 c) = f(c_0) + \frac{\partial f}{\partial c}\Delta c + \frac{\partial f}{\partial \vec{\nabla}c}\vec{\nabla}c + \frac{\partial f}{\partial \nabla^2 c}\nabla^2 c$$

$$+ \frac{1}{2}\left[\frac{\partial^2 f}{\partial c^2}(\Delta c)^2 + \frac{\partial^2 f}{\partial(\vec{\nabla}c)^2}(\vec{\nabla}c)^2 + 2\frac{\partial^2 f}{\partial c\,\partial\vec{\nabla}c}\Delta c\,\vec{\nabla}c\right]. \qquad (16.4)$$

Equation 16.4 may look peculiar because of its derivatives with respect to composition derivatives, such as the derivative with respect to the composition gradient. A detailed concentration profile including all features of $c(x,y,z)$ contains this information, and contributions to $f(c)$ could be assigned to the different features in the composition profile. Such a detailed $c(x,y,z)$ is usually not practical, and it is usually impractical to separate

---

[2] This implies a continuum theory, which is fine for the present arguments. Of course, at the atomic level the material will show statistical fluctuations in local chemical environments, e.g., some atoms in a solid solution may have five solute neighbors, and another may have seven. These considerations are important, but are beyond the present scope.

$c(x, y, z)$ into its different features. Instead we replace $f\big(c(x, y, z)\big)$ with $f(c_0 + \Delta c, \overrightarrow{\nabla} c, \nabla^2 c)$, and we allow some independence between the contributions to $f(c)$ from the local composition and from the spatial derivatives of composition.

It is necessary to go to second order in composition derivatives to consider an instability of free energy against an *infinitesimally* small fluctuation. The first-order terms, linear in Δc and $\overrightarrow{\nabla} c$, must be zero. If there were a linear term, the free energy could be lowered more and more with an increasingly large Δc for the fluctuation, so the fluctuation would not be infinitesimal and a Taylor series approach would not make sense. To discount the gradient term, $\overrightarrow{\nabla} c$, first note that it is a vector. The sign of the gradient (or the sign of dc/dx in one dimension) switches with direction, which is not realistic for a solid solution. Also, as explained at the end of Section 16.2.1, the magnitude $|\overrightarrow{\nabla} c|$ exists over a range of distances that is inversely proportional to the gradient itself, so F cannot depend on $|\overrightarrow{\nabla} c|$ either. The same argument pertains to the last term of Eq. 16.4. This simplifies the result

$$f(c_0 + \Delta c, \overrightarrow{\nabla} c, \nabla^2 c) = f(c_0) + \frac{1}{2}\frac{\partial^2 f}{\partial c^2}(\Delta c)^2 + \frac{1}{2}\frac{\partial^2 f}{\partial(\overrightarrow{\nabla} c)^2}(\overrightarrow{\nabla} c)^2 + \frac{\partial f}{\partial \nabla^2 c}\nabla^2 c. \qquad (16.5)$$

At this point we reduce the problem to one dimension, a practical necessity to avoid numerical computation. Although not fully justified, this has become a standard part of the explanation of spinodal decomposition. The total free energy, F_T, is an integral of the free energy per unit volume, f, over the volume of the material (now the length in x times the area, A)

$$F_T = A \int \left[f(c_0) + \frac{1}{2}\frac{\partial^2 f}{\partial c^2}(\Delta c)^2 + \frac{1}{2}\frac{\partial^2 f}{\partial\left(\frac{\partial c}{\partial x}\right)^2}\left(\frac{\partial c}{\partial x}\right)^2 + \frac{\partial f}{\partial\left(\frac{\partial^2 c}{\partial x^2}\right)}\left(\frac{\partial^2 c}{\partial x^2}\right) \right] dx. \qquad (16.6)$$

The fourth term in brackets (with the $\partial^2 c/\partial x^2$) came from the Laplacian. It can be recast into the form of the third term (with the $(\partial c/\partial x)^2$), which came from the gradient. To do so, integrate by parts following the prescription

$$\int U dV = UV - \int V dU. \qquad (16.7)$$

Identifying dV so that V is a gradient

$$dV = \frac{\partial^2 c}{\partial x^2}dx, \qquad \text{so} \qquad V = \frac{\partial c}{\partial x}, \qquad (16.8)$$

$$U = \frac{\partial f}{\partial\left(\frac{\partial^2 c}{\partial x^2}\right)}, \qquad \text{so} \qquad dU = \left[\frac{\partial}{\partial c}\left(\frac{\partial f}{\partial\left(\frac{\partial^2 c}{\partial x^2}\right)}\right)\frac{dc}{dx} \right] dx. \qquad (16.9)$$

Substituting into Eq. 16.7 (and noting that $c = c(x)$, so $dc/dx = \partial c/\partial x$) gives

$$\int \frac{\partial f}{\partial\left(\frac{\partial^2 c}{\partial x^2}\right)}\left(\frac{\partial^2 c}{\partial x^2}\right)dx = \frac{\partial f}{\partial\left(\frac{\partial^2 c}{\partial x^2}\right)}\frac{\partial c}{\partial x} - \int \left[\frac{\partial}{\partial c}\left(\frac{\partial f}{\partial\left(\frac{\partial^2 c}{\partial x^2}\right)}\right)\right]\left(\frac{\partial c}{\partial x}\right)^2 dx. \qquad (16.10)$$

The first term on the RHS of Eq. 16.10 is to be evaluated at the ends of the material. An important point is that this term does not scale with the volume of the material, whereas the last term does increase with the volume (range of x). The first term can be neglected for bulk materials. Substituting Eq. 16.10 (without the first term on its RHS) into 16.6

$$F_T = A \int f(c_0) + \frac{1}{2} \frac{\partial^2 f}{\partial c^2} (\Delta c)^2$$
$$+ \frac{1}{2} \left[\frac{\partial^2 f}{\partial \left(\frac{\partial c}{\partial x} \right)^2} - 2 \frac{\partial}{\partial c} \left(\frac{\partial f}{\partial \left(\frac{\partial^2 c}{\partial x^2} \right)} \right) \right] \left(\frac{\partial c}{\partial x} \right)^2 \, dx. \tag{16.11}$$

The first line in Eq. 16.11 is familiar from Sections 2.8.2 and 10.6. The second line is new. Notice that it is proportional to $(\partial c/\partial x)^2$, or the square of the gradient in 3D. This is seen more clearly with definitions

$$F_T = A \int f(c_0) + \frac{1}{2} \frac{\partial^2 f}{\partial c^2} (\Delta c)^2 + \frac{1}{2} \left[\kappa_2 - 2 \frac{\partial \kappa_1}{\partial c} \right] \left(\frac{\partial c}{\partial x} \right)^2 \, dx, \tag{16.12}$$

with the new parameters evaluated at the alloy composition c_0

$$\kappa_1 = \frac{\partial f}{\partial \left(\frac{\partial^2 c}{\partial x^2} \right)}, \tag{16.13}$$

$$\kappa_2 = \frac{\partial^2 f}{\partial \left(\frac{\partial c}{\partial x} \right)^2}, \tag{16.14}$$

$$\kappa = \frac{1}{2} \kappa_2 - \frac{\partial \kappa_1}{\partial c}. \tag{16.15}$$

Equation 16.12 simplifies to

$$F_T = A \int f(c_0) + \frac{1}{2} \frac{\partial^2 f}{\partial c^2} (\Delta c)^2 + \kappa \left(\frac{\partial c}{\partial x} \right)^2 \, dx. \tag{16.16}$$

This is sometimes written, with an obvious definition, as

$$F_T = A \int f(c) + \kappa \left(\frac{\partial c}{\partial x} \right)^2 \, dx, \tag{16.17}$$

and in three dimensions it is

$$F_T = \int f(c) + \kappa \left(\overrightarrow{\nabla c} \right)^2 \, dV. \tag{16.18}$$

Equation 16.18, which contains the square gradient energy, is called the "Cahn–Hilliard equation."

16.2.3 Effects of the Square of the Composition Gradient

The composition gradient describes something like a fuzzy surface. If the surface were sharp in composition, the gradient would be large. A more gradual variation in composition gives a smaller gradient. Since it is the square of the gradient that is important, it is beneficial energetically to spread the composition variation over a wider range of distance.[3]

Suppose the solid solution is stable, as is the case at high temperatures. A penalty is paid when the fuzzy surface gets sharper. The quantity in the square brackets in Eq. 16.12 acts like a spring constant, where the free energy increases quadratically with the composition gradient.[4] Note that the second term from the ordinary chemical free energy, $\partial^2 f/\partial c^2$, also acts like a restoring force at high temperature. At the critical temperature where $\partial^2 f/\partial c^2$ changes sign, this ordinary chemical restoring force is lost. On the other hand, the term in square brackets and the gradient energy remain positive, so sharp gradients are disfavored at all temperatures. The net effect of the favorable unmixing term and the unfavorable gradient energy term is to set an optimal spatial scale for the unmixing. The problem becomes one of kinetics. The gradient energy promotes a longer spatial scale, but diffusion occurs more efficiently over smaller spatial scales. With an increasingly larger volume free energy below the critical temperature, which drives diffusion, the characteristic modulation wavelength for the composition fluctuations becomes shorter at lower temperatures. This is discussed in more detail after a mathematical excursion into the calculus of variations.

16.2.4 Preview of the Cahn Approach to Spinodal Decomposition

There is a lot in what follows, and some of the steps may be both unfamiliar and subtle. The steps summarized in Box 16.1 chart the path.

Box 16.1   **Summary of the Cahn Approach to Spinodal Decomposition**

The following approach to spinodal decomposition is that of J.W. Cahn [275], with added explanation. Other references include [267], [276–278]. The Cahn approach, although largely in one dimension, is a classic. It includes these steps:

- Develop a free energy density $f(x, c(x), dc/dx)$ that depends on average composition, and on composition fluctuations in the material through two terms:
  - the compositional effect on the free energy, d^2f/dc^2,
  - the square of the composition gradient, $(dc/dx)^2$.
- Use calculus of variations (developed below as a mathematical excursion) to minimize $f(x, c(x), dc/dx)$ over the volume of material, obtaining a condition for the optimal compositional profile, the "Euler equation."
- Use the method of Lagrange multipliers to conserve chemical composition. This brings a constraint equation and a constant μ, which is the chemical potential.
- Use the gradient of the chemical potential to obtain a diffusive flux with Fick's first law, where flux is proportional to $\overrightarrow{\nabla}\mu$.

---

[3] As we saw in the discussion of Fig. 16.3, an energy proportional to the magnitude of the gradient would have no such preference.

[4] The constant κ_1 is the composition dependence of how the free energy depends on the lumpiness of composition. The constant κ_2 is more obviously like a spring constant.

- Take the divergence of Fick's first law to get an equation for the kinetics of composition change, dc/dt.
- Recast the composition fluctuation $\delta c(x)$ as a Fourier transform.
- Identify the wavevectors, $2\pi/\lambda$, of the compositional fluctuations that can grow, and identify the fastest-growing compositional wavelength, λ.
- Recognize that the lattice parameter of the crystal changes with chemical unmixing, and add elastic energy to the free energy. This suppresses the temperature of unmixing.

16.3 Constrained Minimization of the Free Energy

We seek to minimize the F_T of Eq. 16.12, and seek the condition when F_T becomes unstable against the formation of a small concentration fluctuation. The difficulty is that we need to conserve atoms, and an arbitrary increase in composition in one location must be balanced by a decrease in composition elsewhere. This is not so simple as shown in Fig. 16.1, since we could have more regions with small $-\Delta c$, and fewer regions with very large $+\Delta c$. We seek to minimize F_T subject to the constraint of constant alloy composition

$$A \int_{-L/2}^{L/2} c(x)\, dx = A\, L\, c_0, \tag{16.19}$$

$$\int_{-L/2}^{L/2} \left[c(x) - c_0 \right] dx = 0. \tag{16.20}$$

Minimizing an integral with a constraint equation over the same limits requires a pair of results from a topic of mathematics called "calculus of variations." We next make a brief excursion through this topic.

16.3.1 Calculus of Variations

Suppose we have a function f that depends on x, $c(x)$, and dc/dx. When f is integrated over a range of x

$$F_T = \int_{-L/2}^{L/2} f\left(x, c(x), \frac{dc}{dx}\right) dx, \tag{16.21}$$

the integral has different values, depending on the specific profile of $c(x)$. This $c(x)$ also sets the derivative dc/dx, since c depends only on x. We seek the concentration profile $C(x)$ that gives an extremum of F_T (which may be a maximum or minimum of F_T). Consider a composition profile $c(x)$ that is quite close to this desired $C(x)$, and define a small "variation," $\delta c(x)$ as

$$\delta c(x) \equiv C(x) - c(x). \tag{16.22}$$

Compared with the optimal profile $C(x)$ that minimizes F_T, this nearby profile $c(x)$ causes a change in the integrand at any point x

$$\delta f = f\left(x, C(x), \frac{dC}{dx}\right) - f\left(x, c(x), \frac{dc}{dx}\right). \tag{16.23}$$

Equation 16.23 shows that changes in f originate from the change of $C(x)$ to $c(x)$, and from the change of dC/dx to dc/dx. However, x does not change, so $\delta x = 0$ (and the same range of integration is always used). For small variations $\delta c(x)$, we expect

$$\delta f = \frac{\partial f}{\partial c}\delta c + \frac{\partial f}{\partial \frac{dc}{dx}} \delta\frac{dc}{dx}. \tag{16.24}$$

We make an important observation about the last variation in Eq. 16.24

$$\delta\frac{dc}{dx} = \frac{dC}{dx} - \frac{dc}{dx} = \frac{d}{dx}(C - c) = \frac{d}{dx}\delta c. \tag{16.25}$$

The operator δc commutes with the operator d/dx, and Eq. 16.24 becomes

$$\delta f = \frac{\partial f}{\partial c}\delta c + \frac{\partial f}{\partial \frac{dc}{dx}} \frac{d}{dx}\delta c. \tag{16.26}$$

The condition that F_T is an extremum (minimum, we hope) is simple

$$0 = \int_{-L/2}^{L/2} \delta f\left(x, c(x), \frac{dc}{dx}\right) dx. \tag{16.27}$$

This is an analog to the familiar case that $df = 0$ when seeking an extremum of a function such as $f(x)$.[5] Using Eq. 16.26 in this integral

$$0 = \int_{-L/2}^{L/2} \left[\frac{\partial f}{\partial c}\delta c + \frac{\partial f}{\partial \frac{dc}{dx}} \frac{d}{dx}\delta c\right] dx. \tag{16.28}$$

There is one more simplification needed before the main result. The second term in the square brackets of Eq. 16.28 is integrated by parts to give

$$\frac{\partial f}{\partial \frac{dc}{dx}} \frac{d}{dx}\delta c = \left(\frac{\partial f}{\partial \frac{dc}{dx}} \delta c\right)_{-L/2}^{L/2} - \int_{-L/2}^{L/2}\left(\frac{d}{dx}\frac{\partial f}{\partial \frac{dc}{dx}}\right)\delta c\ dx. \tag{16.29}$$

The simplification is that the first term on the RHS of Eq. 16.29 can be neglected. One argument is that the composition is fixed at the limits, so this term vanishes. A second argument is that this constant term becomes negligible compared to the second term that grows with L. Equation 16.28 becomes

$$0 = \int_{-L/2}^{L/2} \left[\frac{\partial f}{\partial c} - \frac{d}{dx}\frac{\partial f}{\partial \frac{dc}{dx}}\right]\delta c\ dx. \tag{16.30}$$

---

[5] Because f is a function of the functions $c(x)$ and dc/dx, we call f a "functional," instead of a function.

The final trick is an observation. Our $\delta c = \delta c(x)$ is arbitrary. If the term in square brackets in Eq. 16.30 were not zero everywhere, we could pick a $\delta c(x)$ that would make the total integral nonzero. This gives the main result, the "Euler equation" of variational calculus

$$\frac{\partial f}{\partial c} - \frac{\mathrm{d}}{\mathrm{d}x}\frac{\partial f}{\partial \frac{\mathrm{d}c}{\mathrm{d}x}} = 0. \tag{16.31}$$

This Euler equation is the condition for F_T to have a minimum value if we are free to select the composition profile.

16.3.2 Constraint of Constant Composition

We return to the constraint of constant alloy composition, Eq. 16.20, which is the condition of conservation of atoms. This condition is over the same range of x as the extremum problem that gave the Euler equation. We therefore bring in the constraint of solute conservation

$$\int_{-L/2}^{L/2} \delta c \, \mathrm{d}x = 0 \tag{16.32}$$

by adding $N\mu\,\delta c$ to the integrand of Eq. 16.27, which is over the same limits of integration. Here $N\mu$ is a constant called a "Lagrange multiplier," and is yet to be determined. Because Eq. 16.32 equals 0, when $N\mu\,\delta c$ is subtracted from the integrand of Eq. 16.27, it remains zero, of course. The arguments that follow Eq. 16.27 remain the same up to Eq. 16.30,[6] which becomes

$$0 = \int_{-L/2}^{L/2} \left[\frac{\partial f}{\partial c} - \frac{\mathrm{d}}{\mathrm{d}x}\frac{\partial f}{\partial \frac{\mathrm{d}c}{\mathrm{d}x}} - N\mu \right] \delta c \, \mathrm{d}x. \tag{16.33}$$

Unfortunately, we cannot use the previous argument in passing from Eq. 16.30 to Eq. 16.31 because δc is no longer arbitrary owing to Eq. 16.32. Although we lose this degree of freedom in selecting a δc, the new $N\mu$ is an adjustable parameter that can be tuned so the integrand vanishes. It turns out that μ is the chemical potential (Eq. 1.24). It is selected, as usual, to conserve atoms. With the definition

$$K = f - N\mu, \tag{16.34}$$

it is straightforward to show by substitution for f that the Euler equation of 16.31 becomes

$$\frac{\partial K}{\partial c} - \frac{\mathrm{d}}{\mathrm{d}x}\frac{\partial K}{\partial \frac{\mathrm{d}c}{\mathrm{d}x}} = 0. \tag{16.35}$$

---

[6] We could have directly added Eq. 16.32 to 16.30 for the same result.

16.3.3 Minimizing a Constrained Free Energy Functional

Using the free energy per unit volume from Eq. 16.16

$$f = f(c_0) + \frac{1}{2}\frac{\partial^2 f}{\partial c^2}(\Delta c)^2 + \kappa\left(\frac{\partial c}{\partial x}\right)^2, \tag{16.36}$$

and using this definition of $f(c)$ (as in Eq. 16.17)

$$f(c) \equiv f(c_0) + \frac{1}{2}\frac{\partial^2 f}{\partial c^2}(\Delta c)^2, \tag{16.37}$$

after substitution into Eq. 16.35

$$\frac{\partial}{\partial c}\left[f(c) + \kappa\left(\frac{\partial c}{\partial x}\right)^2 - N\mu\delta c\right] - \frac{d}{dx}\frac{\partial}{\partial \frac{dc}{dx}}\left[f(c) + \kappa\left(\frac{\partial c}{\partial x}\right)^2 - N\mu\delta c\right] = 0. \tag{16.38}$$

Recognizing that $\partial/\partial c\ \delta c = \partial/\partial c\ (c - c_0) = 1$

$$\frac{\partial f}{\partial c} - N\mu - \frac{d}{dx}2\kappa\frac{\partial c}{\partial x} = 0, \tag{16.39}$$

$$\frac{\partial f}{\partial c} - 2\kappa\frac{\partial^2 c}{\partial x^2} = N\mu. \tag{16.40}$$

Note that when $\kappa = 0$, we have $\mu = (1/N)\,\partial f/\partial c$, which is the usual form of the chemical potential. In equilibrium, Eq. 16.40 is also a constant. We recognize μ as a chemical potential that now includes the effects of the square gradient energy.

    These free energy arguments suggest how the square gradient energy may alter the unmixing phase diagram of Fig. 2.18b. In essence, the last term in Eq. 16.36 is an extra cost of unmixing, so it is expected that the two-phase region might be diminished in size. The problem is more subtle for two reasons, however. Composition gradients are infinitesimally small at the upper critical temperature for the composition $c = 0.5$, and the infinitesimally small contribution to the free energy can be neglected. More generally, by setting a very long spatial scale for the unmixing, the composition gradient and the square gradient energy can be suppressed. Allowing for equilibrium allows for infinitely long spatial scales, so the square gradient energy has no effect on the equilibrium phase diagram of Fig. 2.18b or Fig. 16.5a. Both the miscibility line and the chemical spinodal remain unchanged. On the other hand, the unmixing process would require an infinite time if the unmixing were over infinite distances. The real situation for unmixing becomes one of kinetics, and we approach it with the diffusion equation. It is in the kinetics where the square gradient energy becomes important in a continuous phase transformation.

16.4 The Diffusion Equation

16.4.1 Diffusion Driven by a Gradient in Chemical Potential

The standard derivation of the diffusion equation in Section 3.2 involves random mass transport between adjacent slices of material. The idea is that the atoms jump in random directions, so if one slice has a higher concentration than a second slice next to it, there will be a net transport of atoms from the first slice to the second. A gradient in composition drives the flux of atoms.

In general, though, the jump frequencies of atoms are biased by their chemical potential, as suggested by Fig. 5.11, for example. The gradient in chemical potential follows the gradient in composition when there are no chemical interactions between atoms. On the other hand, Eqs. 10.89 to 10.92 showed how the diffusion flux is modified when there are chemical preferences. We take a similar approach here, using the gradient of the chemical potential[7] (including the squared gradient energy) to drive the diffusive flux, $\vec{J}$

$$\vec{J} = -M\,\overrightarrow{\nabla\mu},\tag{16.41}$$

where M is a mobility coefficient. Using the chemical potential of Eq. 16.40

$$J = -M\frac{\partial}{\partial x}\left(\frac{\partial f}{\partial c} - 2\kappa\frac{\partial^2 c}{\partial x^2}\right),\tag{16.42}$$

$$J = -M\left(\frac{\partial^2 f}{\partial c^2}\frac{dc}{dx} - 2\kappa\frac{\partial^3 c}{\partial x^3}\right),\tag{16.43}$$

where κ is the gradient energy coefficient, now assumed independent of concentration. Without derivation, the three-dimensional form of Eq. 16.42 is

$$\vec{J} = -M\,\mathbf{grad}\left(\frac{\partial f}{\partial c} - 2\kappa\nabla^2 c\right).\tag{16.44}$$

Solving for fluxes is typically less useful than solving for the time evolution of the concentration profile. We therefore use the continuity equation

$$\frac{\partial c}{\partial t} = -\frac{\partial}{\partial x}J,\tag{16.45}$$

to make the transformation of Eq. 16.43

$$\frac{\partial c}{\partial t} = M\left(\frac{\partial^2 f}{\partial c^2}\frac{\partial^2 c}{\partial x^2} - 2\kappa\frac{\partial^4 c}{\partial x^4}\right).\tag{16.46}$$

This is analogous to the three-dimensional form obtained by applying the three-dimensional continuity equation

$$\frac{\partial c}{\partial t} = -\vec{\nabla}\cdot\vec{J},\tag{16.47}$$

to Eq. 16.44

$$\frac{\partial\delta c}{\partial t} = \vec{\nabla}\cdot\left[M\,\mathbf{grad}\left(\frac{\partial f}{\partial\,\delta c} - 2\kappa\nabla^2\delta c\right)\right],\tag{16.48}$$

with the definition $\delta c(\vec{r}, t) \equiv c(\vec{r}) - c_0$.

---

[7] When lattice sites are conserved and populated fully, the gradients of chemical potential can be replaced by a single gradient in the difference of chemical potentials between the two species A and B in solution.

16.4.2 Fourier Transform Solution

Equation 16.46 has a notoriously inconvenient form, but it is possible to make progress if the concentration fluctuation δc is small. The approach is to work with a Fourier transform expression for the concentration

$$\delta c(x,t) = c(x,t) - c_0 = \int_{-\infty}^{\infty} A(Q,t)\, e^{iQx}\, dQ, \qquad (16.49)$$

where Q is a wavevector, inversely related to the wavelength λ as $Q \equiv 2\pi/\lambda$. The amplitude for the concentration modulation, A, is assumed to be an exponential function of time, and at any time it is the inverse Fourier transform of the concentration fluctuation

$$A(Q,t) = \frac{1}{2\pi}\, e^{+R(Q)t} \int_{-\infty}^{\infty} \delta c(x,t)\, e^{-iQx}\, dx, \qquad (16.50)$$

where $R(Q)$ is a growth rate for the wavevector Q. When we substitute Eq. 16.49 into our modified diffusion equation Eq. 16.46, we note that $\exp(iQx)$ is the only function of x, and $A(Q,t)$ is the only function of t. After dividing by $c(x) - c_0$ in the form of Eq. 16.49

$$R(Q) = M\left[\frac{\partial^2 f}{\partial c^2}(-1)Q^2 - 2\kappa Q^4\right]. \qquad (16.51)$$

To grow a concentration modulation, $R(Q)$ should be positive – if $R(Q)$ is negative, the concentration modulations will decay with time. If κ were zero, so the gradient energy were unimportant, we would be back to the case of unmixing used for obtaining the unmixing phase diagram in Fig. 2.18b, because a concentration modulation would grow whenever $\partial^2 f/\partial c^2$ is negative. For finite κ (which is always positive), the gradient energy tends to suppress unmixing, so we need a larger free energy driving force for unmixing. This requires that $\partial^2 f/\partial c^2$ is negative,[8] and its term in Eq. 16.51 must be larger than the gradient energy term, $2\kappa Q^4$. The critical condition is when the two terms are equal in Eq. 16.51. This condition defines a critical wavevector Q_c

$$Q_c = \left(-\frac{1}{2\kappa}\frac{\partial^2 f}{\partial c^2}\right)^{\frac{1}{2}}. \qquad (16.52)$$

When $Q < Q_c$, a composition modulation can grow because the gradient energy is not large enough to suppress the modulation.

Another important step is to find the wavevector Q_{max} that grows most rapidly. The most rapidly growing wavelength should set the spatial scale for the early stages of spinodal decomposition. Differentiating Eq. 16.51

$$\frac{dR(Q)}{dQ} = 0 = M\left[-2\frac{\partial^2 f}{\partial c^2}Q_{max} - 8\kappa Q_{max}^3\right], \qquad (16.53)$$

$$Q_{max} = \left(-\frac{1}{4\kappa}\frac{\partial^2 f}{\partial c^2}\right)^{\frac{1}{2}}. \qquad (16.54)$$

---

[8] Of course, if $\partial^2 f/\partial c^2$ were positive, all small-amplitude concentration fluctuations would decay exponentially with time, and the system would be stable. The alloy cannot unmix with infinitesimal fluctuations if $\partial^2 f/\partial c^2 > 0$.

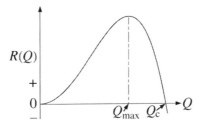

Figure 16.4 Amplification factor $R(Q)$, having the form $Q^2 - Q^4$ of Eq. 16.51. The value Q_c corresponds to $R = 0$, and Q_{max} gives the maximum R.

From Eqs. 16.52 and 16.54, $Q_{max} = Q_c/\sqrt{2}$. Figure 16.4 shows the form of the amplification factor, $R(Q)$, vs. Q. For wavenumbers less than Q_c, i.e., longer wavelengths, $R(Q)$ is positive and the amplitude $A(Q, t)$ grows exponentially with time. The most rapid growth is for modulations with Q_{max}. This selective amplification implies that, shortly after a quench into the unstable region of the phase diagram, the Fourier spectrum will be dominated by a single component. It suggests a real-space concentration profile characterized by an approximately periodic, three-dimensional interconnected network of regions of high and low solute concentrations.

Estimating Q_{max} or $R(Q_{max})$ is challenging. The conventional thinking is that the wavelength λ for maximum growth is of order 10 nm. The characteristic rate is something like 10^{-3} s, but this obviously depends on the atomic mobility or diffusion coefficient. Spinodal decomposition is known to occur in some alloys such as Co–Cu, but it is not so easily identified because its characteristically short time and spatial scales make it a challenge to demonstrate by experiment. Today we believe that unmixing by spinodal decomposition is much less common than unmixing by nucleation and growth. However, only recently have methods become routine for characterizing chemical composition at dimensions less than 10 nm, and new methods of rapid quenching may better prepare an alloy for unmixing by spinodal decomposition.

16.5 Effects of Elastic Energy on Spinodal Decomposition

16.5.1 Thermodynamic Effects of Elastic Energy

In general, atoms in an alloy have different sizes. A concentration fluctuation therefore creates an elastic strain field in the material. For nucleation, we saw that the positive elastic energy from a misfitting spherical nucleus must be added to the work of formation, sometimes markedly changing the kinetics. The strain energy generated by growing concentration fluctuations plays an important role in the spinodal decomposition of crystalline solids, suppressing the temperature of spinodal decomposition.

The elastic strain energy for an arbitrary composition fluctuation can be found in much the same way as was done in solving the diffusion equation by Fourier transform methods.

By Fourier transforming the composition, computing the elastic energy of each Fourier component, and utilizing the fact that the Fourier components do not interact, one obtains a general elastic energy contribution

$$E_{el} = \frac{Y}{1-v}\,\eta^2 \int_{vol} (\delta c)^2 \, d^3\vec{r}, \tag{16.55}$$

where Y is Young's modulus, v is Poisson's ratio and η is the fractional change of lattice parameter with composition. The $\delta c \equiv c(\vec{r}) - c_0$ depends only on the amplitude of the concentration variation, not on the wavelength. To include the elastic energy in the continuum theory of spinodal decomposition, the Cahn–Hilliard free energy equation (Eq. 16.18) must be modified as

$$F = \int_{vol} \left[f(c) + \frac{Y}{1-v}\,\eta^2 (\delta c)^2 + \kappa\,(\overrightarrow{\nabla c})^2 \right] d^3\vec{r}. \tag{16.56}$$

It is a straightforward exercise to show that the limit of stability of a solid solution is no longer $\partial^2 f / \partial c^2 = 0$ but

$$\frac{\partial^2 f}{\partial c^2} + \frac{2\eta^2\,Y}{1-v} = 0. \tag{16.57}$$

A more complete thermodynamic picture of spinodal decomposition can now be illustrated with Fig. 16.5a. The heavy solid line shows the equilibrium miscibility gap (cf. Fig. 2.7). Below the equilibrium phase boundary is a dashed curve corresponding to the locus of points such that $\partial^2 f / \partial c^2 = 0$. This curve is called the "chemical spinodal."

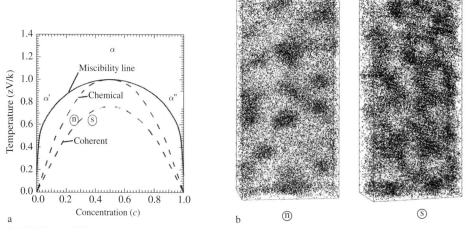

Figure 16.5   (a) Phase diagram of a binary system with a miscibility gap (solid curve). Also shown are the chemical spinodal where $\partial^2 f / \partial c^2 = 0$ and the coherent spinodal for which $\partial^2 f / \partial c^2 + 2Y\eta^2/(1-v) = 0$. (b) Atomic probe tomography maps of Cr atoms in alloys of Fe–27at.%Cr and Fe–38at.%Cr after annealing at a temperature of 85% of the peak temperature of the coherent spinodal, approximately as indicated in the phase diagram of part a. Image "n" shows characteristics of nucleation and growth, whereas "s" shows characteristics of spinodal decomposition. Horizontal width is 8 nm. Reprinted from [279], with permission from Elsevier.

The lower curve, the "coherent spinodal," is the limit of stability with elastic contributions included. It is found by displacing the chemical spinodal downward by an amount equal to the volume integral of $2\eta^2 Y/(1 - v)$ in Eq. 16.57. This contribution can be large in many solid solutions.

Between the solid curve for the miscibility line and the dashed line for the coherent spinodal in Fig. 16.5a, the solid solution is unstable against unmixing by nucleation and growth (the region labeled "metastable" in Fig. 16.2). Figure 16.5b shows the distribution of Cr atoms in two Fe–Cr alloys that were prepared as solid solutions, then annealed to induce unmixing. Their chemical compositions were on the two sides of the spinodal line, as marked approximately by the "n" and "s" in the phase diagram of Fig. 16.5. Notice the different morphologies of these partly unmixed alloys – the alloy "n" (having a composition between the miscibility and spinodal lines) shows characteristics of nucleation and growth, and the alloy "s" (having a composition within the assessed spinodal) shows characteristics of continuous unmixing [279].

16.5.2 Kinetic Effects of Elastic Energy

The elastic energy from unmixing atoms of different sizes also alters the kinetics of the phase transformation. With Eq. 16.56, the linearized form of the diffusion equation (16.48) has a solution in Fourier space given by

$$A(Q, t) = A(Q, t = 0) \exp\left[-MQ^2 \left(\frac{\partial^2 f}{\partial c^2} + 2\eta^2 \frac{Y}{1 - v} + 2\kappa Q^2\right) t\right], \tag{16.58}$$

where the "amplification factor" $R(Q)$ in the exponential is

$$R(Q) = -MQ^2 \left(\frac{\partial^2 f}{\partial c^2} + 2\eta^2 \frac{Y}{1 - v} + 2\kappa Q^2\right). \tag{16.59}$$

In comparison with Eqs. 16.50 and 16.51, the new positive term with the elastic constants decreases the growth rates of concentration waves. The positive elastic energy contribution slows the phase transformation. As with Eq. 16.53, setting $dR/dQ = 0$ gives the wavevector Q_{max} for most rapid growth

$$Q_{\mathrm{max}}^2 = -\frac{1}{4\kappa} \left(\frac{\partial^2 f}{\partial c^2} + 2\eta^2 \frac{Y}{1 - v}\right), \tag{16.60}$$

with the amplification factor

$$R_{\mathrm{max}} = 2\kappa M Q_{\mathrm{max}}^4. \tag{16.61}$$

If the elastic energy plays an important role in the growth rate (as may be expected when the atom sizes differ substantially and η^2 is large), the orientation dependence of the Young's modulus, Y is important. Spinodal decomposition in anisotropic systems is characterized by preferential growth of concentration waves along elastically soft directions [115]. For example, Fig. 6.12c for a Zener ratio $A > 1$ shows that elastic distortions along $\langle 1\,0\,0 \rangle$ cost less energy. We expect spinodal unmixing to be fastest with Q-vectors along the $\langle 1\,0\,0 \rangle$ directions when $A > 1$.

Problems

16.1 Explain why the chemical free energy cannot depend linearly on the magnitude of the composition gradient, i.e., $|\vec{\nabla}c|$.

16.2 (**a**) What are the units of κ in Eq. 16.18?

   (**b**) Calculate an approximate value of κ for an equiatomic alloy that undergoes spinodal unmixing. Suppose the wavelength, λ, for $Q_c = 2\pi/\lambda$ is 10 nm. Suppose the critical temperature for unmixing is 1000 K, and the spinodal decomposition occurs at 500 K.

   (*Hint*: Consider the chemical free energy expressions such as Eqs. 2.28–2.31 and their critical temperature.)

16.3 At a particular time, suppose a concentration fluctuation has the form

$$\delta c(x) = \frac{\delta_0}{2}\left[e^{iQx} + e^{-iQx}\right]. \tag{16.62}$$

With this expression, derive the growth rate $R(Q)$ of Eq. 16.51.

(*Hint*: When using Eq. 16.62 in Eq. 16.50, one term will diverge if the limits are taken at infinity. On physical grounds, you can argue that for c in units of atoms per unit length, the divergent integral should be N, the number of atoms. Fortunately, this N is the same for any value of Q.)

16.4 In the point approximation, the phase diagram for continuous unmixing can have the same shape as the Gorsky–Bragg–Williams order parameter. Specifically, after a 90 degree rotation, the phase boundary for the unmixing phase diagram from $0 < c < 0.5$ in Fig. 2.17 can have a shape identical to $L(T)$ in the ordering diagram of Fig. 2.21.

   (**a**) Setting $e_{AA} = e_{BB}$ in Eq. 2.40, use this energy $E(c)$ and configurational entropy S_{conf} of Eq. 2.44 to find the condition of $dF/dc = 0$. Why is this pertinent to the common tangent construction?

   (**b**) Using the substitution $L' = 1 - 2c$, obtain Eq. 2.55. What does this substitution tell you about how to reconfigure the curve of Fig. 2.17 so it overlaps with the $L(T)$ curve of Fig. 2.21?

   (*Hint*: Remember to change the sign of V for unmixing and ordering.)

   (**c**) If the condition $e_{AA} = e_{BB}$ were removed, explain the change in the unmixing phase diagram. Is this relevant to the ordering problem? Why or why not?

   (**d**) Can you think of a physical relevance for the dashed curve of Fig. 2.18b on an ordering phase diagram?

17

Chapters 12–14 discussed phase transformations where melting and precipitation occurred as distinct regions with distinct interfaces. The specific surface energy σ, multiplied by the area of interface, was essential to the thermodynamics and kinetics. Chapter 16 explained the concept of a square gradient energy $\kappa(\overrightarrow{\nabla c})^2$. When integrated over volume, $\kappa(\overrightarrow{\nabla c})^2$ served much like a surface energy. This second approach is conceptually more general than the first because it is possible to obtain a sharp interface and a specific surface energy σ in the limit as the concentration gradient becomes very steep. "Phase field theory," described in this chapter, works with the composition profile as a field in space, and identifies the equilibrium states or time evolution of composition gradients and interfaces [211, 212, 280]. The phase field method can be used for problems other than chemical unmixing and hence its name – the different *phases* in a material are represented as *fields* in space. The idea is illustrated in Fig. 17.1, which shows a one-dimensional trace through a polycrystalline microstructure, and how a set of order parameters $\{\eta_j\}$ vary along the trace.

This chapter first describes the phase field concept in more detail, and mentions its use for solidification problems, where it has had considerable success. An example of a free energy function is presented, and the different equations for the evolution of

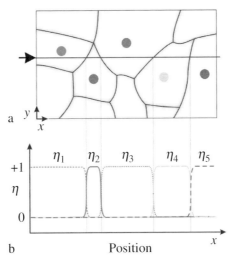

Figure 17.1 Illustrative treatment of grain boundary interfaces with the phase field method. (**a**) Grain structure of polycrystalline material, with each crystallite described by an order parameter η_j. (**b**) Dependence of order parameters on position. After image by Cenna, Wikipedia (2012).

a conserved order parameter (e.g., composition) and a nonconserved order parameter (e.g., spin orientation) are discussed. The Ginzburg–Landau equation is presented, and its volume free energy is used (as the Landau potential) to calculate chemical unmixing near the critical temperature. The structure of an interface, especially its width, is analyzed for the typical case of an antiphase domain boundary, and the dynamics of domain growth are described. Much of the interest in phase field methods stems from advances in computing. Numerical methods for integrating partial differential equations in a phase field model can be more convenient than handling the boundary conditions at moving interfaces between phases. Such methods are beyond the scope of the present chapter, however.

17.1 Spatial Distribution of Phases and Interfaces

17.1.1 Phases as Distinct Volumes with Abrupt Interfaces

To this point we have considered a phase as a region of material that is distinct, compact, and with spatially uniform atom arrangements. A phase is therefore a distinguishable region with uniform properties, located in a well-defined volume. This is a useful definition, and consistent with our experience that we can clearly distinguish a solid from a liquid, for example. For a simple binary alloy, the solid and liquid phases had separate free energy functions, $F_s(c, T)$ and $F_l(c, T)$. At a fixed temperature, there was a specific composition for the solid and a specific composition for the liquid given by the common tangent rule (Fig. 2.5). This distinctness of the two phases was relaxed somewhat when we considered chemical unmixing on an Ising lattice (Fig. 2.17). The composition was allowed to be a continuous variable, and at low temperatures the unmixing was understood with a single $F(c, T)$ curve having minima at two different compositions. As the temperature increased to T_c, however, the distinctness of the two unmixed phases became infinitesimally small.

One way to follow the growth of a microstructure in a material is by tracking the motions of interfaces, such as the velocity of the solidification front during solidification, or the motions of other interfaces in solids (e.g., Sections 11.6 and 14.6). If the two phases are distinct, their interfaces are abrupt and their trajectories are easy to identify.

17.1.2 Phases as Field Quantities with Diffuse Interface

In contrast, this chapter assumes diffuse interfaces between phases. Chapter 16 gave physical examples of such "interfaces" in spinodal decomposition. The chemical composition varied continuously between different regions, and the free energy had a contribution from the square of the composition gradient. We now generalize this approach, where interfaces between phases have state variables that are intermediate between those of the two adjacent phases. New terms in the free energy include the squared gradients of the field variables. The method can be extended to complex materials (Section 9.3) with multiple field variables and multiple gradients.

We first parameterize the phases throughout a material with a continuous number. This number is much like the order parameter, L, used to describe chemical long-range order in Section 2.9. Our interest at the time was finding the thermodynamic value of $L(T)$, giving the order as a function of temperature. There is no reason why L cannot also vary with position, so we can make a spatial field of the LRO parameter in the material, $L(\vec{r})$. In an alloy with domains as shown in Fig. 7.3, L could switch from, say, $+1$ to -1 across the antiphase domain boundary in the middle of the figure. What is new is that the order parameter is now allowed to take intermediate values in the transition region between the two domains.

In "phase field theory" we do the same for phases with the variable $\eta(\vec{r})$. For example, we assign $\eta = 0.0$ for regions of space that are liquid, and $\eta = 1.0$ for regions of space that are solid. We also allow the phase variable $\eta(\vec{r})$ to take intermediate values at the transition between liquid and solid. The free energy is no longer two functions, $F_s(c, T)$ and $F_l(c, T)$ that are confined to regions of solid or liquid, but is now one functional of the function η, i.e., $F(\eta, c, T)$ that extends over all space. For solidification, this free energy functional must be compatible with the materials physics by having two minima at $\eta = 0$ and $\eta = 1$, and no minimum at, for example, $\eta = 0.5$. An additional minimum could produce an additional phase that we do not want, or it might even violate the Gibbs phase rule of Section 2.5.

Figure 17.2a shows a $F(\eta, c, T')$ that is acceptable for solidification. At this temperature, T', the functional F has a minimum for the liquid ($\eta = 0$) at large c, and a minimum for the solid ($\eta = 1$) at small c. (Evidently the B-rich solid melts at a lower temperature than the A-rich solid.) Free energy versus composition curves for three different values of η are shown in Fig. 17.2b. The common tangent intersects only the curves for $\eta = 0$ and $\eta = 1$, and misses the curve for $\eta = 0.5$. Equilibrium is predicted between one solid phase and one liquid phase, and no additional phase is predicted with this free energy functional.

Because $F(\eta, c, T')$ of Fig. 17.2a is continuous in η and c, a transition between two minima of F occurs across a region of space with intermediate values of η and c.

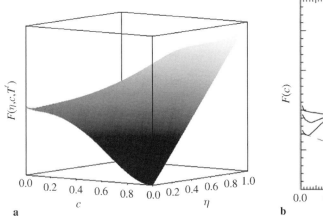
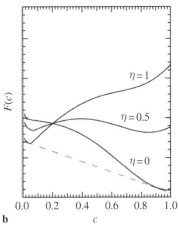

Figure 17.2 (a) Typical function $F(\eta, c, T')$ at fixed T' for both solid and liquid phases, and (b) three curves $F(c, T')$ for different η. The dashed curve, the common tangent, intersects the curves for $\eta = 0$ and $\eta = 1$.

The interface between the solid and liquid phases is therefore diffuse. This is not physically accurate, of course.[1] In other cases, interfaces really are physically diffuse, and the phase field method gives physical information about how a diffuse interface varies in chemical order or in chemical composition. In other situations when a diffuse interface is not physically realistic, the phase field method can still be useful for computational modeling, especially if it predicts an interface that is narrow in width.

Phase field calculations are well suited for numerical simulations of solidification dynamics, and this is an active field of research. Some advances in this field are well beyond the scope of this text, such as the adaptive mesh refinement techniques that allow accurate treatment of small structures. Furthermore, the governing equations are often specific to particular problems. Nevertheless, simulation results such as those in Fig. 13.6 are informative, and Section 13.1 explained some of their features.

17.2 Order Parameters as Field Quantities

17.2.1 Kinetics of Conserved and Nonconserved Order Parameters

Unlike chemical composition, many order parameters are not conserved quantities. Magnetic order parameters are not conserved, for example, because a spin can change orientation without a corresponding change in the other spins. The magnetic alignment can change in one region without compensation in another region. In general, an ordered domain of one type can grow at the expense of its neighbor because at the domain boundary, an atom, a pair of atoms, or a spin can change domain affiliation independently of its neighbors. The conservation or nonconservation of field variables must be consistent with the kinetic mechanism. Three important mechanisms are listed in Table 17.1, and illustrated in Fig. 17.3. They are important mechanisms for changing atoms or spins in kinetic Monte Carlo simulations. The three differ in their timescales, however, with the vacancy mechanism usually being slowest and Glauber dynamics fastest.

Assuming transition probabilities with characteristics of detailed balance (described in Section 5.6.1), all three kinetic mechanisms of Table 17.1 eventually bring a system to a state of equilibrium that is set by temperature and interaction energies. However, when the number of nearest-neighbor bonds that are changed by a kinetic step differs from the coordination number z per moving atom, the state of equilibrium differs from the expected state of thermodynamic equilibrium (see Section 23.3.3 [online]).

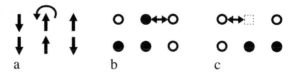

Figure 17.3 Kinetic mechanisms. (**a**) Glauber. (**b**) Kawasaki. (**c**) Vacancy.

[1] Sometimes it can be accurate in an average sense, however.

Table 17.1 Kinetic mechanisms				
Name	Mechanism	1nn bonds changed	Conservative	Efficiency
Glauber	Flip spins individually	z	No	High
Kawasaki	Interchange pair of spins/atoms	$2(z-1)$ for 1nn, $2z$ for distant pairs	Yes	Medium
Vacancy	Interchange atom and vacancy	$z-1$	Yes	Low

For their evolution in time, conserved and nonconserved order parameters obey different governing equations [281]. Details vary, but some general features are:

- The evolution of a conservative parameter, c for example, was developed in differential form in Eq. 16.48. In general form it is[2]

$$\frac{\partial c}{\partial t} = \vec{\nabla} \cdot \left[M_c \, \mathbf{grad}\left(\frac{\partial F_T}{\partial c} \right) \right]. \tag{17.1}$$

- The conservation of solute imposes a constraint on Eq. 17.1, typically requiring Lagrange multipliers as in Section 16.3.2. Nonconserved order parameters avoid the need for such Lagrange multipliers.
- Nonconserved order parameters do not use a diffusion equation (e.g., Eq. 17.1) for their time evolution. For nonconserved order parameters, usually the kinetics are proportional to the steepest gradient in total free energy, and the field variables follow this path directly. The evolution of a nonconserved parameter, η for example, is simply proportional to a mobility times the effect of η on the free energy

$$\frac{\partial \eta}{\partial t} = -M_\eta \, \frac{\partial F_T}{\partial \eta}. \tag{17.2}$$

In a phase field model that includes both c and η, both $\partial c/\partial t$ and $\partial \eta/\partial t$ must be integrated simultaneously to follow the time evolution of the system because they are coupled through $f(\eta, c, T)$ in Eq. 17.10. Equations governing the time evolution of conservative fields tend to be more complicated than equations for nonconserved ones, i.e., they have higher-order spatial derivatives. Nevertheless, both the Cahn–Hilliard equation 16.18 and the Ginzburg–Landau equation 17.3 include contributions from the squared gradient of field variables, allowing phase field models to include interfaces and other heterogeneities.

17.2.2 Ginzburg–Landau Equation

The starting point for phase field models is often the Ginzburg–Landau equation

$$f(\eta) = A\eta^2 + B\eta^4 + C(\overrightarrow{\nabla \eta})^2, \tag{17.3}$$

---

[2] Equation 17.1 is a diffusion equation, where $\partial c/\partial t$ equals the (negative) divergence of flux. The flux, as usual, is M_c times the (negative) gradient of the chemical potential, $\partial F_T/\partial c$.

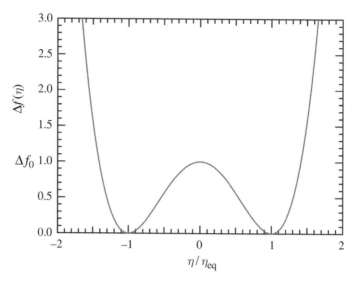

Figure 17.4 Normalized Landau potential of Eq. 17.4.

which expresses the free energy density in terms of an order parameter $\eta(\vec{r})$ and its gradient. The volume free energy density $\Delta f(\eta)$ (Eq. 17.3 without the square gradient term) is a "Landau potential." It is conveniently written in normalized form as

$$\Delta f(\eta) = \Delta f_0 \left[1 - 2\left(\frac{\eta}{\eta_{eq}}\right)^2 + \left(\frac{\eta}{\eta_{eq}}\right)^4 \right]. \tag{17.4}$$

This normalized Landau potential, shown in Fig. 17.4, has two minima at zero for $\eta = \pm\eta_{eq}$, and a local maximum of $+\Delta f_0$ at $\eta = 0$.

17.2.3 An Unmixing Phase Diagram with Landau Theory

We revisit the chemical unmixing phase diagram, using the volume part of Eq. 17.3 (without the square gradient term). In Section 2.7, expressions for the configurational energy $E(c)$ and a configurational entropy term $-TS(c)$ were put in competition at different temperatures. At the critical temperature for unmixing, T_c, there was a precise cancellation of the curvatures of these two contributions to the free energy at $c = 1/2$.

Here we do the same sort of analysis with the Landau potential, using it as a difference in free energy, ΔG, between the unmixed alloy and the solid solution. (Individual terms for energy and entropy are not considered.) The solid solution is stable at high temperatures, where the curvature, $\partial^2\Delta G/\partial c^2$, must be positive. This curvature is zero at T_c, and is negative below T_c. The simplest way to get this behavior is with a linear temperature dependence of the curvature, so it goes from positive to negative as ΔT goes through 0 at T_c, where $\Delta T \equiv T - T_c$

$$\frac{\partial^2 \Delta G}{\partial c^2} \equiv \kappa_2 \frac{\Delta T}{T_c}. \tag{17.5}$$

The κ_2 is assumed to be a positive constant near $\Delta T = 0$ and $c = 1/2$.

A ΔG with the form of Fig. 17.4 is constructed by adding a quartic term to Eq. 17.5

$$\Delta G(c, \Delta T) = \kappa_4 (\Delta c)^4 + \kappa_2 \frac{\Delta T}{T_c} (\Delta c)^2, \tag{17.6}$$

where $\Delta c \equiv c - 1/2$, so $\Delta c = 0$ for a solid solution of $c = 1/2$. The κ_4 is also assumed to be a positive constant. In a Taylor series treatment[3] it would be $\kappa_4 = 1/4! \, \partial^4 \Delta G / \partial c^4$, evaluated at T_c. Equation 17.6 does not have terms that are linear or cubic in Δc. In the subsequent step of finding a common tangent, a linear term is unimportant because taking a derivative with respect to Δc gives a constant offset. Addressing the cubic term brings deeper questions that we defer to Section 18.4.2, but Eq. 17.6 is useful without it.

To find the common tangent, take the derivative of Eq. 17.6 with respect to Δc (noting that $d(\Delta c) = dc$). The ΔG is symmetric about $\Delta c = 0$, so the compositions with zero derivative are on a common tangent, and give the phase boundary

$$0 = \frac{\partial \Delta G}{\partial c} = 4 \kappa_4 (\Delta c)^3 + 2 \kappa_2 \frac{\Delta T}{T_c} (\Delta c), \tag{17.7}$$

$$\Delta c = \sqrt{-\frac{2 \kappa_2}{4 \kappa_4} \frac{\Delta T}{T_c}}, \tag{17.8}$$

$$\Delta c = \pm \sqrt{\frac{\kappa_2}{2 \kappa_4} \left(1 - \frac{T}{T_c} \right)}. \tag{17.9}$$

Equations 17.8 and 17.9 are valid only below T_c, where $\Delta T < 0$. They give a parabolic shape for the unmixing phase diagram, with $\Delta c \propto \sqrt{|\Delta T|}$. At larger Δc, this is not quite the shape of the unmixing phase diagram of Fig. 2.17.[4] Nevertheless, this Landau theory offers good value for the effort required, especially for behavior near critical conditions (e.g., just below T_c when Δc is small). It can predict other physical behavior, as in Section 19.3, where the displacement behavior of Eq. 19.39 is similar to the unmixing behavior of Eq. 17.9.

17.2.4 Extensions to Multiple Fields

The function $F(\eta, c, T)$ of Fig. 17.2 depends on three field variables that can have some independence. This figure shows how the volume part of the free energy varies with η and c. There are also energy costs for sharp variations in these quantities – gradients in both composition and phase fields will have such terms, giving a total free energy of the form

$$F_T(\eta, c, T) = \int_{vol} \kappa_c (\overrightarrow{\nabla c})^2 + \kappa_\eta (\overrightarrow{\nabla \eta})^2 + f(\eta, c, T) \, dV. \tag{17.10}$$

For minimizing $F_T(\eta, c, T)$ with the Euler equation, the square gradients of c and η are not treated in the same way because, as described in Section 17.2.1, one is conservative and the other is not.

---

[3] It is possible to obtain Eq. 17.6 from a Taylor series expansion of the free energy about $\Delta T = 0$ and $\Delta c = 0$ (see Section 18.3).

[4] Near T_c, however, both approaches predict the critical exponent of 1/2 for how Δc depends on ΔT, so both are classical theories in the context of Section 27.4 [online].

Other field variables can be included in a more complete model of a material. Magnetic fields, which are vector fields, vary with η, c, and T. Strain fields, which are tensor fields, also follow variations in the other field variables, and may alter these other fields when the strain energy is important. In a complex microstructure, these spatial variations in the different types of fields generally do not track each other in straightforward ways. Locations with the largest square gradients $(\overrightarrow{\nabla \eta_j})^2$ of different fields $\{\eta_j\}$ may not be co-located. Furthermore, the energetic relationships between the different field quantities, such as a magnetoelastic energy, are not necessarily linear in the field quantities.

The governing equations for the kinetic evolution, or even the equilibrium structures, are therefore not practical for analytical solutions. Calculating the evolution of a complex microstructure invariably requires computer simulations. This is an active topic of current research.

17.3 Domain Boundary Structure

17.3.1 General Features of Domain Boundaries

The Ginzburg–Landau equation can be used to calculate an equilibrium structure, and its evolution with time. Here the Ginzburg–Landau equation is used to calculate the width of a flat interface between ordered domains. The time evolution of a three-dimensional interface is described in Section 17.4.

Chemical order is a good example of a field with a nonconserved order parameter. Figure 7.3 shows two ordered domains with an antiphase domain boundary (APDB) between them. By interchanging pairs of A- and B-atoms near the boundary, one domain can grow at the expense of the other, so the areas of the two domains are not conserved. These dynamics of boundary motion do not require long-range diffusion, unlike spinodal decomposition in Chapter 16 (which also required a Lagrange multiplier to ensure solute conservation).

The boundary in Fig. 7.3 is abrupt, but in general there will be some disorder around the boundary, at least at intermediate temperatures. This disorder is to be understood in an average sense – the atoms and their local configurations at the boundaries are, of course, discrete. Figure 17.5 shows this situation, where a domain on the left has an order parameter of –0.8 and the domain on the right has an order parameter of +0.8. The order parameter undergoes a transition across the APDB. The goal of this section is to calculate the equilibrium width and profile of the transition region.

Away from the boundary, the order parameters in the two domains have the equilibrium values $|\eta_0|$ for the temperature of interest.[5] The sign of the order parameter is useful for identifying the domain, but it does not alter the free energy.[6] The volume free

---

[5] The temperature for $L = \pm 0.8$ would be approximately $T = 0.78\,T_c$, based on Fig. 7.5.
[6] For example, a chessboard with its black square on the upper left has the same number of black–white pairs as a rotated chessboard with its white square at the upper left.

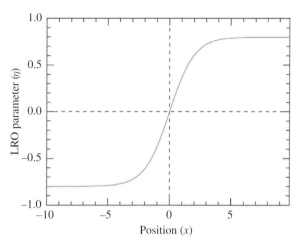

Figure 17.5   Variation of order parameter across antiphase domain boundary at zero position.

energy therefore includes only even powers of the order parameter. The Ginzburg–Landau equation (17.3) has the right form. What is of interest is the cost in free energy of the interface, not the volume free energy $F_T(\eta_0)$ at the equilibrium value of η_0. This is given by the difference

$$\Delta F_{APDB} = F_T(\eta(\vec{r})) - F_T(\eta_0), \tag{17.11}$$

$$\Delta F_{APDB} = \int_{vol} \left(f(\eta(\vec{r})) + \kappa \, (\overrightarrow{\nabla\eta})^2 \right) - f(\eta_0) \, dV, \tag{17.12}$$

$$\Delta F_{APDB} = \int_{-\infty}^{\infty} \Delta f(\eta(x)) + \kappa \left(\frac{d\eta}{dx} \right)^2 \, dx, \tag{17.13}$$

where the last equation was a transition to one dimension, and

$$\Delta f(\eta(x)) \equiv f(\eta(x)) - f(\eta_0). \tag{17.14}$$

17.3.2 Finding the Optimal Boundary Profile

Minimizing a free energy that depends on a field quantity and its gradient, i.e., $f(\eta, \overrightarrow{\nabla\eta})$, requires the Euler equation from variational calculus as described in Section 16.3.1. The Euler equation for our antiphase domain boundary problem, the equivalent of Eq. 16.31, is

$$\frac{\partial f_i}{\partial \eta} - \frac{d}{dx} \frac{\partial f_i}{\partial \frac{d\eta}{dx}} = 0, \tag{17.15}$$

where f_i is the integrand of Eq. 17.13. Following the prescribed steps from the calculus of variations, a minimum of ΔF_{APDB} is found by applying Eq. 17.15 to the integrand of Eq. 17.13, giving a condition on $\Delta f(\eta)$

$$\frac{\partial \Delta f(\eta)}{\partial \eta} - \frac{\mathrm{d}}{\mathrm{d}x} 2\kappa \frac{\mathrm{d}\eta}{\mathrm{d}x} = 0, \tag{17.16}$$

$$\frac{\partial \Delta f(\eta)}{\partial \eta} - 2\kappa \frac{\mathrm{d}^2 \eta}{\mathrm{d}x^2} = 0, \tag{17.17}$$

$$\frac{\partial \Delta f(\eta)}{\partial \eta} = 2\kappa \frac{\mathrm{d}^2 \eta}{\mathrm{d}x^2}. \tag{17.18}$$

An expression for $\Delta f(\eta)$ is found by integrating over η

$$\Delta f(\eta) = 2\kappa \int \frac{\mathrm{d}^2 \eta}{\mathrm{d}x^2} \, \mathrm{d}\eta. \tag{17.19}$$

Here it is more convenient to change to integration over x

$$\Delta f(\eta) = 2\kappa \int \frac{\mathrm{d}^2 \eta}{\mathrm{d}x^2} \frac{\mathrm{d}\eta}{\mathrm{d}x} \, \mathrm{d}x. \tag{17.20}$$

The following relationship is useful for rewriting the integrand of Eq. 17.20

$$\frac{\mathrm{d}}{\mathrm{d}x} \left(\frac{\mathrm{d}\eta}{\mathrm{d}x} \right)^2 = 2 \frac{\mathrm{d}\eta}{\mathrm{d}x} \frac{\mathrm{d}^2 \eta}{\mathrm{d}x^2}, \tag{17.21}$$

so Eq. 17.20 becomes

$$\Delta f(\eta) = \kappa \int \left[\frac{\mathrm{d}}{\mathrm{d}x} \left(\frac{\mathrm{d}\eta}{\mathrm{d}x} \right)^2 \right] \mathrm{d}x, \tag{17.22}$$

$$\Delta f(\eta) = \kappa \left(\frac{\mathrm{d}\eta}{\mathrm{d}x} \right)^2, \tag{17.23}$$

and

$$\frac{\mathrm{d}\eta}{\mathrm{d}x} = \sqrt{\frac{\Delta f(\eta)}{\kappa}}. \tag{17.24}$$

Equation 17.23 is used to rewrite our expression for the minimum change in free energy caused by the presence of an antiphase domain boundary (Eq. 17.13)

$$\Delta F_{\mathrm{APDB}} = 2 \int_{-\infty}^{\infty} \Delta f(\eta) \, \mathrm{d}x. \tag{17.25}$$

It is more convenient to integrate over the range of the order parameter η, instead of the infinite range of x

$$\Delta F_{\mathrm{APDB}} = \int_{-\eta_{\mathrm{eq}}}^{\eta_{\mathrm{eq}}} \Delta f(\eta) \frac{\mathrm{d}x}{\mathrm{d}\eta} \, \mathrm{d}\eta, \tag{17.26}$$

and using Eq. 17.24

$$\Delta F_{\mathrm{APDB}} = \int_{-\eta_{\mathrm{eq}}}^{\eta_{\mathrm{eq}}} \Delta f(\eta) \sqrt{\frac{\kappa}{\Delta f(\eta)}} \, \mathrm{d}\eta, \tag{17.27}$$

$$\Delta F_{\mathrm{APDB}} = \int_{-\eta_{\mathrm{eq}}}^{\eta_{\mathrm{eq}}} \sqrt{\kappa \, \Delta f(\eta)} \, \mathrm{d}\eta. \tag{17.28}$$

Not surprisingly, the free energy cost of an APDB increases with κ, which gives the strength of the square gradient term. The square-root dependence on κ reflects the possibility of altering the width of the boundary for different κ, as shown next.

17.3.3 Boundary Profile with Ginzburg–Landau Potential

We now use the Landau potential, Eq. 17.4, to specify the volume contribution to the free energy, $\Delta f(\eta)$, in Eq. 17.28. Using it in Eq. 17.24 allows us to get an expression for the characteristic width of the APDB

$$\frac{d\eta}{dx} = \sqrt{\frac{\Delta f_0}{\kappa}} \left[1 - 2\left(\frac{\eta}{\eta_{eq}}\right)^2 + \left(\frac{\eta}{\eta_{eq}}\right)^4 \right]^{1/2}. \tag{17.29}$$

In the limit of small η

$$\frac{d\eta}{dx} = \sqrt{\frac{\Delta f_0}{\kappa}}, \tag{17.30}$$

$$dx = \sqrt{\frac{\kappa}{\Delta f_0}}\, d\eta, \tag{17.31}$$

$$\bar{x} = \sqrt{\frac{\kappa}{\Delta f_0}}\, \eta_{eq}, \tag{17.32}$$

where $\bar{x}$ is a characteristic width, obtained by integrating Eq. 17.31 over the range of $-\eta_{eq} < \eta < +\eta_{eq}$. The characteristic width of the APDB, $\bar{x}$, depends on the relative magnitude of the square gradient energy through κ, and the magnitude of the volume free energy through Δf_0.

A complete solution for the shape of the APDB can be obtained by rearranging Eq. 17.29

$$\int \frac{1}{\left[1 - 2\left(\frac{\eta}{\eta_{eq}}\right)^2 + \left(\frac{\eta}{\eta_{eq}}\right)^4 \right]^{1/2}}\, d\eta = \sqrt{\frac{\Delta f_0}{\kappa}} \int dx. \tag{17.33}$$

The integral on the right is trivial, but the scary integral evaluates to a clean analytic result

$$\eta_{eq} \tanh^{-1}\left(\frac{\eta}{\eta_{eq}}\right) = \sqrt{\frac{\Delta f_0}{\kappa}}\, x, \tag{17.34}$$

which gives immediately the profile of the antiphase domain boundary, $\eta(x)$

$$\eta(x) = \eta_{eq} \tanh\left(\frac{1}{\eta_{eq}} \sqrt{\frac{\Delta f_0}{\kappa}}\, x\right). \tag{17.35}$$

This is the profile shown in Fig. 17.5 (see also [282]). This profile is physically reasonable. The width increases with the square gradient energy (larger κ), because a wider interface minimizes the gradient energy cost. On the other hand, the difference in volume free energy (the Δf_0, without the square gradient term) favors as much volume as possible with the full equilibrium value of η_{eq}, and hence favors a narrow interfacial region. (Finally, for

the same gradient $d\eta/dx$, the interface width increases in proportion to the value of η_{eq}.)
Section 21.9.2 applies the same reasoning to the width of an interface between magnetic
domains, another case of a nonconserved order parameter.

17.4 Domain Boundary Kinetics

17.4.1 Nonconservative Dynamics

Chapter 16 developed the kinetic theory of spinodal decomposition with a formulation
analogous to what we will use here for the kinetics of domain interface motion with phase
field theory. The important difference is that chemical composition is a conserved quantity,
whereas ordered domains are not. In both cases, the mechanism involves A- and B-atoms
exchanging positions,[7] and the overall chemical composition is unchanged. On the other
hand, the interchange of neighboring A- and B-atoms between the α- and β-sublattices
does not conserve the sizes of the ordered domains. Over time, one domain grows at the
expense of the other, reducing the area of domain boundaries to minimize the interface
energy. Actually, the "conservative dynamics" of spinodal decomposition is the more
difficult problem – the constraint of constant composition required a Lagrange multiplier
in Section 16.3.2, which became a chemical potential (i.e., the free energy per atom). The
kinetics were driven by atom diffusion down the gradient of the chemical potential, so a
diffusion equation had to be adapted for the kinetics of spinodal decomposition.

   For the "nonconserved dynamics" of antiphase domain boundaries, we impose no
constraint on ordered domains that requires a Lagrange multiplier. We work with the
Euler equation (17.15) directly, minimizing Eq. 17.13. The three-dimensional analog of
Eq. 17.17 is

$$\frac{\delta \Delta F_{\text{APDB}}}{\delta \eta} = \frac{\partial \Delta f}{\partial \eta} - 2\kappa \nabla^2 \eta. \tag{17.36}$$

The previous section used this equation in one dimension, setting it equal to zero to identify
the minimum of ΔF_{APDB} for a one-dimensional APDB.

17.4.2 Stability of a Flat Interface

For kinetics, however, a one-dimensional model has limited value because there can be
no curvature in one dimension.[8] All interfaces in one dimension are flat, and therefore
stable. It is, nevertheless, useful to confirm that a flat APDB does not move. For this case
we set to zero the time dependence for the relaxation of ΔF_{APDB}. As discussed previously

---

[7] Atom movement is likely through a vacancy mechanism, which brings the issues of Chapter 10.
[8] A curved interface is described by a functional dependence of one coordinate on another. This cannot be done
   with one coordinate, of course.

in Section 5.6.2, for developing a kinetic theory the essential idea is that the free energy decreases with time in proportion to the mobility, M

$$\frac{\partial \eta}{\partial t} = -M \left(\frac{\delta \Delta F_{\mathrm{APDB}}}{\delta \eta} \right). \tag{17.37}$$

For our one-dimensional problem (which also pertains to a flat interface in three dimensions), we use results from Section 17.3.2 to minimize the variation of F_{APDB} with respect to the variation in η

$$\frac{\partial \eta}{\partial t} = -M \left(\frac{\partial f_i}{\partial \eta} - \frac{\mathrm{d}}{\mathrm{d}x} \frac{\partial f_i}{\partial \frac{\mathrm{d}\eta}{\mathrm{d}x}} \right) = 0, \tag{17.38}$$

$$\frac{\partial \eta}{\partial t} = -M 2\kappa \int_{-\infty}^{\infty} \left(\frac{\mathrm{d}\eta}{\mathrm{d}x} \right)^2 \mathrm{d}x = 0, \tag{17.39}$$

where Eq. 17.39 was obtained from results of Eqs. 17.15 and 17.23. The result is zero. This is correct because in one dimension the free energy does not depend on the position of the APDB. In three dimensions, this corresponds to a flat interface, which has no thermodynamic tendency to move forward or backward.

17.4.3 Curvature and Growth Kinetics

To study dynamics, we need to work in two or three dimensions. In both cases, we need to align a variable such as x along the direction normal to the APDB, and determine the velocity of the interface along this direction. The natural coordinate system follows the shape of the three-dimensional APDB.[9] The text in Box 17.1 provides a mathematical result for the Laplacian of Eq. 17.36 using a curvilinear coordinate $\vec{\chi}$ that is perpendicular to the APDB. This allows us to use some of our prior results from one-dimensional analysis of an APDB along x to develop the Allen–Cahn theory of APDB dynamics [283]. Returning to Eq. 17.36, using the result from Eq. 17.55

$$\frac{\delta \Delta F_{\mathrm{APDB}}}{\delta \eta} = \left[\frac{\partial \Delta f}{\partial \eta} - 2\kappa \frac{\partial^2 \eta}{\partial \chi^2} \right] - 2\kappa \frac{\partial \eta}{\partial \chi} \vec{\nabla} \cdot \vec{\chi}. \tag{17.40}$$

The first two terms in square brackets in Eq. 17.40 pertain to a flat interface with a normal along $\vec{\chi}$. By Eq. 17.17 this bracketed quantity is zero, so

$$\frac{\delta \Delta F_{\mathrm{APDB}}}{\delta \eta} = -2\kappa \frac{\partial \eta}{\partial \chi} \vec{\nabla} \cdot \vec{\chi}. \tag{17.41}$$

Finally, we substitute into our kinetic equation (17.37) to obtain

$$\frac{\partial \eta}{\partial t} = -M'(K_1 + K_2) \frac{\mathrm{d}\eta}{\mathrm{d}\chi}, \tag{17.42}$$

---

[9] For a spherical domain, for example, it is natural to use a spherical coordinate system with r normal to the surface of the domain to analyze the effect of curvature on growth velocity.

where $K_1 + K_2$ is the mean curvature from the two orthogonal directions on the surface of the APDB. It is related to the divergence as

$$K_1 + K_2 = -\vec{\nabla} \cdot \vec{\chi}. \tag{17.43}$$

A velocity of APDB motion, v, is defined as

$$v \equiv -\frac{\frac{d\eta}{dt}}{\frac{d\eta}{d\chi}}, \tag{17.44}$$

giving the simple expression for curvature-driven interface motion[10]

$$v = M'(K_1 + K_2). \tag{17.45}$$

This seems at first an intuitive equation, stating that a small domain of high curvature will shrink most rapidly. Less intuitive is the definition of a mean curvature in three dimensions, as discussed next.

Box 17.1   **Some Results in Curvilinear Coordinates**

In any curvilinear coordinate system with the coordinates l, m, n, we can calculate incremental lengths along the coordinate directions with the Pythagorean theorem and incremental lengths in a Cartesian system. For example, along the l-direction we define

$$h_l \equiv \sqrt{\left(\frac{\partial x}{\partial l}\right)^2 + \left(\frac{\partial y}{\partial l}\right)^2 + \left(\frac{\partial z}{\partial l}\right)^2}, \tag{17.46}$$

so for a small change of dl in l, the incremental change of distance is $ds = h_l dl$. There are similar expressions for h_m and h_n. In curvilinear coordinates we can likewise adapt expressions for the gradient and divergence that are familiar from Cartesian coordinates

$$\vec{\nabla}\eta = \frac{1}{h_l}\frac{\partial \eta}{\partial l}\hat{l} + \frac{1}{h_m}\frac{\partial \eta}{\partial m}\hat{m} + \frac{1}{h_n}\frac{\partial \eta}{\partial n}\hat{n}, \tag{17.47}$$

$$\vec{\nabla} \cdot \vec{n} = \frac{1}{h_l h_m h_n}\left(\frac{\partial}{\partial l}(n_l h_m h_n) + \frac{\partial}{\partial m}(n_m h_l h_n) + \frac{\partial}{\partial n}(n_n h_l h_m)\right), \tag{17.48}$$

where $\hat{l}, \hat{m}, \hat{n}$, are unit vectors along the coordinates l, m, n. The Laplacian can be written as

$$\nabla^2 \eta = \vec{\nabla} \cdot \vec{\nabla}\eta, \tag{17.49}$$

$$\nabla^2 \eta = \frac{1}{h_l h_m h_n}\left[\frac{\partial}{\partial l}\left(\frac{h_m h_n}{h_l}\frac{\partial \eta}{\partial l}\right) + \frac{\partial}{\partial m}\left(\frac{h_l h_n}{h_m}\frac{\partial \eta}{\partial m}\right) + \frac{\partial}{\partial n}\left(\frac{h_l h_m}{h_n}\frac{\partial \eta}{\partial n}\right)\right]. \tag{17.50}$$

As a check, for a Cartesian system $h_l = h_m = h_n = 1$, and Eqs. 17.47–17.50 assume their familiar forms.

---

[10] It should be noted that the $d\eta/d\chi$ is for a fixed time, and $d\eta/dt$ is for a fixed APDB orientation. Since both are subject to change during microstructural evolution, the velocity v is an instantaneous one.

For our APDB geometry we need a curvilinear coordinate system with one coordinate pointing normal to the surface of the boundary, which we choose as l. The other two coordinates are in the plane of the boundary, so the Laplacian depends only on the coordinate l

$$\nabla^2 \eta = \frac{1}{h_l h_m h_n} \frac{\partial}{\partial l} \left(\frac{h_m h_n}{h_l} \frac{\partial \eta}{\partial l} \right).$$

(17.51)

For consistency of notation with our one-dimensional problem of the APDB, we arrange the distance $d\chi$ as normal to the surface of the APDB. That is, $d\chi = h_l dl$ and the Laplacian is

$$\nabla^2 \eta = \frac{1}{h_m h_n} \frac{\partial}{\partial \chi} \left(h_m h_n \frac{\partial \eta}{\partial \chi} \right).$$

(17.52)

The differentiation is worked as usual, and again with $d\chi = h_l dl$

$$\nabla^2 \eta = \frac{\partial^2 \eta}{\partial \chi^2} + \frac{\partial \eta}{\partial \chi} \frac{1}{h_l h_m h_n} \frac{\partial}{\partial l} (h_m h_n).$$

(17.53)

Likewise, the divergence of a vector along the l direction from Eq. 17.48 is

$$\vec{\nabla} \cdot \vec{\chi} = \frac{1}{h_l h_m h_n} \frac{\partial}{\partial l} (h_m h_n).$$

(17.54)

Recognizing Eq. 17.54 in Eq. 17.53, we obtain the result

$$\nabla^2 \eta = \frac{\partial^2 \eta}{\partial \chi^2} + \frac{\partial \eta}{\partial \chi} \vec{\nabla} \cdot \vec{\chi}.$$

(17.55)

17.4.4 Stability of Curved Boundaries with Saddle Points

Domain evolution in three dimensions is not simple because the interface velocity depends on the mean curvature in Eq. 17.45. Perhaps our intuition is based on growth and shrinkage of spherical particles as in nucleation, for example. For a sphere, in all directions on its surface the curvatures are equal, so $K_1 = K_2$, simplifying the problem. For a saddle surface, on the other hand, the two principal curvatures have opposite signs. It is therefore possible to have $K_1 = -K_2$, and zero velocity of the domain boundary. What is interesting in three dimensions (and not in two) are topological structures such as shown in Fig. 17.6. All points on this surface are saddle points having zero mean curvature. Such a microstructure may have a high free energy, but it is reasonably stable against growth. Monte Carlo simulations of ordered domains in the form of Fig. 17.6 show that small thermal fluctuations of the surface are restored, and the structure is metastable. A large fluctuation that causes two nodes to merge will lead to collapse of the entire structure, however, as shown in Fig. 17.7.

Section 14.6 described the coarsening of precipitate microstructures, but considered only cases where the phase fraction of the precipitates was small. At higher concentrations above a "percolation threshold," the precipitate phase makes a connected network all the way across a large volume of material. For comparison, for the solute atoms in a random

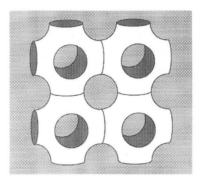

Figure 17.6 Schwarz P-surface, a periodic minimal surface with zero mean curvature everywhere.

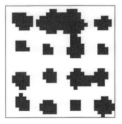

Figure 17.7 Initiation of collapse of a periodic minimal surface in a Monte Carlo simulation, caused by coalescence of units in the upper center. Image shows one plane of atoms on a cut through a three-dimensional structure.

solid solution in an fcc crystal, the percolation threshold is $c = 0.198$, and for a bcc crystal it is $c = 0.245$ [284]. At higher concentrations, especially around equiatomic compositions, it is therefore unrealistic to model unmixed regions as spheres. In alloys of higher concentration, the interfaces between the unmixed phases will have more complicated curvatures, and some of the arguments for the energy of APDBs should apply to this unmixing problem even though the composition is a conserved quantity. Notice again the structures of Fig. 1.7b for chemical unmixing. Many of the points on the isoconcentration surface appear to be at saddle points of the interface, or have at least partial cancellation of their principal curvatures. These topological features of the boundary structure likely play a role in setting the kinetics of coarsening. The lower driving force for moving saddle surfaces may cause these regions of the interface to be relatively persistent in an evolving microstructure, and therefore prominent features of the microstructure.

Problems

17.1 How would you expect the free energy curves in Fig. 17.2 to vary with temperature? If there is no chemical unmixing in the liquid above the melting temperature, discuss how $\left(\partial^2 F(\eta, c, T)/\partial c^2 \right)_T$ may vary with η.

17.2 Explain why the free energy cost of an APDB (cf. 17.28) varies as $\sqrt{\kappa}$.

17.3 Consider a small sphere of an antiphase domain inside an ordered domain.

   (**a**) Using expressions for the Laplacian and divergence in spherical coordinates, develop the equations analogous to Eqs. 17.37 and 17.41 to explain the shrinkage of the antiphase domain with time.

   (**b**) What is the functional form for how the radius of the shrinking domain changes with time?

17.4 Consider the thermal stability of a system having the Landau potential of Fig. 17.4.

   (**a**) Suppose the temperature is low, so the mean-squared displacement, $\langle x^2 \rangle - \langle x \rangle^2$, is small (perhaps 0.2). What is $\langle x \rangle$ in equilibrium? (Show your reasoning or calculation.)

   (**b**) Now suppose the temperature is high, so $\langle x^2 \rangle - \langle x \rangle^2$ is large (perhaps 2). What is $\langle x \rangle$ in equilibrium? (Show your reasoning or calculation.)

   (*Hint*: See Fig. 19.16.)

18

Method of Concentration Waves and Chemical Ordering

This chapter analyzes the thermodynamic stability of "static concentration waves." The idea is that an ordered structure can be described as a variation of chemical composition from site to site on a crystal, and this variation can be written as a wave, with crests denoting B-atoms and troughs the A-atoms, for example. The wave does not propagate, so it is called a "static" concentration wave. Another important difference from conventional waves is that the atom sites are exactly on the tops of crests or at the bottoms of troughs, so we do not consider the intermediate phases of the concentration wave, at least not in our first examples.

A convenient feature of this approach is that an ordered structure can be described by a single wavevector, $\vec{k}_0$, or a small set of related wavevectors, $\{\vec{k}_{0j}\}$. The disordered solid solution has zero amplitude of the concentration wave, so the amplitude of the concentration wave, η, serves as a long-range order parameter. The analysis of the free energy and ordered structure in k-space, rather than in real space, is much like that used for the Fourier transform solution to the diffusion equation for spinodal decomposition in Section 16.4.2. Unlike unmixing, however, concentration waves accommodate the symmetry of the ordered structure, and how this differs from the high temperature solid solution.

This chapter begins with a review of how periodic structures in real space are described by wavevectors in k-space, and then explains the "star" of the wavevector of an ordered structure. A key step for phase transitions is writing the free energy in terms of the amplitudes of static concentration waves. If a second-order phase transition is possible (i.e., the ordered structure evolves from the disordered with infinitesimal amplitude at the critical temperature), the demand for translational symmetry of the free energy sets a condition for the vectors in the star of the ordered structure known as the "Landau–Lifshitz criterion." Chapter 18 ends with a more general formulation of the free energy in terms of static concentration waves, which is an important example of how Fourier transform methods can treat long-range interactions in materials thermodynamics. These methods of static concentration waves were developed by Armen G. Khachaturyan [285–287] and are explained in his book and review article [267, 288].

18.1 Structure in Real Space and Reciprocal Space

18.1.1 First Brillouin Zone

Here is a brief review of reciprocal space. Figure 18.1 shows how a row of atoms responds to waves of different wavevector, k. Perhaps the simplest case is $k = 0$ near the middle of

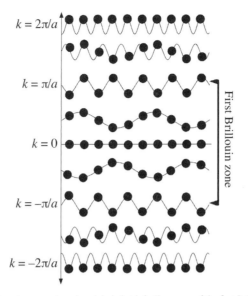

$k = 2\pi/a$

$k = \pi/a$

$k = 0$

$k = -\pi/a$

$k = -2\pi/a$

First Brillouin zone

Figure 18.1   Atom displacements for different wavevectors, k, as labeled at left. The range of the first Brillouin zone, $-\pi/a < k < \pi/a$, is identified at right. Please identify the three waves that give zero relative displacements of atoms, the two waves that give the identical wave of intermediate atom displacements, and check the effect of changing the sign of k.

the figure, which gives a wave of infinite wavelength ($k = 2\pi/\lambda$, where λ is wavelength). For $k = 0$ there are no relative displacements of adjacent atoms, and no wave. Waves of very small k correspond to macroscopic wavelengths, such as sound waves in a solid. The two waves adjacent to the null wave of $k = 0$ in Fig. 18.1 show a typical sinusoidal pattern of atom displacements. They have k of opposite sign, however, so they are mutually out of phase, but are otherwise similar.

When k becomes larger, so the atom displacement pattern has a smaller λ, an issue arises when λ exceeds twice the interatomic spacing, i.e., $\lambda = 2a$, i.e., when $k = \pm2\pi/2a = \pm\pi/a$. As shown in Fig. 18.1, the condition $k = \pm\pi/a$ gives the atom displacement pattern with the shortest wavelength that can be sustained on a one-dimensional chain. What happens at twice this value, $k = 2\pi/a$, is shown at the top and bottom of Fig. 18.1. Here the wavelength is matched exactly to the atom periodicity, so the result is no mutual displacement between adjacent atoms. There is no wave, and the picture is equivalent to that for $k = 0$.[1]

A more subtle point is that for k from π/a to $2\pi/a$, the wave patterns repeat in reverse order from the sequence of k from 0 to π/a. Note carefully how the atom displacement pattern in the second wave from the top in Fig. 18.1 is identical to the wave two below it. Of course only the physical atom displacements are important, and not the wavevector parameter itself. A wide range of k therefore gives a redundant description of the atom displacements. All wavelike atom displacements can be accounted for with the

[1] Sometimes these large k-vectors are used to account for momentum transfers with the center of mass of the crystal, but these issues do not arise in what follows.

wavevectors where $-\pi/a < k < +\pi/a$, a region called the "first Brillouin zone." For all analyses with concentration waves of atoms, we will work within the first Brillouin zone. Importantly, we consider two wavevectors such as $k = +\pi/a$ and $k = -\pi/a$ to be identical because they give the same physical effects.

18.1.2 Reciprocal Lattice

For a crystal, the Brillouin zone plays a role in k-space that parallels the role played by the unit cell in r-space. We can tile an entire crystal in r-space by covering it with translated unit cells, but the content is set by that of one unit cell. Likewise we can tile k-space with Brillouin zones, but the structure is set by the content of the first Brillouin zone.[2] A characteristic of the k-space structure of a crystal is that the origin at $k = 0$ is repeated at intervals of $k = 2\pi/a$ in k-space, as demonstrated with Fig. 18.1. These points separated by multiples of $2\pi/a$ comprise the "reciprocal lattice" of the crystal. Each point of the reciprocal lattice is a wavevector, but a special one that gives a periodicity in the crystal.

A reciprocal lattice is generated from a set of primitive translation vectors in k-space, which have units of length^{-1}. For comparison, a crystal lattice vector in real space, $\vec{R}$, is generated with the primitive lattice translation vectors $\{\vec{a}_1, \vec{a}_2, \vec{a}_3\}$

$$\vec{R} = m\vec{a}_1 + n\vec{a}_2 + o\vec{a}_3, \tag{18.1}$$

where $\{m, n, o\}$ are integers. In the same way, a reciprocal lattice vector, $\vec{g}$, is generated with its own primitive translation vectors $\{\vec{a}_1^*, \vec{a}_2^*, \vec{a}_3^*\}$

$$\vec{g} = h\vec{a}_1^* + k\vec{a}_2^* + l\vec{a}_3^*, \tag{18.2}$$

where $\{h, k, l\}$ are integers. The reciprocal primitive translation vectors $\{\vec{a}_1^*, \vec{a}_2^*, \vec{a}_3^*\}$ are obtained from the real primitive translation vectors $\{\vec{a}_1, \vec{a}_2, \vec{a}_3\}$ by these definitions

$$\vec{a}_1^* = 2\pi \frac{\vec{a}_2 \times \vec{a}_3}{\vec{a}_1 \cdot [\vec{a}_2 \times \vec{a}_3]}, \tag{18.3}$$

$$\vec{a}_2^* = 2\pi \frac{\vec{a}_3 \times \vec{a}_1}{\vec{a}_1 \cdot [\vec{a}_2 \times \vec{a}_3]}, \tag{18.4}$$

$$\vec{a}_3^* = 2\pi \frac{\vec{a}_1 \times \vec{a}_3}{\vec{a}_1 \cdot [\vec{a}_2 \times \vec{a}_3]}. \tag{18.5}$$

The problem is particularly easy for simple cubic crystals, for which $\{\vec{a}_1, \vec{a}_2, \vec{a}_3\} = \{a\hat{x}, a\hat{y}, a\hat{z}\}$. The numerators in Eqs. 18.3 to 18.5 are a^2 (with a direction), and the denominators are a^3 (without a direction)

$$\vec{a}_1^* = \frac{2\pi}{a}\hat{x}, \tag{18.6}$$

$$\vec{a}_2^* = \frac{2\pi}{a}\hat{y}, \tag{18.7}$$

$$\vec{a}_3^* = \frac{2\pi}{a}\hat{z}. \tag{18.8}$$

---

[2] This analogy works very well in one dimension, but for higher dimensions the higher Brillouin zones do not have the same shape, although their volumes are the same.

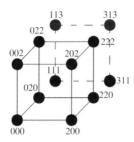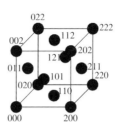

Figure 18.2   Indexed bcc and fcc Cartesian unit cells.

For a simple cubic lattice, inner products of real and reciprocal lattice translation vectors benefit from the relationship

$$\vec{a}_i \cdot \vec{a}_j^* = 2\pi \delta_{ij}, \tag{18.9}$$

where the Kronecker delta δ_{ij} equals 1 when $i = j$, and is zero otherwise. Exponentials involving lattice translation vectors and reciprocal lattice translation vectors become quite simple. Using Eqs. 18.1 and 18.2 for a simple cubic crystal

$$e^{-i\vec{g}\cdot\vec{R}} = e^{-i(h\vec{a}_1^* + k\vec{a}_2^* + l\vec{a}_3^*)\cdot(m\vec{a}_1 + n\vec{a}_2 + o\vec{a}_3)}, \tag{18.10}$$

$$e^{-i\vec{g}\cdot\vec{R}} = e^{-i2\pi(mh + nk + ol)}, \tag{18.11}$$

$$e^{-i\vec{g}\cdot\vec{R}} = e^{-i2\pi \text{ integer}} = +1, \tag{18.12}$$

where the sum of products of integers, $mh + nk + ol$, is an integer. For the periodicities of atom concentrations, Eq. 18.12 means that all wavevectors of Eq. 18.2, denoted $\vec{g} = (h, k, l)$, are equivalent. There is a perfect correspondence to X-ray diffraction. The diffracted wave, evaluated by summing exponential phase factors over atoms at all $\{m, n, o\}$, will be strong whenever the diffraction vector $\vec{\Delta k}$ is equal to $\vec{g}$, a vector of the reciprocal lattice. (This is the "Laue condition" for diffraction.)

Not all reciprocal lattices are so obvious as the simple cubic reciprocal lattice of the simple cubic lattice. A few correspondences between reciprocal lattices and real lattices are worth knowing. An important one is that the reciprocal lattice of a body-centered cubic (bcc) lattice is a face-centered cubic (fcc) lattice, and vice versa. The correspondence is illustrated in Fig. 18.2, indexed with Cartesian axes in units of cube edge $2a$ to avoid half-integral indices.[3] To obtain the allowed diffractions of the bcc crystal, look at its reciprocal lattice, which is fcc. Allowed diffractions can be read from the indices in Fig. 18.2, for example[4]

   bcc diffractions: (110), (200), (211), (220), (222) ...
   (bcc structure factor rule: the three indices sum to an even integer)
   fcc diffractions: (111), (200), (220), (311), (222) ...
   (fcc structure factor rule: the three indices are all even or all odd integers).

---

[3] However, to use the indexed fcc unit cell in Fig. 18.2 as the reciprocal of the bcc, the edge of the bcc unit cell should be a, rather than $2a$.

[4] The missing diffraction peaks, such as (100) in both fcc and bcc, or (110) in fcc, are consequences of using translation vectors of a simple cubic lattice for reference, instead of using the nonorthogonal primitive lattice vectors of the fcc and bcc lattices. The Cartesian axes of simple cubic are more convenient, however, so we are accustomed to using "structure factor rules," which are derived for a cubic lattice with basis vectors.

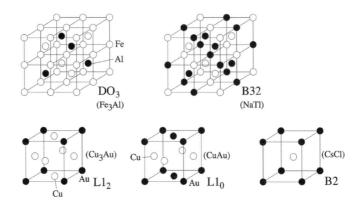

Five ordered structures on bcc and fcc parent lattices. Names of the structures in "strukturbericht" designation are DO_3, B32, $L1_2$, $L1_0$, and B2. The chemical formulae are for "prototypes" of compounds that take these structures.

18.1.3 Ordered Structures Described by Concentration Waves

Numerous ordered structures are found in nature.[5] Figure 18.3 shows five ordered structures on fcc or bcc parent lattices. The atom positions for four of them are listed in Table 18.1. The number of atom position vectors, $\vec{r}_k$, in the unit cell can be as few as two for a binary alloy (and in fact there are two $\vec{r}_k$ for the B2 structure), but if the standard cubic unit cell requires more atoms (such as four for the fcc cell), the ordered structure will require information on atom occupancy for at least this number of sites. These $\{\vec{r}_k\}$ are basis vectors of the ordered unit cell and give real space (r-space) information about which atom is placed where.

It is convenient to describe an ordered structure by specifying a wavevector in k-space because a wavevector can account for the periodic alternation of the different atoms in the structure. A good way to measure the wavevector for an ordered structure is by X-ray diffractometry, which reveals additional wavevectors as "superlattice" diffractions in the diffraction pattern. These are the wavevectors for the static concentration waves of the ordered structure. Analogously, the wavevector can be obtained by performing a Fourier transform of the chemical modulation in the ordered structure. In what follows we go the other way, and pick a wavevector first. We then find the ordered structure that it generates. Perhaps not surprisingly, low-order wavevectors such as (100) and (111) generate simple ordered structures such as $L1_0$ or B2, for example. Wavevectors such as (3,4,5) generate ordered structures with longer-range periodicities that are less common in nature.

A k-space description of an ordered structure starts with the selection of a low-order wavevector, $\vec{k}$, and a wave $e^{-i\vec{k}\cdot\vec{r}}$ is constructed at $\vec{r}$ throughout the real space crystal. The wave is evaluated only at specific atom sites, however, and the wave has no meaning between the atom sites. It is therefore often acceptable to use a complex exponential for a concentration wave, provided it always evaluates to ± 1 at all sites of the crystal.[6]

---

[5] All simple structures that can be constructed with stoichiometric ratios such as 1:1 or 1:3 on fcc or bcc lattices are in fact found in nature.

[6] It is sometimes also appropriate to add the complex conjugate to a wave to ensure that the concentration is real.

Table 18.1 Atom positions in selected ordered structures			
B2	L1$_0$	L1$_2$	D0$_3$
Cs: {0, 0, 0 }	Au: {0, 0, 0}	Au: {0, 0, 0}	Fe: {0, 0, 0}
Cl: a\{1/2, 1/2, 1/2\}	Au: a\{1/2, 1/2, 0\}	Cu: a\{1/2, 1/2, 0\}	Fe: a\{1, 0, 0\}
	Cu: a\{1/2, 0, 1/2\}	Cu: a\{1/2, 0, 1/2\}	Fe: a\{0, 1, 0\}
	Cu: a\{0, 1/2, 1/2\}	Cu: a\{0, 1/2, 1/2\}	Fe: a\{1, 1, 0\}
			Al: a\{1/2, 1/2, 1/2\}
			Fe: a\{3/2, 1/2, 1/2\}
			Fe: a\{1/2, 3/2, 1/2\}
			Al: a\{3/2, 3/2, 1/2\}
			Fe: a\{0, 0, 1\}
			Fe: a\{1, 0, 1\}
			Fe: a\{0, 1, 1\}
			Fe: a\{1, 1, 1\}
			Fe: a\{1/2, 1/2, 3/2\}
			Al: a\{3/2, 1/2, 3/2\}
			Al: a\{1/2, 3/2, 3/2\}
			Fe: a\{3/2, 3/2, 3/2\}

For all $\vec{r}_k$ in Table 18.1, this is easier than it may sound at first, because it uses tricks like the one from Eq. 18.14 to Eq. 18.16

$$e^{-i\vec{k}\cdot\vec{r}_k} = e^{-i\frac{2\pi}{a}(h\hat{x}+k\hat{y}+l\hat{z})\cdot(r_{k,x}\hat{x}+r_{k,y}\hat{y}+r_{k,z}\hat{z})}, \tag{18.13}$$

$$e^{-i\vec{k}\cdot\vec{r}_k} = e^{-i\pi \text{ integer}}, \tag{18.14}$$

$$e^{-i\vec{k}\cdot\vec{r}_k} = +1 \text{ if integer is even,} \quad \text{OR} \tag{18.15}$$

$$e^{-i\vec{k}\cdot\vec{r}_k} = -1 \text{ if integer is odd.} \tag{18.16}$$

In more detail, the exponential in Eq. 18.13 includes

$$\frac{2\pi}{a}(h\hat{x} + k\hat{y} + l\hat{z}) \cdot (r_{k,x}\hat{x} + r_{k,y}\hat{y} + r_{k,z}\hat{z}) = \frac{2\pi}{a}(hr_{k,x} + kr_{k,y} + lr_{k,z}). \tag{18.17}$$

Equation 18.14 will be true if

$$h = \frac{a}{r_{k,x}}\frac{\text{integer}}{2}, \tag{18.18}$$

$$k = \frac{a}{r_{k,y}}\frac{\text{integer}}{2}, \tag{18.19}$$

$$l = \frac{a}{r_{k,z}}\frac{\text{integer}}{2}. \tag{18.20}$$

There are other ways to ensure that Eq. 18.14 is true (so that $e^{-i\vec{k}\cdot\vec{r}_k} = \pm1$, and the composition is real), but the standard practice in this chapter will be to work with a standard

Table 18.2 Concentration waves with selected k-vectors

B2 (111)	L1$_0$ (001)	L1$_2$ (001)+(010)+(100)
$c + \frac{\eta}{2}\, e^{-i\frac{2\pi}{a}(111)\cdot\vec{r}_l}$	$c + \frac{\eta}{2}\, e^{-i\frac{2\pi}{a}(001)\cdot\vec{r}_l}$	$c + \frac{\eta}{4}\left[e^{-i\frac{2\pi}{a}(001)\cdot\vec{r}_l} + e^{-i\frac{2\pi}{a}(010)\cdot\vec{r}_l} + e^{-i\frac{2\pi}{a}(100)\cdot\vec{r}_l} \right]$
Cs: $c + \frac{\eta}{2}$	Au: $c + \frac{\eta}{2}$	Au: $(+1+1+1)\frac{\eta}{4} = c + \frac{3}{4}\eta$
Cl: $c - \frac{\eta}{2}$	Au: $c + \frac{\eta}{2}$	Cu: $(+1-1-1)\frac{\eta}{4} = c - \frac{1}{4}\eta$
	Cu: $c - \frac{\eta}{2}$	Cu: $(-1+1-1)\frac{\eta}{4} = c - \frac{1}{4}\eta$
	Cu: $c - \frac{\eta}{2}$	Cu: $(-1-1+1)\frac{\eta}{4} = c - \frac{1}{4}\eta$

cubic unit cell of edge length a, and Cartesian components $r_{k,x}$, $r_{k,x}$, $r_{k,x}$, that are rational fractions of a. For example, for the B2 structure in Table 18.1, $\vec{r}_{2,x} = a/2$, so conveniently, $h = $ integer.

Examples of concentration waves used to describe B2 and L1$_0$ ordered structures are

$$c_{B2}(\vec{r}) = c + \frac{\eta}{2}\, e^{-i\frac{2\pi}{a}(111)\cdot\vec{r}_k}, \tag{18.21}$$

$$c_{L1_0}(\vec{r}) = c + \frac{\eta}{2}\, e^{-i\frac{2\pi}{a}(001)\cdot\vec{r}_k}, \tag{18.22}$$

These two equations were evaluated at all the $\{\vec{r}_l\}$ of the B2 and L1$_0$ ordered structures listed in Table 18.1, and results are listed in Table 18.2. Note in Table 18.2 that the crests and troughs of the real part of the concentration waves (giving the $\pm\eta$) are at the sites of the two chemical species in the ordered structure.

The amplitudes of the concentration waves are given by the parameter η, which is an order parameter. When $\eta = 0$, the chemical composition is simply c throughout the crystal, and there is no modulation in composition. When $\eta = 1$, the structures shown in Fig. 18.3 are generated. When $0 < \eta < 1$, the concentration wave has intermediate amplitude. As was shown in Fig. 2.21, the order parameter is zero above a critical temperature, but L or η grows larger with temperatures below T_c. Beforc calculating thermodynamic free energies, we develop further the ideas of crystal symmetry in k-space, and identify the minimum number of $\vec{k}$-vectors that define an ordered structure.

18.2 Symmetry and the Star

The symmetry elements of a crystal typically include rotations, reflections, and inversions. When the atom arrangements in an fcc crystal are rotated by $90°$ about a Cartesian axis, for example, the same types of atoms are at the same positions in space. An fcc crystal is "invariant" under such a $90°$ rotation, for example. On the other hand, if there is a periodic arrangement of atoms along the (001) direction as in the L1$_0$ structure shown in Fig. 18.3, a $90°$ rotation about the x-axis will change the atom arrangement into one with periodicity along $\hat{y}$. The formation of the L1$_0$ structure has caused the crystal to lose an element of

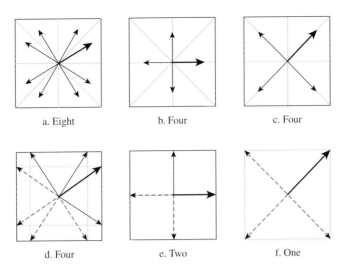

a. Eight   b. Four   c. Four

d. Four   e. Two   f. One

Figure 18.4   First Brillouin zones of square lattice (bounded by squares). Lines of reflective symmetry (through center) are indicated in gray. The initial vector is bold; the others are obtained from it by symmetry operations. Horizontal or vertical gray lines are reciprocal lattice vectors $(2\pi/a)\hat{x}$ and $(2\pi/a)\hat{y}$. Dashed lines denote wavevectors removed from the star because they create identical atomic structures as other wavevectors. (**a**) Arbitrary wavevector; star contains eight vectors. (**b**) Wavevector along symmetry line; star contains four vectors. (**c**) Wavevector along symmetry line; star contains four vectors. (**d**) Arbitrary wavevector orientation, touching special point at zone boundary; star contains four vectors. (**e**) Wavevector along symmetry line, touching special point at zone boundary; star contains two vectors. (**f**) Wavevector along symmetry line, touching special point at corner of zone boundary; star contains one vector (there are reciprocal lattice vectors on all four sides of the square).

rotational symmetry possessed by the parent fcc lattice. *Chemical ordering causes a loss of symmetry.*[7]

To be systematic, we collect all the *unique* wavevectors that generate the ordered structure, and we call this collection the "star." Given one wavevector, the other wavevectors in the star are generated from it by use of the point symmetry operations of the parent lattice. An example is shown in Fig. 18.4a, where symmetry operations on a typical vector on a two-dimensional square lattice generate seven other vectors. There are eight vectors in the star of Fig. 18.4a. When the initial vector lies on a line of symmetry, as in Figs. 18.4b and c, the reflection symmetry operation across the line generates the same vector. All unnecessary vectors must be eliminated from the star (giving four in Figs. 18.4b,c).[8] Generating the star for these three cases of Fig. 18.4a–c was fairly intuitive.

Translational symmetry is a distinguishing feature of crystals, allowing all atoms to be accounted for by lattice translations of a unit cell. These translations must be considered in k-space, as we saw in the discussion of the Brillouin zone in Section 18.1.1 and

---

[7] This sounds nonintuitive to many people because ordered structures display prominent and beautiful symmetries. Nevertheless, the atom patterning in an ordered structure typically causes a loss of translational symmetry between neighboring atoms.

[8] Otherwise the number of vectors in the star will not be unique – numerous equivalent vectors can be created by adding reciprocal lattice vectors, as shown next.

Fig. 18.1. We typically use only wavevectors in a bounded region of k-space (the first Brillouin zone). Otherwise we will generate identical atom arrangements with wavevectors that have different values of $\vec{k}$. This same issue arises for ordered structures when their wavevectors are at special points in the Brillouin zone, especially edges and corners. Consider Fig. 18.4d, in which the vectors of Fig. 18.4a were extended to the edges of the Brillouin zone. The gray lines are reciprocal lattice vectors of length $2\pi/a$, which span the Brillouin zone. Whenever two wavevectors are separated by a reciprocal lattice vector, they specify identical atom arrangements, as for example in Fig. 18.1 for $k = -\pi/a$ and $k = +\pi/a$ (separated by $2\pi/a$), and for $k = -2\pi/a$, $k = 0$, and $k = +2\pi/a$. We therefore say that wavevectors connected by the gray reciprocal lattice vectors in Fig. 18.4 are the same vector. Only one of them is included in our collection called the star. This eliminates half of the wavevectors in Figs. 18.4d and e, and eliminates three of the four wavevectors in Fig. 18.4f, which is a very special situation with only one vector in the star. These special points come up frequently for ordered structures because ordering wavevectors are usually along directions of symmetry of the parent lattice, and their lengths are often submultiples of the spacings in the parent lattice.

Now enumerate systematically the stars for the ordering wavevectors presented in Table 18.2.

- For the (111) wavevector of the B2 structure, first list all eight variants along different directions:

$$(111), (11\bar{1}), (1\bar{1}1), (1\bar{1}\bar{1}), (\bar{1}11), (\bar{1}1\bar{1}), (\bar{1}\bar{1}1), (\bar{1}\bar{1}\bar{1}).$$

Systematically subtracting the other seven vectors from the (111) gives

$$(002), (020), (022), (200), (202), (220), (222).$$

Every one of these is a reciprocal lattice vector of the parent bcc lattice, so all eight variants of the (111) wavevector generate the same ordered structure. There is only one vector in the star for the B2 ordered structure, which we select as (111).

- For the (001) wavevector of the $L1_0$ structure, the six variants are:

$$(001) = (00\bar{1}), (010) = (0\bar{1}0), (100) = (\bar{1}00).$$

The equality symbols mean that the two vectors differ from each other by an (002)-type vector, which is a reciprocal lattice vector of the parent fcc lattice. Therefore half of the six vectors are eliminated from the star. Systematically subtracting the remaining two vectors from the remaining (001) gives

$$(0\bar{1}1), (\bar{1}01).$$

The parent lattice of the $L1_0$ is fcc, for which (110)-type vectors are not reciprocal lattice vectors. We conclude that there are three vectors in the star for $L1_0$ ordering:

$$(001), (010), (100).$$

- The $L1_2$ structure involves the same (001)-type vectors and the same fcc parent lattice as does the $L1_0$, and therefore has the same star:

$$(001), (010), (100).$$

18.3 The Free Energy in *k*-Space with Concentration Waves

18.3.1 Using Concentration Waves to Construct a Free Energy Functional

Compare the free energies of an alloy with and without a concentration wave. A concentration wave in an A–B alloy sets the probability for a B-atom on a site $\vec{r}$ as

$$n(\vec{r}) = c + \Delta(\vec{r}). \tag{18.23}$$

Without the concentration wave, $n(\vec{r}) = c$. The difference in free energy,

$$F_{\text{ord}}(c + \Delta(\vec{r})) - F_{\text{dis}}(c) = \Delta F(\Delta(\vec{r})), \tag{18.24}$$

is a functional[9] of the concentration wave, $\Delta(\vec{r})$.

Our effort in Section 18.3 is to analyze the thermodynamics of second-order phase transitions, for which we focus on small order parameters near the critical temperature. We therefore perform a Taylor series expansion around the disordered free energy. The difference in free energy between an alloy with a small concentration wave and a disordered solid solution with none is

$$\Delta F(\Delta(\vec{r})) = F_{\text{ord}} - F_{\text{dis}}, \tag{18.25}$$

$$\Delta F(\Delta(\vec{r})) = \sum_{\vec{r}} A(\vec{r})\, \Delta(\vec{r}) + \frac{1}{2!} \sum_{\vec{r}} \sum_{\vec{r}'} B(\vec{r}, \vec{r}')\, \Delta(\vec{r})\Delta(\vec{r}')$$

$$+ \frac{1}{3!} \sum_{\vec{r}} \sum_{\vec{r}'} \sum_{\vec{r}''} C(\vec{r}, \vec{r}', \vec{r}'')\, \Delta(\vec{r})\Delta(\vec{r}')\Delta(\vec{r}'')$$

$$+ \frac{1}{4!} \sum_{\vec{r}} \sum_{\vec{r}'} \sum_{\vec{r}''} \sum_{\vec{r}'''} D(\vec{r}, \vec{r}', \vec{r}'', \vec{r}''')\, \Delta(\vec{r})\Delta(\vec{r}')\Delta(\vec{r}'')\Delta(\vec{r}''')$$

$$+ \cdots, \tag{18.26}$$

where the energy parameters are

$$A(\vec{r}) = \left(\frac{\delta F}{\delta \Delta(\vec{r})} \right)_{\Delta=0}, \tag{18.27}$$

$$B(\vec{r}, \vec{r}') = \left(\frac{\delta^2 F}{\delta \Delta(\vec{r})\, \delta \Delta(\vec{r}')} \right)_{\Delta=0}, \tag{18.28}$$

$$C(\vec{r}, \vec{r}', \vec{r}'') = \left(\frac{\delta^3 F}{\delta \Delta(\vec{r})\, \delta \Delta(\vec{r}')\, \delta \Delta(\vec{r}'')} \right)_{\Delta=0}, \tag{18.29}$$

$$D(\vec{r}, \vec{r}', \vec{r}'', \vec{r}''') = \left(\frac{\delta^4 F}{\delta \Delta(\vec{r})\, \delta \Delta(\vec{r}')\, \delta \Delta(\vec{r}'')\, \delta \Delta(\vec{r}''')} \right)_{\Delta=0}. \tag{18.30}$$

---

[9] Again, a functional is a function of a function. Here the free energy function depends on the concentration wave, which is itself a function.

Near the critical temperature for ordering, $A(\vec{r})$ must vanish, or an increasingly large concentration wave (of appropriate sign) would be increasingly favorable. If thermodynamics favors a large concentration wave, the temperature is not near the critical temperature of a second-order phase transition.

In an infinite crystal, a function of two positions, specifically $B(\vec{r}, \vec{r}')$, can depend only on the distance between $\vec{r}$ and $\vec{r}'$, i.e., $\vec{r} - \vec{r}'$.[10] Neglecting for a moment the higher-order terms gives

$$\Delta F(\Delta(\vec{r})) = \frac{1}{2!} \sum_{\vec{r}} \sum_{\vec{r}'} B(\vec{r} - \vec{r}')\, \Delta(\vec{r})\Delta(\vec{r}'). \tag{18.31}$$

The form for the concentration wave

$$\Delta(\vec{r}) = \eta\, e^{-i\vec{k}\cdot\vec{r}} \tag{18.32}$$

has served us well, but we still need to be careful that the concentration is real. We know that $\Delta(\vec{r})$ will be real if $\vec{k}$ is a vector of the reciprocal lattice vector of the parent crystal, since then $e^{-i\vec{k}\cdot\vec{r}} = e^{-i2\pi\,\text{integer}} = +1$ for all $\vec{r}$ (this includes $\vec{k} = 0$). It is also fine to have shorter $\vec{k}$ if $e^{-i\vec{k}\cdot\vec{r}} = e^{-i\pi\,\text{integer}} = -1$, and this will occur for the ordering wavevector. There can be problems with arbitrary $\vec{k}$, however. We call on the complex conjugate $\Delta^*(\vec{r}) = \eta^*\, e^{+i\vec{k}\cdot\vec{r}}$ to ensure reality. Our plan is to first use the following form of $\Delta F(\Delta(\vec{r}))$ for a discussion of possible types of phase transitions at arbitrary $\vec{k}$

$$\Delta F(\Delta(\vec{r})) = \frac{1}{2!} \sum_{\vec{r}} \sum_{\vec{r}'} B(\vec{r} - \vec{r}')\, \Delta(\vec{r})\Delta^*(\vec{r}'), \tag{18.33}$$

$$\Delta F(\Delta(\vec{r})) = \frac{1}{2!} \sum_{\vec{r}} \sum_{\vec{r}'} B(\vec{r} - \vec{r}')\, \eta\, e^{-i\vec{k}\cdot\vec{r}} \eta^*\, e^{+i\vec{k}\cdot\vec{r}'}. \tag{18.34}$$

Later, however, we will focus on the problem for one ordering wavevector $\vec{k}$, for which the exponentials are ± 1, allowing us to avoid the complex conjugates, making it easier to work with the higher-order terms with $C(\vec{r}, \vec{r}')$ and $D(\vec{r}, \vec{r}', \vec{r}'')$ in Eq. 18.26.

Define $\vec{R} = \vec{r} - \vec{r}'$, so $\vec{r}' = \vec{r} - \vec{R}$, and Eq. 18.31 becomes

$$\Delta F(\Delta(\vec{r})) = \frac{1}{2!} \sum_{\vec{r}} \sum_{\vec{R}} B(\vec{R})\, \Delta(\vec{r})\Delta(\vec{r} - \vec{R}), \tag{18.35}$$

where all sums are over an infinite crystal, so shifts of the limits of the sum are not important when changing variables. The form of Eq. 18.34 is

$$\Delta F(\Delta(\vec{r})) = \frac{1}{2!} \sum_{\vec{r}} \sum_{\vec{R}} B(\vec{R})\, \eta\, e^{-i\vec{k}\cdot\vec{r}} \eta^*\, e^{+i\vec{k}\cdot(\vec{r}-\vec{R})}. \tag{18.36}$$

Recognizing $e^{-i\vec{k}\cdot\vec{r}} e^{+i\vec{k}\cdot\vec{r}} = +1$, the first sum over N atoms becomes simply N

$$\Delta F(\Delta(\vec{r})) = \eta^*\eta \frac{1}{2!} \sum_{\vec{r}} (+1) \sum_{\vec{R}} B(\vec{R})\, e^{-i\vec{k}\cdot\vec{R}}, \tag{18.37}$$

$$\Delta F(\Delta(\vec{r})) = \frac{N}{2} |\eta|^2\, b(\vec{k}). \tag{18.38}$$

---

[10] The absolute value of $\vec{r}$ cannot have physical importance because the origin could be anywhere.

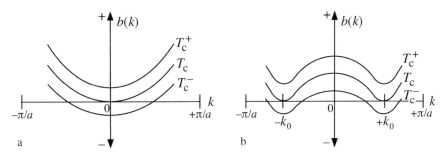

The function $b(k)$ near the critical temperature, T_c. The origin with $b(0) = 0$ is labeled with "0." Temperatures T_c^+ and T_c^- are just above and below T_c, respectively. Note the zero crossing of the lowest point on $b(k)$ for the two cases: (**a**) $b(k)$ for unmixing alloy, with lowest point at $k = 0$; (**b**) $b(k)$ for ordering alloy, with lowest point at $k = \pm k_0$.

where the important quantity $b(\vec{k})$ is defined as the Fourier transform of $B(\vec{R})$

$$b(\vec{k}) = \sum_{\vec{R}} B(\vec{R})\, \mathrm{e}^{-i\vec{k}\cdot\vec{R}}. \tag{18.39}$$

18.3.2 Physical Interpretation of $b(\vec{k})$

For small η, Eq. 18.38 shows how $b(\vec{k})$ contains all information about how the free energy depends on atom arrangements in the alloy. The $b(\vec{k})$ is a k-space function that depends on periodicities of concentration modulations. The minima of $b(\vec{k})$ give the chemical periodicities that minimize ΔF, and are therefore favored thermodynamically. We now consider a few possibilities for the shapes of $b(\vec{k})$ and their temperature dependences.

In Chapter 16 we analyzed the case for alloy instability against long-wave modulations, which give chemical unmixing of the A- and B-atoms. Figure 18.5a shows how we expect $b(\vec{k})$ to change with temperature for an unmixing alloy. At high temperatures above the critical temperature, T_c, the disordered solid solution is stable, so $\Delta F(\Delta(\vec{r})) = F_{\mathrm{ord}} - F_{\mathrm{dis}}$ of Eq. 18.38 is positive.[11] Since $|\eta|^2$ is positive, above T_c the $b(\vec{k})$ is therefore positive for all $\vec{k}$, such as the temperature just above T_c labeled T_c^+ in Fig. 18.5a. (This is also true for an ordering alloy above its critical temperature in Fig. 18.5b for the same reason.) For unmixing, the minimum in $b(\vec{k})$ is at $k = 0$, and with decreasing temperature this minimum comes down to touch zero at the critical temperature T_c. Below the critical temperature, such as at the temperature T_c^- in Fig. 18.5a, $b(\vec{k})$ is negative for some region of $\vec{k}$ around $k = 0$. Spinodal decomposition typically occurs at temperatures well below T_c, where there is some region in k-space with negative $b(\vec{k})$. The thermodynamic preference is for smaller k to minimize the gradient energy and strain energy in the alloy. With deeper undercooling, the $b(\vec{k})$ is increasingly negative over a wider range of k. With larger k the atoms need to move only short distances, so the wavevector for fastest unmixing will tend to become larger with increased undercooling. Kinetics tends to favor a chemical modulation with the largest k, whereas the thermodynamics favor smaller k.

---

[11] For unmixing, F_{ord} pertains to an alloy with a nonzero concentration wave of $k = 0$.

Figure 18.5b shows how we expect $b(\vec{k})$ to change with temperature for an ordering alloy. There are minima in $b(\vec{k})$ for all wavevectors in the star of the ordering wavevector, $\vec{k}_0$, although only two are shown in this one-dimensional k-space. At T_c^+, again $b(\vec{k}_0) > 0$ for all $\vec{k}_0$, so a concentration wave at T_c^+ will cause an increase in the alloy free energy, consistent with Eq. 18.38. At T_c^-, concentration waves of wavevector $\pm\vec{k}_0$ give negative $b(\vec{k}_0)$, so nonzero values of η will be expected in equilibrium because this lowers the alloy free energy in Eq. 18.38.

18.3.3 Special Points

There are "special points" in k-space where we expect $b(\vec{k})$ to be an extremum. There are several such points for a crystal structure of high symmetry, and they are candidate wavevectors for ordering in the crystal. Their full enumeration is beyond the scope of this text, but it is not hard to understand the essential idea of where to look for them. We seek a point $\vec{k}_0$ that gives an extremum of $b(\vec{k})$

$$\frac{\partial}{\partial \vec{k}} b(\vec{k})\bigg|_{\vec{k}_0} = 0 \tag{18.40}$$

for any direction of $\vec{k}$. The places to look for such a condition are along lines of symmetry, such as the lines of reflection in Fig. 18.4. The value of $b(\vec{k})$ must be the same when reflected across such a line of symmetry. The line therefore defines a valley (or mountain) in $b(\vec{k})$. To ensure an extremum, look at the very "special points" where there is a crossing of two lines of symmetry. A minimum exists at the crossing of two valleys. Of course there could also be a maximum or a saddle point at these special points, but these points are candidates for minima of $b(\vec{k})$. Not all ordered structures occur at special points. Nevertheless, when ordering occurs at special points, the wavevector tends to lock to the periodicity of the parent crystal, and the wavevector remains independent of temperature.

18.4 Symmetry Invariance of Free Energy and Landau–Lifshitz Rule for Second-Order Phase Transitions

We now discuss whether an ordering transition can occur as a continuous second-order phase transition, or if it must be a discontinuous first-order transition. The difference is similar to the difference between spinodal decomposition and nucleation and growth. Recall that for nucleation and growth we used two separate free energy curves, one for the new phase and one for the parent phase. Although there is no discontinuity in the total free energy function at the critical temperature T_c, there is a discontinuity in the entropy and energy. These cause a discontinuity in the first derivative of the total free energy with respect to temperature (cf. Fig. 1.2). This discontinuity could be eliminated if there were a continuous variation of structure from the parent phase into the new phase. For spinodal decomposition, our analysis predicted an infinitesimal amount of unmixing at T_c, so a continuous transformation seems possible. With no discontinuity in the first derivative of the free energy, we say that spinodal decomposition is a "second order" transition, whereas nucleation and growth is "first order." This section develops an elegant rule to predict if there must be a discontinuity in order parameter at T_c.

18.4.1 Concentration Waves and Lattice Translations

Consider the free energy in the presence of only one type of concentration wave. That is, assume we already know the type of ordered structure that forms from the disordered solid solution. Our interest is in η at temperatures near the critical temperature, so the picture is much like that of Fig. 18.5b. We focus on the wavevector $\vec{k}_0$ for which $b(\vec{k}_0) = 0$ when the temperature falls to T_c. The $b(\vec{k}_0)$ will vanish for all vectors in the star of $\vec{k}_0$, so after substituting the concentration wave expression Eq. 18.32 into the free energy expression of Eq. 18.26

$$F - F_{\text{dis}} = \frac{N}{2} \sum_{j}^{\text{star}} b(\vec{k}_{0j}) \, |\eta(\vec{k}_{0j})|^2$$

$$+ \frac{1}{3!} \sum_{j_1, j_2, j_3}^{\text{star}} C(\vec{k}_{0j_1}, \vec{k}_{0j_2}, \vec{k}_{0j_3}) \, \eta(\vec{k}_{0j_1}) \eta(\vec{k}_{0j_2}) \eta(\vec{k}_{0j_3})$$

$$+ \frac{1}{4!} \sum_{j_1, j_2, j_3, j_4}^{\text{star}} D(\vec{k}_{0j_1}, \vec{k}_{0j_2}, \vec{k}_{0j_3}, \vec{k}_{0j_4}) \, \eta(\vec{k}_{0j_1}) \eta(\vec{k}_{0j_2}) \eta(\vec{k}_{0j_3}) \eta(\vec{k}_{0j_4}) + \cdots, \quad (18.41)$$

where $C(\vec{k}_{0j_1}, \vec{k}_{0j_2}, \vec{k}_{0j_3})$ is related to $C(\vec{r})$ of Eq. 18.29 by summing over all $\vec{r}$, $\vec{r}'$, and $\vec{r}''$ (and includes the exponentials). The $D(\vec{k}_{0j_1}, \vec{k}_{0j_2}, \vec{k}_{0j_3}, \vec{k}_{0j_4})$ is obtained similarly, with sums over four spatial coordinates.

A fundamental physical requirement is that F is invariant after a lattice translation $\vec{T}$, which shifts the concentration wave

$$F(\Delta(\vec{r})) = F(\Delta(\vec{r} + \vec{T})). \tag{18.42}$$

This $\vec{T}$ is a translation of the disordered solid solution because the temperature is assumed to be close to the critical temperature, where η is infinitesimal or zero. Rewriting $\Delta(\vec{r})$ of Eq. 18.32, and again with a translation by $\vec{T}$

$$\Delta(\vec{r}) = \eta \, e^{-i\vec{k}_0 \cdot \vec{r}}, \tag{18.43}$$

$$\Delta(\vec{r} + \vec{T}) = \eta \, e^{-i\vec{k}_0 \cdot (\vec{r} + \vec{T})}, \tag{18.44}$$

$$\Delta(\vec{r} + \vec{T}) = \left(\eta \, e^{-i\vec{k}_0 \cdot \vec{T}} \right) e^{-i\vec{k}_0 \cdot \vec{r}}. \tag{18.45}$$

Comparing Eqs. 18.43 and 18.45 shows how to account for the lattice translation by a simple substitution

$$\eta \rightarrow \eta \, e^{-i\vec{k}_0 \cdot \vec{T}}, \tag{18.46}$$

which is easily done for the free energy expression of Eq. 18.41

$$F - F_{\text{dis}} = \frac{N}{2} \sum_{j}^{\text{star}} b(\vec{k}_{0j}) \, |\eta(\vec{k}_{0j})|^2$$

$$+ \frac{1}{3!} \sum_{j_1 j_2 j_3}^{\text{star}} C(\vec{k}_{0j_1}, \vec{k}_{0j_2}, \vec{k}_{0j_3}) \, e^{-i(\vec{k}_{0j_1} + \vec{k}_{0j_2} + \vec{k}_{0j_3}) \cdot \vec{T}} \, \eta(\vec{k}_{0j_1}) \eta(\vec{k}_{0j_2}) \eta(\vec{k}_{0j_3})$$

$$+ \frac{1}{4!} \sum_{j_1 j_2 j_3 j_4}^{\text{star}} D(\vec{k}_{0j_1}, \vec{k}_{0j_2}, \vec{k}_{0j_3}, \vec{k}_{0j_4}) \, e^{-i(\vec{k}_{0j_1} + \vec{k}_{0j_2} + \vec{k}_{0j_3} + \vec{k}_{0j_4}) \cdot \vec{T}}$$

$$\times \, \eta(\vec{k}_{0j_1}) \eta(\vec{k}_{0j_2}) \eta(\vec{k}_{0j_3}) \eta(\vec{k}_{0j_4}) + \cdots . \tag{18.47}$$

18.4.2 The Landau–Lifshitz Criterion

Because crystals are invariant against lattice translations, Eqs. 18.41 and 18.47 must be equal. At T_c, $b(\vec{k}_0) = 0$, so we examine the higher-order terms in Eq. 18.47, both at T_c and near T_c. Two important possibilities are:

I. $\{\ e^{-i(\vec{k}_{0j_1} + \vec{k}_{0j_2} + \vec{k}_{0j_3}) \cdot \vec{T}} = 1$

    AND

    $e^{-i(\vec{k}_{0j_1} + \vec{k}_{0j_2} + \vec{k}_{0j_3} + \vec{k}_{0j_4}) \cdot \vec{T}} = 1\ \}$

  OR

II. $\{\ C(\vec{k}_{0j_1}, \vec{k}_{0j_2}, \vec{k}_{0j_3}) = 0.$

    AND

    $D(\vec{k}_{0j_1}, \vec{k}_{0j_2}, \vec{k}_{0j_3}, \vec{k}_{0j_4}) = 0 \text{ or } > 0\ \}$.

Possibility I is the same as

I. $\{\ \vec{k}_{0j_1} + \vec{k}_{0j_2} + \vec{k}_{0j_3} = \vec{g},$

    AND

    $\vec{k}_{0j_1} + \vec{k}_{0j_2} + \vec{k}_{0j_3} + \vec{k}_{0j_4} = \vec{g}\ \},$

where $\vec{g}$ is a vector of the reciprocal lattice of the disordered alloy.

Possibility I gives, with $b(\vec{k}_0) = 0$ at T_c,

$$F - F_{\text{dis}} = \frac{1}{3!} \sum_{j_1 j_2 j_3}^{\text{star}} C(\vec{k}_{0j_1}, \vec{k}_{0j_2}, \vec{k}_{0j_3}) \, \eta(\vec{k}_{0j_1}) \eta(\vec{k}_{0j_2}) \eta(\vec{k}_{0j_3})$$

$$+ \frac{1}{4!} \sum_{j_1 j_2 j_3 j_4}^{\text{star}} D(\vec{k}_{0j_1}, \vec{k}_{0j_2}, \vec{k}_{0j_3}, \vec{k}_{0j_4}) \eta(\vec{k}_{0j_1}) \eta(\vec{k}_{0j_2}) \eta(\vec{k}_{0j_3}) \eta(\vec{k}_{0j_4}) + \cdots , \tag{18.48}$$

when

$$\vec{k}_{0j_1} + \vec{k}_{0j_2} + \vec{k}_{0j_3} = \vec{g}, \text{ AND } \vec{k}_{0j_1} + \vec{k}_{0j_2} + \vec{k}_{0j_3} + \vec{k}_{0j_4} = \vec{g}.$$

Possibility II gives, with $b(\vec{k}_0) = 0$ at T_c,

$$F - F_{\text{dis}} = 0 \tag{18.49}$$

when

$$\vec{k}_{0j_1} + \vec{k}_{0j_2} + \vec{k}_{0j_3} \neq \vec{g}, \text{ AND } \vec{k}_{0j_1} + \vec{k}_{0j_2} + \vec{k}_{0j_3} + \vec{k}_{0j_4} \neq \vec{g}.$$

The following logic is subtle, but beautiful. The required condition for a second-order phase transition is that there is no discontinuity in the order parameter at T_c. The equilibrium amplitude of the concentration wave $\eta(\vec{k}_0)$ must go to zero continuously – the equilibrium free energy is unstable against an infinitesimal concentration wave at a temperature infinitesimally below T_c, not a finite concentration wave.

When Possibility I is true, we can choose the signs of the η so that the third-order term (with $C(\vec{k}_{0j_1}, \vec{k}_{0j_2}, \vec{k}_{0j_3})$) in Eq. 18.48 is negative. If so, we can choose larger and larger η to minimize further the free energy. The free energy is minimized by finite concentration waves, not infinitesimal ones. Possibility I therefore forbids a second-order phase transition.

When Possibility II is true, the free energy can be minimized by infinitesimal concentration waves at temperatures just below T_c. Possibility II does not forbid a second-order phase transition. We arrive at the Landau–Lifshitz criterion for a second-order phase transition (Box 18.1).

Box 18.1   **Landau–Lifshitz Criterion**

The Landau–Lifshitz criterion allows a second-order phase transition if

$$\vec{k}_{0j_1} + \vec{k}_{0j_2} + \vec{k}_{0j_3} \neq \vec{g}, \tag{18.50}$$

where $\{\vec{k}_{0j}\}$ are wavevectors in the star of the ordered structure, and $\vec{g}$ is a reciprocal lattice vector of the disordered alloy. Unless Eq. 18.50 is true, the phase transition must be first order.

This criterion evolved from work by Landau in 1937 [289] using the key idea that the free energy must have the same symmetry as the atom distribution. A group theoretical analysis, presented in the classic text by Landau and Lifshitz [290], shows that for a second-order phase transition to be possible, it must be impossible to construct a third-order term in the free energy of the disordered phase using the characteristic functions of the group of the ordered phase. In a periodic crystal, the characteristic functions are those of Bloch's theorem (Section 6.3). For the ordered structure, the phase factors of the Bloch functions have wavevectors that form the star of the ordered structure. Although group representation theory gives a general criterion for the possibility of a second-order phase transition, the method of concentration waves gives an explicit construction to test if the cubic term must be nonzero. The concentration wave approach allows for quick identification of first-order phase transitions in a number of phase transitions in materials, as shown next.

18.4.3 Applications of the Landau–Lifshitz Criterion

Crystallization from the Melt

A liquid is isotropic, so a crystal can grow with its precise atomic periodicity in any direction. Its star has vectors of all orientations. Figure 18.6 shows how it is possible to construct a reciprocal lattice vector with three wavevectors from this star of the ordered (crystal) phase. Crystallization and melting must be first-order phase transitions.

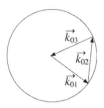

A reciprocal lattice vector, $2\pi/a[0, 0, 0]$, constructed from three wavevectors of the disordered liquid phase.

L1$_2$ Structure

The star of the L1$_2$ structure is $\{(1, 0, 0), (0, 1, 0), (0, 0, 1)\}$. The sum $(1, 0, 0) + (0, 1, 0) + (0, 0, 1) = (1, 1, 1)$ gives a vector of the fcc reciprocal lattice. The L1$_2$ structure must form by a first-order phase transition.

B2 Structure

The star of the B2 structure is $\{(1, 1, 1)\}$. Sums such as $(1, \bar{1}, 1) + (1, 1, 1) + (1, 1, 1) = (3, 1, 3)$ do not give any vectors of the bcc reciprocal lattice. (It takes a little effort to try all combinations of the ± 1. Note, however, that the sum of three odd integers gives an odd integer. Therefore, all three components of (h, k, l) are odd, so they cannot sum to an even integer as required for a bcc reciprocal lattice vector.) The B2 structure may form by a second-order phase transition.

Spinodal Decomposition

Unmixing is isotropic, so spinodal decomposition is similar to the earlier example of crystallization. In this case, however, the thermodynamic unmixing occurs with $\vec{k} = 0$, and of course $0 + 0 + 0 = 0$, a reciprocal lattice vector. Unmixing is therefore first order in a proper thermodynamic treatment.

Nevertheless, spinodal unmixing can be continuous, showing characteristics of a second-order phase transition. It can start with infinitesimal concentration waves having wavevectors that are small, but not zero. Although Fig. 18.6 would seem to forbid a second-order transition, the difference with crystallization is that there is an infinite number of possible lengths for the unmixing wavevector in spinodal decomposition. The chance of finding three that meet the criteria of Fig. 18.6 is therefore infinitesimal. The kinetic process of spinodal decomposition can show characteristics of a second-order phase transition.

18.5 Thermodynamics of Ordering in the Mean Field Approximation with Long-Range Interactions

18.5.1 Thermodynamic Formulation

The present section develops a thermodynamic theory of ordering, and calculates the critical temperature for long-range order. This was done previously in Sections 2.9 and 7.6, but the present approach with concentration waves allows an important extension beyond the first nearest-neighbor bond energies to account for interatomic interactions over all nearest-neighbor shells. Since concentration waves are parameterized by periodicities, and not real space positions, the effort to include distant nearest-neighbor pair energies is essentially the same as the effort to include first-nearest-neighbor pair energies. Unfortunately, the method of concentration waves does not extend easily beyond the point approximation, and short-range order parameters are not included in the analysis. With only long-range order parameters, the entropy is at the same level as for the Gorsky–Bragg–Williams approximation.

Assume a lattice site is occupied by either an A-atom or a B-atom, never both, and the site is never empty. The occupancy variable for the site at $\vec{r}$ is $c(\vec{r})$, defined as

$c(\vec{r}) = 0$, if $\vec{r}$ is occupied by an A-atom,
$c(\vec{r}) = 1$, if $\vec{r}$ is occupied by a B-atom.

An ensemble average of $c(\vec{r})$, an average occupancy of B-atoms, is calculated as

$$\langle c(\vec{r}) \rangle = \frac{0\,e^{-\frac{e_A}{k_B T}} + 1\,e^{-\frac{e_B}{k_B T}}}{e^{-\frac{e_A}{k_B T}} + e^{-\frac{e_B}{k_B T}}}, \tag{18.51}$$

$$\langle c(\vec{r}) \rangle = \frac{1}{e^{-\frac{e_A - e_B}{k_B T}} + 1}. \tag{18.52}$$

The denominator of Eq. 18.51 is the $Z_{1\text{site}}$ of Eq. 2.21. Notice how Eq. 18.52 is essentially the same as the Fermi–Dirac distribution for electrons, which arises from the analogous assumption that each electronic state can accommodate either zero or one electron.

The e_A and e_B have the same meaning as in Eq. 2.20, but here we sum the chemical bond energies from all surrounding atoms, not just the z-atoms in the first nearest neighbor shell

$$e_A = \sum_{\vec{r}'} \left[(1 - c(\vec{r}'))\, e_{AA}(\vec{r} - \vec{r}') + c(\vec{r}')\, e_{AB}(\vec{r} - \vec{r}') \right], \tag{18.53}$$

$$e_B = \sum_{\vec{r}'} \left[(1 - c(\vec{r}'))\, e_{AB}(\vec{r} - \vec{r}') + c(\vec{r}')\, e_{BB}(\vec{r} - \vec{r}') \right], \tag{18.54}$$

$$e_B - e_A = \sum_{\vec{r}'} \Big(e_{AB}(\vec{r} - \vec{r}') - e_{AA}(\vec{r} - \vec{r}')$$

$$+ c(\vec{r}') \big[e_{AA}(\vec{r} - \vec{r}') + e_{BB}(\vec{r} - \vec{r}') - 2e_{AB}(\vec{r} - \vec{r}') \big] \Big). \tag{18.55}$$

As in Eq. 2.32 we define a quantity V, but it is not confined to the first nearest neighbor shell. Figure 18.7 depicts how the central atom at $\vec{r}$ has bonds of various strengths to surrounding atoms at $\vec{r}'$.

$$V(\vec{r} - \vec{r}') \equiv \left[e_{AA}(\vec{r} - \vec{r}') + e_{BB}(\vec{r} - \vec{r}') - 2e_{AB}(\vec{r} - \vec{r}') \right]/4. \tag{18.56}$$

Substituting Eq. 18.56 into 18.55, Eq. 18.52 becomes

$$\langle c(\vec{r}) \rangle = \frac{1}{\exp\left(\frac{\sum_{\vec{r}'} [e_{AB}(\vec{r} - \vec{r}') - e_{AA}(\vec{r} - \vec{r}')]}{k_B T} \right) \exp\left(\frac{\sum_{\vec{r}'} 4V(\vec{r} - \vec{r}')\, c(\vec{r}')}{k_B T} \right) + 1}. \tag{18.57}$$

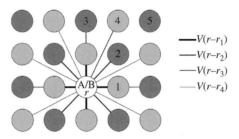

Geometry for ordering problem with arbitrary-range interactions. Modulation of gray shading of neighboring sites represents a small concentration wave, $\Delta(\vec{r}')$.

18.5.2 Small-Amplitude Concentration Waves

We consider an alloy that either has uniform composition, c_0, or may contain a small concentration wave $\Delta(\vec{r})$. We write $c(\vec{r}')$ at the neighboring sites $\{\vec{r}'\}$ as

$$c(\vec{r}') = c_0 + \Delta(\vec{r}'). \tag{18.58}$$

For convenience in what follows, define E_0 and δ to recast Eq. 18.57 into this compact form

$$\langle c(\vec{r}) \rangle = \frac{1}{E_0\, e^{\delta} + 1}, \tag{18.59}$$

which separates factors involving the concentration at the neighboring atom sites into one for constant composition c_0, and another for the small concentration wave $\Delta(\vec{r}')$

$$E_0 = \exp\!\left(\frac{\sum_{\vec{r}'}[e_{AB}(\vec{r}-\vec{r}') - e_{AA}(\vec{r}-\vec{r}')]}{k_B T}\right) \exp\!\left(\frac{\sum_{\vec{r}'} 4V(\vec{r}-\vec{r}')\, c_0}{k_B T}\right), \tag{18.60}$$

$$\delta = \frac{4}{k_B T} \sum_{\vec{r}'} V(\vec{r}-\vec{r}')\, \Delta(\vec{r}'). \tag{18.61}$$

As a reference state we will use a random solid solution, for which there are no concentration waves, i.e., $\Delta(\vec{r}) = 0$ and $\delta = 0$, and Eq. 18.59 becomes

$$c_0 = \frac{1}{E_0 + 1}, \tag{18.62}$$

$$E_0 = \frac{1 - c_0}{c_0}. \tag{18.63}$$

To make further progress, we assume the concentration wave amplitudes are very small, such as near T_c. Therefore δ is very small, so we expand e^{δ} in Eq. 18.59, then use Eq. 18.63

$$\langle c(\vec{r}) \rangle = \frac{1}{E_0\,(1+\delta) + 1}, \tag{18.64}$$

$$\langle c(\vec{r}) \rangle = \frac{1}{\left(\frac{1-c_0}{c_0}\right)(1+\delta) + 1}, \tag{18.65}$$

$$\langle c(\vec{r}) \rangle = \frac{c_0}{1 + \delta(1 - c_0)}. \tag{18.66}$$

For small δ

$$\langle c(\vec{r}) \rangle = c_0 \left[1 - \delta(1 - c_0) \right], \tag{18.67}$$

$$\langle c(\vec{r}) \rangle = c_0 - c_0(1 - c_0)\,\delta. \tag{18.68}$$

Substituting from Eq. 18.61, using Eq. 18.58, recognizing that $c_0 \simeq c$ for small Δ, and recognizing that the ensemble average $\langle \Delta(\vec{r}) \rangle = \Delta(\vec{r})$

$$\Delta(\vec{r}) = -c(1 - c)\frac{4}{k_B T} \sum_{\vec{r}'} V(\vec{r} - \vec{r}')\Delta(\vec{r}'), \tag{18.69}$$

when $\Delta(\vec{r})$ is small. Equation 18.69 is a linearized form of Eq. 18.57, valid when $\Delta(\vec{r})$ is small in equilibrium.

18.5.3 Critical Temperature

Our assumption of very small $\Delta(\vec{r})$ is true near a second-order critical temperature, and Eq. 18.69 can be used to obtain T_c. Different ordered structures will have different values of Δ at different neighboring sites. We want to put this variation into $V(\vec{r})$ to define a $V'(\vec{r})$, and obtain a simple expression for T_c

$$T_c = -c(1 - c)\frac{4}{k_B} \sum_{\vec{r}} V'(\vec{r}). \tag{18.70}$$

In what follows, we will transform this $\sum_{\vec{r}} V'(\vec{r})$ to $\tilde{v}(\vec{k}_0)$, and show a systematic way to go from V to V' for an ordered structure with the star $\vec{k}_0$.

Right now, however, we can readily get the critical temperature for an ordering alloy with first-neighbor interactions but a general composition c, which is a more general result than Eq. 2.58. This case includes the chessboard and B2 structures where the first neighbors are the other species of atom, so $\Delta(\vec{r}_1) = -\Delta(0)$. There are z identical first-nearest-neighbors, so[12]

$$T_c = -c(1 - c)\frac{4z}{k_B}(-V_1). \tag{18.71}$$

For the equiatomic alloy with $c = 0.5$,

$$T_c = \frac{zV_1}{k_B}, \tag{18.72}$$

which agrees with Eq. 2.58. Equation 18.69 is much more general, however, since it allows pairwise interatomic interactions of arbitrary range. It does require information about the ordered structure, and this structural information is mathematically convoluted with the interaction energy in Eq. 18.69.

---

[12] It is not necessary to use concentration waves to obtain Eq. 18.70, but it is necessary to use an order parameter to quantify the difference of atom concentrations on the two sublattices. Our $\Delta(\vec{r})$ does this quite well.

18.5.4 Formulation in k-Space

The method of concentration waves can be used to recast Eq. 18.69 into an interesting form in k-space. The result will still be confined to small-amplitude concentration waves, but it will be easier to determine the critical temperature. We start by Fourier transformation of the quantities $V(\vec{r})$ and $\Delta(\vec{r})$

$$\tilde{v}(\vec{k}) = \sum_{\vec{r}} e^{-i\vec{k}\cdot\vec{r}}\, V(\vec{r}), \tag{18.73}$$

$$\tilde{\delta}(\vec{k}) = \sum_{\vec{r}} e^{-i\vec{k}\cdot\vec{r}}\, \Delta(\vec{r}), \tag{18.74}$$

and the convolution in Eq. 18.69 becomes a multiplication in k-space

$$\tilde{\delta}(\vec{k}) = -\frac{4\,c(1-c)}{k_B T}\, \tilde{v}(\vec{k})\, \tilde{\delta}(\vec{k}). \tag{18.75}$$

The critical temperature can be obtained for the concentration wave with the lowest value of $\tilde{v}(\vec{k})$, here denoted $\tilde{v}(\vec{k_0})$, since this wave is first to achieve $b = 0$ upon cooling, as shown in Fig. 18.5. (Once this concentration wave of $\vec{k_0}$ has grown, the alloy is no longer a disordered solid solution, and it is often not possible to consider instabilities by other small concentration waves.) Since Eq. 18.75 pertains to small concentration waves, which exist near T_c, the temperature defined by Eq. 18.75 is T_c

$$T_c = -\frac{4\,c(1-c)}{k_B}\, \tilde{v}(\vec{k_0}). \tag{18.76}$$

Consider the critical temperature for B2 ordering on the bcc lattice, which has a star containing one vector, $\vec{k_0} = 2\pi/a(1,1,1)$. The $\tilde{v}(\vec{k_0})$ of Eq. 18.73 is evaluated in two steps. First, the exponential phase factors for the concentration wave are evaluated at the nearest-neighbor distances. The distance vectors for the bcc structure are

$$\vec{r}_1 = a\left(\frac{1}{2}, \frac{1}{2}, \frac{1}{2}\right), \quad \vec{r}_2 = a(1,0,0), \quad \vec{r}_3 = a(1,1,0), \quad \vec{r}_4 = a\left(1\frac{1}{2}, \frac{1}{2}, \frac{1}{2}\right), \ldots \tag{18.77}$$

giving the phase factors

$$e^{-i2\pi(111)\cdot\vec{r}_{1j}} = -1, \quad e^{-i2\pi(111)\cdot\vec{r}_{2j}} = +1, \quad e^{-i2\pi(111)\cdot\vec{r}_{3j}} = +1, \quad e^{-i2\pi(111)\cdot\vec{r}_{4j}} = -1, \ldots \tag{18.78}$$

Second, the $V(\vec{r})$ is decomposed into terms for the different nearest-neighbor shells, each with its own V_j (where j denotes the jth shell), and where the interchange potential for the jth nearest-neighbor distance is defined by Eq. 18.56. This V_j is multiplied by the number of sites in the jth shell, which is the sequence of numbers $\{8, 6, 12, 24, 8, 6, \ldots\}$ for the bcc structure. Finally, the product of the sign of Eq. 18.78, the interchange potential V_j, and the number of sites in the bcc shells gives

$$\tilde{v}(\vec{k_0}) = -8V_1 + 6V_2 + 12V_3 - 24V_4 + 8V_5 + \cdots. \tag{18.79}$$

Equation 18.79 for $\tilde{v}(\vec{k_0})$ can be used in Eq. 18.76 to obtain T_c. As expected, a strong, positive V_1 favors B2 order for equiatomic alloys. Other possibilities such as a strong, positive V_4 might promote B2 order, too, but other concentration waves would have to be considered – perhaps the B2 structure is stable, but is not the most stable structure for the alloy with strong V_4 (see Section 22.1 [online]).

Equations 18.69 and 18.75 are valid at temperatures infinitesimally below T_c. At temperatures above T_c, these equations have the trivial solution $\Delta(\vec{r}) = 0$ because the solid solution is stable and concentration waves do not form. Far below T_c^-, the assumptions of Section 18.5.2 are not valid, so Eqs. 18.69 and 18.75 are inaccurate. We then need to use the full, nonlinear expression of Eq. 18.57 which, like Eq. 2.55, is transcendental and has no simple analytical form. The k-space version of Eq. 18.57

$$c_0 + \sum_{j}^{\text{star}} \tilde{\delta}(\vec{k}_j) = \frac{1}{\exp\left(\frac{V_0 c_0 + \sum_{j}^{\text{star}} \tilde{v}(\vec{k}_j)\,\tilde{\delta}(\vec{k}_j)}{k_B T}\right) + 1}, \tag{18.80}$$

is simpler than Eq. 18.57, but it is still without an analytical solution for $\tilde{\delta}(\vec{k}_j)$.

Problems

18.1 An ordered structure with the wavevector $\vec{k}_0 = (\frac{1}{2}\ \frac{1}{2})$ forms on a square lattice.
    (**a**) List the vectors in the star of this ordered structure.
    (**b**) Draw this ordered structure and give it a name. Show the directions of your Cartesian axes x, y.
    (**c**) Can this ordered structure form by a second-order phase transition? (For credit, show some work or reasoning.)

18.2 (**a**) Explain in words why the Landau–Lifshitz criterion for second-order phase transitions uses reciprocal lattice vectors $\vec{g}$ of the disordered alloy, rather than the ordered alloy. In your answer, explain why the free energy for ordering should have the translational periodicity of the disordered phase.
    (**b**) Third- and fourth-order terms in a Taylor series are usually neglected. Why are they important in the development of the Landau–Lifshitz criterion?

18.3 For a random alloy of solute concentration c, the probability of finding n solute atoms in a nearest-neighbor shell having N sites is the binomial probability

$$P(N, n, c) = \frac{N!}{(N-n)!\, n!}\, c^n\, (1-c)^{N-n}. \tag{18.81}$$

    (**a**) Explain why.
    (**b**) For a bcc alloy with solute concentration c, show that the average chemical interaction from the solutes in the 1nn shell is $8cV_1$ by evaluating the sum

$$\overline{V}_1 = \sum_{n=0}^{8} P(8, n, c)\, nV_1 = 8cV_1. \tag{18.82}$$

Such an average interaction was used for the solid solution in Sections 2.7–2.9.

(c) By a similar method, evaluate the variance

$$\langle n^2 \rangle - \langle n \rangle^2. \tag{18.83}$$

(d) For bonding effects from the 1nn and 2nn shells of the bcc structure, we expect

$$\sum_{n=0}^{8} \sum_{m=0}^{6} P(8,n,c)P(6,m,c)(nV_1 + mV_2) = 8cV_1 + 6cV_2. \tag{18.84}$$

Now if $V_1 = V_2 = V$, each of the 14 sites of the 1nn and 2nn shells makes the same contribution to the chemical bonding. Show that

$$\sum_{n=0}^{8} \sum_{m=0}^{6} P(8,n,c)P(6,m,c)(nV + mV) = \sum_{j=0}^{14} P(14,j,c)jV. \tag{18.85}$$

18.4 For the B2 structure:
   (a) Evaluate the concentration wave at positions of the first through fourth neighbors of an A-atom at the origin.
   (b) What is the potential $\tilde{v}(\vec{k}_0)$ through fourth neighbors, where $\vec{k}_0$ is the ordering wavevector?

18.5 An ordered structure with the wavevector $\vec{k}_0 = (0\ 0\ \frac{1}{2})$ forms on a simple cubic lattice.
   (a) Draw this ordered structure on a $2 \times 2 \times 2$ supercell of a simple cubic parent lattice (total of 27 atom sites). Show the directions of the Cartesian axes x, y, z.
   (b) List all vectors in the star of $\vec{k}_0$.
   (c) Can this ordered structure form by a second-order phase transition? (For credit, show some work or reasoning.)
   (d) The $L1_0$ structure with $\vec{k}_0 = (0\ 0\ 1)$ must form by a first-order transition, as shown in the text. For interest, draw one unit cell of this $L1_0$ structure with the same orientation as in part a.

18.6 Figure 18.8 shows the fcc-based ordered structure of CuPt ($L1_1$ structure). Determine if this ordered structure can form by a second-order phase transition.
   (*Hint*: Reference [288] is useful for this problem, especially pages 1–64.)

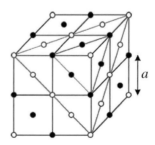

The $L1_1$ structure of CuPt.

18.7 The concentration wave $\Delta(\vec{r})$ is sometimes written as a complex exponential plus its complex conjugate

$$\Delta(\vec{r}) = \frac{1}{2}\left[\eta\,e^{-i\vec{k}\cdot\vec{r}} + \eta^*\,e^{+i\vec{k}\cdot\vec{r}}\right]. \qquad (18.86)$$

Show that this expression for $\Delta(\vec{r})$, when used in Eq. 18.33, gives a result like that of Eq. 18.38.

19

Diffusionless Transformations

Part III has so far described melting, solidification, precipitation, and unmixing phase transformations. For alloys, these all require the diffusion of atoms over moderate distances, for which continuum diffusion equations provide much of the essential behavior. Ordering transformations also require atom movements, although the configurational order can change after relatively few jumps of independent atoms.

This chapter describes diffusionless transformations, in which the atoms in a crystal move cooperatively and nearly simultaneously, distorting the crystal into a new shape. The transformation is very fast, and does not occur by independent jumps of individual atoms. The martensite transformation is the most famous diffusionless transformation, owing to its importance in steel metallurgy. In a martensitic transformation the change in crystal structure occurs by shears and dilatations, and the atom displacements accommodate the shape of the new crystal. The atoms do not move with independent degrees of freedom, so the change in configurational entropy is negligible or small. The entropy change from a martensitic transformation is primarily vibrational (sometimes with electronic entropy, or magnetic entropy for many iron alloys). Twinning is another type of diffusionless transformation where a crystal transforms into a different variant of the same type of crystal.[1]

This chapter begins with a review of dislocations, and how their glide motions can give crystallographic shear. Some macroscopic and microscopic features of martensite are then described, followed by a two-dimensional analog for a crystallographic theory that predicts the martensite "habit plane" (the orientation of a martensite plate in its parent crystal). Displacive phase transitions are explained more formally with Landau theories having anharmonic potentials and vibrational entropy. (These descriptions of second-order and first-order displacive transitions find use again for ferroelectric transitions in Section 21.8.) Finally phonons are discussed, from the viewpoint of soft modes in bcc structures which may be a step along the path of a martensitic transformation, and from the viewpoint of the vibrational entropy of the transformation. A fully satisfying explanation of martensitic transformations is not offered, however, since it is not yet available.

19.1 Dislocations, Mechanisms, and Twinning

Within a crystal formed by a diffusionless transformation, all unit cells had undergone the same distortion. Somehow, the atoms at the initiation point communicate their selection

---

[1] The new twin phase and its parent have the same atomic structure and have the same volume free energy, so twinning is not a "true" phase transition.

of new orientation to neighboring unit cells. This communication is expected to be no faster than the speed of sound, and probably slower. Transmitting the shear to surrounding regions by sound waves has some plausibility, but the atom displacements in the shear process are much larger than typical atom vibrations. (Giant, nonlinear vibrations are discussed in Section 19.4, however.) The agent of change in diffusionless transformations is probably the dislocation. Dislocations carry large displacements, and can move at a good fraction of the speed of sound. The structure, energy, and dynamics of dislocations are now reviewed, and their role in shear transformations is explained.

19.1.1 Structure of a Dislocation

A dislocation is the only type of line defect in a solid. A large body of knowledge has formed around dislocations because their movement is the elementary mechanism of plastic deformation of many crystalline materials. An "edge dislocation" is the easiest to illustrate. In Fig. 19.1a, notice how an extra half-plane of atoms has been inserted in the upper half of the simple cubic crystal. This extra half-plane terminates at the "core" of the edge dislocation line. On the figure is drawn a circuit of five atom spacings, up, right, down, and left. This circuit, called a "Burgers circuit," does not close perfectly when it encloses the dislocation core. (It does close in a perfect crystal, of course. It also closes perfectly in a dislocated crystal if the circuit does not enclose the dislocation core.) The vector from the end to the start of the circuit is defined as the "Burgers vector" of the dislocation, $\vec{b}$.

Dislocations are characterized by their Burgers vector and the direction of their dislocation line. In Fig. 19.1a $\vec{b} = a\hat{x}$, and the dislocation line is along $\hat{y}$. This is a pure "edge dislocation." The other type of a "pure" dislocation has its Burgers vector parallel to the dislocation line. It is a "screw dislocation," and is illustrated in Fig. 19.1b. Around the core of a screw dislocation, the crystal planes form a helix. After completing a Burgers circuit in the x–y plane in Fig. 19.1b, the vector from end to start lies along $\hat{z}$. For a screw dislocation, $\vec{b}$ is parallel to the line of the dislocation in the center of the cylinder in Fig. 19.1b. Finally,

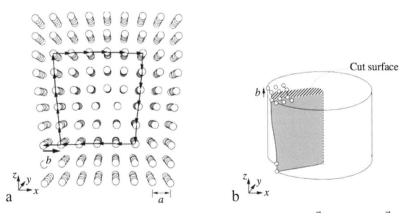

Figure 19.1   (**a**) Edge dislocation in a cubic crystal. Dislocation line is parallel to $\hat{y}$, the Burgers vector $\vec{b} = a\langle 100 \rangle$, and $\vec{b}$ is perpendicular to the dislocation line. (**b**) Screw dislocation in a cylinder of cubic crystal. Dislocation line is parallel to $\hat{z}$, $\vec{b} = a\langle 001 \rangle$, and $\vec{b}$ is parallel to the dislocation line.

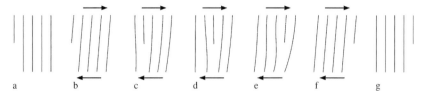

Figure 19.2 (**a**) Initial state of crystal without applied stress. (**b–f**) Applied stress indicated by arrows, and glide of dislocation from left to right. (**g**) Final state of crystal without applied stress.

the magnitude of the Burgers vector b is the "strength" of the dislocation – dislocations with larger b cause larger crystalline distortions.

In general, dislocations are neither pure edge nor pure screw dislocations, but rather have their Burgers vectors at an intermediate angle to their line direction. These are called "mixed dislocations." The $\vec{b}$ for a mixed dislocation is the same everywhere along the line of the dislocation, but the character changes when the dislocation line is curved or kinked, as it usually is.

19.1.2 Dislocation Glide

Figure 19.2 shows how an edge dislocation moves under stress in a process called "glide." The applied shear stress drives the edge dislocation from left to right along a "glide plane" that is halfway down the crystal. Each step occurs as the distorted bonds on one plane are broken at the glide plane, and a half-plane of the lower crystal joins to the half-plane at the dislocation core. The dislocation is preserved in this process, at least until it leaves the surface of the crystal. When the stress is removed, comparing the macroscopic shapes of Fig. 19.2a and 19.2g shows that the crystal has undergone some plastic shear consistent with the applied stress. The alternative of breaking all bonds in the glide plane simultaneously, and doing the shear in one step from Fig. 19.2b to 19.2f, requires stresses that are far larger. Dislocation glide is usually the mechanism for the plastic deformation of crystals.

19.1.3 Elastic Energy of a Dislocation (Self-Energy)

A dislocation generates large elastic strains in the surrounding crystal, as is evident from Fig. 19.1. The strain in the material in the dislocation core (usually considered to be a cylinder of radius $5|\vec{b}|$) is so large that its excess energy cannot be accurately regarded as elastic energy. Because the atom configurations in the core are noncrystalline, sometimes the "core energy" is estimated from the latent heat of melting. Outside the core, however, it is reasonable to calculate the energy by linear elasticity theory. It turns out that this total elastic energy in the surrounding crystal is typically an order-of-magnitude larger than the energy of the core. Approximately, the energy for making a unit length of dislocation line is equal to the elastic energy per unit length of the dislocation.

Figure 19.1b can be used to obtain the energy of a dislocation, using a cut-and-shear process. The dislocation line is located at the edge of the cut, and the Burgers vector is the vector of the shear displacement across the cut. The cut itself requires no energy because the atoms across the cut are later reconnected after the dislocation is made. The energy needed to make the dislocation is the energy to make the shear across the cut surface.

Think of the cut crystal as a spring. An elastic restoring force opposes the shear, and this restoring force is proportional to the shear times the shear modulus, G. The distance of displacement across the cut is b. The elastic energy stored in the crystal is obtained by integrating the force over the distance, x, of shear

$$E_{\text{elas}} \propto \int_{0}^{1\text{cm}} \int_{0}^{b} G\,x\,\mathrm{d}x\,\mathrm{d}z, \tag{19.1}$$

$$E_{\text{elas}} = Gb^2K \ [\text{J/cm}], \tag{19.2}$$

for 1 cm of dislocation line along $\hat{z}$. Here K is a geometrical constant that depends on the size and shape of the crystal (and somewhat on the dislocation character). Neglecting the smaller core energy, the energy cost of creating a unit length of edge dislocation is the E_{elas} of Eq. 19.2.

19.1.4 Stacking Faults in fcc Crystals

Because the self-energy of a dislocation increases approximately as b^2 (Eq. 19.2), dislocations prefer small Burgers vectors. For example, two edge dislocations of $b = a$ accommodate two terminating half-planes, as does one edge dislocation of $b = 2a$, but the energy of the latter is twice as large (i.e., $1 \times (2a)^2 > 2 \times a^2$). Big dislocations therefore decompose into smaller ones by "dissociation reactions," and the two smaller dislocations move apart to reduce the overlap of their elastic fields. The lower limit to the Burgers vector is set by the requirement that the atoms must match positions across the cut in the crystal. This lower limit is typically the distance between nearest-neighbor atoms. Smaller Burgers vectors occur in fcc and hcp crystals, however.

A special dissociation reaction occurs for dislocations on close-packed {111} planes in fcc crystals. Figure 19.3 shows how the stacking of these close-packed planes determines

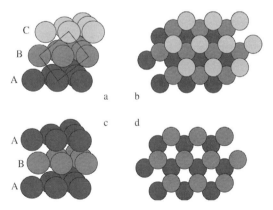

Figure 19.3 (**a**) fcc stacking of close-packed (111) planes; perspective view of three layers, with the cubic face marked with the square. (**b**) Stacking of the three types of (111) planes seen from above. The next layer will be an A-layer, and will locate directly above the dark A-layer at the bottom. (**c**) hcp stacking of close-packed (0001) basal planes; perspective view of three layers. (**d**) hcp stacking of the two types of close-packed planes seen from above. The next layer will be an A-layer, and will locate directly above the dark A-layer at the bottom.

whether the crystal is fcc or hcp. The "perfect dislocation" in the fcc crystal has a Burgers vector of the nearest-neighbor separation, $\vec{b} = 1/2[110]$. The shifts between the adjacent layers of the fcc structure are smaller than this, however, and these small shifts can be obtained by creating a "stacking fault" in the fcc crystal. Specifically, assume that the ABCABCABC stacking of the fcc crystal is interrupted by a small shift of a {111} plane as: ABCAB|ABCABC. Note the error in stacking from placing an A-layer to the immediate right of a B-layer. The structure is still close packed, but there is a narrow region of hcp stacking (...AB|AB...). This region of hcp crystal need not extend to the edge of the crystal, however. The hcp region can be bounded by a "Shockley partial" dislocation, which has a Burgers vector equal to the shift between an A- and a B-layer. This shift is a vector of the type $a/6\langle 112\rangle$.

Consider a specific dissociation reaction when a perfect dislocation decomposes into two Shockley partial dislocations. Note that the total Burgers vectors are equal before and after this reaction[2]

$$a/2[110] \rightarrow a/6[121] + a/6[21\bar{1}]. \tag{19.3}$$

The energy, proportional to the square of the Burgers vector, is smaller for the two Shockley partials on the right than the single perfect dislocation on the left. This is verified by calculating the energies with Eq. 19.2

$$\frac{E_{\text{perfect}}}{E_{\text{2partials}}} = \frac{KG\frac{a^2}{4}\left(1^2 + 1^2 + 0\right)}{2KG\frac{a^2}{36}\left(1^2 + 2^2 + 1^2\right)} = \frac{3}{2}. \tag{19.4}$$

It is energetically favorable for a perfect dislocation in an fcc crystal to split into two Shockley partial dislocations, which move apart to reduce the elastic energy. There is, however, a thin region of hcp crystal between these two Shockley partials (the stacking fault), and the stacking fault energy tends to keep the partials from getting too far apart. Equilibrium separations of Shockley partial dislocations offer a way to determine the stacking fault energy of fcc crystals. This stacking fault energy is approximately (or at least conceptually) related to the free energy difference between the fcc and hcp crystal structures.

19.1.5 Twinning

Figure 19.4 illustrates twinning in an fcc crystal by small displacements of close-packed (111) planes. The original crystal is shown in Fig. 19.4a, with its planes in three different vertical registries indicated by shading and labels A, B, C.[3] Figure 19.4b shows three sequential slips of upper planes of the crystal, performed one layer at a time, and the new registrations of the planes. The magnitudes of the slips are in fact just what is obtained by passing a Shockley partial dislocation across a (111) plane, which is a shift of $a/6\,[1\bar{2}1]$, for

---

[2] This "conservation of Burgers vector" is equivalent to the fact that a dislocation line cannot terminate in the middle of a crystal, but must extend to the surface or form a loop.

[3] The atoms in the different registries are not all in the plane of the paper, but they project onto the positions in the figure.

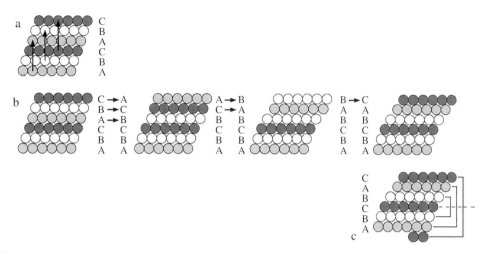

Figure 19.4 (**a**) Side view of stacking sequence of close-packed planes of an fcc crystal. The vertical arrows indicate that A-layers are directly above A-layers, the B directly above B, and C above C. (**b**) Slip of crystal planes above the middle C-layer as a block, followed by slip above the original A (now B), followed by slip above original B (now A). The result is a twin at the upper half of the crystal. (**c**) Reflection symmetry across the dashed twinning plane. See also Fig. 20.4.

example. Three of these give a net shift of $3 \times a/6\,[1\bar{2}1] = a/2\,[101] + a[0\bar{1}0]$. The first vector is a nearest-neighbor vector in the fcc structure, and the second is a translation of the standard fcc unit cell. In other words, these three shifts by a Shockley partial dislocation return an atom to its original vertical registry. (An atom in a C-registry is returned to a C-registry, for example.) These shifts are depicted by the short horizontal arrows in Fig. 19.4b.

Take a look at the systematics of using the short arrows with the notation for the ABC stacking in Fig. 19.4b – each shift requires re-registering *all* planes above the shift, not just the first plane. Finally, after all shifts shown in Fig. 19.4b, the stacking of ABC above the first plane of shift is transformed to CBA (the grayscale sequence is reversed). Figure 19.4c shows that there is a reflection symmetry across the twinning plane, so there are matched planes of A–A, B–B, and C–C planes equidistant from the twinning plane. It is also possible to consider the twinning plane as a grain boundary between two fcc crystals of different orientation.

The relationship between twinning and moving a Shockley partial dislocation on every close-packed plane suggests a process for a twinning called the "pole mechanism." This mechanism includes two dislocations: a screw dislocation that serves as the pole, and a working Shockley dislocation that wraps around the pole and spirals upwards or downwards plane-by-plane. The pole mechanism is depicted in Fig. 19.5. The screw dislocation could have the Burgers vector of a perfect dislocation, or some character that includes the $a/3[111]$ displacement between the close-packed planes. As the working Shockley partial dislocation wraps around the pole, it moves up by one close-packed plane for each wrap. At the same time, the left segment of this dislocation can move down the pole causing appropriate shifts of the lower planes.

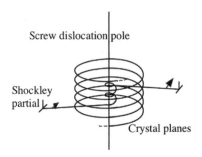

Figure 19.5 A pole mechanism for twinning, constructed with two perpendicular dislocations. The large helix denotes atomic planes intersected by the screw dislocation (the pole). The partial dislocation follows these planes, and has made about two wraps around the vertical pole. One segment spiraled up, the other down.

An issue with the pole mechanism as shown in Fig. 19.5 is that the two segments of the partial dislocation must pass over each other in the first wrap, and this involves significant elastic energy. The two segments probably do not remain straight lines that overlap at all radii simultaneously, but the energy is not fully understood. One possible solution to the problem uses an out-of-plane orientation for one of the segments. The macroscopic shapes of twins seem to be symmetrical, however, and the two counter-rotating segments offer a symmetric transformation product. Another solution is that prior deformation in the material has induced internal stresses that can help the passage of the dislocation segments over each other. A third solution is that the partial dislocation segments move at a very high speed, and their elastic field does not keep up with them. In this case there is a "relativistic" change to dislocation self energies that alters the elastic interaction between the two segments, making their overlap less unfavorable.

Because twinning gives a reorientation of a crystal, twinning can sometimes be driven by applied stress. The deformation of tin is perhaps the most famous. When a bar of tin is bent or deformed at room temperature, it emits audible high-frequency sounds as twinning occurs, known as the "cry of tin." As a mechanism for plastic deformation, twinning often competes with the process of dislocation glide. Twinning tends to occur during the deformation of bcc iron at low temperatures, for example. Twinning is sometimes favored at high deformation rates, but this is material specific.

Phase transformations between fcc and hcp structures have many similarities to the twinning process just considered for fcc crystals. Problem 19.1 works through this case, with reliance on the handy registry notation of Fig. 19.4b and the shifts of letters using the right arrows.

19.2 Martensite

19.2.1 Features of Martensite

A plain carbon steel (for example Fe–1 at.%C) at a temperature of 950 °C has the fcc phase known as "austenite," as discussed in Section 14.3 with the phase diagram of Fig. 14.7. The

Fe atoms occupy the sites of an fcc lattice, and the interstitial C atoms are in octahedral sites such as the site in the center of the standard fcc unit cell. After rapidly quenching the steel to room temperature, it transforms into a body-centered tetragonal (bct) phase called "martensite." Because of the technological importance of martensitic steels, which offer high strength and toughness at modest cost, martensitic phase transformations have been studied extensively for many years. Nevertheless, many aspects of martensitic transformations, including the basic mechanism of atom displacements, remain frustratingly unclear. Martensitic transformations also occur in Fe–Ni, Au–Cd, Ti–Nb, In–Tl, Cu–Zn, Cu–Al, and many other alloys. They occur in superconductors, e.g. V_3Si and Nb_3Sn, and in ceramic systems such as ZrO_2 and $BaTiO_3$. Eight characteristic features of martensite are [244]:

1. Martensitic transformations are "diffusionless." Individual atoms move by less than one interatomic distance and the product phase is formed by a cooperative motion of many atoms. One consequence is that the local chemical composition cannot change during a martensitic transformation, and all local atomic configurations change in the same way [291, 292]. The transformation therefore causes a negligible change in the configurational entropy of the alloy.

2. Martensite usually forms as thin plates in the austenite phase, but other morphologies such as needles and laths can occur. For a given composition, the plates lie on distinct crystallographic planes in the parent phase called the "habit plane." Martensite plates grow at velocities of a fraction of the speed of sound, and growth is terminated when the plate encounters a grain boundary, the specimen surface or another martensite plate. The formation of a martensite platelet can trigger the formation of other plates. During cooling, martensite often forms in a series of bursts, as shown in Fig. 19.6a. On the other hand, the reverse transformation upon heating, as shown in Fig. 19.6b, is more continuous. The transformation temperatures of the forward and reverse martensitic transformation in Fig. 19.6 are very different, showing a large hysteresis for the transformation. This is associated with delayed nucleation of martensite in iron alloys, and this hysteresis often makes it difficult to deduce thermodynamic properties for the martensite transformation.

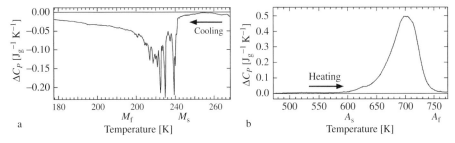

Figure 19.6 Calorimetry measurements of martensitic transformation in $Fe_{71}Ni_{29}$. (**a**) Heat release upon cooling, where the transformation occurs in bursts. The area in the exothermic peaks is approximately proportional to the amount of new phase that has formed. (**b**) Heat absorption upon heating. The reverse transformation from α to γ occurs continuously, but at a much higher temperature than for cooling in panel a.

Figure 19.7 Sequence of martensite plate formation. Growth of plates is terminated at grain boundaries and at other martensite plates. Reproduced, with permission, from [244].

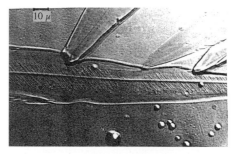

Figure 19.8 Martensite plates in an almost completely transformed Fe–Ni specimen. Image width is 0.14 mm [293, 294]. Reprinted from [293], with permission from Elsevier.

Figure 19.9 Intersecting martensite plates in an Fe–32%Ni specimen. The straight line running through the center of the plates is known as a midrib. The fine lines running across the midrib are twins. From [294], courtesy of W.A. Leslie, U.S. Steel Co.

Figure 19.7 is a schematic diagram of martensite plate formation; notice the successively smaller and smaller plates as more of the specimen is transformed. Figure 19.8 shows an almost fully transformed Fe–Ni alloy. Martensite initially forms as very thin plates that subsequently thicken during the rapid growth stage. The thickening process is inferred from optical and electron microscopy, which reveal the presence of midribs running along the center line of the plate. Midribs, as seen in Fig. 19.9, are thought to be the initial martensite plate.

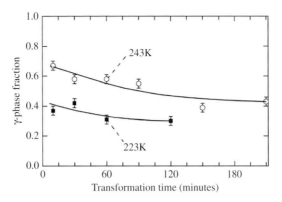

Figure 19.10 Diffractometry measurements of the remaining austenite as martensite forms in $Fe_{71}Ni_{29}$ at different times. Most of the transformation occurs quickly (from 1.0 to the fraction on the y-axis) but some transformation continues after times of hours.

3. Upon cooling, the volume fraction of martensite is a function of temperature. The first martensite plates form at the "martensite start" temperature, M_s, whereas the specimen is completely transformed for temperatures below M_f, the "martensite finish" temperature. The volume fraction may or may not change with time. In the "isothermal" case, the volume fraction of martensite increases with time. For "athermal" martensites, the volume fraction of the product phase changes almost instantaneously at any temperature in the range $M_s > T > M_f$, and holding at temperature does not change the amount of martensite. Some isothermal character to the martensitic transformation in $Fe_{71}Ni_{29}$ is shown in Fig. 19.10, but most of the transformation occurs promptly upon cooling to the low temperature, so this transformation is mostly athermal.

4. There is a crystallographic relationship between the martensite and austenite crystals. In plain carbon steels, Kurdjumov and Sachs [295] determined that the (111) plane in the fcc austenite phase lies parallel to the (011) plane in the bct martensite. These two adjacent planes are the close-packed planes in each structure. In addition, the $[\bar{1}01]$ direction in austenite, which lies in the (111) plane, is oriented parallel to the $[\bar{1}\bar{1}1]$ direction in the martensite. Using the standard notation of "γ" representing the fcc austenite and "α'" denoting the bct martensite, the Kurdjumov–Sachs relationship is

$$(111)_\gamma \parallel (011)_{\alpha'} \qquad [\bar{1}01]_\gamma \parallel [\bar{1}\bar{1}1]_{\alpha'}. \tag{19.5}$$

In Fe–Ni alloys with Ni content greater than 28wt.% Ni, the orientation relationship found by Nishiyama [296] is

$$(111)_\gamma \parallel (011)_{\alpha'} \qquad [\bar{1}\bar{1}2]_\gamma \parallel [0\bar{1}1]_{\alpha'}. \tag{19.6}$$

Other martensitic transformations have different orientation relationships, and these need not involve low-index planes and directions [297].

5. The macroscopic distortion caused by the formation of martensite is a homogeneous shear. Figure 19.11a shows a series of straight parallel lines that were inscribed across the surface of the parent phase. An illustration of the surface relief phenomenon,

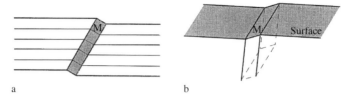

a   b

Figure 19.11 (**a**) View normal to the surface of a specimen with a martensite plate. Parallel lines inscribed on the surface remain straight, but a macroscopic shear due to the transformation is observed. (**b**) Surface upheaval due to the plate intersecting the surface of the sample. From [244].

depicted in Fig. 19.11b, shows the subsurface region. After a martensite plate forms and intersects the sample surface, the lines remain straight in the martensite (labeled M), but make a distinct angle with respect to the lines of the austenite. The habit plane, i.e., the plane separating the parent and martensite phases, is one of zero *macroscopic* distortion. The inside of the martensite plate is sheared or twinned, however.

6. Martensite transformations are often reversible in that the martensite transforms back to austenite upon heating. The reverse reaction starts at a temperature called A_s, and the transformation finishes at the temperature A_f. There is generally thermal hysteresis, so $A_s > M_f$ and $A_f > M_s$. The magnitude of the hysteresis can be quite large, as shown in Fig. 19.6. In a study of the Cu–Al–Ni system, Kurdjumov and Khandros [298] (see also Tong and Wayman [299]) demonstrated the microscopic reversibility of martensite reactions. The first plate to form at M_s during cooling is the last one remaining upon heating. Also, the last plate to appear at M_f is the first plate to retransform at the temperature A_s.

7. An applied stress can promote the martensitic transformation. For any temperature in the range $M_s > T > M_f$, martensite formation is usually increased by plastic strain. Plastic deformation can also cause some formation of martensite plates at temperatures above M_s. The highest temperature at which the martensite can form during deformation is M_d. In contrast, deforming the austenite at temperatures above M_d often inhibits the martensite transformation. For paramagnetic austenite and ferromagnetic martensite, a high magnetic field can also promote the transformation.

8. Suppose austenite is quenched to a temperature below M_s but above M_f, and held at this intermediate temperature for a time. After the holding time, the temperature is lowered again. Martensite does not form immediately. After the holding time, the amount of martensite at lower temperatures is less than the amount obtained by quenching directly to the lower temperatures. This holding treatment is called "stabilization" of austenite [300].

19.2.2 Transformation Mechanisms

Despite the enormous technological importance of martensite in steels, there is no consensus on the atomic-level mechanism (or mechanisms) of the fcc-to-bct martensite transformation in iron alloys. Nevertheless, some detailed models have been proposed,

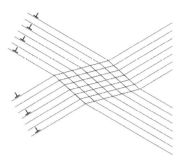

Figure 19.12   Arrays of dislocations in an fcc crystal that could produce a near-bcc structure in their zone of intersection. The lines are different {111} planes in the fcc crystal. The top array would produce a band of hcp crystal, and the lower array may be described as a faulted fcc crystal. Notice that the clockwise shear caused by the first array causes a larger deflection of the lower array of dislocations.

and much discussion is framed in terms of these models. One important model pertains to hexagonal martensite, and has similarities to the twinning transformation shown in Fig. 19.4. Problem 19.1 develops some details of how an fcc-to-hcp transformation can occur in almost the same way as twinning, except only half as many Shockley partial dislocations move along the glide planes (i.e., a Shockley dislocation glides across every other fcc close-packed plane). The formation of this hexagonal ϵ'-martensite occurs in some iron alloys, and this simple mechanism of a coordinated glide of Shockley partial dislocations is believed to be the mechanism of transformation. Notice that a clockwise shear is induced by the twinning mechanism in Fig. 19.4, and a smaller shear occurs with the fcc-to-hcp transformation.

One shear is insufficient to transform an fcc crystal to bcc. Geometrically, the transformation from fcc to bcc requires shears of different orientations in three dimensions. An important proposed mechanism has the euphonious[4] name "Bogers–Burgers double shear." The mechanism of Bogers–Burgers double shear [301], extended to a detailed dislocation mechanism as the Olson–Cohen model [302], includes the same set of partial dislocations that form hcp crystal from fcc, i.e., identical Shockley partial dislocations gliding across alternate close-packed (111) planes. Intersecting this region of hcp crystal (either sequentially or simultaneously) is another array of partial dislocations, although they are spaced on every third glide plane. Figure 19.12 gives the general idea (crystallographic details are given in [302, 303]), by showing two arrays of partial dislocations that cause clockwise shear. The zone of intersection is the transformed bcc martensite. Depicting intersecting bands of glide planes is helpful to show the geometrical relationships between the dislocation arrays, but there is no geometrical reason why the dislocation arrays need to intersect as bands (and in fact a band of hcp phase is not found after martensitic transformations in many materials). The dislocation arrays may begin near the surfaces of the transformed particle, although the formation of dislocations inside a bulk material will require an energy barrier. Another feature of a dislocation mechanism is that the atom shifts should be the same on all planes, and should not be the full displacement

[4] Agreeable to the ear.

of a Shockley partial dislocation on just every second or every third plane. The atoms in the intervening planes must therefore "shuffle" into position, either after the passage of the partial dislocations or simultaneously. These questions are difficult to address by experimental measurement because they are at such a small scale and occur quickly.

19.2.3 Crystallographic Theory of Martensite

An important feature of a martensitic transformation is how the individual plates of martensite are aligned along special crystallographic planes in the parent austenite crystal. The martensite plates lie on a "habit plane," which minimizes the elastic energy in the material. If martensite plates formed without regard to shape or orientation, the transformation strains shown in Fig. 19.13 would be so large that they could create elastic energies well beyond the difference in free energy of the fcc and bct structures. The approach to finding the habit plane is therefore to orient martensite plates for the condition for minimum strain energy. (There is still elastic energy around martensite plates, but it is then less than the free energy of the phase transition.) The geometrical requirements for the habit plane are described here for two dimensions. This two-dimensional problem is incomplete, but the details are far less complicated than the full three-dimensional crystallographic theory of martensite [304–308].

Consider the differences in unit cell of the austenite and martensite, shown in Fig. 19.13. The gray unit cell in Fig. 19.13a is not a cube. It requires a major flattening and expansion to become a bct (nearly bcc) crystal, as shown by the difference between Figs. 19.13b and c. A direct transformation in shape would generate stresses in the surrounding austenite larger than the 10 GPa in Table 8.1. This is too extreme. Nature has found a way to do the martensitic transformation with minimal elastic energy by using a geometrical trick.

The trick is explained with Fig. 19.14. The steps include the "Bain strain," highlighted with an arrow, which is the only step in Fig. 19.14 that involves the atoms in the unit cell. The other steps are macroscopic ones to ensure that the transformed region fits well in the austenite. These steps include slip at the bottom of Fig. 19.14c, which was designed to ensure that the transformed band is macroscopically flat. Finally, there is a rotation of the transformed band, and reinsertion into the original austenite in Fig. 19.14c. Notice the perfect registry of atomic planes across the two interfaces where the transformed zone contacts the austenite. There is, however, a shear of the austenite itself. This distortion is small if the width of the transformed region is small, and martensite forms as plates to minimize this distortion.

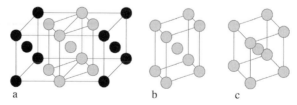

a   b   c

Figure 19.13   (a) Two fcc unit cells, with bct cell in gray. (b) Isolated bct cell from part a. (c) Bain distortion to transform bct cell into bcc cell.

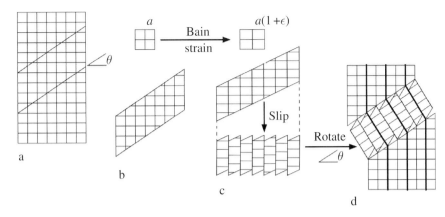

Figure 19.14 (**a**) Original austenite crystal, with pre-selected region to be transformed. (**b**) Pre-selected region and its unit cell. (**c**) Bain strain of martensitic transformation for unit cell and the region. The slip shown at bottom generates a new surface of appropriate length to reconnect to the austenite. (**d**) Rotation of band, which now fits with atomic registry in the original austenite. Dark lines show the registry.

The habit plane in Fig. 19.14a cannot be chosen arbitrarily, and in fact the habit plane changes with the Bain strain. Suppose this strain is ϵ in the horizontal direction of the figure. The unit cell is widened from a to $a(1 + \epsilon)$, as shown in Fig. 19.14c, so the transformed region is too wide to fit in the original austenite. The process of slip is the trick to make it fit again. The slip over the angle θ shortens the surface that needs to fit, and fitting is achieved if

$$1 + \epsilon = \frac{1}{\cos \theta}, \tag{19.7}$$

$$\theta = \arccos\left(\frac{1}{1 + \epsilon}\right). \tag{19.8}$$

For our two-dimensional problem, this θ gives the habit plane of the martensite plate. The thick lines in Fig. 19.14d show its success in achieving the alignment of atom planes across the interface.

We have assured the alignment of atom planes along the y-direction, but in two dimensions we also need to consider atom matching in the x-direction. For this we need another degree of freedom. There was one subtly arbitrary decision in drawing Fig. 19.14c – the slip was chosen to give a horizontal surface because this gave a length match to the habit plane by Eq. 19.8. If the Bain strain were smaller, however, we could have worked with the same transformed region, but ensured matching along the austenite interface by performing a smaller amount of slip than at the bottom of Fig. 19.14c. This extra degree of freedom makes it possible to preserve crystal registry along the austenite interface for different Bain strains. In three dimensions we have similar considerations, with additional degrees of freedom and additional constraints of atom matching. In three dimensions, however, the Bain strain forces the selection of only one habit plane, and this can be calculated geometrically with the crystallographic theory of martensite [304–308].

There is an alternative to the slip process shown in Fig. 19.14 – twinning (as described in Section 19.1.5, but with a different dislocation mechanism than Fig. 19.4). Again, the

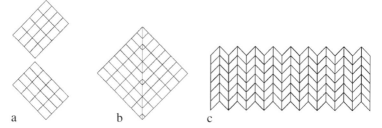

Figure 19.15 (**a**) Two variants of a rectangular phase that are transformation products from a parent square lattice. (**b**) Mutual arrangement of the two variants to minimize shear strains on a macroscale. Note the square geometry of the composite. (**c**) Plate composed of multiple units of twins, with vertical boundaries corresponding to the vertical line shown in panel b.

idea is that twinning on a fine scale allows length matching across the habit plane and atomic registry. The idea is shown in Fig. 19.15. Suppose a square lattice has the two variants of rectangular transformation products shown in Fig. 19.15a. Either one of them will cause substantial strains when reinserted into a square lattice. Together, however, they can make the composite square shown in Fig. 19.15b. The shear components of the strain are eliminated. If the volume per atom were to remain the same in the square and rectangular crystals, the composite of Fig. 19.15b would fit without strain into the parent square crystal. Usually there is some atomic volume mismatch after the transformation, and the formation of a plate structure to minimize the distortion. Such a plate is illustrated in Fig. 19.15c. It is often composed of twin boundaries between the variants of martensite. Although equal fractions of the twins are shown in Fig. 19.15c, it is possible to change the fractions of the twins by changing their widths. This degree of freedom can accommodate a range of transformation strains, and optimizing the fractions of twins is analogous to the selection of the shear at the bottom of Fig. 19.14c.

19.3 Landau Theory of Displacive Phase Transitions

19.3.1 Displacive Instability

A theory for displacive phase transitions can be developed using the "Landau potential"

$$\phi(x) = -\kappa_2 x^2 + \kappa_4 x^4, \tag{19.9}$$

where κ_2 and κ_4 are positive constants that are assumed independent of temperature. Even powers of the displacement coordinate x are appropriate when the atoms in the crystal have inversion symmetry, as is often the case. This potential is graphed in Fig. 19.16a. There are two minima of $\phi(x)$ at

$$x_{min} = \pm\sqrt{\frac{\kappa_2}{2\kappa_4}} \tag{19.10}$$

(easily found by setting to zero the first derivative of Eq. 19.9).

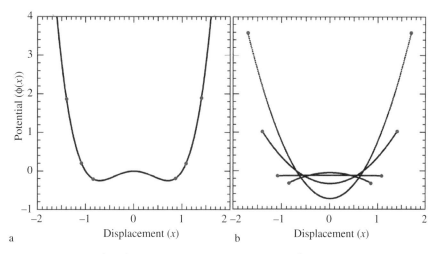

Figure 19.16 (**a**) Graph of the function $-x^2 + x^4$. (**b**) Least-squares fits of a parabola $a + bx^2$ to the curve of part a. The different ranges of the fits are indicated in both graphs by shaded end points.

For a physical picture in what follows, x can be considered the displacement of an atom off a symmetric position in the center of a unit cell. At low temperatures, the average thermal displacement $\langle x \rangle$ is expected to be near one of the minima of $\phi(x)$, causing a distortion of the unit cell. Neither the atomic potential nor the surrounding atomic structure has inversion symmetry in $\pm x$ at low temperatures.

At high temperatures, however, the bump at the bottom of the Landau potential in Fig. 19.16a is small in height compared with the thermal energy. The potential of Eq. 19.9 is then sampled as an even function of x, so at high temperatures, the thermal average displacement is $\langle x \rangle = 0$. At temperatures above T_c, the interesting variable is $\langle x^2 \rangle$, which increases with T. A phase transition is expected upon cooling below T_c, where the symmetry of this high-temperature phase is lost, and the crystal takes on a distorted configuration with two values of $\langle x \rangle$, reaching $\langle x \rangle = \pm x_{min}$ at $T = 0$.

In a big approximation that will be handy for calculating the vibrational entropy, we fit the potential of Fig. 19.16a to a harmonic potential. Typical results are shown in Fig. 19.16b. The quadratic potential depends on the range in x used for the fit, as indicated by the gray endpoints in Figs. 19.16a,b. Fits near $x = 0$ emphasize the bump in the middle of the potential, giving a quadratic potential that is concave downwards. With an increasing range of x, the average fit flattens out. When the range of x is large, the average potential is strongly concave upwards.

The average restoring force F_r against the displacement x is

$$F_r = -\frac{d\phi}{dx}, \tag{19.11}$$

$$F_r = -\kappa x. \tag{19.12}$$

At the temperature where the atom displacement is in the range $-1.08 < x < +1.08$ in Fig. 19.16, we have the condition $\kappa = 0$, i.e., the average restoring force vanishes. This pathological vanishing of κ marks an instability. Well-known results from the equation of motion, $ma = F_r$, are

$$m\frac{\mathrm{d}^2x}{\mathrm{d}t^2} = -\kappa\, x, \tag{19.13}$$

$$x(t) = \frac{x_0}{2}\left(\mathrm{e}^{+i\omega t} + \mathrm{e}^{-i\omega t}\right) = x_0\,\cos\omega\, t, \tag{19.14}$$

$$\omega = \sqrt{\frac{\kappa}{m}}. \tag{19.15}$$

When κ is positive, as at high temperatures, the frequency ω is positive, and normal oscillations occur. At low temperatures, where κ is negative, Eq. 19.15 shows that $\omega = iw$ is imaginary, so

$$x(t) = \frac{x_0}{2}\left(\mathrm{e}^{-wt} + \mathrm{e}^{+wt}\right), \tag{19.16}$$

and e^{+wt} and $x(t)$ diverge with time. An imaginary frequency is the signature of a mechanical instability, where the structure does not oscillate about its original value, but makes a displacive transition into a new structure.

19.3.2 Free Energy

To develop a thermodynamic free energy, $F = E - TS$, with the potential energy of Eq. 19.9, a kinetic energy and an entropy are added. These additional terms come from atom vibrations

$$F(T) = \frac{1}{2}m\left\langle\frac{\mathrm{d}^2x}{\mathrm{d}t^2}\right\rangle - \kappa_{2'}\langle x^2\rangle + \kappa_{4'}\langle x^4\rangle - T\,S_{\mathrm{vib}}. \tag{19.17}$$

The substitution $x^2 \rightarrow \langle x^2\rangle$, where $\langle\ \rangle$ denotes a thermal average, is a change in how we use the Landau potential at high temperatures where $\langle x\rangle = 0$, but the mean-squared displacement $\langle x^2\rangle$ is nonzero.

A simple vibrational entropy from the Einstein model is given by Eq. 7.51, repeated here for a three-dimensional material

$$S_{\mathrm{vib},\omega}(T) \simeq 3\,k_\mathrm{B}\left[\ln\!\left(\frac{k_\mathrm{B}T}{\hbar\omega}\right) + 1\right]. \tag{19.18}$$

The dependence of ω on $\langle x^2\rangle$ is parameterized as a change from the frequency ω_0 at T_c (here the subscript "0" designates the quantity at the critical temperature). It might seem that this frequency is zero, since the system is unstable against a soft mode at T_c, but the vibrational entropy is dominated by other vibrational modes that soften normally with increased temperature. These other modes are not part of the instablility that gives the mechanism of the phase transition.

For the entropy the characteristic frequency is parameterized as

$$\omega^2 = \omega_0^2 - \gamma_2 \left(\langle x^2 \rangle - \langle x_0^2 \rangle \right), \tag{19.19}$$

$$\omega^2 = \omega_0^2 - \gamma_2 \Delta^2 = \omega_0^2 \left(1 - \frac{\gamma_2 \Delta^2}{\omega_0^2} \right), \tag{19.20}$$

where we defined $\Delta^2 \equiv \langle x^2 \rangle - \langle x_0^2 \rangle$. We will assume that ω is not much different from ω_0.

Assuming there is a critical temperature, we seek the change in free energy with $\langle x^2 \rangle$ above T_c. The kinetic energy terms cancel when taking differences of Eq. 19.17, since they depend only on T (see Section 9.7.1). Above T_c we assume a stable parabolic potential as in Fig. 19.16b, so the fourth-order term, $\kappa_4 x^4$, is ignored explicitly (although its effects are implicitly included in γ_2)

$$\Delta F(T) \equiv F(T) - F(T_c), \tag{19.21}$$

$$\Delta F(T) = +\kappa_{2'} \langle x^2 \rangle - \kappa_{2'} \langle x_0^2 \rangle - T\, 3k_B \left[\ln\left(\frac{k_B T}{\hbar \omega} \right) + 1 \right] + T\, 3k_B \left[\ln\left(\frac{k_B T}{\hbar \omega_0} \right) + 1 \right], \tag{19.22}$$

$$\Delta F(T) = \kappa_{2'} \langle x^2 - x_0^2 \rangle - T\, 3k_B \, \ln\left(\frac{\omega_0}{\omega} \right), \tag{19.23}$$

$$\Delta F(T) = \kappa_{2'} \Delta^2 - T\, \frac{3}{2} k_B \, \ln\left(\frac{\omega_0^2}{\omega^2} \right), \tag{19.24}$$

$$\Delta F(T) = \kappa_{2'} \Delta^2 - T\, \frac{3}{2} k_B \, \ln\left(1 + \frac{\gamma_2 \Delta^2}{\omega_0^2} \right), \tag{19.25}$$

$$\Delta F(T) = \kappa_{2'} \Delta^2 - T\, \frac{3}{2} k_B \frac{\gamma_2 \Delta^2}{\omega_0^2} \quad \text{for } T > T_c. \tag{19.26}$$

The approximation $\ln(1 + \delta) \simeq \delta$ was made in Eq. 19.25, and ΔF is normalized to the number of atoms, N.

19.3.3 Critical Temperature and Equilibrium Displacement

Both terms in Eq. 19.26 include Δ^2, which increases with T. The second term changes more rapidly, however, and for $\Delta F = 0$ at T_c the critical temperature is

$$T_c = +\frac{2\kappa_{2'} \, \omega_0^2}{3k_B \, \gamma_2}. \tag{19.27}$$

Equation 19.27 is the analog of Eq. 18.72 for developing chemical order in an alloy, which was $T_c = zV/k_B$. In both cases, the critical temperature is proportional to the strength of the potential that drives the breaking of symmetry – the development of either chemical order or the sensitivity of the potential to atom displacements. The vibrational entropy seen in Eq. 19.27 is $3k_B \ln(1 + \gamma_2 \Delta^2/\omega_0^2) \simeq 3k_B \gamma_2 \Delta^2/\omega_0^2$. If the mean-squared displacement $\langle x^2 \rangle$ causes most vibrations to decrease in frequency rapidly in Eq. 19.19 (i.e., γ_2 is large), the vibrational entropy increases rapidly. This causes larger displacements to be favored at finite temperature. A large γ_2 therefore suppresses the critical temperature for the displacive phase transition, as does a small $\kappa_{2'}$ that parameterizes the potential energy.

Below the critical temperature, the mean displacement $\langle x \rangle$ is no longer zero. The shape of the Landau potential cannot be ignored, so the free energy is

$$F(\langle x \rangle, T) = -\kappa_2 \langle x \rangle^2 + \kappa_4 \langle x \rangle^4 - TS(\langle x \rangle). \tag{19.28}$$

At the temperature $T < T_c$ we seek the equilibrium value of $\langle x \rangle$

$$\frac{dF}{d\langle x \rangle} = 0, \tag{19.29}$$

$$0 = -2\kappa_2 + 4\kappa_4 \langle x \rangle^2 - T \frac{1}{\langle x \rangle} \frac{dS(\langle x \rangle)}{d\langle x \rangle}, \tag{19.30}$$

$$\langle x \rangle = \pm \sqrt{\frac{T \frac{1}{\langle x \rangle} \frac{dS}{d\langle x \rangle} + 2\kappa_2}{4\kappa_4}}. \tag{19.31}$$

It is easy to see that at $T = 0$, Eq. 19.31 gives $\langle x \rangle = \pm \sqrt{\kappa_2/2\kappa_4}$, consistent with Eq. 19.10.

The value of $\langle x \rangle$ at intermediate temperatures below T_c depends on how the vibrational entropy varies with $\langle x \rangle$. This is not a simple problem because many different vibrational modes are present, and the individual modes depend differently on $\langle x \rangle$. In general, we expect vibrational frequencies to increase as the temperature is lowered, following the increase in $\langle x \rangle$ as the system occupies lower parts of the Landau potential. For convenience we assume the vibrational modes behave like Eq. 19.19, but now $\langle x \rangle$ increases with decreasing temperature

$$\omega^2 = \omega_0^2 + \gamma_2 \left(\langle x \rangle^2 - \langle x_0 \rangle^2 \right), \tag{19.32}$$

and since the displacement at T_c, $\langle x_0 \rangle$ is zero

$$\omega = \sqrt{\omega_0^2 + \gamma_2 \langle x \rangle^2}, \tag{19.33}$$

$$S(\langle x \rangle) = 3k_B \left[\ln \frac{k_B T}{\hbar \omega} + 1 \right], \tag{19.34}$$

$$S(\langle x \rangle) = 3k_B \ln \left[\left(\frac{k_B T}{\hbar \omega_0} \right) \left(1 + \frac{\gamma_2 \langle x \rangle^2}{\omega_0^2} \right)^{-1/2} \right] + 3k_B, \tag{19.35}$$

$$S(\langle x \rangle) = 3k_B \left[\ln \left(\frac{k_B T}{\hbar \omega_0} \right) - \frac{\gamma_2 \langle x \rangle^2}{2\omega_0^2} + 1 \right], \tag{19.36}$$

$$\frac{dS(\langle x \rangle)}{d\langle x \rangle} = -\frac{3k_B \gamma_2 \langle x \rangle}{\omega_0^2}. \tag{19.37}$$

The approximation $\ln(1 + \delta)^{-1/2} \simeq -\delta/2$ was used to obtain Eq. 19.36. Substituting Eq 19.37 into Eq. 19.31

$$\langle x \rangle = \pm \sqrt{\frac{-\frac{3k_B \gamma_2 \langle x \rangle}{\omega_0^2} T + 2\kappa_2}{4\kappa_4}}. \tag{19.38}$$

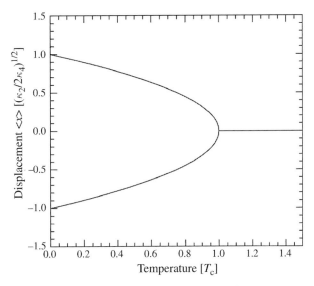

Figure 19.17 Temperature dependence of displacement parameter $\langle x \rangle$ above and below T_c from Eq. 19.39.

Assuming $\langle x \rangle$ is stable below T_c and the system undergoes small amplitude oscillations, we can approximate $\kappa_2 \langle x \rangle^2 \simeq \kappa_{2'} \langle x^2 \rangle$, so Eq. 19.27 for T_c can be written with κ_2 in place of $\kappa_{2'}$, and Eq. 19.38 is written in terms of T_c as

$$\langle x \rangle = \pm \sqrt{\frac{\kappa_2}{2\kappa_4}\left(1 - \frac{T}{T_c}\right)}. \tag{19.39}$$

Equation 19.39 is graphed in Fig. 19.17. The $\langle x \rangle$ goes to zero continuously at T_c, so this Landau theory predicts the displacive transition is second order in free energy.[5] Although Fig. 19.17 agrees with Eq. 19.10 for the displacement at $T = 0$, the slope of $\langle x \rangle(T)$ is not satisfactory at the lowest temperatures. The heat capacity of a material goes to zero at $T = 0$, and the curve in Fig. 19.17 must flatten at low temperatures. The high-temperature limit of Eq. 19.18 was used for the vibrational entropy. It is convenient, but not appropriate at low temperatures.

Because we considered one coordinate, x, the breaking of symmetry below T_c gives two values for $\langle x \rangle$ (plus and minus in Eq. 19.39). In three-dimensional crystals there are frequently more symmetrically equivalent variants of displacements below T_c. Inside a solid material, these different variants can interact through strain energy. This leads to microstructural complexity like that discussed in Section 19.2.3. Another potentially misleading feature of working with one coordinate, x, is that there are many different types of vibrational modes in a real material. The modes that give the structural instability need not be the same modes that are important for the vibrational entropy.

---

[5] Many displacive transitions are first order, but the Landau formalism can be adapted to handle them, as described in Section 19.3.4.

19.3.4 First-Order Landau Theory

We can adapt the Landau theory to accommodate first-order phase transitions. The quadratic and quartic terms in Eq. 19.9 have opposing signs, giving the structure at the bottom of the free energy function in Fig. 19.16. It was the positive quartic term that gave the positive potential at large displacements, essential for stability at high temperatures. The trick to adapting to a first-order phase transition is the addition of a positive sextic term in the displacement, and switching the signs of the quartic and quadratic terms. The potential remains an even function of displacement, as appropriate when the high-temperature phase is symmetric in the displacement coordinate, x.

Potential Energy and Low-Temperature Displacements

A Landau potential suitable for first-order phase transitions is

$$\phi(x) = +\kappa_2 x^2 - \kappa_4 x^4 + \kappa_6 x^6. \tag{19.40}$$

With three independent constants, $\{\kappa_2, \kappa_4, \kappa_6\}$, the function $\phi(x)$ can take on a number of shapes, but some essential ones are shown in Fig. 19.18. At small values of x the quadratic component dominates, and at large values of x the sextic component dominates. The shape at intermediate values of x depends on the relative strength of the quartic component, and a typical variation is shown in Fig. 19.18. For small values of κ_4 there is only the one minimum at $x = 0$, but this potential is not useful in the following model of a phase transition. A larger value of κ_4 is interesting, giving two more minima in the potential that are separated by energy barriers from the minimum at $x = 0$. This is especially evident for $\kappa_4 = 2$ in Fig. 19.18.

The values of x at the minima can be found by setting $d\phi(x)/dx = 0$, from which we obtain the quadratic equation

$$3\kappa_6 \chi^2 - 2\kappa_4 \chi + \kappa_2 = 0, \tag{19.41}$$

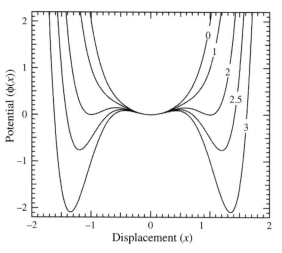

Figure 19.18 Graph of the function $\phi(x) = +1x^2 - \kappa_4 x^4 + 1x^6$ for $\kappa_4 = \{0, 1, 2, 2.5, 3\}$ as labeled.

making the substitution $\chi = x^2$. Using the quadratic formula for a standard solution for χ

$$\chi = \frac{\kappa_4 \pm \kappa_4 \sqrt{1 - 3\kappa_2\kappa_6(\kappa_4)^{-2}}}{3\kappa_6}. \tag{19.42}$$

We are interested in the two deep minima in Fig. 19.18. These minima are away from $x = 0$ where the term $\kappa_2 x^2$ is important, and we also assume that κ_4 is relatively large. We therefore ignore the second term in the radical in Eq. 19.42

$$\chi \simeq \frac{\kappa_4 \pm \kappa_4}{3\kappa_6}. \tag{19.43}$$

We have no interest in the trivial solution $\chi = 0$, so we take the positive sign and obtain for x

$$x \simeq \pm\sqrt{\frac{2\kappa_4}{3\kappa_6}}. \tag{19.44}$$

For the larger values of κ_4 in Fig. 19.18, Eq. 19.44 is accurate to a few percent.

Free Energy Minimization and Critical Temperature

For first-order transitions, the Landau free energy is like that of Eq. 19.22, with a sextic term and the switch of signs of the quadratic and quartic terms described above

$$F(T) = \frac{1}{2}m\left\langle\frac{d^2x}{dt^2}\right\rangle + \kappa_2\langle x\rangle^2 - \kappa_4\langle x\rangle^4 + \kappa_6\langle x\rangle^6 - T S_{vib}(x). \tag{19.45}$$

The vibrational entropy was discussed in Section 19.3.2, where it was handled approximately by assuming three Einstein modes per atom, with a frequency that increases with temperature as parameterized by γ_2. The decrease of vibrational entropy with $\langle x\rangle$ disfavors this displacement at higher temperatures.

Anticipating a first-order phase transition, we seek the condition of equal free energies for a phase with finite $\langle x\rangle$ and a phase with $\langle x\rangle = 0$

$$F(T_c, \langle x\rangle_c) = F(T_c, \langle x\rangle = 0), \tag{19.46}$$

where we expect $\langle x\rangle_c$ to jump discontinuously from 0 to a finite value as the temperature is reduced below T_c. The difference between these two free energies is

$$\Delta F(T) = +\kappa_2\langle x\rangle^2 - \kappa_4\langle x\rangle^4 + \kappa_6\langle x\rangle^6 + T\frac{3}{2}k_B \ln\left(1 + \frac{\gamma_2\langle x\rangle^2}{\omega_0^2}\right), \tag{19.47}$$

$$\Delta F(T) = \left(\kappa_2 + T\frac{3}{2}k_B\frac{\gamma_2}{\omega_0^2}\right)\langle x\rangle^2 - \kappa_4\langle x\rangle^4 + \kappa_6\langle x\rangle^6. \tag{19.48}$$

We seek two conditions for $\Delta F(T_c)$ at the critical temperature:

- First, $\Delta F(T_c) = 0$ for the finite displacement $\langle x\rangle_c$. (The two phases are in equilibrium at T_c.)
- Second, this $\langle x\rangle_c$ corresponds to a minimum of $\Delta F(T)$, so $d\Delta F(T)/dx|_{\langle x\rangle_0} = 0$. (Otherwise a different value of $\langle x\rangle$ would be more stable.)

To find the displacements $\langle x \rangle_c$ for the minima, set $d\Delta F/d\langle x \rangle = 0$ and consider only the dominant quartic and sextic terms. The minima of Eq. 19.44 are recovered

$$\langle x \rangle_0 = \pm \sqrt{\frac{2\kappa_4}{3\kappa_6}}. \tag{19.49}$$

In cases where the order parameter is not strongly dependent on temperature, this $\langle x \rangle_0$ of Eq. 19.49 can be used to obtain the critical temperature. The deep minima of Fig. 19.18 are dominated by the negative quartic and positive sextic parts of the potential, but the quadratic part is important for the temperature dependence and the magnitude of $\Delta F(T, \langle x \rangle)$. Substituting Eq. 19.49 into Eq. 19.48, setting $\Delta F(T_c) = 0$, and after some rearrangements

$$T_c = \frac{\omega_0^2}{3k_B\gamma_2}\left(\frac{4\kappa_4^2}{9\kappa_6} - 2\kappa_2\right). \tag{19.50}$$

Compared to Eq. 19.27 for the second-order transitions, this T_c shows a similar role of vibrational entropy through ω_0^2/γ_2. The coefficients of the potential also set the T_c in a similar way, although the sign of κ_2 is switched for first- and second-order Landau potentials. If there is a change of $\langle x \rangle$ with temperature, as is expected, the value of T_c in Eq. 19.50 can be semiquantitative.

This idealized first-order phase transition used only one coordinate, x, to describe all characteristics of the structure, the transformation path, and the free energy.[6] In a real material it might be true that one coordinate dominates a particular characteristic of the phase transition, but quantifying different characteristics requires different coordinates. A displacement coordinate (or coordinates) is needed to transform the structure from its initial state to its final state. The deformation path from one structure to another must accommodate the change in this structural coordinate, but the deformation path need not be a linear one through only one coordinate – other coordinates may become involved at different stages of the transformation. Finally, the vibrational entropy depends on coordinates of numerous vibrational modes in the material. These modes are likely to change in frequency and polarization with changes in the structural coordinates. This is sometimes modeled by isolating the coordinates of the structural change, and evaluating the vibrational entropy at various amplitudes of the structural transition. (This was done in Vineyard's analysis of the vacancy jump, Section 10.7.) The identification of coordinates for structure, transformation, and entropy may be unclear if the phase transformation occurs rapidly.

---

[6] It is sometimes argued that a "soft mode," i.e., a vibrational mode with a frequency that goes to zero, is responsible for both the large displacements and the thermodynamics of a displacive phase transformation. Although a harmonic mode of zero frequency would provide large amplitudes of displacement, the effective restoring forces will probably change when the displacements are large, so the mode is unlikely to be truly harmonic. Second, a mode with large atomic displacements enhances the entropy of the phase in which it forms, not the product phase. Finally, a single mode contributes an entropy of $k_B \ln(k_B T/\hbar\omega')$. This diverges as the frequency ω' goes to zero, but only logarithmically. In a continuum of modes this divergence does not cause an entropy catastrophe.

19.4 Crystal Instabilities and Phonons

19.4.1 Criteria for Crystal Stability

A stable crystal provides restoring forces against shear and dilatation strains that would change its shape. In other words, there must be a positive cost in elastic energy when the crystal is distorted. An analysis of the elastic energy gives the "Born stability criteria," and for a cubic crystal the conditions on the elastic constants are [73, 309]

$$C_{44} > 0, \ C_{11} + 2C_{12} > 0, \ C_{11} - C_{12} > 0. \tag{19.51}$$

Each of these stability criteria corresponds to a different mode of deformation. For example, from $B = (C_{11}+2C_{12})/3$ (Eq. 6.118), we see that the second inequality demands a positive bulk modulus, $B > 0$. A negative bulk modulus causes a catastrophic instability – the crystal reduces its elastic energy by expanding, so its atoms come apart. The other two criteria involve instabilities against shear distortions.

Of particular interest in the study of martensite is the stability with respect to small-amplitude shears on (110)-type planes in $\langle 110 \rangle$ directions. If $C_{11} - C_{12}$ vanishes at a given temperature, the crystal shears spontaneously in a "soft mode instability." The earliest explanations of martensitic transformations [310, 311] were based on soft mode instabilities. Later, Clapp [312] wrote the free energy of a crystalline solid as

$$F = F_0 + \frac{1}{2} \sum_{i=1}^{6} \sum_{j=1}^{6} C_{ij}\epsilon_i\epsilon_j + \frac{1}{3!} \sum_{i=1}^{6} \sum_{j=1}^{6} \sum_{k=1}^{6} C_{ijk}\epsilon_i\epsilon_j\epsilon_k + \cdots, \tag{19.52}$$

where F_0 is the free energy of the strain-free material, and Voigt notation for strains ϵ is used, as in Section 6.7.3.[7] Stability with respect to strain is governed by the sign of each of the six eigenvalues of the 6×6 matrix of second derivatives

$$\frac{\partial^2 F}{\partial \epsilon_i \partial \epsilon_j} = F_{ij} \qquad (i,j = 1,6), \tag{19.53}$$

much like the analysis of the Hessian matrix of Section 9.6.3. If a particular eigenvalue vanishes at a given temperature, the material is unstable with respect to the deformation mode of the associated eigenvector. A stability analysis from a free energy with only the first summation in Eq. 19.52 will reproduce the inequalities of Eq. 19.51. To account for a phase transformation that depends on temperature, the stability against a lattice distortion must be temperature dependent. This was the role of vibrational entropy in the Landau theory of Section 19.3.

For a crystal with harmonic vibrations, another stability criterion is that all vibrational frequencies must be real.[8] For the time dependence of a typical vibration, as in Eq. 19.16, an imaginary $\omega = iw$ allows for amplification with time, $\exp(+wt)$. This is, of course, an

---

[7] Here $\epsilon_1 = \epsilon_{11}, \epsilon_2 = \epsilon_{22}, \epsilon_3 = \epsilon_{33}, \epsilon_4 = \epsilon_{23}, \epsilon_5 = \epsilon_{31}$ and $\epsilon_6 = \epsilon_{12}$.
[8] For use in Eq. 26.12 [online], $\omega^2_{\vec{kj}} > 0 \ \forall \ \vec{kj}$.

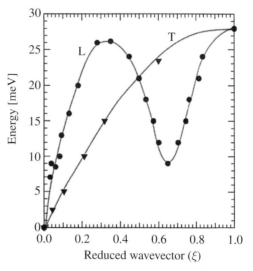

Figure 19.19 Phonon dispersion curves for bcc Ti along [111] at elevated temperature. The parameter ξ gives the magnitude of the wavevector as ξ[111]. The transverse dispersion, labeled "T," has a normal stiffening with wavevector, but the longitudinal dispersion, labeled "L," has a large softening near the wavevector of 2/3[111] which depends on temperature. After [313–315].

instability of the atom movements. At long wavelengths, this instability can be equivalent to the Born stability criteria of Eq. 19.51. Vibrational instabilities can exist at short wavelengths, too [316].

19.4.2 Soft Phonons in bcc Structures

In general, the bcc structure is weak against longitudinal vibrations with $2/3 \langle 111 \rangle$ wavevectors. All bcc elements show a dip in frequency of their longitudinal phonon dispersions near this wavevector, or at least an anomaly.[9] The softening of [111] phonons in bcc Ti is shown in Fig. 19.19. The dip at $2/3 \langle 111 \rangle$ is prominent. At temperatures near a transition to the hexagonal phase, these longitudinal modes dip towards zero frequency, becoming unstable. This behavior could originate with a temperature-dependent change of second-nearest-neighbor force constants [317]. Note, the bcc structure is not stable with first-nearest-neighbor (1nn) central forces alone.

The soft atom displacements in the bcc structure are shown in Fig. 19.20, which is one unit cell of the bcc structure balanced on its corner. Three arrows show atom displacements upwards, and one downwards. Three atoms are marked with horizontal lines, signifying that they are stationary. Planes of equivalent atoms extend out of the plane of the paper.

---

[9] In contrast, phonon dispersions of fcc crystals tend to increase monotonically to the edge of the Brillouin zone. Many bcc crystals also have phonon dispersions that soften along the transverse [110] branch, but the specific wavevector of these soft modes varies for different alloys.

Figure 19.20 Large model of one unit cell of a bcc crystal (Atomium in Brussels), with arrows showing atom displacements in a soft 2/3[111] longitudinal phonon.

This is a longitudinal phonon mode of wavevector 2/3[111].[10] What is interesting about this mode is that its displacements can collapse the bcc structure. Viewed from above, the three atoms with up arrows and the three with the horizontal lines make a hexagonal arrangement. If the mode shown in Fig. 19.20 becomes large in amplitude, these six atoms will briefly form a hexagonal plane. A crystal structure with these six atoms on the same plane is the ω-phase, and "ω-phase collapse" is an instability of the bcc structure. The diffusionless ω-phase transformation in Zr is probably driven by this phonon, and is a type of martensitic transformation.

The atom displacements in a 2/3[111] phonon are shown in more detail in Fig. 19.21. The horizontal lines in the figure identify planes of atoms with the same displacements. Again, notice that pairs of them move together, although these pairs are separated by planes that are stationary. Note also that the vertical chains of first-nearest neighbors along $\langle 111 \rangle$ directions tend to move together in Fig. 19.21. With large atom displacements, however, the interatomic force constants are not expected to behave as for small displacements, i.e., the restoring forces will be nonlinear functions of displacement. The mechanics of how a soft mode causes structural changes is not yet well understood.

19.4.3 Phonons and Entropy

Many bcc metal crystals are unstable energetically at low temperatures, where a close-packed structure is preferred. In pure elements, the bcc–fcc transformation cannot cause any change in configurational entropy because the shear and shuffle processes of atom movements create the product phase deterministically. Owing to the weakness of bcc structures against various distortions, Friedel suggested some time ago that soft transverse modes stabilize the bcc structure at elevated temperatures [318]. Other considerations of entropy in martensitic transformations and twinning were discussed by Cahn [319].

---

[10] For comparison, the shortest wavelength would have the stationary three atoms moving upwards, so the repeat distance would be two planes instead of three as labeled in the figure (hence the 2/3[111] designation for the soft mode).

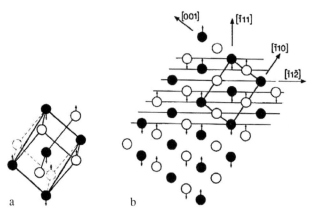

Figure 19.21 (**a**) Standard bcc unit cell (plus two centered atoms from adjacent cells), oriented on end and almost rotated into the position for correspondence to panel b. (**b**) Atoms in two stacked (110) planes, with white atoms in the layer above the black. Arrows show displacements of the atoms in a longitudinal phonon of $\vec{k} = 2/3\,[\bar{1}11]$, with wavefronts shown as horizontal lines. Notice how close-packed columns along $[\bar{1}11]$ move as units (as do entire $(11\bar{2})$ planes). A transverse mode with $\vec{k} = 1/3\,[11\bar{2}]$ and a $[\bar{1}11]$ polarization has the same atom displacements. After [313]. Reprinted by permission of the publisher (Taylor & Francis Ltd, www.tandfonline.com).

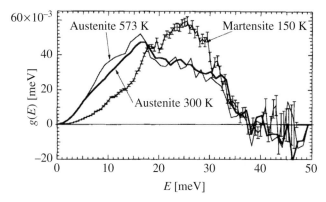

Figure 19.22 Phonon DOS curves of NiTi at different temperatures and different phases as labeled. Courtesy Taylor & Francis Ltd. (www.tandfonline.com), *Philos. Mag. B* [323].

    Planes *et al.* performed a number of calorimetric studies of the heats of martensitic transformations in shape-memory alloys, which tend to have good thermodynamic reversibility. From the measured heats from Cu–Al–Zn, Cu–Al–Ni, and Cu–Al–Be alloys, ΔH, and the temperatures of transformation, the entropy was obtained as $\Delta S = \Delta H/T$. A typical value of ΔS was $0.2\,k_B$/atom [320–322]. They argued that the martensitic transformation temperature depends on the presence of low-energy modes in the high-temperature phase that help stabilize it, and the transformation is promoted when these modes are depopulated at lower temperatures.

    The martensite-to-austenite transformation in NiTi has a latent heat of $0.5\pm0.05\,k_B$/atom, as measured by calorimetry [323]. The phonon DOS curves of the austenite and martensite of Fig. 19.22 were obtained from inelastic neutron scattering. Note how the transverse modes of austenite are prominent at low energies around 10 meV. From these data, with Eq. 7.49 it was found that the vibrational entropy of the martensitic transformation was quite close to the full entropy measured by calorimetry. From simulations and the measured DOS, the vibrational entropy of austenite was found to be larger than that of martensite because the transverse acoustic (TA(2)) and longitudinal acoustic modes of austenite are softer than those of martensite. Simulations suggest that this originates with a large and negative 1nn transverse force constant, which may also be responsible for the soft modes involved in the mechanism of the martensitic transformation.

Problems

19.1 A phase transformation from fcc to hcp may occur by a mechanism similar to that of twinning in fcc structures.

    (a) Show that the ABAB stacking of the hcp structure is generated by the passage of a Shockley partial dislocation on alternate close-packed planes. You can do this conveniently by using the registry notation of letters and arrows written for the three cases of slip in Fig. 19.4b.

    (b) A pole mechanism may be designed for this fcc → hcp transformation, and it will look much like that of Fig. 19.5. Describe the differences of the pole mechanism from that for twinning in fcc crystals.

    (c) Twinning is possible in hcp crystals. Is it possible to make a twin in an hcp structure by our pole mechanism and close-packed planes? Why or why not?

19.2 **(a)** Calculate the angle of shear for twinning in an fcc crystal.

    (b) Calculate the angle of shear for an hcp to fcc transition as described in Problem 19.1a.

19.3 In many steels the bct martensite phase (α') is ferromagnetic, and the fcc austenite phase (γ) is not. In a magnetic field H, the difference in magnetization M can lower the free energy of the martensite with respect to austenite by the amount $M_{\alpha'} H$, where $M_{\alpha'}$ is 2.2 μ_B/atom (the Bohr magneton $\mu_B = 0.92 \times 10^{-24}$ J/T), whereas M_γ is nearly 0.

    (a) Derive a form of the Clapeyron–Clausius equation that predicts how much the temperature of γ/α' equilibrium is shifted by a magnetic field.

    (b) For a magnetic field of 10 T, how much of a change is expected in the temperature of γ/α' equilibrium? Assume the entropy of the austenite phase is 0.5 k_B/atom greater than that of martensite.

    (c) Do you expect an effect of the magnetic field on the nucleation rate of martensite? Why or why not?

19.4 Suppose the vibrational frequency ω did not depend on the displacement as in Eq. 19.19, but instead as

$$\omega^2 = \omega_0^2 - \gamma_4 \left(\langle x^4 \rangle - \langle x_0^4 \rangle \right). \tag{19.54}$$

How will the critical temperature, T_c, differ from Eq. 19.27? Why?

19.5 The martensitic transformation in Fig. 19.14 included a slip process in part c that allowed refitting the transformed region with crystallographic registry in part d. An alternative to slip is twinning. The modification to the diagram of Fig. 19.14 is that the Bain strain from b to c would have some unit cells with tetragonality along x and others along y. (They are all along x in Fig. 19.14c.)

(a) By alternating unit cells with the two tetragonalities, using the same fraction of each variant (1/2), construct a transformed region that can fit back in the original crystal as in Fig. 19.14d.

(b) Is the angle of the habit plane θ the same as if you had performed a shear process instead of twinning?

19.6 Using the free energy expression for the first-order Landau theory of a displacive transition in Eq. 19.48, show that the change with temperature in the minimum of the free energy at $\langle x \rangle_0$ is approximately

$$\Delta F = k_{\mathrm{B}}(T - T_{\mathrm{c}}) \frac{\gamma_2 \, \kappa_4}{\omega_0^2 \, \kappa_6}. \tag{19.55}$$

20

Thermodynamics of Nanomaterials

Nanostructured materials are of widespread interest in science, engineering, and technology. For the purpose of thermodynamics, it is useful to define nanomaterials as materials with structural features of approximately 10 nm or smaller, i.e., tens of atoms across. Unique physical properties of nanomaterials originate from one or two of their essential features:

- Nanomaterials have high surface-to-volume ratios, and a large fraction of atoms located at, or near, surfaces.
- Nanomaterials confine electrons, phonons, excitons, or polarons to relatively small volumes, altering their energies. Analogously, the confinement of structural defects such as dislocations or internal interfaces alters defect energies, too.

   Some practical questions are if a nanostructure can be synthesized, or if a nanostructure is adequately stable at a modest temperature. More basic questions are how the thermodynamics of nanostructured materials differs from conventional bulk materials. In short, the energy is increased by the surfaces, interfaces, or composition gradients in nanostructures. A nanostructure generally has a higher entropy than bulk material, however, and at finite temperature the entropy contribution to the free energy can help to offset the higher internal energy term in the free energy $F = E - TS$.

   Sections 11.1 and 11.5 covered important aspects of surface energy, including surface relaxation and reconstruction processes that are driven by chemical energy. This chapter discusses the structure of nanomaterials, the thermodynamics of interfaces in nanostructures, electron states in nanostructures, and the entropy of nanostructures. The unique properties of nanomaterials are of interest in electronics, mechanics, optics, magnetics, chemistry, biology, and medicine. An overview of the properties of nanomaterials is well beyond the scope of this text (see [324]), but some relevant surface phenomena were presented in Chapter 11, and magnetic nanocrystals are discussed in Section 20.6.

20.1 Energies of Atoms at Grain Boundaries in Nanocrystals

20.1.1 Dimensional Scaling and Its Validity

If two microstructures are "self-similar" (meaning that they look the same when scaled in size by uniform enlargement), any measure of surface area per unit volume decreases as

$1/d$, where d is the distance used for scaling. In general, the grain boundary energy per unit volume scales as $1/d$. However, the widths of grain boundaries alter this geometrical relationship at small length scales. As a rule of thumb, for grain boundaries of 2.5 atoms in width, a polycrystalline metal with a crystallite size of 10 nm will have 10% of its atoms in grain boundaries. At smaller scales, however, the fraction of atoms at edges between grain surfaces begins to increase rapidly. At crystal sizes of 2 to 3 nm, approximately 10% of all grain boundary atoms are located at corners or triple junctions of grain boundaries. Below 2 nm, however, few atoms remain in the crystal interior. It is probably better to describe consolidated nanocrystals below 2 nm as amorphous. Furthermore, an amorphization transition may occur at this size or somewhat larger.

Most crystal orientations cannot accommodate special grain boundary orientations as in Fig. 11.8c (remember, there are crystal orientation angles out of the plane of the paper, and the orientation of the boundary itself can be varied). In general, there is significant structural disorder in grain boundaries, with atoms having unfavorable local coordinations, neighbor distances, and bond angles. These all make positive contributions to the grain boundary energy, and for nanocrystals there can be a substantial fraction of atoms in these unfavorable grain boundary sites. Figure 20.1 shows the fraction of atoms located at and near grain boundaries for some bcc and fcc polycrystals. The widths of the grain boundaries corresponding to these curves were 1.0 nm for the bcc crystals and 0.5 nm for the fcc crystals.

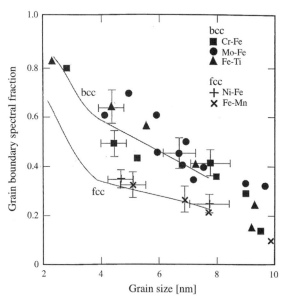

Figure 20.1 Fraction of atoms at and near grain boundaries in consolidated nanocrystals versus the characteristic grain size of the nanocrystals. The experimental data were fit with a model with a variable width of the grain boundary region between crystals, and the best fits for fcc and bcc crystals are shown as solid lines. Reprinted from [325] with the permission of AIP Publishing.

20.1.2 Comparison of Grain Boundary Structure and Amorphous Structure

There are similarities between atom positions in an amorphous phase and inside a grain boundary region, but the two are different. The atoms in a grain boundary are constrained to accommodate two rigid, neighboring crystals. Computer simulations show that their local neighbor coordination is less optimal than what they can achieve in an amorphous structure. As a rule of thumb, the enthalpy per mole of a grain boundary region exceeds that of an amorphous phase by perhaps a factor of 2 [158]. (This assumes of course that the boundary is one without special orientation, as is generally the case.) For example, using a grain boundary energy of $1 \, J/m^2$ for a material with a grain size of 10 nm gives about 2 kJ/mol for transition metal densities, consistent with calorimetry measurements [326]. If this enthalpy is assigned to the 10% of atoms in grain boundaries in this 10 nm material, these atoms have an enthalpy of 20 kJ/mol. To make a comparison to an amorphous phase, in general the crystallization of an amorphous phase has a similar heat release as crystallization from a liquid. If melting is at 1500 K, and using the $0.8 \, k_B$/atom of Section 12.5, the amorphous phase has a higher enthalpy than the polycrystalline material by about 10 kJ/mol. This is about a factor of 2 lower than the molar enthalpy of the grain boundary region.

The idea that the atoms in grain boundaries have a higher molar enthalpy than those in an amorphous phase has been used to explain an amorphization transition in a material subjected to mechanical attrition, a process that reduces the grain size by high-energy milling. When the grain size of NiTi was reduced to about 5 nm, the material transformed to an amorphous phase rather than undergo further reduction of its grain size [327]. This is a low-temperature phenomenon, driven by enthalpy alone. It is expected to occur at a grain size for which

$$f(d)E_{gb} + [1 - f(d)]E_{xtl} = E_{am}, \tag{20.1}$$

where $f(d)$ is the fraction of atoms in the grain boundary for a crystallite size of d, and the molar energies E refer to the grain boundary, crystal, and amorphous phase.[1] For NiTi, this occurred at $d = 5$ nm when f was approximately 0.5.

20.2 Gibbs–Thomson Effect

20.2.1 Chemical Potential

Because smaller particles have larger surface-to-volume ratios, we expect a less favorable change in free energy when we add an atom to a small particle than a large one. In other words, we expect a higher chemical potential for atoms in and near small particles.

---

[1] The enthalpy of chemical disorder in NiTi may play a role in this phase transformation if it differs substantially in the crystalline and amorphous phases.

This increase in chemical potential with decreasing particle radius is called the "Gibbs–Thomson effect."

Consider the equilibrium between two phases when one nucleates as a small particle in the other. Homogeneous nucleation from the melt is a convenient example. For small crystals, the difference in free energy of solid and liquid, ΔG (which is zero at the melting temperature for a big solid), has an additional term for a small particle equal to $A\sigma$, which is the area of interface between the two phases times the specific surface energy. This surface energy was added to the bulk free energy when analyzing the nucleation problem in Eq. 4.2, repeated here

$$\Delta G(T) = V\,\Delta G_V(T) + A\,\sigma, \tag{20.2}$$

where ΔG_V is the Gibbs free energy per unit volume of forming the new phase from the parent phase, and V is the volume of the new phase.

To work with Eq. 20.2, it is traditional to assume a spherical particle. This is convenient in what follows, where we seek the effect on ΔG from the radius of curvature, r. Spherical particles were considered in Section 4.2.2, for which the ΔG with bulk and surface energies is graphed in Fig. 4.3. Figure 4.3 shows that for an undercooling with a specific ΔG_V (which is negative), the positive surface energy is cancelled when the radius is $r = -3\sigma/\Delta G_V$. This is not quite the result needed for equilibrium, however – Fig. 4.3 shows that the solid particle of $r = -3\sigma/\Delta G_V$ will grow continuously to reduce its free energy further as atoms move from the liquid to the solid nucleus.

At equilibrium the chemical potential is equal for atoms in the nanoparticle and in the adjacent liquid (otherwise atoms would move from one to the other). Instead of finding the radius where the surface energy equals the volume free energy per atom, we ask instead, "How does the difference in free energy between nanoparticle and matrix, ΔG, change by moving another atom from the matrix to the nanoparticle?" From the definition of chemical potential (Eq. 1.24), we obtain its difference between nanoparticle and matrix[2]

$$\Delta\mu = \left(\frac{\partial \Delta G}{\partial N}\right)_{T,P}. \tag{20.3}$$

Setting $\Delta\mu = 0$ is the condition of no preference for atoms moving between the nanoparticle and the adjacent matrix because this movement across the interface does not alter ΔG. This was the condition for the critical radius in nucleation theory of Section 4.2. The radius for this equilibrium is therefore the critical radius shown in Fig. 4.3, i.e., $R^* = -2\sigma/\Delta G_V$, where the curve of $\Delta G(R)$ has zero first derivative (Eq. 4.4). The equilibrium is an unstable one, with a smaller particle shrinking and a larger one growing, as explained for nucleation in Section 4.2.2. This equilibrium for a nanoparticle occurs when the increase in ΔG is $\Delta G_V = -2\sigma/r$. This can be converted to a chemical potential through the factor ρ, the

---

[2] When the free energy scales with the number of atoms, the chemical potential is simply the total free energy divided by the number of atoms, i.e., $\mu = G/N$. This simple relationship works when all contributions to G scale with N. Surface energy does not scale with N, however, so it is necessary to calculate $\partial\Delta G(N)/\partial N$ instead.

density in units of atoms/volume (or moles/volume). For a general radius, the bias in chemical potential caused by the surface energy is

$$\Delta\mu = \frac{2\sigma}{r\rho}. \tag{20.4}$$

This is the Gibbs–Thomson effect.

20.2.2 Melting of Nanoparticles

The Gibbs–Thomson effect explains how small crystals have lower melting temperatures than large crystals. At the melting temperature (which now depends on the radius of curvature of the solid, r), the free energies of the solid and liquid are equal. With all quantities on a molar basis

$$0 = F_l(T_m(r)) - F_s(T_m(r)), \tag{20.5}$$

$$0 = E_l - T_m(r)S_l - E_s - \frac{2\sigma}{r\rho} + T_m(r)S_s, \tag{20.6}$$

$$T_m(r) = \frac{E_l - E_s - \frac{2\sigma}{r\rho}}{S_l - S_s}. \tag{20.7}$$

Assume that S_s does not change with curvature. Recognizing $T_m(r = \infty) = \Delta E/\Delta S$ from Eq. 1.4, and using the definition of the latent heat $L = (S_l - S_s)T_m(r = \infty)$ from Eq. 1.5, Eq. 20.7 becomes

$$T_m(r) = \left[1 - \frac{2\sigma}{rL\rho}\right]T_m(r = \infty). \tag{20.8}$$

To estimate how much the Gibbs–Thomson effect alters the melting temperature of small particles, assume $L = 10^4$ J/mol, $\sigma = 2$ J/m^2, $\rho = 0.15$ mol/cm^3, so

$$T_m(r) \sim \left[1 - \frac{3}{r}\right]T_m(r = \infty), \tag{20.9}$$

where r is in nm. Clearly r must be quite small if there is to be a substantial effect on T_m. If $r = 10$ nm, for example, T_m decreases to 0.7 of its bulk value for the approximate parameters given here. Although the present analysis is not reliable for very small nanoparticles (for example, the concept of curvature is inappropriate for a surface of several atoms across), this analysis suggests that the atomic structures of isolated nanocrystals having dimensions of a few nanometers may not be stable at modest temperatures.

Experimental measurements of the melting temperatures of nanocrystalline gold particles are shown in Fig. 20.2. The gold nanoparticles were deposited on two different substrates (silica and graphite), and the particle–substrate interaction alters the melting temperature. In both cases, the smaller particles have suppressed melting temperatures.

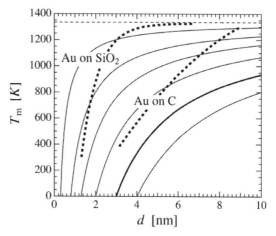

Figure 20.2 Graph of Eq. 20.9 (dark solid), with curves of similar form. Thick dashed curves are fits to melting temperatures T_m of Au nanoparticles of different diameters d, on two substrates: silica and highly oriented pyrolytic graphite [328]. T_m of bulk Au is 1336 K (horizontal dashed line).

20.2.3 Phase Boundaries

Because the Gibbs–Thomson effect alters the melting temperature (suppressing T_m in all known cases), it should not be surprising that phase boundaries on a T–c phase diagram are altered when one or more of the phases is nanostructured. Please refer to Fig. 4.6a, which was used to understand nucleation, but in some ways is even more appropriate for nanomaterials.[3] For a nanostructured β-phase, the contribution ΔG^* that is shown as elevating $F_\beta(c)$ in Fig. 4.6a is now obtained from the $\Delta\mu$ of Eq. 20.4. The curve $F_\alpha(c)$ in Fig. 4.6a is depicted as unchanged, but for many nanomaterials this curve is also shifted upwards by an amount that generally differs from the shift of $F_\beta(c)$. The enlargement of the composition range of the α-phase, and change in composition of the new β-phase, can be immediately understood with the aid of Fig. 4.6a (more explanation is given in Section 4.4.1). Finally, Fig. 4.6b indicates that phases may appear or disappear in phase diagrams of nanostructured materials. As was the case for melting, however, these large effects are not expected to be significant unless material dimensions are several nanometers or so.

20.3 Atomic Structures of Nanocrystals

20.3.1 Thermal Instabilities of Nanoparticles

Small particles can undergo thermodynamic fluctuations in their size and shape. A standard expression, with derivation discussed in Problem 20.4, is that the fractional spread in particle energy $\sigma_E/\langle E\rangle$ is

---

[3] The radius of a nucleus, which is a thermal fluctuation, is not usually taken as a fixed quantity; a radius for a nanoparticle can be specified.

$$\frac{\sigma_E}{\langle E \rangle} = \frac{\sqrt{k_B T^2 \, C_V}}{\langle E \rangle}, \tag{20.10}$$

where C_V is the heat capacity and $\langle E \rangle$ is the average energy. The approximate trend is that both C_V and $\langle E \rangle$ scale with the number of atoms, N, so the fractional spread in particle energy scales as $N^{-1/2}$. Smaller particles are therefore expected to have larger fluctuations in size, but even a 3 nm particle would have only a ~2% fluctuation in diameter.

Isolated nanoparticles may undergo dynamical rearrangements, where they switch between a set of configurations at finite temperature. Reasons for unstable particle shapes include:

• The activation barriers between favored configurations are low for small particles, so transitions could be driven by temperature.
• The particles have premelted surface layers, so their surface atoms can reconfigure quickly. Premelting of surfaces occurs at temperatures below T_m, and this premelting might be suppressed considerably for nanocrystals. Even if not, the curves of Fig. 20.2 show that small particles are liquid at modest temperatures, and hence unstable in their shapes.

Enthalpy fluctuations can occur as a nanoparticle distorts from a sphere to an ellipsoid, for example, and undergoes a small increase in surface-to-volume ratio. It is risky to apply such continuum models to nanoparticles, of course, and other types of enthalpy fluctuations are possible. For example, Fig. 20.3 shows clusters of 144 gold atoms with very different structures that are formed under somewhat similar synthesis conditions [329]. Figure 20.3a shows an icosahedral nanocrystal quite unlike the fcc nanocrystal shown in 20.3b. There is only a small difference in free energy of these two different structures, but fluctuations between these two structures are probably not driven by temperature.

Kinetic processes that occur during the synthesis of nanocrystals can control the atomic structure. Figure 20.4 shows a high-resolution image of a Pt nanocrystal [330] that nucleated in a solution with chloroplatinic acid as a precursor [331]. Peptide molecules in the solution helped control the size and shape of the nanocrystals when the precursor was being reduced to metallic platinum. These biomolecules undergo preferential binding to different crystallographic surfaces of Pt crystals, restricting the growth of these surfaces. The coverage of biomolecules offers some control over the shapes and facets of the

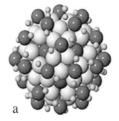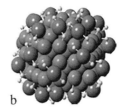

a   b

Figure 20.3 Structural models consistent with pair distribution functions from X-ray diffraction measurements on clusters of Au_{144} [329]. (**a**) Icosahedral structure of Au_{144} cluster. Dark and light gray spheres denote Au atoms at different radii. Notice the five-fold rings of darker Au atoms. (**b**) fcc structure of Au_{144} cluster. Small light spheres denote possible hydrocarbon ligands.

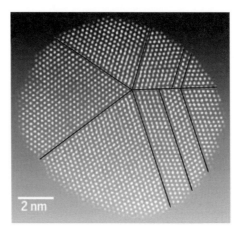

Figure 20.4   Image of a Pt nanocrystal obtained from an aberration-corrected scanning transmission electron microscope, showing a high density of twins. Lines denote twin boundaries. Reprinted by permission from *Nature* [330] (2013).

nanocrystals. Stacking faults and twins in fcc crystals were discussed in Chapter 19.1. Figure 20.4 shows nine twins within this single 11 nm nanocrystal. This is a high density of defects. Perhaps these numerous defects form during synthesis because the metal atoms in nanocrystals already have higher chemical potentials.

20.3.2 Chemical Configurations within Alloy Nanocrystals

Chemical Unmixing in Core–Shell Nanostructures

Section 4.7 discussed surface segregation of chemical species in alloy clusters, and implications for nucleation. Surface chemical segregation can occur in the synthesis of nanoparticulate alloys. Such a chemical segregation can produce "core-shell" nanostructures, where the surface layers have compositions that differ from the particle cores.[4] Motivated in part by the use of nanoparticles as catalysts for gas-phase chemical reactions, surface segregation in nanoparticles has become an important topic. It is often possible to understand why a binary alloy will form a core–shell nanostructure by using a T–c phase diagram, especially with additional information about surface energies. For binary alloys, the phase diagram should show a strong tendency for chemical unmixing, such as a miscibility gap that extends to high temperatures. Surface energy can motivate the unmixing to occur to the outer parts of the nanoparticle. In general, the outer layers of the nanoparticle will be enriched in the chemical species that has the lowest average surface energy. For similar elements and crystal structures, the surface energy tends to increase with the melting temperature, much like the grain boundary energy discussed in Section 11.4.2. For example, the Au–Pt phase diagram shows strong chemical unmixing, and Pt has a higher melting temperature than Au. We therefore expect it is possible to

---

[4] Other core–shell nanoparticles can be synthesized by the oxidation of surface layers. The oxide layer may passivate the surface against further oxidation, at least for some time.

make core–shell nanoparticles with Pt in the core and Au in the outer shell. If coarsening can be suppressed, by isolating the nanoparticles perhaps, this core–shell structure should be stable to relatively high temperatures.

Chemical Order within Nanocrystals

Advances in aberration-corrected transmission electron microscopes, with computational methods for tomographic 3D reconstructions from images at many tilts, now enable detailed determinations of all atoms in nanoparticles. Jianwei Miao's group has identified the positions of 6569 iron atoms and 16 627 platinum atoms in a Pt–Fe nanoparticle of approximately 8 nm diameter, shown in Fig. 20.5a [332]. This particle had been heated at 600 °C to induce chemical order, and a measure of chemical short-range order was used to identify the different ordered domains. It is expected that this Pt–Fe alloy should have primarily $L1_2$ chemical order, and indeed most of the atoms are in the large $L1_2$ domains shown in Fig. 20.5b. The $L1_2$ order is imperfect, however, and these domains contain many antisite defects. There are some different domains in the core of the nanoparticle, including an fcc Pt–Fe solid solution and a thin region of $L1_0$ chemical order. Both these iron-rich domains help accommodate some of the extra iron in this nanoparticle of composition $Pt_{0.7168}Fe_{0.2832}$. During their formation by pyrolysis, Pt–Fe nanoparticles tend to develop a Pt-rich core [333], and vestiges of this core seem to persist after considerable chemical order was induced in the nanoparticle of Fig. 20.5.

Figure 20.5 Pt (gray) and Fe (darker) atom positions in $Pt_{0.72}Fe_{0.28}$ nanoparticle of approximately 8 nm diameter from 3D tomographic imaging with an aberration-corrected transmission electron microscope [334]. (**a**) Outer atoms in the nanoparticle. (**b**) Exploded view of ordered domains of Pt and Fe within the nanoparticle. Reprinted by permission from *Nature* [332] (2013).

20.4 Electron Energies in Nanomaterials

20.4.1 Energies of Free Electrons in Nanostructures

Most of the thermodynamic energy of nanostructures is associated with the atomic structure local to the interfaces. However, the reduction in size of the nanostructure brings important changes to the electronic energy owing to the confinement of electrons in small volumes. The energy of a gas of free electrons in three dimensions increases with electron density as $E \sim (N/V)^{2/3}$ (Section 6.3), where N is the number of electrons in a box of volume V. This free electron model proves useful for understanding some properties of metals, such as heat capacity and bulk modulus. A material with nanoscale dimensions does not have a substantially different density of free electrons, however, so to a first approximation the energy of its free electrons and their properties are not altered. The confinement of electrons to a smaller box changes the numbers and energy levels of the electrons, however, and this redistribution of electron energy levels can alter electronic properties such as transport and optical behavior. In some cases the elastic constants and bonding energy are changed, too.

Consider a large three-dimensional cube of material, and reduce its size one dimension at a time, as shown across Fig. 20.6a. For a free electron gas, changing these dimensions can be understood with a simple argument based on the number of available states in k-space, $N(k)$. For free electrons in a three-dimensional box, this was given in Eq. 6.65 as

$$N(k) = \frac{1}{8} \frac{4\pi}{3} k^3 \frac{1}{\left(\frac{\pi}{L}\right)^3}. \tag{20.11}$$

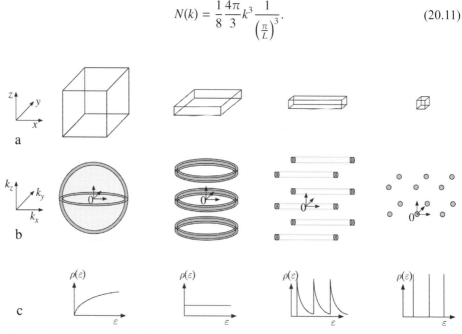

Figure 20.6   **(a)** Reducing the dimensionality of a box for free electrons from (left to right) 3, 2, 1, 0. **(b)** The shells in k-space for each dimensionality are at the edges, shaded with dark gray. **(c)** The electron densities of states, $\rho(\varepsilon)$, for structures of each dimensionality.

Here $N(k)$ is the total number of quantum states within the volume of a sphere of radius k.[5] This $N(k)$ goes with the volume of a three-dimensional sphere as k^3. For d spatial dimensions, the number of states in a region bounded by k is

$$N(k) \sim k^d. \tag{20.12}$$

A density of states in k is the number of states in a specific shell, ring, or point in 3, 2, or 1 dimensions, shown as the thin gray volumes in Fig. 20.6b (and indicated by the arc in Fig. 6.4)

$$\frac{\mathrm{d}N(k)}{\mathrm{d}k} \sim k^{d-1}. \tag{20.13}$$

We want a density of states in ε, so convert to energy as

$$\frac{\mathrm{d}N}{\mathrm{d}\varepsilon}\frac{\mathrm{d}\varepsilon}{\mathrm{d}k} \sim k^{d-1}. \tag{20.14}$$

Using the relationship for free electrons (which have only kinetic energy, $p^2/2m$)

$$\varepsilon = \frac{\hbar^2 k^2}{2m}, \tag{20.15}$$

$$\frac{\mathrm{d}\varepsilon}{\mathrm{d}k} \sim k \sim \sqrt{\varepsilon}, \tag{20.16}$$

we see that the shape of the electronic density of states $\rho(\varepsilon)$ depends on dimensionality as

$$\frac{\mathrm{d}N}{\mathrm{d}\varepsilon} \equiv \rho(\varepsilon) \sim \varepsilon^{(d-2)/2}, \tag{20.17}$$

for $d \geq 1$. For $d = 3, 2, 1$ the electronic density of states has the characteristic shapes $\rho(\varepsilon) \sim \sqrt{\varepsilon}$, $\rho(\varepsilon) \sim 1$, $\rho(\varepsilon) \sim 1/\sqrt{\varepsilon}$, respectively. These cases are shown in Fig. 20.6c.

The directions in k-space corresponding to the narrow dimension(s) of the boxes give discrete energy separations in Fig. 20.6c. These gaps are typical of the wide spacings of energy levels in a small box.[6] Around these points are spectral shapes that originate from the continuum of free electron states along the long dimension(s) of the box. Especially at low temperatures, the electronic excitation spectrum of a single-wall carbon nanotube looks much like the $\rho(\varepsilon)$ for the long rod in Fig. 20.6c. The case for $d = 0$ must be treated separately because there is no dimension for free electron behavior. This case of isolated nano-dots is essentially like that of an atom, which has discrete energy levels, as shown in Fig. 20.6c.

20.4.2 Tight-Binding Model in Two Dimensions

The electron states in two and three dimensions require two concepts beyond those of the tight-binding model for a linear chain in Section 6.3.2:

- The k variable is now a vector, $\vec{k}$, and the energy $\epsilon(\vec{k})$ varies with the direction of $\vec{k}$.

---

[5] $N(k)$ is obtained as the total volume of k-space divided by the volume $(\pi/L)^3$ which accommodates a single quantum state (see Fig. 6.4).

[6] For example, an electron wavefunction can make short, high-energy oscillations along the thin dimension of a plate.

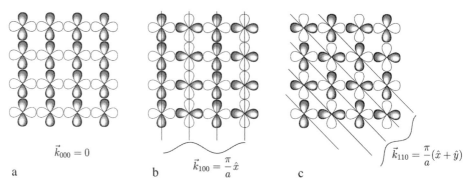

$$\vec{k}_{000} = 0$$

a

$$\vec{k}_{100} = \frac{\pi}{a}\hat{x}$$

b

$$\vec{k}_{110} = \frac{\pi}{a}(\hat{x} + \hat{y})$$

c

Figure 20.7 Atomic orbitals on a square lattice: (**a**) all with the same phase; (**b**) phases alternate along $\hat{x}$; (**c**) phases alternate along diagonal $\hat{x} + \hat{y}$.

- The number of k-points increases with $|\vec{k}|$. From the volumes of shells in 3D it can be shown that there are very few k-points in an interval near $k = 0$, compared to the number of k-points in an interval at larger k. In 2D and 3D, there is a suppression of the high density of states at small ϵ, shown for $\rho(\epsilon)$ in Figs. 6.3c and 20.6c.

The first point is illustrated with the example in Fig. 20.7. Here is a square lattice, and the bonding involves more complicated atomic wavefunctions (orbitals) placed at each site.[7] Importantly, the phase varies inside each atomic orbital. For reference, the phases of the white lobes are +1, whereas those of the dark lobes are –1. Figure 20.7a shows the simple case where there are no phase changes between the orbitals at the different atom sites. All the orbitals are drawn with identical phase, which can be a phase factor of simply +1 at every atom. This is consistent with $\vec{k} = 0$, for which $e^{i\vec{k}\cdot\vec{r}_i} = e^{i0} = +1$ for any atom site $\vec{r}_i$. This is the low-energy configuration of orbitals, since every connection between lobes is bonding (every white lobe connects to a white lobe, and every black connects to a black).

On the other hand, Fig. 20.7c is the highest energy configuration, since every connection between lobes is antibonding (white to black). The case of Fig. 20.7b is intermediate, since all vertical lobes connect to lobes of the same phase and are bonding, whereas all horizontal lobes are antibonding. Please take a moment to look at the phase relationships between the different atomic orbitals in Figs. 20.7b and c. Note how the phase is depicted at the bottom of the figures with wave crests and troughs for wavevectors $\vec{k}_{100}$ and $\vec{k}_{110}$.

To formalize the situation of Fig. 20.7, calculate a total energy from the four bonds to each atom. An atom at position 0 has bonds left, right, below, and above, so the bonds to the four neighbors have the net energy (cf. Eq. 6.57 for the linear chain)

$$\epsilon(\vec{k}) = \langle -a\hat{x}|\mathcal{H}|0\rangle + \langle a\hat{x}|\mathcal{H}|0\rangle + \langle -a\hat{y}|\mathcal{H}|0\rangle + \langle a\hat{y}|\mathcal{H}|0\rangle, \qquad (20.18)$$

$$\epsilon(\vec{k}) = \left(e^{i\vec{k}\cdot(-a\hat{x})} + e^{i\vec{k}\cdot a\hat{x}}\right)\langle a|\mathcal{H}|0\rangle + \left(e^{i\vec{k}\cdot(-a\hat{y})} + e^{i\vec{k}\cdot a\hat{y}}\right)\langle a|\mathcal{H}|0\rangle, \qquad (20.19)$$

---

[7] These look like d-orbitals of symmetry $x^2 - y^2$. For each orbital, along $\hat{x}$ its phase has the opposite sign as along $\hat{y}$. Incidentally, the same result of Eq. 20.21 is obtained when s-orbitals are at the atom sites.

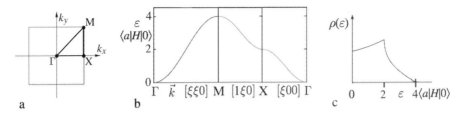

Figure 20.8 (**a**) Special points in a section of k-space for square lattice. (**b**) $\epsilon(\vec{k})$ along the [100] direction and [110] direction for square lattice of Fig. 20.7. (**c**) Density of states of electrons for square lattice.

$$\epsilon(\vec{k}) = \left(e^{-ik_x a} + e^{ik_x a}\right)\langle a|\mathcal{H}|0\rangle + \left(e^{-ik_y a} + e^{ik_y a}\right)\langle a|\mathcal{H}|0\rangle, \tag{20.20}$$

$$\epsilon(\vec{k}) = 2\left(\cos(k_x a) + \cos(k_y a)\right)\langle a|\mathcal{H}|0\rangle. \tag{20.21}$$

where Eq. 20.19 was obtained by using Bloch's theorem.

The energy $\epsilon(\vec{k})$ of Eq. 20.21 depends on the direction of $\vec{k} = k_x\hat{x} + k_y\hat{y}$. This is awkward to plot because $\epsilon(\vec{k})$ is a three-dimensional set of data (two dimensions of $\vec{k}$ and one of ϵ). Figure 20.8a is a map to interpret the standard plots. It is a sector of k-space around $k = 0$, a special point that is always labeled Γ. The four sides of the big square are where $k_x = \pm\pi/a$ and $k_y = \pm\pi/a$. A special point on this square is labeled X (on the x-axis). The corner, labeled M, is where $\vec{k} = \pi/a\hat{x} + \pi/a\hat{y}$. (Many other arcane letters for special points in k-space were defined in 1936 [335].) Figure 20.8b shows one-dimensional graphs of $\epsilon(\vec{k})$ along special directions of $\vec{k}$, specifically along the three dark lines in Fig. 20.8a. There are three panels within Fig. 20.8b, for these three different directions in k-space. The energy varies along $\vec{k}$ as specified by Eq. 20.21, divided by 2 to avoid double-counting of bonds, and shifted so the lowest energy is zero.

20.4.3 Density of States in Two Dimensions

The second difference between band structures in 2D (or 3D) and 1D is in the counting of the different energies for the electron density of states (DOS), $\rho(\epsilon)$. The important concept is that each allowed $\vec{k}$ vector gives an energy $\epsilon(\vec{k})$. In the 1D chain it was simple to map an interval in k into an interval in ϵ, giving the DOS of Fig. 6.3c. This is harder to do in 2D and 3D. Here we describe an algorithm that steps through a broad range of k_x and k_y in equally spaced intervals. It covers the full area of the square of Fig. 20.8a.[8] For each pair $\{k_x, k_y\}$ it calculates $\epsilon(\vec{k})$ with Eq. 20.21. This generates many different values of ϵ varying from a minimum (which we shift to 0 for reference) to a maximum of $4\langle a|\mathcal{H}|0\rangle$. Binning these values into equal intervals of $\Delta\epsilon$ gives the histogram of Fig. 20.8c. This is the electron DOS for our tight-binding model on a 2D square lattice. It is approximately constant at small ϵ, as expected from the dimensional argument of Eq. 20.17.

There is a surprise in the DOS of Fig. 20.8c at $\epsilon = 2\langle a|\mathcal{H}|0\rangle$. The DOS has a singularity in its first derivative, called a "Van Hove singularity." The analysis of such features can

---

[8] Actually, owing to symmetry, it need cover only the area in the dark triangle.

take us rather far afield, but a hint of their origin is found by looking at Figs. 20.7b and c. For the phases of the orbitals, Fig. 20.7b shows the shortest wavelength that is possible along $\hat{x}$, with alternate orbitals having fully opposite phase. This corresponds to neighbors along $\hat{x}$ differing in phase by π, so $e^{i\pi} = -1$ and the phase factor switches sign between neighbors. Increasing the value of k_x can only decrease this variation of phase. In fact, if this phase difference between neighbors increases to 2π, we return to a constant phase for all atoms (as in Fig. 20.7a) since $e^{i2\pi} = +1$. At this Van Hove singularity, we have reached the top of the [100] band in the right panel of Fig. 20.8b. Another subtlety is that the energy can go up to $\epsilon = 2\langle a|\mathcal{H}|0\rangle$ with k_x (or k_y) alone, but larger energies are possible only with both together. However, for some values of k_x the contribution from k_y may have already reached its maximum. The largest value of $\epsilon = 4\langle a|\mathcal{H}|0\rangle$ is possible only when both x and y contributions are maximum, and there is only one $\vec{k}$ where this occurs ($k_x = \pi/a, k_y = \pi/a$). This single $\vec{k}$ contributes little to the histogram, so the DOS, $\rho(\epsilon)$, is infinitesimal at this maximum ϵ.

20.5 Entropy of Nanomaterials

Entropic Stabilization

The configurational entropy from nanostructural degrees of freedom, and especially the vibrational entropy from these new degrees of freedom, tends to stabilize a nanomaterial at finite temperatures. Consider, for example, a nanostructured material with a grain boundary enthalpy of $+3$ kJ/mole [326, 336]. At a temperature of 500 K, an entropy increase of $+0.75\,k_B$/atom over the bulk material would null the free energy difference between the nanostructured material and the bulk material, i.e., $\Delta F = \Delta E - T\Delta S = 0$. Although such a thermodynamic stabilization has not yet been demonstrated,[9] entropic effects should alter the kinetic stability of nanomaterials.

20.5.1 Configurational Entropy from Microstructural Degrees of Freedom

Grouping Degrees of Freedom

Consolidated nanoparticles can be configured in many ways, so we expect some configurational entropy if equivalent configurations can be explored. Entropies of thermodynamic importance require the number Ω of $S = k_B \ln \Omega$ (Fig. 7.1) to be large. Specifically, the degrees of freedom for nanostructural configurations need to be comparable with the large number of degrees of freedom at the atomic scale. Consolidating n degrees of freedom into 1 reduces S by a factor of n, so mesoscopic and microscopic structures in materials generally have too few degrees of freedom to compete with entropic contributions from atomistic quantities. Nevertheless, what is lost in the consolidation can be partly recovered

---

[9] If the nanostructure were favored thermodynamically, the material might find a different crystal structure possessing some of the local atomic arrangements of the nanocrystal.

by the degrees of freedom in position and orientation of the consolidated nanocrystals. As a consolidated nanocrystal is heated so its atoms can diffuse, the material can explore some degrees of freedom of the nanostructure. Increased atom mobility at higher temperatures leads to crystal growth, of course, but this requires specific types of atomic motions. There are more ways that the nanostructured degrees of freedom can be explored. These microstructural degrees of freedom provide a configurational entropy that we now estimate.

Positions and Orientations of Crystals

Consider the following algorithm for constructing a polycrystalline material. Cover space with a grid of N points separated by an atomic diameter. Pick randomly x points on the grid, where $x \ll N$. Use each of these x points as a nucleation site for an individual crystal, which grows outwards until it impinges on a neighboring crystal. The algorithmic grid does not pertain to crystallographic orientations, only centers, so there are additional degrees of freedom to select the orientations of the individual crystals. The Ω of Eq. 7.1 is a product of two factors, Ω_{cent} for the number of ways of picking the centers of the x crystals, and Ω_{orient} for the number of ways to orient the crystals

$$S_{\text{cf,nano}} = k_B \ln \left(\Omega_{\text{cent}} \Omega_{\text{orient}} \right) . \tag{20.22}$$

To evaluate the numbers of positions and numbers of orientations, we recognize that chemical bonding sets the final position of each atom. If an atom in a relaxed structure is briefly displaced by a small fraction of an atomic diameter, it will move back into its original position. For both position and orientation, the smallest grid separations for combinatorics are therefore approximately an atomic distance.

Evaluating Ω_{cent} is the same calculation for distributing atoms of a solid solution on an Ising lattice; the result of Eq. 2.41 is adapted for our new variables

$$\Omega_{\text{cent}} = \frac{N!}{x! \, (N - x)!} . \tag{20.23}$$

The calculation of Ω_{orient} is a bit different. For simplicity, assume each crystal contains the same number of atoms, N/x. Assume further that all crystals are spheres, with radius

$$r = \left(\frac{N}{x} \frac{3}{4\pi} \right)^{1/3} . \tag{20.24}$$

The units of distance are now atom diameters. The number of atoms on the surface is simply $(4\pi r^2)$. This is the number for orientations of the z-axis for one crystal.[10] To complete the orientation analysis, we allow for rotations of the crystallographic coordinate system about the z-axis. The number of orientations is approximately the number of atoms around the diameter of one spherical crystal, $(2\pi r)$. The number of ways to orient one crystal is

---

[10] There are some orientations equivalent by symmetries such as mirror symmetry. These give a numerical correction that proves unimportant in what follows. Similarly, adjacent crystals may lock into orientations with a finer grid than considered here, but this numerical correction is also ignored.

$$\Omega_{\text{orient1}} = \left(4\pi\, r^2\right)\left(2\pi\, r\right), \tag{20.25}$$

$$\Omega_{\text{orient1}} = 6\pi\, \frac{N}{x}, \tag{20.26}$$

which scales with the number of atoms in the crystal, N/x. The x crystals can all have different mutual orientations, however, giving a larger number of independent orientations for all crystals

$$\Omega_{\text{orient}} = \left(6\pi\, \frac{N}{x}\right)^x, \tag{20.27}$$

and giving for Eq. 20.22

$$S_{\text{cf,nano}} = k_B \ln\left[\frac{N!}{x!\,(N-x)!}\left(6\pi\, \frac{N}{x}\right)^x\right]. \tag{20.28}$$

We define the ratio $f \equiv x/N$, which is analogous to our definition of concentration c in the alloy problem, and we seek the entropy per atom in the nanostructure. Using the Stirling approximation with manipulations similar to those following Eq. 2.41, and ignoring a term with $x \ln(6\pi)$

$$S_{\text{cf,nano}} = -k_B\left[(1-f)\ln(1-f) + 2f \ln f\right], \tag{20.29}$$

which is close to the entropy of mixing of Eq. 2.31.[11] Since f is small

$$S_{\text{cf,nano}} = k_B\left[f - 2f \ln f\right]. \tag{20.30}$$

This configurational entropy of our model nanocrystalline material is graphed in Fig. 20.9, in terms of the radius and in terms of the volume of an individual nanocrystal in a polycrystalline material. This contribution is of order $0.1\, k_B$/atom when the nanocrystals contain of order 100 atoms, but it could be larger if there are more microstructural features than considered here. For nanocrystals containing more than 10 000 atoms, however, the microstructural configurational entropy is probably negligible.

Grain Boundary Region

Within a grain boundary of 1 nm in width [325, 326], the atoms must accommodate the edges of the bounding crystals, and are not positioned as arbitrarily as in a glass, as discussed in Section 20.1.2. Using an entropy of fusion of $\sim 0.8\, k_B$/atom would be a generous upper bound for the excess entropy of the grain boundary region. When normalizing by the number of atoms in the material, however, this must be reduced by the fraction of atoms in the grain boundary. Very approximately, a nanocrystalline material with a crystallite size of 10 nm has a fraction of 0.1 of its atoms in grain boundaries, so the entropy per atom in the nanocrystalline material may be approximately $0.08\, k_B$/atom from this contribution. Although this estimate may be too large, the grain boundary disorder

---

[11] If we guessed that the entropy of x groups of N/x atoms is reduced by a factor of x/N compared with N individual atoms, we would have obtained Eq. 20.29 without the factor of 2 that comes from orientational disorder.

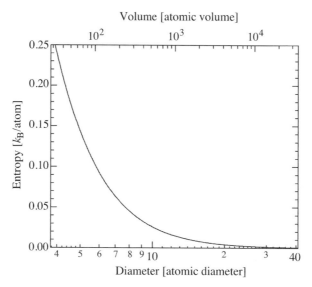

Figure 20.9 Configurational entropy of polycrystalline nanostructure from Eq. 20.30.

probably makes at least as large a contribution to the configurational entropy as the microstructural degrees of freedom discussed in Section 20.5.1.

20.5.2 Vibrational Entropy of Nanostructures

Low-Energy Vibrations of Nanostructures

The difference in vibrational entropy between nanostructured and bulk materials is usually larger than the difference in configurational entropy. Experimental measurements show a number of differences in the phonon DOS of nanocrystalline and bulk materials. In general, the prominent peaks in the DOS (van Hove singularities, see Section 26.1.2 [online]) are the same for nanocrystals and large crystals [337–340]. Although Fig. 20.10 shows that the DOS is broadened with smaller crystallite size, the energies of the van Hove singularities do not shift substantially [341]. This indicates that the interatomic forces are the same for most atoms in both materials. The energies of the van Hove singularities change with crystal structure and with pressure, but these changes are similar for both nanocrystalline and bulk materials [342].

Figure 20.10 (and 20.11) shows that the phonon DOS of the nanocrystalline material has more spectral weight at energies below about 20 meV, and further analysis of the data of Fig. 20.10 showed that this enhancement scales inversely with the crystallite size. This enhancement of the DOS at low energies has been found in all studies of compacted nanocrystals [338–341]; [343–345]. The trend with crystal size implies that the extra spectral weight at low energies originates from degrees of freedom in the microstructure. These may include vibrations that localize at grain boundary regions or propagate along them, motions of the nanocrystals with respect to each other, or some combinations of these dynamics. The transfer of phonon spectral weight to lower energies for smaller nanocrystals causes the vibrational entropy to increase as much as $0.3\,k_B$/atom.

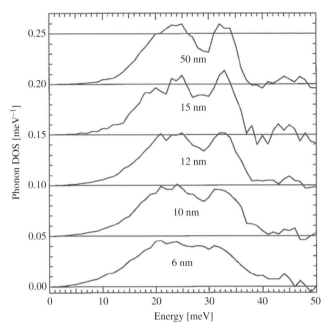

Figure 20.10   Phonon DOS curves from nanocrystalline fcc Ni_3Fe prepared by high-energy ball milling, followed by subsequent annealing to induce grain growth to sizes as labeled. Reprinted with permission from [341]. Copyright (1998) by the American Physical Society.

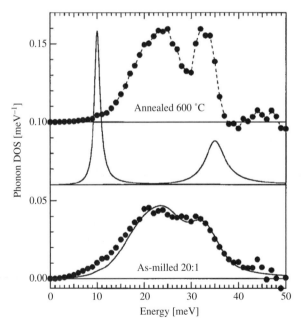

Figure 20.11   The characteristic shape of the damped harmonic oscillator function is shown in the center for modes at 10 and 35 meV having the quality factor $Q = 7$. Solid curve at bottom is the "convolution" of the damped harmonic oscillator function for $Q = 7$ with the experimental DOS from the sample annealed at 600 °C (points at top). Points at bottom are experimental DOS curve from 6 nm Ni–Fe material. Reprinted with permission from [341]. Copyright (1998) by the American Physical Society.

In a challenging experiment, the ^{57}Fe partial phonon DOS curves were measured from isolated nanocrystals on a Si substrate [346] (also [347]). The nanocrystals of 8–10 nm diameter consisted of bcc Fe cores, surrounded by oxide shells. At phonon energies below 5 to 10 meV, exponents n in the expression $g(\varepsilon) \propto \varepsilon^n$ were found to be 1.33, 1.84, and 1.86 for three different samples, showing a clear low-dimensional behavior in the partial DOS of the Fe.[12] This tendency toward 2D vibrational behavior is likely due to the large number of low-coordinated atoms at the surface or interface. Perhaps it is more appropriate to consider these low-energy vibrations as eigenmodes of small particles [348–350].

Broadening of Spectral Features of Nanocrystalline Materials

The second salient feature of the phonon DOS of nanocrystalline materials is a broadening of all features of the DOS, which seems independent of the enhancement of the spectrum at low energies. One plausible explanation is that the broadening originates with the "damping" of the vibrational modes. Such behavior can originate with anharmonic interatomic potentials, as may perhaps exist for atoms in irregular sites in grain boundaries. If the broadening originates with damping, the spectral shape of a damped harmonic oscillator can be used to broaden the phonon DOS of bulk material to match that of the nanocrystalline. The adjustable parameter is Q, the "quality factor" of the oscillator (i.e., the number of cycles for which the fraction e^{-1} of the phonons will remain). For spectra measured to date on consolidated nanocrystals, a damped harmonic oscillator model has given values of Q as low as 5. This high damping may contribute to the low thermal conductivity of nanostructured materials, but it may not affect the entropy significantly. A second plausible explanation is that the spectral broadening originates with a distribution of harmonic modes. If true, the higher vibrational frequencies would suppress the vibrational entropy.

20.6 Magnetic Nanoparticles

For magnetism as described in Section 21.2, there are two types of electron exchange interactions that are important – intraatomic exchange and interatomic exchange. The intraatomic exchange is summarized in Hund's rule, stating that equivalent electron levels in an atom are filled by electrons with the same spin. This intraatomic exchange is very strong, of order 1 eV, and much larger than thermal energies. It is responsible for atoms having robust magnetic moments, even at high temperatures.

The interatomic exchange is much weaker and more variable. It couples the magnetic moments between different atoms, aligning them at low temperatures. In a magnetic Curie transition it is the alignment of magnetic moments that is disrupted by temperature, not the moments of the individual atoms. The strength of this interatomic exchange, $J(\vec{r})$,

---

[12] When the sound velocity is independent of wavelength, in three dimensions $n = 2$, and in two dimensions $n = 1$.

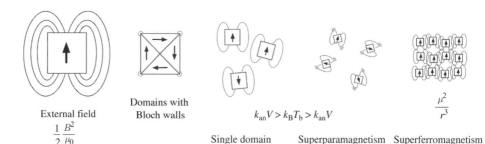

Single domain   Superparamagnetism   Superferromagnetism

Figure 20.12   Magnetizations of bulk crystals to the left, and smaller crystals to the right. See text for descriptions.

depends on the types of atoms and their local configurations. It is therefore not surprising that nanostructured materials, with their large fraction of atoms at and near interfaces, have different $\{J\}$ and Curie temperatures than bulk materials. The general expectation is that surface atoms experience a lower interatomic exchange force, and nanocrystals or very thin films (a few atomic layers) have lower Curie temperatures than bulk material. The problem is not easy to analyze, however, since the exchange interactions are difficult to calculate and there may also be changes in magnetic moments near the surfaces of nanoparticles.

A large magnetic crystal tends to form magnetic domains, as shown in the second image of Fig. 20.12 (and shown for a ferroelectric material in Fig. 21.10). The basic idea is that a single domain has a magnetic field that extends into the space around the crystal, and the energy of this magnetic field B goes as B^2 times the volume. By forming magnetic domains as in Fig. 20.12, the magnetic flux is confined within the material, and the energy in the external field is much smaller. The penalty is the formation of boundaries between the domains. These "Bloch walls," described in Section 21.9.2, contain magnetic moments that are not aligned optimally, and a surface energy is associated with the Bloch wall. For smaller magnetic crystals, the external field energy is smaller, but the surface energy in Bloch walls is relatively large. Below a minimum size, sometimes around 1 μm, it is favorable to have a single magnetic domain within the crystal, and no Bloch wall.

The magnetic moments at individual atoms have the same interactions as in bulk material, but there are fewer of them in a small particle, assumed a single domain. Sometimes thermal energy can reorient the direction of magnetization of the entire domain. This process depends on the "magnetocrystalline anisotropy," i.e., how favorable is the alignment of the magnetization along a crystallographic direction.[13] A ferromagnet made of amorphous material and shaped as a sphere tends to be isotropic in the energy of its magnetization, so relatively little thermal energy is required to reorient its magnetization. On the other hand, "hard" permanent magnets tend to have strong preferences for aligning magnetic moments along favored crystallographic directions. The total energy required to misalign the magnetic moment away from its preferred crystallographic axis scales with the number of electron spins in the crystal. The magnetocrystalline anisotropy energy is

[13] Magnetocrystalline anisotropy requires a coupling of electron spins to the electron distribution in the material, such as by spin–orbit coupling.

$E_{an} = \kappa_{an} V$, where κ_{an} is a material constant and V is the volume of the crystal. This E_{an} is an activation barrier for flipping the magnetization, so this reorientation occurs at the rate

$$\Gamma(T) = \Gamma_0 \, e^{-\kappa_{an} V/k_B T}. \tag{20.31}$$

Approximately, there is a "blocking" temperature $T_b = \kappa_{an} V/k_B$ below which the total magnetic moment of a single-domain nanocrystal remains aligned in one direction. Above T_b the spins remain aligned within each nanocrystal, but the spin orientation of a single nanocrystal fluctuates rapidly (and independently of its neighbors). The size of the crystal where this occurs is typically a few to tens of nanometers. The state above T_b is sometimes called "superparamagnetic," since each nanocrystal maintains a large net spin, but temperature causes misalignment between the net spins of the different nanocrystals (see Fig. 20.12).

Suppose isolated superparamagnetic crystals are brought close together and allowed to interact magnetically. At intermediate separations between particles of order 1 nm, the magnetic interactions between the particles may allow them to have a cooperative mutual alignment. The blocking temperature is raised, and it is sometimes useful to describe the behavior as ferromagnetic, where the interacting spins are the magnetic moments of the crystals themselves. This behavior is sometimes called "superferromagnetic" [351, 352].[14] When all particles are in close contact, bulk ferromagnetism is recovered, at least in principle.

Problems

20.1 The second row in Fig. 20.6 shows a continuous density of states in k-space for the large cube, but a stacked set of plates for a box of two dimensions, and an array of rods for a box of one dimension. Explain the electronic DOS for the latter two structures by using the solution of the Schrödinger equation for a particle in a box of one or two dimensions, and explain how the separation of the plates or rods in k-space changes with the size of the box.

20.2 Suppose a random, equiatomic A–B binary alloy on a simple cubic lattice transforms into a molecular crystal, where each site on a (different) simple cubic lattice contains one $A_1 B_1$ molecule. Assume that each molecule can have six orientations, and assume its orientation is independent of its neighbors.

  What is the change in configurational entropy for this transformation?

20.3 Explain how the Van Hove singularity in Fig. 20.8c relates to Fig. 20.8a. In particular, explain what happens to the number of states contributing to the DOS when k_x or k_y equals π/a. Show how this condition corresponds to drawing a circle within the square of Fig. 20.8a.

---

[14] With disorder in nanoparticle shapes or positions, a collective spin glass behavior is possible, too [353].

20.4 Thermodynamic fluctuations are expected in small systems. Suppose the fluctuation involves one degree of freedom in the particle shape or size. If the particle is small, the energy cost for a change in this degree of freedom may be of order $k_B T$, so fluctuations are expected.

   This argument can be made more precise for energy fluctuations, leading to Eq. 20.10 in the text. Recall that the thermal average of the mean-squared energy, $\langle E^2 \rangle$, is

$$\langle E^2 \rangle = \frac{\sum_j \varepsilon_j^2 \, e^{-\varepsilon_j/k_B T}}{\mathcal{Z}}, \tag{20.32}$$

where $\mathcal{Z}$ is the partition function. Here we assume that the ε_j are energy states of the particle, and the $\{\varepsilon_j\}$ do not change with temperature.

   (a) Write expressions for $\mathcal{Z}$ and $\langle E \rangle$.

   (b) The root-mean-squared energy is σ_E, where $\sigma_E^2 = \langle E^2 \rangle - \langle E \rangle^2$, and we seek to calculate it from the expressions above. The *Big Trick* is to take the derivative of $\langle E \rangle$ with respect to temperature. You will need to take the derivative of $\mathcal{Z}$ in the process (which is in the correct expression for $\langle E \rangle$ from part a).

   (c) Now rearrange your result from part b into the form

$$\langle E^2 \rangle + \langle E \rangle^2 = k_B T^2 \frac{\partial \langle E \rangle}{\partial T}. \tag{20.33}$$

   (d) Using the relationship $C_V = \partial \langle E \rangle / \partial T$, obtain Eq. 20.10 in the text.

20.5 Equation 20.23 for Ω_{cent} was derived by assuming that all crystals nucleated at the same time, so any point on a grain boundary is equidistant to its two closest centers. Now assume that nucleation occurs over a range of times from zero to the time when most of the crystals impinge. Estimate the change in configurational entropy from this new effect.

20.6 Magnetic nanoparticles
   (a) (Surface-to-volume ratio) Below a critical diameter, a small magnetic particle will be a single domain. Calculate the critical size.
      (*Hints*: You may assume that the magnetic energy is proportional to the volume of the domain, whereas the energy of the domain wall scales with the area of the wall. You may assume any convenient shape for the particle and the domains.)
   (b) (Interactions) When large numbers of small magnetic nanoparticles are brought together into a group, it is observed that the macroscopic magnetic properties of the bulk material at room temperature depend strongly on the distance of separation between the nanoparticles. Explain why this is true.

20.7 **(a)** When electrons are added to a free electron gas of different dimensions (3, 2, 1, 0), explain the differences in the change of electron energy.
   (b) How do you expect the bulk modulus to change with electron density in these different dimensions?

21

Magnetic and Electronic Phase Transitions

The biggest magnetic effects in materials originate with unpaired electron spins and their alignments.[1] The electron spins interact with each other, and patterns and structures develop in their orientations at low temperatures. With temperature, pressure, and magnetic field, these spatial patterns of electron spins are altered, and several trends can be understood by thermodynamics.

Much of this chapter describes how magnetic structures change with temperature. The emphasis is on magnetic moments localized to individual atoms, as may arise from unpaired $3d$ electrons at an iron atom, for example. The strong *intra*atomic exchange interaction gives an atom a robust magnetic moment, but the magnetic moments at adjacent iron atoms interact through *inter*atomic exchange interactions. Interatomic exchange interactions are typically weaker, with energies comparable to thermal energies. Interatomic exchange is analogous to the chemical energy preferences of pairs of atoms in a binary alloy that develops chemical order. The critical temperature of chemical ordering, T_c, corresponds to the Curie temperature for a magnetic transition, T_C, and short-range chemical order above T_c finds an analog in the "Curie–Weiss law" for paramagnetic susceptibility above T_C. For chemical ordering the atom species are discrete types, whereas magnetic moments can vary in strength and direction as vector quantities. This extra freedom allows for diverse magnetic structures, including antiferromagnetism, ferrimagnetism, frustrated structures, and spin glasses. (Superparamagnetism and superferromagnetism were described in Section 20.6, and a quantum magnetic phenomenon is described in Section 28.7 [online].) The vectorial character of spin interactions can also give rise to localized spin structures such as skyrmions (Section 21.6).

Pairs of ions with opposite charge tend to move towards each other to minimize their Coulomb energy, creating an electric dipole. An electromechanical phase transition can occur when the energy for a displacement of positive and negative ions in a unit cell is comparable to thermal energies. This ferroelectric transition has some similarities to the ferromagnetic transition, but is described with the Landau theory of Chapter 19.

Domains were discussed in Chapter 19 for martensitic transformations, and explained as a way to reduce the energy in the elastic field surrounding a transformed particle.

---

[1] The orbital angular momentum of an atomic electron is largely "quenched" when the potential energy for the electron follows the symmetry of the crystal structure. Nevertheless, some spin–orbit coupling alters the magnetic moments at transition metals, and causes magnetocrystalline anisotropy (an energetic preference for spin alignments along crystallographic directions). In rare earth elements, the orbital angular momentum couples to the electron spin to make a total spin for use in the thermodynamics of spin alignments in this chapter.

Analogously, domains in ferroelectric and ferromagnetic materials can reduce the energy in surrounding elastic and magnetic fields. Finally, the width of a boundary between two magnetic domains is estimated with an energetics argument.

21.1 Overview of Magnetic and Electronic Phase Transitions

21.1.1 Some Magnetic Structures and Phase Transitions

Curie Transition

When a ferromagnetic material such as iron is heated above its Curie temperature, it seems "nonmagnetic," but it is actually "paramagnetic." Above the Curie temperature the iron atoms still possess magnetic moments, but they fluctuate in orientation. Since the moments lose their long-range order and no longer point in the same direction, the iron appears nonmagnetic. If one could take a snapshot of the local arrangements of spins, however, one would see short-range order in the spin alignments in the paramagnetic phase. This ferromagnetic to paramagnetic "Curie transition" is much like the order–disorder transition of Section 2.9.[2] The magnetic response of a material below and above T_C can be altered by applied magnetic fields, however, but there is no good analogy of this for chemical ordering.

At low temperatures, however, a ferromagnet responds differently to temperature than an ordered alloy. Spins are able to orient in multiple directions, whereas a lattice site can be occupied by only an A-atom or a B-atom. In general, the lowest energy excitation of a spin system is not a single misoriented spin, but is a wavelike spin reorientation across the entire crystal. Spin waves are analogous to vibrational modes in a crystal.[3] Possible two-dimensional spin wave structures on a one-dimensional chain, which may propagate from left to right, are

$$\uparrow \nearrow \uparrow \nwarrow \uparrow \nearrow \uparrow \nwarrow \uparrow \nearrow \uparrow \nwarrow \uparrow \nearrow \uparrow \nwarrow \uparrow \nearrow \uparrow \nwarrow \uparrow \nearrow \uparrow \nwarrow \uparrow \nearrow, \text{ or}$$

$$\uparrow \nearrow \rightarrow \searrow \downarrow \swarrow \leftarrow \nwarrow \uparrow \nearrow \rightarrow \searrow \downarrow \swarrow \leftarrow \nwarrow \uparrow \nearrow \rightarrow \searrow \downarrow \swarrow \leftarrow \nwarrow.$$

Néel Transition

Antiferromagnetism is another type of spin structure that is quite common in transition metal oxides. It has alternating orientations for neighboring spins, for example a chain of $\uparrow \downarrow \uparrow \downarrow \uparrow \downarrow \uparrow \downarrow$, as opposed to a ferromagnet with its aligned spins as $\uparrow \uparrow \uparrow \uparrow \uparrow \uparrow \uparrow \uparrow$

---

[2] Because solute atoms are conserved, however, our previous analysis considered "right" and "wrong" duos of atom pairs. For spins there is no such conservation condition, and to some extent the mean field calculation of the Curie temperature is even easier.

[3] When their minimum energies of excitation are quantized as $\hbar\omega$, these two types of excitations are called "magnons" and "phonons."

or ↓ ↓ ↓ ↓ ↓ ↓ ↓ ↓ . Above a "Néel temperature," the antiferromagnetic structure is lost, and paramagnetism develops in an analogous way to that above the Curie temperature of a ferromagnet. Unlike ferromagnetism, antiferromagnetism causes only subtle changes in bulk magnetic properties. Its discovery required microscopic probes of spin alignments, such as polarized neutron diffractometry.

Another spin structure is a "ferrimagnet," which is similar to an antiferromagnet, but some of the spins are of different size or abundance, for example ⇑ ↓ ⇑ ↓ ⇑ ↓ ⇑ ↓ . There is a net magnetic moment in a ferrimagnet at low temperature. Magnetite, or lodestone, is perhaps the most familiar ferrimagnet. The properties of lodestone have been recognized for two millennia, and it has been used in magnetic compasses for about a millennium.

Frustration

For antiferromagnets and ferrimagnets it is not assured that each atomic spin will correspond properly to a crystal site. There is a natural geometric pattern for spins in an antiferromagnet, and if this is incompatible with the crystal structure, the magnetic structure can be "frustrated." For example, try to construct an antiferromagnet on a triangular lattice, with neighboring sites having opposite spins. A difficulty arises as soon as two opposite spins are placed as neighbors on a triangle

The first two sites have opposite spins, as is favored by a first-neighbor antiferromagnetic interaction, but what about the third? Consider the two possibilities

The first is unsatisfactory because it places two ↓ spins together, but the second places two ↑-spins as first neighbors. There is no geometrical solution to the problem of constructing an antiferromagnet on a triangular lattice – the condition is called "frustration." Frustration has a significant effect on the free energy of an antiferromagnet. A frustrated system often has difficulty finding its lowest-energy ground state, and the ground state may not be unique. A ground state for first-neighbor antiferromagnetic interactions on a triangular lattice is shown in Fig. 21.1a. It is obviously a compromise for spin orientations, but all first-neighbor orientations are favorable.

The Kagome lattice is related to the triangular lattice as shown in Fig. 21.1b. With first-neighbor interactions, the antiferromagnetic Kagome lattice has multiple ground states, such as those shown in Figs. 21.1b, c, d. The multiple ways to configure spins on the central hexagon and the neighboring hexagons gives a configurational entropy at 0 K. It also leads to interesting quantum dynamics at 0 K, since the spin system can be considered a superposition of different ground states. A discussion of quantum excitations on such spin structures is presented in Section 28.7 [online].

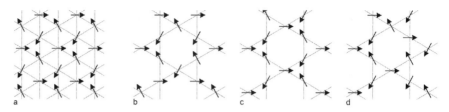

Figure 21.1 (**a**) A ground state for a Heisenberg antiferromagnet on a triangular lattice. Note horizontal rows of equivalent spin orientations, giving long-range order. (**b, c, d**) Three ground states for an antiferromagnet on a Kagome lattice. Panel b has the same spin orientations as panel a, but is missing 1/3 of the sites. Panels c and d are ground states with spin orientations different from panel b, but with equivalent energies from first-neighbor spin alignments.

Spin-Glass Transition

A "spin glass" is roughly analogous to a glassy structure of atoms. A spin glass can occur for the atomic spins of an amorphous material if there are some antiferromagnetic interactions. Spin glasses can also occur in crystalline alloys if solute atoms alter the interatomic exchange interactions. A model for a spin glass usually includes some randomness in the signs and magnitudes of interactions between neighboring spins. Frustration is a general feature of spin glasses, and the spin system typically has many spatial structures that are similar in energy. Some are accessible with few reorientations of spins, others with many. The larger reorientations become more improbable at lower temperatures, and cooling below the spin-glass transition temperature causes the magnetic response to become sluggish. Above the transition temperature the system acts increasingly like a paramagnet. The magnetic susceptibility (i.e., the change in spin alignment under a small change in applied magnetic field) is a maximum at the spin-glass transition temperature. This peak in response to an applied magnetic field is a signature of the spin-glass transition. Higher magnetic fields reduce the number of favorable magnetic states, and suppress the magnetic susceptibility at the spin-glass transition temperature.

21.1.2　Electronic Transitions

Verwey Transition

A peculiar phase transition occurs in magnetite, Fe_3O_4, at a temperature of 122 K. Magnetite is an insulator below this temperature, but an electronic conductor above. Electron spectroscopies show that Fe^{2+} and Fe^{3+} ions sit on specific crystallographic sites below 122 K, but there is evidence that these sites have an average valence between 2 and 3 at higher temperatures. The "Verwey transition" at 122 K includes a charge localization transition, where the extra electron of the Fe^{2+} becomes mobile above this temperature and contributes to electrical conductivity. Although "Verwey transition" has come to mean a charge localization transition, the situation in magnetite is not fully understood. A structural distortion accompanies the transition, and the interplays between the distortion, charge, and magnetism are complicated.

Metal–Insulator Transition

By the arguments of Section 6.3, we generally expect a crystal of atoms to have electronic bands that range from bonding to antibonding. The picture includes a Fermi surface with mobile electrons and metallic behavior. Only when bands are completely full or completely empty do we expect immobile electrons or holes that cannot find unoccupied states just across the Fermi surface. Metallic behavior is not so universal as this picture suggests, however, owing to how electrons interact with each other over various distances. Over short distances an electron is expected to see the potential energy from the ion core of its atom, giving a tendency for the electron to localize at its atom. A characteristic size for localization would be the Bohr radius, $r_B \equiv \hbar^2/(m^*e^2)$ (adapted with m^* for an effective electron mass). There is a second length scale that comes from the screening of the positive ion core owing to the presence of other electrons in the crystal. This screening length is tricky to calculate, and it is a function of the k-vectors of the electrons and the shape of the potential. A simple model is that of a screened Coulomb potential with the form $-e/r \exp(-r/r_{sc})$, where the screening length r_{sc} depends on the density of free electrons, but also on their k-vectors. Details are challenging, but here is the key argument. With a high density of free electrons, the screening length r_{sc} is small. When $r_{sc} < \kappa r_B$, the electron does not see the attractive ion core, becomes unbound from the atom, and adds to the free electron density. (Here κ is a numerical constant that depends on the details of the model.) With more free electrons the screening becomes more effective at other atoms, r_{sc} decreases further, and the conductivity increases suddenly. This is called a "metal–insulator transition," causing metallic behavior in some metal oxides upon heating.

Ferroelectric Transition

Ferroelectric phase transitions are named by analogy to ferromagnetic phase transitions. Below a critical temperature, a ferroelectric material develops an electric dipole moment from asymmetrical displacements of the positive and negative ions in its unit cell. Above the critical temperature, the material will on average have a symmetrical unit cell, and no net dipole moment (sometimes called the paraelectric phase, in analogy to paramagnetism). In $BaTiO_3$, a "displacive" ferroelectric, the Ti ion is centered in an octahedron of O ions at high temperatures, but below the critical temperature the Ti^{4+} and Ba^{2+} ions in all unit cells undergo the same small displacement with respect to the O^{2-}. In "order–disorder" ferroelectrics, all unit cells are distorted at high temperatures, but are aligned along random directions. Upon cooling, these unit cell distortions become aligned along one direction, leading to a macroscopic dipole moment. Finally, "relaxor ferroelectrics" such as $PbMg_{1/3}Nb_{2/3}O_3$ (PMN) have polar nanodomains that are misoriented at high temperatures, but become aligned at low temperatures. Compared with displacive ferroelectrics, relaxor ferroelectrics exhibit a more gradual change in properties across the ferroelectric

Table 21.1 Nomenclature for some electronic and magnetic phase transitions				
	Structure			
Entity	Parallel ($k = 0$)	Antiparallel ($k \sim \pi/a$)	Disordered high T	Glass low T
Spin	Ferromagnet	Antiferromagnet (Ferri-)	Paramagnet	Spin glass
	Curie	Néel		
Charge		Verwey		
Electric	Ferroelectric	Antiferroelectric	Paraelectric	
dipole				Burns

Curie temperature, sometimes argued to be a type of glass transition. By analogy to
frustrated interactions in glasses, solid solutions of lead zirconium titanate (PZT) can
change with chemical composition from ferroelectric to antiferroelectric behavior.[4]

21.1.3 Complex Materials

Table 21.1 summarizes the magnetic and electronic phase transitions discussed so far, and
superparamagnetism or even superconductivity could be added to this list. Likewise, a
complementary table could be prepared for chemical phase transitions such as unmixing
and ordering, precipitation, and martensitic transformations. It is often interesting to
consider materials with phenomena from both tables, such as materials that have a
magnetic or electronic structure that changes with atomic structure.

Phase transitions are driven by discontinuities in the free energy of a material, or in
one of the derivatives of the free energy. So far we have been considering the change in
free energy to be dominated by one phenomenon or another, but there can be more than one
important contribution to the free energy at a phase transition. For example, the martensitic
transformation from fcc to bcc iron often involves a change from a paramagnetic to
ferromagnetic phase. Although a magnetic contribution can be added to the total free
energy of the martensitic transition, it may be easier to treat the magnetic effects separately.
For example, by studying the thermal shifts of phase boundaries over a range of magnetic
fields, the magnetic contribution to the free energy, $H \Delta M$, can be related to the contribution
$T \Delta S$ for the martensitic transformation.

A new term, "complex material," has come to mean a material with interdependent
properties that all contribute to the free energy. The concept of a complex material was
discussed in Section 17.2.4, where coupled fields of composition, strain, and magnetic
fields were added to the free energy. Electronic and spin degrees of freedom allow for
new types of complexity. All these properties change at a phase transition. Working
backwards from observation, however, the problem is more challenging. Just because
multiple properties are observed to change at a phase transition, it is not obvious how
or if these properties are interdependent, or if they all make significant contributions to the
free energy. The field of complex materials is an active field of research today.

---

[4] Intermediate compositions of PZT are of wide interest for engineering applications as sensors or displacement
transducers, and PZT is available as a machinable ceramic.

21.2 Exchange Interactions

21.2.1 Exchange in the Heisenberg Model

For the Heisenberg model of magnetism, the energy of a spin system is

$$E = -\frac{1}{2}\sum_i^N \sum_j^N J(\vec{r}_i - \vec{r}_j)\, \vec{s}_i \cdot \vec{s}_j, \tag{21.1}$$

and $J(\vec{r}_i - \vec{r}_j)$ is the "exchange energy."[5] In Eq. 21.1 the spin orientation at site i is taken as a vector, $\vec{s}_i$, and pairwise energies depend on the relative orientation of the spins at the two sites i and j. This Heisenberg model of magnetism may be appropriate when the spins are localized on atomic sites (the $\vec{r}_i$ and $\vec{r}_j$ in Eq. 21.1).[6] The spins are often dimensionless, so J has units of energy. (The factor 1/2, as usual, corrects for the error in running the sums to count contributions from both the i'–j' and the j'–i' pairs of spins, which are of course the same pair.) For magnets, J is expected to originate from "quantum mechanical exchange," but J is often treated as a model parameter.

21.2.2 Quantum Mechanical Exchange

The second, and arguably more authentic, usage of the word "exchange" in magnetism is quantum mechanical. It is the physical origin of a spin-dependent interaction, reflecting the Pauli principle for electron–electron interactions. For two electrons, the total wavefunction must be antisymmetric under the interchange of the spatial or spin coordinates. The exchange energy arises from a change to the repulsive Coulomb energy between electrons when two (or more) electrons have an antisymmetric total wavefunction.

As a preliminary to identifying the exchange energy, first consider the electrostatic energy of the Coulomb interaction of two electron densities

$$E_C = +\int_{\vec{r}_1}\int_{\vec{r}_2} |\psi_\alpha(\vec{r}_1)|^2 \frac{e^2}{|\vec{r}_1 - \vec{r}_2|} |\psi_\beta(\vec{r}_2)|^2\, d^3\vec{r}_1 d^3\vec{r}_2, \tag{21.2}$$

where the charge distributions of the two electrons are $|\psi_\alpha(\vec{r}_1)|^2$ and $|\psi_\beta(\vec{r}_2)|^2$, so small differential volumes of charge, $d^3\vec{r}_1$ and $d^3\vec{r}_2$, have a repulsive Coulomb interaction that goes as the inverse distance between them. This is as expected from classical electrostatics. For efficiency we switch to Dirac notation, but with subscripts 1 and 2 to denote each individual electron

$$E_C = +\langle\alpha_1|\langle\beta_2|\frac{e^2}{|\vec{r}_1 - \vec{r}_2|}|\alpha_1\rangle|\beta_2\rangle. \tag{21.3}$$

---

[5] This usage of the word "exchange" is similar to that of the "exchange energy" in the alloy problem, illustrated in Fig. 2.15.
[6] The Heisenberg model is not appropriate for understanding contributions to magnetism from free electrons.

Now we demand that the total wavefunction is antisymmetric in interchange of electrons between coordinates. For example, this two-electron wavefunction (using the "Hartree" form[7] of a product of one-electron wavefunctions)

$$|\text{Har}_{1,2}\rangle = |\alpha_1\rangle|\beta_2\rangle\big(\uparrow_1\downarrow_2 - \uparrow_2\downarrow_1\big). \tag{21.4}$$

is antisymmetric because interchanging the two electrons has the effect of multiplying $|\text{Har}_{1,2}\rangle$ by -1; watching the subscripts, $(\uparrow_1\downarrow_2 - \uparrow_2\downarrow_1) = -1(\uparrow_2\downarrow_1 - \uparrow_1\downarrow_2)$. If the spins of the two electrons are the same, however, the antisymmetry must be in the spatial coordinates, and the spatial factor of Eq. 21.4 is not satisfactory. We use instead the Hartree–Fock form for the spatial wavefunction of two electrons

$$|\text{HF}_{1,2}\rangle = \frac{1}{\sqrt{2}}\Big[|\alpha_1\rangle|\beta_2\rangle - |\alpha_2\rangle|\beta_1\rangle\Big], \tag{21.5}$$

which ensures that the wavefunction changes sign upon interchange of the two electrons. If this new form of Eq. 21.5 is used in the expression for the Coulomb energy, unlike Eq. 21.2 it gives four separate integrals

$$E_C = \langle \text{HF}_{1,2}|\frac{e^2}{|\vec{r}_1 - \vec{r}_2|}|\text{HF}_{1,2}\rangle, \tag{21.6}$$

$$2E_C = + \langle\alpha_1|\langle\beta_2|\frac{e^2}{|\vec{r}_1 - \vec{r}_2|}|\alpha_1\rangle|\beta_2\rangle - \langle\alpha_1|\langle\beta_2|\frac{e^2}{|\vec{r}_1 - \vec{r}_2|}|\alpha_2\rangle|\beta_1\rangle$$
$$- \langle\alpha_2|\langle\beta_1|\frac{e^2}{|\vec{r}_1 - \vec{r}_2|}|\alpha_1\rangle|\beta_2\rangle + \langle\alpha_2|\langle\beta_1|\frac{e^2}{|\vec{r}_1 - \vec{r}_2|}|\alpha_2\rangle|\beta_1\rangle, \tag{21.7}$$

but pairs of them evaluate to the same result because they involve only a relabeling of coordinates (integrals 1,4 and 2,3)

$$E_C = \langle\alpha_1|\langle\beta_2|\frac{e^2}{|\vec{r}_1 - \vec{r}_2|}|\alpha_1\rangle|\beta_2\rangle - \langle\alpha_1|\langle\beta_2|\frac{e^2}{|\vec{r}_1 - \vec{r}_2|}|\alpha_2\rangle|\beta_1\rangle, \tag{21.8}$$

or as explicit integrals

$$E_C = + \int_{\vec{r}_1}\int_{\vec{r}_2} |\psi_\alpha(\vec{r}_1)|^2 \frac{e^2}{|\vec{r}_1 - \vec{r}_2|}|\psi_\beta(\vec{r}_2)|^2 \, d^3\vec{r}_1 d^3\vec{r}_2$$
$$- \int_{\vec{r}_1}\int_{\vec{r}_2} \psi_\alpha^*(\vec{r}_1)\psi_\beta^*(\vec{r}_2)\frac{e^2}{|\vec{r}_1 - \vec{r}_2|}\psi_\alpha(\vec{r}_2)\psi_\beta(\vec{r}_1) \, d^3\vec{r}_1 d^3\vec{r}_2. \tag{21.9}$$

The first term in Eq. 21.9 is the same as Eq. 21.2, coming from the repulsive Coulomb energy of the charge densities of the two electrons.

The second term is new. It has a negative sign, reducing the Coulomb energy, although it is Coulombic in origin. Look carefully at the coordinates of the two wavefunctions in Eq. 21.9 – the two electrons are exchanged between ψ_α^* and ψ_α (and between ψ_β^* and ψ_β). This second term, the "exchange energy," appears only when the two spins are the same, so the two-electron wavefunction is antisymmetric in the spatial coordinates of the electrons. A physical interpretation of the exchange energy is that when the two electrons have the

[7] To be consistent with the Hartree–Fock wavefunction $|\text{HF}_{1,2}\rangle$ of Eq. 21.5, the symmetric spatial part should really be written as $[|\alpha_1\rangle|\beta_2\rangle + |\alpha_2\rangle|\beta_1\rangle]/\sqrt{2}$, but this leads to the same results in what follows.

same spin, the electrons are correlated in time so they tend to stay out of each other's way. The instantaneous correlations of their positions reduces their overlap and reduces the Coulomb energy. This is intuitively consistent with a blunt statement of the Pauli principle – two electrons of the same spin will avoid each other, and hence reduce their Coulomb repulsion energy.

21.2.3 Exchange Hole

It is useful to look at the electron densities in the Coulomb and exchange terms of Eq. 21.9

$$\rho_{\rm C} = |\psi_\alpha(\vec{r}_1)|^2 \, |\psi_\beta(\vec{r}_2)|^2, \tag{21.10}$$

$$\rho_{\rm ex} = -\psi_\alpha^*(\vec{r}_1)\psi_\beta^*(\vec{r}_2) \, \psi_\alpha(\vec{r}_2)\psi_\beta(\vec{r}_1). \tag{21.11}$$

We choose ψ_α^* and ψ_β^* to be orthogonal, so integrating Eq. 21.11 over all positions of the electrons gives zero. Suppose, however, that the electrons are instantaneously close together. A proper consideration of the correlations in electron motion requires a time variable, which raises the level of difficulty, but we can get insight by a change in spatial coordinate $\vec{r}_2 = \vec{r}_1 + \vec{\delta r}$ so

$$\rho_{\rm ex}(\vec{\delta r}) = -\psi_\alpha^*(\vec{r}_1)\psi_\beta^*(\vec{r}_1 + \vec{\delta r})\psi_\alpha(\vec{r}_1 + \vec{\delta r})\psi_\beta(\vec{r}_1). \tag{21.12}$$

We are interested in the behavior as $\vec{\delta r}$ becomes small, so the electrons become very close together. One approach is to approximate $\psi_\alpha(\vec{r}_1 + \vec{\delta r}) \simeq \psi_\alpha(\vec{r}_1) + 1/2\,\vec{\nabla}\psi_\alpha\,\vec{\delta r} + \cdots$. For small $\vec{\delta r}$

$$\rho_{\rm ex}(\vec{\delta r} \to 0) = -\int_{\delta r} \psi_\alpha^*(\vec{r}_1)\psi_\beta^*(\vec{r}_1)\psi_\alpha(\vec{r}_1)\psi_\beta(\vec{r}_1) \, {\rm d}^3\vec{\delta r} + O\vec{\delta r}, \tag{21.13}$$

$$\rho_{\rm ex}(\vec{\delta r} \to 0) \simeq -|\psi_\alpha(\vec{r}_1)|^2 \, |\psi_\beta(\vec{r}_2)|^2, \tag{21.14}$$

where $O\vec{\delta r}$ means terms proportional to $\vec{\delta r}$ (containing $\vec{\nabla}\psi_\alpha\,\vec{\delta r}$, for example), which we neglect for simplicity. For small $\vec{\delta r}$, the negative exchange density $\rho_{\rm ex}$ of Eq. 21.14 cancels the usual charge density $\rho_{\rm C}$ of Eq. 21.10, and there is no charge overlap. We deduce that each electron carries with it an "exchange hole," which excludes electrons of the same spin. More detailed work shows that this exchange hole is generally small on the spatial scale of the wavefunction.

21.2.4 Exchange and Shapes of Wavefunctions

Since electrons of like spin tend to avoid each other at any one time, their spatial wavefunctions can spread out a bit and interpenetrate each other more. Such spreading reduces the electron kinetic energy. Exchange therefore favors electrons of like spin in nearby orbitals (giving rise to "Hund's rule," described in the next section). In general, the exchange interaction polarizes the electron density, giving more electrons of up spin than down spin. The electron wavefunctions have shapes that differ for up and down spin. The altered shapes of the wavefunctions alter the potential energy (as in Eq. 21.2), which may further alter the shapes of the electron wavefunctions.

Consider a magnetic iron atom, with more $3d\uparrow$ than $3d\downarrow$ electrons. The $3d$ states have relatively high energy and angular momentum, so they are relatively far from the iron nucleus. The $1s$ electrons are closer to the nucleus. The exchange interactions between the $3d\uparrow$ electrons and the $1s\uparrow$ electrons allow the $1s\uparrow$ electrons to avoid the more abundant $3d\uparrow$ electrons. The shapes of the $1s\uparrow$ wavefunction expand outwards a bit, compared with the $1s\downarrow$ wavefunction. This is seen experimentally as an abundance of $1s\downarrow$ electrons at the iron nucleus,[8] and a magnetization at the nucleus that is opposite in direction from the lattice magnetization from the $3d\uparrow$ electrons.

21.2.5 Hund's Rule and Stoner Criterion for Ferromagnetism

The previous sections explained how exchange interaction can lower the energy of electrons of parallel spins, and a consequence is "Hund's rule" for filling the electron orbitals at an atom. Suppose the electron states are degenerate (have the same energy), such as the five $3d$-levels at a transition metal atom (when it is an isolated atom, not necessarily when in a solid). Hund's rule says that all states of the same spin are filled with electrons first, before any states of the opposite spin are filled.

The five $3d$-levels at an iron atom are broadened into bands when the iron atoms are in a solid. Both types of electron spins are present in the bonding states, so there is an energy competition with Hund's rule. As the bands are filled, do the electrons go into the bonding states regardless of spin, or do they fill antibonding states of the same spin as expected by Hund's rule? This depends on the width in energy of the band splitting. If the bonding–antibonding splitting is small, Hund's rule dominates and we expect ferromagnetism. This idea is formulated below as the "Stoner criterion" for band ferromagnetism. It is based on the electron density of states at the Fermi level. If this is high, then it does not cost much bonding energy to move a significant number of electrons into states of the majority spin ("spin up," or $\uparrow$, by convention). The more electrons of majority spin, the more favorable the energy from exchange interactions, and the $\uparrow$-spin band is shifted even more to lower energy, and the material becomes ferromagnetic. This competition between bonding energy and exchange energy allows a straightforward calculation of the criterion for ferromagnetism as follows.

The density of electron states $\rho(\epsilon)$ has units of electron states per eV on a per atom basis. Figure 21.2a shows how an energy range of δE at the Fermi level accounts for the number of electron states $\delta n = \rho(\epsilon_F)\,\delta E$. The total (positive) change in bonding energy of shifting these δn electrons by δE is $\Delta E_b = \delta E\,\delta n$, or

$$\Delta E_b = \frac{(\delta n)^2}{\rho(\epsilon_F)}. \tag{21.15}$$

---

[8] There is a hyperfine interaction, seen easily by ^{57}Fe Mössbauer spectrometry, where nuclear energy levels are altered by the electron density at the nucleus. Near the nucleus, nonrelativistic electrons have wavefunctions that go as r^l, where the angular momentum quantum number $l = 0$ for s-electrons. The electrons at the nucleus are primarily s-electrons. When a magnetic field is applied to the material, the lattice magnetization increases, but the hyperfine magnetic field decreases. This is explained by the "core polarization" effect described here.

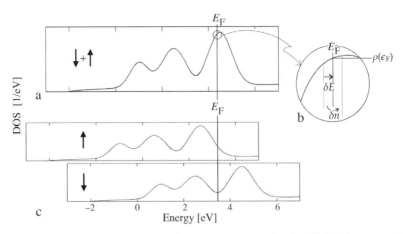

Figure 21.2 **(a)** Electronic DOS of nonmagnetic bcc Fe, showing the Fermi energy at a peak in the DOS. **(b)** Enlargement of the electron DOS at the Fermi energy, showing how the shift of δn electrons corresponds to the energy shift and DOS as $\delta E\, \rho(\epsilon_F)$. **(c)** An approximate spin-polarized electron DOS of bcc Fe. The DOS curves for $\uparrow$ and $\downarrow$ spin have approximately the same shape, but the $\uparrow$-spin states are shifted to lower energy.

This unfavorable bonding energy has competition from the favorable exchange energy. For no spin imbalance (and hence no ferromagnetism), there are $n/2$ $\uparrow$-spin electrons and $n/2$ $\downarrow$-spin electrons, where n is the total number of electrons. The exchange interactions are only between electrons of like spin, so without polarization the exchange energy has $\frac{1}{2}\,(n/2)^2$ contributions from each spin type. (The prefactor of $\frac{1}{2}$ corrects for the overcounting of electron pairs in both directions.) The exchange energy between a pair of parallel electron spins is assumed to be $-I$, which is negative.

We seek the difference in exchange energy ΔE_{ex} between the polarized and unpolarized conditions. In the polarized condition there are $n/2 + \delta n$ $\uparrow$-spin electrons, and $n/2 - \delta n$ $\downarrow$-spin electrons. The unpolarized case has two spin states with $n/2$ electrons each:

$$\Delta E_{\mathrm{ex}} = \frac{1}{2}\left(\frac{n}{2}+\delta n\right)^2 I + \frac{1}{2}\left(\frac{n}{2}-\delta n\right)^2 I - 2\frac{1}{2}\left(\frac{n}{2}\right)^2 I, \tag{21.16}$$

$$\Delta E_{\mathrm{ex}} = \delta n^2\, I. \tag{21.17}$$

Comparing these two energy contributions, the condition for ferromagnetism is

$$\Delta E_{\mathrm{ex}} > \Delta E_{\mathrm{b}}, \tag{21.18}$$

$$\delta n^2 I > \frac{\delta n^2}{\rho(\epsilon_F)}, \tag{21.19}$$

$$I\,\rho(\epsilon_F) > 1. \tag{21.20}$$

Equation 21.20 is the Stoner criterion for ferromagnetism. It occurs when the exchange interactions are large or there is a high density of states at the Fermi level. Within reason, this condition can predict the ferromagnetism of transition metals. Approximately, for bcc iron $I = 1$ eV and $\rho(\epsilon_F) = 1.5$/eV, consistent with the Stoner criterion.

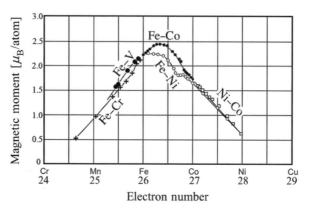

Average magnetic moment per atom of 3d transition metal alloys. The alloys were Fe–V, Fe–Cr, Fe–Co, Fe–Ni, and Ni–Co is added. After [354].

21.2.6 Slater–Pauling Curve

Perhaps surprisingly, the rigid band model of Section 6.5.3 can sometimes explain rather delicate behavior, such as how the magnetic moments of $3d$ transition metals change with alloying. Because bcc Fe is ferromagnetic, the states for ↑-spin and ↓-spin electrons are separated in energy. The ↑-spin states are lower in energy, and more of them are occupied, as shown in Fig. 21.2c, but both DOS curves are nearly the same shape. When bcc Fe is alloyed with Co, which donates electrons to the bands of both spins, the ↑-spin states fill more rapidly because of their higher DOS at E_F (approximately as shown in Fig. 21.2c). Adding more ↑-spins than ↓-spins gives an increase to the magnetic moments in the alloy. At an intermediate composition, around $c_{Co} = 0.3$, the filling of the ↑-spin states moves the Fermi level up the ↑-spin DOS curve to where there are fewer states. More of the additional electrons then go into the ↓-spin states. The magnetic moments therefore reach a maximum and then decrease with c_{Co}, as shown in Fig. 21.3. These plots of magnetization versus electron-to-atom ratio for Fe–Co and other $3d$ transition metal alloys are the "Slater–Pauling curve." It is reasonably explained by the filling of "rigid" electron bands that do not change their shape with alloying or band filling.

21.3 Thermodynamics of Ferromagnetism

The word "exchange" was first assigned to J in the Heisenberg model (Eq. 21.1), and we return to this meaning of exchange to understand the thermodynamics of magnetism. This J originates with quantum mechanical exchange, which is a feature of Coulomb interactions between electrons of like spins. For ferromagnetic metals such as Fe, Ni, and Co, the J in the Heisenberg model couples local magnetic moments at atoms through delocalized electrons in the band structure. There are many interesting trees in this forest of wavefunctions and their interactions, but to explain quantum mechanical exchange, the analysis of Section 21.2 did not venture beyond electron–electron interactions on the same atom. This is intraatomic exchange. For iron, Fig. 21.2c showed an intraatomic exchange of order 1 eV. In units of $k_B T$, this corresponds to a temperature of order 10^4 K, well beyond temperatures of concern to us here. We therefore assume that the magnetic moments at atoms are robust, and do not change with temperature.

Intraatomic exchange gives the magnitude of $\vec{s}_i$ in Eq. 21.1, which is the unpaired spin at the atom i. The high strength of the intraatomic exchange allows the assumption that the magnitude of $\vec{s}_i$ is unchanged with temperature. Aligning these magnetic moments within groups of atoms is a separate problem, however. The important interatomic interactions between neighboring spins of different alignments are parameterized by the J in the Heisenberg model. The alignment energies are of order $k_B T$ and are central to the thermodynamics of magnetism, such as understanding the Curie transition in ferromagnets. The interatomic exchange between local atomic spins involves an exchange interaction with an intermediary, such as conduction electrons in metals [355–357] or oxygen atoms in ionic crystals [358–361]. We bypass these interesting trees to avoid getting lost in the forest.

21.3.1 Local Moment Ising Problem

Exchange and Local Moments

A central feature of chemical ordering in an alloy (Section 2.9.1) is that the energy cost to interchange an A-atom with a B-atom on the other sublattice depends on the degree of chemical order, given by the long-range order parameter, L. Figure 2.19 conveys the idea. At high temperatures where $L = 0$, there is no change in energy when swapping a pair $\{A\alpha, B\beta\}$ to obtain $\{B\alpha, A\beta\}$. At low temperatures where the sublattices α and β are populated primarily by one species of atoms (so $L \neq 0$), the energy difference is large. In equilibrium, the probability of a pair of atoms being on the correct sublattice depends on all the other atoms, which control L. Chemical ordering is a cooperative phase transition.

Here we treat the problem of ferromagnetism in essentially the same way, using an Ising lattice with local magnetic moments. Two spin orientations[9] are assumed, and the energy to flip a single spin depends on the alignment of spins at the neighboring sites. Below a Curie temperature, magnetic order develops cooperatively, and the material has a bulk magnetic moment. Table 21.2 compares terms in the problems of chemical ordering and magnetic ordering.

For the alloy problem the solute concentrations are conserved, and we enforced atom conservation by interchanging A-atoms and B-atoms in pairs. For the Ising problem for ferromagnetism, a ↑-spin changes to ↓-spin without such conservation. An additional consequence is that sublattices are not required for ferromagnetism. In analogy to the alloy problem, however, we first calculate the energy cost of a flipped spin by considering the spin alignments at its neighboring sites.

---

[9] It might be argued that two spin orientations are consistent with a local spin of $\pm 1/2$, but this quantum mechanics justification is naïve.

Table 21.2 State and thermodynamic variables for chemical ordering and ferromagnetism

Variable	Ordering (Equiatomic)	Ferromagnetism
Numbers on sites	$R \equiv N_A^\alpha = N_B^\beta$, $W \equiv N_B^\alpha = N_A^\beta$	$N^\uparrow$, $N^\downarrow$
Order parameter	$L \equiv (R - W)/(N/2)$	$L \equiv (N^\uparrow - N^\downarrow)/N$
Energies	e_{AB} , $(e_{AA} + e_{BB})/2$	$-J$, $+J$
Thermodynamic differences	$4V \equiv e_{AA} + e_{BB} - 2e_{AB}$	$4J$

Alignment Energies

In the Ising model, the energy of a $\uparrow$-spin depends on the orientations of its z first-nearest neighbors (1nn), and similarly for a $\downarrow$-spin

$$e_\uparrow = -J(n_\uparrow - n_\downarrow), \tag{21.21}$$

$$e_\downarrow = -J(n_\downarrow - n_\uparrow) = J(n_\uparrow - n_\downarrow), \tag{21.22}$$

where there are $n_\uparrow$ neighbors with $\uparrow$-spin, $n_\downarrow$ with $\downarrow$-spin, and $n_\uparrow + n_\downarrow = z$. Here J is positive, so two spins of like sign have an energetically favorable exchange interaction.[10] Because of the difficulty in knowing the numbers $n_\uparrow$ and $n_\downarrow$ around each individual atom, $n_\uparrow$ and $n_\downarrow$ are approximated as their average over all sites. This is the "point approximation" discussed in Sections 2.9.2 and 7.2. For magnetic thermodynamics, an applied magnetic field, B, can be included, giving a favorable energy $-\mu B$ for a $\uparrow$-spin and an unfavorable $+\mu B$ for a $\downarrow$-spin (where a spin has a magnetic moment, μ)

$$e_\uparrow = -Jz\left(\frac{N_\uparrow}{N} - \frac{N_\downarrow}{N}\right) - \mu B, \tag{21.23}$$

$$e_\downarrow = +Jz\left(\frac{N_\uparrow}{N} - \frac{N_\downarrow}{N}\right) + \mu B. \tag{21.24}$$

The equations of the alloy problem for comparison are Eqs. 2.51 and 2.52, which are more complicated because of solute conservation. The alloy problem also includes two sublattices, so instead of the definition in Eq. 2.50, the long-range order parameter is now defined as

$$L \equiv \frac{N_\uparrow - N_\downarrow}{N} . \tag{21.25}$$

As before, this L can range over $-1 < L < 1$, since $N_\uparrow$ and $N_\downarrow$ can range from 0 to N. In terms of L, the local energies are

$$e_\uparrow = -JzL - \mu B, \tag{21.26}$$

$$e_\downarrow = +JzL + \mu B. \tag{21.27}$$

---

[10] A comparison with Eqs. 2.19 and 2.20 is useful.

21.3.2 Critical Temperature (Curie Temperature)

At a finite temperature, the relative probabilities of a ↑-spin and a ↓-spin are their Boltzmann factors

$$\frac{N_\uparrow}{N_\downarrow} = \frac{e^{-e_\uparrow/k_B T}}{e^{-e_\downarrow/k_B T}} = \frac{e^{(JzL+\mu B)/k_B T}}{e^{-(JzL+\mu B)/k_B T}}, \tag{21.28}$$

$$\frac{N_\uparrow}{N_\downarrow} = e^{2(JzL+\mu B)/k_B T}. \tag{21.29}$$

With Eq. 21.25 (and $N_\uparrow + N_\downarrow = N$), the ratio $N_\uparrow/N_\downarrow$ is

$$\frac{1+L}{1-L} = e^{2(JzL+\mu B)/k_B T}, \tag{21.30}$$

reminiscent of Eq. 2.55 for the chemical ordering problem. This is the practical end of a calculation of $L(T)$, but sometimes it is convenient to use the obscure identity

$$\tanh^{-1} x = \frac{1}{2} \ln\left(\frac{1+x}{1-x}\right) \tag{21.31}$$

to present a more compact expression

$$L = \tanh\left(\frac{JzL+\mu B}{k_B T}\right). \tag{21.32}$$

To obtain the critical temperature for ordering, we use the same trick as for Eq. 2.58. At the critical temperature, L will be infinitesimally small, so Eq. 21.30 can be linearized. After taking the logarithms of both sides of Eq. 21.30

$$L = \frac{JzL+\mu B}{k_B T}. \tag{21.33}$$

When there is no applied magnetic field ($B = 0$) there is a unique critical temperature

$$T_C = \frac{Jz}{k_B}, \tag{21.34}$$

which compares directly to Eq. 2.58 for the alloy problem. (The uppercase C in the subscript denotes Curie.) Equation 21.34 predicts the critical condition when the thermal energy $k_B T$ equals the exchange energy over the first nearest neighbor shell, zJ. An applied magnetic field stabilizes magnetic order at higher temperatures, as discussed next.

21.3.3 High-Temperature Magnetic Susceptibility

Equation 21.30 shows that an applied magnetic field ($B \neq 0$) will always induce some magnetic order, even at the highest temperatures. Assuming L and B are small, however, Eq. 21.33 can be used for $T > T_C$. Substituting T_C from Eq. 21.34 into Eq. 21.33

$$L = \frac{T_C L}{T} + \frac{\mu B}{k_B T}, \qquad (21.35)$$

$$\frac{L}{\mu B} = \frac{1}{k_B(T - T_C)}, \qquad (21.36)$$

$$\frac{M}{\mu^2 N B} = \frac{1}{k_B(T - T_C)}, \qquad (21.37)$$

where the magnetization M is defined as the number of unpaired spins times the magnetic moment per spin, μ (i.e., $M \equiv \mu L N$). Another definition is the magnetic susceptibility, $\chi \equiv M/B$, which is the change in magnetization per applied field

$$\chi(T) = \frac{\mu^2 N}{k_B(T - T_C)}. \qquad (21.38)$$

Equation 21.38 is the "Curie–Weiss law" for ferromagnets above the critical temperature. It is often used to fit experimental data of $\chi(T)$. The fits usually give values of T_C that are slightly larger than the zero-field result, but the local moment Ising model is generally successful at high temperatures.

21.3.4 Low-Temperature Behavior (Classical)

Magnetization in the low-temperature limit can be calculated easily with Eq. 21.30. Inverting it and setting $B = 0$ for cooperative ferromagnetism

$$\frac{1 - L}{1 + L} = e^{-2\frac{JzL}{k_B T}}. \qquad (21.39)$$

Substituting the result of Eq. 21.34, $T_C = zJ/k_B$, and because $L \simeq 1$, it is safe to approximate $1 + L \simeq 2$ and $LT_C \simeq T_C$

$$\frac{1 - L}{2} = e^{-2\frac{T_C}{T}}, \qquad (21.40)$$

$$L = 1 - 2e^{-2\frac{T_C}{T}}. \qquad (21.41)$$

This low-temperature result is straightforward to test experimentally, but it proves inaccurate. The problem is that, while the high-temperature behavior can be approximated with classical statistical mechanics of independent spins, the low-temperature behavior requires attention to the quantum mechanics of excitations of a periodic system, known as "spin waves."

21.4 Spin Waves

The Heisenberg model for a linear chain of spins has a low-temperature behavior that is more interesting than flipped individual spins. The energy of a flipped spin is $+4J$ in a first-neighbor model, where $-J$ is the exchange interaction energy for like spins, and the spin has two neighbors of opposite spin. This energy proves to be higher than the energy of a spin wave with long wavelength, as shown next.[11]

There are differences in the treatment of Section 6.3 for the energy of bands $\epsilon(k)$ and the proper treatment of spin waves, but there are so many similarities that the full argument is not repeated here. The first difference is that a spin wave is not static and it has the form $\exp[-i(kx - \omega_k t)]$, with wavevector k and energy $\hbar\omega_k = \epsilon_k$.[12] Second, the spin wave needs another dimension in which to oscillate: y-directed spins in a chain along x need to twist into the z-direction, or perhaps rotate their tips in an x–z plane. (The latter are sometimes described as the motion of wheat stalks in a light wind – groups of wheat stalks rotate around their vertical axes, and there is a characteristic distance between stalks tilted at the same angle at the same time.) The vectorial nature of spin waves, and the vectorial character of a magnetic torque, lead to separate equations of motion for the spins along different coordinates. Nevertheless, the energy has the same form as in Eq. 6.59 (shown in Fig. 6.3)

$$\hbar\omega_k = -J\cos(\vec{k} \cdot \vec{a}), \tag{21.42}$$

or by setting the zero of energy at $k = 0$ where all spins are aligned

$$\hbar\omega_k = J[1 - \cos(\vec{k} \cdot \vec{a})]. \tag{21.43}$$

This result for a linear chain would need a numerical factor like $z/2$ to account for multiple neighbors on a 2D or 3D lattice. For an isotropic spin wave of small k (and low energy), Eq. 21.43 becomes

$$\hbar\omega_k = J\frac{1}{2}(ka)^2 \equiv \epsilon_k, \tag{21.44}$$

$$k = \sqrt{\frac{2\epsilon_k}{Ja^2}}. \tag{21.45}$$

To calculate the numbers of spin waves at low T, we need their energy spectrum. The problem is similar to that of a free electron gas, described in Section 6.4.1 with the help of Fig. 6.4. The quantized spin waves must match the boundaries of a box of length L, but for consistency with Eq. 21.43 the box is assumed a periodic one. This has an impact on the number of spin waves in k-space, modifying the result of Eq. 6.65 for a static box of length L, where a function like $\sin(\pi x/L)$ satisfies the boundary conditions at both 0 and L. This is not so for a periodic box because $\sin(\pi x/L)$ has opposite slopes at 0 and L. Only even integers are allowed for the k-vectors, such as $\sin(2\pi x/L)$ and $\sin(4\pi x/L)$. This halves the number of k-points along x, and similarly for y and z. The density of spin wave states as a function of k for a periodic box is 1/8 that of Eq. 6.66

$$\frac{dN}{dk} = \frac{1}{16}\pi k^2 \left(\frac{L}{\pi}\right)^3 = \frac{1}{2}\pi k^2 \left(\frac{L}{2\pi}\right)^3, \tag{21.46}$$

---

[11] Another point is that a single flipped spin is not an eigenstate of a periodic structure. Recall from Section 6.3 that the eigenstates of a Hamiltonian with a periodic potential must satisfy Bloch's theorem, so they have a phase factor of $\exp(-ikx)$.

[12] With increasing t, the same value of $(kx - \omega_k t)$ is obtained at larger x, so this wave propagates to increasing x.

which is a general result for quantized waves in a 3D periodic box. Using Eq. 21.44 to substitute for both k and dk in Eq. 21.46 gives the energy spectrum for spin waves, which has the same $\sqrt{\epsilon}$ dependence as was found for the free electron density of states[13] (Eq. 6.72)

$$\rho_{sw}(\epsilon)\, d\epsilon = \frac{1}{8\sqrt{2}\pi^2}\left(\frac{L}{a\sqrt{J}}\right)^3 \sqrt{\epsilon}\, d\epsilon, \tag{21.47}$$

$$\rho_{sw}(\epsilon)\, d\epsilon = K\sqrt{\epsilon}\, d\epsilon, \quad \text{where} \quad K \equiv \frac{1}{8\sqrt{2}\pi^2}\left(\frac{L}{a\sqrt{J}}\right)^3. \tag{21.48}$$

At low temperatures only a few spin waves are expected, so the spin waves are treated as independent. Their numbers are calculated as boson particles of zero chemical potential, as for phonons. The total number of spin waves versus temperature, $n_{sw}(T)$, is a thermally weighted average

$$n_{sw}(T) = \int_0^\infty \frac{1}{\exp\left(\frac{\epsilon}{k_B T}\right) - 1}\, \rho_{sw}(\epsilon)\, d\epsilon, \tag{21.49}$$

i.e., the number of states in the energy interval $\rho_{sw}(\epsilon)\, d\epsilon$ times the boson thermal occupancy of a state of energy ϵ. With Eq. 21.48 for $\rho_{sw}(\epsilon)$

$$n_{sw}(T) = K\int_0^\infty \frac{1}{\exp\left(\frac{\epsilon}{k_B T}\right) - 1}\, \sqrt{\epsilon}\, d\epsilon. \tag{21.50}$$

We use a standard method of converting the integral to a dimensionless one, so all important physical variables appear in a prefactor outside the integral. Making the substitutions $x \equiv \epsilon/k_B T$, $\epsilon = x k_B T$, and $d\epsilon = k_B T\, dx$

$$n_{sw}(T) = K\int_0^\infty \frac{1}{e^x - 1}\, \sqrt{k_B T}\, \sqrt{x}\, k_B T\, dx, \tag{21.51}$$

$$n_{sw}(T) = K\,(k_B T)^{3/2}\int_0^\infty \frac{1}{e^x - 1}\, \sqrt{x}\, dx. \tag{21.52}$$

The dimensionless definite integral can be evaluated numerically (or by evaluating a Riemann zeta function), and has the value 2.315. Bringing K from Eq. 21.48 and evaluating constants

$$n_{sw}(T) = 0.02073 \left(\frac{L}{a}\right)^3 \left(\frac{k_B T}{J}\right)^{3/2}. \tag{21.53}$$

Such numerical accuracy is not warranted, since we need to multiply J by a coordination number for at least one nearest-neighbor shell.

Equation 21.53 shows that at low temperatures, the number of spin waves increases with temperature as $T^{3/2}$, so at low temperatures the decrease in magnetization goes as

---

[13] Both spin wave and free electron energies scale as k^2 – the energy of a free electron is $(\hbar^2/2m)\,k^2$, compared with $(Ja^2/2)\,k^2$ for spin waves (Eq. 21.44).

$T^{3/2}$. Each quantized spin wave (called a "magnon") accounts for one flipped spin (the minimum change in the spin system), so

$$M(T) = [N - n_{\mathrm{sw}}(T)]\mu. \tag{21.54}$$

In terms of the magnetization per unit volume

$$M_{\mathrm{V}}(T) = M_{\mathrm{V}}(0) - \kappa_{\mathrm{M}} \left(\frac{T}{T_{\mathrm{C}}}\right)^{3/2}, \tag{21.55}$$

where $J \sim T_{\mathrm{C}}$ from Eq. 21.34, and κ_{M} is a material constant. This low-temperature result is known as the "Bloch $T^{3/2}$ law" for ferromagnets. It is different from the Ising lattice result for the order parameter at low temperatures, Eq. 21.41, which assumed that individual spins are flipped at specific sites.

21.5 Thermodynamics of Antiferromagnetism

21.5.1 Néel Transition

The Néel temperature for an antiferromagnet can be calculated with an Ising model in much the same way as for a ferromagnet. There are several possibilities for the spin structure of an antiferromagnet, however, and we select only one structure, shown in 2D as

$$
\begin{array}{cccccccccc}
\uparrow & \downarrow & \uparrow & \downarrow & \uparrow & \downarrow & \uparrow & \downarrow & \uparrow & \downarrow \\
\downarrow & \uparrow & \downarrow & \uparrow & \downarrow & \uparrow & \downarrow & \uparrow & \downarrow & \uparrow \\
\uparrow & \downarrow & \uparrow & \downarrow & \uparrow & \downarrow & \uparrow & \downarrow & \uparrow & \downarrow \\
\downarrow & \uparrow & \downarrow & \uparrow & \downarrow & \uparrow & \downarrow & \uparrow & \downarrow & \uparrow \\
\uparrow & \downarrow & \uparrow & \downarrow & \uparrow & \downarrow & \uparrow & \downarrow & \uparrow & \downarrow \\
\downarrow & \uparrow & \downarrow & \uparrow & \downarrow & \uparrow & \downarrow & \uparrow & \downarrow & \uparrow \\
\end{array}
$$

This is the same chessboard structure as the ordered alloy of Fig. 2.19a, and is composed of two interpenetrating square sublattices, oriented diagonally on the page. Following Fig. 2.19a, one sublattice, occupied by $\uparrow$-spins, is denoted α, and the other with $\downarrow$-spins is denoted β. On this structure, all first-nearest neighbors (1nn) of a spin are located on the other sublattice. With perfect antiferromagnetic order, all $\uparrow$-spins have four 1nn $\downarrow$-spins, and vice versa. This is favorable for a positive 1nn exchange interaction, J, using a sign convention consistent with Eq. 21.1. The energies for the $\uparrow$-spin and $\downarrow$-spin, analogous to Eqs. 21.21 and 21.22, are therefore

$$e_\uparrow = J(n_\uparrow - n_\downarrow), \tag{21.56}$$

$$e_\downarrow = J(n_\downarrow - n_\uparrow), \tag{21.57}$$

where there are $n_\uparrow$ neighbors with $\uparrow$-spin and $n_\downarrow$ with $\downarrow$-spin as before. The signs are switched compared with the ferromagnet problem of Eqs. 21.21 and 21.22.

Like our analysis of the ferromagnet, we use again the point approximation (also called the "mean field," or "Gorsky–Bragg–Williams" approximation). The number of ↑-spin first neighbors is $n_\uparrow$, and the LRO parameter is defined as

$$n_\uparrow = \frac{zN_\uparrow}{\frac{N}{2}}, \tag{21.58}$$

$$L \equiv \frac{N_\uparrow^\alpha - N_\downarrow^\alpha}{\frac{N}{2}}. \tag{21.59}$$

Since the α-sublattice (↑-spin) accounts for half of all sites, this LRO parameter differs by a factor of 2 from the definition Eq. 21.25 for ferromagnetism, but is actually the same as Eq. 2.50 for the problem of chemical ordering. Again recall that for chemical ordering of an alloy there was a strict conservation of solute, which ensured symmetry in occupancies of the two sublattices, e.g., $N_\uparrow^\alpha = N_\downarrow^\beta$ in the present notation. There is no conservation of spin, but there is symmetry of the sublattices, at least in the absence of an applied magnetic field. A statistical average over the equivalent sublattices ensures $N_\uparrow^\alpha = N_\downarrow^\beta$ for the antiferromagnet. Continuing in the point approximation, the equivalents of Eqs. 21.26 and 21.27 are

$$e_\uparrow^\alpha = +JzL, \tag{21.60}$$

$$e_\downarrow^\beta = +JzL, \tag{21.61}$$

$$e_\uparrow^\beta = -JzL, \tag{21.62}$$

$$e_\downarrow^\alpha = -JzL. \tag{21.63}$$

If we calculate the ratio of $e_\uparrow^\alpha$ to $e_\uparrow^\beta$ by the ratio of their Boltzmann factors as in Eq. 21.28, the equations are the same as for the ferromagnet, giving a result analogous to Eq. 21.30

$$\frac{1+L}{1-L} = e^{2(-JzL)/k_BT}. \tag{21.64}$$

The Néel temperature therefore has the same form as the Curie temperature, recognizing the change in sign of J for the 1nn exchange energy (and again $B = 0$)

$$T_N = \frac{-Jz}{k_B}. \tag{21.65}$$

21.5.2 Antiferromagnetic Susceptibility

At low temperatures, the magnetic susceptibility of an antiferromagnet depends on the orientation of the applied field with respect to the spin directions. If the magnetic field is along the direction of either ↑-spins or ↓-spins (they are equivalent), a change in magnetization requires flips of spins, which becomes energetically more difficult when there are fewer disordered spins at low temperatures. On the other hand, a magnetic field perpendicular to the spin direction causes the spins to tilt towards the direction of the applied field. This requires less energy, and hence the perpendicular susceptibility of an antiferromagnet is larger than the parallel susceptibility, and is less dependent on temperature.

At low temperatures, magnetic disorder from flips of individual spins is not expected for the same reason as for the ferromagnet – spin waves are the lowest-energy excitations of an antiferromagnet. (Also, the periodicity of spin waves allows them to be eigenstates of a periodic solid.) There is, however, a fundamental difference between spin waves in ferromagnets and antiferromagnets, originating with the nature of magnetic forces between adjacent spins. A spin creates a magnetic field along the direction of its orientation, assumed the z-direction. A spin at site a feels a magnetic torque from its neighboring spins along the x-direction at $a - 1$ and $a + 1$ through the magnetic cross-product relation $(\vec{\tau} = \vec{\mu} \times \vec{B})$. Its angular momentum $\hbar \vec{s}_a$ changes in proportion to this torque

$$\hbar \frac{d\vec{s}_a}{dt} = 2J \left(\vec{s}_a \times \vec{s}_{a-1} + \vec{s}_a \times \vec{s}_{a+1} \right). \tag{21.66}$$

This is the basic relationship from which we could have developed Eq. 21.43. Without working the details, however, we can see quickly that an antiferromagnet, with its opposite signs for spins at a and $a + 1$, will have different magnetic torques than a ferromagnet. On the other hand, the y-component of the spin at the same site a gives the same x-component of torque at a as for the ferromagnet. Compared with the ferromagnet, there is a difference in sign between the local and neighboring contributions to the torque. Factors e^{ika} and e^{-ika} for the energy contributions in the equivalent of Eq. 6.58 have opposite signs, giving an energy that goes as $\sin(ka)$, unlike $\cos(ka)$ for the ferromagnetic spin wave.

At small k, the dispersion relations $\epsilon(k)$ for the antiferromagnetic spin waves are linear in k, whereas the ferromagnetic spin waves are quadratic in k. The population of spin waves with temperature is therefore different, so instead of the "Bloch $T^{3/2}$ law" for ferromagnets, for antiferromagnets the statistical mechanics causes the internal magnetization to decrease with temperature as T^3. The picture of antiferromagnetic spin waves being similar to phonons is not necessarily reliable near $k = 0$. Anisotropy in the magnetic interactions can cause a gap so that a finite energy, ϵ_g, is required to excite even long-wavelength spin waves in an antiferromagnet. In this case, temperature will activate spin waves with a probability $e^{-\epsilon_g/k_B T}$, giving an exponential decrease of magnetization with temperature.

At temperatures above T_N, the antiferromagnet has lost its long-range magnetic order, and has the magnetic susceptibility of a paramagnetic phase. Instead of the Curie–Weiss behavior of Eq. 21.38, however, the antiferromagnetic susceptibility above the Néel temperature is

$$\chi(T) = \frac{\mu^2 N}{k_B(T + T_N)}. \tag{21.67}$$

21.6 Dzyaloshinskii–Moriya Interactions and Skyrmions

21.6.1 D-M Vector Interaction

Our models of exchange have so far considered energies proportional to the alignment of spins as $\vec{s}_i \cdot \vec{s}_j$, or spin alignments in applied fields as $\vec{s}_i \cdot \vec{B}$. The Dzyaloshinskii–Moriya

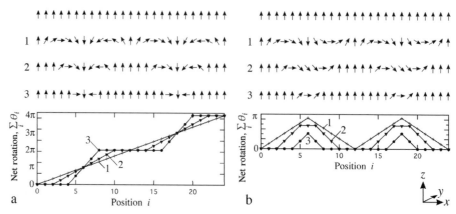

Figure 21.4   (a) Ferromagnetic chain drawn above chains 1, 2, 3 that develop skyrmion defects. (b) Ferromagnetic chain above three general domain boundaries. For both a and b, graphs at bottom are the cumulative rotation angles, $\sum_i \theta_i$, for the three chains above them (starting at the spin at $i = 0$ in the chain).

(D-M)[14] interaction [362–364] is fundamentally different, with an energy proportional to the cross-product of spin vectors

$$H_{\text{DM}} = -\vec{D}_{ij} \cdot \left(\vec{s}_i \times \vec{s}_j \right) . \tag{21.68}$$

Evidently $\vec{D}_{i,j} = -\vec{D}_{j,i}$ to account for the reversed direction of the cross product when the vectors are taken in reverse order. The magnitude of $\vec{D}$ is usually small. It originates with a spin–orbit coupling between atom sites, and is spatially asymmetric.[15]

Even if the energy of the Dzyaloshinskii–Moriya interaction is smaller than the energy of the Heisenberg exchange interaction (Eq. 21.1), it can create interesting magnetic structures and magnetic phase transitions. Consider a row of spins with a ferromagnetic exchange interaction J, plus a Dzyaloshinskii–Moriya interaction with magnitude D. Assume only first-nearest-neighbor interactions. The energy is

$$E = \sum_i^N \left[-J_{i,i+1}\, \vec{s}_i \cdot \vec{s}_{i+1} - \vec{D}_{i,i+1} \cdot \vec{s}_i \times \vec{s}_{i+1} \right]. \tag{21.69}$$

We seek the ground state of the 1D chain oriented along $\hat{x}$ at the top of Fig. 21.4, with the spins confined to lie in the x–z-plane (coordinate system is in lower right). The spins are free to make an angle θ with respect to their neighbors. This angle would be zero if $D = 0$, and the ground state would be aligned spins as expected for a ferromagnet. For nonzero $\vec{D}_{i,i+1} = D\hat{y}$, however, Eq. 21.69 becomes

$$E = \sum_i^N \left[-J\, s^2 \cos\theta - D\hat{y} \cdot s^2 \sin\theta\, \hat{y} \right], \tag{21.70}$$

$$E = -N\, s^2 \left[J \cos\theta + D \sin\theta \right], \tag{21.71}$$

---

[14] Pronunciations: jell - o' - shin - ski, mo - re' - ya, skurm - e' - un .

[15] In essence, two spins interact through a spin–orbit interaction with an intervening atom, but this atom does not lie along the line between the two spins. A vector direction for the Dzyaloshinskii–Moriya interaction can be justified with the line between the spins and the displacement of the intervening atom off the midpoint of this line.

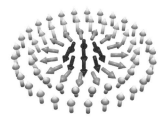

Figure 21.5 Spin configuration of a hedgehog skyrmion in a 2D spin array. Spins away from the center are $\uparrow$-spins. Image courtesy of F. Büttner.

where θ is the rotation of $\vec{s}_{i+1}$ with respect to $\vec{s}_i$. The $\theta_{\min}$ giving the lowest E is found where the derivative with respect to θ is zero

$$\frac{\mathrm{d}E}{\mathrm{d}\theta} = 0 = -N\,s^2\left[-J\sin\theta_{\min} + D\cos\theta_{\min}\right], \tag{21.72}$$

$$\frac{\sin\theta_{\min}}{\cos\theta_{\min}} = \frac{D}{J}, \tag{21.73}$$

$$\theta_{\min} = \tan^{-1}\left(\frac{D}{J}\right) \simeq \frac{D}{J}, \tag{21.74}$$

where the approximation is reasonable for $D < J$ and small θ. A chiral spin structure is predicted to be the ground state. The clockwise rotation of Fig. 21.4a.1 is favored for positive D, and the anticlockwise rotation is unfavorable.

21.6.2 Topological Protection

There is a fundamental difference between the spin structures of Figs. 21.4a.1 and 21.4b.1 that is most evident around spin number 6 in the two chains. Consider the cumulative rotation down the 1D chain, $\theta = \sum_i \theta_i$. This is graphed at the bottoms of Figs. 21.4a and b. The total θ increases steadily in Fig. 21.4a, but goes up and down in Fig. 21.4b. These two structures respond very differently to an applied field, or any tendency to grow a region of $\uparrow$-spins at the expense of $\downarrow$-spins. This is shown by the spin structures numbered 2 and 3 in Figs. 21.4a and b. For Fig. 21.4b all spins can be gradually moved to the $\uparrow$-spin orientation by small, continuous changes, and all inverted spins can be eliminated continuously. This is not possible for the spin rotation structure in Fig. 21.4a. Enlarging the regions of $\uparrow$-spins leads to two tight, knot-like zones that cannot be reduced without a discontinuous rotation of the inverted spin, or by propagation of the defects out of the end of the chain. The spin defects in Fig. 21.4a are "topologically protected," and are expected to have some robustness against thermal disorder or applied magnetic fields.

Building on the image of Fig. 21.4a, a topologically protected defect in a two-dimensional spin structure is depicted in Fig. 21.5. This defect is called a "skyrmion," specifically a "hedgehog skyrmion." (A vortex skyrmion is also possible in two dimensions.) A line through its center gives a 2π rotation of spins, so it has topological protection.[16]

---

[16] The array of spins around it are all pointing up, with rotation $\theta = 0$, so if hedgehog skyrmions are brought together, it is not appropriate to add their spin rotations as for the linear chain of Fig. 21.4a.

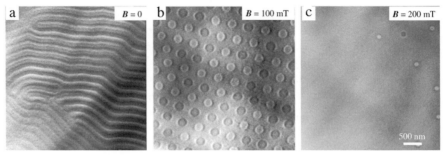

Figure 21.6 Lorentz TEM images observed for Sc-doped barium ferrite showing changes in magnetic domain structure caused by increasing magnetic field. (**a**) Stripe domains at zero field. (**b**) Skyrmion lattice at 100 mT. (**c**) Onset of ferromagnetism at 200 mT. From [365].

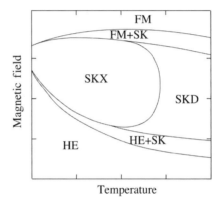

Figure 21.7 Possible B–T phase diagram for spin structures including skyrmion crystal (SKX), disordered skryrmions (SKD), helical phase (HE), and ferromagnetic phase (FM).

21.6.3 Magnetic Phase Diagram

In three dimensions it is possible to stack the structures of Fig. 21.5 into vertical tubes. Figure 21.6 shows images made through a material with a skyrmion phase at intermediate magnetic field. The low-field phase is an array of striped magnetic domains with alternating spins, and a high applied magnetic field gives a ferromagnetic structure. The skyrmions form at intermediate fields, and Fig. 21.6b shows tubular structures oriented end-on, as they extend out of the plane of the paper. A hexagonal skyrmion crystal structure is evident.

Magnetic structures can change their configurations with temperature, pressure, or applied magnetic field. Skyrmions have some freedom to behave as independent particles, or quasiparticles, giving the material a source of configurational entropy. An approximate B–H phase diagram is shown in Fig. 21.7. This was drawn assuming that the skyrmion phases have higher entropy than the phase with helical spins or the ferromagnetic phase, and hence an increased range of stability at higher temperatures. Figure 21.7 has some similarities to the phase diagram of $Fe_{0.5}Co_{0.5}Si_1$ [366], but Fig. 21.7 includes a disordering of the skyrmion crystal at higher temperatures. The phase boundaries between the skyrmion phase and the helical or ferromagnetic phases are characteristic of first-order phase transitions, since the structures are discontinuously different. In other cases such as MnSi, there is a conical spin structure that forms at low temperatures and intermediate magnetic fields. The skyrmion crystal develops inside the conical phase region at higher temperatures [367]. Other spin structures such as worm-like short tubes may appear, especially when T or B brings the material near a phase boundary.

21.7 Thermodynamics of Ionic Crystals

Here we consider an ionic material in an electrostatic potential $\Phi(\vec{r})$. We show that the electric charges are, by themselves, less interesting than the dipole moments formed with positive and negative charges. The thermally averaged orientation of an isolated dipole in an electric field is calculated, but cooperative effects and ferroelectric transitions are described in Section 21.8.

The charge distribution in the material is $\rho(\vec{r})$, and the electrostatic energy is

$$E_{\text{elec}} = \int_{-\infty}^{\infty} \rho(\vec{r}) \, \Phi(\vec{r}) \, d^3 r. \tag{21.75}$$

Expanding $\Phi(\vec{r})$ in a Taylor series[17] about a selected $\vec{r}_0$

$$\Phi(\vec{r}_0 + \vec{r}) = \Phi(\vec{r}_0) + \vec{r} \cdot \vec{\nabla}\Phi|_{\vec{r}_0} + \cdots, \tag{21.76}$$

where the next term would be a quadrupole term that requires tensor notation. Using the electric field $\vec{E} = -\vec{\nabla}\Phi$, Eq. 21.75 becomes

$$E_{\text{elec}} = \int_{-\infty}^{\infty} \rho(\vec{r}) \left[\Phi(\vec{r}_0) - \vec{r} \cdot \vec{E}(\vec{r}_0) \right] d^3 r. \tag{21.77}$$

The total charge Q and the electric dipole moment $\vec{\tau}$ are defined as

$$Q \equiv \int_{-\infty}^{\infty} \rho(\vec{r}) \, d^3 r, \quad \vec{\tau} \equiv \int_{-\infty}^{\infty} \vec{r} \rho(\vec{r}) \, d^3 r, \tag{21.78}$$

and are useful to simplify the electrostatic energy

$$E_{\text{elec}} = Q \, \Phi(\vec{r}_0) - \vec{\tau} \cdot \vec{E}(\vec{r}_0). \tag{21.79}$$

The material is assumed homogeneous, and the spatial coordinates were integrated away, so the energy E_{elec} does not vary with the position or momentum coordinates that we have been using to calculate a partition function. There is a degree of freedom in the angle θ for the orientation of $\vec{\tau}$ along the direction of $\vec{E}$, where

$$\vec{\tau} \cdot \vec{E}(\vec{r}_0) = \tau E_0 \cos\theta. \tag{21.80}$$

---

[17] Assume $\vec{r}_0$ is inside the material.

Integrating the Boltzmann probability over differential solid angles $d\Omega$ gives the partition function, Z

$$Z = \int_{4\pi} \exp\left(-\frac{E_{\text{elec}}}{k_B T}\right) d\Omega = \int_{4\pi} \exp\left(-\frac{Q\Phi(\vec{r}_0) - \tau E_0 \cos\theta}{k_B T}\right) d\Omega, \tag{21.81}$$

$$Z = \exp\left(-\frac{Q\Phi(\vec{r}_0)}{k_B T}\right) \int_{4\pi} \exp\left(\frac{\tau E_0 \cos\theta}{k_B T}\right) d\Omega, \tag{21.82}$$

$$Z = \exp\left(-\frac{Q\Phi(\vec{r}_0)}{k_B T}\right) 2\pi \int_0^\pi \exp\left(\frac{\tau E_0 \cos\theta}{k_B T}\right) \sin\theta\, d\theta. \tag{21.83}$$

Changing variables to $\xi = \tau E_0/k_B T$ and $\eta = \cos\theta$, we work an integral of the form I to obtain Z

$$I = 2\pi \int_{-1}^{+1} e^{\xi\eta} d\eta = 2\pi \frac{e^{+\xi} - e^{-\xi}}{\xi}, \tag{21.84}$$

$$Z = 4\pi \exp\left(-\frac{Q\Phi(\vec{r}_0)}{k_B T}\right)\left(\frac{\sinh\xi}{\xi}\right). \tag{21.85}$$

The partition function is now a product of a factor from the total charge, and a second factor from the dipole moment. The first factor is not interesting for phase transitions in materials, assuming that the ions keep their valence as a function of temperature and applied field. The product $Q\Phi(\vec{r}_0)$ is therefore constant, and does not depend on configurations of atoms. Furthermore, we generally expect a material to be neutral, with net charge $Q = 0$, making this term even less interesting.

The energy can be calculated from Eq. 1.18 as

$$E = k_B T^2 \frac{\partial}{\partial T} \ln Z, \tag{21.86}$$

$$E_{\text{elec}} = -\tau E_0 \left[\text{ctanh}\left(\frac{\tau E_0}{k_B T}\right) - \frac{k_B T}{\tau E_0}\right]. \tag{21.87}$$

Define the Langevin function $L(\xi) \equiv \text{ctanh}(\xi) - 1/\xi$ to simplify this expression

$$E_{\text{elec}} = -\tau E_0 L\left(\frac{\tau E_0}{k_B T}\right). \tag{21.88}$$

The characteristic dipole moment of a material is small, and it is difficult to apply high voltages across a material without electrical breakdown, so usually $\tau E_0 \ll k_B T$. For small values of $\tau E_0/k_B T$, the Langevin function is approximately $L(\xi) \simeq \xi/3$. In this "high-temperature limit," which usually includes room temperature, the electrostatic energy is

$$E_{\text{elec}}(T) = -\frac{(\tau E_0)^2}{3k_B T}. \tag{21.89}$$

This E_{elec} is monotonic with temperature, as may be expected for isolated dipoles without cooperative interactions.

The entropy can be calculated from 21.85 as

$$S_{\text{elec}}(T) = k_B \ln\left(\frac{4\pi \sinh\xi}{\xi}\right) + k_B \xi \ln\xi. \tag{21.90}$$

At high temperatures (small ξ) the entropy per dipole is

$$S_{\text{elec}}(T_{\text{hi}}) = k_{\text{B}} \ln(4\pi),\qquad(21.91)$$

as each dipole samples all angles of orientation. At low temperatures the entropy has the form

$$S_{\text{elec}}(T_{\text{lo}}) = -k_{\text{B}} \ln\left(\frac{\tau E_0}{k_{\text{B}} T}\right).\qquad(21.92)$$

Equation 21.92 has a logarithmic singularity at low temperatures (and a negative sign). This is cause for alarm. The problem is the unreliability of classical statistical mechanics at low temperatures, originating with the continuous orientation averaging of the Boltzmann factors in Eq. 21.84. This is usually not an issue for electric fields in materials, since $T \gg \tau E_0/k_{\text{B}}$, as explained after Eq. 21.88. As for the analogous magnetic problem of spins in magnetic fields, however, at low temperatures quantum behavior must be treated properly. A partition function for a magnetic spin is best prepared as a sum of discrete Boltzmann factors for each quantized spin orientation. Likewise, the Langevin function is replaced with a Brillouin function that accounts for the finite spin orientations, but this is beyond the scope of this book.

Return to the point of factoring the partition function in Eq. 21.85. The multipole expansion of the potential $\Phi(\vec{r})$ (Eq. 21.76) gave new quantities for the charge distribution in the material. The monopole moment (total charge) was separated from the dipole moment, accounting for the two factors Z of Eq. 21.85. This allows a rigorous statement that the charges in ionic crystals cannot bring a structural phase transition in a uniform electrostatic potential. For there to be an energy preference for a specific configuration of ions, it is necessary to have a gradient in the potential (an electric field) in the material, and a dipole moment.

21.8 Ferroelectric Transition

Section 19.3 developed the Landau theory for a displacive phase transition, obtaining a second-order phase transition with a displacement $\langle x \rangle(T)$ graphed in Fig. 19.17, or a first-order phase transition with a potential as shown in Fig. 19.18. This is essentially what is needed to predict the "Curie temperature," T_{C}, of a ferroelectric to paraelectric phase transition.

In a ferroelectric phase, each unit cell has a dipole moment where a cation is not surrounded symmetrically by anions. The electric dipole moment of a unit cell, $\vec{\tau}_{\text{e}}$, is

$$\vec{\tau}_{\text{e}} = \int_{\text{cell}} \vec{r}\,\rho(\vec{r})\,\mathrm{d}^3\vec{r},\qquad(21.93)$$

where $\rho(\vec{r})$ is the electric charge density, a scalar field that is typically negative near anions such as O^{2-} and positive near cations such as Ba^{2+} or Ti^{4+}. The origin of the coordinate system is centered in the unit cell so that

$$0 = \int_{\text{cell}} \vec{r}\,\mathrm{d}^3\vec{r}.\qquad(21.94)$$

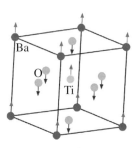

Figure 21.8 BaTiO$_3$ with perovskite structure, showing displacements of O^{2-}, Ti^{4+}, and Ba^{2+} ions in a soft phonon that leads to a tetragonal distortion of the cubic unit cell.

With charge neutrality in the unit cell, the "strength" of the ferroelectric polarization is directly proportional to asymmetrical displacements of the cations with respect to the anions. The dipole moment $\vec{\tau}_e$ of Eq. 21.93 vanishes when $\rho(\vec{r})$ is symmetrical.[18] The $\vec{\tau}_e$ of the unit cell and the electric polarization of the crystal are proportional to the atom displacements. These vary with temperature, as shown for a second-order transition in Fig. 19.17, for example.

With these preliminaries out of the way, consider the symmetric unit cell of Fig. 21.8. This compound, BaTiO$_3$, has a cubic structure at high temperature. For calculating $\vec{\tau}_e$ of Eq. 21.93, and for parameterizing the displacement used in Landau theory, the origin of the coordinate system should be placed at the center of the unit cell of Fig. 21.8, since there is inversion symmetry for the charges about the center. Below a critical temperature of 380 K, BaTiO$_3$ undergoes a first-order ferroelectric transition, as shown by the arrows in Fig. 21.8. (The ion displacements are not all identical, and there is a net tetragonal distortion to the unit cell.) The polarization is proportional to the displacement, and the critical temperature of Eq. 19.50 can describe the essential behavior of both polarization and displacement. As expected in a first-order transition, the polarization and tetragonality make sudden jumps upon cooling below T_c, but both increase somewhat upon greater cooling. Some classic experimental data are shown in Fig. 21.9 [368], where X-ray diffraction was used to detect changes in the shape of the unit cell. Below a temperature of about 115 °C, the cubic unit cell takes on a tetragonal distortion, with the two a-axes shrinking and the c-axis expanding. Another structural transformation occurs below about 0 °C, and the shape of the unit cell changes again below about –90 °C. These are all first-order phase transitions, although the cubic-to-tetragonal transformation looks somewhat like second order.

Under an applied electric field, either static or oscillatory, the anions and cations in a ferroelectric material experience Coulomb forces in opposite directions. Recall that displacements of atoms in opposite directions are described by optical phonons, where there is no change in the center of mass of the unit cell, but the different atoms move against each other. An important concept is that an electric field, if tuned to the frequency of an optical phonon, can induce resonance of the ions in the unit cell. These oscillating ions then act as small antennas that generate electromagnetic radiation of the same frequency,

---

[18] With equal charges on both sides of center, $\rho(\vec{r}) = \rho(-\vec{r})$, so the integral of Eq. 21.93 vanishes by symmetry.

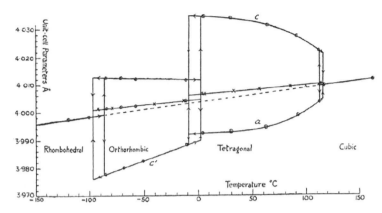

Figure 21.9 Temperature dependence of unit cell parameters of BaTiO$_3$, measured by X-ray diffractometry. From [368].

which in turn helps drive the oscillations of ions in other unit cells.[19] This coupled system of a photon plus phonon is called a "polariton." It can exist only for transverse optical phonons because these generate dipole oscillations that radiate along the direction of the phonon k-vector. An emitted transverse photon has the proper polarization to sustain the ion oscillations. Light moves quickly compared with phonons, so the wavelengths of polaritons are large, and the more so for lower frequencies. One model of ferroelectric behavior is a polariton of zero frequency and infinite wavelength. This viewpoint is most appropriate for a crystal that is unstable against ion displacements, as described by Landau theory in Section 19.3.

21.9 Domains

21.9.1 Minimizing the Energy in External Fields

The ferroelectric phase, and ferromagnetic phases as well, generally do not have one orientation within a piece of material. Instead they form "domains" that have different orientations of polarization. An idealized domain structure is shown in Fig. 21.10. This structure would be compatible with a cubic crystal that has three x-, y-, and z-variants for ferromagnetism, for example. This domain structure succeeds in minimizing the energy in the external magnetic field – the magnetic flux makes a set of loops inside the material, so only small amounts of flux are expected to escape at corners and edges. Domain structures are rarely as perfect as shown in Fig. 21.10. Nevertheless, domains of finer spatial scale are generally more efficient in minimizing the external energy in a magnetic field. The same reasoning pertains to the electric fields of ferroelectrics, and the strain fields of martensite, as discussed in Section 19.2.3.

---

[19] If the photon frequency is a bit above the frequency of the optical phonon, the ion motions lag out of phase with respect to the photon. The ion radiation cancels out the photon, and it does not propagate.

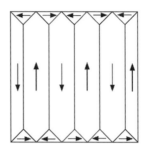

Domain structure compatible with the tetragonal distortion of BaTiO$_3$, or with ferromagnetic domains.

21.9.2 Bloch Walls

There is an energy penalty to form a domain wall, however. In the case of ferromagnetic domains, the domain walls frequently have widths that span tens of unit cells or more. In ferromagnetic "Bloch walls," the magnetic moments do not reverse abruptly at the boundary between domains, but instead make a gradual rotation from unit cell to unit cell. The width of a Bloch wall is a balance between the low energy for a small rotation of adjacent magnetic moments (favoring a wide Bloch wall), versus the unfavorable "anisotropy energy" associated with having the rotated magnetic moments pointing away from the preferred crystallographic axes (favoring a narrow Bloch wall).

Section 17.3 calculated the width of an antiphase domain boundary (APDB), and like a Bloch wall between magnetic domains, the APDB is a structure in a system with a nonconserved order parameter. The phase field approach for the width of the APDB between ordered domains is therefore essentially the same as for the width of a Bloch wall between two magnetic domains.

The physical phenomenon of a squared gradient energy for composition, $\kappa(\overrightarrow{\nabla c})^2$, is analogous to an exchange energy of misaligned spins. The restoring force between two spins misaligned by the angle ϕ is proportional to ϕ, and across a Bloch wall over which N spins are rotated gradually by the angle π from $\uparrow$ to $\downarrow$, we expect $\phi = \pi/N$. The exchange energy cost of the misaligned spins in the Bloch wall is $NJs^2\phi^2$, where J is the exchange energy of Eq. 21.1. There is also an energy penalty for misorienting a spin with respect to a favorable crystallographic direction of the magnetic crystal, the "anisotropy energy." The anisotropy energy is analogous to the bulk free energy of the ordered domains, and it scales with the total number of spins in the Bloch wall, since these spins are not favorably aligned crystallographically. We write the anisotropy energy penalty as KNa^3, where a is the interspin distance, and is used to give traditional units of volume. Equation 17.32 for the APDB width, repeated here, is

$$\bar{x}_{\mathrm{APDB}} = \sqrt{\frac{\kappa}{\Delta f_0}}\, \eta_{\mathrm{eq}}, \tag{21.95}$$

where $\bar{x}$ is a characteristic width. With the replacements of analogous variables, the width of a Bloch wall is

$$\bar{x}_{\mathrm{Bloch}} = \sqrt{\frac{\pi^2 Js^2}{K}}. \tag{21.96}$$

Problems

21.1 Consider the ground state of the Heisenberg antiferromagnet on a triangular lattice.

    (a) By reversing the spins, draw an alternative ground state to Fig. 21.1. Are there other variants of the ground state? If so, describe them.

    (b) How much more energy per spin do you expect for this triangular antiferromagnet compared with an antiferromagnet on a square lattice?

21.2 Start with the definitions $\sinh x = (e^x - e^{-x})/2$ and $\cosh x = (e^x + e^{-x})/2$.

    (a) Derive Eq. 21.31.

    (b) Use Eq. 21.31 to obtain Eq. 21.32 from Eq. 21.30.

21.3 Equation 21.32 for the long-range order $L(T)$ in the Gorsky–Bragg–Williams approximation contains $L(T)$ inside the tanh() function. An iterative solution is possible with an algorithm that evaluates $L(T) = \tanh(L'(T)/T)$ at each step, where $L'(T)$ is the function $L(T)$ from the previous step. Write a loop to implement this algorithm, iterating on an initial function $L'_0(T)$ over the range $0 < T < 2$ (which proves to be in units of T_C).

    (a) If the initial $L'_0(T) = 0.0$, show that this trivial solution is retained after 100 steps.

    (b) If the initial $L'_0(T) = 1.0$, graph $L(T)$ after 1, 10, 100, and 1000 steps.

    (c) Show how an initial $L'_0(T < 1.0) = 1.0$ and $L'_0(T \geq 1.0) = 0.0$ approaches the final shape more rapidly than the constant 1.0 used for part (b) after 1, 10, 100, and 1000 steps, although for $T < 1.0$ they are the same.

21.4 The method used from Eqs. 21.44 through 21.53 can be used for other problems.

    (a) Use this method with the $\epsilon_k = \sin(\vec{k} \cdot \vec{a})$ relationship of the antiferromagnet spin wave dispersion (versus $\epsilon_k = 1 - \cos(\vec{k} \cdot \vec{a})$ for the ferromagnet) to show that the spin wave density increases as T^3 in a simple antiferromagnet.

    (b) Use this method to calculate the heat capacity of a free electron gas in a periodic box. To do so, use $\epsilon_k = \hbar^2 k^2/(2m_e)$, and allow two spin states per volume interval in k-space. You will need to calculate first the average energy $\langle \epsilon \rangle$, not just the total number of excitations as for the spin wave problems. Next, differentiate the result with respect to T to get the heat capacity.

21.5 For a ferromagnet in the mean field approximation, show that the magnetization near T_C goes as $\sqrt{T_C - T}$. (This " critical exponent" of 1/2 proves to be too large, as discussed in Chapter 27 [online].)

21.6 An applied magnetic field B makes a natural change to $\epsilon_\uparrow$ and $\epsilon_\downarrow$ in Eqs. 21.26 and 21.27. Is there an equivalent effect in the chemical ordering problem when adding interstitial atoms that have chemical interactions with the ordering species? What about an applied electric field for chemical ordering in an ionic compound? (*Hint*: Consider conserved versus nonconserved quantities in the problems of chemical ordering and ferromagnetism.)

21.7 Section 21.7 showed some problems at low temperatures with the classical thermodynamics of the alignment of a dipole moment in a field. The essence of the

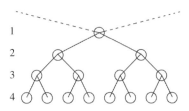

Figure 21.11 A Cayley tree with three levels from the top. Notice that there are no loops in a Cayley tree, so it is often used in models of phase transitions in materials, too. This diagram could be a simplification for a honeycomb lattice.

low-temperature behavior can be understood with the behavior of a "two-level system," for which there are two states of a system at energies 0 and ϵ_1. Use this partition function for a two-level system

$$Z_{2lvl} = e^{-\frac{0}{k_B T}} + e^{-\frac{\epsilon_1}{k_B T}}. \tag{21.97}$$

to answer the following questions:
(a) Calculate the energy and entropy of the two-level system versus temperature.
(b) What is occupancy of the upper level as a function of temperature?
(c) What is the reason for the suppressed entropy at temperatures below approximately ϵ_1/k_B?

21.8 Peruse Reference [369]. This 1982 paper uses some of the methods developed in Reference [370], but you need not read this 1980 paper in detail.
(a) Do your best to explain Equation 9 in the 1980 paper. At least explain why the combinatoric factors alternate between the numerator and denominator.
(b) For the 1982 paper, briefly explain how much of an improvement is seen by going to larger clusters in (i) the critical temperatures, (ii) the phase boundaries.

21.9 Social networks have structures and phase transitions with similarities to problems on Ising lattices. Figure 21.11 shows a "Cayley tree," where a node at the top is connected to pairs of nodes at the next level below. It has an increasing number of nodes at an increasing distance. A connectivity matrix with elements p_{ij} is defined with i starting at various levels in the tree. The difference $i - j$ is the distance, x_{ij}, between levels. The elements of the matrix $\underline{p}$ include a factor for the strengths of the links between nodes at the levels i and j

$$p_{ij} \propto \exp(-\alpha\, x_{ij}), \tag{21.98}$$

where α parameterizes the strengths of the links.
(a) Show that the matrix $\underline{p}$ is diagonal when α is large.
(b) For what value of α is $\underline{p}$ a random matrix (so that all p_{ij} are equally probable)? The convention is to include the number of links as a factor for p_{ij}.
(c) Suppose there is an opinion at each node, denoted as ↑ or ↓. What characteristics of the opinion field are expected when α is large and positive? A phase transition has been found in this model system, depending on the strengths of the links between nodes [371]. Give a similarity to and a difference from the Curie transition.

Further Reading

Part I

J.J. Hoyt, *Phase Transformations* (McMaster Innovation Press, 2010).

D.A. Porter, K.E. Easterling, and M.Y. Sherif, *Phase Transformations in Metals and Alloys* (Boca Raton, FL: CRC Press, 2009).

P.G. Shewmon, *Diffusion in Solids* (Warrendale, PA: The Minerals, Metals & Materials Society, 1989).

Part II

H. Ibach and H. Lüth, *Solid-State Physics*: *An Introduction to the Principles of Materials Science* Fourth Edn. (Springer, 2009).

C. Kittel and H. Kroemer, *Thermal Physics* Second Edn. (New York: W.H. Freeman, 1980).

C. Kittel, *Introduction to Solid State Physics* Eighth Edn. (or earlier) (New York: Wiley, 2004).

Part III

M.E. Fine, *Phase Transformations in Condensed Systems* (New York: Macmillan, 1964).

J.J. Hoyt, *Phase Transformations* (McMaster Innovation Press, 2010).

P.G. Shewmon, *Transformations in Metals* (New York: McGraw-Hill, 1969).

Chapter 1

C.R. Barrett, W.D. Nix, and A.S. Tetelman, *The Principles of Engineering Materials* (Upper Saddle River, NJ: Prentice-Hall, 1973), Chapters 1–5.

Chapter 2

J.C. Slater, *Introduction to Chemical Physics* (New York: McGraw-Hill, 1939), Chapter 17.

Chapter 3

P.G. Shewmon, *Diffusion in Solids* (Warrendale, PA: TMS, 1989).

Chapter 4

K.F. Kelton and A.L. Greer, *Nucleation in Condensed Matter: Applications in Materials and Biology* (Oxford: Pergamon Press, 2005).

Chapter 5

B. Chalmers, *Physical Metallurgy* (New York: Wiley, 1959).

Chapter 6

A. Cottrell, *Introduction to the Modern Theory of Metals* (London: Institute of Metals, 1988).

D.G. Pettifor, *Bonding and Structure of Molecules and Solids* (Oxford: Clarendon Press, 1995).

Chapter 7

D. de Fontaine, "Configurational Thermodynamics of Solid Solutions," *Solid State Physics* **34**, 73–274 (1979).

B. Fultz, "Vibrational Thermodynamics of Materials," *Progress in Materials Science* **55**, 247–352 (2010).

Chapter 8

J.C. Slater, *Introduction to Chemical Physics* (New York: McGraw-Hill, 1939), Chapter 13.

Chapter 9

L. Onsager, *Proceedings of the New York Academy of Science* **41**, 241 (1945).

Chapter 10

P. Shewmon, *Diffusion in Solids* (Warrendale, PA: The Minerals, Metals & Materials Society, 1989).

G.E. Murch and A.S. Nowick, *Diffusion in Crystalline Solids* (Orlando, FL: Academic Press, 1984).

Chapter 11

J.B. Hudson, *Surface Science* (New York: Wiley, 1992), Chapters 1, 17.

A. Zangwill, *Physics at Surfaces* (Cambridge: Cambridge University Press, 1988), Chapters 1–3, 16.

Chapter 12

D.C. Wallace, *Statistical Physics of Crystals and Liquids* (Singapore: World Scientific, 2002), Chapter 5.

Chapter 13

B. Chalmers, *Physical Metallurgy* (New York: John Wiley & Sons, 1962).

W. Kurz and D.J. Fisher, *Fundamentals of Solidification* (Zurich: Trans Tech, 1989).

J.J. Hoyt, *Phase Transformations* (McMaster Innovation Press, 2010).

Chapter 14

H.I. Aaronson, M. Enomoto, and J.K. Lee, *Mechanisms of Diffusional Phase Transformations in Metals and Alloys* (Boca Raton, FL: CRC Press, 2010).

Chapter 15

J.W. Christian, *Transformations in Metals and Alloys* Second Edn. (Oxford: Pergamon Press, 1975).

Chapter 16

J.W. Cahn and J.E. Hilliard, "Free Energy of a Nonuniform System. 1. Interfacial Free Energy," *Journal of Chemical Physics* **28**, 258 (1958).

J.W. Cahn, "On Spinodal Decomposition," *Acta Metallurgica* **9**, 795 (1961).

Chapter 17

N. Provatas and K. Elder, *Phase Field Methods in Materials Science and Engineering* (Weinheim: Wiley-VCH, 2010).

Chapter 18

A.G. Khachaturyan, "Ordering in Substitutional and Interstitial Solid Solutions," *Progress in Materials Science* **22**, 1–150 (1978).

A.G. Khachaturyan, *Theory of Structural Transformations in Solids* (New York: Wiley-Interscience, 1983).

Chapter 19

P.G. Shewmon, *Transformations in Metals* (New York: McGraw-Hill, 1969), Chapter 8.

Chapter 20

C. Kittel, *Introduction to Solid State Physics* Eighth Edn. (New York: Wiley, 2004), Chapter 18.

H.-E. Schaefer, *Nanoscience* (Heidelberg: Springer, 2010).

Chapter 21

C. Kittel, *Introduction to Solid State Physics* Eighth Edn. (New York: Wiley, 2004).

J.M.D. Coey, *Magnetism and Magnetic Materials* (Cambridge: Cambridge University Press, 2010).

M.E. Lines and A.M. Glass, *Principles and Applications of Ferroelectrics and Related Materials* (Oxford: Oxford University Press, 2001).

References

[1] L. Onsager, *Phys. Rev.* **65**, 117 (1944).

[2] A. Cerezo, J.M. Hyde, M.K. Miller, *et al.*, *Phil. Trans. Roy. Soc. London A* **341**, 313 (1992).

[3] W. Hume-Rothery and G.V. Raynor, *The Structure of Metals and Alloys* (Institute of Metals, London, 1962).

[4] A. Cottrell, *Introduction to the Theory of Metals* (Institute of Metals, London, 1988).

[5] L.S. Darken and R.W. Gurry, *Physical Chemistry of Metals* (McGraw–Hill, New York, 1953), p. 74.

[6] D.G. Pettifor, *Bonding and Structure of Molecules and Solids* (Clarendon Press, Oxford, 1995).

[7] A.R. Miedema, P.F. de Chatel, and F.R. de Boer, *Physica B,C* **100**, 1 (1980).

[8] H. Okamoto, *Desk Handbook Phase Diagrams for Binary Alloys* (ASM International, Materials Park, OH, 2000).

[9] J.W. Gibbs, *Trans. Conn. Acad.* **3**, 108 (1876).

[10] P. Villars, Ed., with H. Okamoto and K. Cenzual, *ASM Alloy Phase Diagram Database* (ASM International, Materials Park, OH, 2006–2013).

[11] C. Kittel, *Thermal Physics* (John Wiley, New York, 1969), Chapter 2.

[12] B.E. Warren, *X-Ray Diffraction* (Dover, Mineola, New York, 1990).

[13] W. Gorsky, *Z. Physik* **50**, 64 (1928).

[14] W.L. Bragg and E.J. Williams, *Proc. Roy. Soc. London A* **145**, 699 (1934).

[15] W.L. Bragg and E.J. Williams, *Proc. Roy. Soc. London A* **151**, 540 (1935). *Ibid.* **152**, 231.

[16] H.A. Bethe, *Proc. Roy. Soc. London A* **150**, 552 (1935).

[17] D. de Fontaine, *Acta Metall.* **23**, 553 (1975).

[18] D.R.F. West and N. Saunders, *Ternary Phase Diagrams in Materials Science*, Third Edn. (Institute of Materials, London, 2002).

[19] N. Saunders and A.P. Miodownik, *CALPHAD (Calculation of Phase Diagrams): A Comprehensive Guide*, Volume 1 (Pergamon Press, Oxford, 1998).

[20] P.E.A. Turchi, A. Gonis, and R.D. Shull, Eds., *CALPHAD and Alloy Thermodynamics* (TMS, Warrendale, PA, 2002).

[21] P.E.A. Turchi, I.A. Abrikosov, B. Burton, *et al.*, *CALPHAD* **31**, 4 (2007).

[22] R. Sarmiento-Perez, T.F.T. Cerqueira, I. Valencia-Jaime, *et al.*, *New J. Phys.* **15**, 115007 (2013).

[23] P.M. Morse and H. Feshbach, *Methods of Theoretical Physics* (McGraw–Hill, New York, 1953), Chapters 5 and 10.

[24] R.O. Simmons and R.W. Balluffi, *Phys. Rev.* **117**, 52 (1960).

[25] R.O. Simmons and R.W. Balluffi, *Phys. Rev.* **125**, 862 (1962).

[26] R.O. Simmons and R.W. Balluffi, *Phys. Rev.* **129**, 1533 (1963).

[27] R. Feder and A.S. Nowick, *Philos. Mag.* **15**, 805 (1967).

[28] H.-E. Schefer, K. Frenner, and R. Würschum, *Phys. Rev. Lett.* **82**, 948 (1999).

[29] J.K. Lee and H.I. Aaronson, *Acta Metall.* **23**, 799 (1975).

[30] U. Dahmen, S. Hagège, F. Faudot, T. Radetic, and E. Johnson, *Philos. Mag.* **84**, 2651 (2004).

[31] J.W. Gibbs, *Trans. Conn. Acad.* **11**, 382 (1873).

[32] M. Volmer and A. Weber, *Z. Phys. Chem.* **119**, 277 (1926).

[33] Z. Farkas, *Z. Phys. Chem.* **A125**, 236 (1927).

[34] R. Becker and W. Döring, *Ann. Phys.* **24**, 1 (1935).

[35] J.B. Zeldovich, *Acta Physicochim.* **18**, 1 (1943).

[36] K.F. Kelton and A.L. Greer, *Nucleation in Condensed Matter: Applications in Materials and Biology* (Pergamon Press, Oxford, 2010).

[37] H. Trinkaus and M.H. Yoo, *Philos. Mag.* **A55**, 269 (1987).

[38] G. Shi, J.H. Seinfeld, and K. Okuyama, *Phys Rev.* **A41**, 2101 (1990).

[39] D.T. Wu, in *Solid State Physics*, Volume 50, H. Ehrenreich and F. Spaepen, Eds. (Academic Press, New York, 1997) p. 37, Section 11.

[40] G.H. Gulliver, *J. Inst. Met.* **9**, 120 (1913).

[41] E. Scheil, *Z. Metallk.* **34**, 70 (1942).

[42] W. Klement, R.H. Willens, and P. Duwez, *Nature* **187**, 869 (1960).

[43] C. Kittel, *Introduction to Solid State Physics* Fourth Edn. (Wiley, New York, 1971), p. 143.

[44] N.I. Medvedeva, Y.N. Gornostyrev, and A.J. Freeman, *Phys. Rev. B* **67**, 134204 (2003).

[45] The Fermi Surface Database www.phys.ufl.edu/fermisurface/.

[46] T.S. Choy, J. Naset, J. Chen, S. Hershfield, and C. Stanton, *Bull. Amer. Phys. Soc.* **45**, 42 (2000).

[47] D.S. Sholl and J.A. Steckel, *Density Functional Theory: A Practical Introduction* (John Wiley and Sons, Hoboken, New Jersey, 2009).

[48] M. Frary, "Anisotropic Elasticity," Wolfram Demonstrations Project (Wolfram Research Inc., 2011).

[49] J.D. Eshelby, *J. Appl. Phys.* **25**, 255 (1954).

[50] J.D. Eshelby, *Solid State Phys.* **3**, 79 (1956).

[51] J.D. Eshelby, *Proc. Roy. Soc. London A* **241**, 376 (1957).

[52] F.R.N. Nabarro, *Proc. Roy. Soc. London A* **175**, 519 (1940).

[53] R. Kikuchi, *Phys. Rev.* **81**, 988 (1951).

[54] R. Kikuchi, *J. Chem. Phys.* **60**, 1071 (1974).

[55] D. de Fontaine, in *Solid State Physics*, Volume 34, H. Ehrenreich, F. Seitz, and D. Turnbull, Eds. (Academic Press, New York, 1979), p. 73.

[56] R.H. Fowler and E.A. Guggenheim, *Proc. Roy. Soc. London A* **174**, 189 (1940).

[57] J.M. Sanchez, F. Ducastelle, and D. Gratias, *Physica A* **128**, 334 (1984).

[58] J.W.D. Connolly and A.R. Williams, *Phys. Rev. B* **27**, 5169 (1983).

[59] R. Kikuchi, *Phys. Rev.* **81**, 988 (1951).

[60] C.M. Van Baal, *Physica* **64**, 571 (1973).

[61] J.M. Sanchez and D. de Fontaine, *Phys. Rev. B* **21**, 216 (1980).

[62] P. Cenedese and R. Kikuchi, *Physica A* **205**, 747 (1994).

[63] W. Nernst and F.A. Lindemann, *Berl. Ber.* 494 (1911).

[64] E.S.R. Gopal, *Specific Heats at Low Temperatures* (Plenum, New York, 1966).

[65] J.P. McCullough and D.W. Scott, Eds. *Experimental Thermodynamics Volume 1: Calorimetry of Non-Reacting Systems* (Plenum, New York, 1968).

[66] B.E. Lang, J. Boerio-Goates, and B.F. Woodfield, *J. Chem. Thermodynamics* **38**, 1655 (2006).

[67] R. Bachmann, F.J. DiSalvo, T.H. Geballe, *et al.*, *Rev. Sci. Instr.* **43**, 205 (1972).

[68] G.R. Stewart, *Rev. Sci. Instr.* **54**, 1 (1983).

[69] J.C. Lashley, M.F. Hundley, A. Migliori, *et al.*, *Cryogenics* **43**, 369 (2003).

[70] C.A. Kennedy, M. Stancescu, R. Marriott, and M.A. White *Cryogenics* **47**, 107 (2007).

[71] A. Einstein, *Ann. Phys.* **22**, 180 (1907).

[72] A.A. Maradudin, E.W. Montroll, G.H. Weiss, and I.P. Ipatova, *Theory of Lattice Dynamics in the Harmonic Approximation* (Academic Press, New York, 1971).

[73] M. Born and K. Wang, *Dynamical Theory of Crystal Lattices* (Clarendon Press, Oxford, 1988).

[74] M.T. Dove, *Introduction to Lattice Dynamics* (Cambridge University Press, Cambridge, 1993).

[75] A. van de Walle and G. Ceder, *Rev. Mod. Phys.* **74**, 11 (2002).

[76] G. Moraitis and F. Gautier, *J. Phys. F: Metal Phys.* **7**, 1421 (1977).

[77] J.A.D. Matthew, R.E. Jones, and V.M. Dwyer, *J. Phys. F: Metal Phys.* **13**, 581 (1983).

[78] A.A.H.J. Waegemaekers and H. Bakker, *Mater. Res. Soc. Symp. Proc.* **21**, 343 (1984).

[79] G.D. Garbulsky and G. Ceder, *Phys. Rev. B* **53**, 8993 (1996).

[80] S. Baer, *J. Phys. C: Solid State Phys.* **16**, 4103 (1983).

[81] J. Mahanty and M. Sachdev, *J. Phys. C* **3**, 773 (1970).

[82] H. Bakker, *Philos. Mag. A* **45**, 213 (1982).

[83] H. Bakker, *Phys. Stat. Solidi B* **109**, 211 (1982).

[84] O. Delaire, T. Swan-Wood, and B. Fultz, *Phys. Rev. Lett.* **93**, 185704 (2004).

[85] B. Fultz and J.M. Howe, *Transmission Electron Microscopy and Diffractometry of Materials* Fourth Edn. (Springer, Heidelberg, 2013).

[86] M.H.F. Sluiter, M. Weinert, and Y. Kawazoe, *Phys. Rev. B* **59**, 4100 (1999).

[87] A. van de Walle and G. Ceder, *Phys. Rev. B* **61**, 5972 (2000).

[88] E.J. Wu, G. Ceder, and A. van de Walle, *Phys. Rev. B* **67**, 134103 (2003).

[89] J.C. Slater, *Introduction to Chemical Physics* (McGraw-Hill, New York, 1939), Chapter 13.

[90] S. Desgreniers, Y.K. Vohra, and A.L. Ruoff, *J. Phys. Chem.* **94**, 1117 (1990).

[91] G. Weck, S. Desgreniers, and P. Loubeyre, *Phys. Rev. Lett.* **102**, 255503 (2009).

[92] R.P. Dias and I.F. Silvera, *Science* **355**, 715 (2017).

[93] M.L. Winterrose, M.S. Lucas, A. F. Yue, *et al.*, *Phys. Rev. Lett.* **102**, 237202 (2009).

[94] X. Tong, X. Xu, B. Fultz, *et al.*, *Phys. Rev. B* **95**, 094306 (2017).

[95] R.B. Schwarz and W.L. Johnson, *Phys. Rev. Lett.* **51**, 415 (1983).

[96] W.L. Johnson, *Prog. Mater. Sci.* **30**, 81 (1986).

[97] L.J. Chen, J.H. Lin, T.L. Lee, *et al.*, *Microsc. Res. Tech.* **40**, 136 (1998).

[98] W.S. Gorsky, *Phys. Zeitschr. Sowjetunion* **8**, 457 (1935).

[99] J.L. Snoek, *Physica* **8**, 711 (1941).

[100] A.H. Cottrell and M.A. Jaswon, *Proc. Roy. Soc.* **A199**, 104 (1949).

[101] T. Vreeland, D.S. Wood, and D.S. Clark, *Trans. Amer. Soc. Metals* **45**, 620 (1953).

[102] T. Holstein, *Ann. Phys.* **8**, 325 (1959). *Ibid* **8**, 342 (1959).

[103] D. Emin and T. Holstein, *Ann. Phys.* **53**, 439 (1969).

[104] I.G. Austin and N.F. Mott, *Adv. Phys.* **18**, 41 (1969).

[105] N. Mott, *J. Non-Cryst. Solids* **1**, 1 (1968).

[106] H. Schmid, *Ferroelectrics* **162**, 317 (1994).

[107] G. Heckmann, *Ergeb. Exakten Naturwiss.* **4**, 100 (1925).

[108] J.F. Nye, *Physical Properties of Crystals: Their Representation by Tensors and Matrices* (Oxford University Press, Oxford, 1985).

[109] L.J. Nagel, B. Fultz, J.L. Robertson, and S. Spooner, *Phys. Rev. B* **55**, 2903 (1997).

[110] M.E. Manley, B. Fultz, and L.J. Nagel, *Philos. Mag. B* **80**, 1167 (2000).

[111] L.J. Nagel, Ph.D. thesis in materials science, California Institute of Technology (1996).

[112] V. Gopalan, V. Dierolf, and D.A. Scrymgeour, *Annu. Rev. Mater. Res.* **37**, 449 (2007).

[113] Z. Zhao, X. Ding, T. Lookman, J. Sun, and E.K.H. Salje, *Adv. Mater.* **25**, 3244 (2013).

[114] L. Tartar, *The General Theory of Homogenization: A Personalized Introduction*, Lecture Notes of the Unione Matematica Italiana **7** (Springer-Verlag, Berlin Heidelberg, 2010).

[115] J.W. Cahn, *Acta Metall.* **10**, 179 (1962).

[116] J.Y. Huh, J.M. Howe, and W.C. Johnson, *Scripta Metall.* **24**, 2007 (1990).

[117] M. Hillert and L.-I. Staffansson, *Acta Chem. Scand.* **24** , 3618 (1970).

[118] B. Sundman and J. Ågren, *J. Phys. Chem. Solids* **42**, 297 (1981).

[119] X.L. Yeh, K. Samwer, and W.L. Johnson, *Appl. Phys. Lett.* **42**, 242 (1983).

[120] Xian-Li Yeh, Ph.D. thesis in applied physics, California Institute of Technology (1987).

[121] R.C. Bowman, C.H. Luo, C.C. Ahn, C.K. Witham, and B. Fultz, *J. Alloys Compounds* **217**, 185 (1995).

[122] K. Samwer, X.L. Yeh, and W.L. Johnson, *J. Non-Cryst. Solids* **61**, 631 (1984).

[123] J.R. Manning, *Acta Metall.* **15**, 817 (1967).

[124] R. Kikuchi and H. Sato, *J. Chem. Phys.* **53**, 2702 (1970).

[125] H. Bakker, *Philos. Mag.* **40**, 525 (1979).

[126] H. Sato and R. Kikuchi, *Acta Metall.* **24**, 797 (1976).

[127] B. Fultz, *J. Chem. Phys.* **87**, 1604 (1987).

[128] R.W. Siegel, *Annu. Rev. Mater. Sci.* **10**, 393 (1980).

[129] Zs. Kajcsos, *Phys. Scripta* **T25**, 26 (1989).

[130] A. Seeger, *J. Phys. F: Metal Phys.* **3**, 248 (1973).

[131] M.J. Puska and R.M. Nieminen, *Rev. Modern Phys.* **66**, 841 (1994).

[132] E. Boronski, *Europhys. Lett.* **75**, 475 (2006).

[133] A.D. Smigelskas and E.O. Kirkendall, *Trans. AIME* **171**, 131 (1947).

[134] G. Martin, *Phys. Rev. B* **30**, 1424 (1984).

[135] G. Martin and P. Bellon, *Solid State Physics*, Volume 50, H. Ehrenreich and F. Spaepen, Eds. (Academic Press, New York, 1996), p. 189.

[136] G. Vineyard, *J. Phys. Chem. Solids* **3**, 121 (1957).

[137] S. Rice, *Phys. Rev.* **112**, 804 (1958).

[138] W.K. Burton, N. Cabrera, and F.C. Frank, *Nature* **163**, 398 (1949).

[139] W.K. Burton, N. Cabrera, and F.C. Frank, *Phil. Trans. Roy. Soc. London* **243**, A 866 (1951).

[140] K.A. Jackson, *J. Cryst. Growth* **24/25**, 130 (1974).

[141] W.M. Lomer and J.F. Nye, *Proc. Roy. Soc. London A* **212**, 576 (1952).

[142] W.T. Read and W. Shockley, *Phys. Rev.* **78**, 275 (1950).

[143] H. Van Swygenhoven, D. Farkas, and A. Caro, *Phys. Rev. B* **62**, 831 (2000).

[144] M. Yuasa, T. Nakazawa, and M. Mabuchi, *J. Phys.: Condens. Matter* **24**, 265703 (2012).

[145] D. Olmsted, S.M. Foiles, and E.A. Holm, *Acta Mater.* **57**, 3694 (2009).

[146] W. Setyawan and R.J. Kurtz, *Scripta Mater.* **66**, 558 (2012).

[147] M.D. Sangid, T. Ezaz, H. Sehitoglu, and I.M. Robertson, *Acta Mater.* **59**, 283 (2011).

[148] D. Udler and D.N. Seidman, *Phys. Rev. B* **54**, R11133 (1996).

[149] S.M. Foiles, *Scripta Mater.* **62**, 231 (2010).

[150] R. Kikuchi and J.W. Cahn *Phys. Rev. B* **21**, 1893 (1980).

[151] R. Pandit, M. Schick, and M. Wortis, *Phys. Rev. B* **26**, 5112 (1982).

[152] J.W. Cahn and R. Kikuchi, *Phys. Rev. B* **36**, 418 (1987).

[153] P.R. Cantwell, M. Tang, S.J. Dillon, *et al.*, *Acta Mater.* **62**, 1 (2014).

[154] T. Frolov, D.L. Olmsted, M. Asta, and Y. Mishin, *Nature Commun.* **4**, 1899 (2013).

[155] T. Frolov, M. Asta, and Y. Mishin, *Phys. Rev. B* **92**, 020103 (2015).

[156] P.R. Cantwell, S. Ma, S.A. Bojarski, G.S. Rohrer, and M.P. Harmer, *Acta Mater.* **106**, 78 (2016).

[157] D.R. Clarke, *J. Amer. Ceramic Soc.* **70**, 15 (1987).

[158] P. Keblinski, S.R. Phillpot, D. Wolf, and H. Gleiter, *Phys. Rev. Lett.* **77**, 2965 (1996).

[159] D. McLean, *Grain Boundaries in Metals* (Clarendon Press, Oxford, 1957). Chapter 5.

[160] R.V. Zucker, D. Chatain, U. Dahmen, S. Hagège, and W.C. Carter, *J. Mater. Sci.* **47**, 8290 (2012).

[161] P.J. Desré and A.R. Yavari, *Phys. Rev. Lett.* **64**, 1533 (1990).

[162] U. Gösele and K.N. Tu, *J. Appl. Phys.* **53**, 3252 (1982).

[163] M.-A. Nicolet and S.S. Lau, in *VLSI Electronics*, Volume 6, N.G. Einspruch and G.B. Larrabee, Eds. (Academic Press, New York, 1983), p. 329.

[164] R. Walser and R. Bené, *Appl. Phys. Lett.* **28**, 624 (1976).

[165] J. Purewal, Ph.D. thesis in materials science, California Institute of Technology (2010).

[166] J.G. Dash *Contemp. Phys.* **89** (1989).

[167] R. Rosenberg, *Phys. Today* **58**, 50 (2005).

[168] Y. Yang, M. Asta, and B.B. Laird, *Phys. Rev. Lett.* **110**, 096102 (2013).

[169] J. Mellenthin, A. Karma, and M. Plapp, *Phys Rev. B* **78**, 184110 (2008).

[170] M. Tang, W.C. Carter, and R.M. Cannon, *Phys. Rev. B* **73**, 024102 (2006).

[171] Y. Yang, D.L.Olmsted, M. Asta, and B.B. Laird, *Acta Mater.* **60**, 4960 (2012).

[172] J.P. Palafox-Hernandeza, B.B.Laird, and M. Asta, *Acta Mater.* **59**, 3137 (2011).

[173] A.A. Minakov, A. Wurm, and C. Schick, *Eur. Phys. J. E* **23**, 4353 (2007).

[174] J. Daeges, H. Gleiter, and J.H. Perepezko, *Phys. Lett. A* **119**, 79 (1986).

[175] S. Takeya, *Appl. Phys. Lett.* **88**, 074103 (2006).

[176] M. Forsblom and G. Grimvall, *Nature* **4**, 388 (2005).

[177] A.B. Belonoshko, N.V. Skorodumova, A. Rosengren, and B. Johansson, *Phys. Rev. B* **73**, 012201 (2006).

[178] H.J. Fecht and W.L. Johnson, *Nature* **334**, 50 (1988).

[179] A.C. Lawson, *Philos. Mag.* **89**, 1757 (2009).

[180] J.H. Rose, J. Ferrante, and J.R. Smith, *Phys. Rev. Lett.* **47**, 675 (1981).

[181] J.H. Rose, J.R. Smith, F. Guinea, and J. Ferrante, *Phys. Rev. B* **29**, 2963 (1984).

[182] X. Tang, C.W. Li, and B. Fultz, *Phys. Rev. B* **82**, 184301 (2010).

[183] M.H.G. Jacobs and R. Schmid-Fetzer, *Phys. Chem. Minerals* **37**, 721 (2010).

[184] P.J. Spencer and the Scientific Group Thermodata Europe (SGTE), *Landolt–Börnstein / New Series Group IV: Physical Chemistry*, Volume 19 (Springer, Heidelberg, 1999).

[185] SGTE Scientific Group Thermodata Europe www.met.kth.se/sgte/.

[186] N. Bock, D. Coffey, and D.C. Wallace, *Phys. Rev. B* **72**, 155120 (2005).

[187] N. Bock, D.C. Wallace, and D. Coffey, *Phys. Rev. B* **73**, 075114 (2006).

[188] M.G. Kresch, M.S. Lucas, O. Delaire, J.Y.Y. Lin, and B. Fultz, *Phys. Rev. B* **77**, 024301 (2008).

[189] M. Forsblom and G. Grimvall, *Phys. Rev. B* **72**, 132204 (2005).

[190] F. Körmann, A. Dick, B. Grabowski, *et al.*, *Phys. Rev. B* **78**, 033102 (2008).

[191] F. Körmann, A. Dick, B. Grabowski, T. Hickel, and J. Neugebauer, *Phys. Rev. B* **85**, 125104 (2012).

[192] D.C. Wallace, *Statistical Physics of Crystals and Liquids: A Guide to Highly Accurate Equations of State* (World Scientific, Singapore, 2003).

[193] F.A. Lindemann, *Phys. Z.* **11**, 609 (1910).

[194] J.J. Gilvarry, *Phys. Rev.* **102**, 308 (1956).

[195] K. Gschneidner, Jr., *Solid State Physics*, Volume 16, F. Seitz and D. Turnbull, Eds. (Academic Press, New York, 1965), p. 275.

[196] W. Kauzmann, *Chem. Rev.* **43**, 219 (1948).

[197] C.A. Angell, K.L. Ngai, G.B. McKenna, P.F. McMillan, and S.W. Martin, *J. Appl. Phys.* **88**, 3113 (2000).

[198] P.G. Debenedetti and F.H. Stillinger, *Nature* **410**, 259 (2001).

[199] M.D. Ediger, *Annu. Rev. Phys. Chem.* **51**, 99 (2000).

[200] W.L. Johnson, M.D. Demetriou, J.S. Harmon, M.L. Lind, and K. Samwer, *MRS Bull.* **32**, 644 (2007).

[201] R. Bohmer, K.L. Ngai, C.A. Angell, and D.J. Plazek, *J. Chem. Phys.* **99**, 4201 (1993).

[202] C.A. Angell, *Science* **267**, 1924 (1995).

[203] H.L. Smith, C.W. Li, A. Hoff, *et al.*, *Nature Phys.* **13**, 900 (2017).

[204] M. Goldstein, *J. Chem. Phys.* **64**, 4767 (1976).

[205] P.D. Gujrati and M. Goldstein, *J. Phys. Chem.* **84**, 859 (1980).

[206] G.P. Johari, *J. Chem. Phys.* **112**, 7518 (2000).

[207] J.M. Kosterlitz and D.J. Thouless, *J. Phys. C: Solid State Phys.* **6**, 1181 (1973).

[208] W. Kurz and D.J. Fisher, *Fundamentals of Solidification* (Trans Tech, Switzerland, 1989).

[209] B. Chalmers, *Physical Metallurgy* (Wiley, New York, 1959).

[210] B. Chalmers, *Principles of Solidification* (John Wiley & Sons, New York, 1964), pp. 118, 119.

[211] A. Karma and W.J. Rappel, *Phys. Rev. E* **57**, 4323 (1998).

[212] W.J. Boettinger, J.A. Warren, C. Beckermann, and A. Karma, *Annu. Rev. Mater. Res.* **32**, 163 (2002).

[213] J.A. Warren and W.J. Boettinger, *Acta Metall. Mater.* **43**, 689 (1995).

[214] J.J. Hoyt, M. Asta, and A. Karma, *Mater. Sci. Eng. Rep.* **41**, 121 (2003).

[215] H.D. Brody and M.C. Flemings, *Trans. AIME*, **236**, 651 (1966).

[216] T.F. Bower, H.D. Brody, and M.C. Flemings, *Trans. AIME*, **236**, 624 (1966).

[217] P.G. Saffman and G. Taylor, *Proc. Roy. Soc.* **245**, 312 (1958).

[218] S.-C. Huang and M.E. Glicksman, *Acta Metall.* **29**, 701 (1981).

[219] J.S. Langer, *Rev. Mod. Phys.* **52**, 1 (1980).

[220] W.W. Mullins and R.F. Sekerka, *J. Appl. Phys.* **34**, 323 (1963).

[221] W.W. Mullins and R.F. Sekerka, *J. Appl. Phys.* **35**, 444 (1964).

[222] G.P. Ivanstov, *Doklady Akad. Nauk SSSR* **58**, 567 (1947).

[223] J.S. Langer and H. Müller-Krumbhaar, *Acta Metall.* **26**, 1681 (1978). *Ibid.* **26**, 1689 (1978). *Ibid.* **26**, 1697 (1978).

[224] J. Lipton, M.E. Glicksman, and W. Kurz, *Mater. Sci. Eng.* **65**, 57 (1984).

[225] M.E. Glicksman, M.B. Koss, and E.A. Winsa, *Phys. Rev. Lett.* **73**, 573 (1994).

[226] A.J. Clarke, D. Tourret, Y. Song, *et al.*, *Acta Mater.* **129**, 203 (2017).

[227] E. Ben-Jacob, N.D. Goldenfield, J.S. Langer, and G. Schon, *Phys. Rev. Lett.* **51**, 1930 (1981). *Ibid.* **29**, 330 (1984).

[228] D.A. Kessler, J. Koplik, and H. Levine, *Phys. Rev. A* **33**, 3352 (1986).

[229] D.A. Kessler and H. Levine, *Acta. Metall.* **36**, 2693 (1988).

[230] S. Gurevich, A. Karma, M. Plapp, and R. Trivedi, *Phys. Rev E* **81**, 011603 (2010).

[231] K. Hono, *Prog. Mater. Sci.* **47**, 621 (2002).

[232] S.C. Wang and M.J. Starink, *Int. Mater. Rev.* **50**, 193 (2005).

[233] I.J. Polmear, *Trans. Metall. Soc. AIME* **230**, 1331 (1964).

[234] J.A. Taylor, B.A. Parker, and I.J. Polmear, *Metall. Sci.* **12**, 478 (1978).

[235] A. Garg and J.M. Howe, *Acta Metall. Mater.* **39**, 1939 (1991).

[236] H.K. Hardy, *J. Inst. Met.* **78**, 169 (1950).

[237] H. Kimura and R. Hashiguti, *Acta Metall.* **9**, 1076 (1961).

[238] R. Banerjee, S. Nag, J. Stechschulte, and H.L. Fraser, *Biomaterials* **25**, 3413 (2004).

[239] M. Bachhav, L. Yao, G.R. Odette, and E.A. Marquis, *J. Nucl. Mater.* **453**, 334 (2014).

[240] R.F. Hehemann, K.R. Kinsman, and H.I. Aaronson, *Metall. Trans.* **3A**, 1077 (1972).

[241] United States Steel Company, *Atlas of Isothermal Transformation Diagrams* (U.S. Steel Company, Pittsburgh, PA, 1951).

[242] J.S. Langer and A.J. Schwartz, *Phys. Rev.* **A21**, 948 (1980).

[243] K. Binder and D. Stauffer, *Adv. Phys.* **25**, 343 (1976).

[244] J.J. Hoyt, *Phase Transformations* (McMaster Innovation Press, Hamilton, ON, 2010).

[245] A.N. Kolmogorov, *Akad. Nauk SSSR, Izv., Ser. Matem.* **355**, 1 (1937).

[246] W.A. Johnson and P.A. Mehl, *Trans. AIME* **135**, 416 (1939).

[247] M. Avrami, *J. Chem. Phys.* **7**, 1103 (1939).

[248] M. Avrami, *J. Chem. Phys.* **8**, 212 (1940).

[249] M. Avrami, *J. Chem. Phys.* **9**, 177 (1941).

[250] J.W. Cahn, *Acta Metall.* **4**, 449 (1956).

[251] I.M. Lifshitz and V.V. Slyozov, *J. Phys. Chem. Solids* **19**, 35 (1961).

[252] C. Wagner, *Z. Electrochem.* **65**, 581 (1961).

[253] R.W. Balluffi, S.M. Allen, and W.C. Carter, *Kinetics of Materials* (Wiley-Interscience, Hoboken, NJ, 2005), Chapter 15 and references therein.

[254] C.E. Krill and L.Q. Chen, *Acta Mater.* **50**, 3057 (2002).

[255] J.W. Cahn, *Acta Metall.* **10**, 1 (1962).

[256] J.C. Baker and J.W. Cahn, in *Solidification*, T.J. Hughel and G.F. Bolling, Eds. (ASM, Metals Park, OH, 1971), p. 23.

[257] M. Hillert and B. Sundman, *Acta Metall.* **24**, 731 (1976).

[258] M. Hillert, *Acta Mater.* **47**, 4481 (1999).

[259] M.J. Aziz and T. Kaplan, *Acta Metall.* **36**, 2335 (1988).

[260] J.W. Cahn, *Acta Metal.*, **10**, 789 (1962).

[261] D.M. Herlach, *Mater. Sci. Eng. R* **12**, 177 (1994).

[262] S. Walder and P.L. Ryder, *Acta Metall. Mater.* **43**, 4007 (1995).

[263] S.L. Sobolev, *Acta Mater.* **61**, 7881 (2013).

[264] H. Humadi, J.J. Hoyt, and N. Provatas, *Phys. Rev. E* **93**, 010801(R) (2016).

[265] G. Liu, G.J. Zhang, X.D. Ding, J. Sun, and K.H. Chen, *Mater. Sci. Eng. A* **344**, 113 (2003).

[266] J.D. Eshelby, *Solid State Physics*, Volume 3 (Academic Press, New York, 1956), p. 79.

[267] A.G. Khachaturyan, *Theory of Structural Transformations in Solids* (Wiley-Interscience, New York, 1983).

[268] F. Bitter, *Phys. Rev.* **37**, 1527 (1931).

[269] M.M. Crum, communication cited in F.R.N. Nabarro, *Proc. Roy. Soc.* A **125**, 519 (1940).

[270] J.W. Cahn and F.C. Larché, *Acta Metall.* **32**, 1915 (1984).

[271] R.B. Schwarz and A.G. Khachaturyan, *Phys. Rev. Lett.* **74**, 2523 (1995).

[272] R.B. Schwarz and A.G. Khachaturyan, *Acta Mater.* **54**, 313 (2006).

[273] S. Luo, W. Luo, J.D. Clewley, T.B. Flanagan, and R.C. Bowman, Jr., *J. Alloys Compounds* **231**, 473 (1995).

[274] C.K. Witham, Ph.D. Thesis in materials science, California Institute of Technology (2000).

[275] J.W. Cahn, *Acta Metall.* **9**, 795 (1961).

[276] J.W. Cahn and J.E. Hilliard, *J. Chem. Phys.* **28**, 258 (1958).

[277] J.W. Cahn and J.E. Hilliard, *J. Chem. Phys.* **31**, 688 (1959).

[278] J.S. Langer, *Ann. Phys.* **65**, 53 (1971).

[279] Wei Xiong, P. Hedström, M. Selleby, *et al.*, *CALPHAD: Computer Coupling of Phase Diagrams and Thermochemistry* **35**, 355 (2011).

[280] J.S. Langer, *Rev. Mod. Phys.* **52**, 1 (1980).

[281] P.C. Hohenberg and B.I. Halperin, *Rev. Mod. Phys.* **49**, 435 (1977).

[282] T. Mohri, in *Alloy Physics*, W. Pfeiler, Ed. (Wiley–VCH, Weinheim, 2007), Chapter 10.

[283] S.M. Allen and J.W. Cahn, *Acta Metall.* **27**, 1085 (1979).

[284] D. Stauffer, *Introduction to Percolation Theory* (Taylor & Francis, London, 1985).

[285] A.G. Khachaturyan, *Phys. Met. Metallog.* **13**, 493 (1962).

[286] A.G. Khachaturyan, *Sov. Phys. Solid State* **5**, 16 (1963).

[287] A.G. Khachaturyan, *Sov. Phys. Solid State* **5**, 548 (1963).

[288] A.G. Khachaturyan, *Prog. Mater. Sci.* **22**, 1-150 (1978).

[289] L.D. Landau *Zh. Eksp. Teor. Fiz.* **7**, 19 (1937). *Ibid* **7**, 627 (1937). Translated and reprinted in L.D. Landau, *Collected Papers*, Volume 1 (Nauka, Moscow, 1969), pp. 234–252.

[290] L.D. Landau and E.M. Lifshitz, *Statistical Physics* (Addison-Wesley, Reading, MA, 1969), Chapters 13, 14.

[291] E.Z. Kaminsky and G.V. Kurdjumov, *Zh. Tekh. Fiz.* **6**, 984 (1936).

[292] G.V. Kurdjumov, V.I. Miretzskii, and T.I. Stelletskaya, *Zh. Tekh. Fiz.* **2**, 1956 (1939).

[293] R.L. Patterson and C.M. Wayman, *Acta Metall.* **14**, 347 (1966).

[294] P.G. Shewmon, *Transformations in Metals* (McGraw-Hill, New York, 1969).

[295] G.V. Kurdjumov and G. Sachs, *Z. Phys.* **64**, 325 (1930).

[296] Z. Nishiyama, *Sci. Rep. Tohoku Univ.* **23**, 637 (1934).

[297] D.S. Lieberman, M.S. Weschler, and T.A. Read, *J. Appl. Phys.* **26**, 473 (1955).

[298] G.V. Kurdjumov and G. Khandros, *Dokl. Nauk. SSSR* **66**, 211 (1949).

[299] H.C. Tong and C.M. Wayman, *Acta Metall.* **23**, 209 (1975).

[300] Z. Nishiyama, *Martensitic Transformation* (Academic Press, New York, 1978).

[301] A.J. Bogers and W.G. Burgers, *Acta Metall.* **12**, 255 (1964).

[302] G.B. Olson and M. Cohen, *J. Less-Common Metals* **28**, 107 (1972).

[303] L. Bracke, L. Kestens, and J. Penning, *Scripta Metall.* **57**, 385 (2007).

[304] M.S. Wechsler, D.S. Lieberman, and T.A. Read, *Trans. AIME* **197**, 1503 (1953).

[305] J.S. Bowles and J.K. Mackenzie, *Acta Metall.* **2**, 129 (1954).

[306] J.K. Mackenzie and J.S. Bowles, *Acta Metall.* **2**, 138 (1954).

[307] J.K. Mackenzie and J.S. Bowles, *Acta Metall.* **5**, 137 (1957).

[308] J.W. Christian, *J. Inst. Metals* **84**, 385 (1956).

[309] M. Born, *Proc. Cambridge Philos. Soc.* **36**, 160 (1940).

[310] C. Zener, *Elasticity and Anelasticity of Metals* (University of Chicago Press, Chicago, 1948).

[311] E.S. Scheil, *Anorg. Allg. Chem.* **207**, 21 (1932).

[312] P.C. Clapp, *Phys. Stat. Sol. B* **57**, 561 (1973).

[313] W. Petry, *Phase Trans.* **31**, 119 (1991).

[314] W. Petry, A. Heiming, J. Trampenau, *et al.*, *Phys. Rev. B* **43**, 10933 (1991).

[315] W. Petry, *J. Phys. IV* **5 C2**, 15 (1995).

[316] G. Grimvall, B. Magyari-Köpe, V. Ozolins, and K.A. Persson, *Rev. Mod. Phys.* **84**, 945 (2012).

[317] J. Trampenau, W. Petry, and C. Herzig, *Phys. Rev. B* **47**, 3132 (1993).

[318] J. Friedel, *J. Phys. Lett. (Paris)* **35**, 59 (1974).

[319] J.W. Cahn, *Prog. Mater. Sci.* **36**, 149 (1992).

[320] L. Mañosa, A. Planes, J. Ortín, and B. Martínez, *Phys. Rev. B* **45**, 7633 (1992).

[321] L. Mañosa, A. Planes, J. Ortín, and B. Martínez, *Phys. Rev. B* **48**, 3611 (1993).

[322] E. Obradó, L. Mañosa, and A. Planes, *Phys. Rev. B* **56**, 20 (1997).

[323] P. Bogdanoff and B. Fultz, *Philos. Mag. B* **81**, 299 (2001).

[324] H.E. Schaefer, *Nanoscience* (Springer, Heidelberg, 2010).

[325] B. Fultz, H. Kuwano, and H. Ouyang, *J. Appl. Phys.* **77**, 3458 (1995).

[326] L.B. Hong and B. Fultz, *J. Appl. Phys.* **79**, 3946 (1996).

[327] K. Yamada and C.C. Koch, *J. Mater. Res.* **8**, 1317 (1993).

[328] M.A. Pushkin, V.I. Troyan, P.V. Borisyuk, V.D. Borman, and V.N. Tronin, *J. Nanosci. Nanotechnol.* **12**, 8676 (2012).

[329] K.M.Ø. Jensen, P. Juhas, M.A. Tofanelli, *et al.*, *Nature Commun.* **7**, 11859 (2016).

[330] C.C. Chen, C. Zhu, E.R. White, *et al.*, *Nature* **496**, 74 (2013).

[331] C.Y. Chiu, Y. Li, L. Ruan, X. Ye, C.B. Murray, and Y. Huang, *Nature Chem.* **3**, 393 (2011).

[332] Y. Yang, C.-C. Chen, M.C. Scott, *et al.*, *Nature* **542**, 75 (2017).

[333] S. Saita and S. Maenosono, *Chem. Mater.* **17**, 6624 (2005).

[334] J. Miao, P. Ercius, and S.J.L. Billinge, *Science* **353**, aaf2157 (2016).

[335] L.P. Bouckaert, R. Smoluchowski, and E.Wigner, *Phys. Rev.* **50**, 58 (1936).

[336] A. Tschöpe and R. Birringer, *Acta Metall. Mater.* **41**, 2791 (1993).

[337] K. Suzuki and K. Sumiyama, *Mater. Trans. JIM* **36**, 188 (1995).

[338] J. Trampenau, K. Bauszuz, W. Petry, and U. Herr, *Nanostruct. Mater.* **6**, 551 (1995).

[339] B. Fultz, J.L. Robertson, T.A. Stephens, L.J. Nagel, and S. Spooner, *J. Appl. Phys.* **79**, 8318 (1996).

[340] H.N. Frase, L.J. Nagel, J.L. Robertson, and B. Fultz, *Philos. Mag. B* **75**, 335 (1997).

[341] H.N. Frase, B. Fultz, and J.L. Robertson, *Phys. Rev. B* **57**, 898 (1998).

[342] A.B. Papandrew, A.F. Yue, B. Fultz, *et al.*, *Phys. Rev. B* **69**, 144301 (2004).

[343] B. Fultz, C.C. Ahn, E.E. Alp, W. Sturhahn, and T.S. Toellner, *Phys. Rev. Lett.* **79**, 937 (1997).

[344] E. Bonetti, L. Pasquini, E. Sampaolesi, A. Deriu, and G. Cicognani, *J. Appl. Phys.* **88**, 4571 (2000).

[345] H.N. Frase, L.J. Nagel, J.L. Robertson, and B. Fultz, in *Chemistry and Physics of Nanostructures and Related Non-Equilibrium Materials*, E. Ma, B. Fultz, R. Shull, J. Morral, and P. Nash, Eds. (TMS, Warrendale, PA, 1997), p. 125.

[346] B.R. Cuenya, A. Naitabdi, J. Croy, *et al.*, *Phys. Rev. B* **76**, 195422 (2007).

[347] B.R. Cuenya, W. Keune, R. Peters, *et al.*, *Phys. Rev. B* **77**, 165410 (2008).

[348] A. Tamura, H. Higeta, and T. Ichinokawa, *J. Phys. C* **15**, 4975 (1982).

[349] A. Tamura and T. Ichinokawa, *J. Phys. C* **16**, 4779 (1983).

[350] A. Tamura, H. Higeta, and T. Ichinokawa, *J. Phys. C* **16**, 1585 (1983).

[351] M.F. Hansen, C.B. Koch, and S. Mørup, *Phys. Rev. B* **62**, 1124 (2000).

[352] S. Bedanta and W. Kleemann, *J. Phys. D Appl. Phys.* **42**, 013001 (2009).

[353] S. Mørup, M.F. Hansen, and C. Frandsen, *Beilstein J. Nanotechnol.* **1**, 182 (2010).

[354] R.M. Bozorth, *Ferromagnetism* (Van Nostrand, New York, 1951).

[355] M.A. Ruderman and C. Kittel, *Phys. Rev.* **96**, 99 (1954).

[356] T. Kasuya, *Prog. Theor. Phys.* **16**, 45 (1956).

[357] K. Yosida, *Phys. Rev.* **106**, 893 (1957).

[358] J.B. Goodenough, *J. Phys. Chem. Solids* **6**, 287 (1958).

[359] J. Kanamori, *J. Phys. Chem. Solids* **10**, 87 (1959).

[360] P.W. Anderson, *Solid State Physics*, Volume 14, F. Seitz and D. Turnbull, Eds. (Academic Press, New York, 1963), p. 99.

[361] J.B. Goodenough, *Scholarpedia* **3**, 7382 (2008).

[362] I. Dzyaloshinskii, *J. Phys. Chem. Solids* **4**, 241 (1958).

[363] T. Moriya, *Phys. Rev. Lett.* **4**, 228 (1960).

[364] T. Moriya, *Phys. Rev.* **120**, 91 (1960).

[365] X. Yu, M. Mostovoy, Y. Tokunaga, W. Zhang, *et al.*, *Proc. Natl. Acad. Sci. USA* **109**, 8856 (2012).

[366] X.Z. Yu, Y. Onose, N. Kanazawa, *et al.*, *Nature* **465**, 901 (2010).

[367] A. Bauer, M Garst, and C. Pfleiderer, *Phys. Rev. Lett.* **110**, 177207 (2013).

[368] H.F. Kay and P. Vousden, *Philos. Mag.* **40**, 1019 (1949).

[369] J.M. Sanchez, D. de Fontaine, and W. Teitler, *Phys. Rev. B* **26**, 1465 (1982).

[370] J.M. Sanchez and D. de Fontaine, *Phys. Rev. B* **21**, 216 (1980).

[371] M. Woloszyn, D. Stauffer, and K. Kulakowski, *Eur. Phys. J. B* **57**, 331 (2007).

Index